THE ISOQUINOLINE ALKALOIDS

CHEMISTRY AND PHARMACOLOGY

This is Volume 25 of
ORGANIC CHEMISTRY
A series of monographs
Editors: ALFRED T. BLOMQUIST and HARRY WASSERMAN

A complete list of the books in this series appears at the end of the volume.

The Isoquinoline Alkaloids

CHEMISTRY AND PHARMACOLOGY

Maurice Shamma

Department of Chemistry
The Pennsylvania State University
University Park, Pennsylvania

1972

ACADEMIC PRESS New York and London

VERLAG CHEMIE GmbH Weinheim/Bergstr.

ACADEMIC PRESS, INC.
111 Fifth Avenue, New York, New York 10003

VERLAG CHEMIE
GmbH Weinheim/Bergstr.

United Kingdom Edition published by
ACADEMIC PRESS, INC. (LONDON) LTD.
24/28 Oval Road, London NW1

LIBRARY OF CONGRESS CATALOG CARD NUMBER: 73-154384

PRINTED IN THE UNITED STATES OF AMERICA

Logical pictures can depict the world

LUDWIG WITTGENSTEIN

CONTENTS

PREFACE

The intent of this work is to present the chemistry, spectroscopy, biogenesis, and pharmacology of the isoquinoline alkaloids in a concise manner. It is also a critical review of the interesting and significant aspects of these alkaloids. It is not a compendium, however, so that no pretense is made of complete coverage of the field such as is found in Manske's "The Alkaloids." The format used is essentially a new one for a book on alkaloids. The chapters include sections on degradation, synthesis, reactions, absolute configuration, biogenesis, pharmacology, and mass spectroscopy, as well as on nuclear magnetic resonance and ultraviolet spectroscopy.

The discussion of the biogenesis of the isoquinolines has been incorporated into each of the chapters, and not relegated to a final chapter, so as to dramatize the view that the most meaningful way to look at each group of natural products is from the standpoint of biogenesis.

Throughout this book infrared absorptions are quoted both in terms of wavelength and frequency. Ultraviolet log e values follow between brackets the λ_{max} values, and those ultraviolet references quoting Holubek and Štrouf refer to the useful compendium entitled "Spectral Data and Physical Constants of Alkaloids" edited by J. Holubek and O. Štrouf, published by the Publishing House of the Czechoslovak Academy of Sciences in Prague. All nuclear magnetic resonance data are given in δ values; they were obtained in deuteriochloroform solution unless otherwise indicated. The term "tetrahydroisoquinoline" refers to the 1,2,3,4-tetrahydro system. It is suggested that henceforth the prefix "nor" in the isoquinoline alkaloids refer solely to the N-nor series.

The decision was made to include in this book the emetine and protostephanine bases. The groups that were excluded belong to the morphine, hasubanonine, Erythrina, Amaryllidaceae, tubulosine, and colchicine series, since these are usually considered apart from the isoquinoline bases. The literature to 1971 has been covered, and some material that appeared since has also been incorporated.

I would like to take this opportunity to express my gratitude to Mrs. Mary Boyer Muldrow and to Drs. V. St. Georgiev, J. L. Moniot, S. W. Scott, J. A. Weiss, C. D. Jones, and R. W. Lagally, as well as to Miss Sarah Ann Gallagher, for commenting on parts of the manuscript.

MAURICE SHAMMA

Chapter 1/THE SIMPLE TETRAHYDROISOQUINOLINES

Occurrence: Cactaceae, Chenopodiaceae, Fumariaceae, Leguminoseae, Nymphaceae, Papaveraceae; and Ranunculaceae

Approximate Number: 30

Some Simple Tetrahydroisoquinolines of Interest:

(+)-Salsoline

(−)-Salsolidine

(+)-Calycotomine

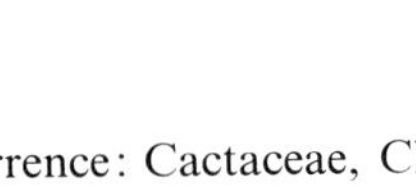

Anhalamine

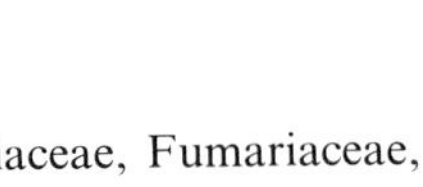

Hydrocotarnine

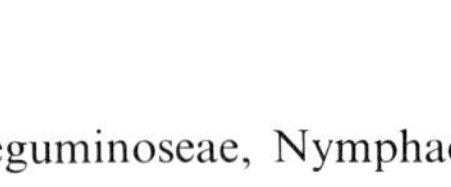

(−)-Anhalonine

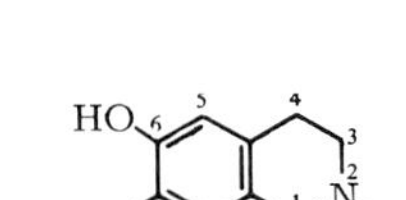

Anhalonidine

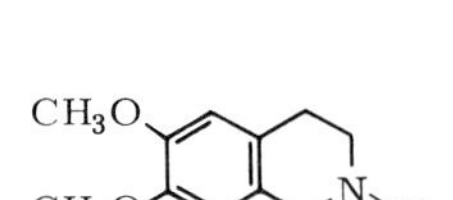

(−)-Lophophorine

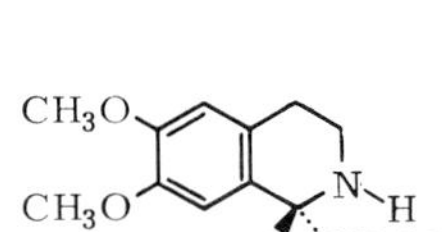

Lophocerine

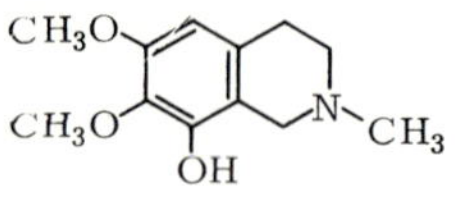
Anhalidine

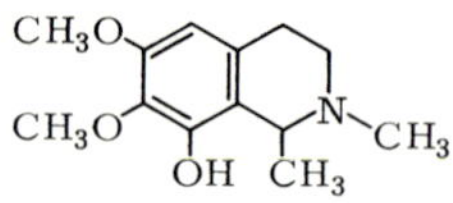
Pellotine

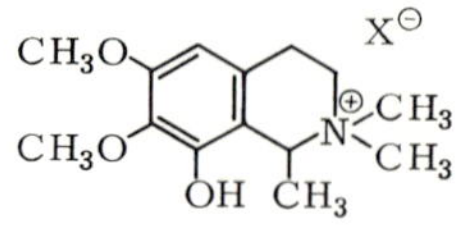
Peyotine

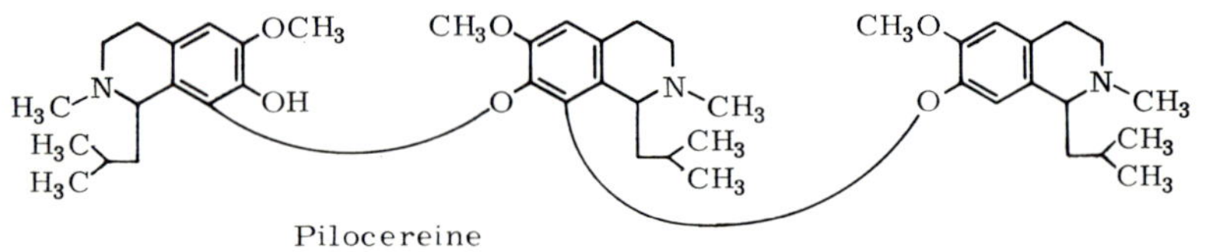
Pilocereine

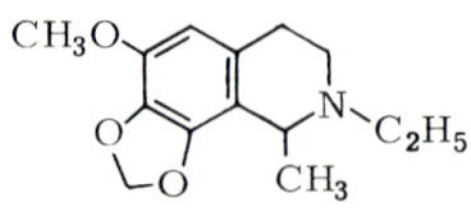
Peyophorine

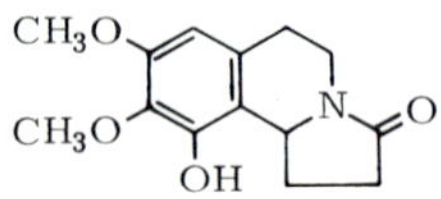
Peyoglutam

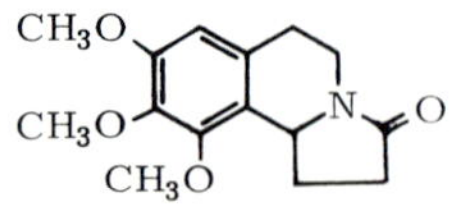
Mescalotam

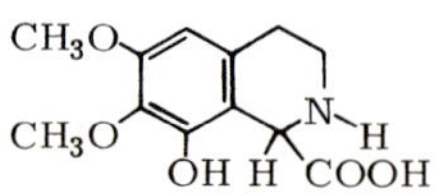
Peyoxylic acid

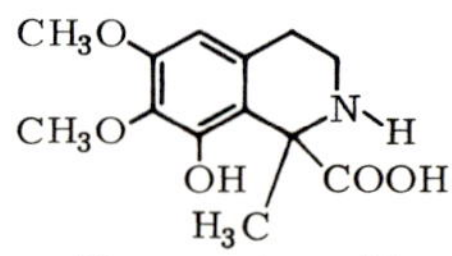
Peyoruvic acid

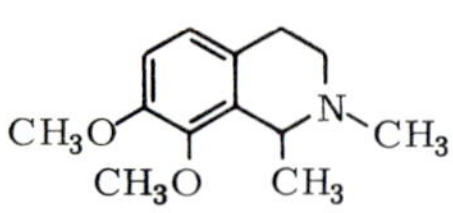
Tepenine

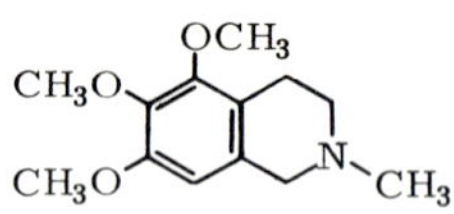
Tehaunine

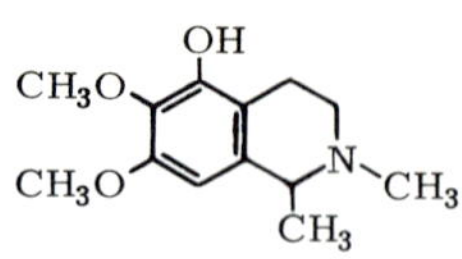
Gigantine

I. Introduction

Many of the simple tetrahydroisoquinolines have been found in the Cactaceae. They include such alkaloids as lophocerine, which incorporates an isobutyl side chain, and its trimer pilocereine. Peyophorine is an uncommon isoquinoline alkaloid since it possesses an *N*-ethyl group, while peyoglutam and mescalotam are lactams rather than amines. Peyoxylic and peyoruvic acids are α-amino acids. The unusual substitution pattern of tepenine should be noted.[1,1a]

Tetrahydroisoquinolines are also present in some members of the Papaveraceae and Fumariaceae, in which case their biogenesis is probably different from that for the Cactaceae alkaloids. Hydrocotarnine, for example, which is found in opium, may very well be formed through the known facile cleavage of the C-1 to C-9 bond of the phthalideisoquinoline base narcotine, although formal proof to this effect is still missing. The biogenesis of the simple tetrahydroisoquinolines obtained from the Cactaceae will be discussed in a later section in this chapter.

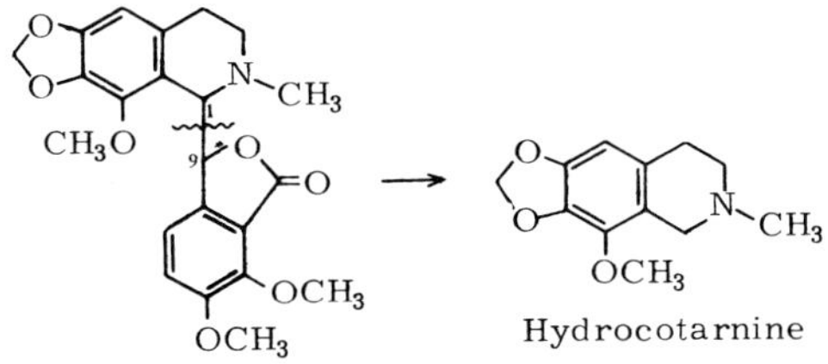

Narcotine
(present in opium)

II. Degradations

Chemical degradation is nowadays seldom used as a tool in the structural elucidation of new simple tetrahydroisoquinolines. Rather, extensive use is made of spectroscopic data, and the conclusions are usually confirmed by synthesis.

A. Lophocerine

A systematic degradation of the cactus alkaloid lophocerine has recently been carried out not for the purposes of structural elucidation, since the structure of the alkaloid was known, but rather in connection with biogenetic studies in which labeled lophocerine had been produced in the plant (Scheme I).[1b]

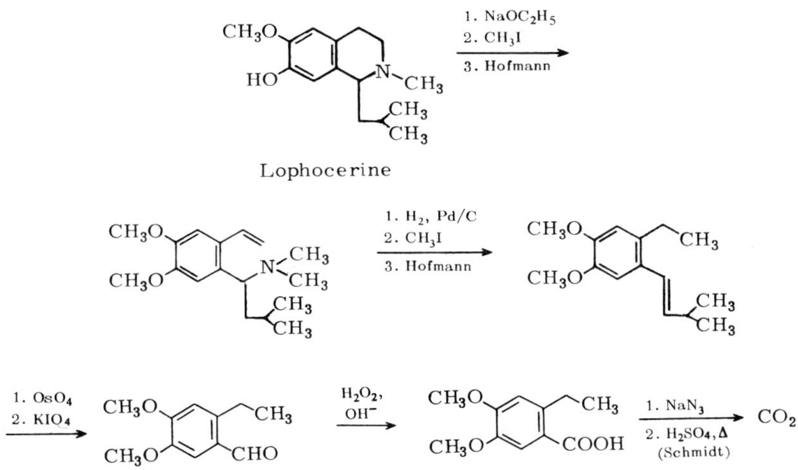

Scheme I

B. Pilocereine

Pilocereine is an interesting trimeric tetrahydroisoquinoline base found in some giant cacti, and its structural elucidation deserves special comment. The alkaloid was originally thought to analyze for $C_{30}H_{44}N_2O_4$. Since permanganate oxidation yielded a mixture of isobutyric and isovaleric acids, and the alkaloid appeared to contain two nitrogen atoms incorporated as *N*-methyl groups, it was assumed at first that pilocereine was simply a dimer of lophocerine so that it was assigned structure **1**.

This view was strengthened by the belief that the alkaloid possessed one phenolic hydroxyl and one diphenyl ether linkage, together with two *O*-methyl groups. Indeed, *O*-ethylpilocereine, presumably **2**, was found to yield four amines, **3**–**6**, upon potassium in liquid ammonia reduction, *a method which cleaves aryl ethers preferentially over alkyl aryl ethers*. These results seemed to vindicate the assignment of expression **1** to the alkaloid.

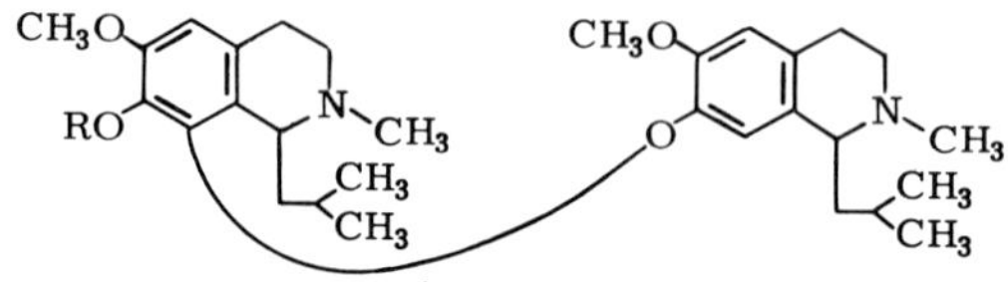

1, R = H
2, R = C_2H_5

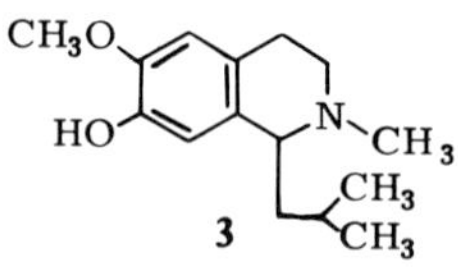

3

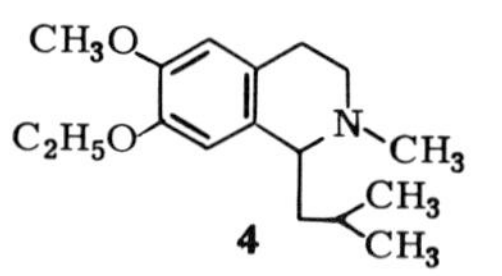

4

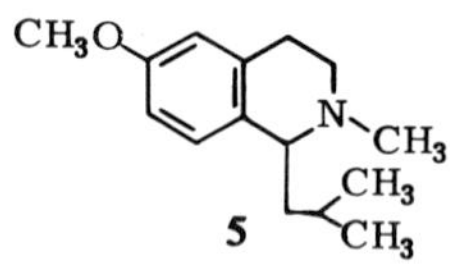

5

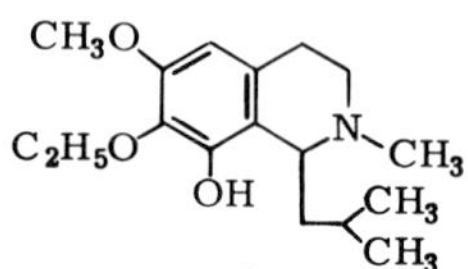

6

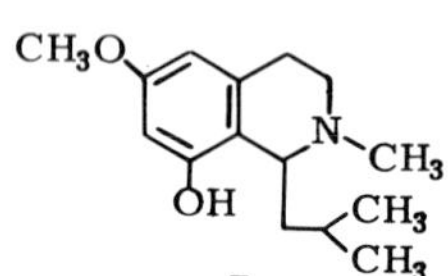

7

With the advent of mass spectrometry, however, it was quickly determined that the molecular ion for *O*-methylpilocereine was at *m/e* 757, which would make the alkaloid trimeric rather than dimeric. Furthermore, analysis of the newly available NMR spectrum of pilocereine acetate showed four aromatic hydrogens, three *O*-methyl and three *N*-methyl groups, together with only one *O*-acetyl and three isobutyl residues.

A reinvestigation of the potassium in liquid ammonia reduction of *O*-ethylpilocereine then showed that the tetrahydroisoquinoline **7** had also been generated, and this product could not be readily accommodated on the basis of structure **1** for pilocereine. These data necessitated the revision of the structural assignment for the alkaloid to expression **8**, which is a trimer of lophocerine.[2]

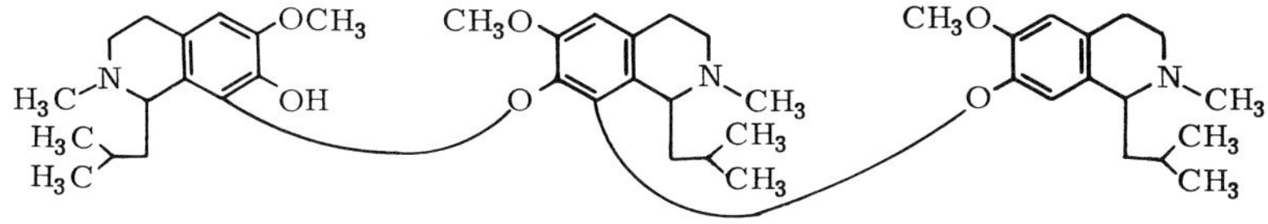

8

Pilocereine

Structure **1** was subsequently assigned to isopilocereine, a cryptophenolic fragment obtained as one of the products of the potassium in liquid ammonia reduction of pilocereine.[2]

The synthesis of pilocereine (**8**), or rather a mixture of pilocereine diastereoisomers, has been achieved through phenolic oxidative coupling of lophocerine. The reaction product was a complex mixture from which material corresponding to pilocereine could be isolated by thin-layer chromatography.[3,4]

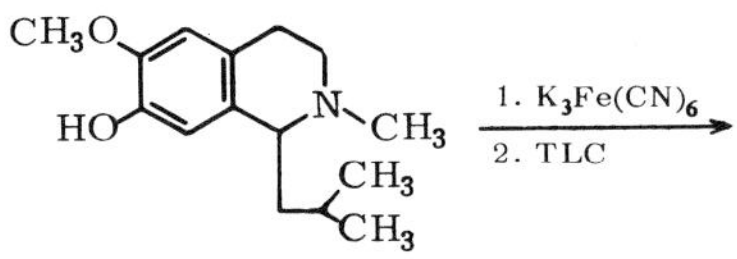

Lophocerine

Pilocereine (**8**)

Piloceredine, which accompanies pilocereine in *Lophocereus schottii* (Engelm.) Britt. et Rose (Cactaceae), is diastereoisomeric with the latter alkaloid.[2]

III. Some Nonbasic Isoquinolines: Peyoglutam, Mescalotam, and Related Alkaloids

The best known alkaloid from the peyote cactus, *Lophophora williamsii* (Lemaire) Coult. (Cactaceae), is mescaline, which corresponds to 3,4,5-trimethoxyphenylethylamine and is a powerful psychotomimetic agent.

The nonbasic fraction from peyote has been investigated by Kapadia and Fales, who isolated from that source 14 new amides, lactams, and imides which can be considered to be alkaloids. Besides such relatively simple species as *N*-acetylmescaline and *N*-formylanhalinine, the lactams peyoglutam and mescalotam were obtained.[5]

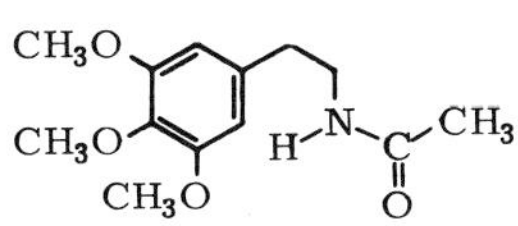

N-Acetylmescaline

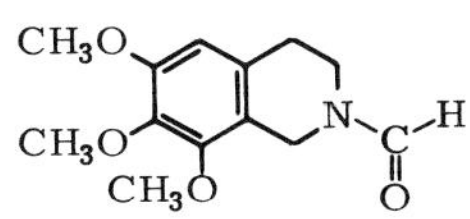

N-Formylanhalinine

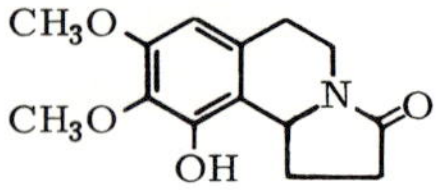

Peyoglutam

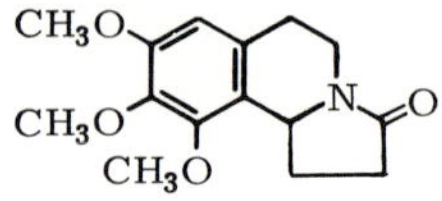

Mescalotam

Mescalotam was synthesized as shown from mescaline hydrochloride and α-oxoglutaric acid, and peyoglutam was prepared by an analogous route.

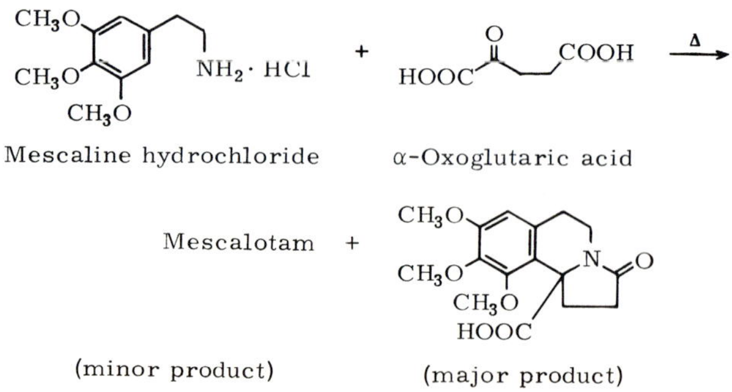

Other interesting compounds obtained from the nonbasic fraction of peyote are the imides **9–11**[5] and the substituted pyrrole peyonine.[6]

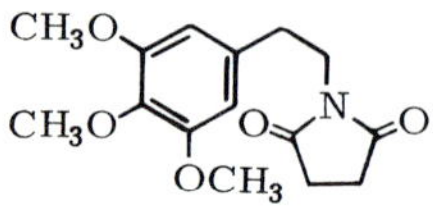

9

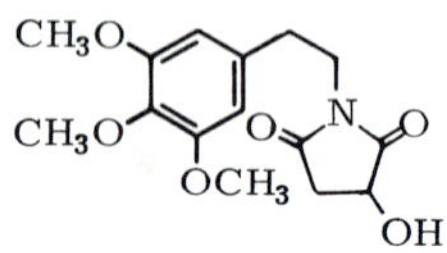

10

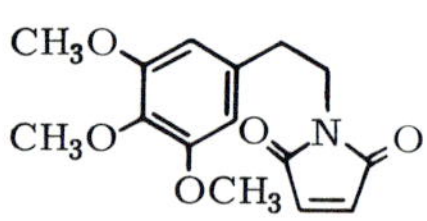

11

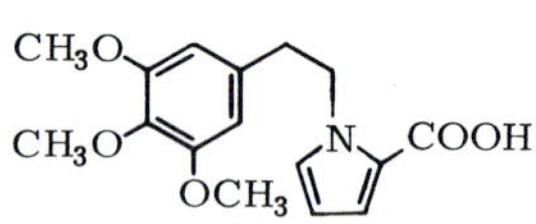

Peyonine

The isolation of the above compounds indicates that mescaline can react in the plant with acids of the citric acid (Krebs) cycle, so that amidic conjugates are formed.

Very recently, Kapadia and co-workers have found that among the peyote isoquinolines there were two new amino acids, peyoxylic and peyoruvic acid. That these are likely precursors for the 6,7,8-trisubstituted tetrahydroisoquinoline alkaloids was suggested by their facile decarboxylation in the plant system.[7]

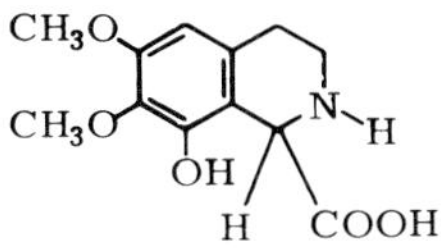

Peyoxylic acid

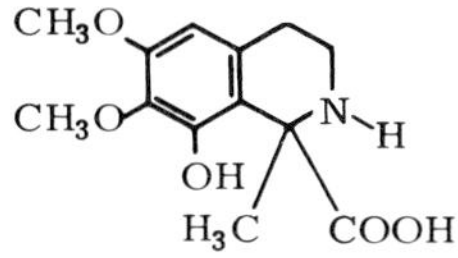

Peyoruvic acid

IV. Methods of Synthesis

A wide variety of methods are available for the preparation of simple tetrahydroisoquinolines, and these are outlined below. Although no fully authenticated 4-hydroxytetrahydroisoquinoline has yet been isolated from plants, it is more than probable that such natural species do exist. Emphasis has been placed, therefore, upon synthetic methods for the preparation of C-4 oxygenated species.

A. The Bischler–Napieralski Cyclization

The Bischler–Napieralski cyclization[8] is one of the methods of choice for the preparation of simple tetrahydroisoquinolines – the amide derived from a substituted phenethylamine being cyclized under the influence of acid to a 3,4-dihydroisoquinoline. Amides with electron-releasing groups on the aromatic ring are readily cyclized, while electron-withdrawing groups hinder the reaction. The following sequence, starting with *N*-acetylmescaline, afforded the methiodide of *O*-methylpellotine.[9]

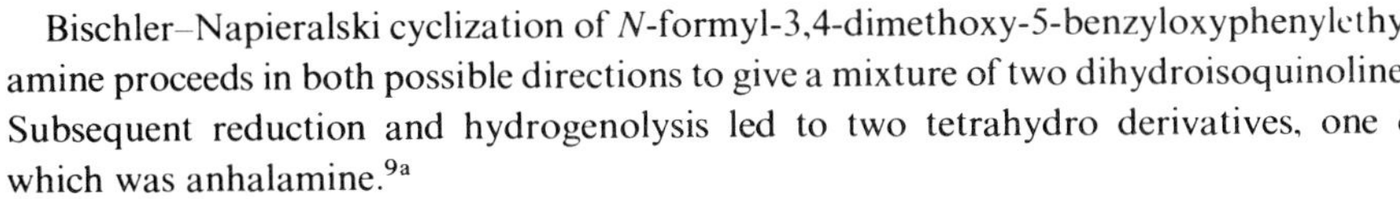

N-Acetylmescaline

O-Methylpellotine methiodide

Bischler–Napieralski cyclization of *N*-formyl-3,4-dimethoxy-5-benzyloxyphenylethylamine proceeds in both possible directions to give a mixture of two dihydroisoquinolines. Subsequent reduction and hydrogenolysis led to two tetrahydro derivatives, one of which was anhalamine.[9a]

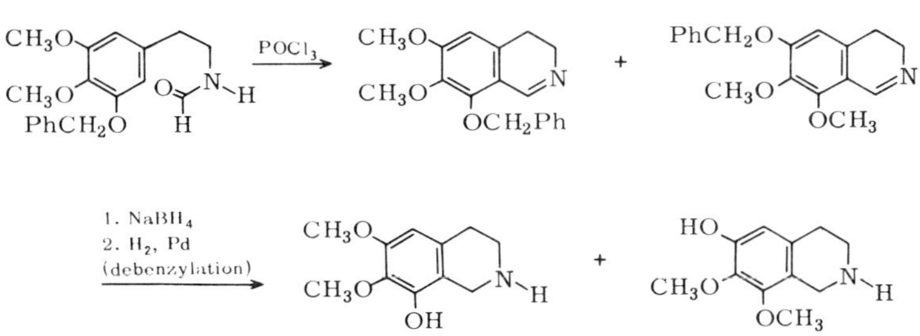

Anhalamine

The Bischler–Napieralski cyclization can be extended to the preparation of C-4 oxygenated isoquinolines. The ethylene ketal of ω-benzamidoacetoveratrone was cyclized with phosphorus pentoxide in pyridine to the imine ketal **12**.[10]

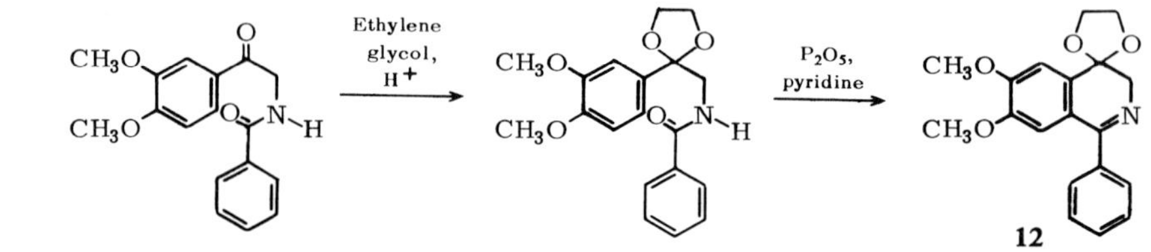

12

ω-Benzamidoacetoveratrone

B. Friedel–Crafts Acylation or Modified Bischler–Napieralski Cyclization

If a urethane instead of an amide is cyclized into the aromatic ring, a lactam is generated which can be reduced to the tetrahydroisoquinoline stage.[11]

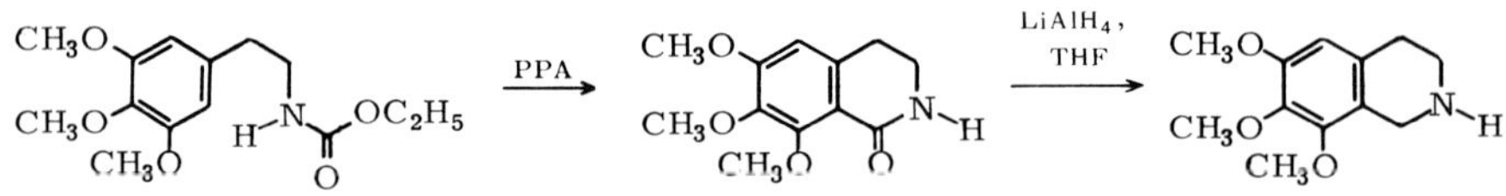

The tetrahydroisoquinoline can then be oxidized to the imine and thus functionalized at C-1:

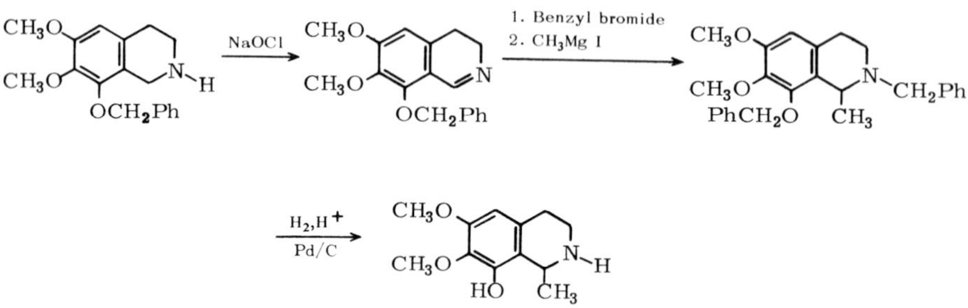

Anhalonidine

C. The Pictet–Spengler Cyclization

A phenethylamine substituted with hydroxyl, methoxyl, or methylenedioxy groups may be cyclized directly to a tetrahydroisoquinoline in the presence of an aldehyde under acid conditions.[12] Using this approach, Schöpf and Bayerle obtained an 83% yield of *O*-norsalsoline by maintaining dopamine and acetaldehyde at pH 5 at room temperature for several days.[13]

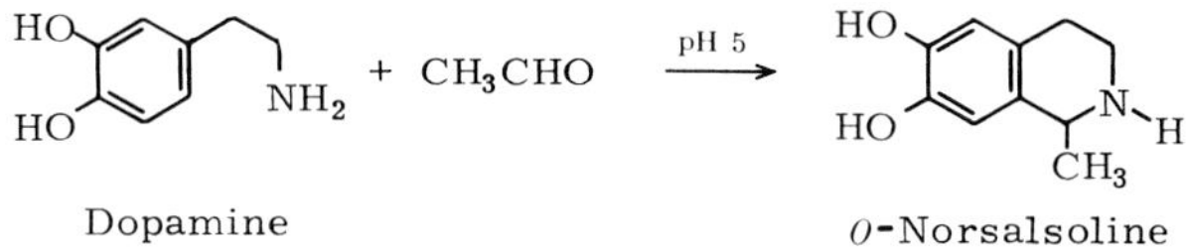

Pictet–Spengler reaction of 3,4-dimethoxy-5-benzyloxyphenylethylamine with formaldehyde led to a mixture of products resulting from cyclization in both possible directions[9a]:

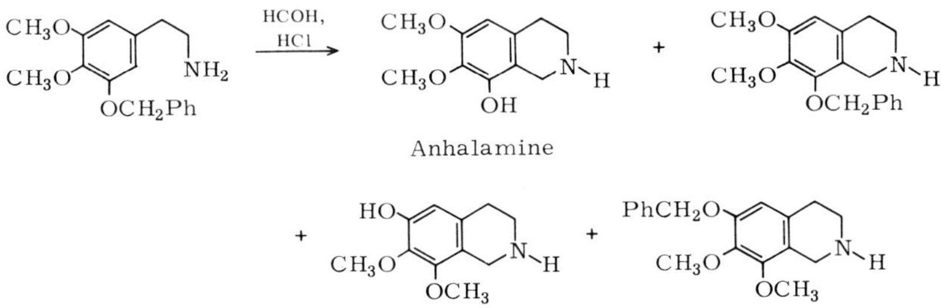

4-Hydroxytetrahydroisoquinolines can also be obtained by this procedure since treatment of the amino alcohol **13** with formaldehyde and formic acid to effect *N*-methylation afforded the bicyclic alcohol **14**, resulting from Pictet–Spengler cyclization.[14]

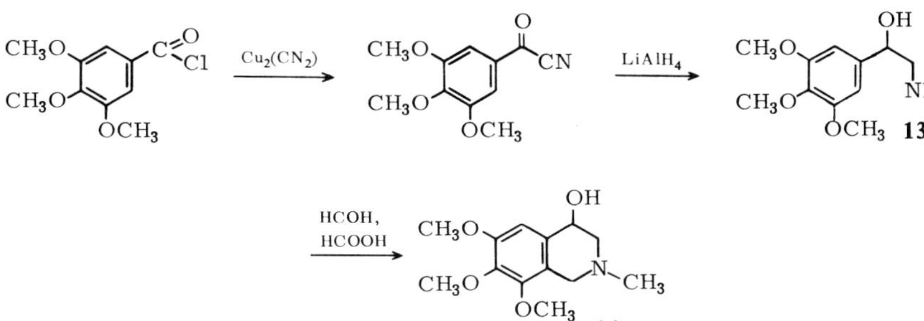

D. Phenolic Cyclization

A Pictet–Spengler type cyclization can occur *without* the use of acid catalysis if an activating phenolic function is present para to the cyclization site. For example, the substituted phenethylamine **15** will condense with acetone simply on heating to provide the tetrahydroisoquinoline **16**.[15]

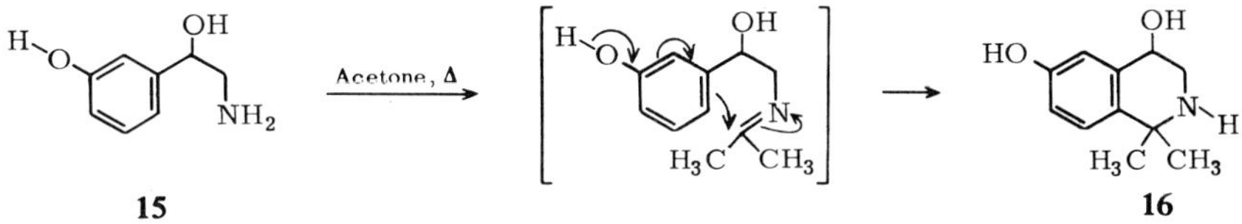

Similarly, condensation of the phenethylamine **17** with pyruvic acid gives the amino acid **18**. Such transformations may have analogies in biogenetic processes.

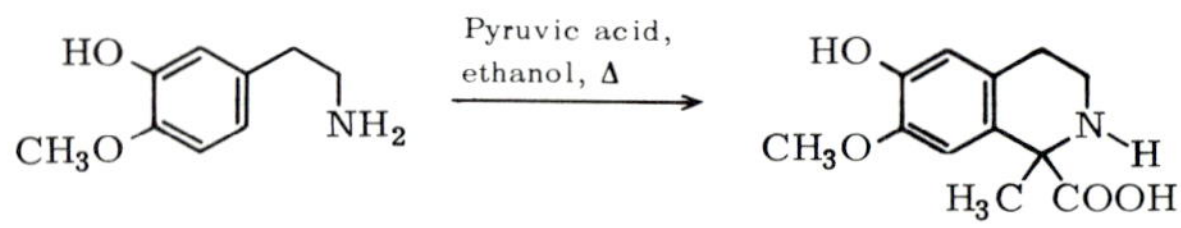

17 **18**

E. The Pomeranz–Fritsch Cyclization[16]

Acid-catalyzed cyclization of a benzalamino acetal results in formation of an isoquinoline derivative. However, extension of the method to the use of ketimines in place of aldimines gives much poorer yields, as experienced in the synthesis of the isoquinoline **19** (Scheme II).

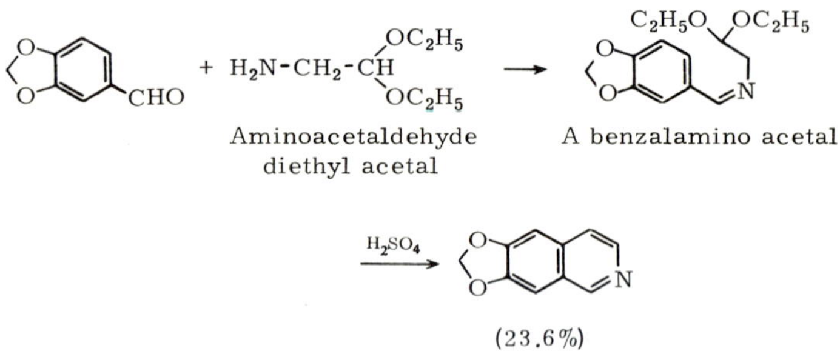

But:

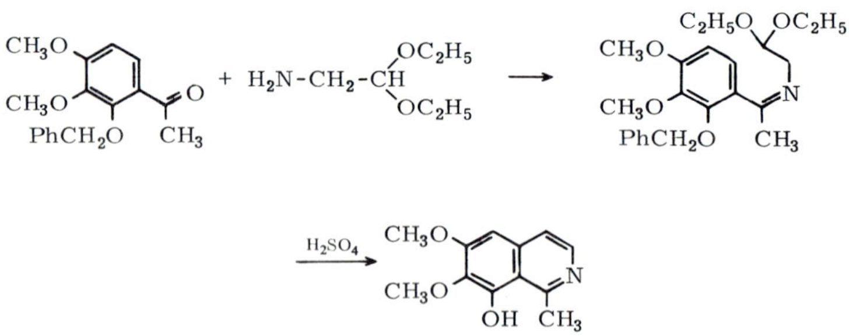

19
(13%)

Scheme II

F. The Bobbitt Modifications of the Pomeranz–Fritsch Cyclization

A variety of reaction schemes have been worked out by Bobbitt and co-workers which allow the synthesis of a broad spectrum of substituted tetrahydroisoquinolines. When an aldehyde such as 3,4,5-trimethoxybenzaldehyde is combined with an equimolar amount of aminoacetaldehyde diethyl acetal in ethanol and the mixture is reduced with Adams catalyst, the product is the secondary amine **20**. If this amine is taken up in 6 *N* hydrochloric acid, allowed to stand at room temperature for 15 hours, and the solution subsequently treated with hydrogen and a palladium catalyst, the product is the tetrahydroisoquinoline **21** which is obtained in high yield.[17]

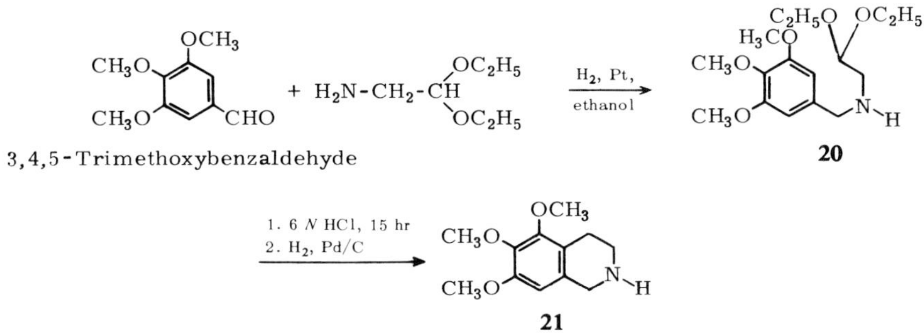

The method can be extended to the preparation of *N*-methyltetrahydroisoquinolines through reductive *N*-methylation of the secondary amine intermediate, e.g., **22**.[18]

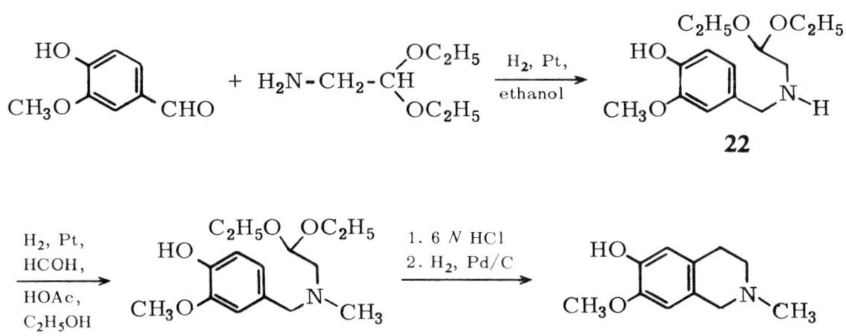

The original imine formed from the condensation of the aromatic aldehyde with aminoacetaldehyde diethyl acetal can be treated with an alkyl Grignard reagent and the resulting secondary amine then cyclized to afford a 1-alkyltetrahydroisoquinoline.[19]

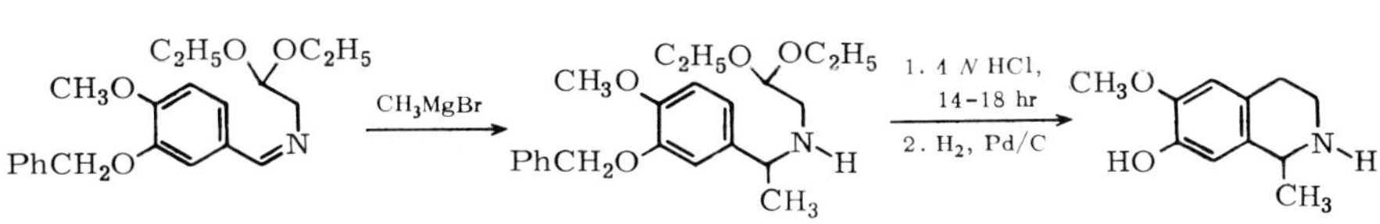

In an ancillary development, it was found that a more direct route to the acetal required for cyclization involves Mannich condensation of the appropriate phenol with formaldehyde and a properly substituted amino acetal.[20]

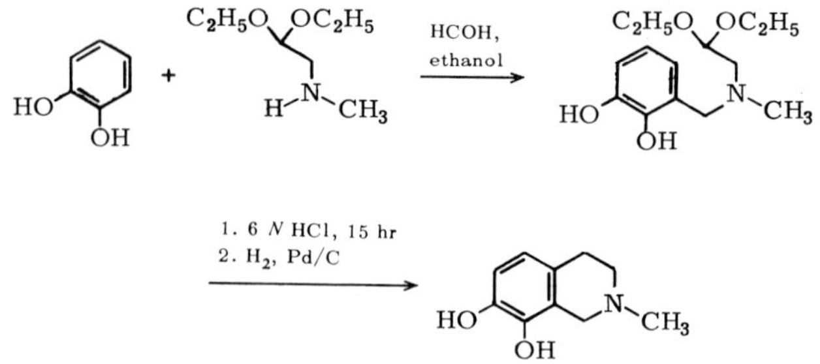

All the above cyclizations proceed through the intermediacy of 4-hydroxytetrahydroisoquinolines which can actually be isolated if hydrogenolysis with palladium on carbon is omitted.[21,22]

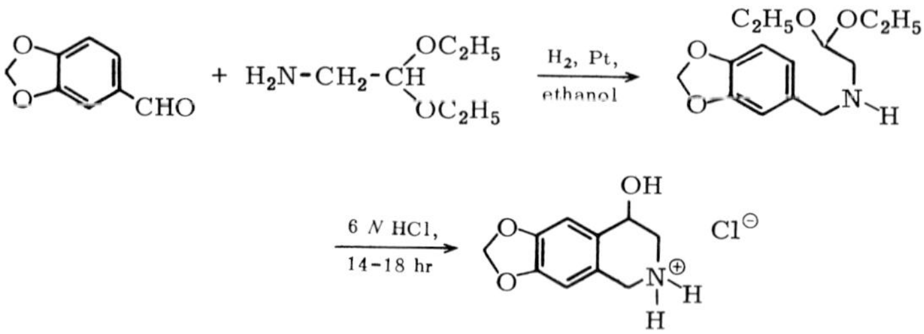

G. *An Approach Using N-Benzylglycine Derivatives*

In this synthesis, the Schiff base obtained from the condensation of 3,4-dimethoxybenzaldehyde with methylamine was reduced with sodium borohydride. The resulting benzylamine was alkylated with ethyl chloroacetate to yield the *N*-benzylglycine derivative **23**. Friedel–Crafts cyclization then furnished the α-amino ketone **24**.[23,24]

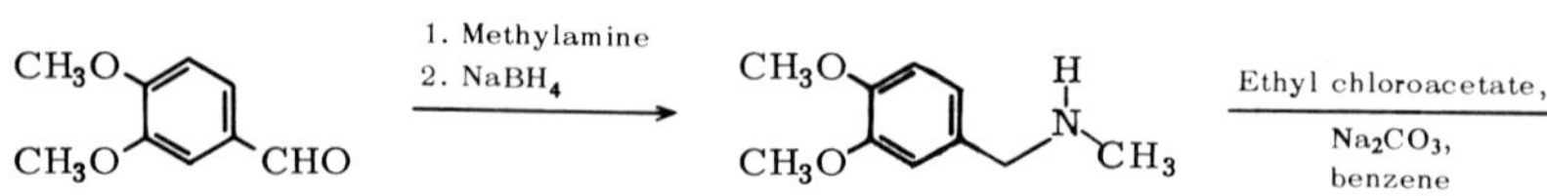

3,4-Dimethoxybenzaldehyde

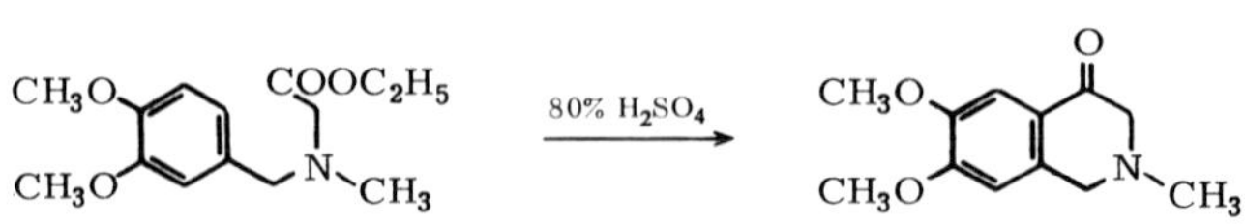

23 **24**

H. The Dieckmann Cyclization

A toluene derivative such as **25** was first transformed to the acetophenone **26** via Friedel–Crafts acylation. Oxidation with hypobromite then generated an *o*-toluic acid which was esterified and treated with *N*-bromosuccinimide.

Reaction of the resulting bromo ester with *N*-benzylglycine ethyl ester gave rise to the amino diester **27** which upon Dieckmann cyclization produced the β-keto ester **28**. The amino ketone **29** was subsequently obtained by simple hydrolysis and decarboxylation (Scheme III).[25]

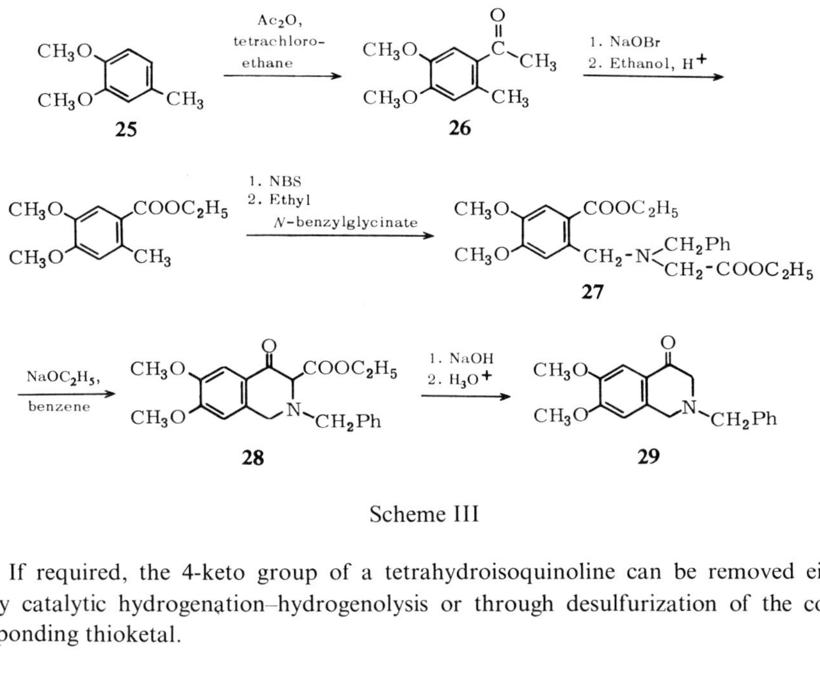

Scheme III

If required, the 4-keto group of a tetrahydroisoquinoline can be removed either by catalytic hydrogenation–hydrogenolysis or through desulfurization of the corresponding thioketal.

I. A Pathway Using Chloromethyl Methyl Ether

When the amide **30** was treated with chloromethyl methyl ether in glacial acetic acid at room temperature, the product was the *N*-benzoyltetrahydroisoquinoline **31**.[26]

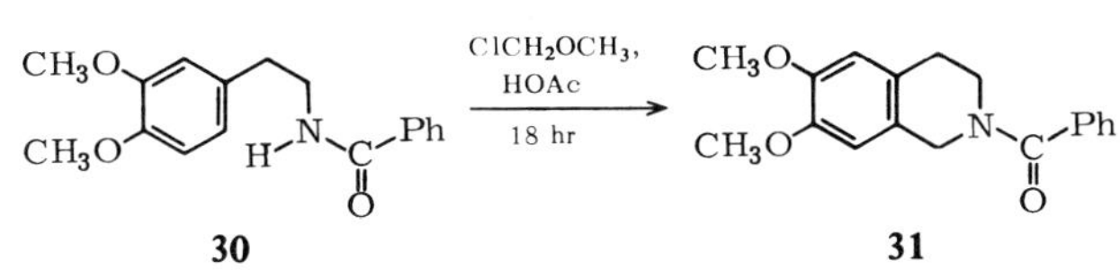

Similarly, reaction of the amide **32** with chloromethyl methyl ether in THF yielded the derivative **33**.

J. The Cyclization of α-Amino Nitriles

α-Amino nitriles, mono- or disubstituted at the α-carbon, can be cyclized to 4-oxotetrahydroisoquinolines in the presence of concentrated sulfuric acid. However, this reaction has not been extended to the preparation of 4-oxotetrahydroisoquinolines unsubstituted at C-3.[27]

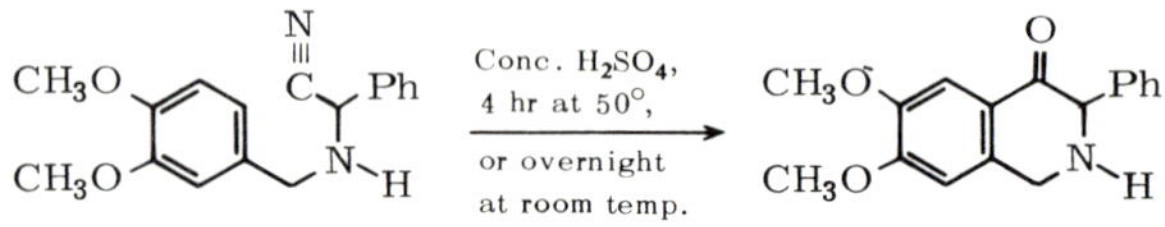

K. Dehydrative Cyclization of Amine Oxides

In formic acid solvent, sulfur dioxide can effect the dehydrative cyclization of substituted *N,N*-dimethylphenethylamine *N*-oxides to yield 2-methyltetrahydroisoquinolines. If the reaction is run in a more nucleophilic solvent such as water, *N*-demethylation rather than cyclization takes place, resulting in the formation of the *N*-methylphenethylamine derivative.[28]

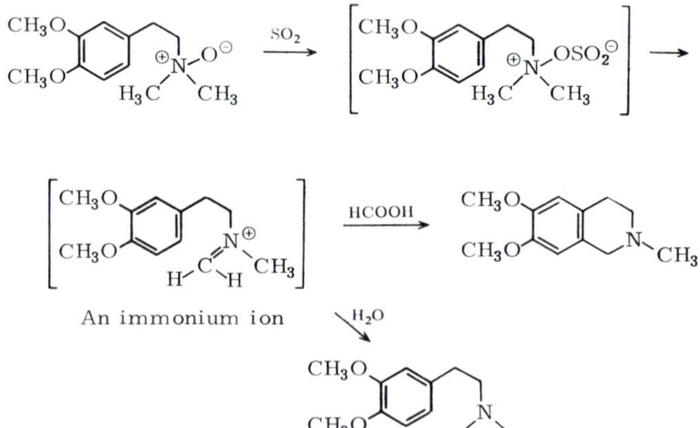

Exactly the same products shown above can be obtained with the identical amine oxide, but using ferrous ion in dilute sulfuric acid and a base such as pyridine. This is a one-electron transfer system which converts the tertiary amine oxide first to the corresponding tertiary amine radical cation. This species gives rise to the same immonium ion as shown in the sulfur dioxide case, which as indicated can either cyclize to the 2-methyltetrahydroisoquinoline or undergo hydrolysis to the *N*-methylphenethylamine derivative.[28a]

Tertiary amine oxides have actually been suggested as intermediates in the metabolic dealkylation of tertiary amines and in the formation of heterocyclic rings during alkaloid biogenesis[29,30]; the above transformations indicate the feasibility of both processes.[31]

L. *Preparations via 3-Isochromanones*

Treatment of homoveratric acid with formaldehyde and mineral acid forms 6,7-dimethoxy-3-isochromanone (**34**). When this lactone is refluxed with benzylamine, the corresponding *N*-benzyllactam **35** is obtained.[32]

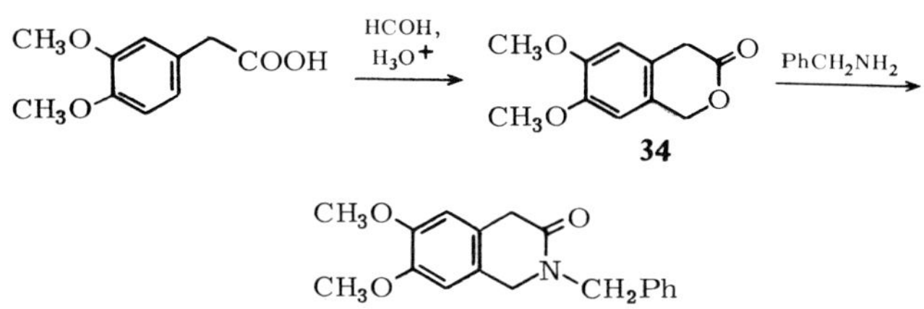

34 **35**

Alternatively, the isochromanone **34** reacts with ethanolic hydrogen bromide to form the bromo ester **36**. Treatment with methylamine then gives the lactam **37**, which can be reduced to the tetrahydroisoquinoline **38**.[32]

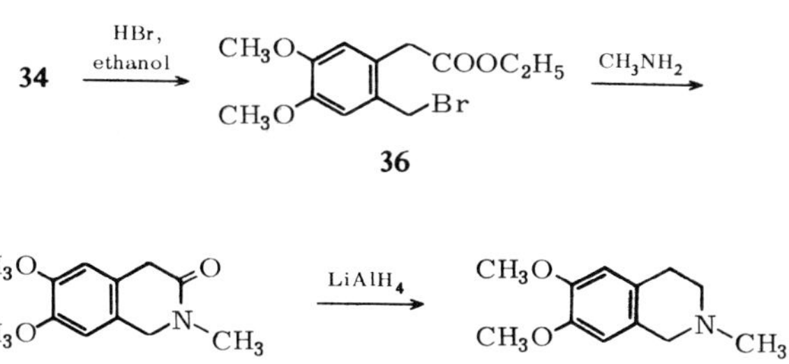

34 **36** **37** **38**

It has also been found that reaction of the isochromanone **34** with hydrazine furnishes the hydrazide **39**. This product upon reaction with hydrochloric acid gives a 76% yield of the lactam **40**.[33]

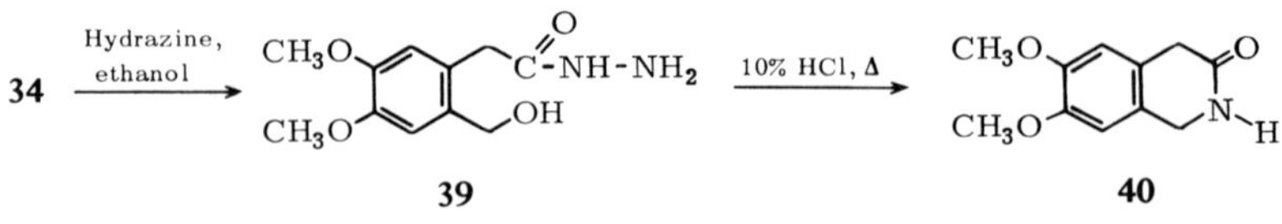

M. *Reductive Decyanation of α-Amino Nitriles*

It has been revealed in a preliminary communication that α-amino nitriles, e.g., **41**, usually obtained from the dehydration of α-amino amides, can readily undergo decyanation to the corresponding amine when treated with sodium borohydride. This approach may allow the synthesis of optically active tetrahydroisoquinolines since racemization does not occur if an asymmetric center is present at C-1.[34]

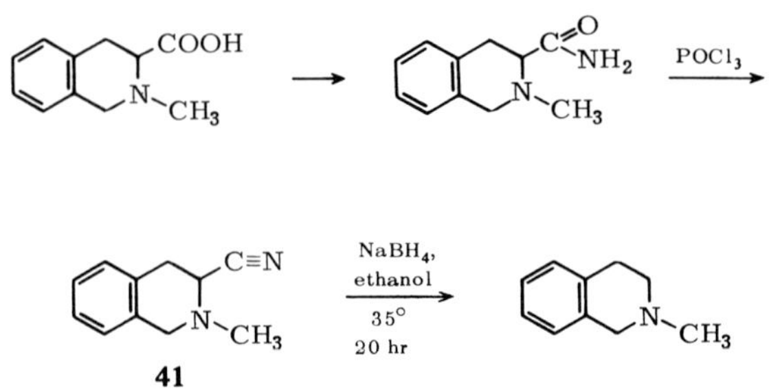

N. *Friedel–Crafts Alkylation*

The latest method for the formation of isoquinolines involves treatment of benzylamine, or a substituted benzylamine, with 2-bromoethanol followed by Friedel–Crafts alkylation using aluminum chloride in hot decalin.[34a]

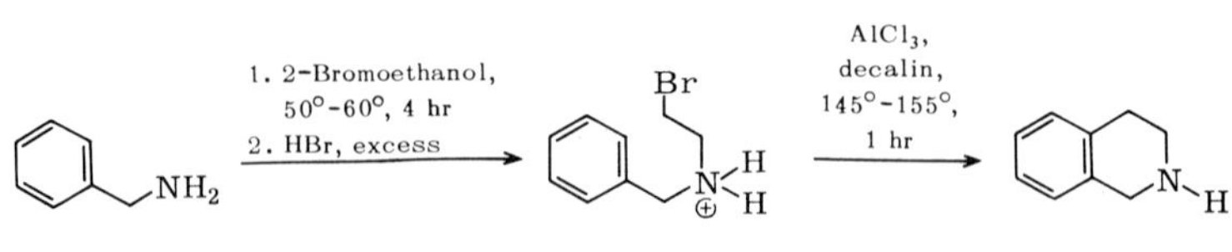

V. The Gigantine Problem

The alkaloid (+)-gigantine, $C_{13}H_{19}NO_3$, isolated from the saguaro cactus *Carnegia gigantea* (Engelm.) (Cactaceae) was given structure **42** on spectral grounds.[35] However, the unequivocal synthesis of the two diastereoisomeric forms of **42** clearly indicated that neither compound corresponded to gigantine, so that the structural assignment for the alkaloid had to be reconsidered.[36]

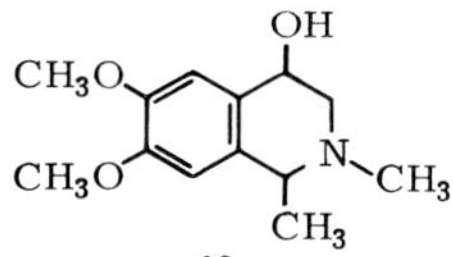

42

The preparation of the two diastereoisomers started with the oxidation of the *N*-benzylamino ketone hydrochloride **43** with mercuric acetate to afford the quaternary base **44**. Reaction with methyl magnesium iodide gave the ketone **45**. Reduction of this material first with sodium borohydride and then with hydrogen and palladium on carbon led to a mixture of two diastereoisomeric secondary amines which could be separated. Reductive *N*-methylation subsequently furnished the desired tertiary amines **42a** and **42b**, neither of which corresponded to the natural base gigantine (Scheme IV).[36]

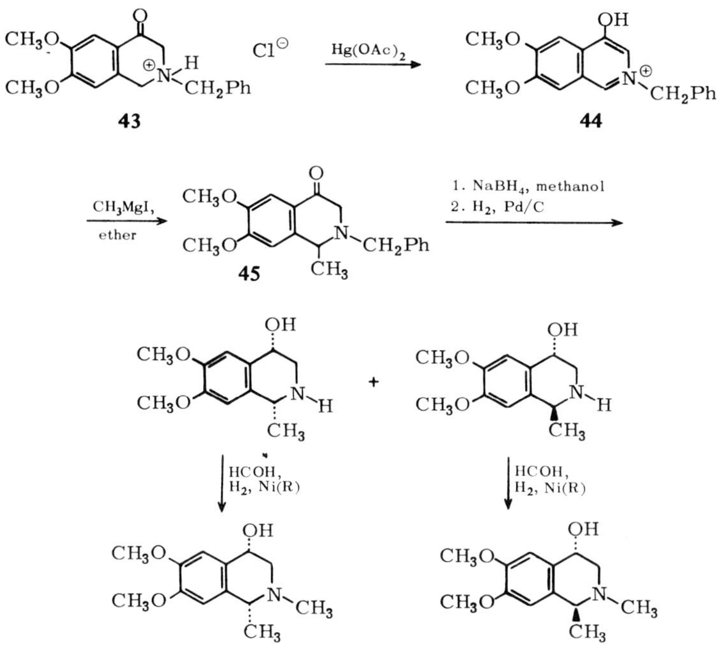

Scheme IV

The synthetic compounds **42a** and **42b** were completely characterized by spectral means. The NMR spectra obtained at 100 MHz are summarized below. It should be noted that the C-1 methyl group in species **42b**, being quasi-axial, comes at $\delta 1.17$ and further upfield than in **42a**, where it is quasi-equatorial and appears at $\delta 1.41$.

The chemical shifts and coupling constants for the C-3 and C-4 protons are so similar for both diastereoisomers that the two compounds must be nearly identical in structure in this region, as denoted by conformations **42aa** and **42bb**.[36]

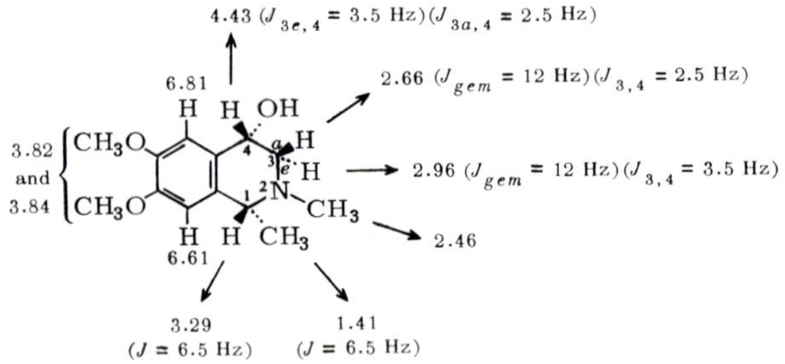

NMR spectral values for **42a**

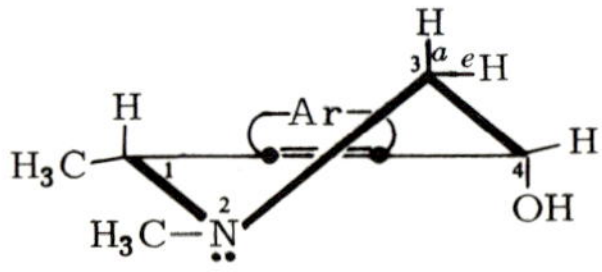

42aa

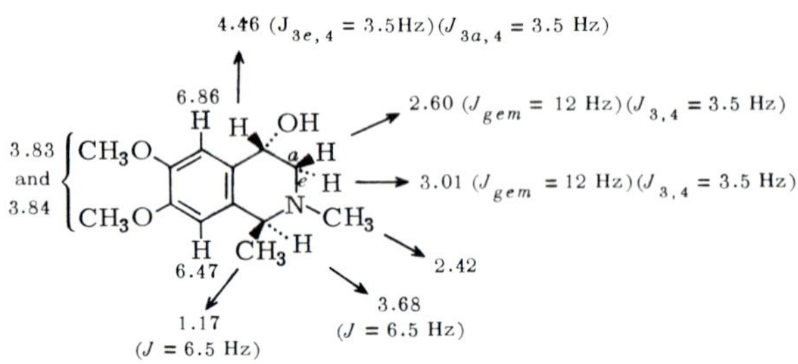

NMR spectral values for **42b**

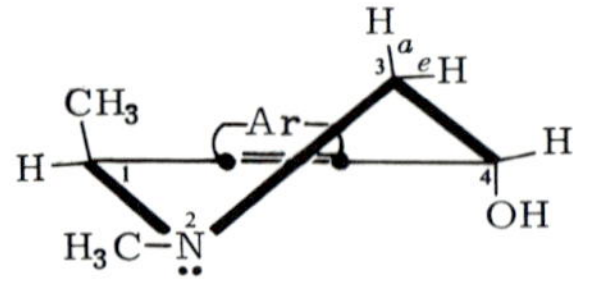

42bb

The correct structure of gigantine was deduced by Kapadia and co-workers, who observed that the alkaloid gives a positive Gibb's test for a free position para to a phenolic hydroxyl.[37] Additionally, in the NMR spectrum of the alkaloid, the $\delta 6.29$ peak assigned to the C-8 proton is shifted upfield by 0.41 ppm upon the addition of base, and a change of this magnitude is characteristic of a proton para to a phenolic group.[37,38] The structure of gigantine was then confirmed by a synthesis of the racemic compound (Scheme V). The required benzyloxy amide **47** could be obtained either from the amide **45a** or from the amine **46**.[39]

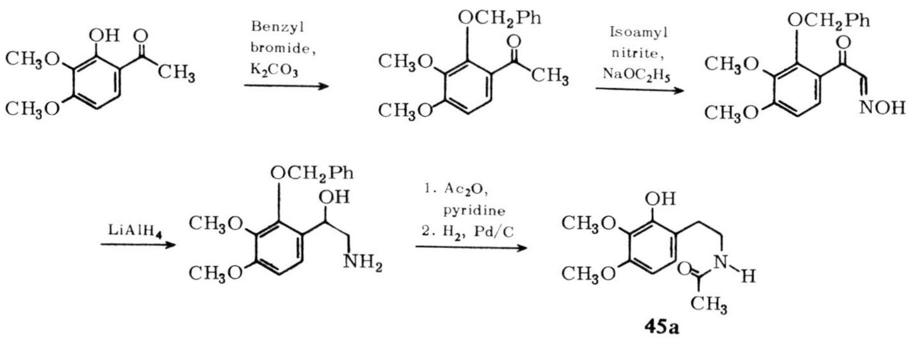

And:

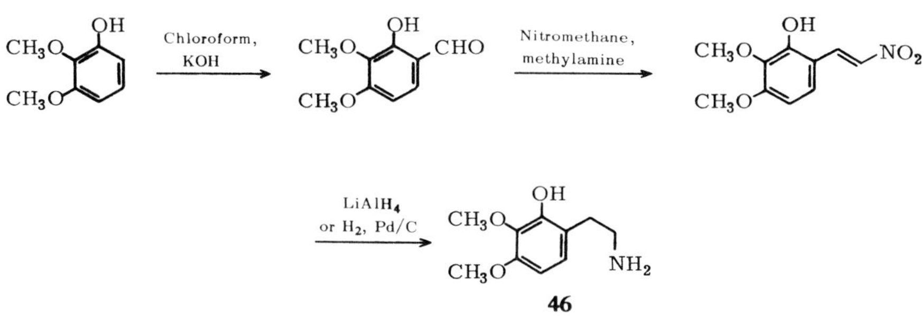

Then:

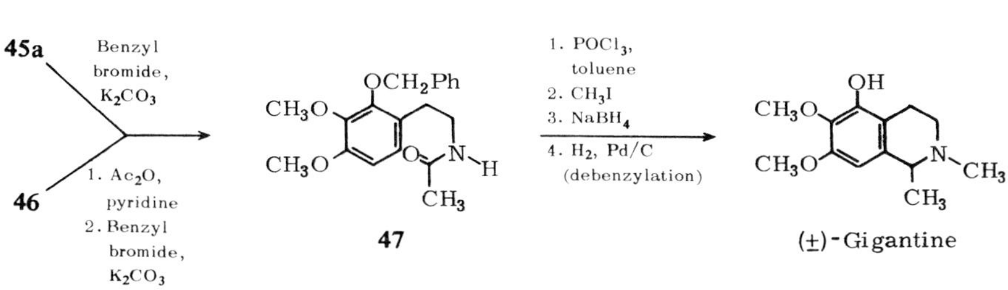

Scheme V

VI. Some Reactions of Tetrahydroisoquinolines and Their Derivatives

A. C-4 Hydroxylation of a Tetrahydroisoquinoline via the Rearrangement of a Quinol Acetate

It has long been known that *p*-toluquinol can be rearranged in acid to 2-methylhydroquinone.[40]

p-Toluquinol

In a related transformation, it was found that the quinol acetate **48** prepared from 6-hydroxytetralin and lead tetraacetate rearranged to 5,7-diacetoxytetralin in the presence of sulfuric acid and acetic anhydride.[41]

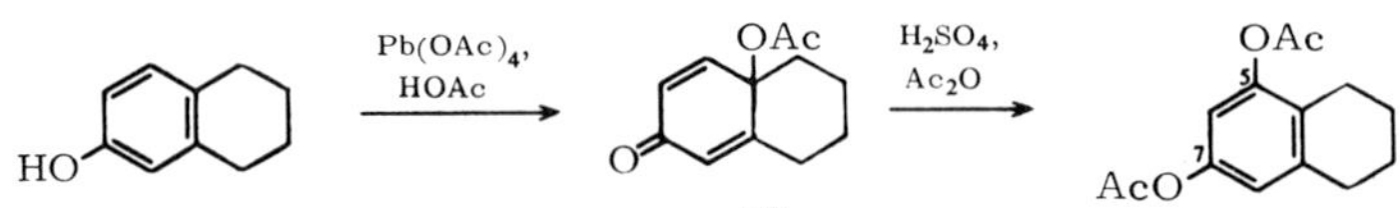

6-Hydroxytetralin **48**

An attempt was made, therefore, to extend the above transformations to the tetrahydroisoquinoline series. Treatment of the alkaloid corypalline with lead tetraacetate provided the corresponding quinol acetate. However, further treatment with sulfuric acid and acetic anhydride unexpectedly formed the tetrahydroisoquinoline **49**.[24]

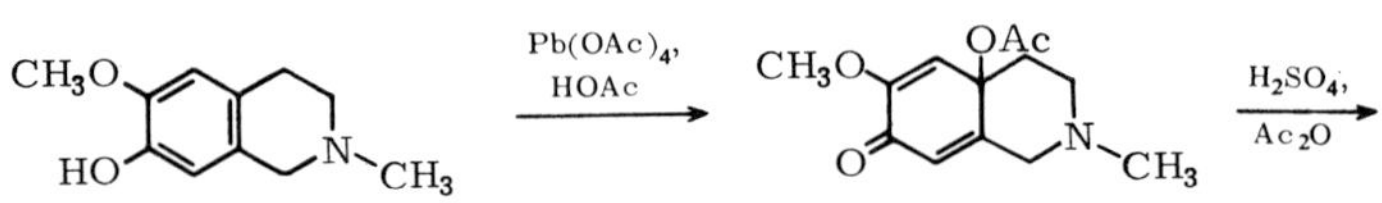

Corypalline

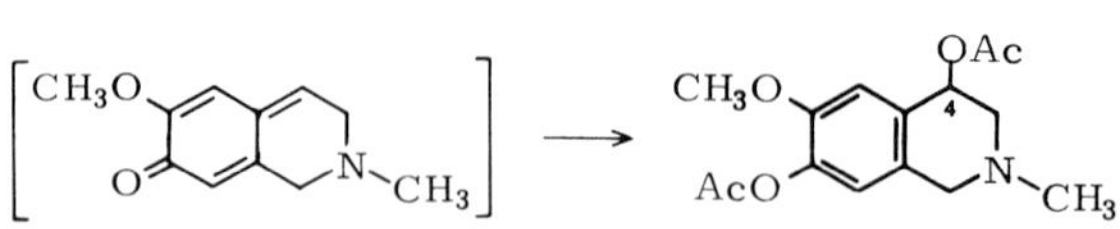

Quinone methide **49**

B. The Reactivity of C-4 Oxygenated Tetrahydroisoquinolines

A C-4 acetoxy group is particularly reactive when a free or potential phenolic group is present at C-7. For example, treatment of the diacetate **49** with potassium hydroxide in methanol at room temperature leads to the phenol **51** through the intermediacy of the quinone methide **50**. No such substitution takes place when a methoxyl rather than an acetoxy group is present at C-7.[42]

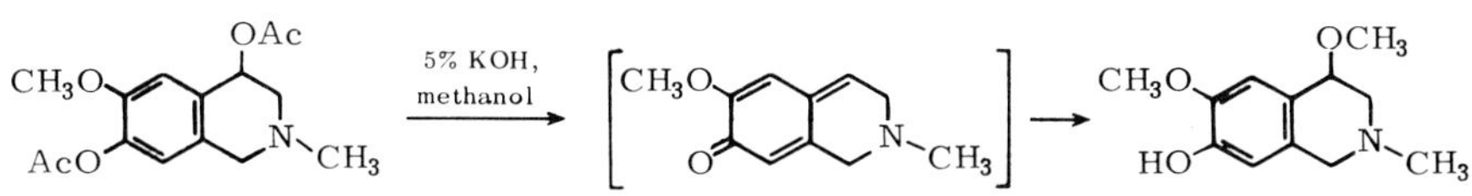

49 **50** **51**

In a similar vein, reaction of diacetate **49** with methylamine and potassium hydroxide provides the diamino phenol **52**. Thiols may also be used as nucleophiles in lieu of alcohols or amines.[43]

49 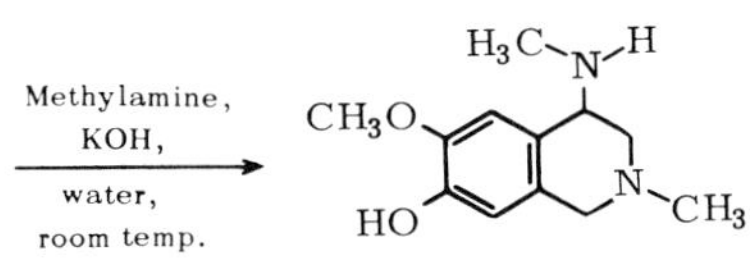

52

The 4,6,7-trimethoxyamide **53** shows unusual reactivity since it undergoes substitution at C-4 when recrystallized from ethanol even though it is devoid of a free phenol at C-7. The product is the 4-ethoxy derivative **54**. Here again the reaction probably proceeds through a quinonoidal intermediate.[26]

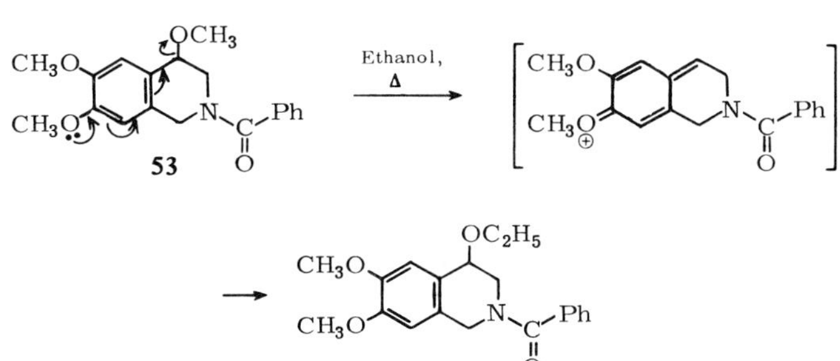

53 **54**

C. C-4 Alkylation of Isoquinoline Systems

A quaternary salt such as **55** can be alkylated in the presence of an aromatic aldehyde under reducing conditions to give a 4-benzyltetrahydroisoquinoline.[44,45]

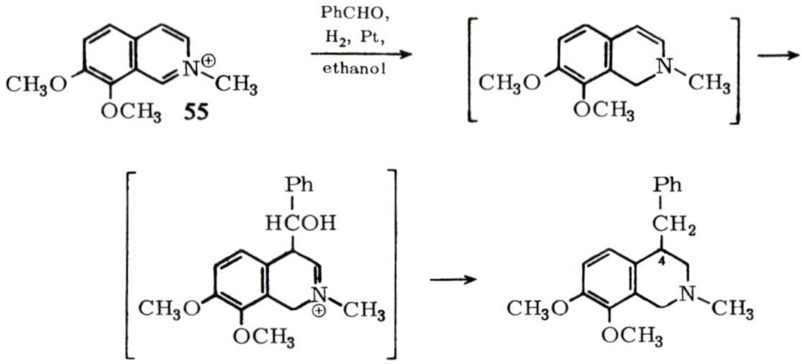

In a related transformation, the 1,2-dihydroisoquinoline **56** has been shown to react in alcoholic hydrochloric acid solution with a variety of aldehydes to form 4-substituted 2-methylisoquinoline salts.[22]

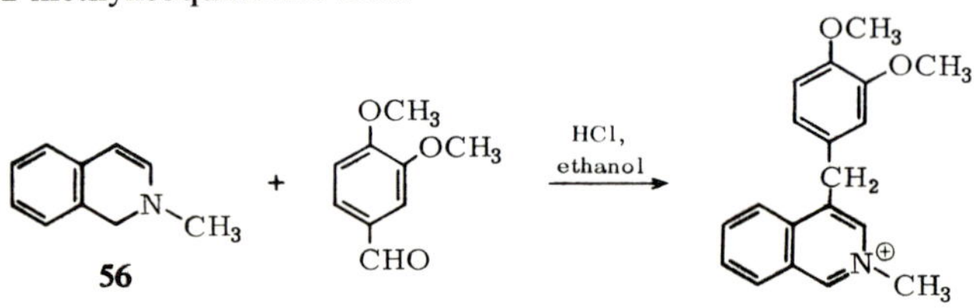

Alternatively, the amino acetal **57** can be treated with an aromatic aldehyde and concentrated hydrochloric acid to afford a high yield of the 4-benzyltetrahydroisoquinoline **58**.

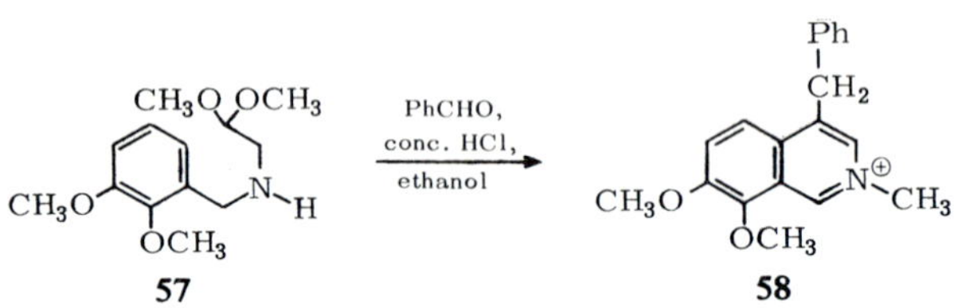

D. Selective Hydrolysis of the Methoxyl Substituents

Studies on the selective hydrolysis of isoquinoline derivatives are due almost exclusively to Brossi and his associates. Specifically substituted isoquinolines which might be difficult to prepare by other means may sometimes be readily obtained by this method. Treatment of the tetrahydroisoquinoline **59** with 20% hydrochloric acid results in hydrolysis of all the ether functions so that the product is the triphenol **60**.[46]

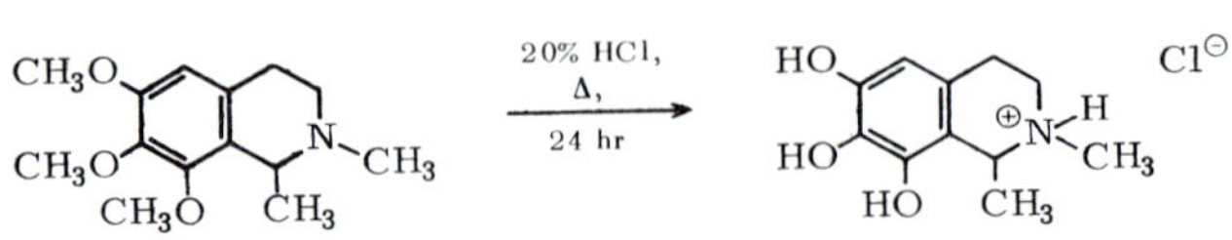

Under identical experimental conditions, however, the analogously substituted dihydroisoquinoline **61** undergoes only partial hydrolysis to the monophenol **62**.[46]

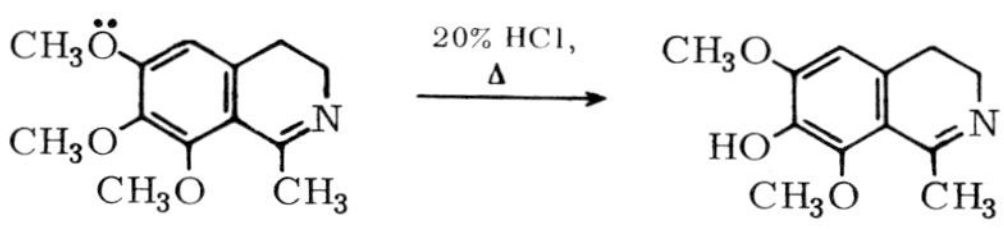

61 **62**

Other examples of the selective hydrolysis of the methoxyl groups of 3,4-dihydroisoquinolines follow[46a]; they indicate that both electronic and steric factors are important in deciding which group will be hydrolyzed.

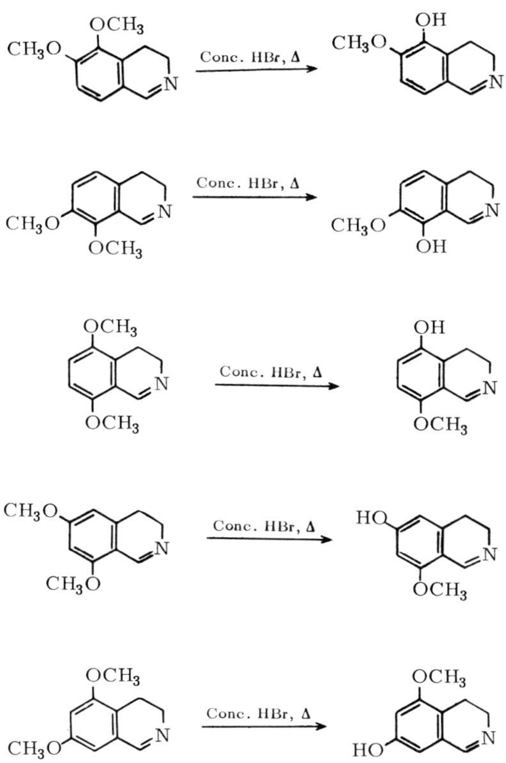

Concentrated hydrobromic acid selectively hydrolyzes the 6,7-dimethoxydihydroisoquinoline **63**. The resulting phenol **64** can then be oxidized or reduced to afford the corresponding isoquinoline or tetrahydroisoquinoline derivatives.[47]

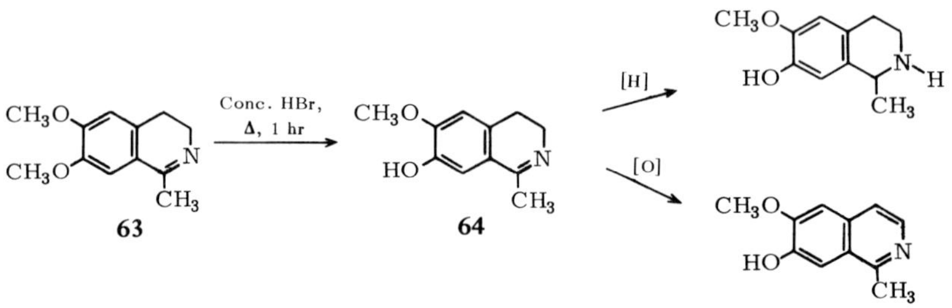

63 64

In like fashion, corypalline can be conveniently prepared from 6,7-dimethoxy-3,4-dihydroisoquinoline by treatment with concentrated hydrobromic acid and catalytic hydrogenation of the imine methiodide intermediate.[47a]

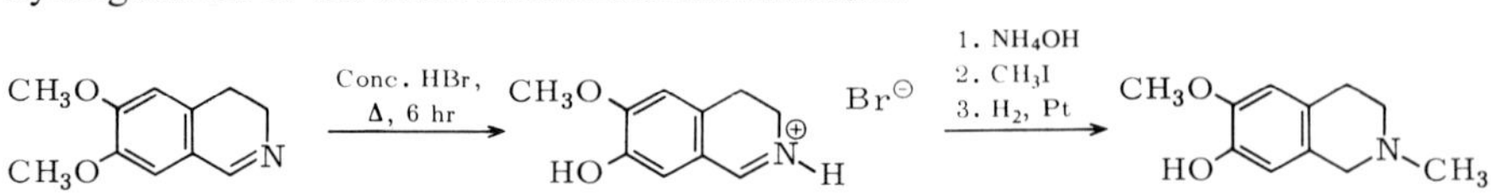

6,7-Dimethoxy-3,4-dihydroisoquinoline

Corypalline, isolated as HCl salt

Preferential hydrolysis can also be performed on 4-ketotetrahydroisoquinolines. Hydrolysis of **65** with hydrobromic acid in acetic acid produced the phenolic ketone **66**, which was then hydrogenolized to a simple tetrahydroisoquinoline.[48]

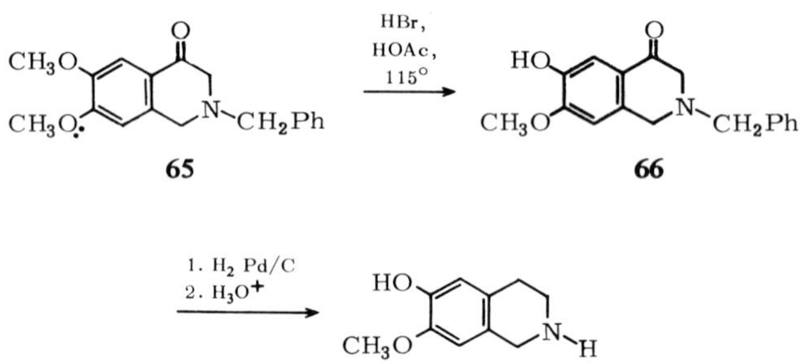

65 66

An interesting case concerns the selective hydrolysis of the lactam **67** with concentrated hydrochloric acid to furnish the phenol **68** through the intermediacy of an oxonium ion.[49]

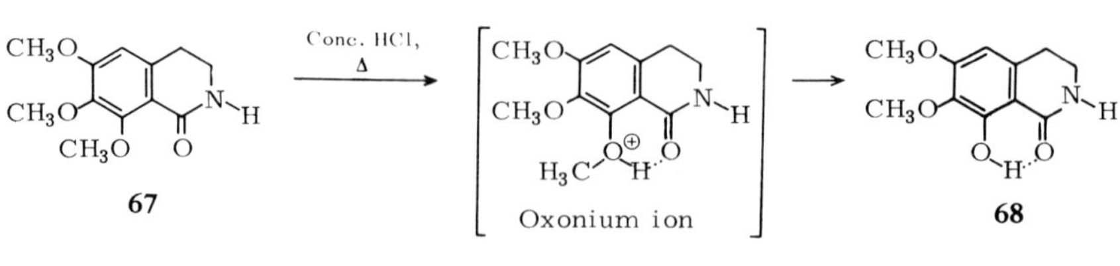

67 68

Selective hydrolysis can be performed even in the absence of a carbonyl or an imine function. When the trimethoxyphenethyltetrahydroisoquinoline **69** was refluxed with

20% hydrochloric acid, the main product was found to be the monophenol indicated.[50] The faster rate of cleavage of a 1,2,3-trimethoxybenzene occurs mainly because of the enhanced basicity of the central methoxyl as it is sterically twisted out of the plane of the benzene ring.[50a]

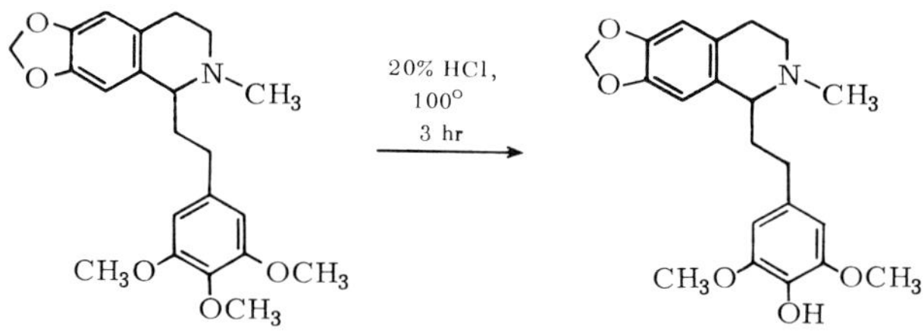

69

A somewhat similar example concerns the simple tetrahydroisoquinoline below[50]:

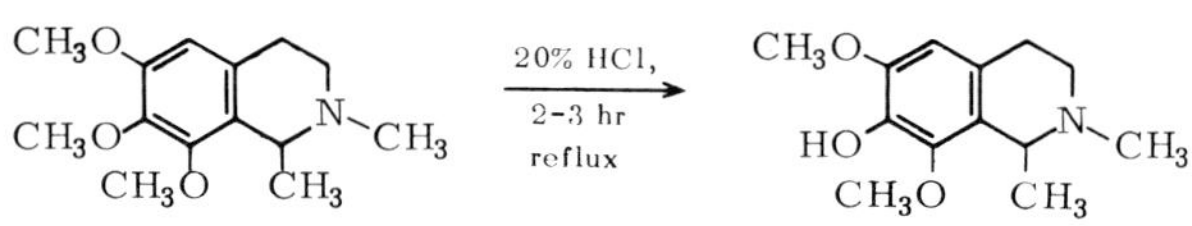

With a longer reflux time, both the C-7 and C-8 methoxyl groups of (±)-*O*-methylanhalonidine can be hydrolyzed, thus allowing a new synthesis of (±)-anhalonine and (±)-lophophorine (Scheme Va).[50b]

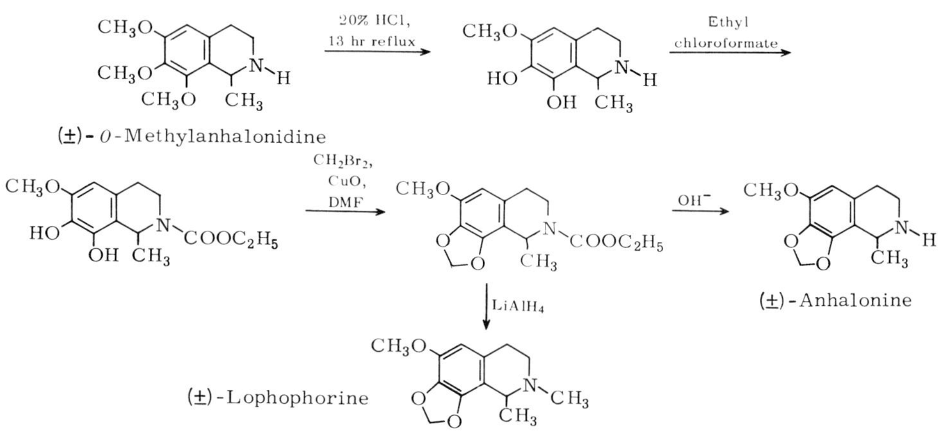

Scheme Va

A case of selective hydrolysis which was further extended to the synthesis of the benzylisoquinoline alkaloid petaline (see Chapter 2, Section VI, F) is the transformation of the simple tetrahydroisoquinoline **70** into a mixture of the monophenolic compounds **71**, **72**, and **73** as well as unreacted starting material (Scheme VI).[51] (For other cases of selective hydrolysis of methoxyl groups see Chapter 2, Section VI, F and Sections VII, B and L; and Chapter 10, Section V, C.)

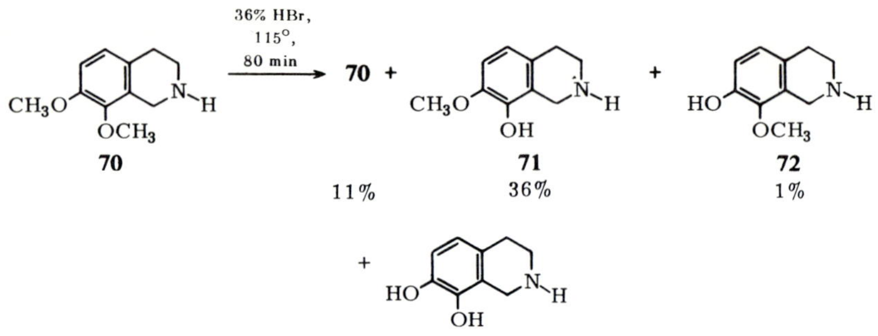

70 **71** **72**

11% 36% 1%

73

5%

Scheme VI

E. Phenolic Oxidative Coupling

Phenolic oxidative coupling has been recognized as an important pathway in biosynthesis, and the formation of pilocereine (see above) involves such oxidation.

When (±)-*N*-methylisosalsoline was oxidized in 3% aqueous ammonium acetate solution at pH 7 with crude horseradish peroxidase and hydrogen peroxide, the dimers **74** and **75** were obtained after methylation of the product with diazomethane.[52] But oxidative coupling of (±)-*N*-methylisosalsoline with 2 moles of potassium ferricyanide and aqueous sodium carbonate yielded only the dimer **74**.[53]

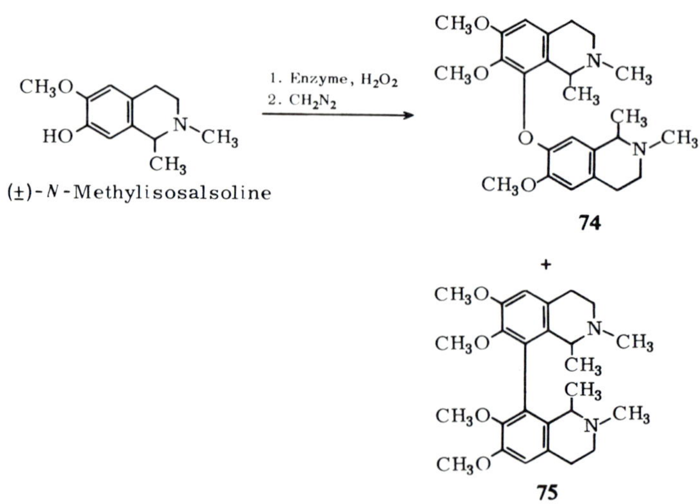

74

\+

75

In a separate study, it has been demonstrated that potassium ferricyanide oxidation of (±)-*N*-methylisosalsoline under somewhat different experimental conditions yields three racemic products, **75a**–**75c**, with **75b** and **75c** being rotational isomers. However, when coupling is induced by electrolytic oxidation, only **75b** is formed. The product

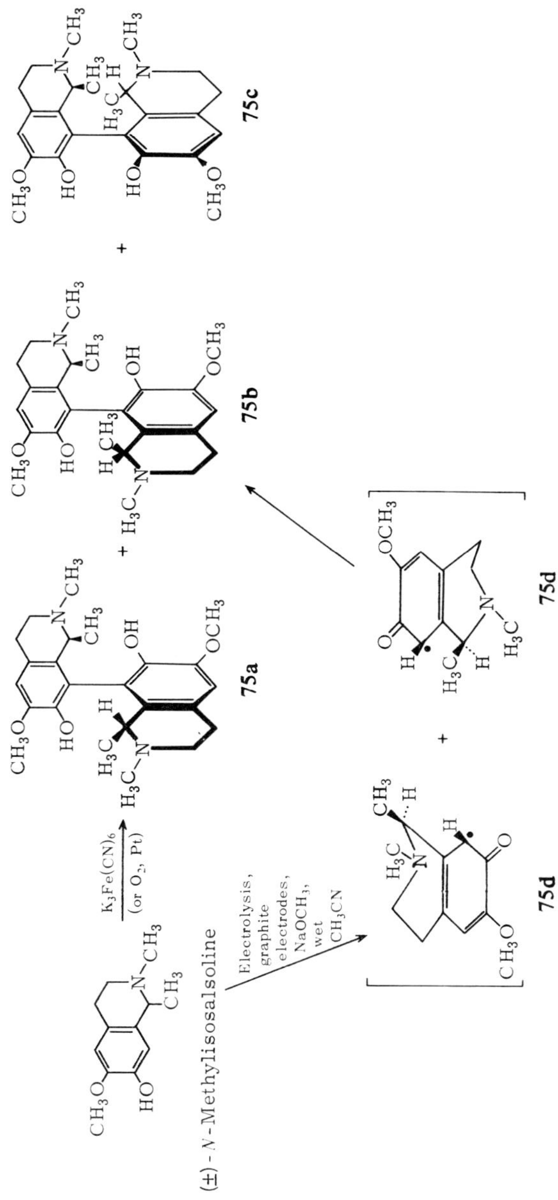
(±)-N-Methylisosalsoline
$K_3Fe(CN)_6$
(or O_2, Pt)
Electrolysis, graphite electrodes, $NaOCH_3$, wet CH_3CN
75a
75b
75c
75d
75d

75b results from the dimerization of two identical moieties, **75d**, on the graphite surface, so that this reaction is both stereoselective and stereospecific.[53a]

Oxidation of (±)-lophocerine with horseradish peroxidase and hydrogen peroxide followed by treatment with diazomethane afforded a diastereoisomeric mixture of the dimer **76**.[52]

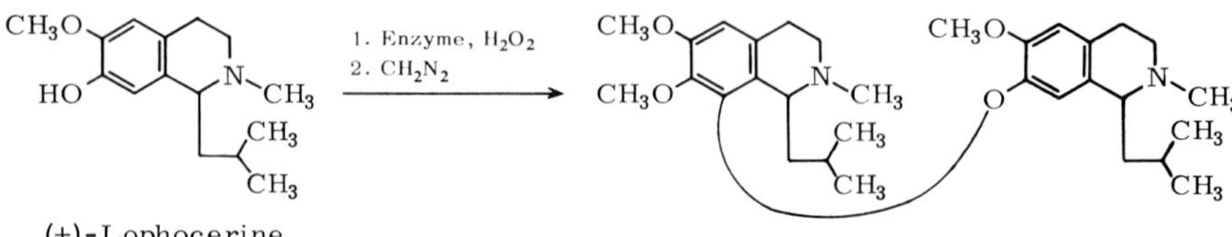

(±)-Lophocerine

76

Electrolytic or photochemical oxidation of the simple tetrahydroisoquinoline alkaloid corypalline has been shown to yield mainly the carbon–carbon dimer **77**.[54] The amount of carbon–oxygen coupling becomes significant only when a bulky substituent is present at C-1.[55]

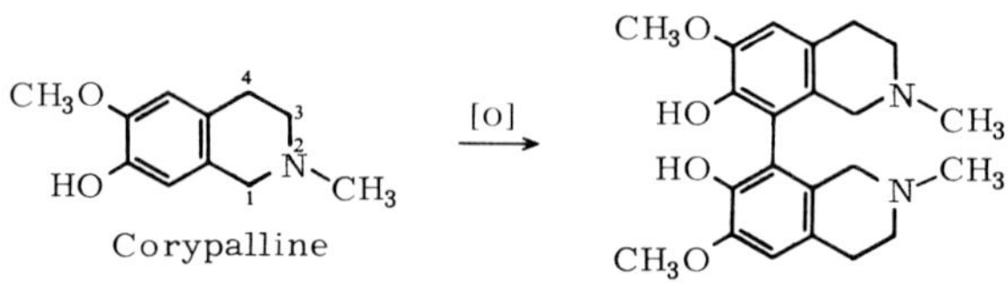

Corypalline

77

It is probable that phenolic coupling may proceed along several routes, the choice of the actual pathway depending largely upon the nature of the mechanism of action of the particular oxidizing agent being used. A complete kinetic study has been carried

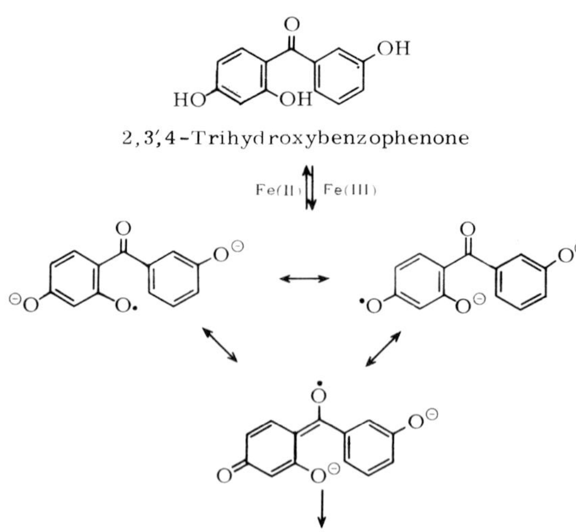

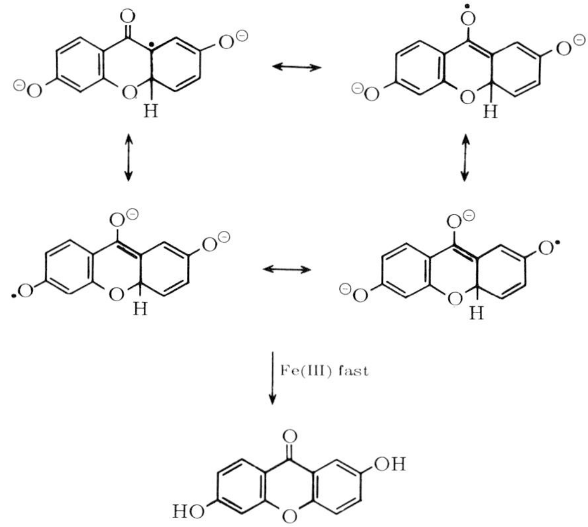

2,6-Dihydroxyxanthone

Scheme VII

out on the highly efficient intramolecular coupling of 2,3′,4-trihydroxybenzophenone to 2,6-dihydroxyxanthone using potassium ferricyanide in aqueous sodium hydroxide.[55a] The oxidation in this case is first order in ferricyanide and is inhibited by the presence of ferrocyanide. The oxidation is also first order in the benzophenone, and a large increase in reaction rate was observed with increasing ionic strength. Specific cations catalyzed the reaction, with $Na^{\oplus}$ being the least and $Cs^{\oplus}$ the most effective. The aromatic substitution mechanism of the type described in Scheme VII is the only process consistent with all the experimental facts (Scheme VII).

F. Acid-Catalyzed Rearrangement of 1-Substituted 1,2-Dihydroisoquinolines

See Chapter 2, Section VII, F.

G. Keto–Enol Tautomerism of Some 3,4-Dihydroisoquinolines

A study of the ultraviolet and fluorescence spectra of several 3,4-dihydroisoquinoline derivatives has shown that all such species investigated having a free hydroxyl group at position 6 undergo tautomerization above pH 4 or 5 to the quinonoidal form **78**, with maximum quinone formation around pH 8. Above pH 8, formation of the anion **79** is favored.[56]

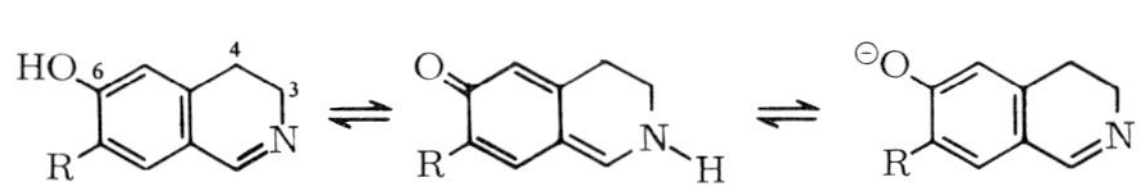

78 **79**

H. Cleavage of Tetrahydroisoquinoline p-Toluenesulfonamides

Secondary amines have sometimes been protected as their *p*-toluenesulfonamide derivatives in the course of alkaloid synthesis. In a number of cases, however, this protective group has proven difficult to remove. Using the *p*-toluenesulfonamides of tetrahydroisoquinolines and 1-benzyltetrahydroisoquinolines as examples, Umezawa and co-workers have shown that cleavage of the protective function can be efficiently performed using reductive conditions coupled with UV irradiation. Specifically, the sulfonamide **80** could be converted to the secondary amine **81** with a high-pressure mercury lamp in an ethanolic medium containing sodium borohydride and sodium carbonate.[57]

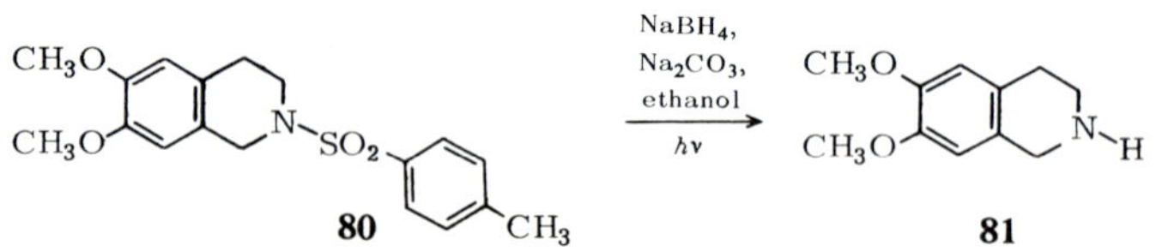

I. Quaternization

N-Alkylation of a C-1 substituted simple tetrahydroisoquinoline gives mostly the trans product with respect to the C-1 substituent, i.e., the major product has the entering alkyl group trans to the substituent at C-1. The illustrations that follow exemplify this point.[57a]

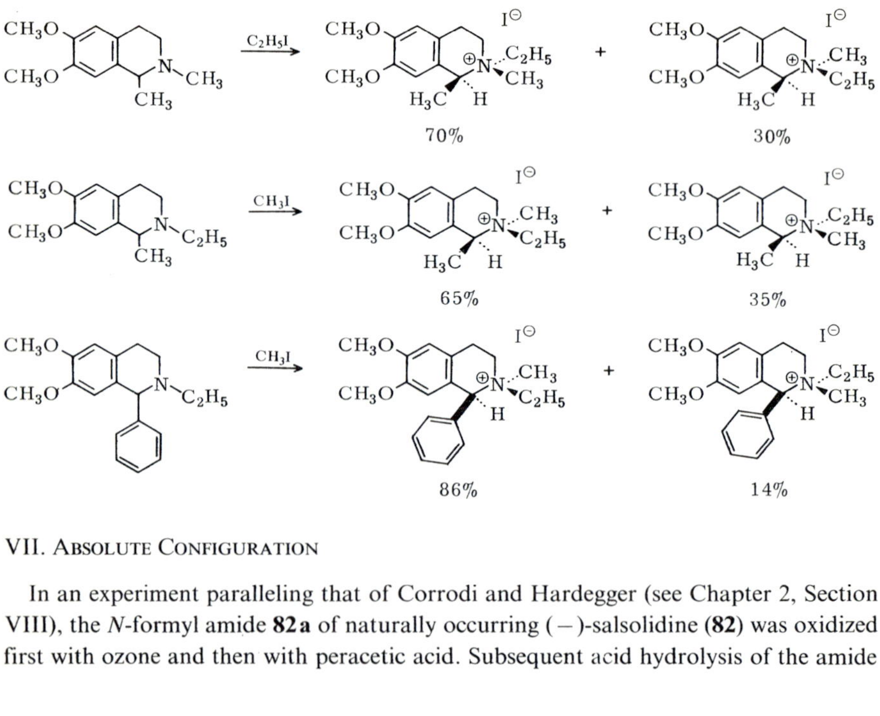

VII. Absolute Configuration

In an experiment paralleling that of Corrodi and Hardegger (see Chapter 2, Section VIII), the *N*-formyl amide **82a** of naturally occurring (−)-salsolidine (**82**) was oxidized first with ozone and then with peracetic acid. Subsequent acid hydrolysis of the amide function provided the previously unknown amino diacid **83** which was prepared from L-alanine as shown. It follows that the stereochemistry of (−)-salsolidine is as indicated in Scheme VIII.[58]

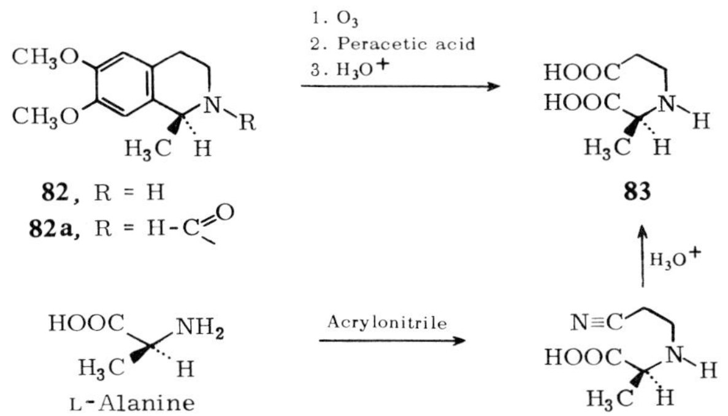

Scheme VIII

The chirality of naturally occurring (+)-calycotomine was elucidated by interrelation with (−)-salsolidine (**82**). Reaction of (+)-calycotomine with *p*-toluenesulfonyl chloride gave rise to the disulfonyl derivative **84**. Reduction with lithium aluminum hydride in boiling tetrahydrofuran generated the levorotatory *p*-toluenesulfonamide **85**. Since the corresponding *p*-toluenesulfonamide from (−)-salsolidine is dextrorotatory, it follows that (+)-calycotomine must have an absolute configuration opposite to that of (−)-salsolidine (**82**).[58]

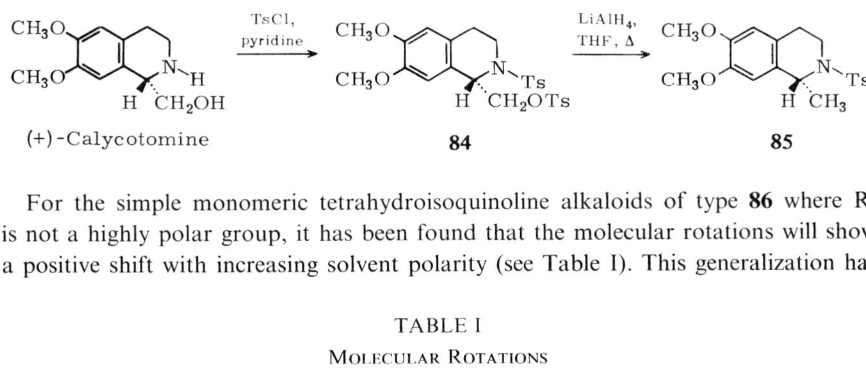

For the simple monomeric tetrahydroisoquinoline alkaloids of type **86** where R′ is not a highly polar group, it has been found that the molecular rotations will show a positive shift with increasing solvent polarity (see Table I). This generalization has

TABLE I

MOLECULAR ROTATIONS

	Increasing polarity →			
	C_6H_6	$CHCl_3$	C_2H_5OH	1 *N* HCl
(−)-Salsolidine	−133°	−	−123°	−57°
(−)-*N*-Ethylsalsolidine	+14°	+18°	+20°	+30°
(−)-Anhalonine	−	−124°	−	−108°
(−)-Lophorine	−	−111°	−	−44°

been used to determine the absolute configurations of the cactus alkaloids (−)-anhalonine and (−)-lophophorine, both of which exhibit such a shift. They, therefore, possess the same *S* configuration as (−)-salsolidine (**82**).[58]

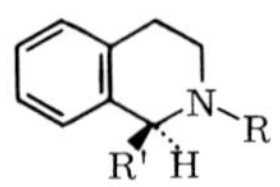

86

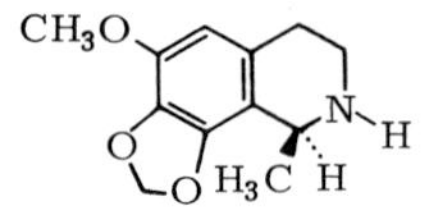

(-)-Anhalonine

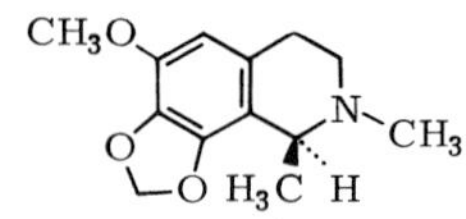

(-)-Lophophorine

An X-ray analysis of (−)-anhalonine hydrobromide and (+)-*O*-methylanhalonidine hydrobromide, besides confirming the absolute configuration of the former salt, revealed that the latter also belongs to the *S* configuration.[50a] The method of molecular rotational differences is inconclusive for (+)-*O*-methylanhalonidine since this minor constituent of the peyote cactus exhibits the following $[\alpha]_D^{25}$ values: $CHCl_3$, +20.6°; CH_3OH, +11.5°; and 1 *N* HCl, +19.3°. The X-ray data were also useful in settling bond lengths, bond angles, and conformations.[50a]

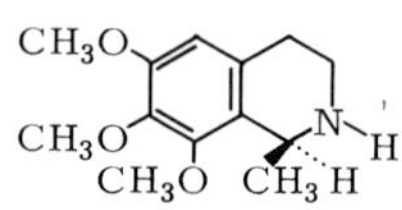

(+)-*O*-Methylanhalonidine

VIII. Biosynthesis

A substantial amount of work has been carried out using labeled precursors to determine the biosynthetic pathway for the formation of mescaline and the simple tetrahydroisoquinolines. Evidence is accumulating that in the Cactaceae, dopamine originating from tyrosine can be *O*-methylated to 4-hydroxy-3-methoxyphenylethylamine which then undergoes oxidation to 4,5-dihydroxy-3-methoxyphenylethylamine (**87**).[59,60] The position of further *O*-methylation of this intermediate is all important in determining the biogenetic fate of the molecule. If *O*-methylation proceeds at the C-4 hydroxyl group, species **88** is formed, which subsequently gives rise to such tetrahydroisoquinolines as anhalonidine, pellotine, anhalamine, and anhalidine. If, on the other hand, *O*-methylation of the diphenol **87** is at the C-5 hydroxyl, then the amine **89** is generated which is efficiently further *O*-methylated to mescaline.[61−63] It should be pointed out that Mannich condensation to form anhalonidine, pellotine, anhalamine, and anhalidine is favored by the presence of the activating phenolic function in **88** ortho to the condensation site (Scheme IX).

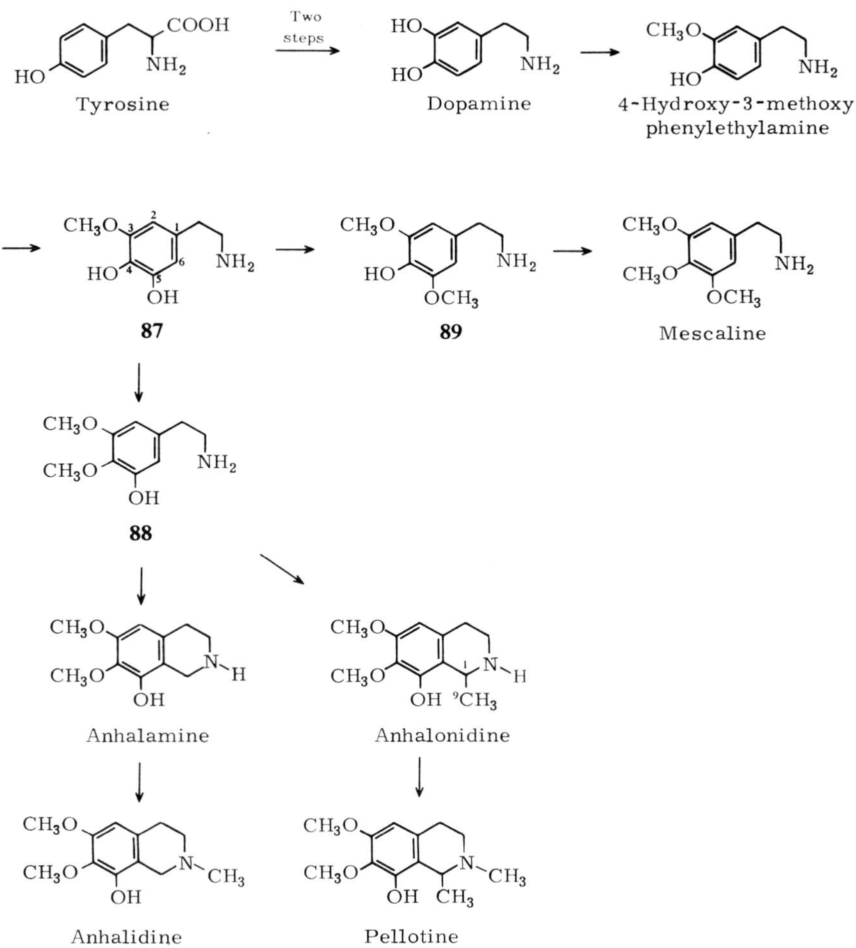

Scheme IX

It first seemed reasonable to assume that the two-carbon unit (C-1 and C-9) in the Cactaceae alkaloids anhalonidine and pellotine was derived from acetic acid. However, administration of acetic acid-1-^{14}C to the peyote cactus, *Lophophora wi''iamsii* (Lemaire) Coult., yielded pellotine with the activity equally divided between C-1 and C-9, suggesting that acetic acid is not a direct precursor ot the two-carbon unit.[59,64,65]

Labeled sodium pyruvate was therefore fed to the peyote cactus, and a relatively high specific incorporation of the C-3 of pyruvate into the C-9 of anhalonidine was found. Consequently, it was suggested that tetrahydroisoquinoline alkaloids having a methyl group at C-1 are formed through the intermediacy of acetyl coenzyme A and the *N*-acetylphenethyl amine **90**.[65]

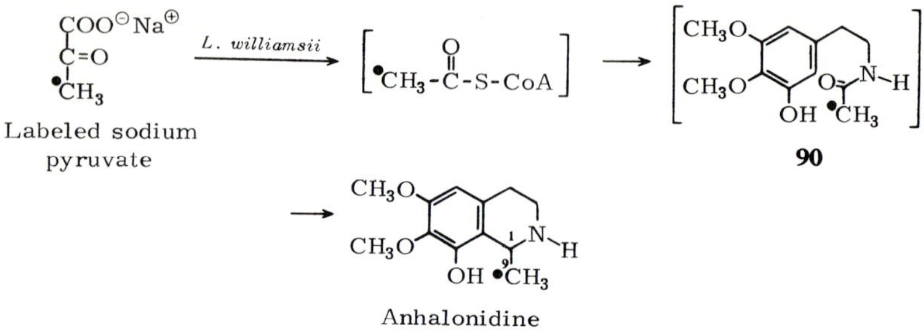

More work is required to establish rigorously the precise biogenetic route involved in the construction of ring B of the various simple tetrahydroisoquinolines. In particular, the possibility of an α-keto acid being the source of the C-1 atom according to the scheme outlined below must not be discounted.[1,5]

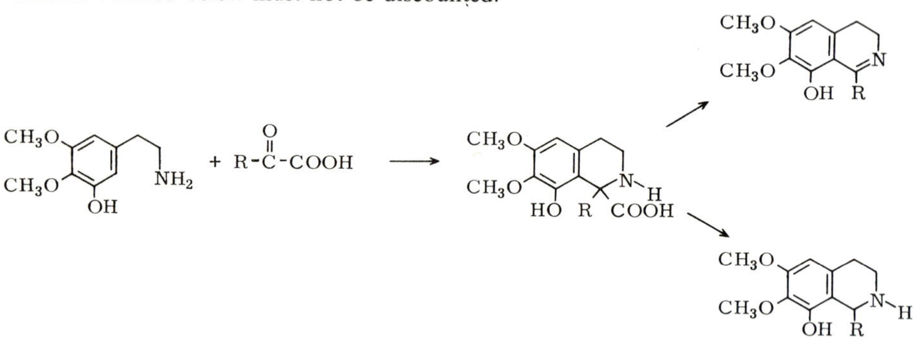

The biosynthesis of lophocerine in the cactus *Lophocereus schottii* (Engelm.) Britt. et Rose (Cactaceae) using labeled precursors has also been studied. Racemic leucine or racemic mevalonic acid lactone can act as precursors for the isoprenoid moiety of the alkaloid while racemic tyrosine can act as the precursor for the phenylethylamine portion of the tetrahydroisoquinoline nucleus.[1b]

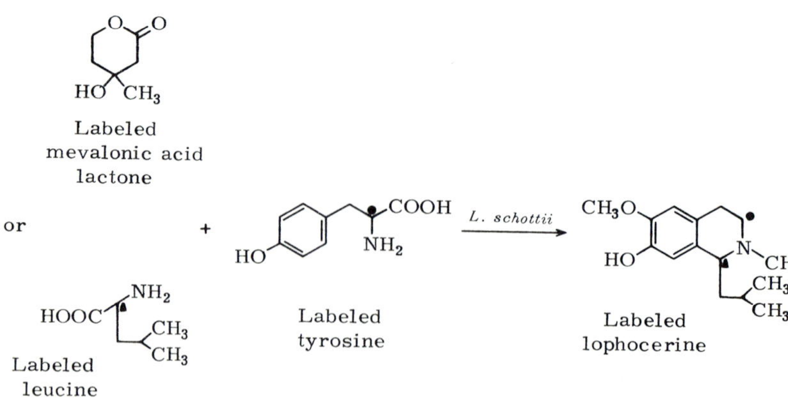

When L-[methyl-^{14}C]methionine was injected into *L. schottii*, lophocerine labeled on the *O*- and *N*-methyl groups was obtained. And when [*N*-methyl-^{14}C]lophocerine was fed to the plant, the pilocereine obtained through phenolic oxidative coupling was labeled on the *N*-methyl groups.[66]

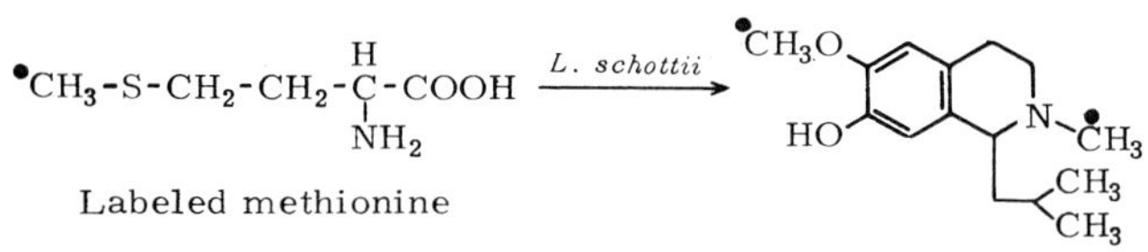

Labeled methionine

Labeled lophocerine

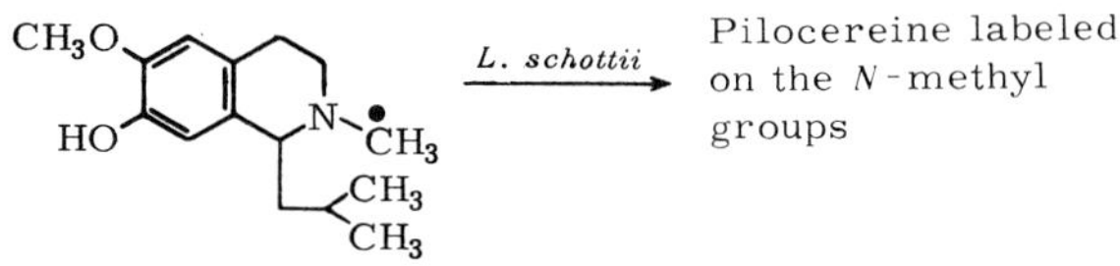

Labeled lophocerine

The isolation of Krebs cycle conjugates of mescaline from the peyote cactus, discussed in Section III, gives further insight into the biogenesis of the tetrahydroisoquinolines.[5,6,9]

IX. Pharmacology

A tetrahydroisoquinoline that has been introduced commercially is debrisoquine (Declinax, Hoffmann–La Roche), which contains a guanidine residue and has hypotensive activity. Analogs of debrisoquine that are also active are **91**, **92**, and **93**.[67]

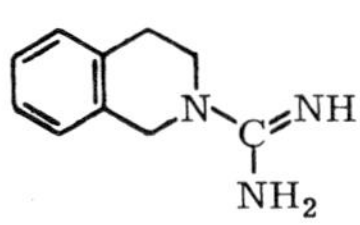

Debrisoquine

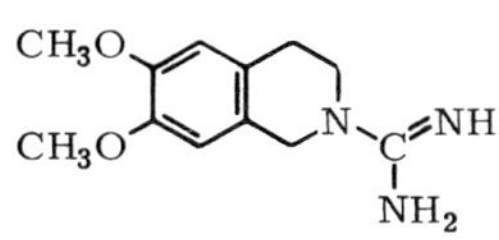

91

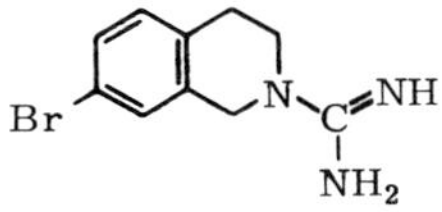

92

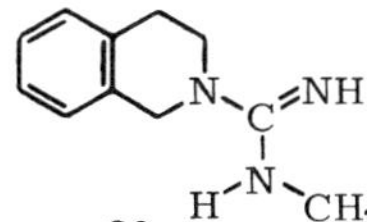

93

Simple tetrahydroisoquinolines have been shown to have both central and peripheral pharmacological effects.[68] They may have central nervous system depressant or stimulant and convulsant properties.[68] They can also cause vasopressor or depressor actions, various effects on the smooth muscle,[68] and lipid-mobilizing activity.[69] The saguaro alkaloid gigantine caused hallucinogenic reactions when tested on squirrel monkeys and cats.[35]

Anhalamine, anhalonidine, anhalidine, and pellotine have little activity as anticonvulsants, tranquilizers or muscle relaxants, and possess no significant hallucinogenic action.[9a]

Bisobrin is a synthetic bistetrahydroisoquinoline that has shown promising clinical use as a fibrinolytic agent. Fibrin forms the frame-work or support of a clot, and dissolution of the fibrin should result in lysis of the clot and restoration of the blood flow.[70]

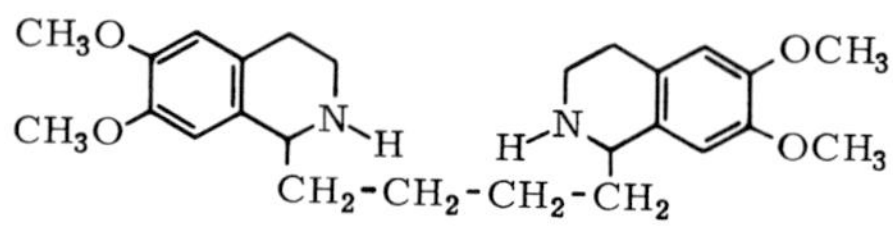

Bisobrin

The synthetic bisonium salt Hedaquinium chloride is used as an external antifungal agent,[70a] while Lodal is an astringent.[70b]

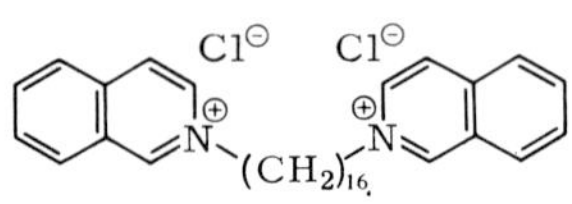

Hedaquinium chloride
(Allen and Hanburys Ltd.)

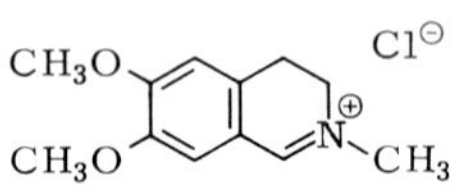

Lodal

X. The Possible Role of Tetrahydroisoquinoline Alkaloids in Alcoholism

Cow adrenal glands are known to be rich in epinephrine and norepinephrine. When these glands were perfused with dilute solutions of acetaldehyde – a metabolic product of ethanol – at 37°, Mannich condensation readily occurred to form tetrahydroisoquinolines[71]:

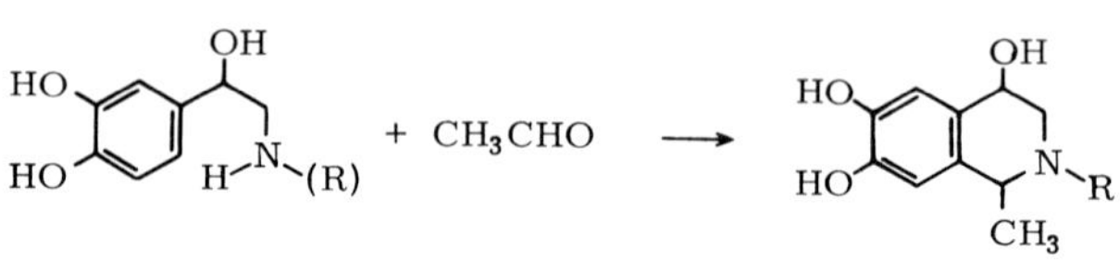

Epinephrine, R = CH_3
Norepinephrine, R = H

R = CH_3 or H

Furthermore, using rat brain stem homogenates, it was found that incubation mixture of dopamine and acetaldehyde, together with the tissue homogenate, gave salsolinol. Small amounts of salsolinol were also formed when the acetaldehyde was replaced by ethanol, so that an alcohol dehydrogenase (ADH) is presumably present in the homogenate[71a]:

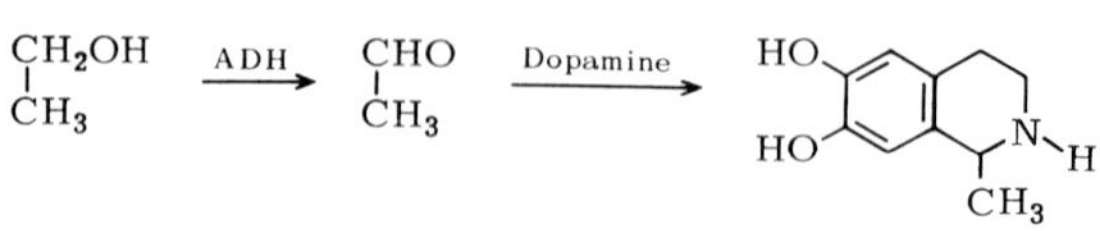

Salsolinol

Tetrahydroisoquinolines formed as a result of alcohol ingestion may, therefore, account for the hyperexcitability, tremulousness, hallucinosis and seizures which are known to occur in some alcoholics when the concentration of alcohol in the blood is falling or is minimal.[71,71a]

(For a related theory of alcohol addiction together with a refutation of this theory, see Chapter 2, Section XI.)

XI. Mass Spectroscopy

A perennial problem in the interpretation of the mass spectra of simple tetrahydroisoquinoline alkaloids is the determination of which peak represents the molecular ion. *O*-Methylpellotine, for example, upon electron impact loses its C-1 substituent very readily so that the base peak is at *m/e* 236. The molecular ion peak is a weak signal at *m/e* 251.

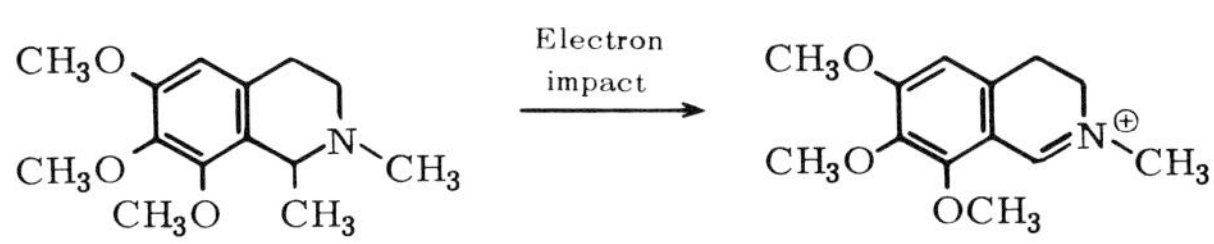

O-Methylpellotine, $M^{\oplus}$ *m/e* 251 — *m/e* 236

To locate the molecular ion, it is desirable to run a chemical ionization mass spectrum of the compound. In this procedure, a mixture of a reactant gas (>99%) and sample (<1%) is submitted to electron bombardment at pressures of about 1 mm. The reactant gas, usually methane, will be extensively ionized to produce such primary ions as CH_4^+, CH_3^+, CH_2^+, and CH^+, and these, because of ion–molecule reactions with methane, will generate the secondary ions CH_5^+, $C_2H_5{}^+$, and $C_2H_3^+$. When the secondary ions collide with a sample molecule, they donate to it a proton or abstract a hydride ion, thus generating quasimolecular ions at *m/e* $(M + 1)^+$ or $(M - 1)^+$, respectively. The important point here is that the sample ions produced will not be energy rich and will possess a relatively long lifetime.[72]

In the chemical ionization mass spectrum of *O*-methylpellotine, the quasi-molecular ion at *m/e* 252 (59%) is accompanied by ions at *m/e* 250 (98%) and 236 (100% base). The peaks at *m/e* 250 and 236 reflect the equally facile cleavage of the C-1 hydrogen and the C-1 methyl group. Chemical ionization can be carried out with a normal electron bombardment source that has been adapted to accommodate the high pumping rate required.

The mass spectra of **42a** and **42b** were quite similar and exhibited a molecular ion at *m/e* 237 which was only 2–3% of the base peak at *m/e* 222. Other intense peaks were at *m/e* 204, 194, and 179.[36]

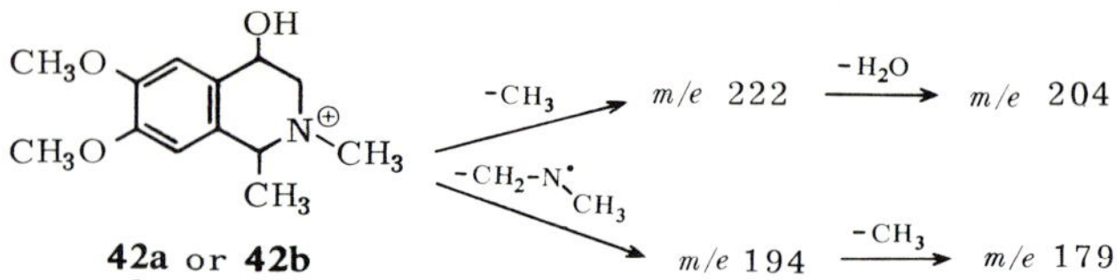

42a or **42b**
$M^{\oplus}$, *m/e* 237

XII. NMR Spectroscopy

The subject of the NMR spectra for simple tetrahydroisoquinolines has already been partially considered in connection with the gigantine problem. Another study has been concerned with the spectra of the 4-hydroxyisoquinolines **94**–**97**. The protons at C-4 consistently appear as asymmetric triplets in the region δ4.8–5.1. For species **94** and **95**, the C-1 protons appear as singlets between δ4.2 and 4.4. However, the situation is somewhat different with compounds **96** and **97**, which are substituted at C-8. The C-1 protons are now nonequivalent and form AB patterns centered at δ4.3.[21]

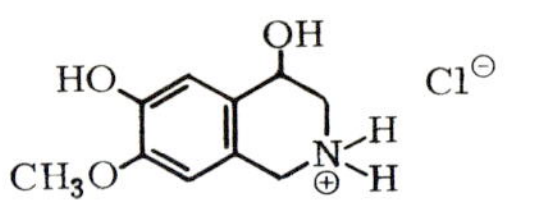

94

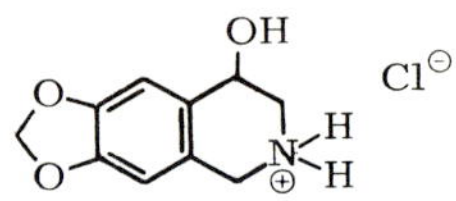

95

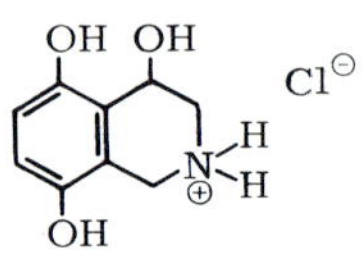

96

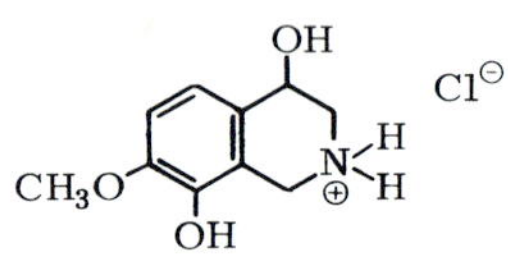

97

An investigation of the NMR spectra of simple tetrahydroisoquinoline hydrochlorides monosubstituted in the aromatic ring, using DMSO-d_6 as solvent, has demonstrated that C-5 and C-8 substituted derivatives can be distinguished from their C-6 or C-7 substituted counterparts by the multiplicity of the signals of the aromatic protons: the C-5 and C-8 oxygenated compounds show a triplet and two doublets, whereas the C-6 or C-7 oxygenated derivatives give rise to a doublet of doublets and two doublets. The aromatic substitution patterns may be further distinguished by the magnitude of the difference between the chemical shifts of the C-1 and C-4 methylene groups. The four examples that follow illustrate some of these points.[72a]

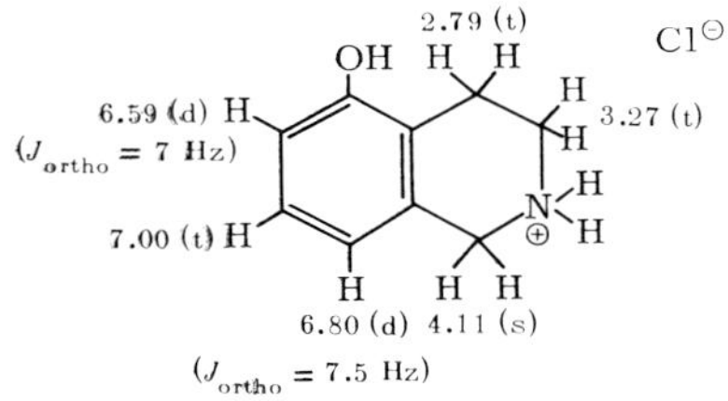

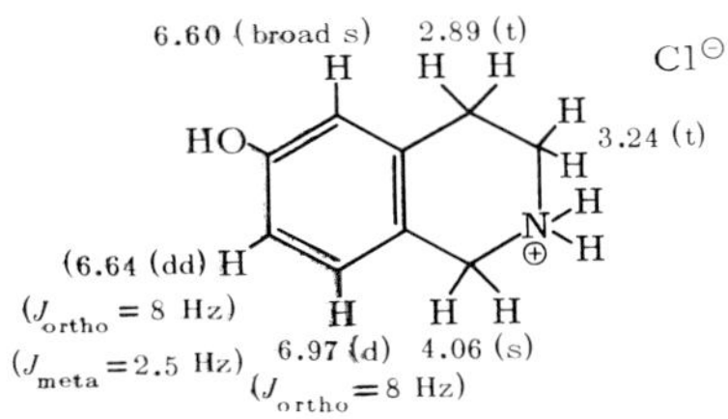

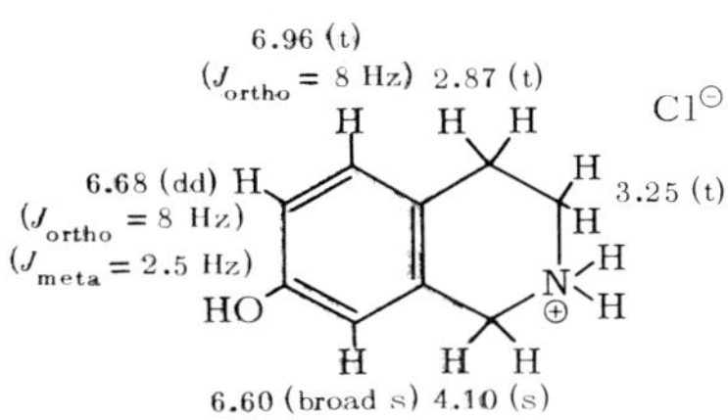

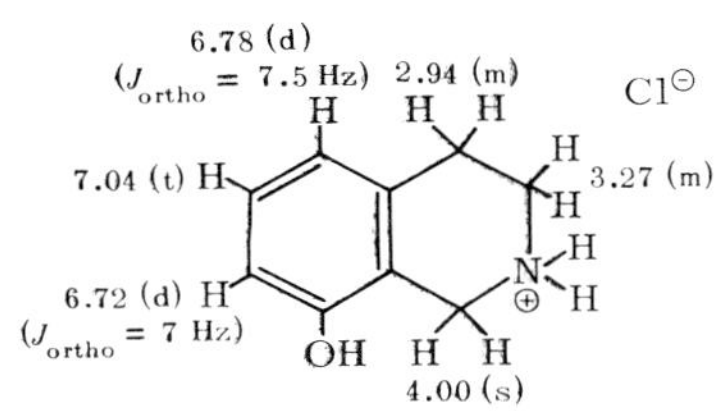

The NMR spectra of *N*-acetylated and C-1-substituted *N*-acetylated 1,2,3,4-tetrahydroisoquinolines clearly show two acetyl resonances of different intensities, pointing to the presence of unequal populations of conformers **98** and **99** at room temperature. This phenomenon is due to the restricted rotation of the *N*-acetyl group. It has been estimated that when R is benzyl, the activation energy for *N*-acetyl rotation may be as high as 20.6 kcal/mole.[73,74]

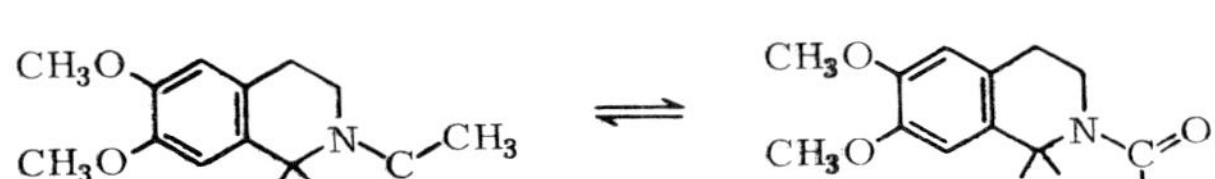

98 **99**

XIII. UV Spectroscopy

Simple tetrahydroisoquinolines substituted at C-6 and C-7 show maxima near 284 mμ. Replacement of methoxyl and hydroxyl substituents by a methylenedioxy group results in a bathochromic shift and an increase in the extinction coefficient, e.g., pellotine hydrochloride vs. lophophorine hydrochloride.

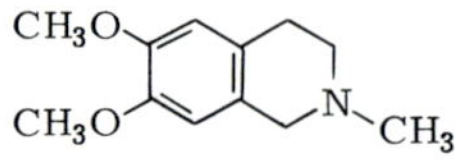

6,7-Dimethoxy-2-methyl-tetrahydroisoquinoline[75] (*O*-Methylcorypalline)

λ_{max}^{EtOH} 235 sh and 285 mμ (3.83 and 3.59)

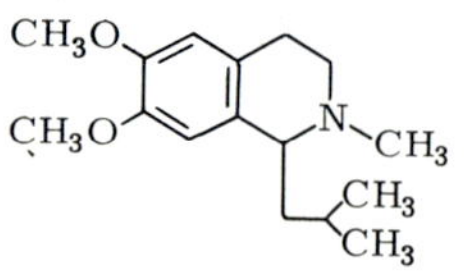

O-Methyllophocerine[76]

λ_{max}^{EtOH} 285 mμ (3.59)

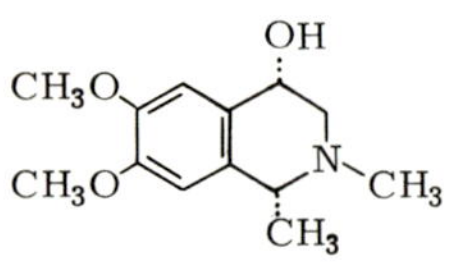

Synthetic compound **42a**[36]

$\lambda_{max}^{2\text{-Propanol}}$ 227–228 sh, 282–283, and 298 sh mμ (3.99, 3.55, and 3.50)

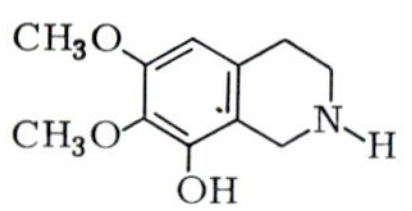

Anhalamine[11]

λ_{max}^{EtOH} 227 sh, 272, and 280 sh mμ (3.98, 2.90, and 2.84)

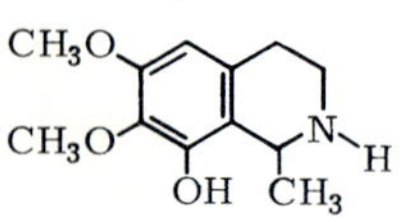

Anhalonidine[11]

λ_{max}^{EtOH} 270 and 278sh mμ (2.87 and 2.81)

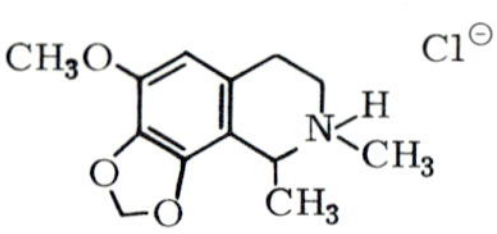

Lophophorine hydrochloride[77]

λ_{max}^{MeOH} 278 mμ (3.01)

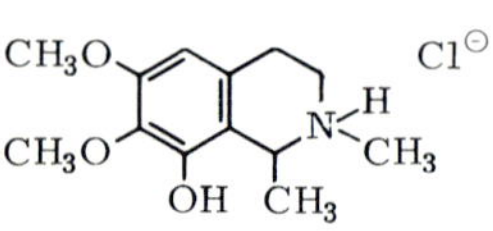

Pellotine hydrochloride[78]

λ_{max}^{MeOH} 272 mμ (2.86)

REFERENCES

1. For a comprehensive review on the peyote alkaloids, see G. J. Kapadia and M. B. E. Fayez, *J. Pharm. Sci.* **59**, 1699 (1970).
1a. Alkaloids of the Cactaceae have been discussed by S. Agurell, *Lloydia* **32**, 206 (1969).
1b. D. G. O'Donovan and H. Horan, *J. Chem. Soc., C* p. 2791 (1968).
2. C. Djerassi, H. W. Brewer, C. Clarke, and L. J. Durham, *J. Amer. Chem. Soc.* **84**, 3210 (1962).
3. J. M. Bobbitt, R. Ebermann, and M. Schubert, *Tetrahedron Lett.* p. 575 (1963).
4. For a slightly different approach, see B. Franck, G. Blaschke, and K. Lewejohann, *Justus Liebigs Ann. Chem.* **685**, 207 (1965).
5. G. J. Kapadia and H. M. Fales, *Chem. Commun.* p. 1688 (1968).
6. G. J. Kapadia and R. J. Highet, *J. Pharm. Sci.* **57**, 191 (1968).
7. G. J. Kapadia, M. B. E. Fayez, Y.N. Vaishnav, H. M. Fales, and G. S. Rao, *Lloydia* **32**, 525 (1969).
8. For a review of the Bischler–Napieralski cyclization, see W. M. Whaley and T. R. Govindachari, *Org. React.* **6**, 74 (1951).
9. E. Späth, *Monatsh. Chem.* **42**, 97 (1921).
9a. A Brossi, F. Schenker, R. Schmidt, R. Banziger, and W. Leimgruber, *Helv. Chim. Acta* **49**, 403 (1966).
10. N. Itoh and S. Sugasawa, *Tetrahedron* **6**, 16 (1959).
11. A. Brossi, F. Schenker, and W. Leimgruber, *Helv. Chim. Acta* **47**, 2089 (1964).
12. For a review of the Pictet–Spengler synthesis of tetrahydroisoquinolines, see W. M. Whaley and T. R. Govindachari, *Org. React.* **6**, 151 (1951).
13. C. Schöpf and H. Bayerle, *Justus Liebigs Ann. Chem.* **513**, 190 (1934).
14. O. M. Friedman, K. N. Parameswaran, and S. Burstein, *J. Med. Chem.* **6**, 227 (1963).
15. T. Kametani, K. Fukumoto, H. Agui, H. Yagi, K. Kigasawa, H. Sugahara, M. Hiiragi, T. Hayasaka, and H. Ishimaru, *J. Chem. Soc., C* p. 112 (1968).
16. For a review of the Pomeranz–Fritsch cyclization, see W. J. Gensler, *Org. React.* **6**, 191 (1951).
17. J. M. Bobbitt, J. M. Kiely, K. L. Khanna, and R. Ebermann, *J. Org. Chem.* **30**, 2247 (1965).
18. J. M. Bobbitt, D. N. Roy, A. Marchand, and C. W. Allen, *J. Org. Chem.* **32**, 2225 (1967).
19. J. M. Bobbitt, A. S. Steinfeld, K. H. Weisgraber, and S. Dutta, *J. Org. Chem.* **34**, 2478 (1969).
20. J. M. Bobbitt and C. P. Dutta, *J. Org. Chem.* **34**, 2001 (1969).
21. J. M. Bobbitt and J. C. Sih, *J. Org. Chem.* **33**, 856 (1968).
22. D. W. Brown, S. F. Dyke, and M. Sainsbury, *Tetrahedron* **25**, 101 (1969).
23. G. Grethe, H. L. Lee, M. Uskoković, and A. Brossi, *J. Org. Chem.* **33**, 491 and 494 (1968).
24. For a related synthesis, see B. Umezawa, O. Hoshino, and Y. Terayama, *Chem. Pharm. Bull.* **16**, 180 (1968).
25. I. G. Hinton and F. G. Mann, *J. Chem. Soc., London* p. 599 (1959).
26. B. Jaques, R. H. L. Deeks, and P. K. J. Shah, *Chem. Commun.* p. 1283 (1969).
27. D. N. Harcourt and R. D. Waigh, *Chem. Commun.* p. 692 (1968); *J. Chem. Soc., C* p. 967 (1971).
28. P. A. Bather, J. R. L. Smith, R. O. C. Norman, and J. S. Sadd, *Chem. Commun.* p. 1116 (1969).
28a. J. R. L. Smith, R. O. C. Norman, and A. G. Rowley, *Chem. Commun.* p. 1238 (1970).
29. D. M. Ziegler and F. H. Pettit, *Biochem. Biophys. Res. Commun.* **15**, 188 (1964).
30. J. P. Ferris, R. D. Gerwe, and G. R. Gapski, *J. Amer. Chem. Soc.* **89**, 5270 (1967).
31. A. R. Battersby, *Proc. Chem. Soc., London* p. 189 (1963).
32. J. Finkelstein and A. Brossi, *J. Heterocycl. Chem.* **4**, 315 (1967).
33. G. Rosen and F. D. Popp, *Can. J. Chem.* **47**, 864 (1969).
34. S. Yamada and H. Akimoto, *Tetrahedron Lett.* p. 3105 (1969).
34a. L. W. Deady, N. Pirzada, and R. D. Topsom, *Chem. Commun.* p. 799 (1971).

35. J. E. Hodgkins, S. D. Brown, and J. L. Massingill, *Tetrahedron Lett.* p. 1321 (1967).
36. G. Grethe, M. Uskoković, T. Williams, and A. Brossi, *Helv. Chim. Acta* **50**, 2397 (1967).
37. G. J. Kapadia, M. B. E. Fayez, M. L. Smith, and G. S. Rao, *Chem. Commun.* p. 856 (1970).
38. R. J. Highet and P. F. Highet, *J. Org. Chem.* **30**, 902 (1965).
39. G. J. Kapadia, G. S. Rao, M. B. E. Fayez, B. K. Chowdhury, and M. L. Sethi, *Chem. Ind. (London)* 1593 (1970). For a different synthetic approach see A. M. Choudhury, *Chem. Ind. (London)* p. 578 (1971).
40. E. Bamberger, *Justus Liebigs Ann. Chem.* **390**, 164 (1912).
41. S. Goodwin and B. Witkop, *J. Amer. Chem. Soc.* **79**, 179 (1957).
42. B. Umezawa, O. Hoshino, and Y. Yamanashi, *Tetrahedron Lett.* p. 933 (1969).
43. O. Hoshino, Y. Yamanashi, and B. Umezawa, *Tetrahedron Lett.* p. 937 (1969). See also B. Umezawa, O. Hoshino, Y. Terayama, K. Ohyama, Y. Yamanashi, T. Inoue, and T. Toshioka, *Chem. Pharm. Bull.* **19**, 2138 (1971), and immediately succeeding papers.
44. R. Grewe, W. Krüger, and E. Vangermain, *Chem. Ber.* **97**, 119 (1964).
45. W. J. Gensler, K. T. Shamasundar, and S. Marburg, *J. Org. Chem.* **33**, 2861 (1968).
46. A. Brossi, M. Baumann, and R. Borer, *Monatsh. Chem.* **96**, 25 (1965).
46a. A. Brossi and S. Teitel, *Helv. Chim. Acta* **53**, 1779 (1970).
47. H. Bruderer and A. Brossi, *Helv. Chim. Acta* **48**, 1945 (1965).
47a. A. Brossi, J. O'Brien, and S. Teitel, *Org. Prep. Proc.* **2**, 281 (1970).
48. G. Grethe, V. Toomé, H. L. Lee, M. Uskoković, and A. Brossi, *J. Org. Chem.* **33**, 504 (1968).
49. A. Brossi, F. Schenker, and W. Leimgruber, *Helv. Chim. Acta* **47**, 2089 (1964).
50. A. Brossi, J. Van Burik, and S. Teitel, *Helv. Chim. Acta* **51**, 1965 (1968); A. Brossi and S. Teitel, *Chem. Commun.* p. 1296 (1970).
50a. C. F. Wilcox, Jr. and M. A. Seager, *J. Org. Chem.* **34**, 2319 (1969); O. Yonemitsu, H. Nakai, Y. Kanoaka, I. L. Karle, and B. Witkop, *J. Amer. Chem. Soc.* **91**, 4591 (1969); R. G. Wilson and D. H. Williams, *J. Chem. Soc., C* p. 2475 (1968).
50b. A. Brossi, J. F. Blount, J. O'Brien, and S. Teitel, *J. Amer. Chem. Soc.* **93**, 6248 (1971).
51. G. Grethe, H. L. Lee, M. R. Uskoković, and A. Brossi, *Helv. Chim. Acta* **53**, 874 (1970).
52. Y. Inubushi, Y. Aoyagi, and M. Matsuo, *Tetrahedron Lett.* p. 2363 (1969).
53. O. Hoshio, H. Hara, M. Wada, and B. Umezawa, *Chem. Pharm. Bull.* **18**, 637 (1970).
53a. J. M. Bobbitt, I. Noguchi, H. Yagi, and K. H. Weisgraber, *J. Amer. Chem. Soc.* **93**, 3551 (1971); and J. M. Bobbitt, H. Yagi, S. Shibuya, and J. T. Stock, *J. Org. Chem.* **36**, 3006 (1971).
54. J. M. Bobbitt, J. T. Stock, A. Marchand, and K. H. Weisgraber, *Chem. Ind.* (*London*) p. 2127 (1966).
55. J. M. Bobbitt, K. H. Weisgraber, A. S. Steinfeld, and S. G. Weiss, *J. Org. Chem.* **35**, 2884 (1970).
55a. G. A. Hamilton and P. D. McDonald, private communication (1970); for a review on oxidative coupling see H. Musso, *in* "Oxidative Coupling of Phenols" (W. I. Taylor and A. R. Battersby, eds.), p. 1. Marcel Dekker, New York, 1967.
56. G. Jonsson, *Acta Chem. Scand.* **20**, 2755 (1966).
57. B. Umezawa, O. Hoshino, and S. Sawaki, *Chem. Pharm. Bull.* **17**, 1120 (1969).
57a. G. Bernáth, J. Kóbor, K. Koczka, L. Radics, and M. Kajtár, *Tetrahedron Lett.* p. 225 (1968).
58. A. R. Battersby and T. P. Edwards, *J. Chem. Soc. London* p. 1214 (1960).
59. A. R. Battersby, R. Binks, and R. Huxtable, *Tetrahedron Lett.* p. 563 (1967).
60. E. Leete, *J. Amer. Chem. Soc.* **88**, 4218 (1966).
61. A. G. Paul, K. L. Khanna, H. Rosenberg, and M. Takido, *Chem. Commun.* p. 838 (1969).
62. J. Lundström and S. Agurell, *Tetrahedron Lett.* p. 3371 (1969).
63. K. L. Khanna, M. Takido, H. Rosenberg, and A. G. Paul, *Lloydia* **32**, 524 (1969); *Phytochemistry* **9**, 1811 (1970).
64. A. R. Battersby, R. Binks, and R. Huxtable, *Tetrahedron Lett.* p. 6111 (1968).

65. E. Leete and J. D. Braunstein, *Tetrahedron Lett.* p. 451 (1969).
66. D. G. O'Donovan and H. Horan, *J. Chem. Soc., C* p. 1737 (1969).
67. O. Schier and A. Marxer, *Progr. Drug Res.* **13**, 112 (1969).
68. A. M. Hjort, E. J. de Beer, and D. W. Fassett, *J. Pharmacol. Exp. Ther.* **62**, 195 (1938); D. W. Fassett and A. M. Hjort, *ibid.* **63**, 253 (1938); A. M. Hjort, E. J. de Beer, J. S. Buck, and L. O. Randall, *ibid.* **76**, 263 (1942).
69. E. Tóth, G. Fassina, and E. S. Soncin, *Arch. Int. Pharmacodyn. Ther.* **169**, 375 (1967).
70. J. M. Schorr, ed., "Chemical Control of Fibrinolysis-Thrombolysis." Wiley, New York, 1970
70a. "The Merck Index," 8th ed., p. 518 (1968).
70b. "The Merck Index," 8th ed., p. 625 (1968).
71. G. Cohen and M. Collins, *Science* **167**, 1749 (1970).
71a. Y. Yamanaka, M. J. Walsh, and V. E. Davis, *Nature* (*London*) **227**, 1143 (1970).
72. H. M. Fales, G. W. A. Milne, and M. L. Vestal, *J. Amer. Chem. Soc.* **91**, 3682 (1969); H. M. Fales, H. A. Lloyd, and G. W. A. Milne, *ibid.* **92**, 1590 (1970).
72a. F. Schenker, R. Schmidt, T. Williams, and A. Brossi, (in press) (1971).
73. G. Fraenkel, M. P. Cava, and D. R. Dalton, *J. Amer. Chem. Soc.* **89**, 329 (1967).
74. D. R. Dalton, K. C. Ramey, H. J. Gisler, Jr., L. J. Lendvay, and A. Abraham, *J. Amer. Chem. Soc.* **91**, 6367 (1969).
75. J. Knabe, *Arch. Pharm.* (*Weinheim*) **292**, 652 (1959).
76. C. Djerassi, J. J. Beereboom, S. P. Marfey, and S. K. Figdor, *J. Amer. Chem. Soc.* **77**, 484 (1955).
77. Holubek and Štrouf, Spectrum No. 451.
78. Holubek and Štrouf, Spectrum No. 464.

Chapter 2 / THE BENZYLISOQUINOLINES

Occurrence: Anonaceae, Berberidaceae, Combretaceae, Hernandiaceae, Lauraceae, Magnoliaceae, Menispermaceae, Monimiaceae, Nymphaceae, Papaveraceae, Ranunculaceae, Rhamnaceae, and Rutaceae

Approximate Number: 32

Some Benzylisoquinolines of Interest:

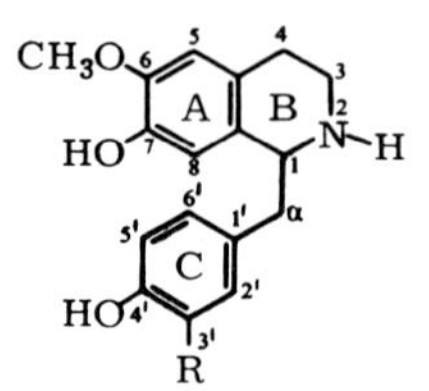

(±)-Coclaurine, R = H

N-Nororientaline,[1] R = OCH_3

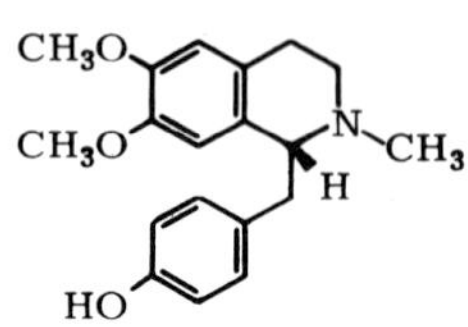

D-(−)-Armepavine

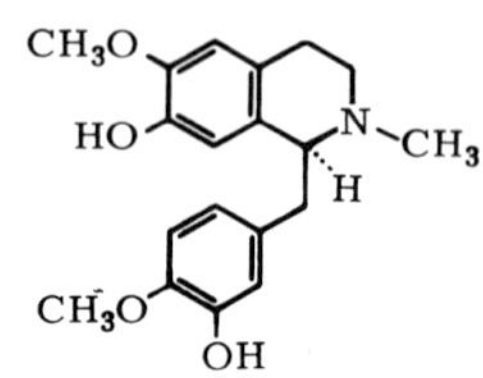

L-(+)-Reticuline

[D-(−)-Reticuline also a natural product]

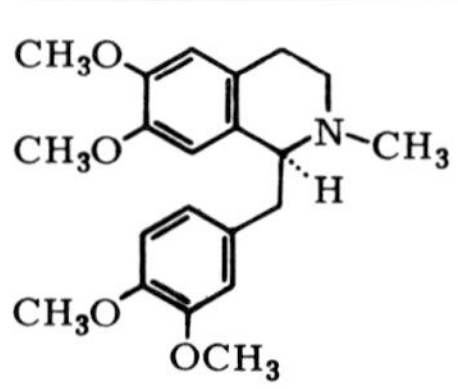

L-(+)-Laudanosine

(*S*-configuration)

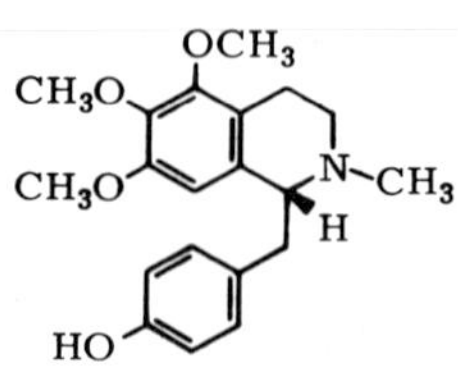

D-(−)-Thalifendlerine

(*R*-configuration)

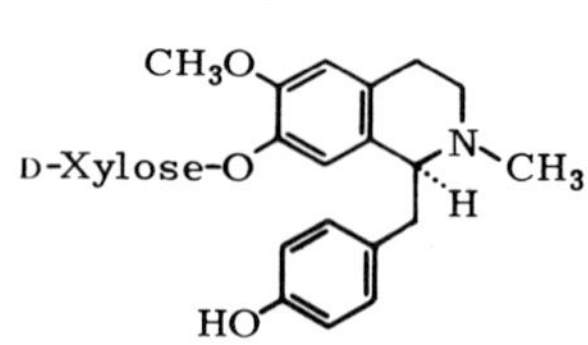

L-(+)-Latericine

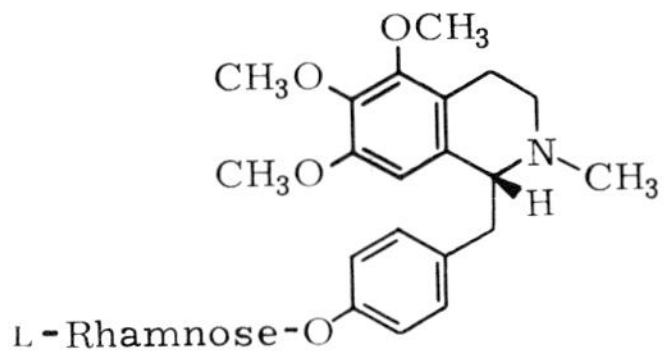

D-(−)-Veronamine

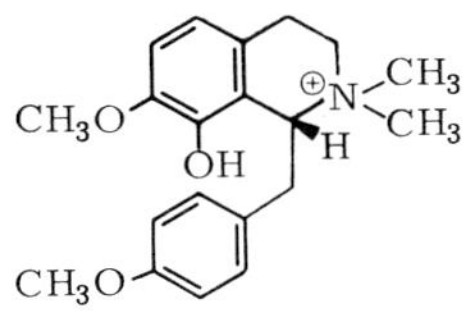

D-(−)-Petaline

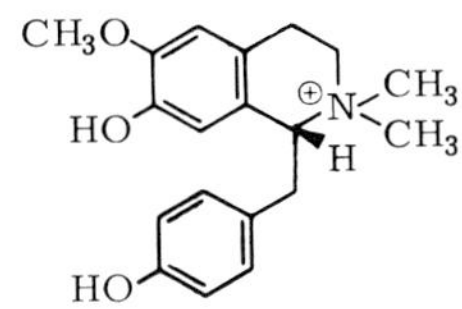

D-(−)-Magnocurarine

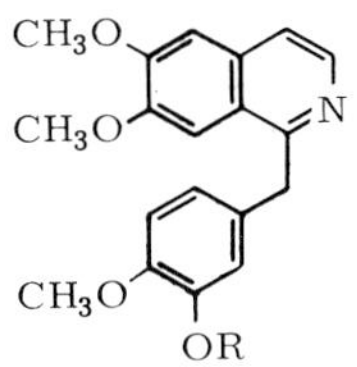

Papaverine, R = CH_3
Palaudine, R = H

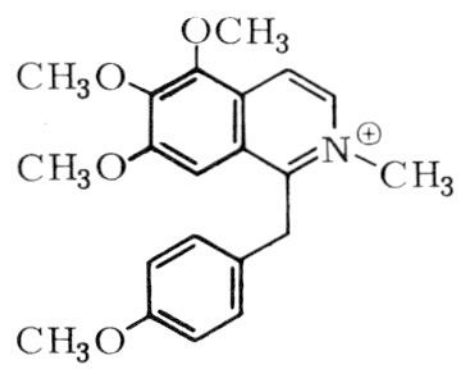

Takatonine

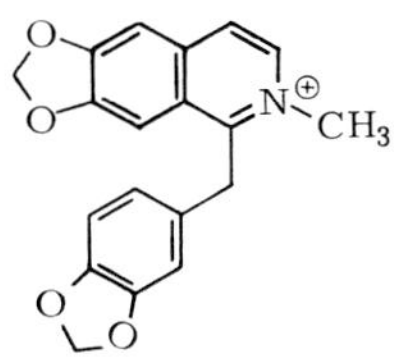

Escholamine

I. Introduction

The benzylisoquinolines occupy a paramount position in alkaloid chemistry because they act as *in vivo* precursors to so many of the other naturally occurring isoquinolines: isoquinolones, pavines, isopavines, bisbenzylisoquinolines, cularines, dibenzopyrrocolines, morphines, cularine-morphine dimers, proaporphines, aporphines, protoberberines, erythrina bases, and others.[2]

The benzylisoquinoline alkaloids are either of the 1,2,3,4-tetrahydro type, such as coclaurine and *N*-nororientaline, or of the completely aromatic type, as in the case of papaverine, palaudine, and escholamine.

Ring A in the benzylisoquinoline alkaloids may possess two or three oxygenated substituents, while ring C has only one or two substituents. Two glycosidic benzylisoquinoline alkaloids are known–latericine and veronamine. The unusual substitution pattern of petaline should be noted since it is very likely that a petaline-like analog is the precursor for the cularine alkaloids. No C-4 hydroxylated benzylisoquinolines have yet been isolated from nature, although it is more than probable that they do exist.

II. Degradation of Laudanosine, a Tetrahydrobenzylisoquinoline Alkaloid

The degradation of the opium alkaloid laudanosine, first isolated in 1871, has recently been restudied by Battersby and his group in connection with biogenetic studies involving labeled precursors. Some of the transformations described in Scheme I were, therefore, carried out on radioactive compounds.[3]

Hofmann degradation of laudanosine methiodide gave the trans-methine **1** in 68% yield and the cis isomer **2** in 9% yield. These compounds were reduced catalytically to the bicyclic base **3** which could also be obtained directly by sodium in liquid ammonia reduction of laudanosine methiodide. *N*-Methylation of **3** followed by another Hofmann degradation furnished the styrene derivative **4**, which yielded formaldehyde upon ozonolysis.

Additionally, methines **1** and **2** were oxidized to veratric acid and the amino acid **5**. The latter material was then converted to the substituted benzoic acid **6** which could be further oxidized or reduced as indicated in Scheme I.[3]

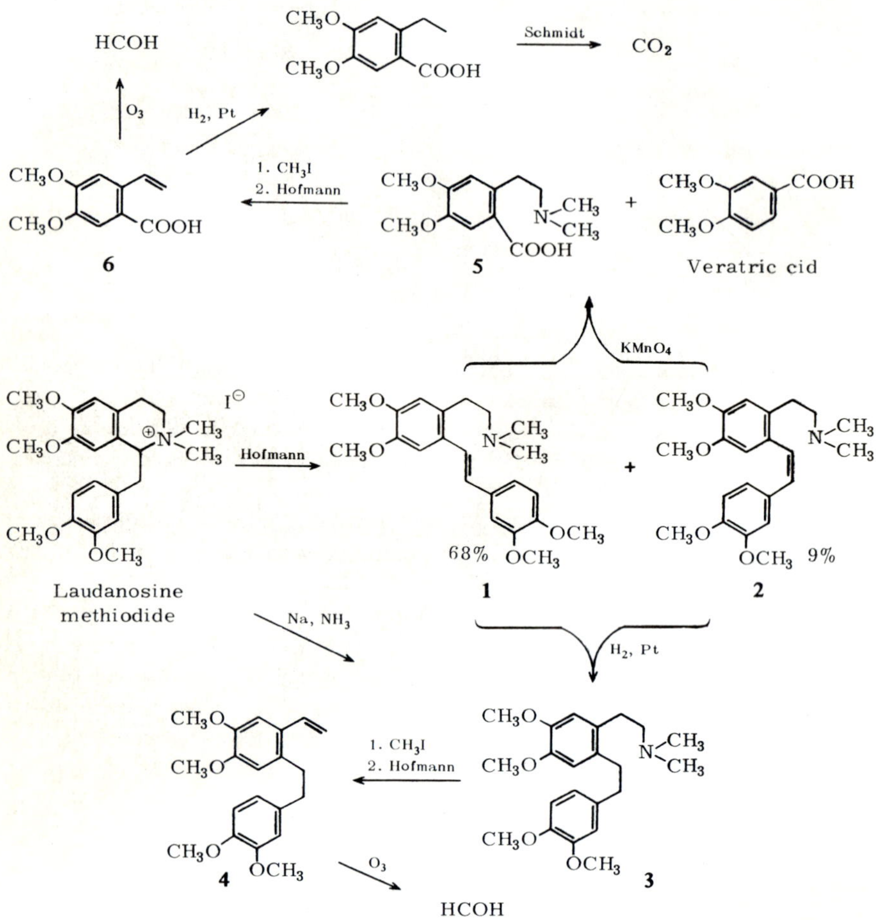

Scheme I

Complete *O*-demethylation of laudanosine has been achieved by refluxing with 48% hydrobromic acid[4] or aluminum chloride.[5] But with 35% hydrochloric acid at 100° for 20 minutes, apparently all four possible monophenolic isomers are formed and have been isolated.[6]

III. Structural Elucidation of Takatonine, a Quaternary Benzylisoquinoline Alkaloid

When a new benzylisoquinoline alkaloid is of the quaternary aromatic type, it is usually reduced to the tetrahydro stage. This transformation can be brought about catalytically, with zinc in acetic acid or with a mixed metal hydride. The tetrahydro derivative is then degraded by the Hofmann sequence.

Takatonine, $C_{21}H_{24}NO_4^+X^-$, is an optically inactive quaternary base found in various Thalictrum species. It incorporates one *N*-methyl and four methoxyl groups and upon reduction with zinc and acetic acid affords tetrahydrotakatonine.[7]

A Hofmann degradation on tetrahydrotakatonine methiodide led to the methine **7** which was oxidized to anisic acid and the amino acid **8**. A further Hofmann sequence on **8** followed by oxidation afforded 3,4,5-trimethoxyphthalic anhydride (Scheme II).

To eliminate the possibility of the third methoxyl group in ring A being situated at C-8, the alternate structure **9** was prepared and found to be different from tetrahydrotakatonine.[8]

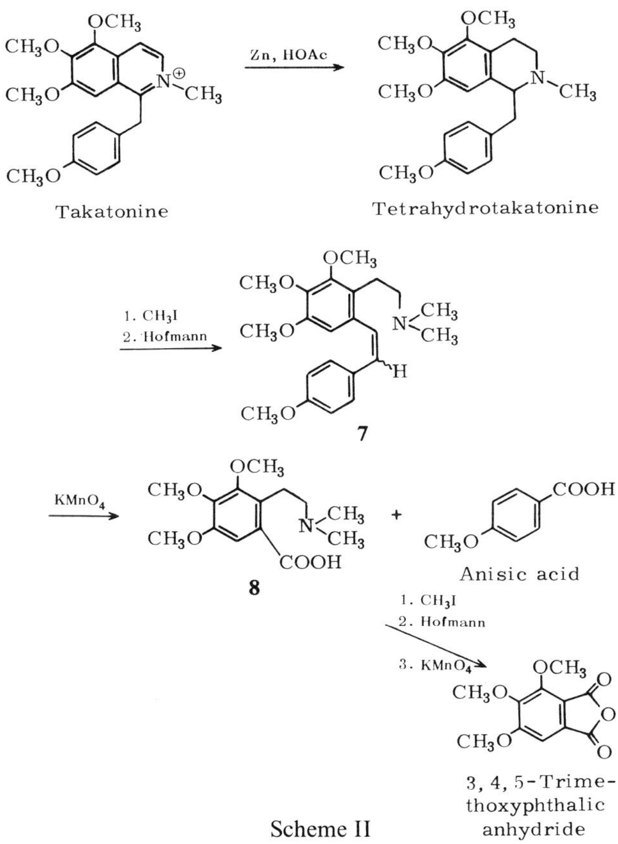

Scheme II

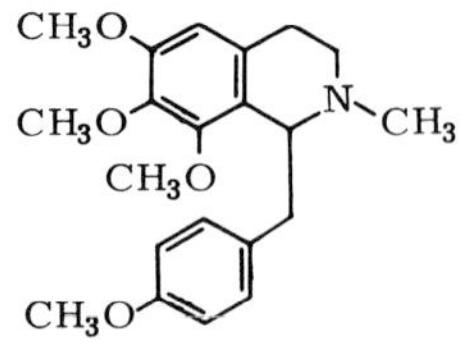

9

IV. Petaline, a Tetrahydrobenzylisoquinoline of Unusual Substitution Pattern

Petaline, $C_{20}H_{26}NO_3^+X^-$, is a quaternary tetrahydrobenzylisoquinoline containing one phenolic hydroxyl and two methoxyl groups. The alkaloid is isolated from extracts of the fresh tuberous roots of *Leontice leontopetalum* Linn. (Berberidaceae), a plant which grows wild in Lebanon and is used there as a folk remedy for grand mal epilepsy.[9]

The alkaloid underwent Hofmann degradation under exceptionally mild conditions, since passage of a solution of the salt through a hydroxide ion-exchange column was sufficient to form the methine **10**.

The mesylate of the methine **10** was then oxidized to produce 3-methanesulfonyloxy-4-methoxyphthalic acid (**11**) and anisic acid. The alternate structure **12** for the methine base was eliminated since it was synthesized and found to be different from **10** (Scheme III).[10,11]

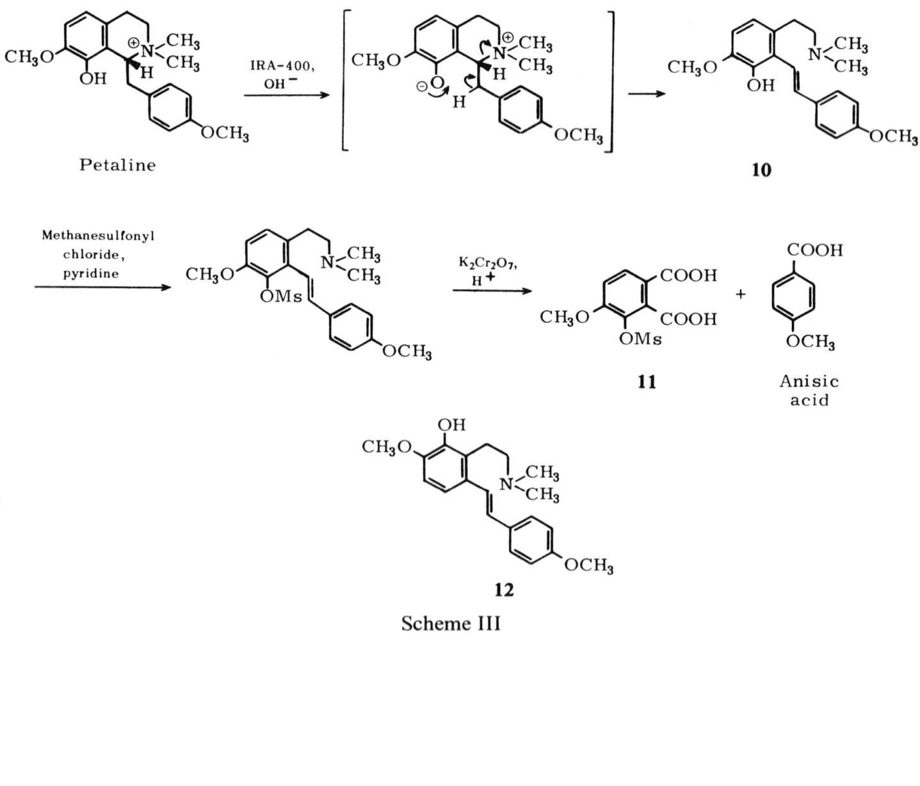

Scheme III

The structure assigned to petaline explains the facile Hofmann degradation that the alkaloid undergoes since the phenolate anion can readily abstract a hydrogen atom beta to the quaternary nitrogen atom as indicated.[9] Petaline is the only known 7,8-disubstituted benzylisoquinoline alkaloid; this rare substitution pattern is also encountered in the simple tetrahydroisoquinoline alkaloid tepenine, as well as in the cularine bases and a few of the protoberberines.

V. Two Glycosidic Tetrahydrobenzylisoquinoline Alkaloids: Latericine and Veronamine

A. *Latericine*

Preininger and co-workers have investigated the structure and stereochemistry of the alkaloid (+)-latericine, $C_{24}H_{29}NO_8$, which was obtained in small yields from a variety of Papaver species.[11]

Hydrolysis of latericine with sulfuric acid gave an aglycone which was characterized as (+)-7-demethylarmepavine, while the sugar moiety corresponded to D-(+)-xylose. *O*-Methylation of latericine with diazomethane and subsequent hydrolysis of the product gave the aglycone L-(+)-*N*,*O*-dimethylcoclaurine, thus fixing the position of the glycoside moiety in the alkaloid (Scheme IV).

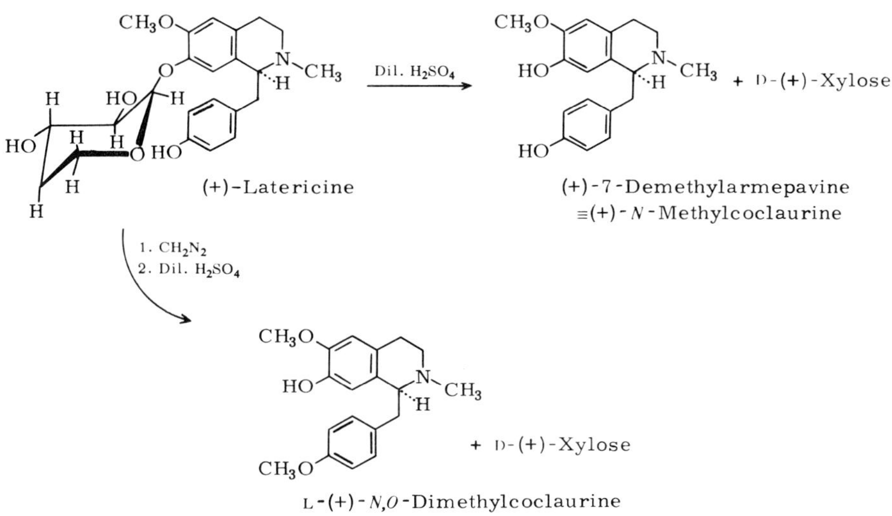

Scheme IV

The anomeric center in latericine was assigned the alpha configuration on the basis of Klyne's rule, which states that the sum of the molecular rotatory contributions of the carbohydrate component and the aglycone is equal to the molecular rotation of the glycoside[12]:

$$[M_D] \text{ Glycoside} = [M_D] \text{ Aglycone} + [M_D] \text{ Carbohydrate}$$

Since in the present case the carbohydrate contribution should be nearly equal to the value for the molecular rotatory contribution of either the α- or the β-methyl-D-xylopyranoside, one can approximate by stating:

$$[M_D] \text{ Latericine} \approx [M_D] \text{ L-(+)-}N,O\text{-Dimethylcoclaurine} + [M_D]\ \alpha\text{- or } \beta\text{-methyl-D-xylopyranoside}$$

The molecular rotations for α- and for β-methyl-D-xylopyranoside are known to be +253° and −108°, respectively, while the molecular rotations found for latericine and L-(+)-*N*,*O*-dimethylcoclaurine are +401° and +245°. Using the molecular rotational contribution for α-methyl-D-xylopyranoside, one obtains +401° ≈ +245° +253°.

This is a closer approximation than can be achieved by using the value for the molecular rotation for β-methyl-D-xylopyranoside. It follows that the anomeric center in latericine must have the alpha configuration as indicated.[11]

B. Veronamine

(−)-Veronamine, $C_{26}H_{35}NO_8$, was isolated from *Thalictrum fendleri* Engelm. (Ranunculaceae) and upon acid hydrolysis yielded (−)-thalifendlerine, previously

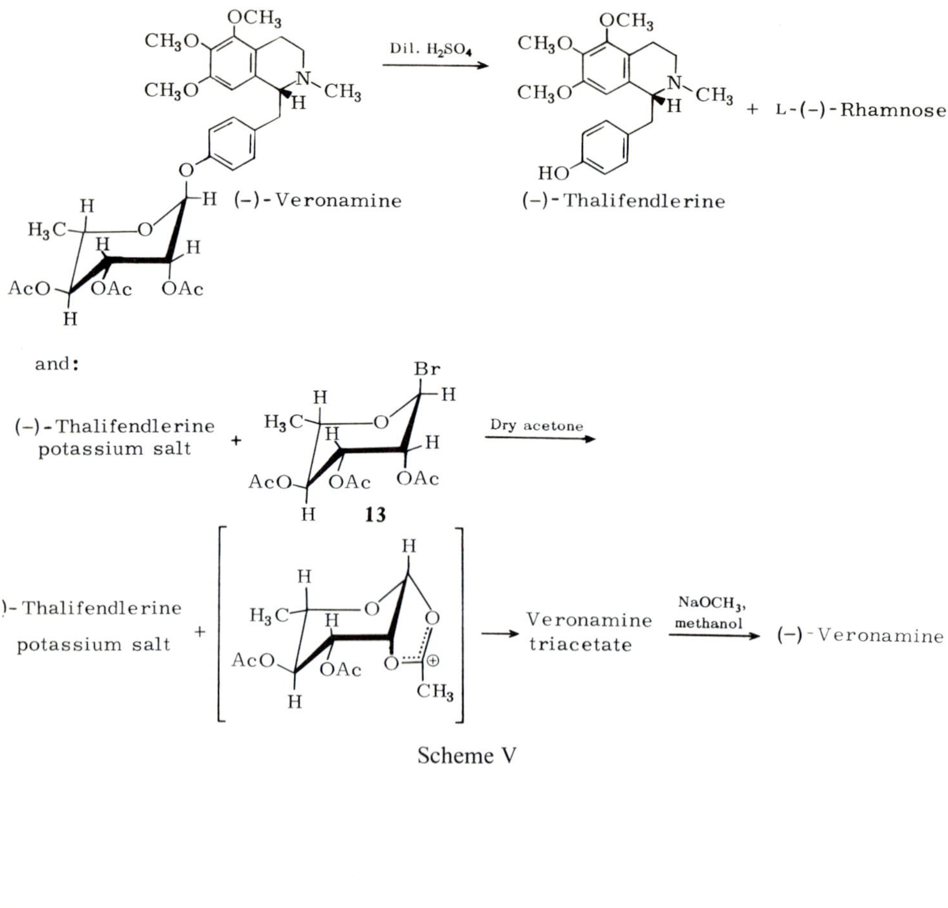

Scheme V

isolated as an alkaloid from the same source, and L-(−)-rhamnose. (−)-Veronamine was then synthesized through the Koenigs-Knorr condensation of the dry potassium salt of (−)-thalifendlerine with 2,3,4-tri-*O*-acetyl-α-L-rhamnopyranosyl bromide (**13**) followed by Zemplén methanolysis to cleave the acetate functions.

It is known that the Koenigs-Knorr reaction usually occurs with Walden inversion at the anomeric center when alkali is employed. However, an acetoxyl neighboring group effect prevails with the rhamnopyranosyl bromide **13**, so that in this instance glycoside formation occurs with overall retention of configuration at the anomeric center which remains alpha (Scheme V).[13]

VI. Methods of Synthesis

A. The Bischler–Napieralski Cyclization

The most common method for the construction of benzylisoquinoline systems involves the use of the Bischler–Napieralski cyclization. The syntheses of takatonine and tetrahydrotakatonine are presented in Scheme VI as examples of such a procedure.[8]

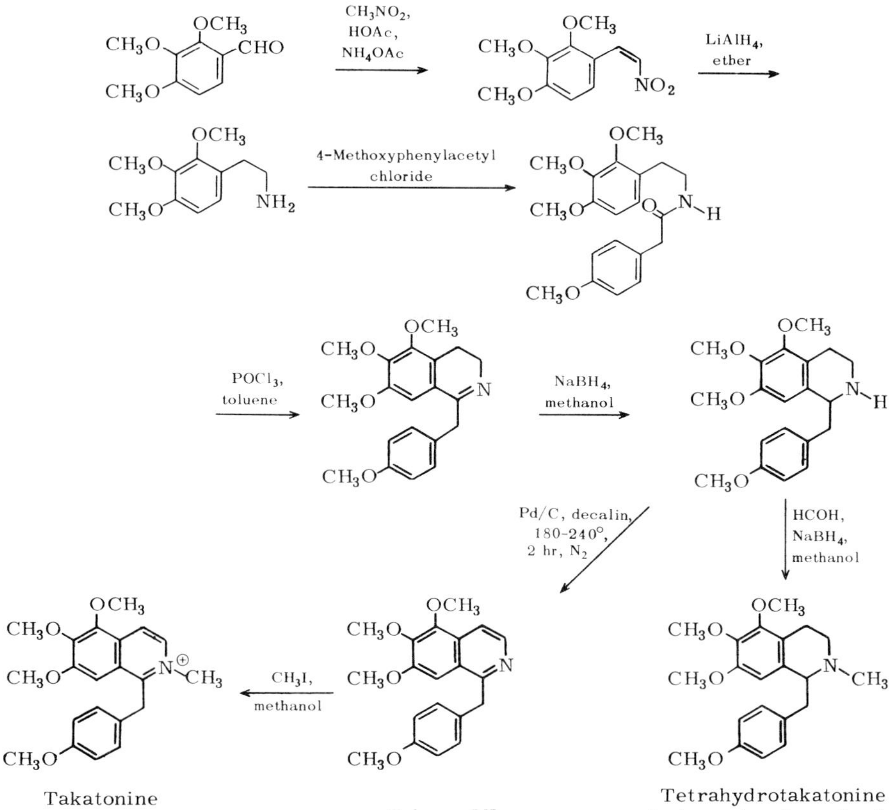

Scheme VI

A worthwhile variation of this approach developed by Teitel and Brossi avoided the use of lithium aluminum hydride in the reduction of the nitrostyrene, as well as the usual and tedious blocking and deblocking of phenolic functions as benzyl ethers. Additionally, acetonitrile was shown to be a superior solvent in effecting the Bischler–Napieralski cyclodehydration with phosphorus oxychloride.[14] In this fashion, (±)-coclaurine was obtained in 61% overall yield from the readily accessible 4-hydroxy-3-methoxyphenylethylamine (**14**) (Scheme VII).

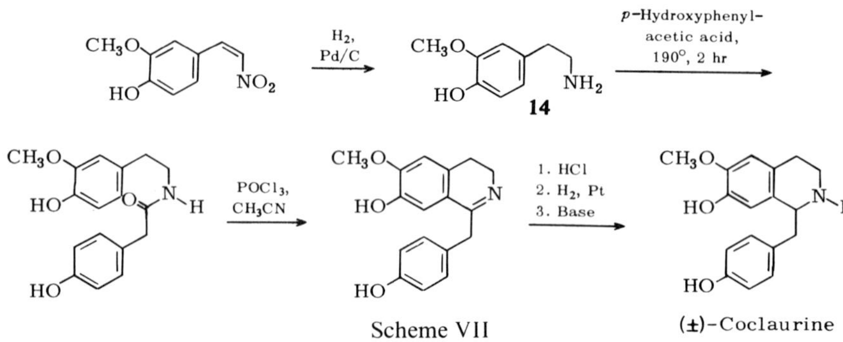

Scheme VII

Another modification of the Bischler–Napieralski approach exists which allows for the direct synthesis of benzylisoquinolines with an aromatic ring B such as papaverine. The nitrostyrene **15** adds the elements of methanol under basic conditions to yield the nitro ether **16**. This material can then be carried through the synthetic sequence shown in Scheme VIII.[15,16,16a]

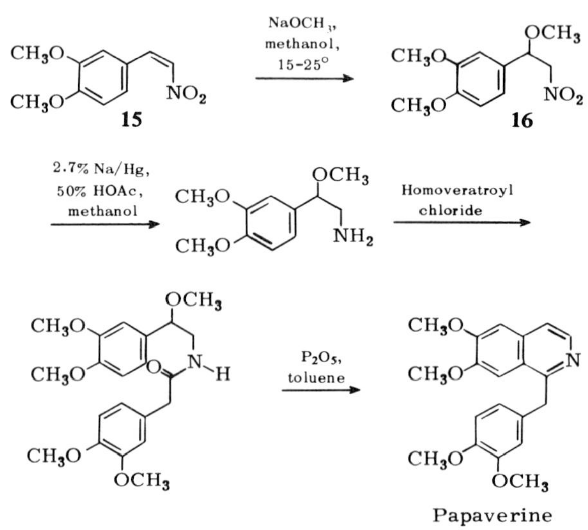

Scheme VIII

B. Reaction with a Grignard Reagent

The unusual substitution pattern of the quaternary base petaline has already been pointed out. The first synthesis of the alkaloid was achieved through the intermediacy of a Grignard reagent. Formylation of the tetrahydroisoquinoline **17** gave the *N*-formyl derivative **18** whose sodium salt was benzylated with benzyl chloride in DMF. Reduction with lithium aluminum hydride followed by dehydrogenation with mercuric acetate led to the 3,4-dihydroisoquinoline salt **19**. Salt **19** in the perchlorate form was then added in small portions to an excess of *p*-methoxybenzylmagnesium chloride in ether. Hydrogenolytic debenzylation of the product afforded the tetrahydrobenzylisoquinoline **20**, and *N*-methylation finally provided (±)-petaline iodide.[17]

The intermediate **20** was also resolved using *O*,*O*′-dibenzoyl-(+)-tartaric acid, and (−)-petaline iodide, identical in all respects with the natural product, was obtained upon *N*-methylation with methyl iodide (Scheme IX).[17]

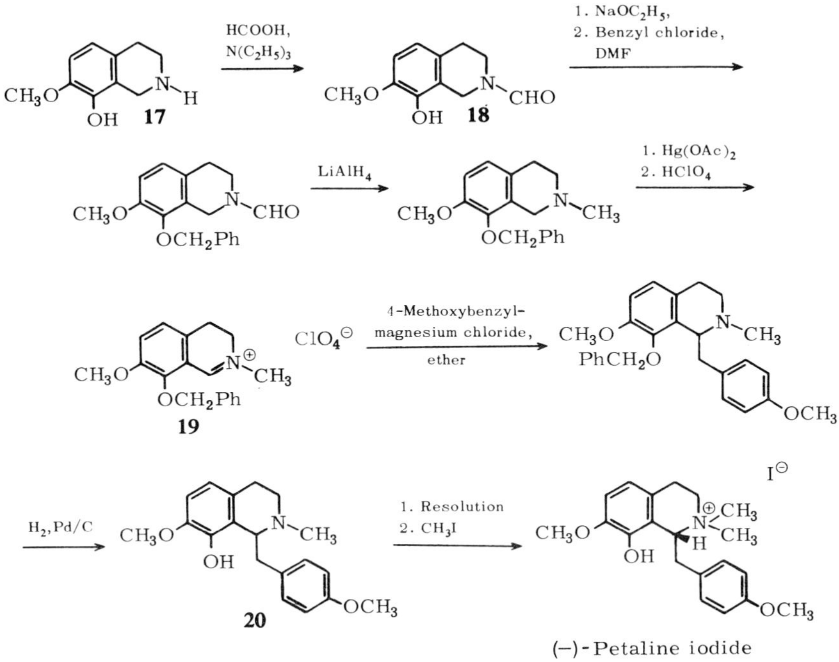

Scheme IX

Other efficient methods for the resolution of 1-substituted tetrahydroisoquinolines[18] involve the use of (+)-camphor-10-sulfonic acid,[19] (2*R* : 3*R*)-2′-nitrotartranilic acid,[20] (−)-quinic acid,[21] and (+)-tartaric acid.[22]

Laudanosine has also been synthesized by the Grignard approach, but in this case it is imperative that the initial 3,4-dimethoxybenzyl chloride from which the Grignard reagent is formed be freshly distilled and of high purity. The organomagnesium reagent does not form satisfactorily otherwise.[23]

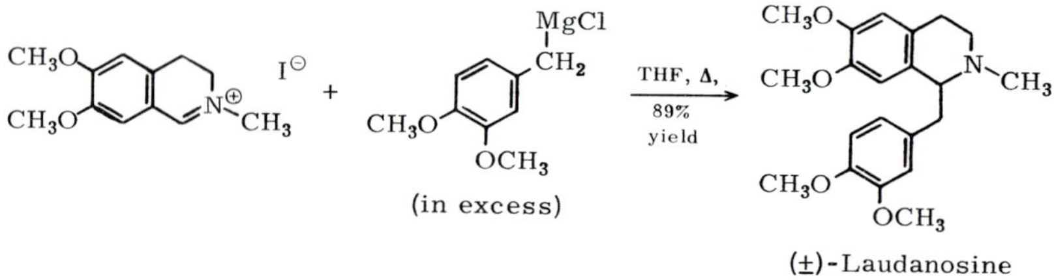

C. *The Use of Reissert Compounds*

Base-induced removal of the C-1 hydrogen of a Reissert compound yields an anion which reacts readily with a benzyl halide.[24] The anion is prepared at room temperature using sodium hydride in DMF.[25,26] Hydrolysis of the alkylated Reissert compound then gives rise to a benzylisoquinoline. An alternate method for generating the Reissert anion is through the use of triethylbenzylammonium hydroxide in an aqueous medium.[27]

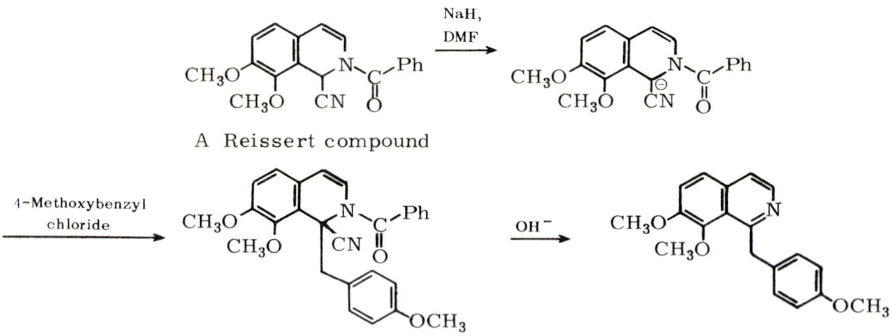

In a variation on this theme, it has also recently been shown that dihydro-Reissert compounds can be used in the preparation of 1,1-disubstituted tetrahydroisoquinolines.[28]

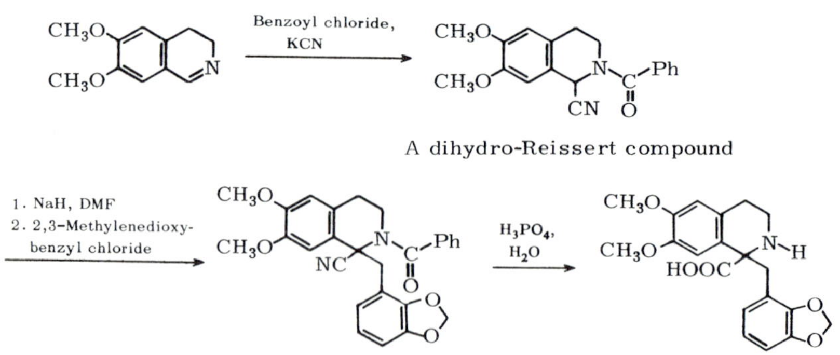

A 1,1-disubstituted tetrahydroisoquinoline

D. *The Pomeranz–Fritsch Cyclization*

This method has been utilized in the preparation of 6,7,8-substituted benzylisoquinolines, but the yields are usually poor. Condensation of the ketone **21** with aminoacetaldehyde diethyl acetal generated the Schiff base **22**. Cyclization of this material with polyphosphoric acid afforded a mixture of the desired benzylisoquinoline **23** and the benz[*e*][1,4]oxazepine **24**.[29,30]

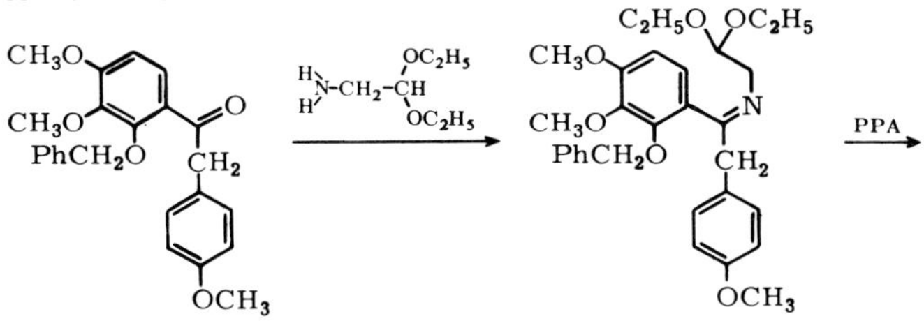

21 **22**

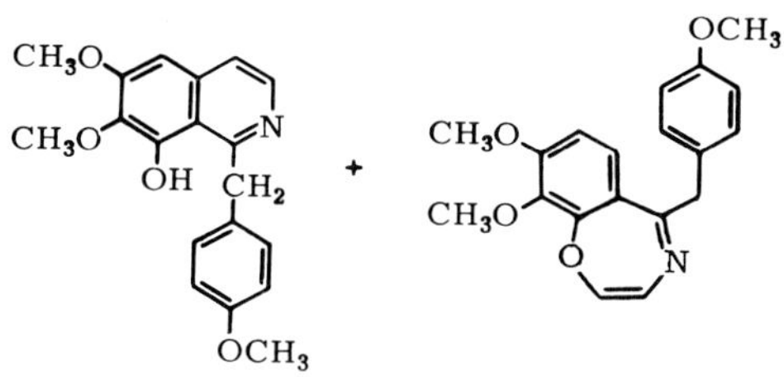

23 **24**

Dalton has extended Bobbitt's modification of the Pomeranz–Fritsch cyclization to the preparation of C-4 oxygenated tetrahydroisoquinolines. Careful Friedel–Crafts acylation of methylenedioxybenzene with the acid chloride of *o*-nitrophenylacetic acid generated the ketone **24a** which was converted to the primary amine **24b** through

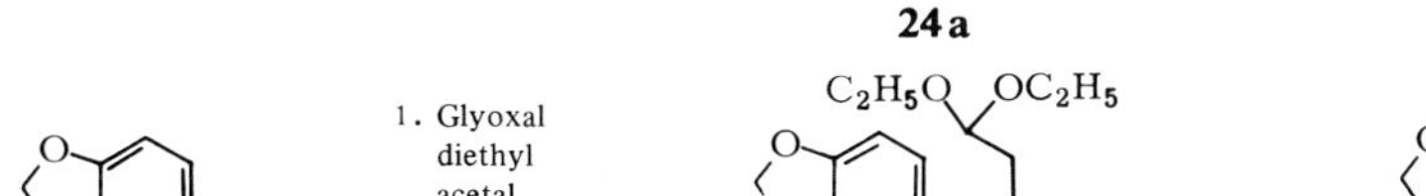

24a

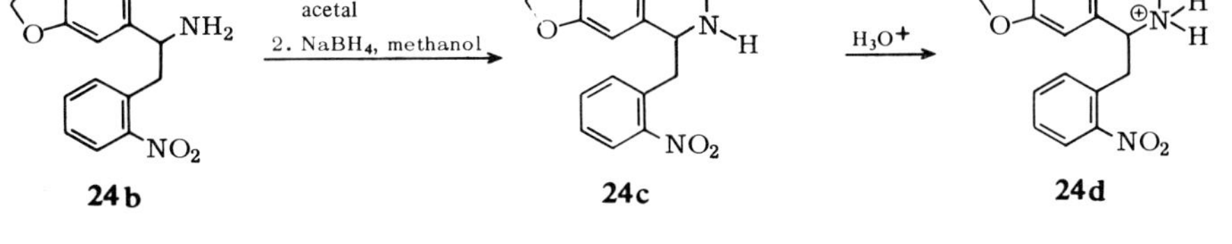

24b **24c** **24d**

diborane reduction of the corresponding oxime acetate. Reaction of the amine **24b** with glyoxal diethyl acetal and reduction with sodium borohydride afforded the acetal **24c** as an oil. Treatment of **24c** with dilute hydrochloric acid followed by removal of the solvent at low temperature yielded the amine hydrochloride **24d** which crystallized spontaneously.[30a]

E. The Pictet–Spengler Condensation

One of the early biogenetic-type syntheses of a tetrahydrobenzylisoquinoline was carried out by Späth and Berger in 1930. Pictet–Spengler condensation of homoveratraldehyde with homoveratrylamine at room temperature in the presence of hydrochloric acid gave (±)-*N*-norlaudanosine.[31]

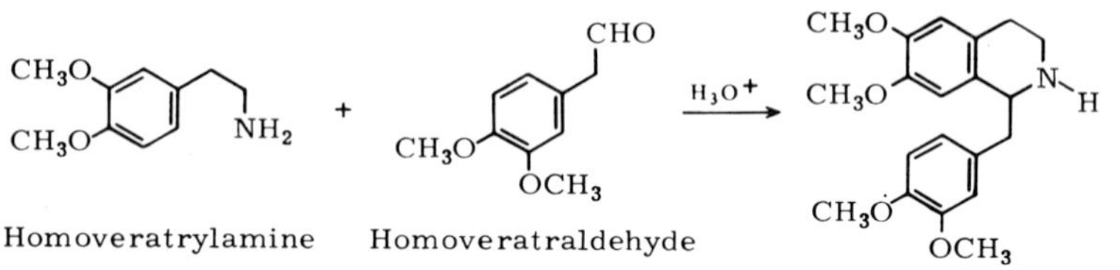

Homoveratrylamine Homoveratraldehyde (±)-*N*-Norlaudanosine

Such a condensation works particularly well when a free phenolic function is available para to the cyclization site. For example, condensation at pH 5 of homoisovanillylamine with homoveratraldehyde followed by Eschweiler–Clarke *N*-methylation gave (±)-pseudolaudanine in 41% yield.[32] Condensation will, in fact, take place even without the use of an acid catalyst (see also Chapter 1, Section IV, D) to afford species **25**.[33]

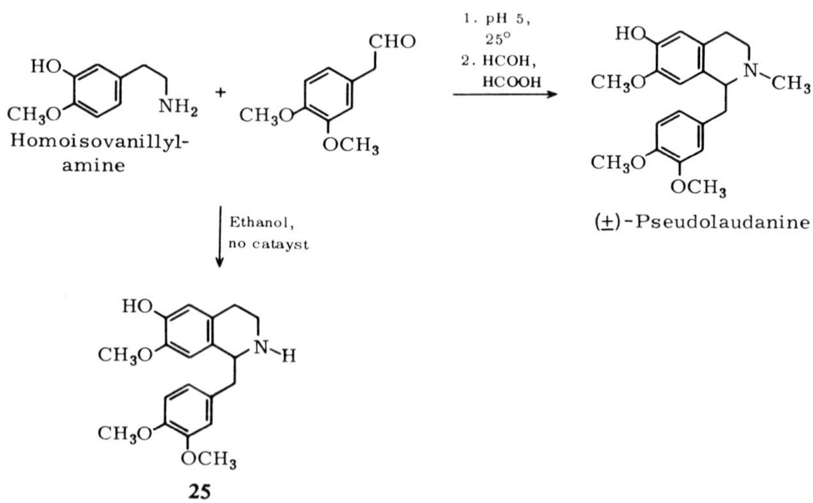

Homoisovanillylamine (±)-Pseudolaudanine

25

The Pictet–Spengler condensation has been used in the third synthesis of the alkaloid petaline. Reaction of 2-bromo-5-hydroxy-4-methoxyphenylethylamine with *p*-methoxyphenylacetaldehyde generated a brominated tetrahydroisoquinoline derivative which was converted to petaline iodide as illustrated below.[33a]

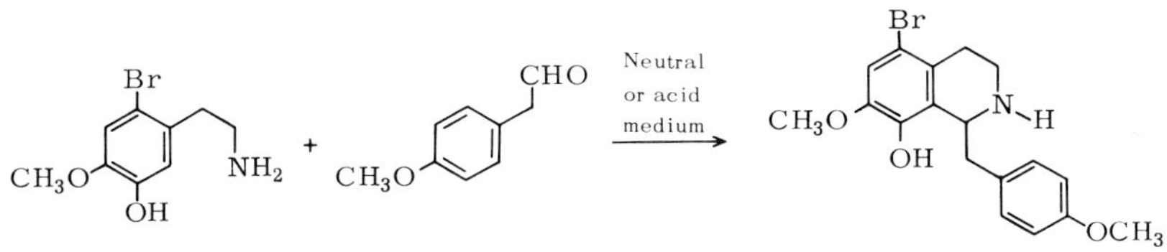

2-Bromo-5-hydroxy-4-methoxyphenylethylamine

p-Methoxyphenylacetaldehyde

1. HCOH, $NaBH_4$
2. Catalytic debromination
3. CH_3I
→ Petaline iodide

F. The Stevens Rearrangement

The base-induced Stevens rearrangement of ammonium salts has been used with a variety of substituted *N*-benzyltetrahydroisoquinolines to produce 1-benzyltetrahydroisoquinolines. The yields may vary all the way from 17 to 85%, and in some of the cases a number of side products accompany the desired tetrahydrobenzylisoquinoline.[34]

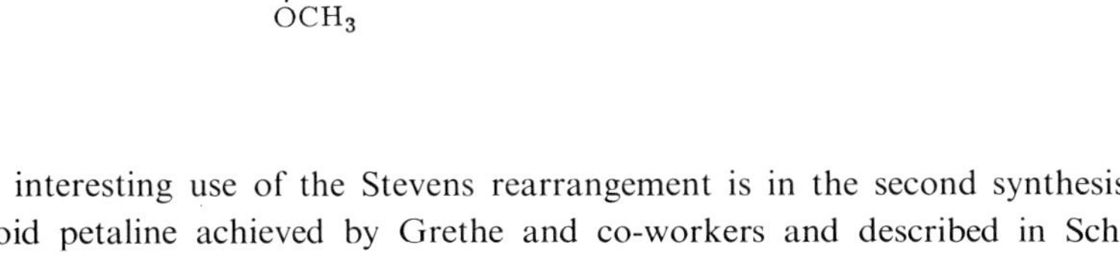

An interesting use of the Stevens rearrangement is in the second synthesis of the alkaloid petaline achieved by Grethe and co-workers and described in Scheme X. In this case the N → C rearrangement proceeded in 85% yield. The selective demethylation of the trimethoxy ether **26** to the diphenol **27** was achieved using 36% HBr.[35]

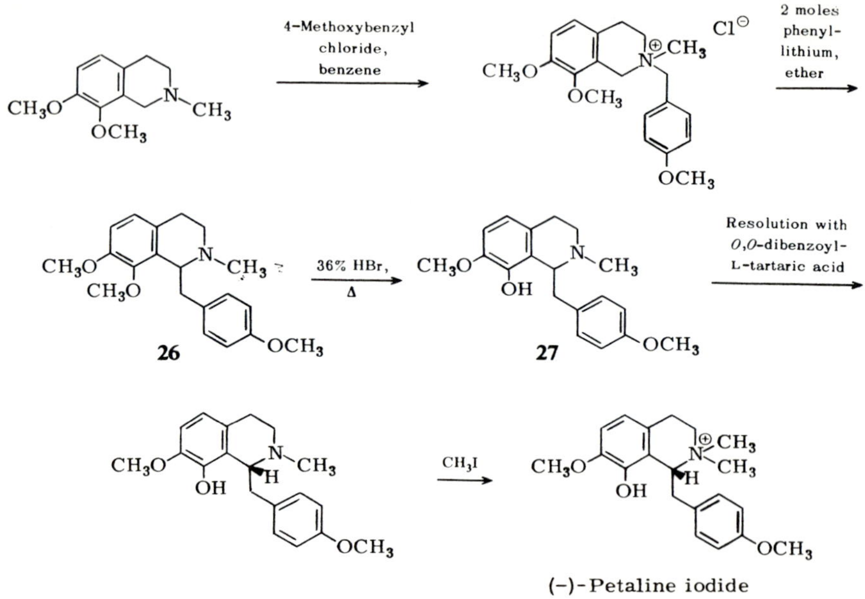

Scheme X

The fact that optically active **26** when selectively *O*-demethylated yielded racemic product suggests a mechanism for the transformation.

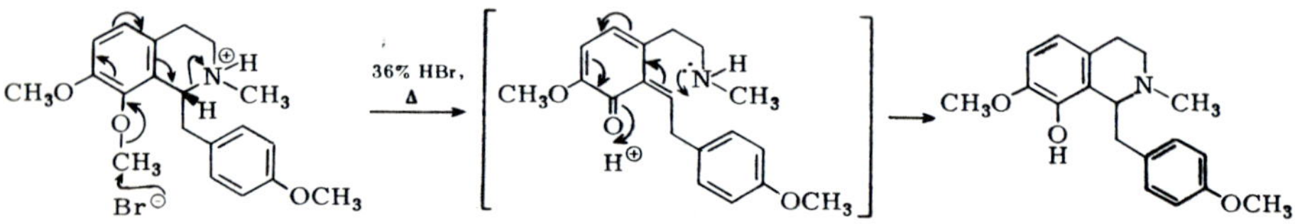

(+)-**26** (±)-**27**

G. The Use of a Peptide Chain

The model sequence delineated below was carried out by Lawton and his co-workers to support the suggestion that the biosynthesis of isoquinoline alkaloids involves peptide chains. Reaction of the peptide analog **28** with the very reactive cyclic carbonic anhydride 5-benzylidene-1,3-dioxolane-2,4-dione gave a pyruvyl derivative which underwent facile acid-catalyzed cyclization at room temperature to the benzylisoquinoline **29**. Hydrolysis with 10% hydrochloric acid then gave the amino acid **30**.

The transformation of **28** to **30** exemplifies the use of a peptide backbone for directing and facilitating chemical or biochemical transformations.[36]

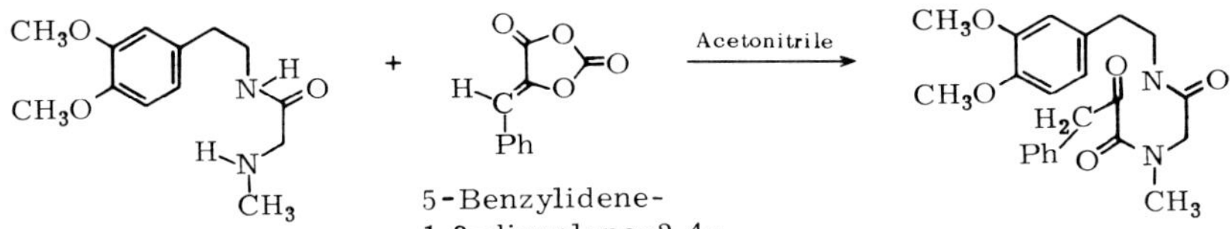

28

5-Benzylidene-1,3-dioxolane-2,4-dione

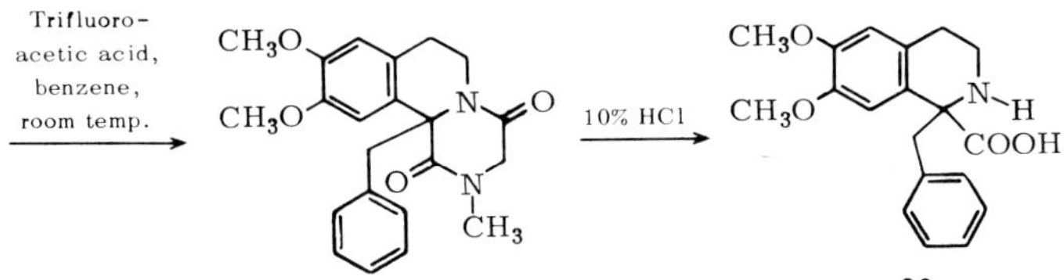

29 **30**

H. Nitrilium Salts

Lora-Tamayo and his co-workers have developed a new synthesis of isoquinolines whereby a substituted phenethyl chloride such as **31** is condensed with homoveratronitrile–stannic chloride complex. The reaction must involve the ring closure of the cationic species **32**.[37]

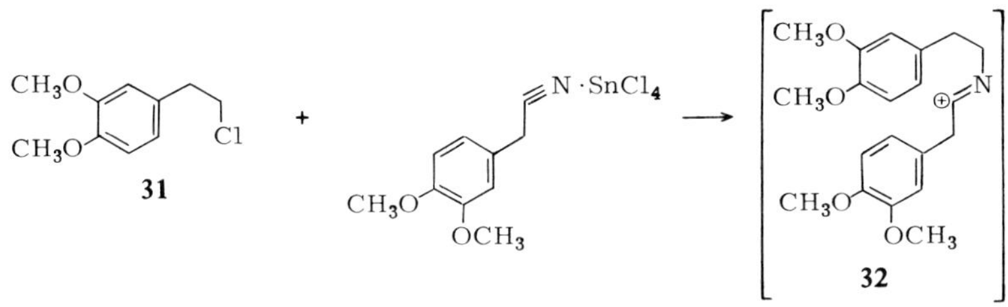

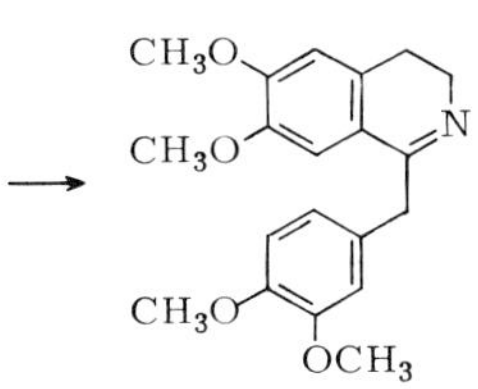

I. The Nitrotoluene Approach

Since a methyl group ortho or para to an aromatic nitro function has acidic properties, it has been found that condensation of 3,4-dihydro-6,7,8-trimethoxyisoquinoline methiodide (**33**) and *p*-nitrotoluene in the presence of sodium hydride gave rise to the tetrahydroisoquinoline **34**. This nitrated benzylisoquinoline was then converted to the phenolic derivative **35** through reduction and diazotization.[38]

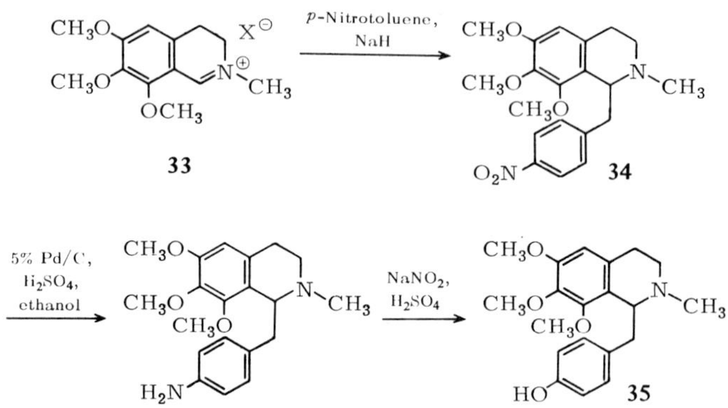

J. The Self-Condensation of Homoveratric Acid

A recent synthesis of a benzylisoquinoline involves simply the self-condensation of homoveratric acid in the presence of polyphosphoric acid to afford a keto acid. Heating this product with ammonium acetate in acetic acid was found to yield a product originally formulated as "3-oxopapaverine,"[39,39a] but which must exist in the tautomeric form shown since the NMR spectrum of the product shows only one set of methylene protons at δ4.40 as a singlet. The old structure for "3-oxopapaverine" would have necessitated the presence of two sets of methylene protons.[39b]

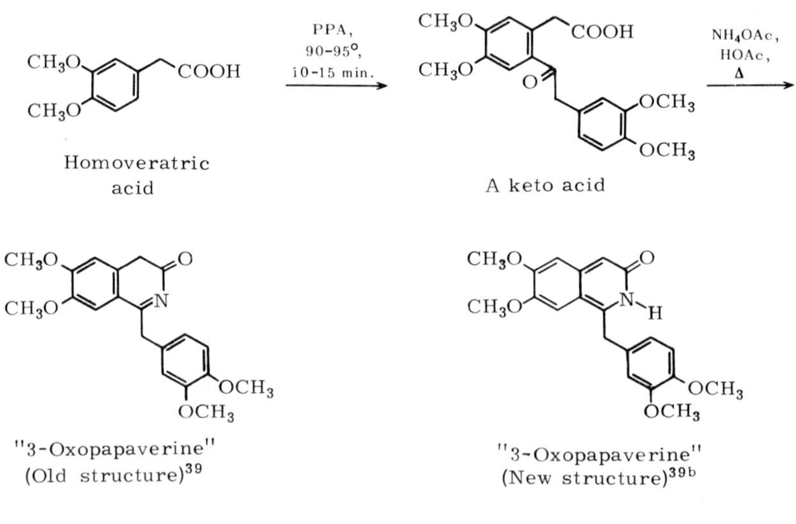

VII. Some Reactions of Benzylisoquinolines

A. Phenolic Oxidative Coupling

The most important reaction that benzylisoquinolines can undergo in plants is phenolic oxidative coupling.[40-42] Such oxidations ultimately lead to proaporphines, aporphines, cularines, bisbenzylisoquinolines, dibenzopyrrocolines, morphines, and other alkaloidal types. Detailed discussions of these transformations will be presented in the succeeding chapters. The first example that will be mentioned here involves the oxidation of $(\pm)$-4′-*O*-methyl-*N*-methylcoclaurine (**36**). Two racemic and diastereoisomeric bisbenzylisoquinolines (**37** and **38**) were produced and separated. The total yield was 15%, a relatively high conversion for such a transformation.[43]

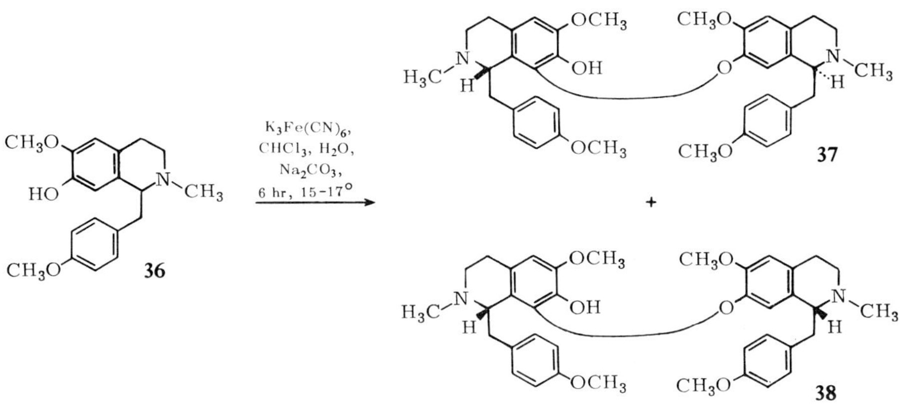

36 **37** **38**

One of the peculiarities of the oxidative coupling of tetrahydrobenzylisoquinolines carried out *in vitro* is that if phenolic groups are present in both rings A and C, the products most often isolated are those resulting from head-to-head coupling, a C-7 hydroxyl group being conspicuously prone to coupling.[23] However, no naturally occurring bisbenzylisoquinoline is known which possesses this specific type of bonding without accompanying tail-to-tail coupling.

Abnormal products have sometimes been isolated from attempts at phenolic oxidative coupling of benzylisoquinolines. Treatment of the phenol **39** with ferric chloride unexpectedly gave species **40** and **41**.[44]

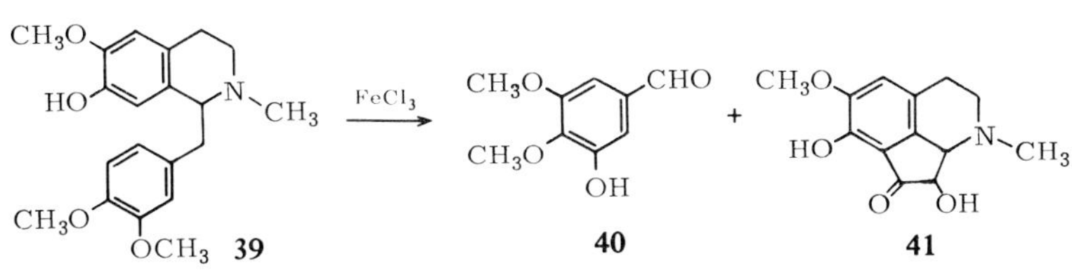

39 **40** **41**

The enzymatic oxidation of $(\pm)$-*N*-norarmepavine at pH 6.5 with crude horseradish peroxidase and hydrogen peroxide gave a complex mixture which included the isoquinolines **42**, **43**, and **44**.[45]

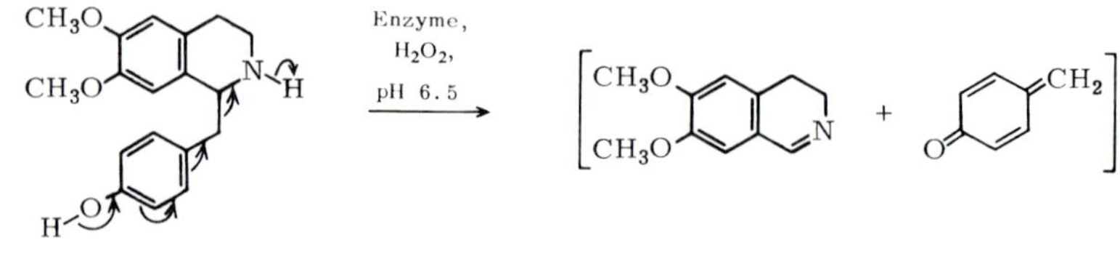

(±)-*N*-Norarmepavine

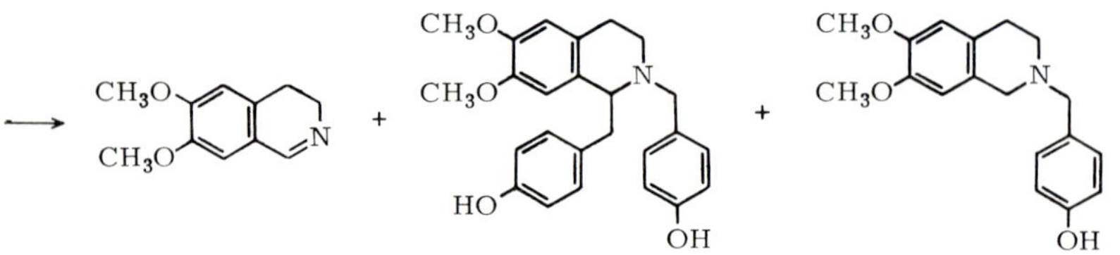

42 **43** **44**

The phenolic oxidative coupling of *N*-methylcoclaurine using homogenized potato peels and hydrogen peroxide has also been investigated. The products were dimers and trimers resulting from carbon–oxygen head-to-tail coupling.[46]

The oxidation of tetrahydrobenzylisoquinolines to provide morphinane derivatives usually proceeds in poor yield. But Franck and co-workers found that satisfactory yields may be obtained by using a manganese dioxide–silica gel system which supplies the substrate to the oxidant at high dilution. The yield is thus improved by cutting down on competing polymerization reactions.[47]

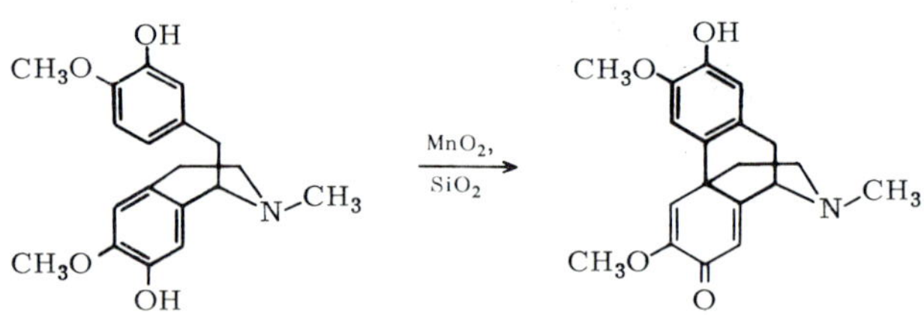

Reticuline Isosalutaridine

B. Oxidation

Laudanosine has been oxidized by means of mercuric acetate and ethylenediaminetetraacetic acid into a mixture of *N*-methylpapaverine salt and 3,4-dihydropapaverine.[48]

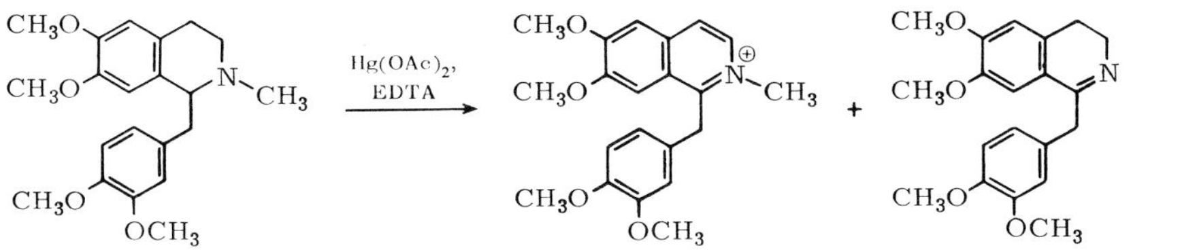

Laudanosine | N-Methylpapaverine salt | 3,4-Dihydropapaverine

The further oxidation of a 3,4-dihydroisoquinoline to an isoquinoline derivative is somewhat difficult to perform, but can be achieved either by the above method or through palladium black dehydrogenation at around 200° in an inert solvent.

Papaverine can be efficiently oxidized with dichromate, permanganate, or selenium dioxide, to papaveraldine. Under milder conditions, using mercuric acetate, papaverinol is generated.[48,49]

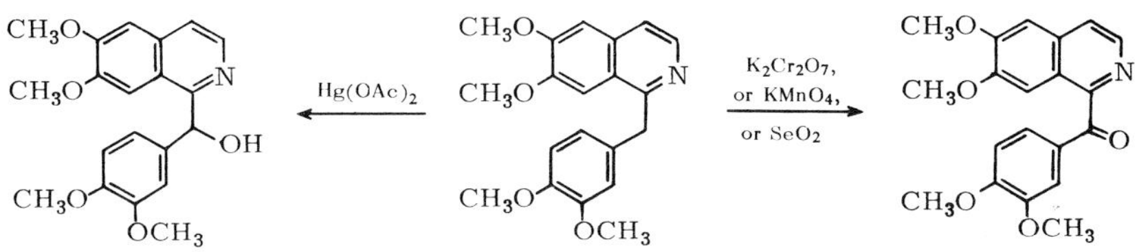

Papaverinol | Papaverine | Papaveraldine

Oxidation at the α-methylene occurs even more readily with 3,4-dihydrobenzylisoquinolines. For example, selective hydrolysis of the imine **45** with 20% hydrochloric acid results in the formation of a free phenolic group at C-7, but the product is also oxidized at the C-α position. The ketone **46** can then be transformed into a tetrahydrobenzylisoquinoline (**47**) by Clemmensen reduction or by zinc in acetic acid.[50]

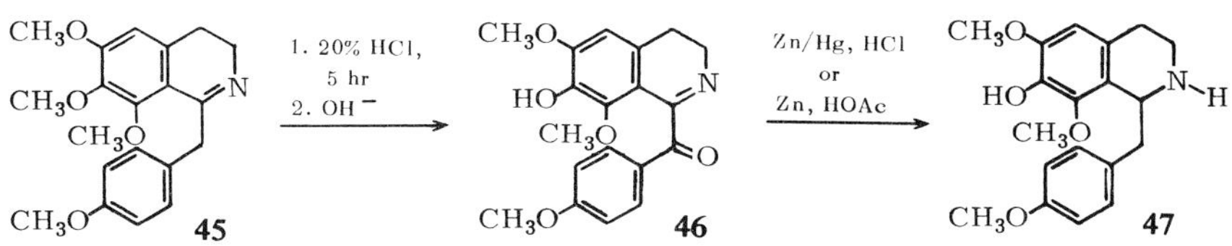

45 | **46** | **47**

Alternatively, Bischler–Napieralski cyclization of the amide **48** followed by hydrolysis of the benzyloxy group has also been shown to give the ketone **46**, oxidation having probably occurred during work-up.[50]

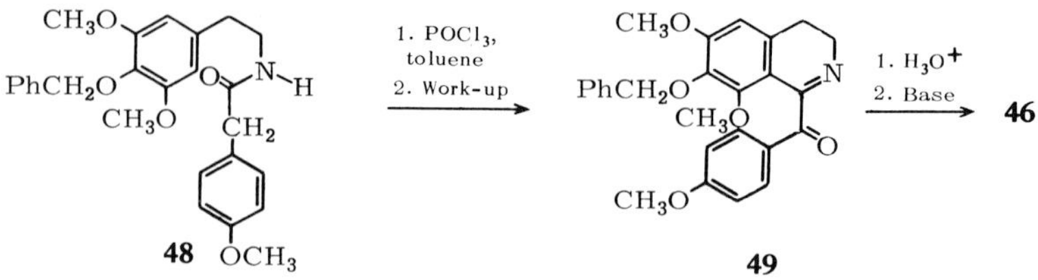

In fact, if a 3,4-dihydroisoquinoline is simply stirred for 1–3 weeks, an imino ketone is produced which may be cleanly reduced to an amino alcohol with sodium borohydride.[51]

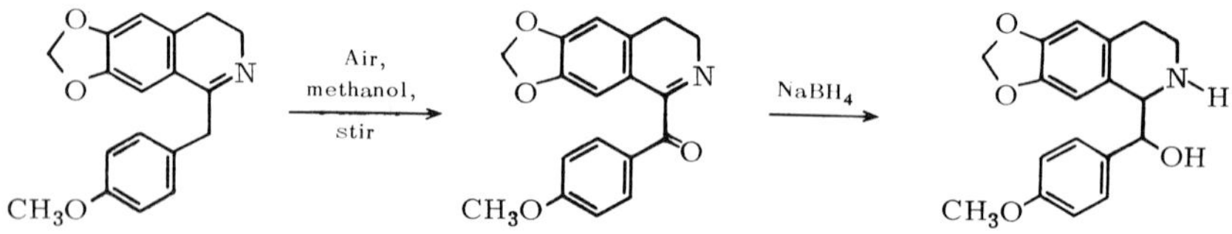

Permanganate oxidation of a tetrahydrobenzylisoquinoline can lead to a bicyclic lactam through cleavage of the C-1 to C-α bond. This transformation has been used to advantage in the structural elucidation of some isoquinolone alkaloids. For instance, oxidation of the Thalictrum alkaloid thalifendlerine of established structure yielded the bicyclic lactam **50** identical in all respects with the new isoquinolone alkaloid *N*-methylthalidaldine.[52]

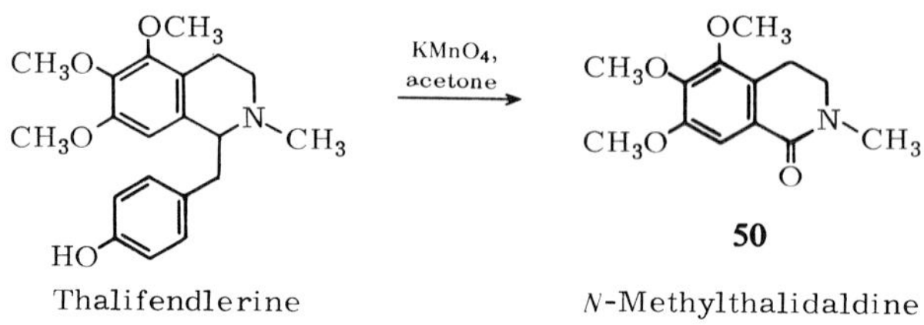

C. Reduction

Papaverine can be easily reduced catalytically to tetrahydropapaverine, but reduction with tin and hydrochloric acid yields two products: tetrahydropapaverine (*N*-norlaudanosine) and pavine (see also Chapter 4, Section I, A).

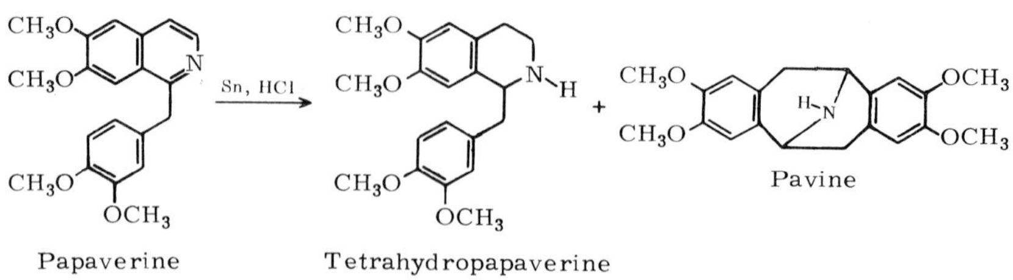

If papaverine methiodide is reduced with lithium aluminum hydride in a nonprotic solvent such as tetrahydrofuran, the product is the 1,2-dihydroisoquinoline **51**.[53] But reduction of the methiodide with sodium borohydride in aqueous methanol is one of the better methods for the production of racemic laudanosine.[54]

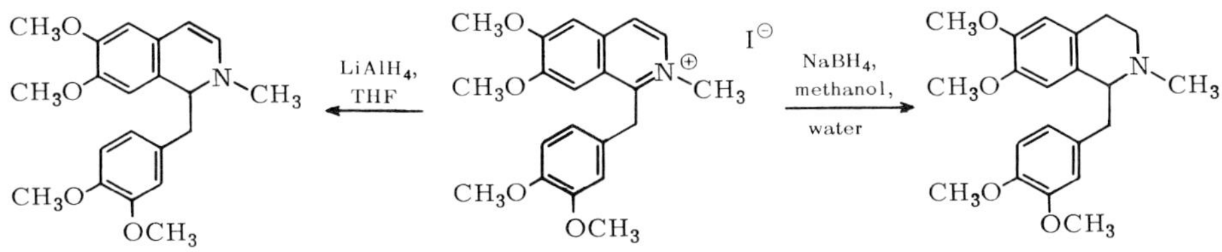

51 Papaverine methiodide (±)-Laudanosine

D. Nitration, Sulfonation, Acetylation, and Bromination

Ring A in papaverine is deactivated and less sensitive to electrophilic attack than ring C, so that the alkaloid can be cleanly nitrated with nitric acid in acetic acid to 6′-nitropapaverine (**52**).[55,56] Papaverine can also be sulfonated at C-6′ with cold sulfuric acid or acetylated at the same position under the influence of acetic anhydride and sulfuric acid. In like fashion, bromination occurs at C-6′.

E. Cleavage of the C-1 to C-α Bond of Benzylisoquinoline Salts

When 6′-nitropapaverine (**52**) was quaternized with methyl iodide and the resulting salt treated with excess potassium borohydride in refluxing aqueous ethanol, fission of the C-1 to C-α bond occurred with formation of 2-methyl-1,2,3,4-tetrahydroisoquinoline and 2-nitro-4,5-dimethoxytoluene. The following mechanism was presented by Neumeyer and co-workers to rationalize the formation of the cleavage products.[56]

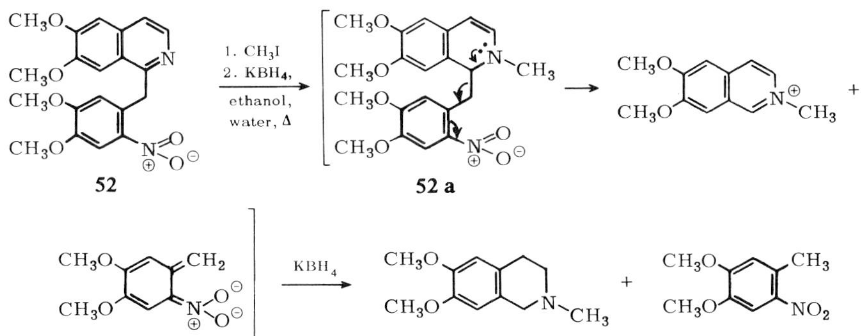

2-Methyl-tetrahydroisoquinoline 2-Nitro-4,5-dimethoxytoluene

Such a cleavage does not occur with the nitrobenzylisoquinoline salt **53**, which is simply reduced to its tetrahydrobenzylisoquinoline analog. The nitro group in the intermediate **52a** is coplanar with ring C and is able to stabilize the formation of the nitrotoluene through resonance. But in salt **53** the nitro group is sterically hindered and forced out of the plane of ring C, so that only reduction to the tetrahydrobenzylisoquinoline derivative takes place.[56]

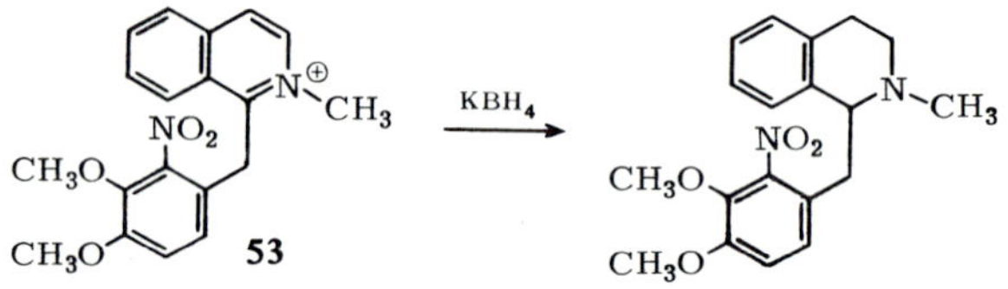

F. Rearrangement of 1,2-Dihydroisoquinolines

An interesting overall 1,3-shift was first observed by Knabe and Kubitz in 1963. When *N*-methyl-1,2-dihydropapaverine (**54**) was heated with dilute acid at 100° for a few minutes, the product was the rearranged immonium ion **55**.[57]

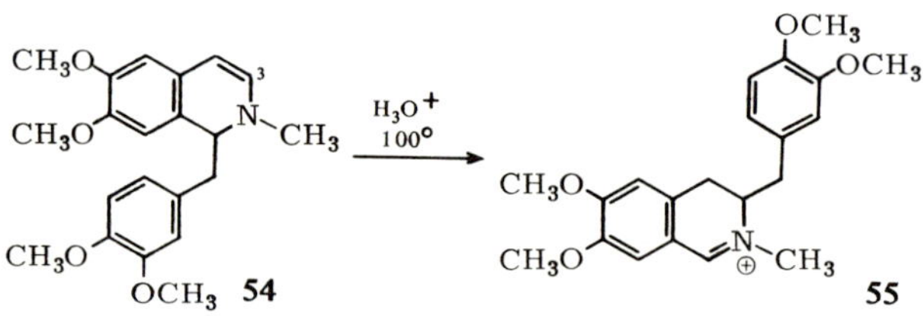

Intermolecular cross migration must have occurred in this instance since heating a mixture of **54** and its tetra-*O*-ethyl homolog in dilute hydrochloric acid gave rise to a mixture of four 3-benzyl-3,4-dihydroisoquinoline salts.[58]

On the other hand, it has been found by Dyke and his associates that rearrangement of a mixture of the amino acetals **56** and **57** leads to only two products, **58** and **59**, indicating that in this case the reaction takes an intramolecular course.[59]

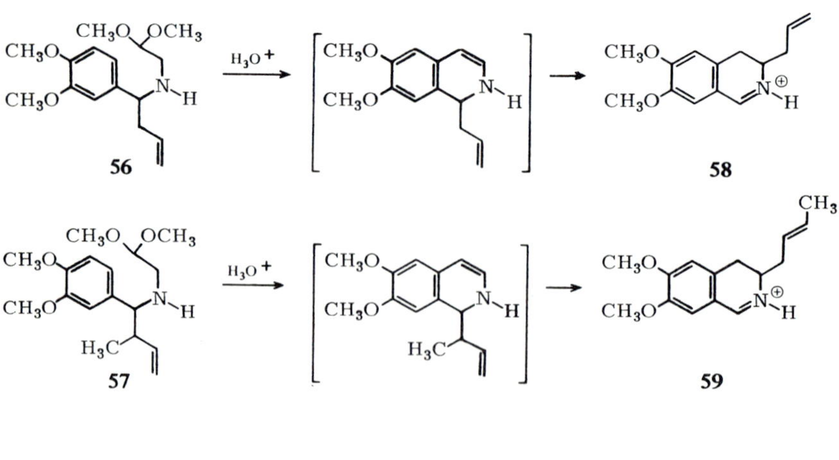

In line with this later finding is the observation that rearrangement of the amino acetal **60** affords an allenic immonium salt, so that in this specific case the reaction has been viewed as a concerted suprafacial [3,3] sigmatropic process.[59]

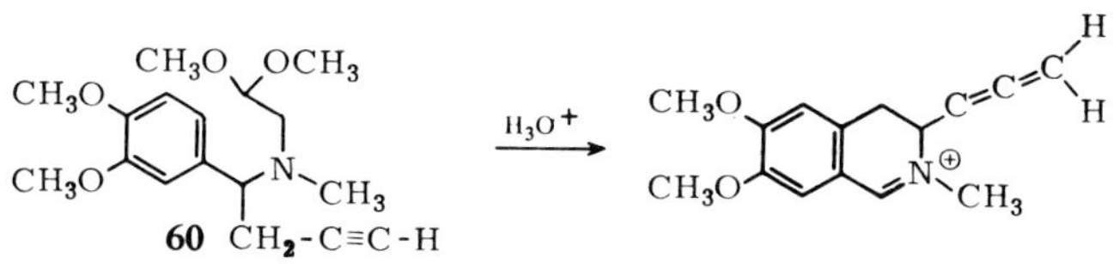

The rearrangement has also been observed with 1,1-disubstituted 1,2-dihydroisoquinolines.[60] When the enamine **61** was treated with dilute acid, the products obtained were the immonium salt **62** and the isoquinoline salt **63** which had been originally utilized in the preparation of the enamine **61**.

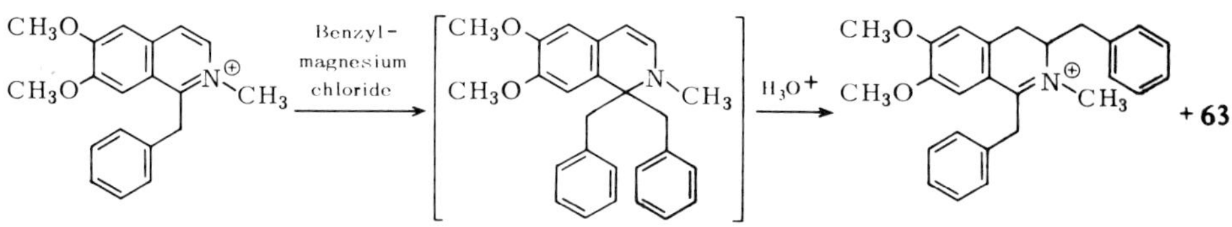

63 **61** **62**

Several recent investigations have been carried out relating to the mechanism of the rearrangement of 1,2-dihydroisoquinolines,[60,60a] but some of the results continue to be apparently contradictory and difficult to reconcile. For example, Knabe and his group have found that[60b,c]:

(i) The yield of rearranged product drops off sharply if the benzylic moiety that migrates is substituted with an electron attracting function such as a bromo or a nitro group in the *ortho* or *para* positions, so that the benzylic moiety does not migrate as a carbanion.

(ii) The rearrangement of the optically active dihydroisoquinoline **63a**, possessing the asymmetric center in the side chain, yielded a racemic rearranged product.

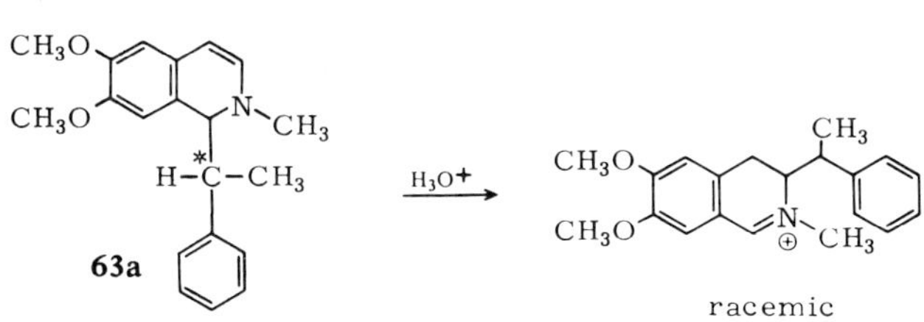

(iii) The rearrangement could be of a free radical nature since there is an initial induction period which is shortened by a nitrogen atmosphere, but is extended by the presence of oxygen. The reaction is also slowed down when free radical traps are present.

(iv) In the rearrangement of *N*-isopropyl-1,2-dihydropapaverine, the symmetrical bis(3,4-dimethoxyphenyl)ethane was isolated, which is presumably formed from free radical precursors.

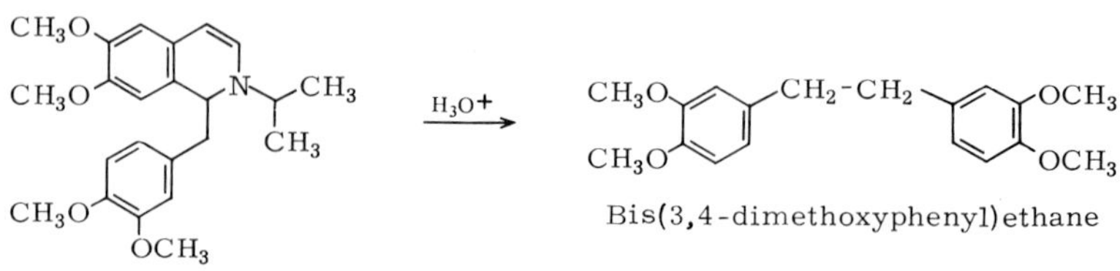

Dyke and his associates, on the other hand, have determined that[60d]:

(i) Rearrangement of **56** in dilute hydrochloric acid is intramolecular and concerted, with no evidence for cross-migration.

(ii) Acid treatment of the optically active amino aldehyde represented below followed by sodium borohydride reduction of the resulting imine yielded an optically active tetrahydroisoquinoline resulting from intramolecular rearrangement.

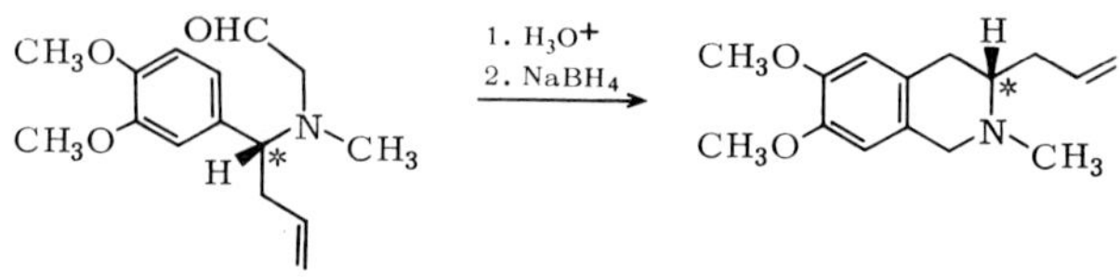

It should be noted in conclusion that the optimum pH for the dihydroisoquinoline rearrangement is 4.2,[60b,c] and that an important side reaction that accompanies the rearrangement is the formation of pavine derivatives[60d] (see Chapter 4). It might just be possible that at low temperatures the rearrangement of 1,2-dihydroisoquinolines is a concerted suprafacial sigmatropic process, while at higher temperatures a free radical mechanism is involved.

G. Condensation at the α-Methylene of Benzylisoquinolines

The α-methylene group in papaverine is activated. Formaldehyde reacts readily at this center to give methylenepapaverine (**64**) which can be hydrogenated to α-methylnorlaudanosine (**65**).[61]

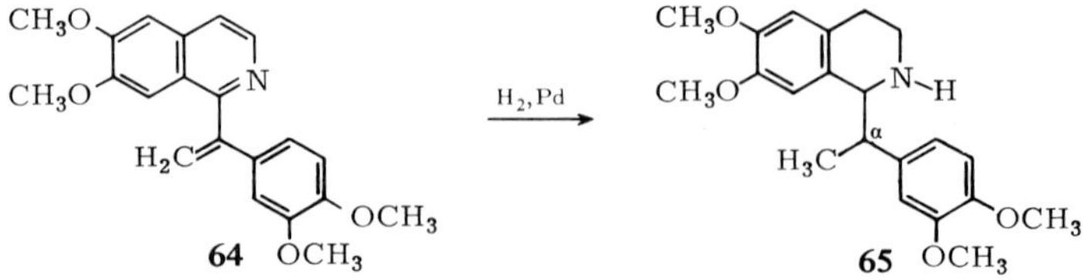

64 **65**

When 6′-acetylpapaverine (**66**) was subjected to hypobromite oxidation, it was possible to isolate the bromoindenone **67**. The latter material could be converted to the indene **68**, the indane **69**, or the indanol **70** depending upon reducing conditions (Scheme XI).[62]

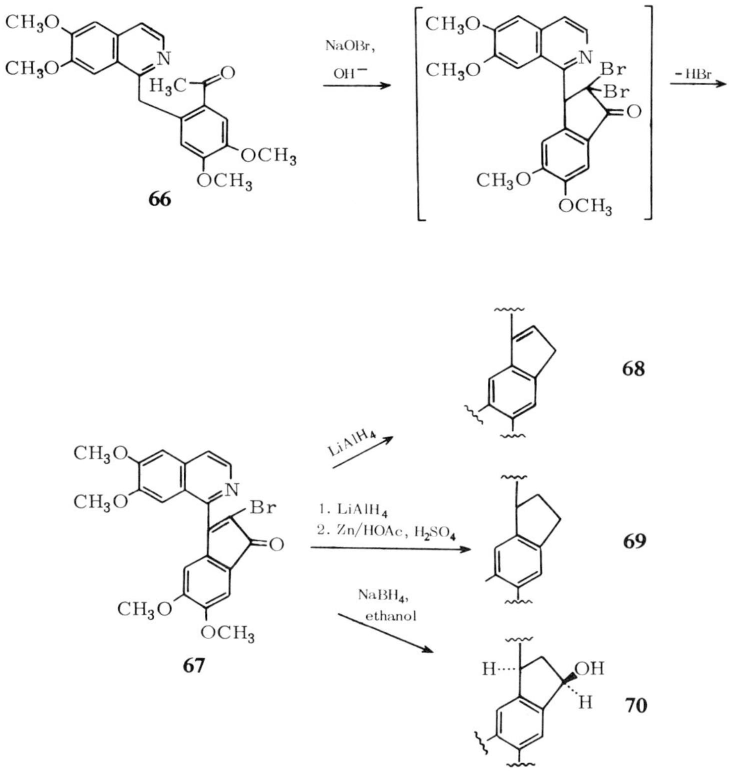

Scheme XI

H. Formation of 6′-Hydroxymethylpapaverine and Methylenedipapaverine

When formaldehyde is added to papaverine in acetic acid with a little hydrochloric acid, condensation occurs at C-6′ rather than at the α-methylene. The product is the crystalline 6′-hydroxymethylpapaverine which can undergo acid-catalyzed condensation with a molecule of papaverine to yield methylenedipapaverine. Otherwise, heating 6′-hydroxymethylpapaverine in a formic acid – formamide solvent leads to norcoralydine.[62a]

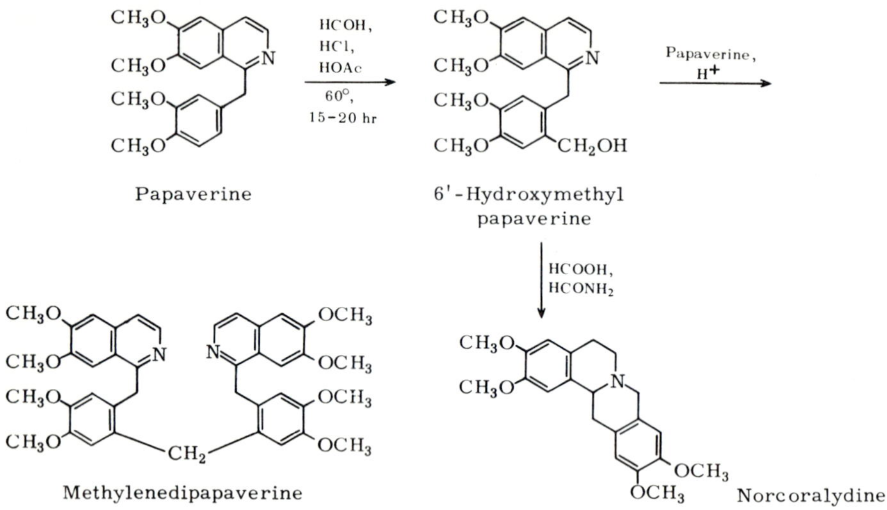

Papaverine

6'-Hydroxymethyl papaverine

Methylenedipapaverine

Norcoralydine

I. Azaberbinones from 6'-Nitropapaverine

Cava and co-workers have shown that iodine oxidation of 6′-nitropapaverine yields a dark red azaberbinone *N*-oxide which upon bisulfite reduction produces a yellow

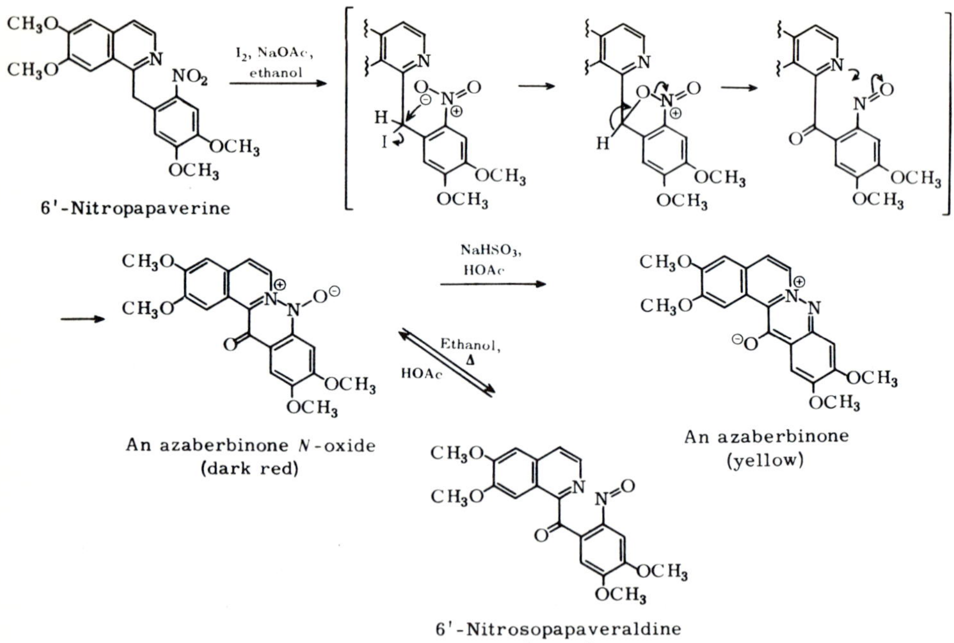

6'-Nitropapaverine

An azaberbinone *N*-oxide (dark red)

An azaberbinone (yellow)

6'-Nitrosopapaveraldine (green)

azaberbinone. Treatment of the dark red azaberbinone *N*-oxide with hot ethanol leads to the green 6′-nitrosopapaveraldine which reverts back to the dark red azaberbinone *N*-oxide on treatment with acetic acid.[62b]

The aforementioned transformations are to be compared with the reaction of 6′-nitropapaverine with methanolic potassium hydroxide which leads to the yellow anthranilopapaverine.[62c]

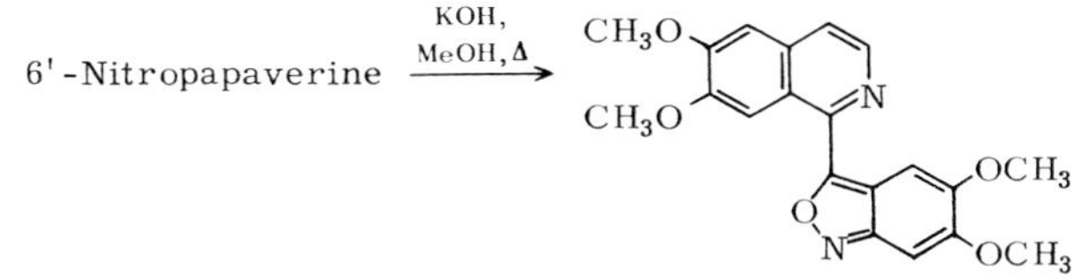

Anthranilopapaverine (yellow)

J. Formation of a Lactam from a 1-Benzoyl-3,4-dihydroisoquinoline

The dark green reaction product generated by heating the dihydroisoquinoline **70a** in acetic anhydride has been formulated as the tetracyclic lactam **70b**. No mechanism was offered for this transformation.[62d]

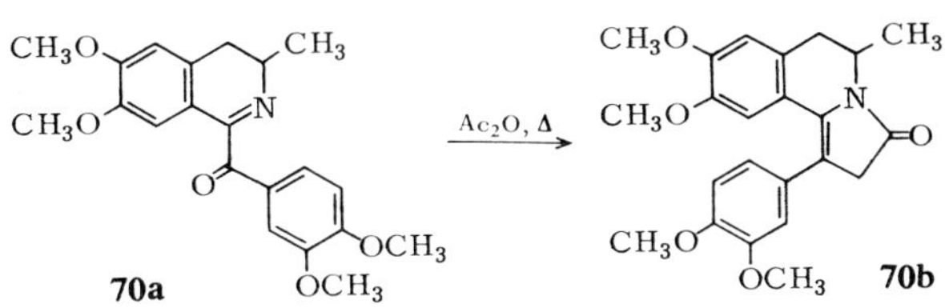

K. Racemization of Tetrahydrobenzylisoquinolines

Optically active tetrahydrobenzylisoquinolines can be racemized with Adams catalyst in ethanol at room temperature, but the presence of phenolic groups slows down the reaction rate. Some of the yields and reaction times are indicated below. It is worth pointing out that the *O*-diphenolic compound **71** could not be racemized. Similar epimerizations have been observed with the tetrahydroprotoberberines.[63]

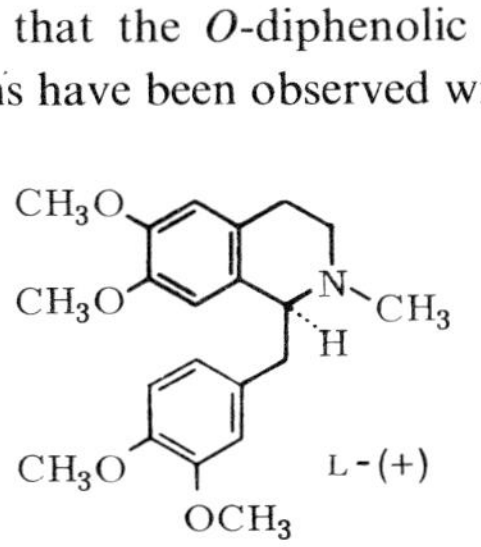

45 hr, 92% yield of racemate

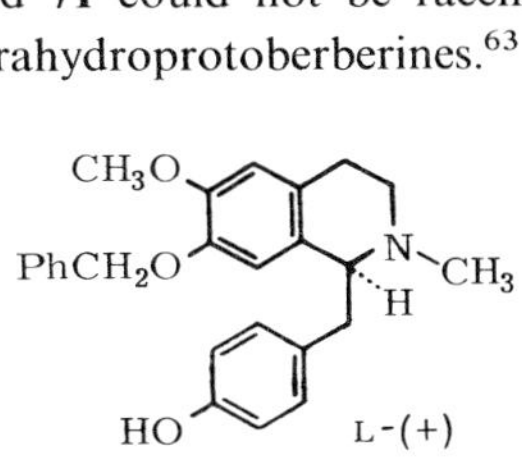

112 hr, 83% yield of racemate

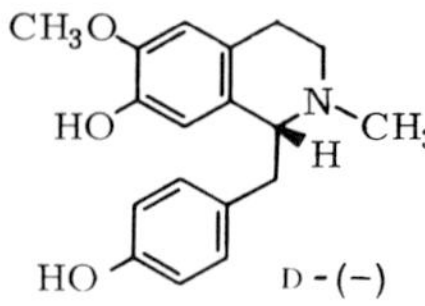

129 hr, 78% yield of racemate

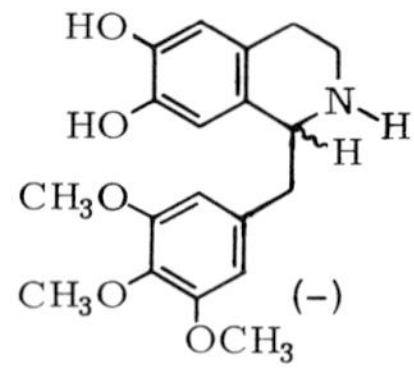

71

No racemization

L. Photolysis

When papaverine is photolyzed in methanol or ethanol, cleavage of the benzyl group occurs with concomitant introduction of the alcohol alkyl residue at C-1.[64]

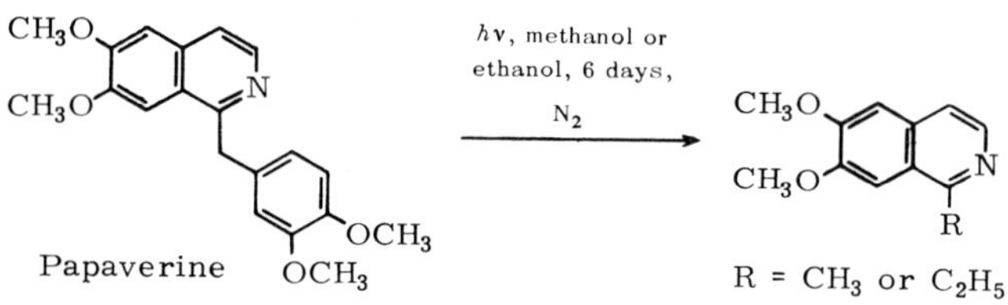

M. Fission of Ring B with an Acid Chloride

If an *N*-methyltetrahydrobenzylisoquinoline is refluxed in the presence of benzoyl chloride, cleavage of ring B occurs through *β*-elimination with formation of a bicyclic amide. Similar reactions have been observed with aporphines and phthalideisoquinolines.[65]

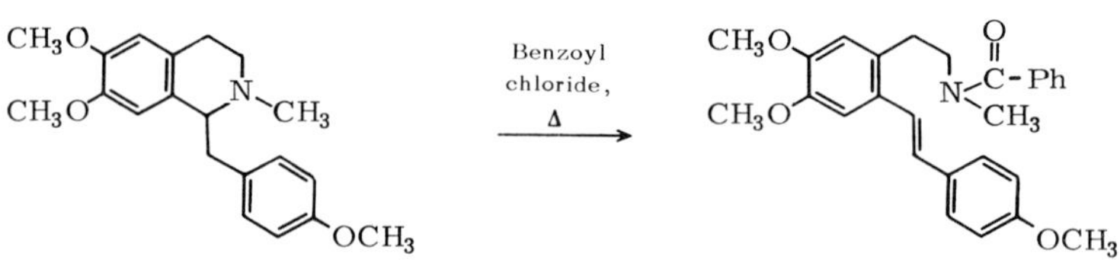

N. Fission of Ring B with Cyanogen Bromide under Solvolytic Conditions

When an *N*-methyltetrahydrobenzylisoquinoline such as laudanosine is treated with cyanogen bromide in an ethanol–chloroform solvent, an S_N2 displacement with nucleophilic attack by ethanol on the intermediate cyano quaternary salt occurs, resulting in cleavage of the C-1 to N-2 bond.[66]

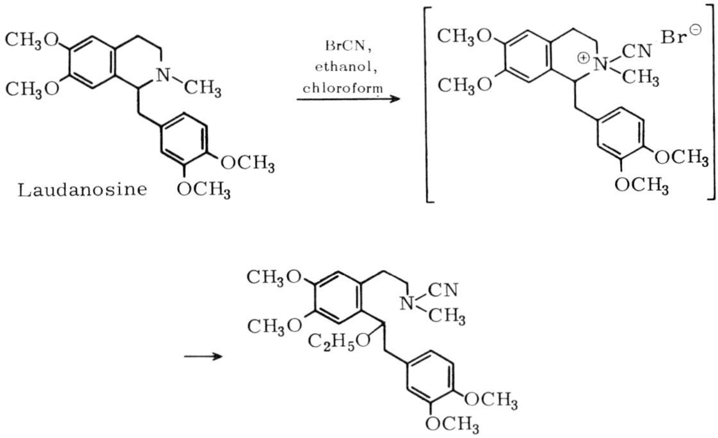

Laudanosine

O. *Selective O-Demethylation of Papaverine and Thalifendlerine*

The course of *O*-demethylation of papaverine with mineral acid involves first the two methoxyl groups on ring C to provide a mixture of the monophenols **72** and **73**, and the diphenol **74**. The next methoxyl group to be hydrolyzed is that at C-7 to form the triphenol **75**. Finally the C-6 methoxyl is affected to give papaveroline.[67]

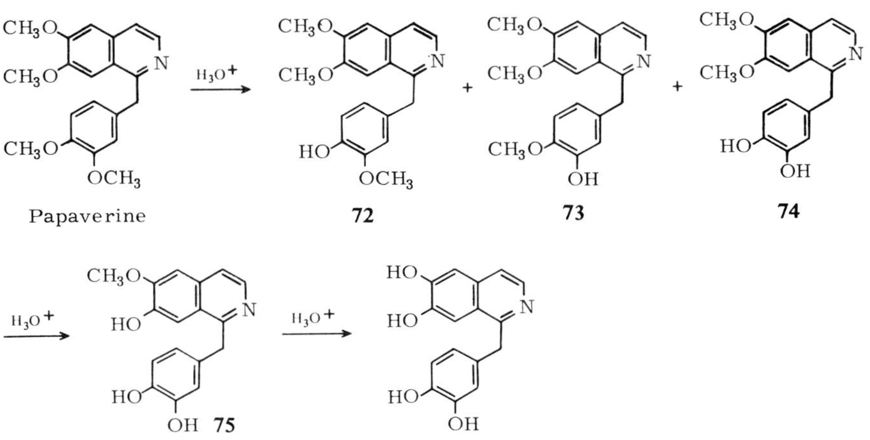

Papaverine **72** **73** **74** **75** Papaveroline

If, on the other hand, a thermal S_N2 process is used, *O*-demethylation proceeds in a different direction. Fusion of papaverine hydrochloride furnished the monophenol **76** and the betaine **77**. When the latter was heated with methyl iodide in a sealed tube, the monophenol **78** was produced.[18]

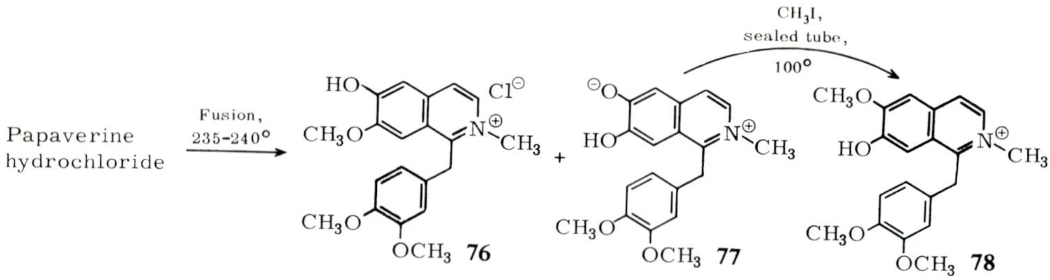

Access to the monophenolic analogs of laudanosine is therefore available since each of the above products can be reduced to the tetrahydro derivative. (See also Chapter 4, Section I, C.)

The preferential cleavage of the middle of three neighboring aromatic methoxyl groups by mineral acids has been explained on steric grounds, and has recently been applied to the tetrahydroisoquinoline alkaloid thalifendlerine.[67]

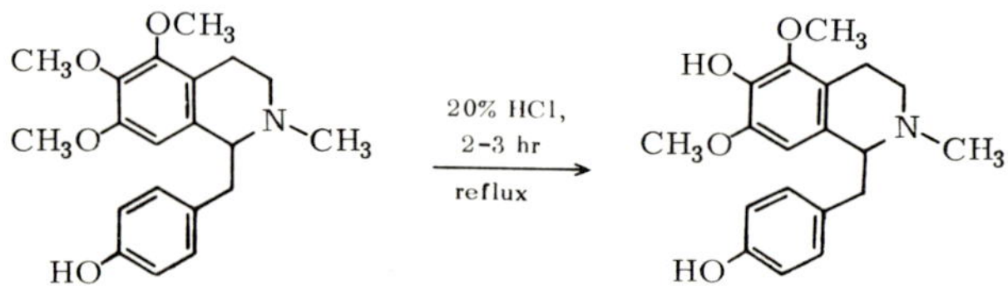

Thalifendlerine

(For other examples of the preferential cleavage of methoxyl ethers see this chapter, Sections VI, F and VII, B; Chapter 1, Section VI, D; and Chapter 10, Section V, C.)

VIII. Absolute Configuration

In a by now classical experiment, Corrodi and Hardegger reported in 1956 that ozonization of (−)-*N*-norlaudanosine afforded *N*-*β*-carboxyethyl-L-aspartic acid of known absolute configuration.[68]

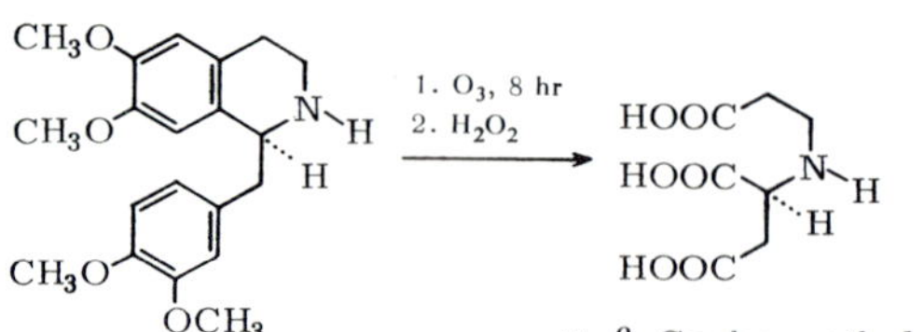

(−)-*N*-Norlaudanosine

N-*β*-Carboxyethyl-L-aspartic acid

The determination of the chirality of the other tetrahydrobenzylisoquinoline alkaloids therefore ultimately revolves around direct or indirect comparison with (−)-*N*-norlaudanosine or one of its derivatives.

The shape of the ORD curve of a benzyltetrahydroisoquinoline near 200−225 mμ can be used to establish its absolute configuration. D-(−)-Laudanidine as well as D-(−)-armepavine give ORD curves with ascending tails below 225 mμ. On the other hand, L-(−)-*N*-norarmepavine shows a descending tail. It is also a fact that the specific rotation at the sodium D line is not a sufficient criterion in this series and that ORD measurements must be carried out before conclusions may be drawn concerning absolute configuration.[69−71]

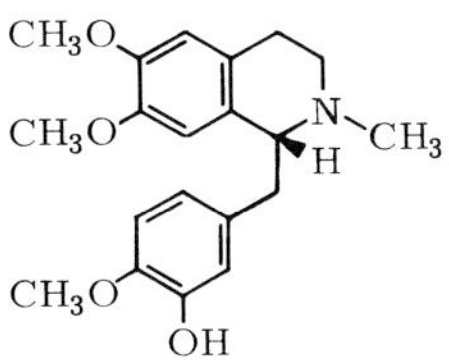

D-(−)-Laudanidine

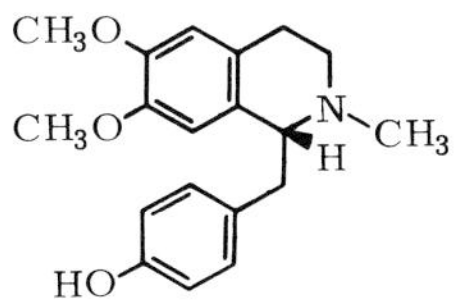

D-(−)-Armepavine (*R* configuration)

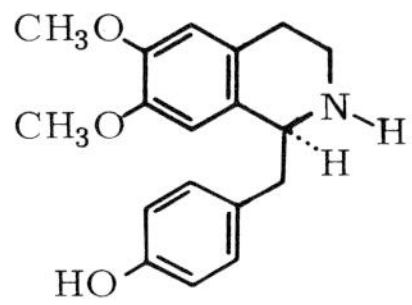

L-(−)-*N*-Norarmepavine (*S* configuration)

Another useful ORD feature is the sign of the Cotton effect between 270 and 290 mμ and between 240 and 255 mμ. Tetrahydrobenzylisoquinolines of the D series show negative Cotton effects at these wavelengths, while the L enantiomers exhibit positive effects. Quaternization of the nitrogen causes no substantial change in the ORD curve.[70,71] Several ORD and CD curves of tetrahydrobenzylisoquinolines have been recorded.[35,35a] Whenever comparing ORD or CD curves, it is imperative that the substitution pattern of the compounds being considered be as similar as possible.

IX. Biosynthesis

When (+)-tyrosine labeled as shown was fed to *Papaver somniferum* L., only carbons 1 and 3 of papaverine were found to be radioactive – the two units derived from tyrosine being incorporated with nearly equal weight.[72]

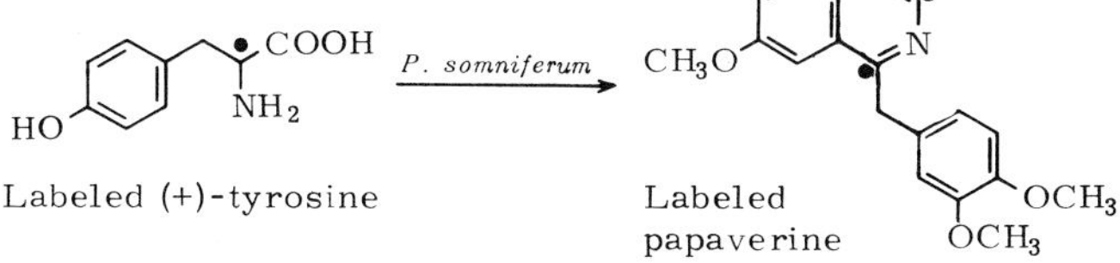

Labeled (+)-tyrosine

Labeled papaverine

Nevertheless, the two aromatic units that originate from tyrosine and that partake in the Pictet–Spengler type condensation leading to the benzylisoquinoline nucleus are not identical.[73,74] After labeled dopamine had been used as a precursor for the phthalide-isoquinoline alkaloid hydrastine, it was found that only one dopamine unit had been incorporated.[75] There is a possibility that the other unit required for the biogenesis of *N*-norlaudanosoline is 3,4-dihydroxyphenylpyruvic acid.[76] Tyrosine is therefore converted to dopa which then affords dopamine as well as 3,4-dihydroxyphenylpyruvic acid.

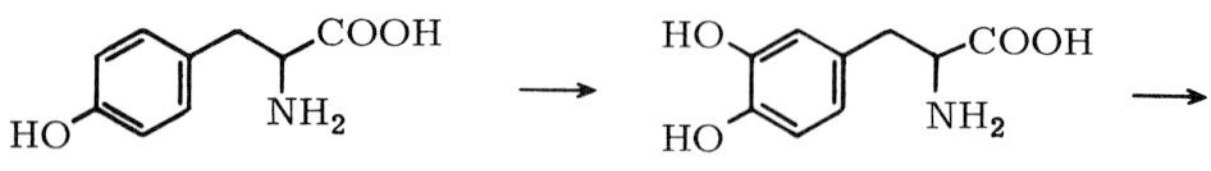

Tyrosine Dopa

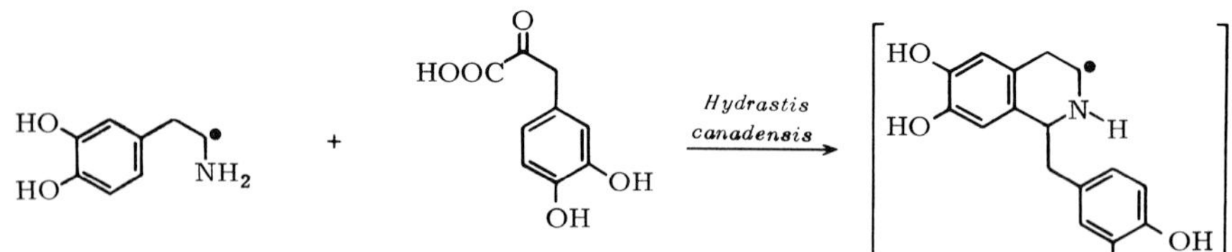

Labeled dopamine 3,4-Dihydroxyphenyl-pyruvic acid *N*-Norlaudanosoline

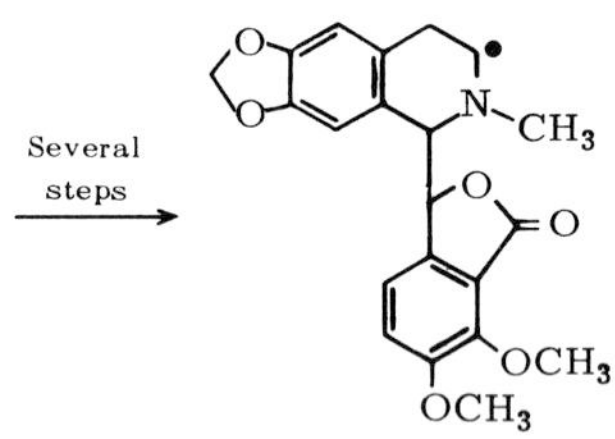

Labeled hydrastine

Labeled (±)-*N*-norlaudanosoline has been shown to be the precursor of reticuline in *P. somniferum*, and reticuline is the precursor for the morphine alkaloids.[77] The reticuline fraction from opium consists of about 60% of the (+)-isomer and 40% of the (−)-isomer, and this ratio results from the fact that the (−)-enantiomer is converted more readily into the morphine bases.[78]

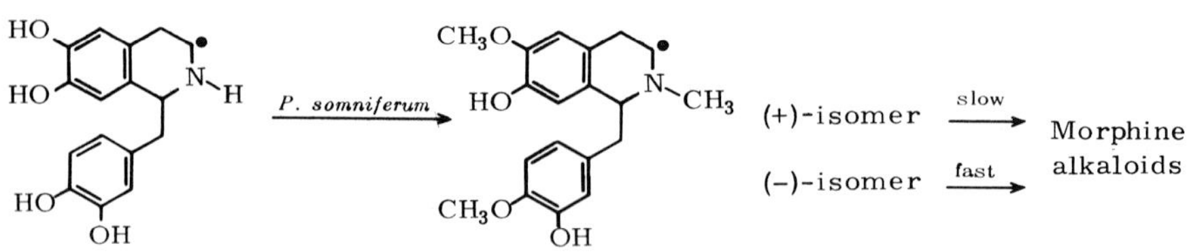

Labeled *N*-norlaudanosoline Labeled reticuline

N-Norlaudanosoline is also the precursor of papaverine in *P. somniferum* since feeding the labeled tetrahydrobenzylisoquinoline base to the plant resulted in 1.5% incorporation into papaverine.[79]

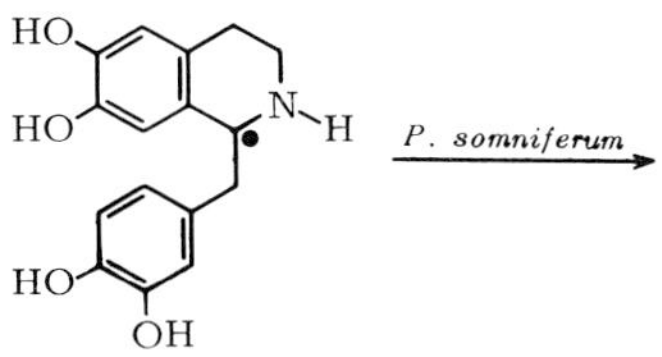

Labeled *N*-norlaudanosoline Labeled papaverine

X. Pharmacology

The main effect of papaverine hydrochoride is relaxation of the smooth muscle.[49] It also has coronary vasodilator properties, but it is not as effective as nitroglycerine in the treatment of angina pectoris. Papaverine hydrochloride has been used in vasospasm accompanying peripheral arterial embolism, pulmonary embolism, and cerebrovascular thrombosis.[80] The drug exhibits a distinct local anesthetic activity. It is not a narcotic and is not addictive, but its side effects are drowsiness, constipation, increased reflex excitability, and gastric distress.

Papaverine glyoxylate salt possesses interesting spasmolytic, vasodilating, and oxygen-sparing properties useful in the treatment of arterial and venous disorders and in cases of disturbed oxidation–reduction metabolic processes.[81] Some tetrahydropapaverine derivatives may be useful as muscle relaxants.[82]

A series of synthetic analogs of papaverine have also found some use in therapy. Dioxyline (Eli Lilly and Co.) is used in the form of its phosphate salt as a coronary and peripheral vasodilator,[82a] while its methylenedioxy analog, 3-methyl-6,7-methylenedioxy-1-piperonylisoquinoline, is a smooth muscle relaxant.[82b] Ethaverine, the tetraethyl homolog of papaverine, is another smooth muscle relaxant.[82c]

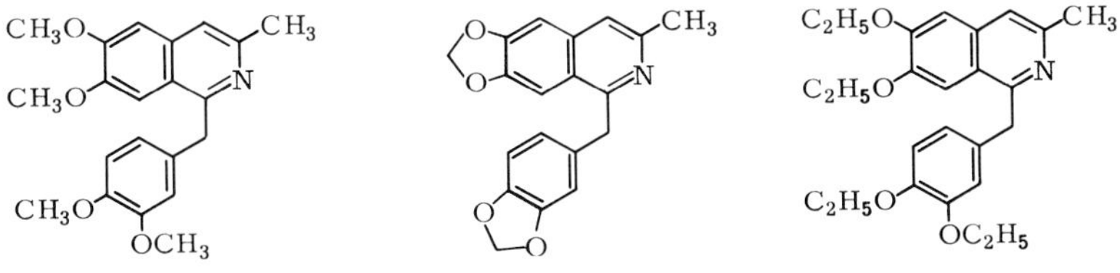

Dioxyline 3-Methyl-6,7-methylenedioxy-1-piperonylisoquinoline Ethaverine

The new alkaloid (+)-demethylcoclaurine, lately isolated from the embryo of the lotus, *Nelumbo nucifera* Gaertn. (Nymphaceae), was also found to relax significantly the smooth muscle and uterine strips.[82d]

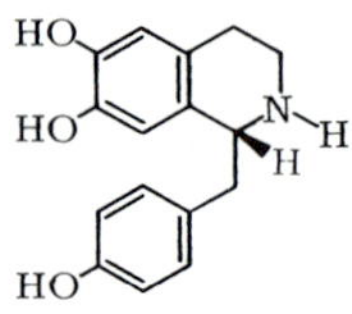

(+)-Demethylcoclaurine

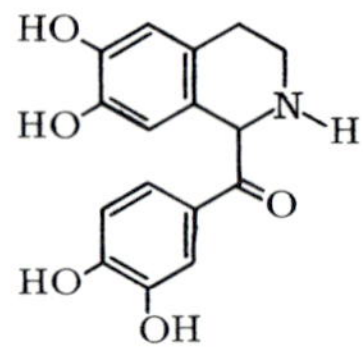

BB 572

The ketonic and tetraphenolic tetrahydrobenzylisoquinoline BB 572, produced by Bristol Laboratories, is a smooth muscle relaxant, a gastric secretion inhibitor, and a tachycardic as well as a β-adrenergic simulator.

N-Norlaudanosoline has been reported to have sympathomimetic activity.[84,85]

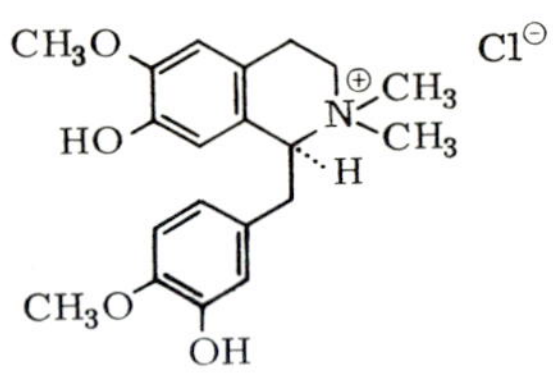

(+)-Tembetarine

The cardiovascular effects of the glycosidic benzylisoquinoline veronamine have been studied. Intravenous injection of graded doses of veronamine to anesthetized dogs produced a dose-related decrease in mean arterial blood pressure and a reduction in pressure of the perfused hindlimb. The duration of the systemic hypotension varied, lasting 1–3 minutes in low doses of veronamine, while a dose of 1.0 mg/kg produced hypotension lasting 5–20 minutes. Bradycardia, which outlasted the systemic hypotension, and reductions in left ventricular *dp/dt* were observed.[85a] (+)-Tembetarine chloride has also shown some hypotensive activity in anesthetized cats and dogs.[85b]

Appropriately substituted benzylisoquinolines such as 1-(3′,4′,5′-trimethoxybenzyl)-6,7-dihydroxytetrahydroisoquinoline are potent β-adrenergic agonists, involved in the mobilization of free fatty acids in adipose tissue.[85c] This same compound is also a potent tracheal relaxation agent, since tracheal relaxation and lipolysis are considered to be β-adrenergic receptor systems.[85d,e] The compound is an effective bronchodilator.[83]

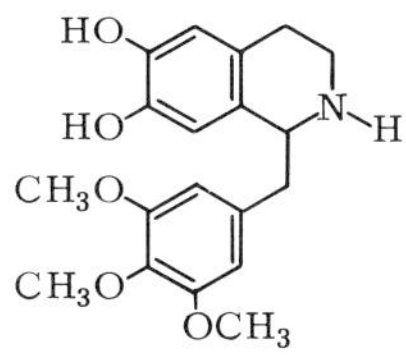

1-(3',4',5'-Trimethoxybenzyl)-6,7-dihydroxytetrahydroisoquinoline (≡Trimethoquinol)

XI. A Possible Biochemical Basis for Alcohol Addiction: A Hypothesis and its Critique

In Chapter 1, Section X, it was mentioned that simple tetrahydroisoquinolines may account for the hallucinosis and seizures that alcoholics undergo when the concentration of alcohol in the blood is receding. The following is a theory for alcohol addiction.

Because of the resemblance of the symptoms occurring upon withdrawal of either alcohol or the opium alkaloids, it seems possible that addiction to these chemicals is very similar, if not identical. The ingestion of ethanol results in an increase in the acetaldehyde concentration in the brain. Acetaldehyde in turn increases the amounts of 3,4-dihydroxyphenylacetaldehyde by inhibiting the action of the enzyme aldehyde dehydrogenase which would normally oxidize this diphenolic aldehyde to 3,4-dihydroxyphenylacetic acid.

With excessive concentrations of 3,4-dihydroxyphenylacetaldehyde in the brain, Pictet–Spengler condensation with dopamine can occur to form the key intermediate *N*-norlaudanosoline. *N*-Norlaudanosoline can then undergo further transformation to morphine. As discussed above in the section on biosynthesis, it is known that *N*-norlaudanosoline is the requisite intermediate in the biosynthesis of morphine in the opium poppy, *P. somniferum*, and a similar transformation may occur in man.

Using rat tissue homogenates with dopamine as a substrate, it was determined that addition of ethanol or acetaldehyde to the substrate and the cofactor for aldehyde dehydrogenase, nicotinamide adenine dinucleotide, gave *N*-norlaudanosoline after incubation. Furthermore, when labeled *N*-norlaudanosoline was administered intravenously to rats, it was found that radioactive morphine-like alkaloids were produced in the animals.[86]

Alcohol dependence might, therefore, result from alcohol-induced inhibition of the oxidation of 3,4-dihydroxyphenylacetaldehyde from dopamine to 3,4-dihydroxyphenylacetic acid. It is also possible that other addictive drugs such as chloral hydrate and paraldehyde may have a similar effect of inhibiting aldehyde dehydrogenase and promoting the synthesis of *N*-norlaudanosoline (Scheme XII).[86]

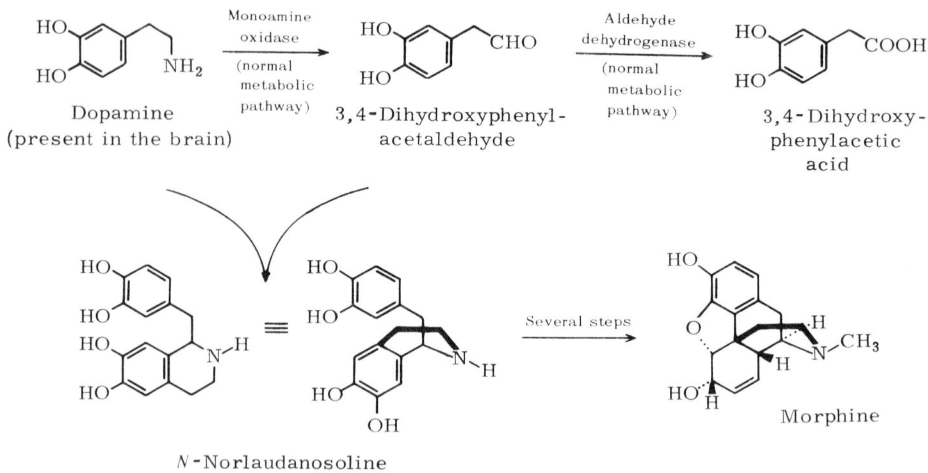

Scheme XII

The above hypothesis has been seriously questioned by Seevers who has noted that the physiological effects of alcohol ingestion are substantially different from those resulting from morphine uptake, and that no specific mutual cross dependence or cross tolerance exists between morphine-like drugs and ethanol. It is a fact that morphine users prefer barbiturates, amphetamines, or cocaine over alcohol which is usually their last choice.[86a]

XII. Mass Spectroscopy

The main cleavage that the benzyltetrahydroisoquinolines undergo is at the C-1 to C-α bond, which is doubly benzylic and also adjacent to the nitrogen atom. In the case of compound **79**, the molecular ion has an intensity of only 0.2% of the base peak, which is at *m/e* 178. An intense peak is also found at *m/e* 163 due to the loss of the methyl group from the base ion. Since the charge prefers to remain on the nitrogen atom, the ion representing the lower half or ring C of the molecule shows up only as a minor peak.[87,88]

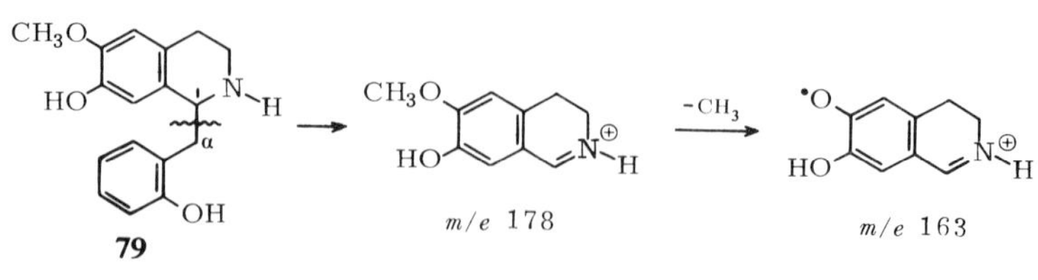

The most important peaks in the mass spectrum of papaverine are due to the M, $M-1$, $M-15$ (base), and $M-31$ ions. The other peaks are minor and of little analytical significance.[89]

XIII. NMR Spectroscopy

Because of steric factors, ring C in tertiary and quaternary tetrahydrobenzylisoquinolines is located close to or slightly on top of ring A. The net result is an upfield shift for the C-8 proton and the C-7 methoxyl. When the nitrogen is secondary, however, ring C lies in the proximity of ring B, and no upfield shift is detected. Two examples illustrate this point.[90-93]

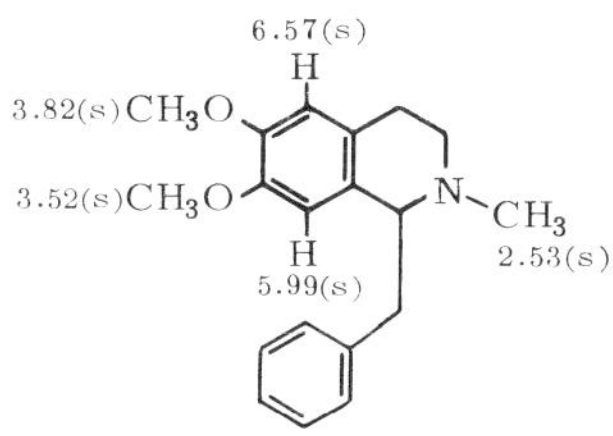

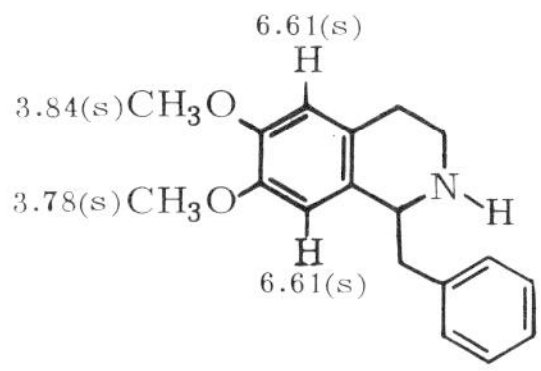

In the case of the alkaloid *O*-methylarmepavine, the following assignments may be made for the NMR peaks[94]:

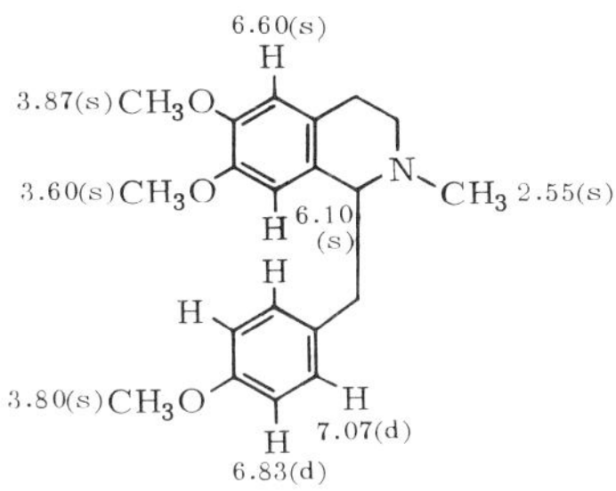

NMR spectral values for *O*-methylarmepavine

Georgiev and Mollov have observed that no upfield shift of the C-8 proton occurs when the C-7 substituent is hydroxyl.[95a]

Working with 7,8,4′-trisubstituted tetrahydrobenzylisoquinolines, it has been noted that the chemical shifts of the individual methoxyl substituents can be identified by recording the spectrum of the compound first in deuteriochloroform and then in DMSO-d_6. The solvent shifts are in the order $C_8\text{-}CH_3O > C_7\text{-}CH_3O > C_{4'}\text{-}CH_3O$. The following example is a case in point.[35]

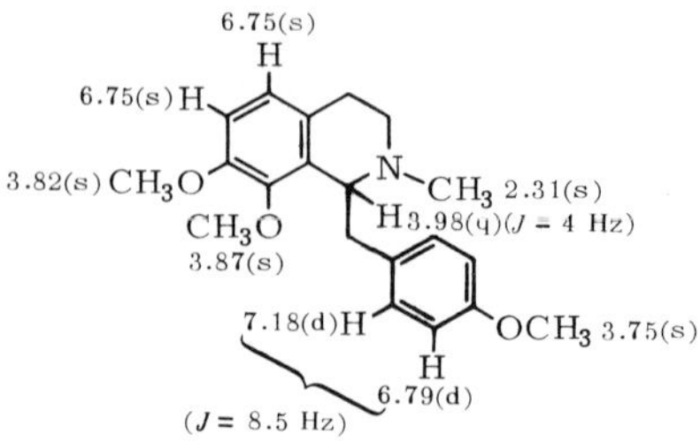

Spectrum taken in $CDCl_3$

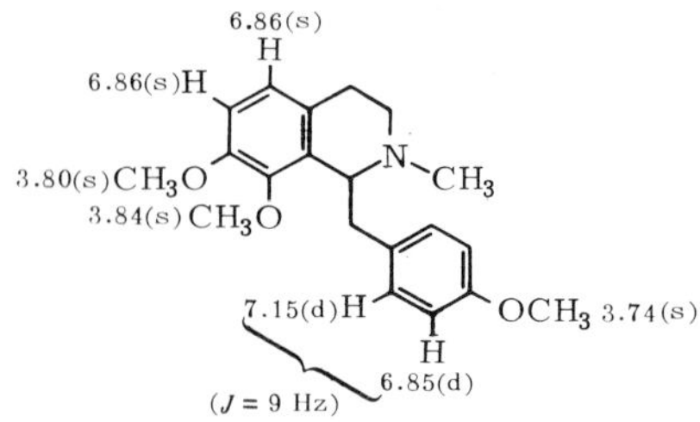

Spectrum taken in DMSO-d_6

The NMR spectrum (solvent unspecified) of the quaternary tetrahydrobenzyl-isoquinoline alkaloid lotusine chloride shows the two *N*-methyl groups with different chemical shifts because of the asymmetry around the nitrogen atom.[82d] The methoxyl as well as the C-8 protons appear appreciably upfield due to the large steric factor involved when the nitrogen atom is quaternary.

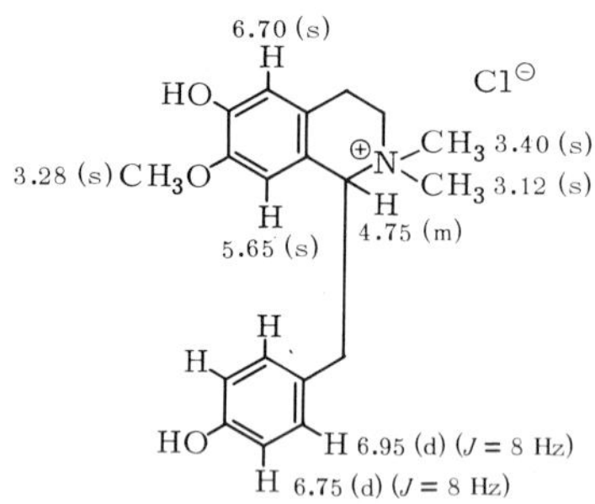

NMR spectral values for lotusine chloride

The NMR spectrum in DMSO-d_6 for the new quaternary aromatic benzylisoquinoline alkaloid *N*-methylpalaudinium chloride, obtained from *Thalictrum polygamum* Muhl. (Ranunculaceae), shows the $\overset{\oplus}{N}-CH_3$ peak downfield at $\delta 4.38$.[94a]

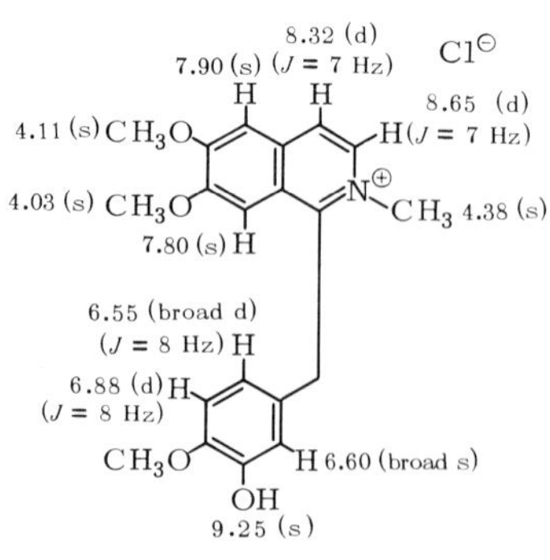

NMR spectral values for *N*-methylpalaudinium chloride

XIV. X-Ray Analysis

An X-ray study of an alkaloid hydrobromide monohydrate isolated from *Alseodaphne archboldiana* (Allen) Kost., a New Guinea tree of the Lauraceae family, revealed the crystal analyzed to be that of D-(+)-coclaurine. The C-6 methoxyl points away from the C-7 hydroxyl, and the C-1 hydrogen is beta. The ring C plane is nearly perpendicular to the mean plane of the isoquinoline ring system. Since the nitrogen is secondary, ring C lies on the nitrogen side, as previously predicted by NMR spectroscopy.[95]

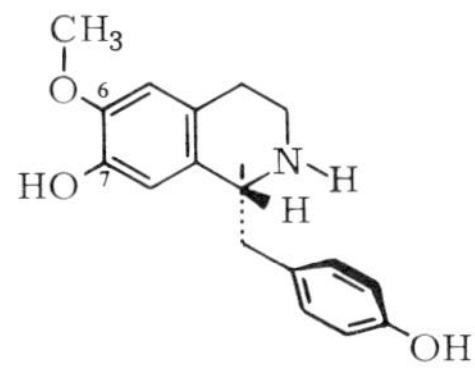

D-(+)-Coclaurine

XV. UV Spectroscopy

Tetrahydrobenzylisoquinolines show a maximum between 280 and 285 mμ. The extinction coefficient of this maximum is increased whenever methoxyl groups are replaced by methylenedioxy[96]; cf. tetrahydropapaverine and tetrahydroescholamine (see also Chapter 1, Section XIII). The spectra of benzylisoquinolines such as papaverine are more complex and are affected by the addition of acid.

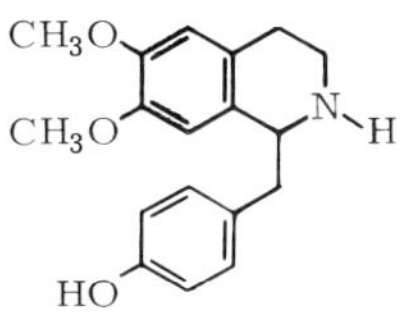

N-Norarmepavine[97]

λ_{max}^{EtOH} 282.5 mμ (3.71)
λ_{min}^{EtOH} 253 mμ (2.67)

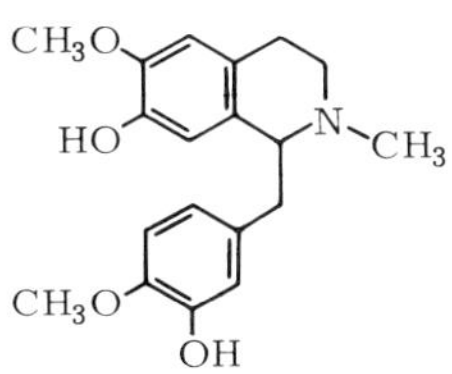

Reticuline[97]

λ_{max}^{EtOH} 285 mμ (3.83)
λ_{min}^{EtOH} 256 mμ (2.79)

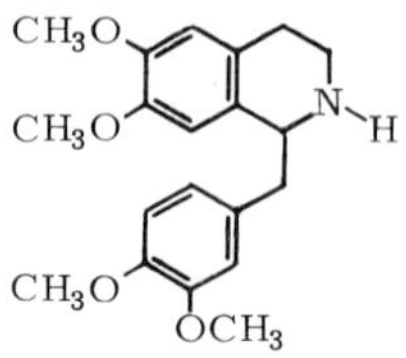

Tetrahydropapaverine[96]
(not yet isolated from nature)

λ_{max}^{EtOH} 225 sh and 282 mμ (4.24 and 3.85)

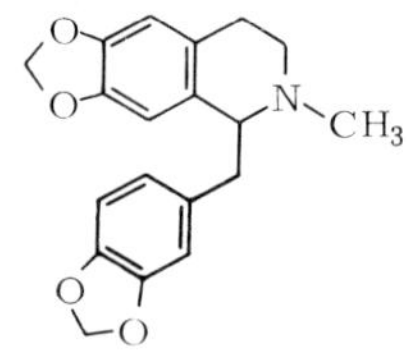

Tetrahydroescholamine[98]
(not yet isolated from nature)

λ_{max}^{EtOH} 237 and 290 mμ (4.1 and 4.1)
λ_{min}^{EtOH} 258 mμ (3.3)

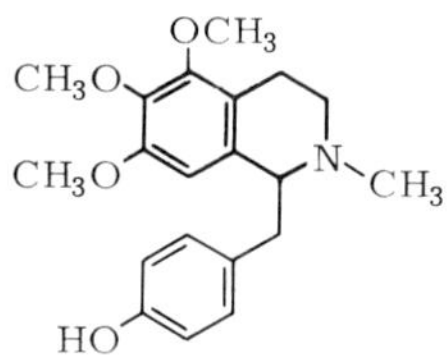

Thalifendlerine[99]

λ_{max}^{EtOH} 282 mμ (3.49)

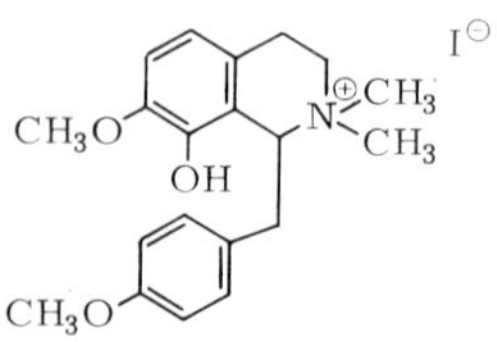

Petaline iodide[71]

λ_{max}^{EtOH} 206 sh, 233 sh, 278, and 284 mμ (4.87, 4.35, 3.59, and 3.59)

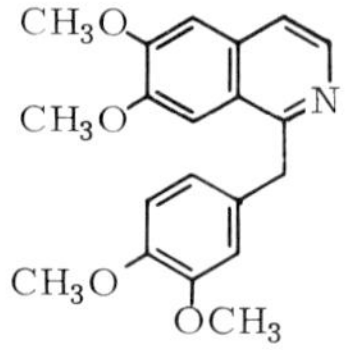

Papaverine[100]

λ_{max}^{EtOH} 239, 279–280, 314, and 327 mμ (4.83, 3.86, 3.60, and 3.67)
λ_{min}^{EtOH} 215, 261, 305, and 319 mμ (4.32, 3.77, 3.42, and 3.54)

Hydrochloride:
λ_{max}^{EtOH} 249–250, 280–282, and 311 mμ (4.69, 380, and 3.82)
λ_{min}^{EtOH} 214, 269, and 294 mμ (4.25, 3.74, and 3.73)

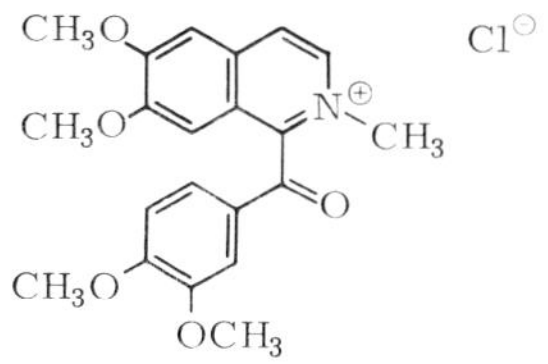

N-Methylxanthaline chloride[101]

λ_{max}^{EtOH} 257 and 330 mμ (4.70 and 4.31)

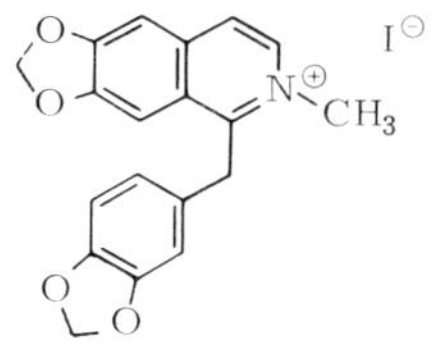

Escholamine iodide[102]

λ_{max}^{MeOH} 253, 292, 314, and 346 mμ (4.73, 3.92, 4.01, and 3.87)

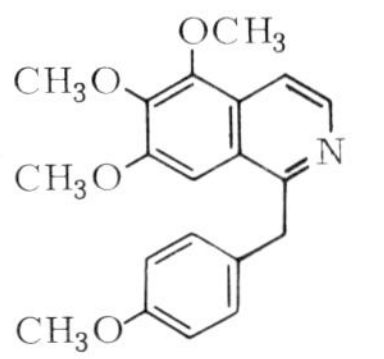

1-(4'-Methoxybenzyl)-5,6,7-trimethoxyisoquinoline[8] (not isolated from nature)

λ_{max}^{MeOH} 241.5, 278, 325–329, and 337 mμ (4.67, 3.70, 3.52, and 3.60)

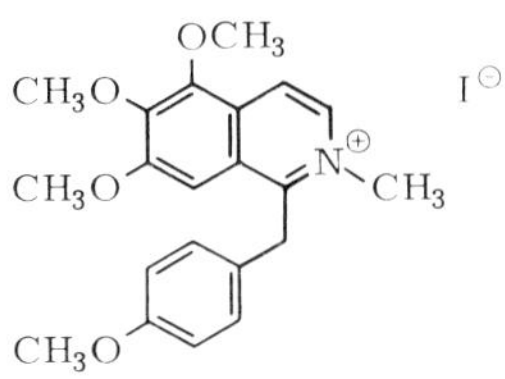

Takatonine iodide[8]

λ_{max}^{MeOH} 265 and 318 mμ (4.61 and 3.69)

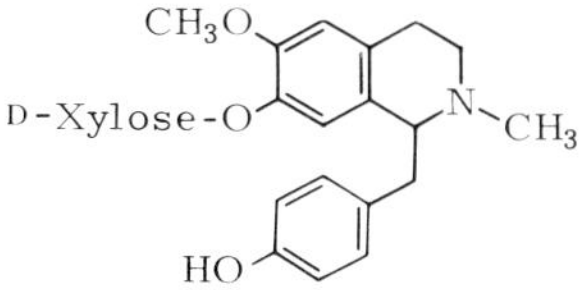

Latericine[103]

λ_{max}^{MeOH} 226 and 282 mμ (4.22 and 3.72)

REFERENCES

1. K. Ito, H. Furukawa, and H. Tanaka, *Chem. Commun.* p. 1076 (1970).
2. For a review on the benzylisoquinoline alkaloids, see V. Deulofeu, J. Comin, and M. J. Vernengo, *in* "The Alkaloids" (R. H. F. Manske, ed.), Vol. 10, p. 401. Academic Press, New York, 1968.
3. A. R. Battersby and B. J. T. Harper, *J. Chem. Soc., London* p. 3526 (1962).
4. C. Schöpf and K. Thierfelder, *Justus Liebigs Ann. Chem.* **497**, 22 (1932).
5. R. Robinson and S. Sugasawa, *J. Chem. Soc., London* p. 789 (1932).
6. A. Burger, Dissertation, Vienna (1927).
7. E. Fujita and T. Tomimatsu, *J. Pharm. Soc. Jap.* **79**, 1082 (1959).
8. S. Kubota, T. Masui, E. Fujita, and S. M. Kupchan, *J. Org. Chem.* **31**, 516 (1966).
9. N. J. McCorkindale, D. S. Magrill, M. Martin-Smith, S. J. Smith, and J. B. Stenlake, *Tetrahedron Lett.* p. 3841 (1964).
10. N. J. McCorkindale, A. W. McCulloch, D. S. Magrill, B. Caddy, M. Martin-Smith, S. J. Smith, and J. B. Stenlake, *Tetrahedron* **25**, 5475 (1969).
11. V. Preininger, A. D. Cross, and F. Šantavý, *Collect. Czech. Chem. Commun.* **31**, 3345 (1966).
12. W. Klyne, *Biochem. J.* **47**, xli (1950).
13. M. Shamma, M. G. Kelly, and Sr. M. A. Podczasy, *Tetrahedron Lett.* p. 4951 (1969); also in press.
14. S. Teitel and A. Brossi, *J. Heterocycl. Chem.* **5**, 825 (1968). The catalytic hydrogenation of β-nitrostyrenes can best be carried out using palladium on carbon as the catalyst and dilute hydrochloric acid as the solvent: D. P. Wagner, A. I. Rachlin, and S. Teitel, *Synthetic Commun.* ***1*** (1), 47 (1971).
15. K. W. Rosenmund, M. Nothnagel, and H. Riesenfeldt, *Chem. Ber.* **60**, 392 (1927).
16. C. Mannich and O. Walther, *Arch. Pharm.* (*Weinheim*) **265**, 1 (1927).
16a. A. Brossi and S. Teitel, *Helv. Chim. Acta* **49**, 1757 (1966).
17. G. Grethe, M. Usković, and A. Brossi, *J. Org. Chem.* **33**, 2500 (1968).
18. B. K. Cassels and V. Deulofeu, *Tetrahedron* **22**, Suppl. 8, Part II, 485 (1966).
19. J. Knabe and P. Horn, *Arch. Pharm.* (*Weinheim*) **300**, 547 (1967).
20. T. A. Montzka, T. L. Pindell, and J. D. Matiskella, *J. Org. Chem.* **33**, 3993 (1968).
21. G. K. Hughes, E. Ritchie, and W. C. Taylor, *Aust. J. Chem.* **6**, 315 (1953).
22. Y. Inubushi, Y. Masaki, S. Matsumoto, and F. Takami, *J. Chem. Soc., C* p. 1547 (1969).
23. M. Shamma and K. F. Foley, unpublished results (1967).
24. J. R. Kershaw and B. C. Uff, *Chem. Commun.* p. 331 (1966).
25. F. D. Popp and J. W. Wefer, *Chem. Commun.* p. 207 (1966).
26. M. P. Cava, M. V. Lakshmikantham, and M. J. Mitchell, *J. Org. Chem.* **34**, 2665 (1969).
27. M. Makosza, *Tetrahedron Lett.* p. 677 (1969).
28. M. Shamma and C. D. Jones, *J. Org. Chem.* **35**, 3119 (1970).
29. T. Kametani, K. Ohkubo, and I. Noguchi, *J. Chem. Soc., C* p. 715 (1966).
30. See also T. Kametani, K. Ohkubo, and S. Takano, *J. Pharm. Soc. Jap.* **89**, 1048 (1969).
30a. D. R. Dalton, S. I. Miller, C. K. Dalton, and J. K. Crelling, *Tetrahedron Lett.* p. 575 (1971).
31. E. Späth and F. Berger, *Chem. Ber.* **63**, 2098 (1930).
32. A. Chatterjee and S. K. Kundu, *Chem. Ber.* **99**, 1764 (1966).
33. T. Kametani, K. Fukumoto, H. Agui, H. Yagi, K. Kigasawa, H. Sugahara, M. Hiiragi, T. Hayasaka, and H. Ishimaru, *J. Chem. Soc., C* p. 112 (1968).
33a. T. Kametani, T. Kobari, and K. Fukumoto, *J. Chem. Soc., C* p. 1796 (1971).
34. G. Grethe, H. L. Lee, and M. R. Usković, *Tetrahedron Lett.* p. 1937 (1969).
35. G. Grethe, H. L. Lee, M. R. Usković, and A. Brossi, *Helv. Chim. Acta* **53**, 874 (1970).
35a. N. S. Bhacca, J. C. Craig, R. H. F. Manske, S. K. Roy, M. Shamma, and W. A. Slusarchyk, *Tetrahedron* **22**, 1467 (1966).

36. G. E. Krejcarek, B. W. Dominy, and R. G. Lawton, *Chem. Commun.* p. 1450 (1968).
37. M. Lora-Tamayo, R. Madroñero, G. G. Muñoz, J. M. Marzal, and M. Stud, *Chem. Ber.* **94**, 199 (1961).
38. T. Kametani, S. Kano, and Y. Watanabe, *J. Pharm. Soc. Jap.* **86**, 417 (1966).
39. G. N. Dorofeenko and V. G. Korobkova, *Zh. Obshch. Khim.* **49**, 249 (1970); *Chem. Abstr.* **73**, 15043x (1970).
39a. For an authoritative and independent study of lactam-lactim equilibria, see D. A. Evans, G. F. Smith, and M. A. Wahid, *J. Chem. Soc., B* p. 590 (1967). For a related synthesis via an isochromanone, see I. W. Elliott, *J. Heterocycl. Chem.* **7**, 1229 (1970). For related work see N. J. McCorkindale and A. W. McCulloch, *Tetrahedron* **27**, 4653 (1971).
39b. M. Shamma and R. P. Volante, unpublished results.
40. D. H. R. Barton and T. Cohen, *Festschr. Prof. Dr. Arthur Stoll siebzigsten Geburtstag 1957* p. 117 (1957).
41. D. H. R. Barton, G. W. Kirby, W. Steglich, and G. M. Thomas, *Proc. Chem. Soc., London* p. 203 (1963).
42. H. Erdtman and C. A. Wachtmeister, *Festschr. Prof. Dr. Arthur Stoll siebzigsten Geburtstag 1957* p. 144 (1957).
43. M. Tomita, Y. Masaki, K. Fujitani, and Y. Sakatani, *Chem. Pharm. Bull.* **16**, 688 (1968).
44. T. Kametani and I. Noguchi, *J. Chem. Soc., C* p. 447 (1968).
45. Y. Inubushi, Y. Aoyagi, and M. Matsuo, *Tetrahedron Lett.* p. 2363 (1969).
46. T. Kametani, H. Nemoto, T. Kobari, and S. Takano, *J. Heterocycl. Chem.* **7**, 181 (1970).
47. B. Franck, G. Dunkelmann, and H. J. Lubs, *Angew. Chem.* **79**, 1066 (1967); *Angew. Chem., Int. Ed. Engl.* **6**, 1075 (1967).
48. J. Knabe, *Arch. Pharm. (Weinheim)* **292**, 416 (1959); **293**, 121 (1960); J. Knabe and G. Grund, *ibid.* **296**, 854 (1963); J. Knabe and H. Rohoff, *Chem. Ber.* **97**, 3452 (1964).
49. A. Burger, *Alkaloids* **4**, 29 (1954).
50. T. Kametani, S. Kano, and T. Kikuchi, *J. Pharm. Soc. Jap.* **86**, 423 (1966).
51. J. A. Weisbach, J. L. Kirkpatrick, E. Macko, and B. Douglas, *J. Med. Chem.* **11**, 752 (1968).
52. M. Shamma and Sr. M. A. Podczasy, *Tetrahedron* **27**, 727 (1971).
53. A. R. Battersby, D. J. LeCount, S. Garratt, and R. I. Thrift, *Tetrahedron* **14**, 46 (1961).
54. B. Witkop and J. B. Patrick, *J. Amer. Chem. Soc.* **75**, 4474 (1953).
55. M. Shamma and W. A. Slusarchyk, unpublished results.
56. J. L. Neumeyer, M. McCarthy, and K. K. Weinhardt, *Tetrahedron Lett.* p. 1095 (1967); and with P. L. Levins, *J. Org. Chem.* **33**, 2890 (1968).
57. J. Knabe and J. Kubitz, *Arch. Pharm. (Weinheim)* **297**, 129 (1964).
58. J. Knabe and K. Detering, *Chem. Ber.* **99**, 2873 (1966).
59. D. W. Brown, S. F. Dyke, R. G. Kinsman, and M. Sainsbury, *Tetrahedron Lett.* p. 1731 (1969).
60. For additional examples of the rearrangement see J. Knabe and K. Detering, *Arch. Pharm. (Weinheim)* **300**, 97 (1967).
60a. Also see W. Wiegrebe and E. Roesel, *Arch. Pharm. (Weinheim)* **302**, 310 (1969); J. Knabe and H.-D. Höltje, *Tetrahedron Lett.* p. 2107 (1969); J. Knabe and H. Powilleit, *Arch. Pharm. (Weinheim)* **303**, 37 (1970); J. Knabe and H.-D. Höltje, *ibid.* p. 404; J. Knabe, W. Krause, H. Powilleit, and K. Sierocks, *Pharmazie* **25**, 313 (1970).
60b. J. Knabe, *Arch. Pharm.* **303**, 64 (1970).
60c. J. Knabe and H. Powilleit, *Arch. Pharm.* **303**, 37 (1970).
60d. M. Sainsbury, S. F. Dyke, D. W. Brown, and R. G. Kinsman, *Tetrahedron* **26**, 5265 (1970).
61. E. Späth and E. Kruta, *Chem. Ber.* **62**, 1024 (1929).
62. W. Wiegrebe, H. Reinhart, H. Budzikiewicz, and U. Krüger, *Tetrahedron* **25**, 2899 (1969).
62a. P. Mathieu and J. Gardent, *C. R. Acad. Sci. Ser. C* **267**, 1416 (1968).
62b. M. P. Cava, M. J. Mitchell, and D. T. Hill, *Chem. Commun.* p. 1601 (1970).
62c. R. Pschorr, *Chem. Ber.* **37**, 1926 (1904).

62d. H. Budzikiewicz and U. Krüger, *Justus Liebigs Ann. Chem.* **737**, 119 (1970).
63. T. Kametani, M. Ihara, and K. Shima, *J. Chem. Soc., C* p. 1619 (1968).
64. F. R. Stermitz, R. Pua, and H. Vyas, *Chem. Commun.* p. 326 (1967).
65. A. L. Mndzhozan, V. A. Mnatsakanyan, and L. S. Arutyunyan, *Arm. Khim. Zh.* **22**, 842 (1969); *Chem. Abstr.* **72**, 21808w (1970).
66. J. D. Albright and L. Goodman, *J. Amer. Chem. Soc.* **91**, 4317 (1969).
67. A. Brossi and S. Teitel, *J. Org. Chem.* **35**, 1684 (1970); *Chem. Commun.* p. 1296 (1970).
68. H. Corrodi and E. Hardegger, *Helv. Chim. Acta* **39**, 889 (1956).
69. J. C. Craig and S. K. Roy, *Tetrahedron* **21**, 401 (1965).
70. A. R. Battersby, I. R. C. Bick, W. Klyne, J. P. Jennings, P. M. Scopes, and M. J. Vernengo, *J. Chem. Soc., London* p. 2239 (1965).
71. J. C. Craig, M. Martin-Smith, S. K. Roy, and J. B. Stenlake, *Tetrahedron* **22**, 1335 (1966).
72. A. R. Battersby and B. J. T. Harper, *J. Chem. Soc., London* p. 3526 (1962).
73. H. Rapoport, N. Levy, and F. R. Stermitz, *J. Amer. Chem. Soc.* **83**, 4298 (1961).
74. J. R. Gear and I. D. Spenser, *Nature (London)* **191**, 1393 (1961).
75. I. D. Spenser and J. R. Gear, *J. Amer. Chem. Soc.* **84**, 1059 (1962).
76. E. Leete and J. B. Murril, *Tetrahedron Lett.* p. 147 (1964).
77. A. R. Battersby, G. W. Evans, R. O. Martin, M. E. Warren, and H. Rapoport, *Tetrahedron Lett.* p. 1275 (1965).
78. E. Brochmann-Hanssen and B. Nielsen, *Tetrahedron Lett.* p. 1271 (1965).
79. A. R. Battersby, R. Binks, R. J. Francis, D. J. McCaldin, and H. Ramuz, *J. Chem. Soc., London* p. 3600 (1964).
80. A. Goth, "Medical Pharmacology," 4th ed., p. 434. Mosby, St. Louis, Missouri, 1968.
81. Laboratoires Houde Patent, *Chem. Abstr.* **73**, 66789c (1970).
82. Allen and Hanburys Ltd., Brit. Pat. 863717 (1961).
82a. "The Merck Index," 8th ed., p. 385 (1968).
82b. "The Merck Index," 8th ed., p. 688 (1968).
82c. "The Merck Index," 8th ed., p. 427 (1968).
82d. H. Koshiyama, H. Ohkuma, H. Kawaguchi, H.-Y. Hsu, and Y.-P. Chen, *Chem. Pharm. Bull.* **18**, 2564 (1970).
83. Y. Iwasawa and A. Kiyomoto, *Jap. J. Pharmacol.* **17**, 143 (1967); *Ann. Allergy* **26**, 504 (1968); *J. Pharm. Soc. Jap.* **65**, 34 (1969).
84. E. Tóth, G. Fassina, and E. Santi Sonsin, *Arch. Int. Pharmacodyn. Ther.* **169**, 375 (1967).
85. R. Santi, M. Ferrari, C. E. Tóth, A. R. Contessa, G. Fassina, A. Bruni, and S. Luciani, *J. Pharm. Pharmacol.* **19**, 45 (1967).
85a. R. A. Hahn, P. N. Patil, J. L. Beal, M. G. Kelly, and M. Shamma, in press.
85b. D. G. Patel, A. Tye, P. N. Patil, A. M. Burkman, and J. L. Beal, *Lloydia* **33**, 36 (1970).
85c. D. R. Feller, R. F. Shonk, and D. D. Miller, *J. Pharm. Pharmacol.* **22**, 869 (1970).
85d. Y. Iwasawa and A. Kiyomoto, *Jap. J. Pharmacol.* **17**, 143 (1967).
85e. E. Yamato, M. Hirakura, and S. Sugasawa, *Tetrahedron* Suppl., **8**, Part I, 129 (1966).
86. V. E. Davis and M. J. Walsh, *Science* **167**, 1005 (1970).
86a. M. H. Seevers, *Science* **170**, 1113 (1970). For an attempt at a rebuttal of Seevers' views, see V. E. Davis and M. J. Walsh, *Science* **170**, 1114 (1970).
87. M. Ohashi, J. M. Wilson, H. Budzikiewicz, M. Shamma, W. A. Slusarchyk, and C. Djerassi, *J. Amer. Chem. Soc.* **85**, 2807 (1963).
88. M. Tomita, H. Furukawa, T. Kikuchi, A. Kato, and T. Ibuka, *Chem. Pharm. Bull.* **14**, 232 (1966).
89. A. Tatematsu and T. Goto, *J. Pharm. Soc. Jap.* **85**, 152 (1965).
90. D. R. Dalton, M. P. Cava, and K. T. Buck, *Tetrahedron Lett.* p. 2687 (1965).
91. M. Tomita, T. Shingu, K. Fujitani, and H. Furukawa, *Chem. Pharm. Bull.* **13**, 921 (1965).
92. H. Furukawa, T. H. Yang, and T. J. Lin, *J. Pharm. Soc. Jap.* **85**, 472 (1965).
93. M. Tomita, T. Shingu, and H. Furukawa, *J. Pharm. Soc. Jap.* **86**, 373 (1966).

94. G. J. Kapadia, N. J. Shah, and R. J. Highet, *J. Pharm. Sci.* **53**, 1140 (1964).
94a. M. Shamma and J. L. Moniot, *J. Pharm. Sci.* **61**, 295 (1972).
95. J. Fridrichsons and A. Mc. L. Mathieson, *Tetrahedron* **24**, 5785 (1968).
95a. V. St. Georgiev and N. M. Mollov, *Compt. rend. Acad. bulg. Sc.* **24**, 1329 (1971).
96. L. Hruban and F. Šantavý, *Collect. Czech. Chem. Commun.* **32**, 3414 (1967).
97. M. Tomita and M. Kozuka, *J. Pharm. Soc. Jap.* **84**, 362 (1964).
98. L. Slaviková and J. Slavík, *Collect. Czech. Chem. Commun.* **31**, 3362 (1966).
99. M. Shamma, M. A. Greenberg, and B. S. Dudock, *Tetrahedron Lett.* p. 3595 (1965).
100. P. M. Oertreicher, C. G. Farmilo, and L. Levi, *Bull. Narcotics* **7**, 42 (1954).
101. J. Kunitomo, E. Yuge, and Y. Nagai, *J. Pharm. Soc. Jap.* **86**, 456 (1966).
102. Holubek and Štrouf, Spectrum No. 436.
103. Holubek and Štrouf, Spectrum No. 154.

Chapter 3 / THE ISOQUINOLONES

Occurrence: Hernandiaceae, Monimiaceae, and Ranunculaceae
Number: 8
Structures:

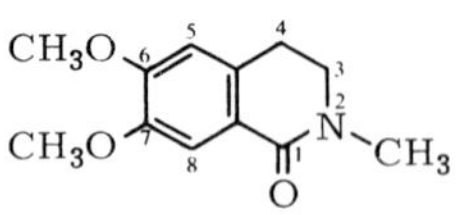

N-Methylcorydaldine

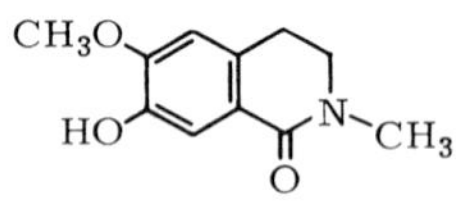

Thalifoline

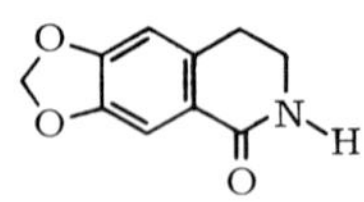

Noroxyhydrastinine

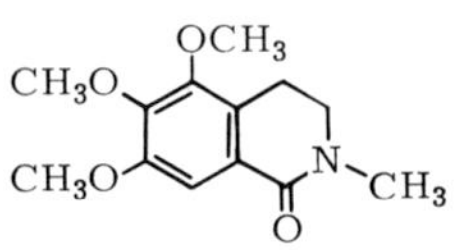

N-Methylthalidaldine

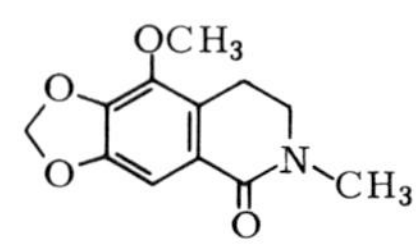

Thalflavine

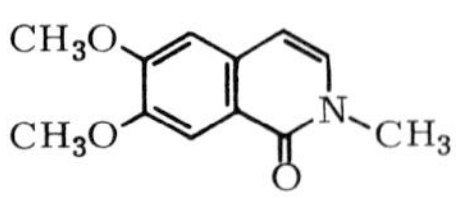

N-Methyl-6,7-dimethoxy-isoquinolone[1,2]

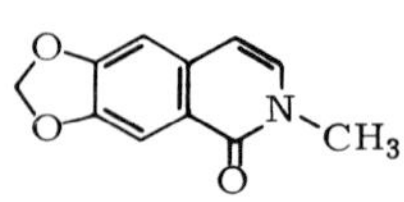

Doryanine

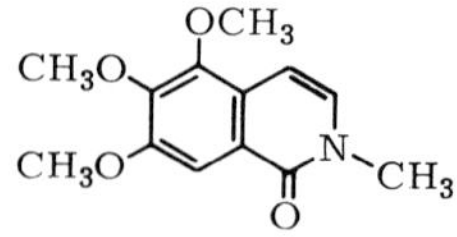

Thalactamine

I. Introduction

The isoquinolones are a small group of alkaloids present in plants only in minor amounts. No biogenetic studies have been conducted on this group, but it is possible that the isoquinolones are derived from the oxidation of benzylisoquinoline precursors.[1,3] The isoquinolones can be subdivided into two categories: those with a totally aromatic nucleus such as doryanine and thalactamine, and those which incorporate a C-3,4 single bond.

II. Structural Elucidation and Synthesis

A. The 3,4-Dihydroisoquinolones

Doskotch and co-workers have carried out a detailed structural study of the isoquinolones noroxyhydrastinine and thalifoline obtained from *Thalictrum minus* L. var. *adiantifolium* Hort. (Ranunculaceae).[4] The IR spectrum of noroxyhydrastinine showed intense peaks at λ_{KBr} 3.15, 3.29, and 5.99 μ (3175, 3040, and 1670 cm^{-1}). The first two peaks were assigned to the associated N–H stretching vibration of a lactam while the latter was consistent with the presence of a δ-lactam carbonyl. A methylenedioxy group was suggested by the presence of absorption bands at 3.57 μ (2800 cm^{-1}) and 10.81 μ (925 cm^{-1}).

The mass spectrum of noroxyhydrastinine was most informative, showing the molecular ion peak at *m/e* 191, the base peak at *m/e* 134, and other intense peaks at *m/e* 162, 104, 76, and 43. The fragmentation pattern in Scheme I was presented to explain the data.

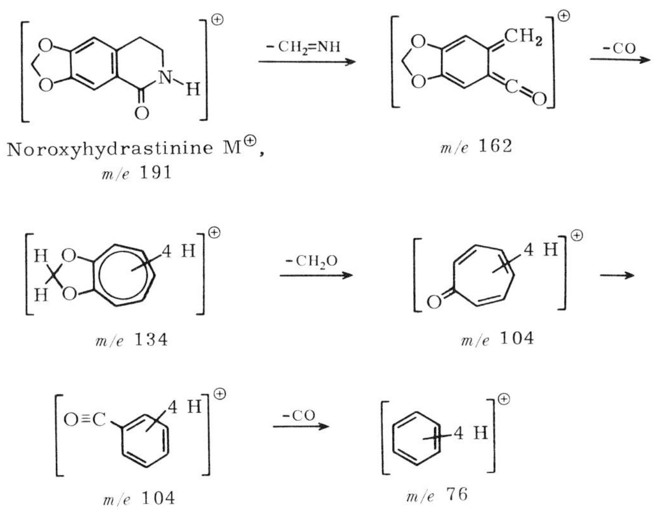

Scheme I

The peak at *m/e* 43 (CONH) is consistent with a lactam structure. More importantly, there were four distinct metastable peaks, at *m/e* 137, 111, 81, and 56; the rationalization for these is given in Table I, where $m = b^2/a$.

TABLE I
METASTABLE PEAKS FOR NOROXYHYDRASTININE

Fragment *a* (m/e)	Fragment *b* (m/e)	Metastable peak *m* (m/e)
191	162	137
162	134	111
134	104	81
104	76	56

Finally, a comparison with the natural product of noroxyhydrastinine prepared by the potassium permanganate oxidation of berberine chloride showed the two materials to be identical.[5] Another synthesis of noroxyhydrastinine relies on the phosphorus oxychloride-induced cyclization of the urethane **1**.[6]

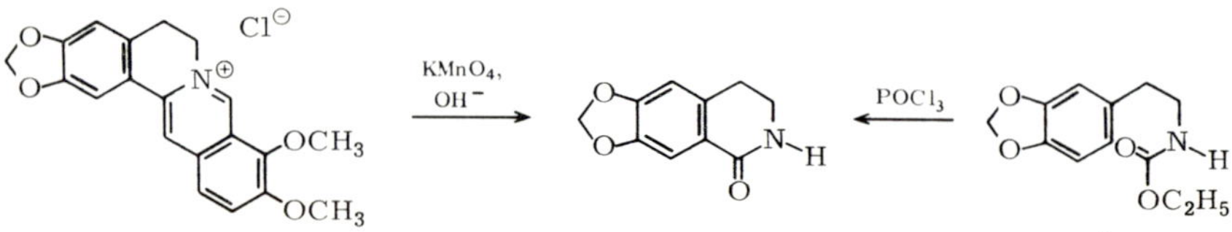

Berberine chloride Noroxyhydrastinine **1**

The structural work on thalifoline closely parallels that for noroxyhydrastinine, except that a synthesis of thalifoline was developed beginning with 3-methoxy-4-benzyloxyphenylethylamine.[4] The intermediate formyl amide was not isolated, but was immediately cyclized to the imine **2**. Quaternization with methyl iodide and ferricyanide oxidation gave thalifoline.

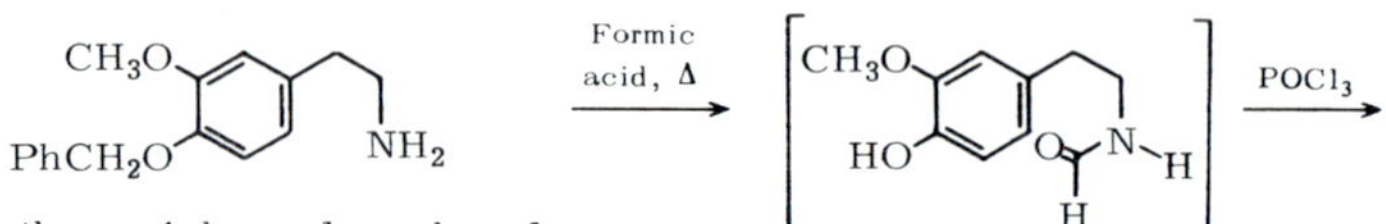

3-Methoxy-4-benzyloxyphenylethylamine A formyl amide

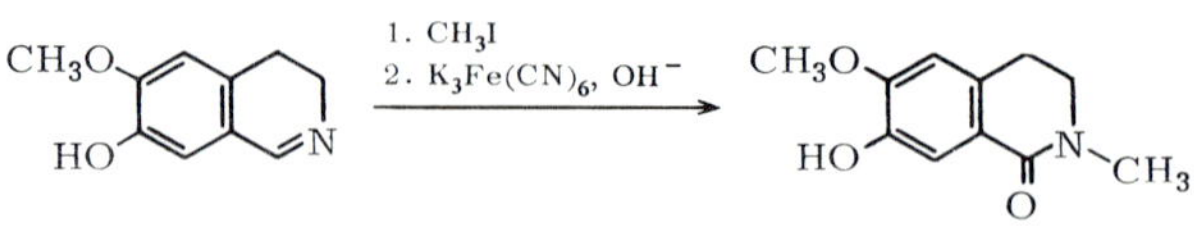

2 Thalifoline

Similarly, the structural elucidation of *N*-methylthalidaldine is based on spectral data, especially NMR, which are summarized below. The conclusions were then corroborated by oxidation of naturally occurring thalifendlerine with permanganate to yield material identical with *N*-methylthalidaldine (see Chapter 2, Section VII, B).

Significantly enough from a biogenetic viewpoint, both thalifendlerine and *N*-methylthalidaldine were found in the same plant, *Thalictrum fendleri* Engelm. *ex* Gray.[7]

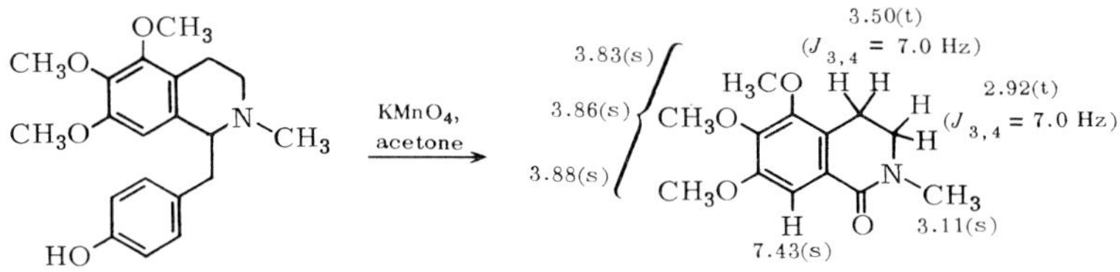

Thalifendlerine — *N*-Methylthalidaldine

B. The Aromatic Isoquinolones

The structural elucidation of doryanine, $C_{11}H_9O_3N$, obtained from *Doryphora sassafras* Endlicher (Monimiaceae), again relies largely upon spectral data, and the conclusions were supported by synthesis.[3]

Doryanine shows a strong carbonyl band at 6.02 μ (1661 cm^{-1}) suggestive of a conjugated carbonyl group. The UV spectrum contained seven maxima in the range from 231 to 338 mμ, a complex pattern which indicated the presence of a highly conjugated chromophore. The NMR spectrum of doryanine was well defined and is summarized here.

The synthesis of the alkaloid was achieved through ferricyanide oxidation of the corresponding isoquinoline salt.[2] An attempted preparation of the alkaloid via dehydrogenation of noroxyhydrastinine was not successful.[2]

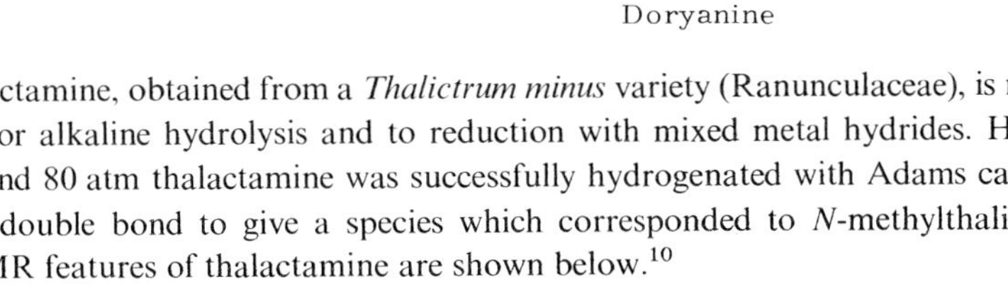

Doryanine

Thalactamine, obtained from a *Thalictrum minus* variety (Ranunculaceae), is resistant to acid or alkaline hydrolysis and to reduction with mixed metal hydrides. However, at 80° and 80 atm thalactamine was successfully hydrogenated with Adams catalyst at the 3,4 double bond to give a species which corresponded to *N*-methylthalidaldine. The NMR features of thalactamine are shown below.[10]

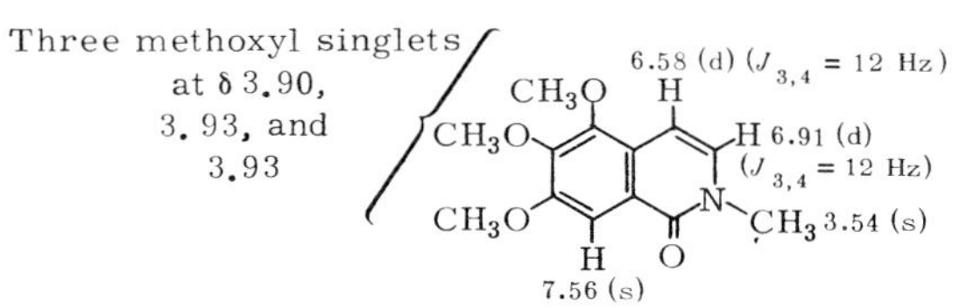

NMR spectral data for thalactamine

III. An Alternate Synthesis of 3,4-Dihydroisoquinolones

Acetylation of the 3,4-dihydroisoquinoline **3** gives rise to the 1-methylene-2-acetylisoquinoline **4**. Ozonization of the latter compound furnishes the *N*-acetyl-3,4-dihydroisoquinolone **5**. Otherwise **4** can be reduced to the acetylated simple tetrahydroisoquinoline **6** or hydrolyzed with mineral acid to the substituted acetophenone **7**.[6]

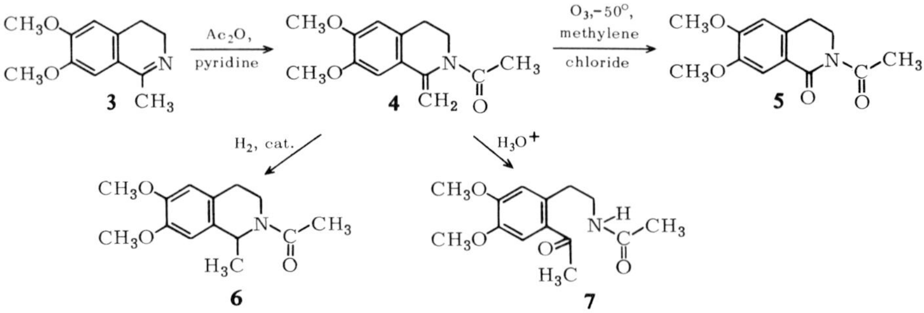

IV. Pharmacology

Corydaldine exhibited analgetic properties in different so-called rheumatic tests, but the results in human pharmacology were disappointing.[8]

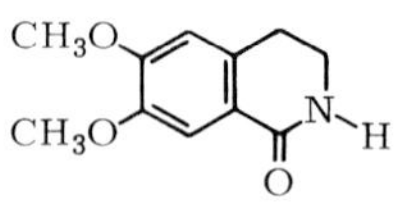

Corydaldine

V. UV Spectroscopy

Noroxyhydrastinine[9]	λ_{max}^{MeOH}	223, 261, and 304 mμ (4.31, 3.58, and 3.67)
Thalifoline[9]	λ_{max}^{MeOH}	224, 261, and 302 mμ (4.41, 3.87, and 3.77)
Doryanine[3]	λ_{max}^{EtOH}	231, 248, 258, 284, 294, 325, and 338 mμ (4.39, 4.44, 4.34, 3.77, 3.83, 3.58, and 3.45)
Thalactamine[10]	λ_{max}	247, 270, 281, and 293 mμ (4.62, 3.60, 3.68, and 3.74)

References

1. M. P. Cava, K. Bessho, B. Douglas, S. Markey, and J. A. Weisbach, *Tetrahedron Lett.* p. 4279 (1966).
2. M. P. Cava and K. T. Buck, *Tetrahedron* **25**, 2795 (1969).

3. S. A. Gharbo, J. L. Beal, R. H. Schlessinger, M. P. Cava, and G. H. Svoboda, *Lloydia* **28**, 237 (1965).
4. R. W. Doskotch, P. L. Schiff, Jr., and J. L. Beal, *Tetrahedron* **25**, 469 (1969).
5. W. H. Perkin, Jr., *J. Chem. Soc., London* **57**, 992 (1890).
6. A. Brossi, J. Würsch, and O. Schnider, *Chimia* **12** 114 (1958).
7. M. Shamma and Sr. M. A. Podczasy, *Tetrahedron* **27**, 727 (1971).
8. A. Brossi, H. Besendorf, L. A. Pirk, and A. Rheiner, Jr., *Med. Chem.* **5**, 281 (1965).
9. R. W. Doskotch, P. L. Schiff, and J. L. Beal, *Lloydia* **32**, 29 (1969).
10. N. M. Mollov and H. B. Dutschewska, *Tetrahedron Lett.* p. 1951 (1969).

Chapter 4 / THE PAVINES AND ISOPAVINES[1]

Occurrence: Lauraceae, Papaveraceae, and Ranunculaceae
Number: 13 Pavines and 7 Isopavines
Structures:

Pavines

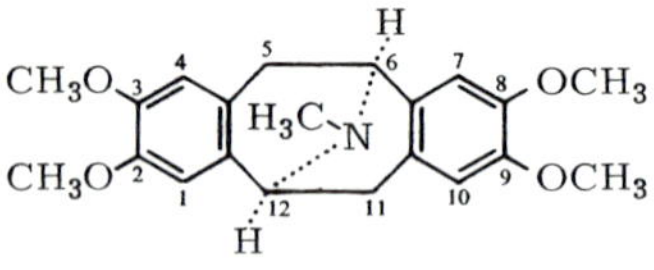

(−)-Argemonine
[also (−)-argemonine *N*-metho salt and (−)-argemonine *N*-oxide]

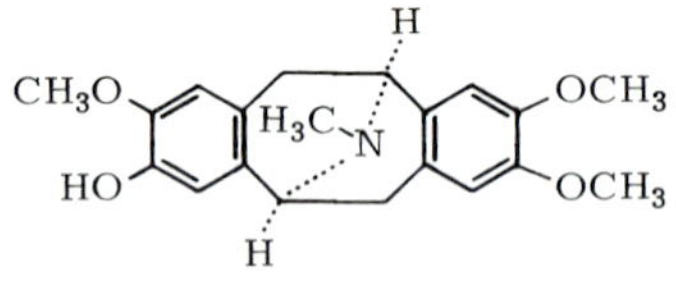

(−)-Norargemonine

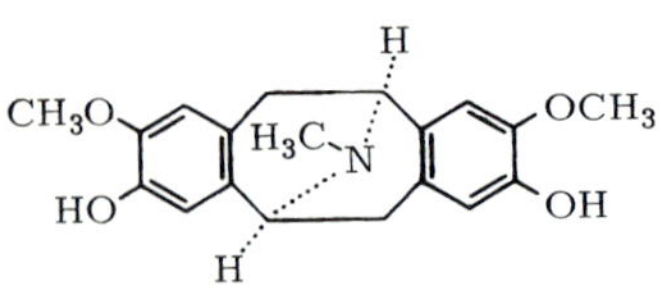

(−)-Bisnorargemonine

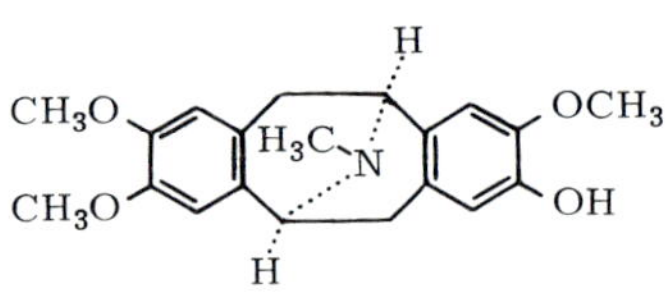

(−)-Isonorargemonine

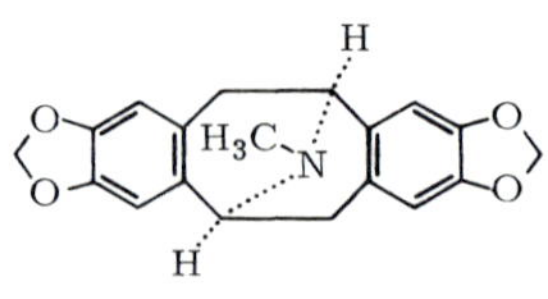

(−)-Eschscholtzine
(also *N*-metho salt: (−)-californidine)

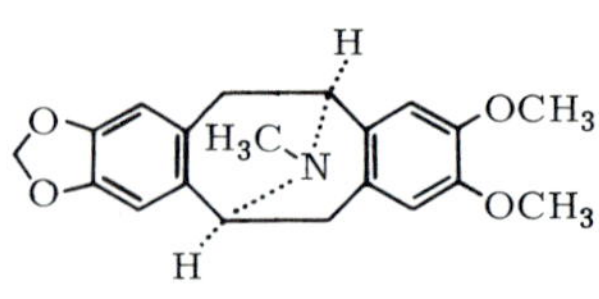

(−)-Eschscholtzidine

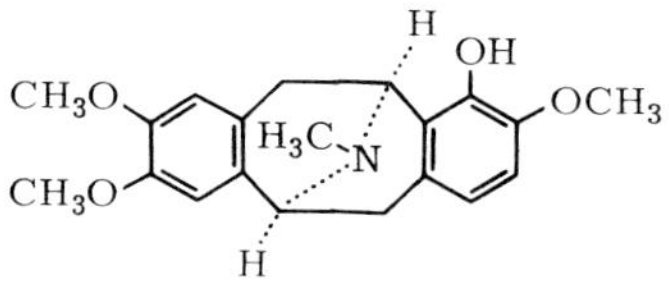

(−)-Platycerine

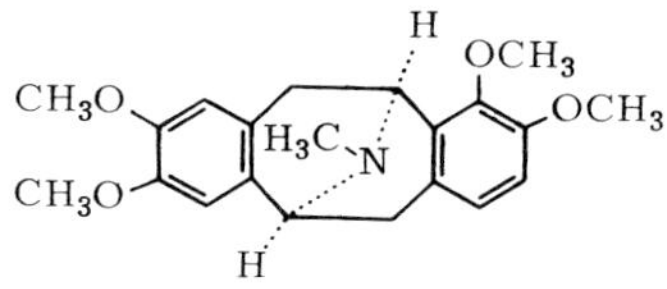

(−)-*O*-Methylplatycerine

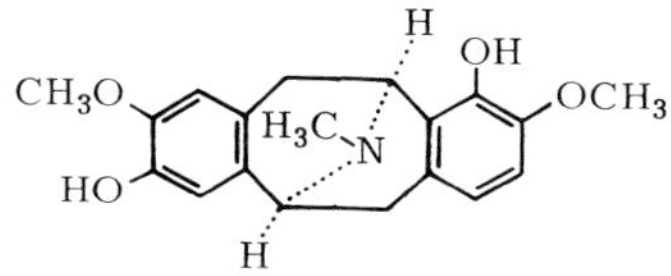

(−)-Munitagine

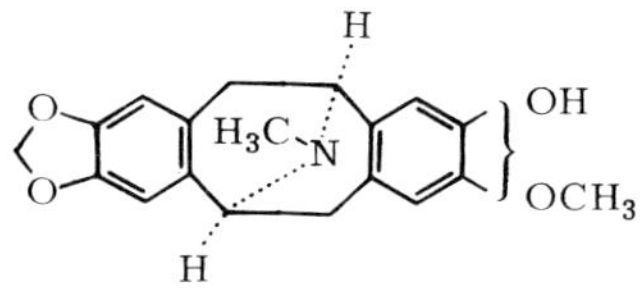

(−)-Caryachine[2]
[also (±)-caryachine]

Isopavines

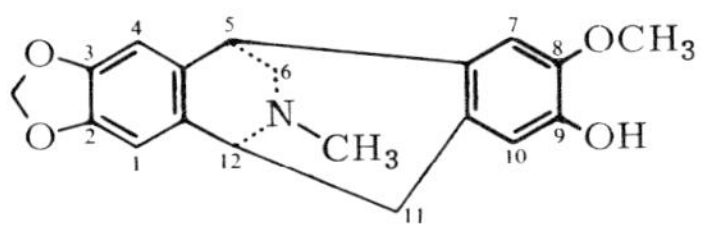

(−)-Amurensine

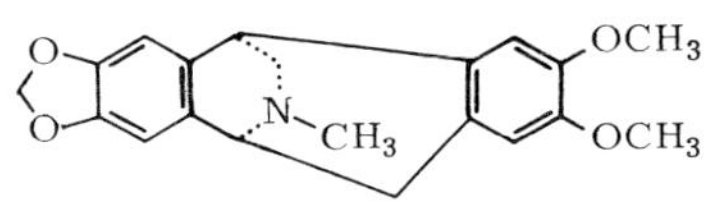

(−)-Amurensinine

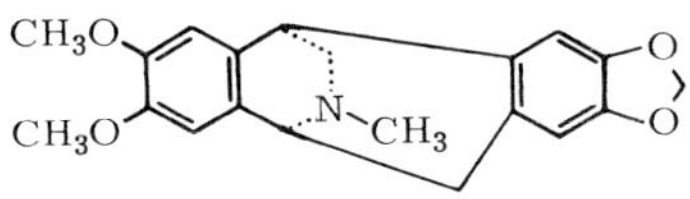

(−)-Reframine[3]
[also *N*-metho salt: (−)-remrefine[4]]

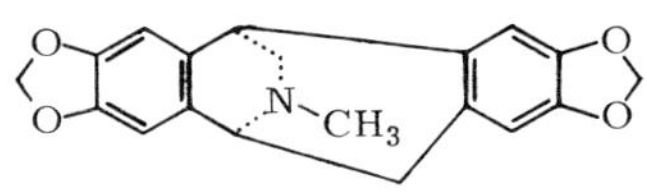

(−)-Reframidine[3]

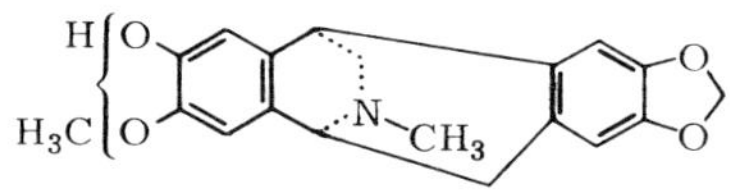

(−)-Reframoline[3]

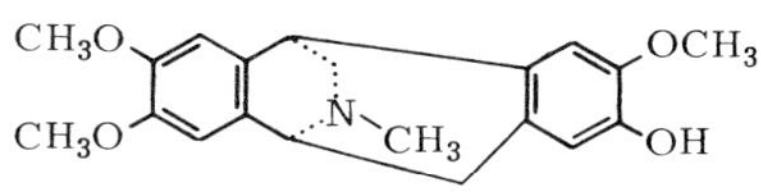

(−)-Thalisopavine

I. Structural Elucidation and Synthesis

A. Pavine

It had been determined before the turn of the century[5] that reduction of the benzylisoquinoline alkaloid papaverine with tin and hydrochloric acid affords tetrahydropapaverine and a crystalline secondary base, pavine, $C_{20}H_{23}O_4N$, whose structure was not elucidated until the 1950s by Schöpf,[6] and by Battersby,[7] and whose isolation has never been reported from a natural source.

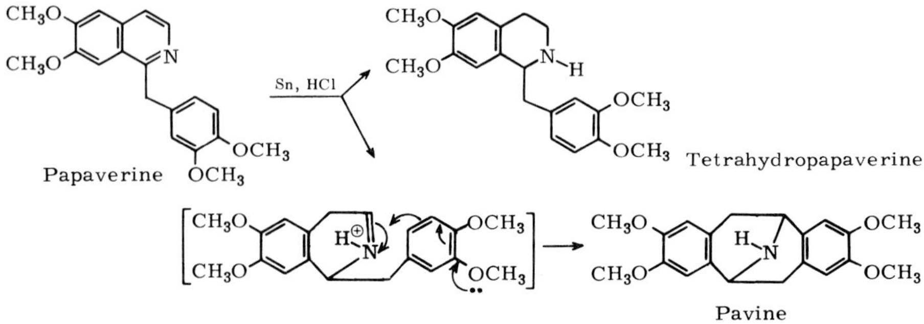

The main features in the degradative evidence can be summarized as shown in Scheme I. *N*-Methylation of pavine afforded *N*-methylpavine whose methiodide upon Hofmann degradation furnished *N*-methylpavine methine. Hydrogenation of the hydrochloride salt gave dihydro-*N*-methylpavine methine, whose methiodide was subjected again to Hofmann conditions. The product was the crystalline dihydropavine bismethine which was devoid of nitrogen.

The foregoing product was reduced catalytically to the dibenzocyclooctadiene **1**. Permanganate oxidation of dihydropavine bismethine gave as the major product the diacid **2** which was compared with authentic material, and as the minor product a ketonic material which was tentatively assigned the α-dione structure **3**.

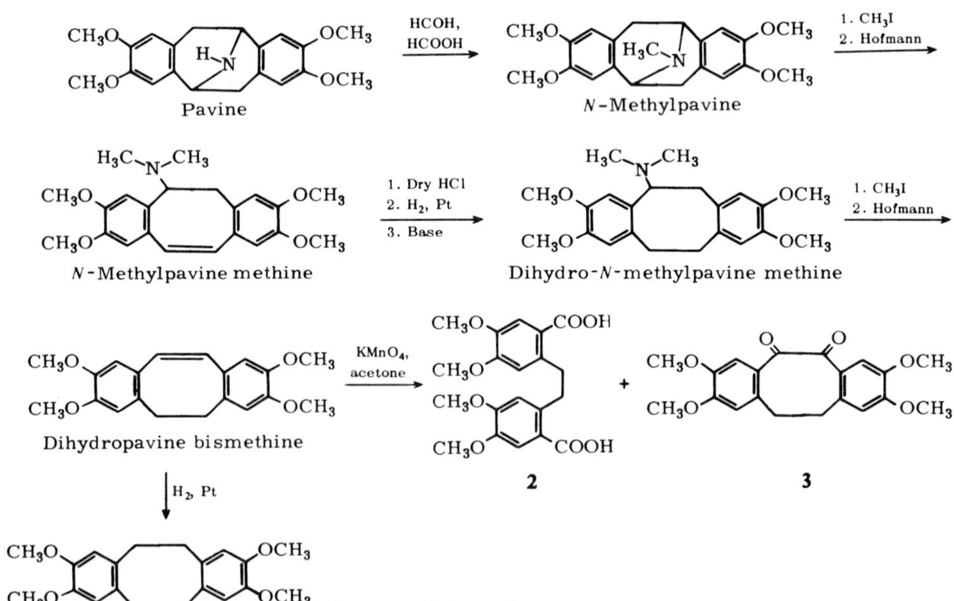

1

Scheme I

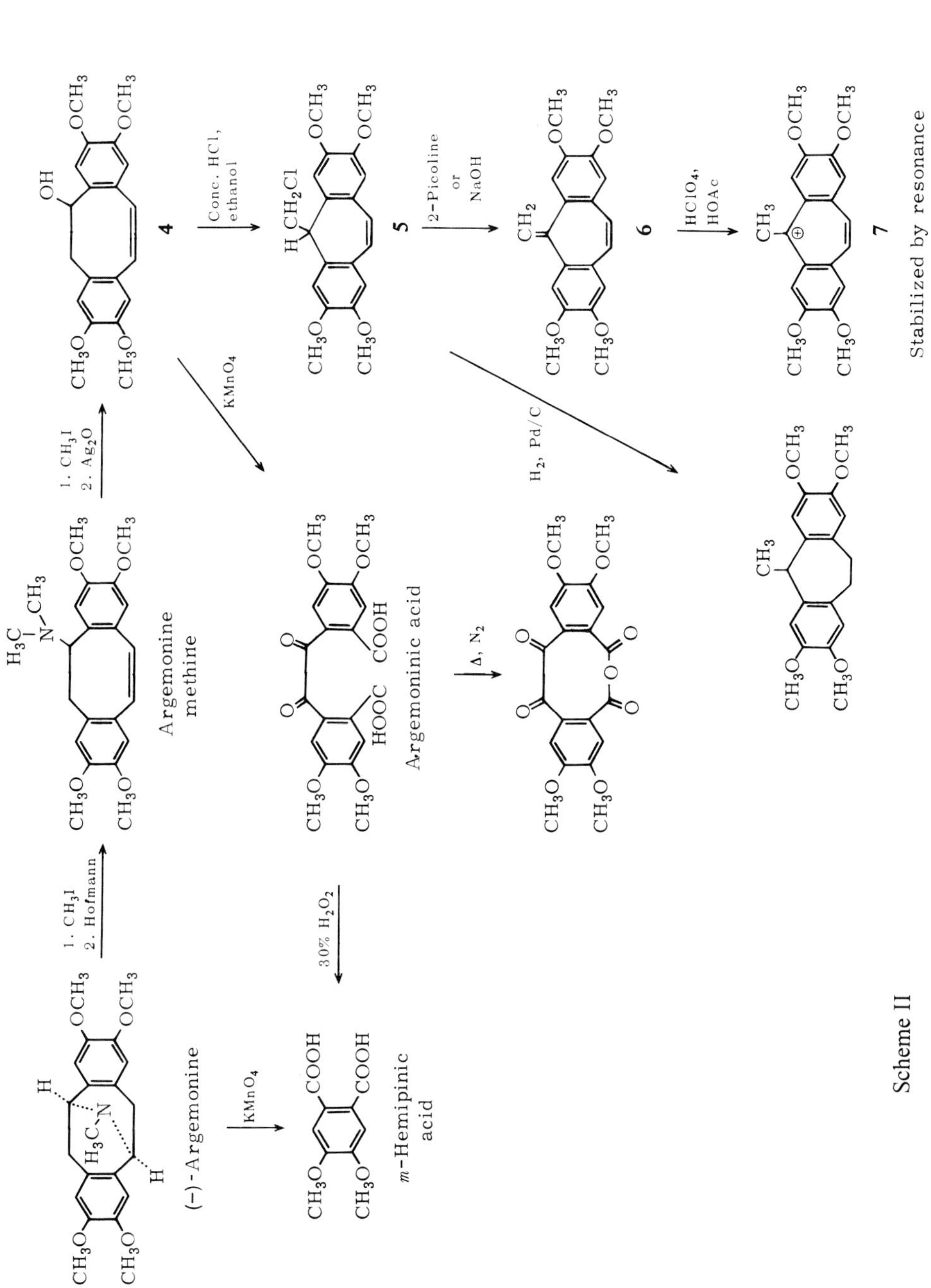
(−)-Argemonine
KMnO4
m-Hemipinic acid
1. CH3I
2. Hofmann
Argemonine methine
1. CH3I
2. Ag2O
4
Conc. HCl, ethanol
5
2-Picoline or NaOH
6
HClO4, HOAc
7
Stabilized by resonance
KMnO4
Argemoninic acid
30% H2O2
Δ, N2
H2, Pd/C

Scheme II

B. Argemonine

In 1963, it was recognized by Stermitz and others that the alkaloid (−)-argemonine, which occurs in a variety of *Argemone* species (Papaveraceae) and had been known for some time, is simply (−)-*N*-methylpavine.[8,9]

The chemistry of argemonine is both similar and complementary to that of pavine. Direct oxidation of the alkaloid with potassium permanganate leads to *m*-hemipinic acid. The methiodide salt of optically active argemonine methine upon treatment with silver oxide gives the alcohol **4**. Reaction of this material with concentrated hydrochloric acid induces a Wagner–Meerwein rearrangement with formation of the chloro derivative **5**. The transformations of this derivative are indicated in Scheme II. Among the interesting products are the cycloheptatriene derivative **6**, which is also obtained from isopavine alkaloids, and the dibenzocycloheptatrienyl cation **7**.

When the alcohol **4** is oxidized with permanganate, argemoninic acid is obtained, which upon heating leads to the corresponding anhydride (Scheme II).[7,10,11]

C. Norargemonine and Isonorargemonine

(−)-Norargemonine, a companion alkaloid to argemonine, is a mono-*O*-demethylargemonine since *O*-methylation with diazomethane yields (−)-argemonine. Because of the symmetry of the molecule, (−)-norargemonine must be one of the two structures below:

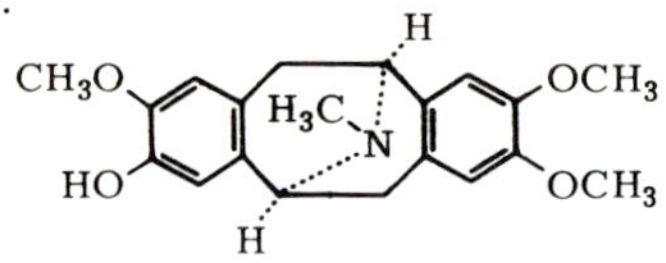

(−)-Norargemonine

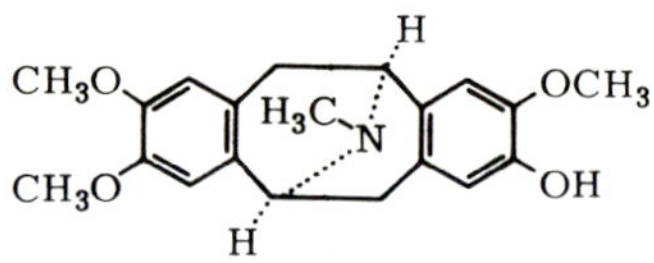

(−)-Isonorargemonine

It was known that when papaverine hydrochloride is heated slightly above its melting point for several minutes, a separable mixture of the phenol betaine **8** and the salt **9** is formed. The salt **10**, which is a position isomer of **9**, is formed when the betaine **8** is heated in a sealed tube with methyl iodide (Scheme III).[12] (See also Chapter 2, Section VII, O.)

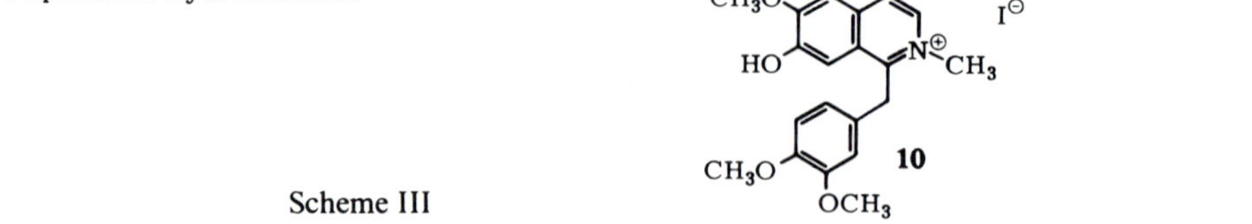

Scheme III

Reduction of the salts **10** and **9** with tin and hydrochloric acid gave mainly the tetrahydrobenzylisoquinolines **11** and **13**, respectively, but also small amounts of the bridged structures **12** and **14**. Compound **12** was then found to correspond to (±)-norargemonine. A few years later it was shown that the alternate structure **14** was the racemic form of the newly isolated alkaloid (−)-isonorargemonine (Scheme IV).[12,13]

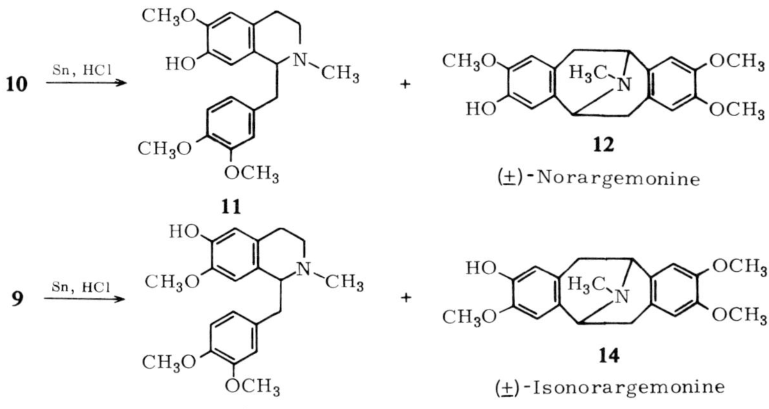

Scheme IV

D. Isopavine

The action of concentrated sulfuric acid on the amino acetal **15** yields the tetracyclic structure called isopavine. As with pavine, the structure of isopavine was confirmed by degradation (Scheme V).[14,15]

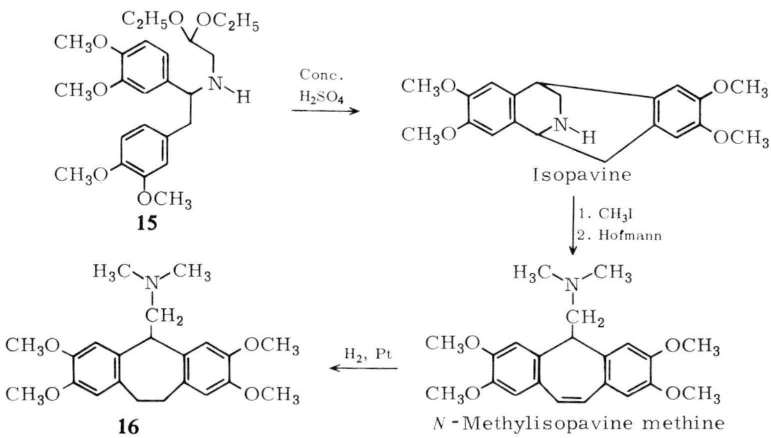

Scheme V

N-Methylisopavine methiodide was smoothly converted by Hofmann elimination to *N*-methylisopavine methine. This product was found to absorb one mole of hydrogen catalytically to give the dibenzocycloheptadiene derivative **16** of established structure. Isopavine, like pavine, has not been found in plant material.

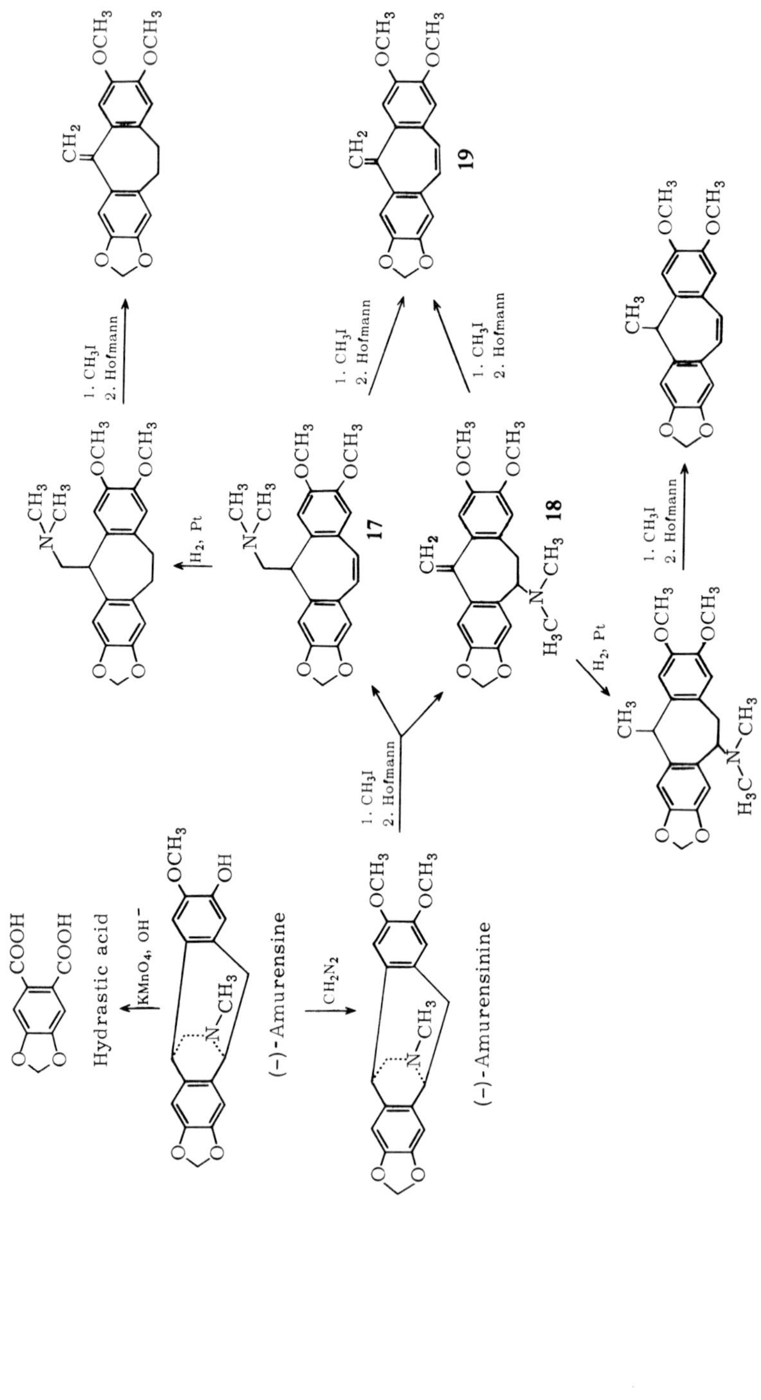
Hydrastic acid
KMnO4, OH−
(−)-Amurensine
CH2N2
(−)-Amurensinine
1. CH3I
2. Hofmann
H2, Pt
17
18
19

Scheme VI

E. *Amurensine and Amurensinine: Two Isopavine Alkaloids*

Šantavý and co-workers have investigated the first two naturally occurring isopavines, (−)-amurensine and (−)-amurensinine. *O*-Methylation of amurensine, $C_{19}H_{19}O_4N$, yields amurensinine, $C_{20}H_{21}O_4N$. Hofmann degradation of the methiodide of amurensinine led to the major product **17** and the minor product **18**. Both of these species could be converted into the cycloheptatriene derivative **19** related to compound **6** obtained from argemonine. The further transformations of the methines **17** and **18** are indicated in Scheme VI. It will be noted that direct oxidation of amurensine with potassium permanganate gave rise to hydrastic acid. The positions of the methoxyl and hydroxyl functions on the other benzenoid ring were derived from NMR data and color tests. Mass spectrometry was also useful in settling the structures of these alkaloids (Section IV).[16]

F. *Thalisopavine, Another Isopavine-Type Alkaloid*

The molecular formula $C_{20}H_{23}O_4N$ was assigned to thalisopavine by Kupchan and co-workers on the basis of elemental analysis and the mass spectrum (M^+ m/e 341) of the alkaloid. NMR spectroscopy showed the presence of one *N*-methyl group and

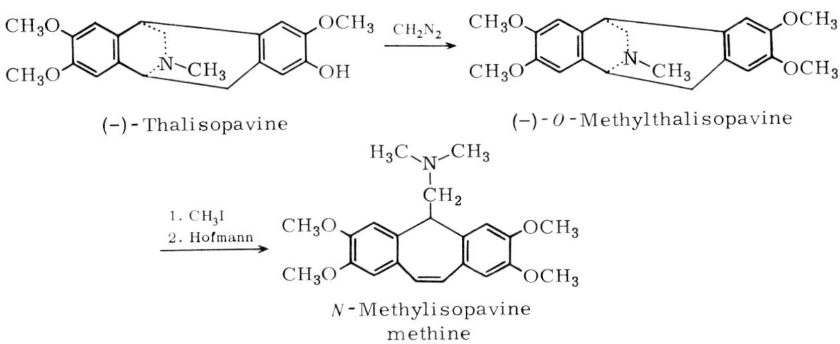

and

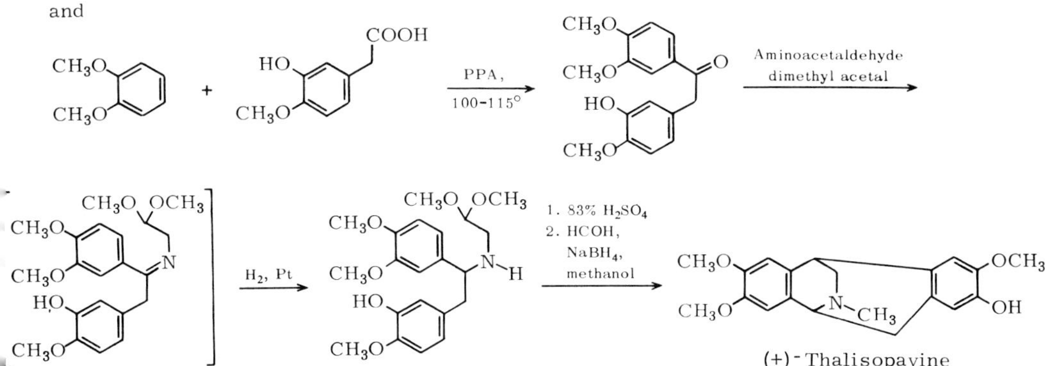

Scheme VII

three methoxyls – the spectrum exhibiting general similarities with that of the known isopavine alkaloid amurensine.

A bathochromic shift was observed in the UV spectrum upon the addition of base, indicative of the presence of a phenolic function. Diazomethane *O*-methylation of thalisopavine produced *O*-methylthalisopavine, the methiodide salt of which was subjected to Hofmann degradation. The methine obtained is optically inactive because of its meso nature, and was shown to be identical with the known *N*-methylisopavine methine derived from isopavine (Scheme VII).[17]

The phenolic group in thalisopavine was tentatively placed at C-9, and this assignment was confirmed by a total synthesis of (±)-thalisopavine whose steps parallel those for isopavine (Scheme VII).

II. Absolute Configuration

Hofmann degradation of the *N*-benzyl quaternary salt of (−)-argemonine yields the methine base **20** in high yield, and hydrogenation and hydrogenolysis over palladized charcoal then generates the secondary base **21**. Ozonization of the *N*-formyl derivative

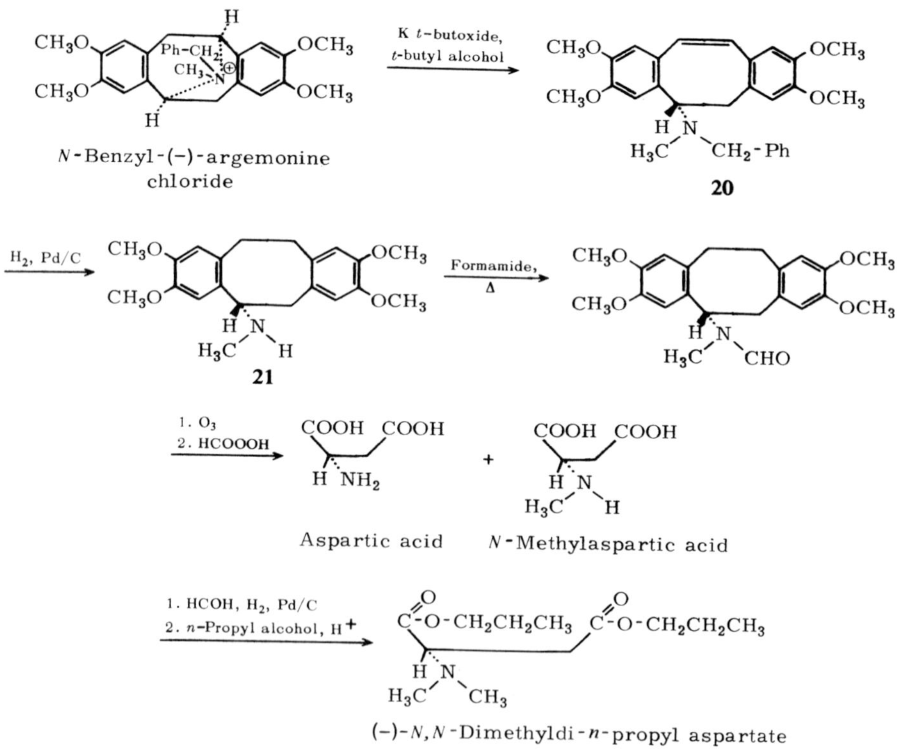

Scheme VIII

of **21** followed by treatment with performic acid leads to aspartic acid and *N*-methylaspartic acid. The mixture of amino acids was not separated but rather *N*-methylated. The resulting *N,N*-dimethylamino acid was immediately converted into its di-*n*-propyl ester. Comparison with an authentic sample of (–)-*N,N*-dimethyldi-*n*-propyl aspartate showed the two materials to be identical. The absolute configuration of naturally occurring (–)-argemonine is shown in Scheme VIII.[18]

An alternate method for the establishment of the absolute configuration of (–)-argemonine relies upon an empirical comparison of the ORD curve of the alkaloid with that of the synthetic tetracyclic base **22**, (+)-Tröger's base, supposedly of known absolute configuration.[19] It has been cogently pointed out, however, that such an empirical approach is ambiguous and can lead to erroneous results. A much more reliable method is a nonempirical analysis of the CD bands of (–)-argemonine based on the exciton theory of the optical rotatory power of dimeric systems.[19a]

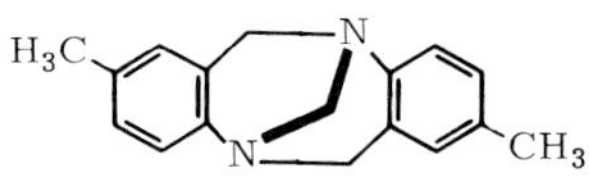

22

The absolute configuration of the isopavines has been determined by the application of the nonempirical aromatic chirality method[19b] to (–)-amurensine in ethanol solution. The first of the two pairs of split CD curves centered around 285 mμ can be correlated to the ^1L_b band. The fact that the first Cotton effect at longer wavelength is negative shows that the chirality of the corresponding benzene transitions is also negative or left-handed as shown here.[19c]

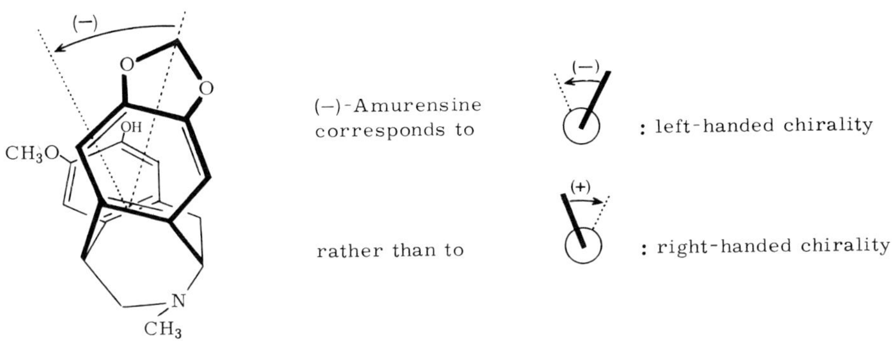

(–)-Amurensine

Similarly, the second pair of Cotton effects, also negative and centered around 240 mμ, which is assignable to interactions between ^1L_a transitions leads to the assignment of the same absolute configuration.[19c] The method is somewhat similar to, but simpler than, the excition treatment of Mason and co-workers.[19a,19d] (For a more detailed discussion of the aromatic chirality method see Chapter 32.)

Argemonine and amurensine, therefore, possess the same absolute configuration, with the nitrogen bridge situated below the mean plane of these molecules. This conclusion can be extended to the other pavines and isopavines since they usually possess in common strong negative specific rotations.[19c]

III. Biosynthesis

The only literature report of investigations in the biosynthesis of argemonine-type alkaloids involves the attempted incorporation of labeled (+)-reticuline into (−)-argemonine using mixed plants of *Argemone mexicana* L. and *A. hispida* Gray. (Papaveraceae).[20] Insignificant uptake of the labeled precursor was indicated, but this result may be assigned simply to the minimal amounts of argemonine present in these plants.[21]

The following tentative sequence has been presented by Stermitz to explain the biosynthesis of argemonine, norargemonine, and bisnorargemonine in *A. munita* and *A. hispida*. (+)-Reticuline has been isolated from both plants and could easily undergo oxidation to the iminium salt **23**. Intramolecular cyclization[7,18] would then generate bisnorargemonine and munitagine. Stepwise *O*-methylation of bisnorargemonine would subsequently lead to norargemonine and argemonine (Scheme IX). All these alkaloids are known to be present in these two plants.[21]

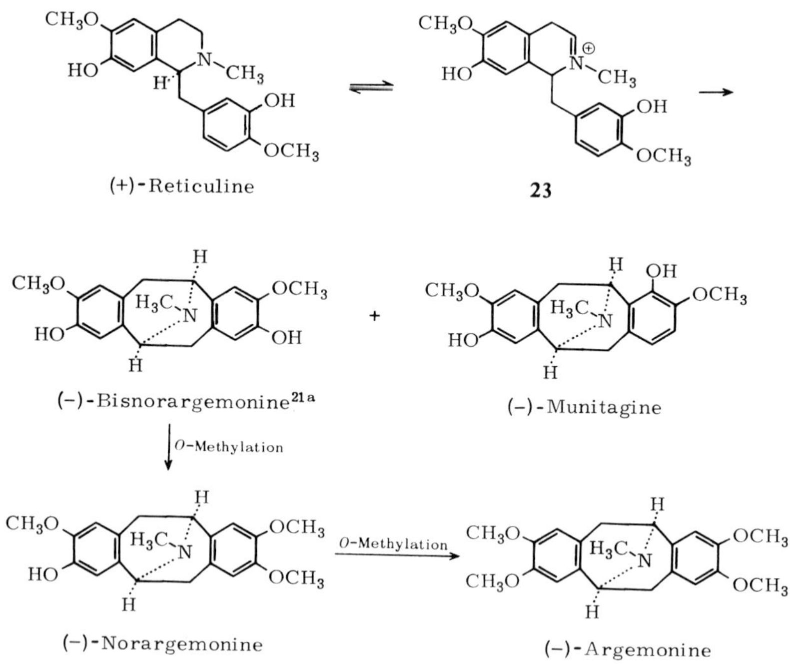

Scheme IX

An alternate thesis has been advanced by Brown, Dyke, *et al.* which is based on the intermediacy of 4-hydroxytetrahydroisoquinolines and explains the formation of pavines as well as isopavines. The 4-hydroxytetrahydroisoquinoline **24** when treated in vitro with acid was found to give the tetracyclic base **25** through displacement of the hydroxyl group. By analogy, therefore, a 4-hydroxytetrahydrobenzylisoquinoline should also cyclize in acid, or in the plant, to give rise to an isopavine structure.[22]

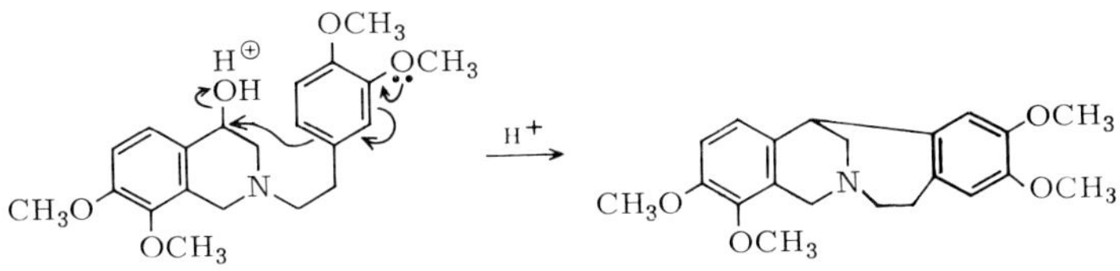

24 **25**

Following the established procedure for obtaining isopavine systems, it was then found that the amino acetal **26** when treated with concentrated hydrochloric acid at room temperature for 5 days gave amurensinine in 24% yield. But it was not possible to isolate the intermediate 4-hydroxytetrahydroisoquinoline **27** in a pure state, probably because it cyclizes very rapidly to the isopavine nucleus.

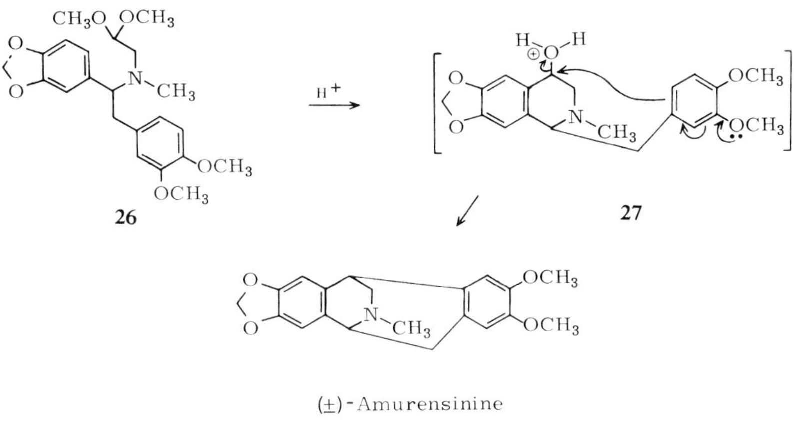

(±)-Amurensinine

By extrapolation, a 4-hydroxynorlaudanosoline analog would also be involved in the biosynthesis of the pavines, first by dehydration to a 1,2-dihydroisoquinoline, followed by double-bond isomerization and intramolecular cyclization (Scheme X). 4-Hydroxytetrahydrobenzylisoquinolines have not yet been isolated from plants, but analogously substituted aporphine and oxoaporphine alkaloids are known.[22]

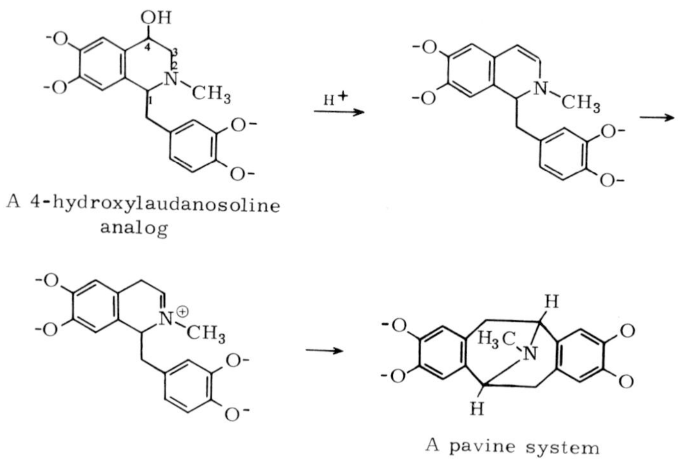

Scheme X

The finding that pavines and isopavines possess the identical absolute configuration lends some added weight to the idea of a 4-hydroxylated tetrahydrobenzylisoquinoline acting as a common precursor to these two groups of alkaloids.[19c]

IV. Mass Spectroscopy

The mass spectrum of argemonine shows a parent peak about one-third the size of the base peak, which is at *m/e* 204. There is also an intense $(M-1)^+$ ion at *m/e* 354 (Scheme XI).[23,24]

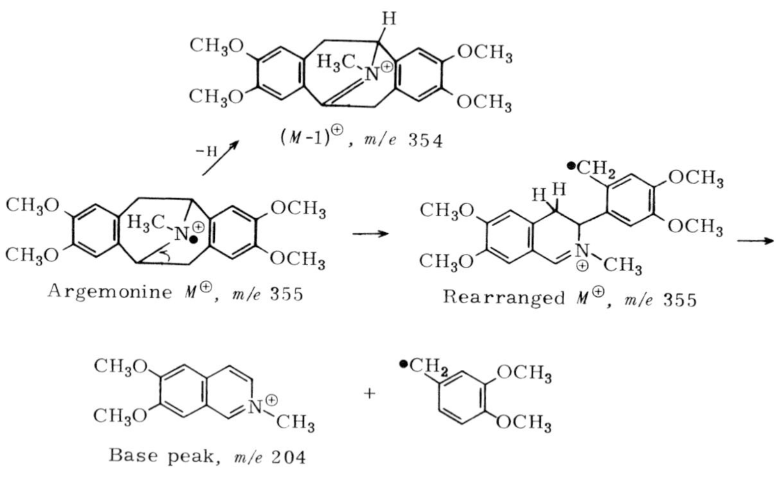

Scheme XI

The mass spectrum for the nonsymmetrical pavine alkaloid eschscholtzidine is slightly more complex. Besides the M^+ and $(M-1)^+$ peaks, there is a peak at *m/e* 204 and a base peak at *m/e* 188. The fragmentation mechanism in Scheme XII has been presented to explain the experimental results.[23]

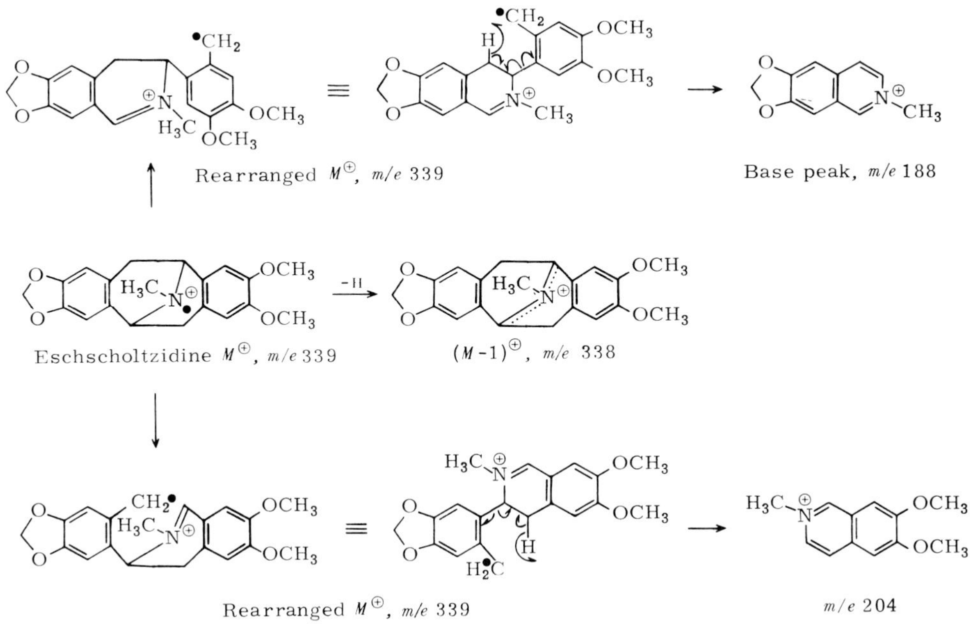

Scheme XII

In the mass spectrum of platycerine, Scheme XIII, the *m/e* 190 peak is only 30% of the intensity of the *m/e* 204 peak. This difference has been attributed to the lesser contribution to stabilization of the *m/e* 190 peak from the *o*-quinonoidal canonical form as compared to the *p*-quinonoidal form of the *m/e* 204 ion.[13]

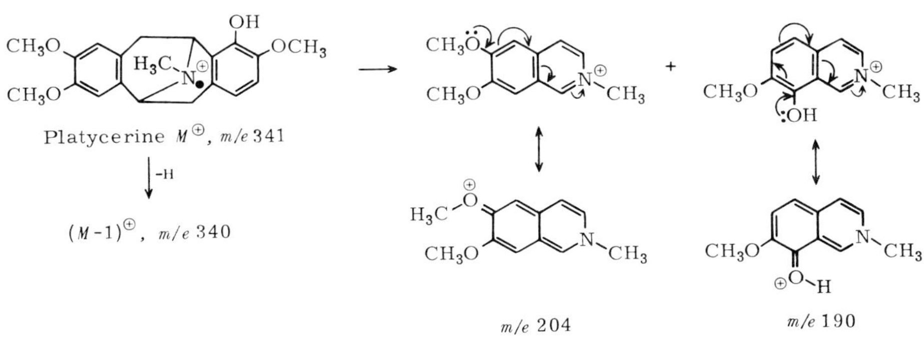

Scheme XIII

The isopavine alkaloid amurensinine shows intense peaks at *m/e* 339, 338, 296, and 188 (base) which point to the cleavage sequence described in Scheme XIV. It should be noted that the $(M-1)^+$ ion can be represented by either of two possible structures. Additionally, the cleavage of the molecular ion is such as to allow differentiation between the substituents on ring A and those on ring D – an item of importance in structural elucidation. Since the molecular ion of an isopavine can readily undergo a retrograde diene condensation with the loss of *m/e* 43 corresponding to a $H_2C{=}N{-}CH_3$ unit, mass spectrometry allows for the differentiation between pavine and isopavines on the basis of the abundance of the $(M-43)^+$ ion. This ion is not normally found in the pavine series.[25]

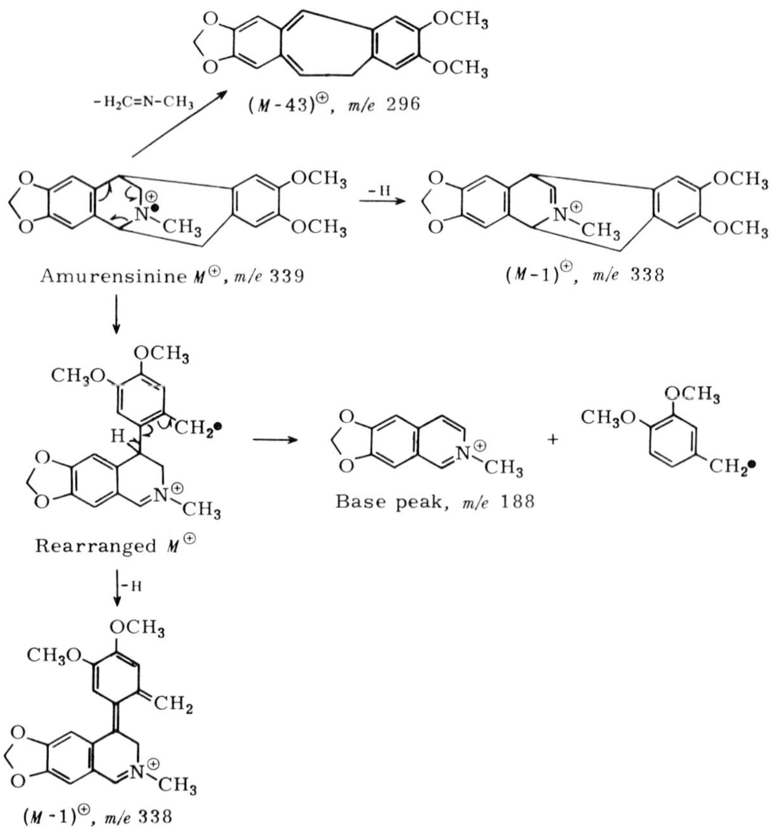

Scheme XIV

V. NMR Spectroscopy

The pavine bases are particularly amenable to NMR spectral analysis, and the spectra for argemonine and eschscholtzine are summarized below. The methylenedioxy absorptions in the spectrum of eschscholtzine are split into two doublets because of the asymmetry of the molecule.[1,12,23]

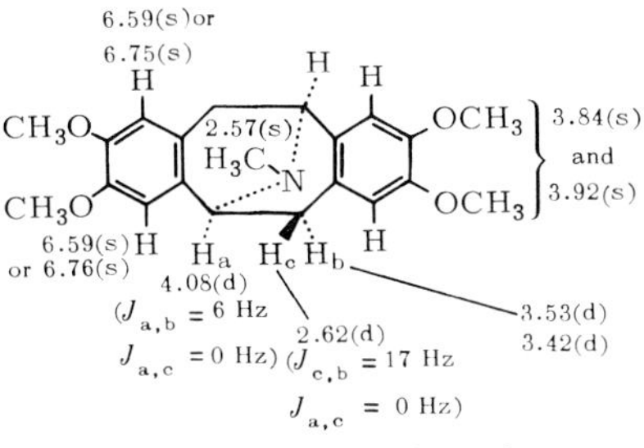

NMR spectral values for argemonine[23]

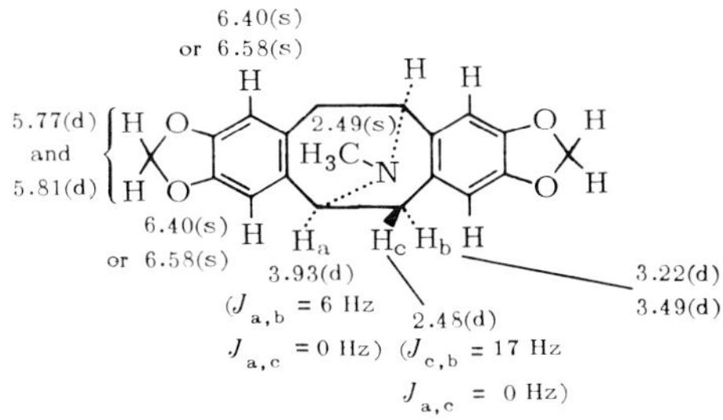

NMR spectral values for eschscholtzine[23]

An interesting aspect of the NMR studies on the diphenolic pavine alkaloid munitagine relates to the determination of the locations of the hydroxyl and methoxyl groups in the molecule. The alkaloid gave a positive Gibb's test, indicating a phenol with a free para position. A negative test with alcoholic ferric chloride solution ruled out the possibility of a catechol (1,2-diphenol) arrangement, while a positive test with Millon's reagent pointed to the presence of a phenol with a free ortho position. Oxidation of *O*,*O*-dimethylmunitagine with potassium permanganate yielded equimolar amounts of hemipinic and *m*-hemipinic anhydrides. These results established the structure of *O*,*O*-dimethylmunitagine as either **29** or **31** (Scheme XV).[21]

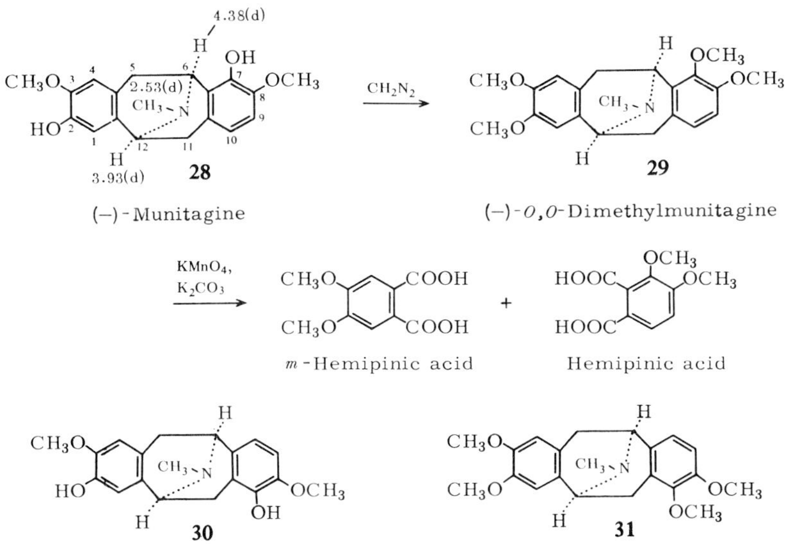

Scheme XV

When the NMR spectrum of munitagine in DMSO-d_6 and base was obtained, it was determined that the C-10 aromatic proton underwent a substantial upfield shift of about 0.7 ppm in relation to the corresponding chemical shift in deuteriochloroform. This proton must therefore lie para to a phenolic group, since protons ortho or meta to a phenol undergo an upfield shift of about half that magnitude. At this stage the spectral results confirmed the conclusions from the Gibb's test, but still did not differentiate between structures **28** and **30** for munitagine.[21]

Analysis of the deuteriochloroform NMR spectrum of munitagine next indicated that the C-12 proton appeared as a doublet at δ 3.93, while the C-6 proton was represented by a doublet further downfield at δ 4.38. The downfield chemical shift of the C-6 hydrogen must be due to the proximity of this hydrogen to an oxygen function, namely the phenolic group at C-7. It follows that munitagine must be represented by **28** and not **30**.[21]

The NMR spectra of several of the isopavine alkaloids have been recorded, but it is more difficult to make specific assignments to some of the peaks.[1,16] The methylenedioxy hydrogens in amurensine and amurensinine again appear as two doublets because of the asymmetry of these species.[16] The *N*-methyl singlet absorptions in amurensine, amurensinine, and thalisopavine come at δ 2.47, 2.50, and 2.48, respectively – very close to the corresponding *N*-methyl peaks for the pavine bases argemonine, eschscholtzine, and munitagine, which are at δ 2.57, 2.49, and 2.53, respectively.

VI. UV Spectroscopy

The UV spectra of the pavine alkaloids are not particularly useful for elucidation of structural details, the maximum absorption appearing as a broad peak between 287 and 295 mμ in ethanol. The cyclohexane spectra are more distinctive: for argemonine and its *O*-methyl derivatives well-defined maxima are found at 284, 288, and 294 mμ.[21] The isopavines usually show a shoulder at 250 mμ in ethanol solution which does not appear in the spectra of the alkaloids of the pavine type. But this criterion alone is insufficient to allow a conclusive differentiation between pavines and isopavines.

Argemonine[23] (*N*-Methylpavine)	λ_{max}^{EtOH} 287 mμ (4.00)
Eschscholtzine[23]	λ_{max}^{EtOH} 295 mμ (4.04)
Californidine iodide[11]	λ_{max}^{MeOH} 292 mμ (4.02)
	λ_{min}^{MeOH} 256 mμ (2.93)
Caryachine[2]	λ_{max}^{EtOH} 291.5 mμ (4.02)
	λ_{min}^{EtOH} 253.5 mμ (2.70)
Platycerine[26]	λ_{max}^{MeOH} 283 mμ (3.83)
Amurensine[16]	λ_{max}^{EtOH} 230, 250 sh, and 290 mμ (4.07, 3.67, and 3.95)
	λ_{min}^{EtOH} 263 mμ (3.29)

Amurensinine[16]	λ_{max}^{EtOH} 230, 250 sh, and 294 mμ (4.07, 3.67, and 3.95)
	λ_{min}^{EtOH} 263 mμ (3.29)
Thalisopavine[17]	λ_{max}^{EtOH} 289 mμ (4.06)

VII. The Differentiation between Pavine and Isopavine Bases

As mentioned above, the isopavine bases show a shoulder at 250 mμ which does not usually appear in the UV spectra of the pavine alkaloids. Additionally, the mass spectra of the isopavines show an $(M-43)^+$ ion which is missing in the pavines.

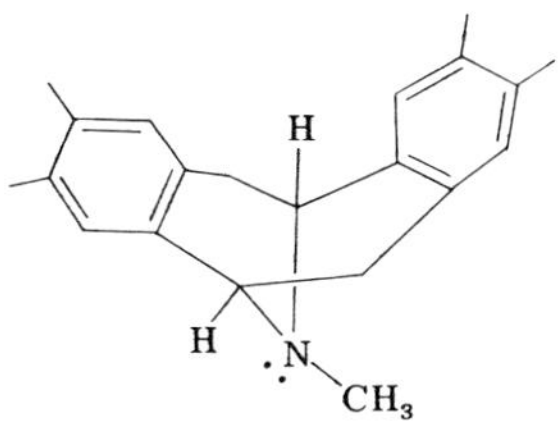

Pavine base

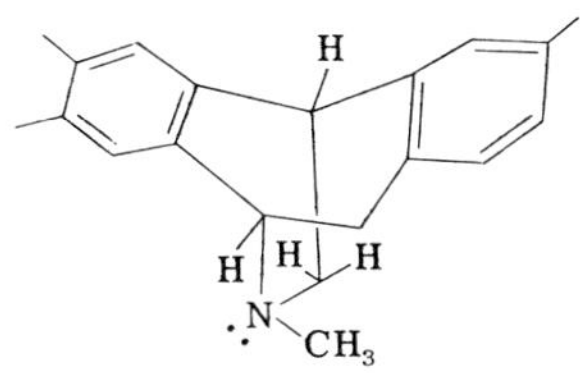

Isopavine base

A third criterion for differentiating is based on the measurement of rates of methiodide formation in acetonitrile solution at 25°. Pavine alkaloids show pseudo-first order rates of *N*-methylation with methyl iodide around 272×10^{-4} sec^{-1}, while the isopavine base amurensine *N*-methylates almost twice as fast, with a rate of 508×10^{-4} sec^{-1}. In the pavine series, the *N*-methyl group is bordered by two methine groups, while in the isopavines the *N*-methyl is adjacent to a methine and a methylene group – the less-hindered isopavines therefore methylate at a faster rate (see Table I).[27]

TABLE I

Rates of *N*-Methylation for the Pavines – Isopavines at 25°[a]

Alkaloid	Rate × 10⁴ sec⁻¹
Argemonine	273
Norargemonine	272
Bisnorargemonine	273
Munitagine	271
Amurensine	508

[a] For a more complete discussion of rates of methiodide formation see Chapter 16, Section V, B.

REFERENCES

1. For a review of the pavine and isopavine alkaloids, see F. Šantavý, *Alkaloids* **12**, 370 (1970).
2. S.-T. Lu and P.-K. Lan, *J. Pharm. Soc. Jap.* **86**, 177 (1966).
3. L. Dolejš and J. Slavík, *Collect. Czech. Chem. Commun.* **33**, 3917 (1968).
4. (−)-Remrefine ≡ (−)-roemrefine. M. S. Yunusov, S. T. Akramov, and S. Yu. Yunusov, *Khim. Prir. Soedin* **4**, 225 (1968); *Chem. Abstr.* **70**, 68581r (1969).
5. G. Goldschmiedt, *Monatsh. Chem.* **19**, 321 (1898).
6. C. Schöpf, *Angew. Chem.* **62**, 453 (1950).
7. A. R. Battersby and R. Binks, *J. Chem. Soc., London* p. 2888 (1955).
8. F. R. Stermitz, S.-Y. Lwo, and G. Kallos, *J. Amer. Chem. Soc.* **85**, 1551 (1963).
9. M. J. Martell, Jr., T. O. Soine, and L. B. Kier, *J. Amer. Chem. Soc.* **85**, 1022 (1963).
10. M. M. Abdel-Monem and T. O. Soine, *J. Pharm. Sci.* **56**, 976 (1967).
11. J. Slavík, L. Slavíkova, and K. Haisová, *Collect. Czech. Chem. Commun.* **32**, 4420 (1967).
12. F. R. Stermitz and J. N. Seiber, *Tetrahedron Lett.* p. 1177 (1966).
13. F. R. Stermitz and K. D. McMurtrey, *J. Org. Chem.* **34**, 555 (1969).
14. E. Waldmann and C. Chwala, *Justus Liebigs Ann. Chem.* **609**, 125 (1957).
15. A. R. Battersby and D. A. Yeowell, *J. Chem. Soc., London* p. 1988 (1958).
16. F. Šantavý, L. Hruban, and M. Maturová, *Collect. Czech. Chem. Commun.* **31**, 4286 (1966).
17. S. M. Kupchan and A. Yoshitake, *J. Org. Chem.* **34**, 1062 (1969).
18. A. C. Baker and A. R. Battersby, *J. Chem. Soc., C* p. 1317 (1967).
19. O. Červinka, A. Fábryová, and V. Novák, *Tetrahedron Lett.* p. 5375 (1966).
19a. S. F. Mason, K. Schofield, R. J. Wells, J. S. Whitehurst, and G. W. Vane, *Tetrahedron Lett.* p. 137 (1967).
19b. R. Harada and K. Nakanishi, submitted for publication.
19c. M. Shamma, J. L. Moniot, W. K. Chan, and K. Nakaniski, *Tetrahedron Lett.* p. 3425 (1971).
19d. S. F. Mason and G. W. Vane, *J. Chem. Soc., London* p. 370 (1966).
20. D. H. R. Barton, R. H. Hesse, and G. W. Kirby, *J. Chem. Soc., London* p. 6379 (1965).
21. F. R. Stermitz and J. N. Seiber, *J. Org. Chem.* **31**, 2925 (1966).
21a. For a synthesis of bisnorargemonine, see C.-H. Chen, T. O. Soine, and K.-H. Lee, *J. Pharm. Sci.* **59**, 1529 (1970).
22. D. W. Brown, S. F. Dyke, G. Hardy, and M. Sainsbury, *Tetrahedron Lett.* p. 1515 (1969).
23. R. H. F. Manske, K. H. Shin, A. R. Battersby, and D. F. Shaw, *Can. J. Chem.* **43**, 2183 (1965).
24. M. Tomita, S.-T. Lu, and T. Ibuka, *J. Pharm. Soc. Jap.* **86**, 414 (1966).
25. L. Dolejš and V. Hanuš, *Collect. Czech. Chem. Commun.* **33**, 600 and 3917 (1968).
26. Holubek and Štrouf, Spectrum No. 210.
27. M. Shamma, C. D. Jones, and J. A. Weiss, *Tetrahedron* **25**, 4347 (1969).

Chapter 5 / THE BISBENZYLISOQUINOLINES

Sources: Anonaceae, Berberidaceae, Hernandiaceae, Lauraceae, Magnoliaceae, Menispermaceae, Monimiaceae, Nymphaeceae, and Ranunculaceae
Approximate Number: Over 100

I. Introduction

The bisbenzylisoquinolines are the largest group of isoquinoline alkaloids.[1,1a,b,c] They are dimeric bases which for purposes of classification can be subdivided into the 21 subgroups depicted. The aromatic substituents may be hydroxyl, methoxyl, or methylenedioxy. When two or more diphenyl ether linkages are present, a large ring is formed which usually incorporates 18 atoms. This ring may also be 16-membered as in the case of the rodiasine type, 19-membered as in the thalicrine type, or even 21-membered as in the thalmine type. Two asymmetric centers are present in the bisbenzylisoquinolines except when one or both of the nitrogen atoms are in the form of an imine.

Bisbenzylisoquinolines are often encountered in plants in the form of complex mixtures, where the individual components may differ in absolute configuration or degree of *N*- or *O*-methylation. An excellent TLC system for separating these components has been described by Döpke which consists of a chloroform–ethyl acetate–methanol system (40 : 40 : 20) on Silica Gel G (Merck) prepared with 0.1 *N* sodium hydroxide.[1d]

Bisbenzylisoquinoline Subgroups

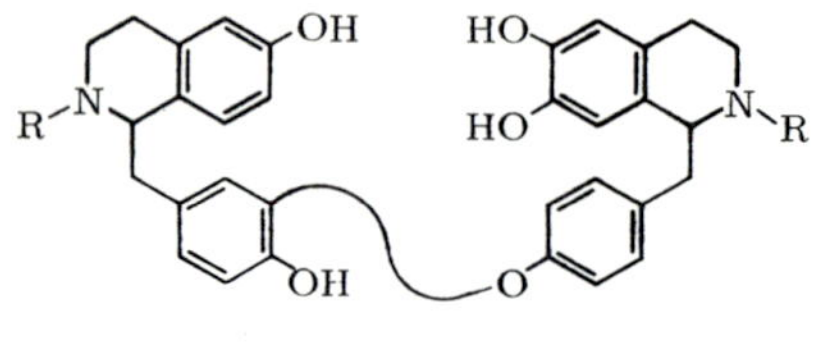

Thalisopine type (?)

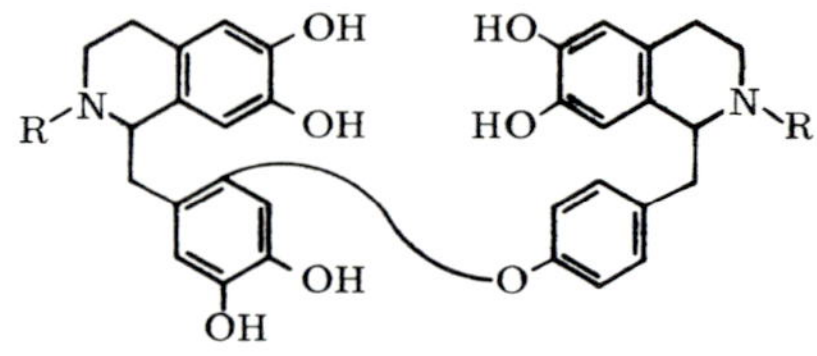

Magnolamine type

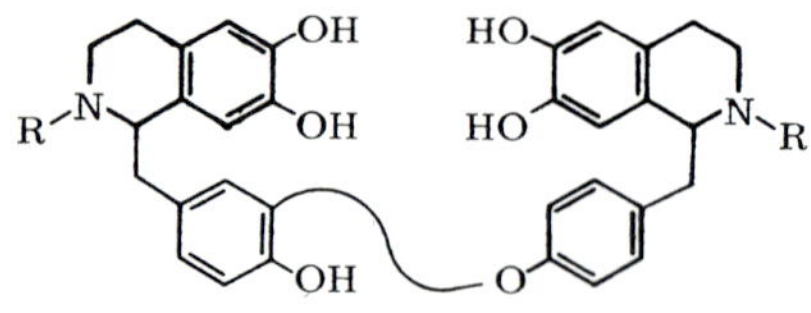

Dauricine type

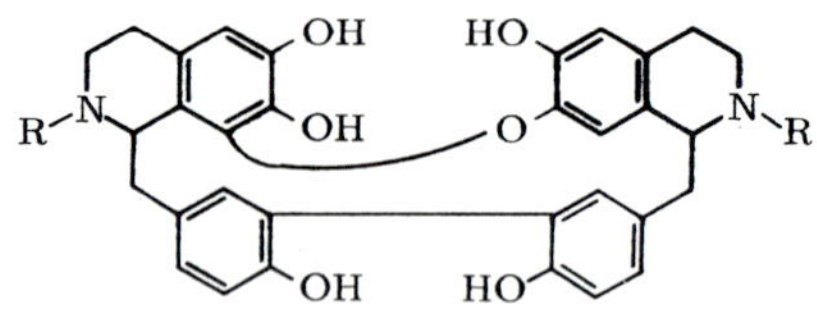

Rodiasine type

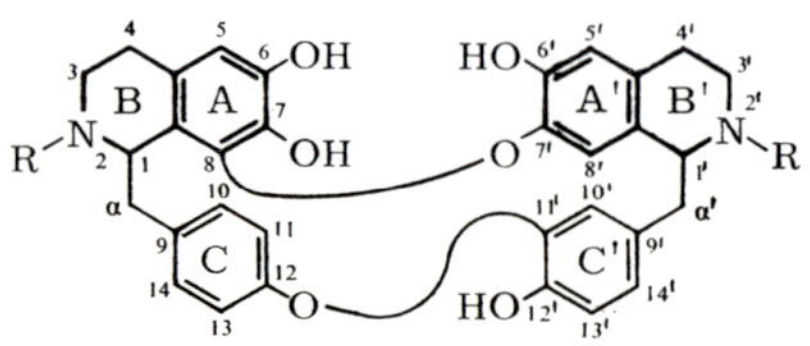

Oxyacanthine type

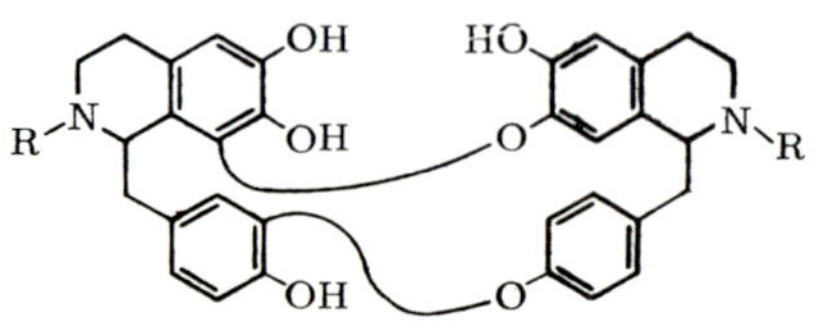

Berbamine type

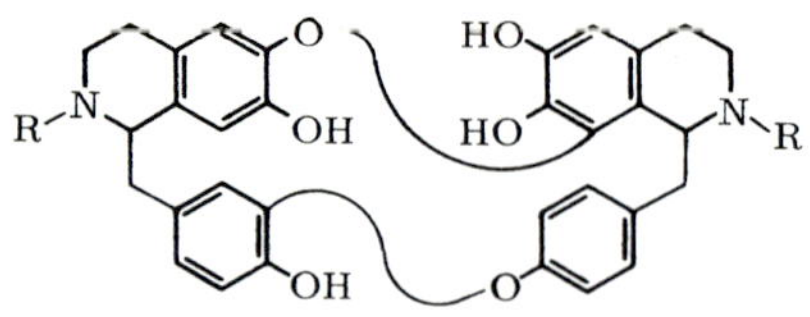

Thalicrine type

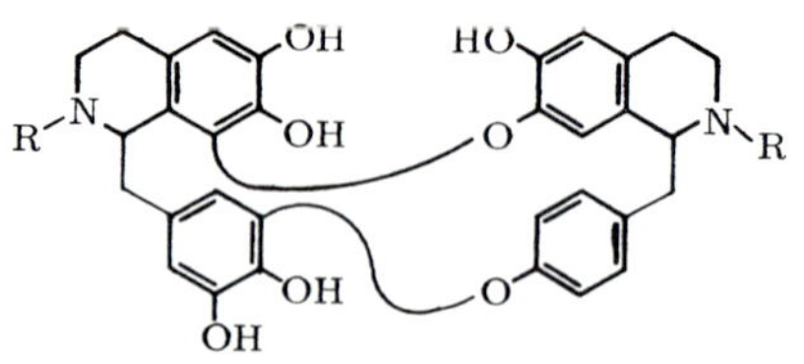

Tenuipine type

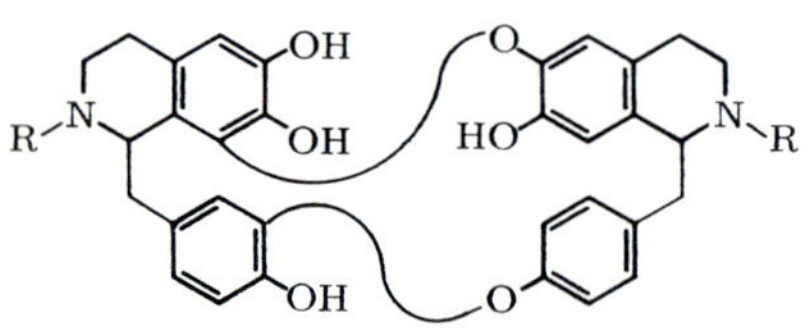

Thalicberine type

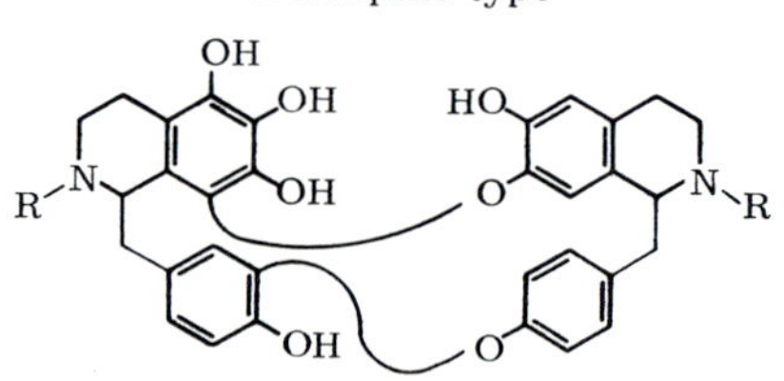

Hernandezine type

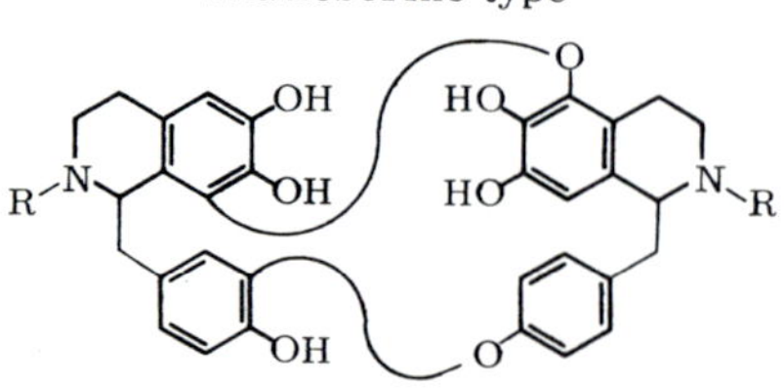

Thalidasine type

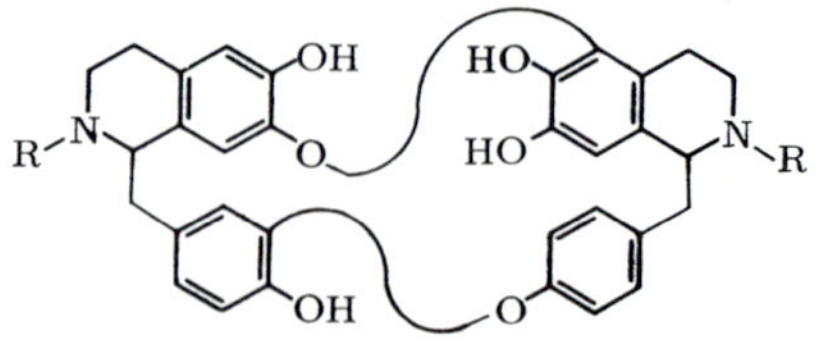

Thalmine type

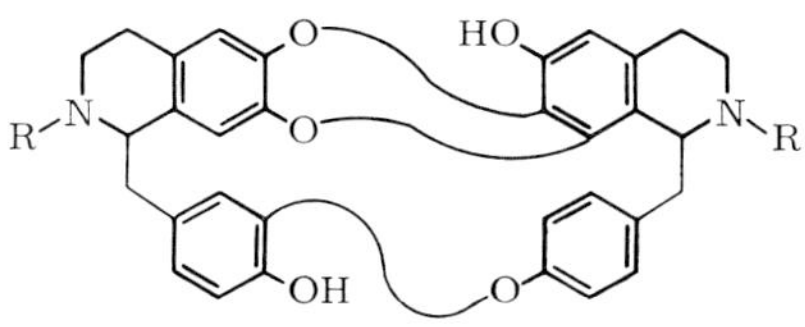

Trilobine-isotrilobine-
micranthine type

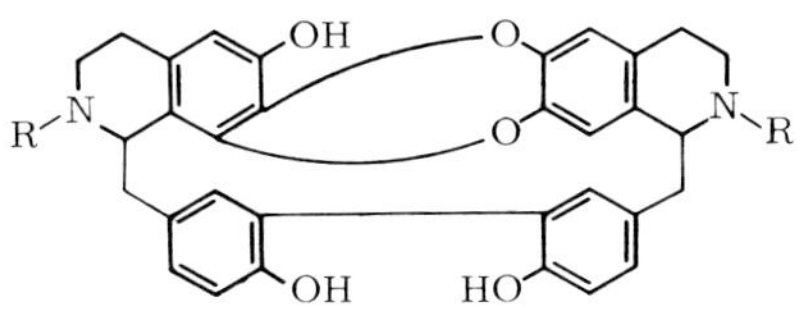

Tiliacorine type

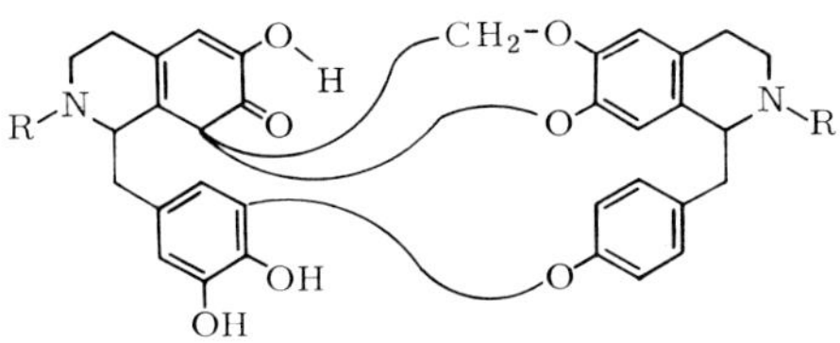

Repanduline type

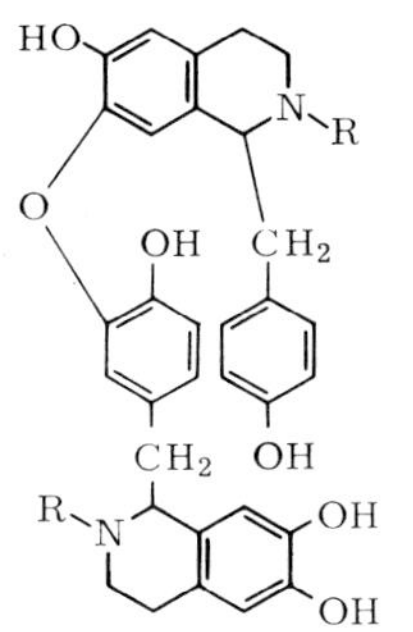

Liensinine type

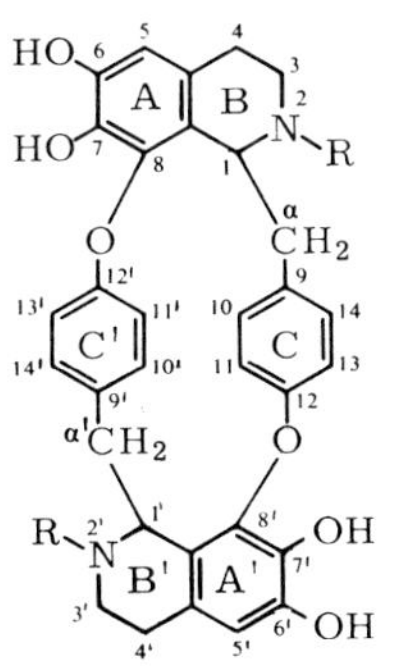

Isochondodendrine
type

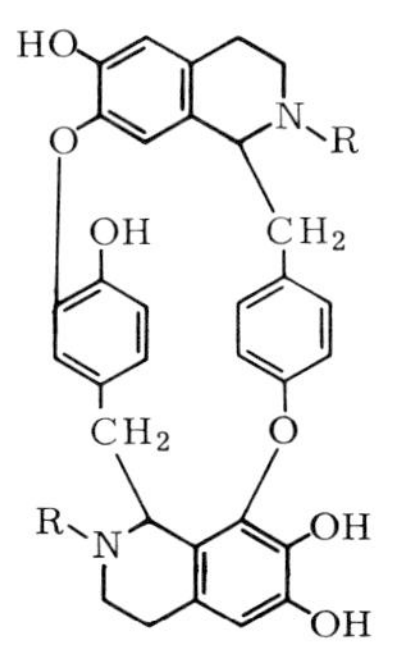

Curine-chondocurine
type

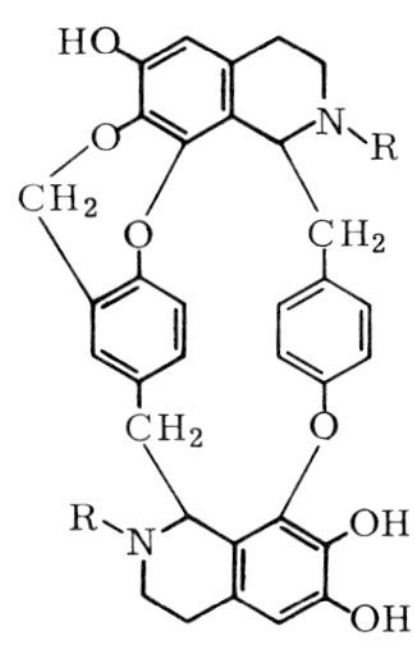

Insularine type

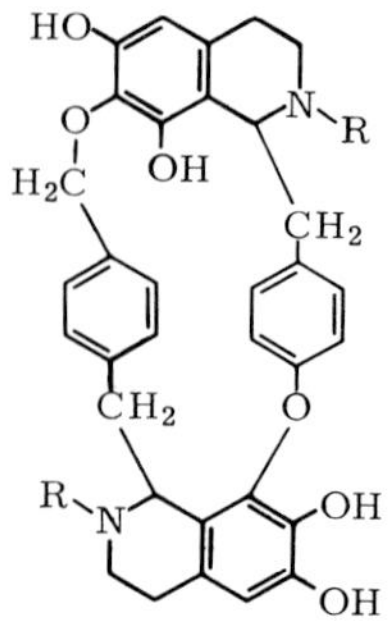

Cissampareine type

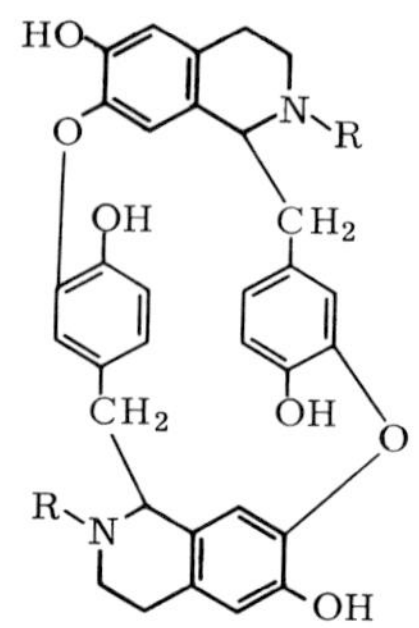

Hayatine type

II. A Classical Degradation of a Bisbenzylisoquinoline Alkaloid: The Structure of Berbamine

The alkaloid (+)-berbamine analyzes for $C_{37}H_{40}O_6N_2$ and possesses three methoxyls, one phenolic hydroxyl, two diphenyl ether linkages, and two *N*-methyl groups as determined by von Bruchhausen and co-workers. Hofmann degradation of *O*-methylberbamine yielded as one of the products the bismethine **1** which upon ozonolysis afforded the dialdehydes **2** and **3**. Dialdehyde **2** was characterized as the diacid **4** (Scheme I).

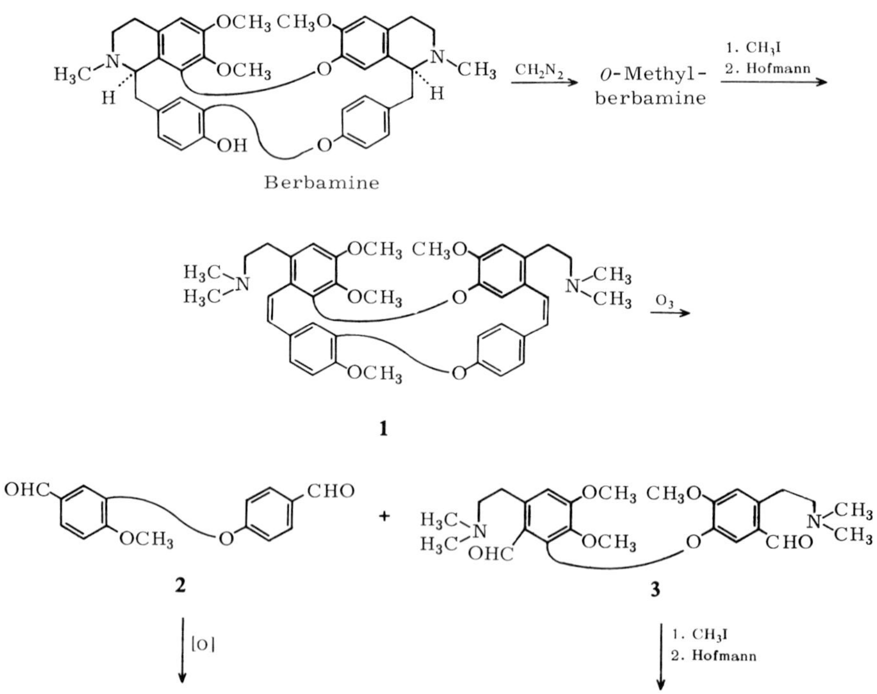

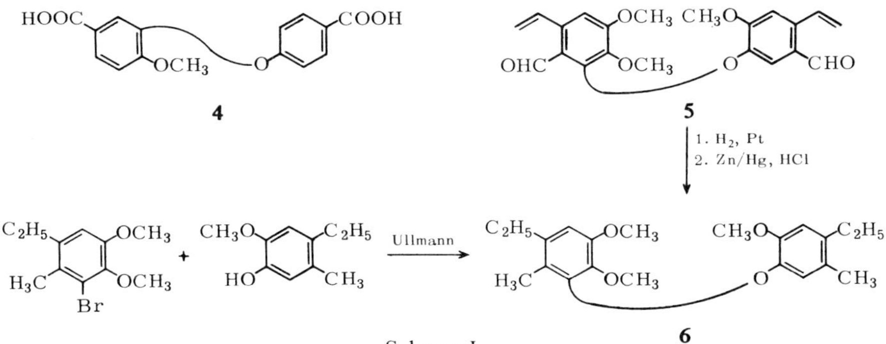

Scheme I

The basic dialdehyde **3** was converted by a second Hofmann degradation into the diene dialdehyde **5**. Catalytic hydrogenation of compound **5** followed by Clemmensen reduction afforded the diphenyl ether **6**, which was unambiguously synthesized using an Ullmann condensation.[2,3]

It can thus be seen that the early degradative work on the bisbenzylisoquinolines relied mainly upon Hofmann degradations and oxidative studies. The position of the phenolic function in berbamine was obtained at a later date utilizing the sodium in liquid ammonia cleavage of *O*-ethylberbamine to generate the benzylisoquinolines **7** and **8**.[4]

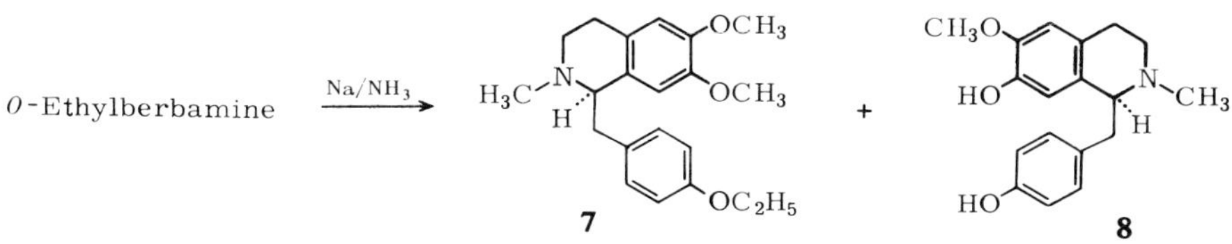

III. Some Comments on the Sodium in Liquid Ammonia Cleavage of Bisbenzylisoquinolines

Sartoretto and Sowa reported in 1937 on the cleavage of diaryl ethers with sodium in liquid ammonia.[5] Their findings can be summarized in the following representative equations*:

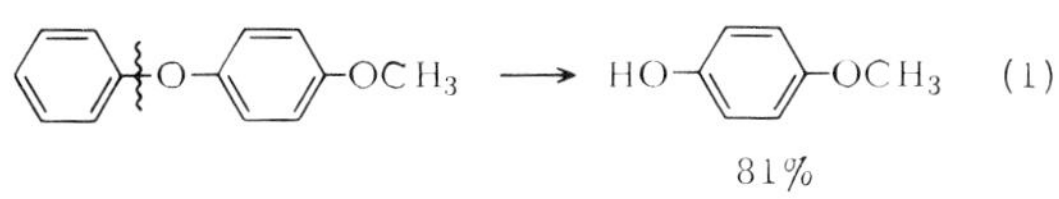

* These data were obtained before the advent of chromatography as a standard laboratory tool and may therefore need some revision.

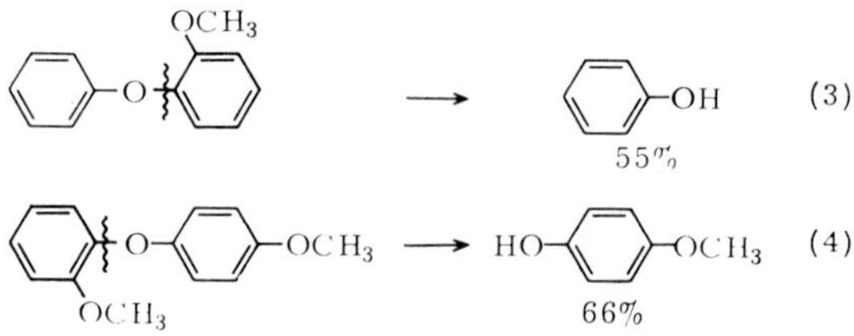

Whenever an *o*-methoxyl group is present, cleavage will occur preponderantly ortho to this methoxyl group, as shown in Eqs. (3) and (4). Fission of the bond between the methoxylated ring and the oxygen of the diphenyl ether occurs in the following order:

$$o\text{-}CH_3O > m\text{-}CH_3O > p\text{-}CH_3O$$

A simplified mechanism for the reductive cleavage of phenyl *m*-methoxylphenyl ether is shown in Scheme II, given that the metal in liquid ammonia system is a rich source of electrons. The successive steps involve addition of an electron to form an anion radical followed by protonation with ammonia acting as the proton source.

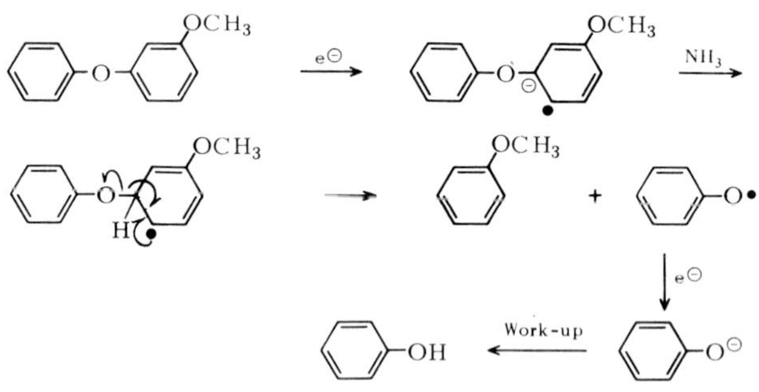

Scheme II

It may be that the phenyl *p*-methoxyphenyl ether cleaves in the opposite direction because the site para to the methoxyl group is already rich in electrons through resonance so that the first electron prefers to add to the unmethoxylated ring as in **9**. When an *o*-methoxyl group is present, the inductive effect of that methoxyl may assist in the addition of the first electron to the methoxylated ring, causing the fission indicated in Eqs. (3) and (4).

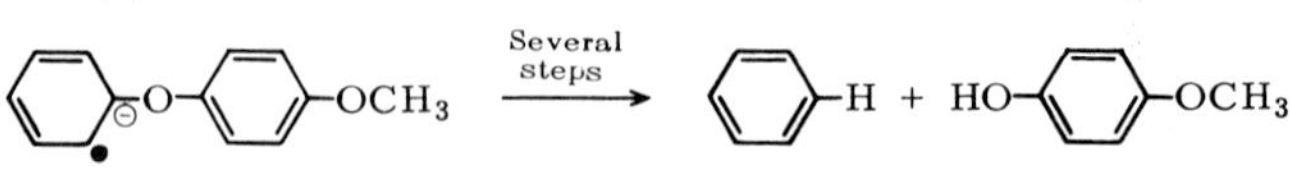

9

The exact mechanism of phenyl ether cleavage is somewhat obscure. An attempt was made to discuss this reaction in terms of molecular parameters estimated from a modified Hückel MO treatment,[6] but more work remains to be done before firm conclusions can be drawn.

The sodium in liquid ammonia reduction of aromatic ethers was first applied successfully to the bisbenzylisoquinolines by Tomita and co-workers.[7] If the alkaloid is phenolic, it is first either *O*-methylated or *O*-ethylated. In a few instances, apparently only two benzylisoquinolines are obtained from the reduction. In most cases, complex mixtures of benzylisoquinolines are produced which must be separated chromatographically.

When a diphenylenedioxide bridge is a feature of the dimeric molecule, as denoted by a bluish coloration with sulfuric acid and a little nitric acid, a two-stage reductive fission of the molecule is required as illustrated for the (+)-isotrilobine case.[8]

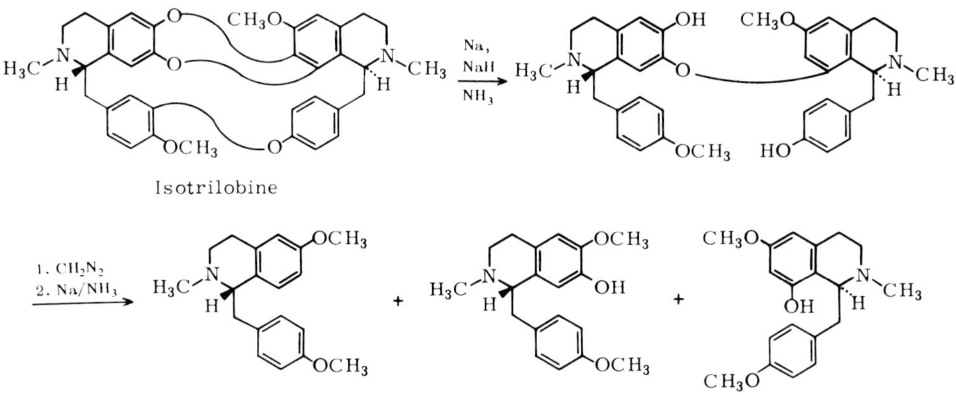

It is important to bear in mind that there is an intrinsic uncertainty in a sodium in liquid ammonia cleavage even when the two resulting isoquinoline units are completely characterized. For example, the structure (**10**) assigned to thalisopine, isolated from

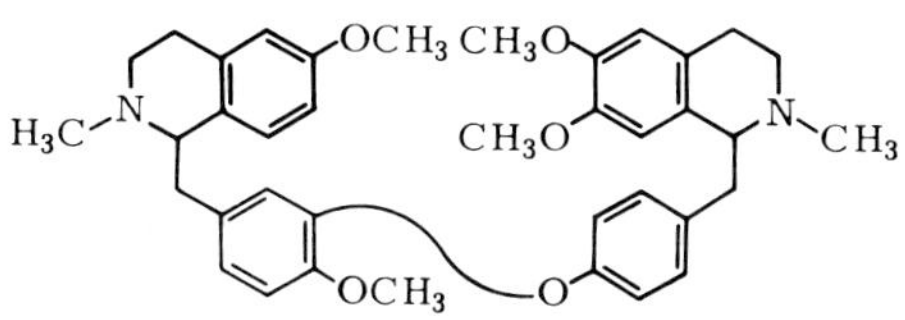

10

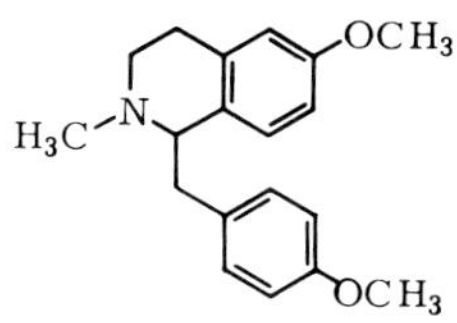

11

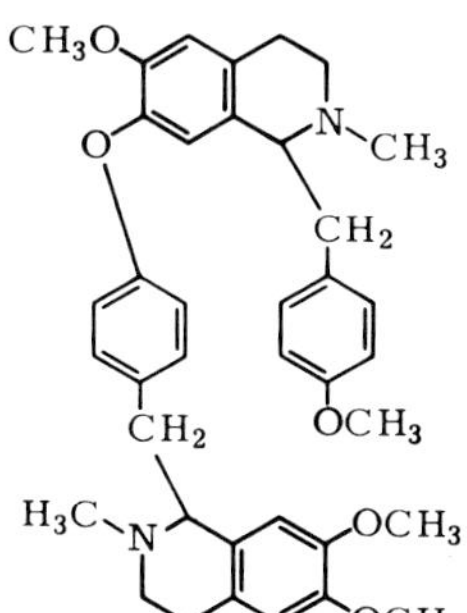

12

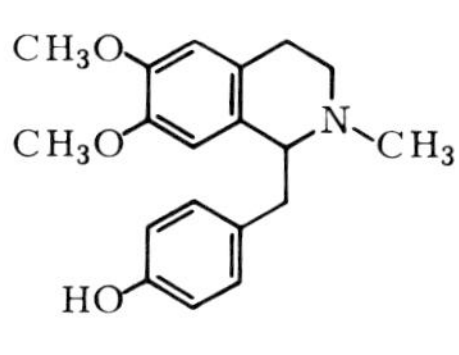

Armepavine

Thalictrum isopyroides C.A. May (Ranunculaceae), depends almost solely upon the isolation of armepavine and the isoquinoline **11** from the sodium in liquid ammonia reduction.[9] It should be noted, however, that **12** is a more valid expression for the alkaloid.[1b] In this context, a reduction using ND_3 can be useful in locating the site of the diphenyl ether linkage.[9a] (See also Chapter 32.)

IV. The Structures of Tubocurarine, Chondocurarine, and Chondocurine

The old structures for the curine-type alkaloids (+)-tubocurarine, (+)-chondocurarine, and (+)-chondocurine, arrived at by classical chemical degradative means, are given below. The first two alkaloids were supposed to differ in the location of the phenolic function in ring A′, while (−)-chondocurine was assumed to be the free base corresponding to (+)-chondocurarine.

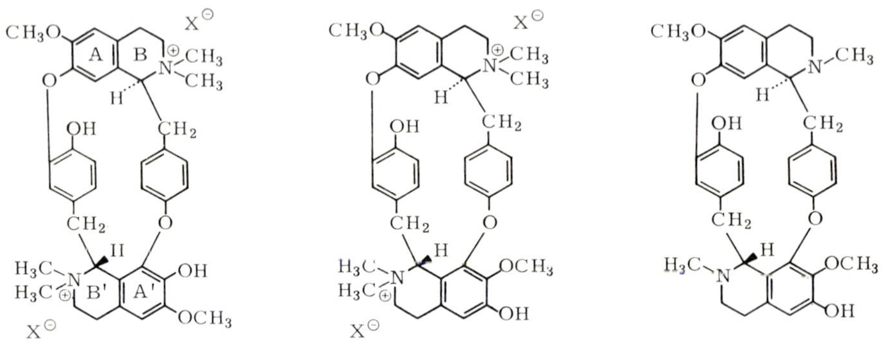

Old structure for (+)-tubocurarine

Old structure for (+)-chondocurarine

Old structure for (+)-chondocurine

It has lately been found, however, that the free base (+)-tubocurine, obtained by thiophenolate *N*-demethylation of (+)-tubocurarine corresponds in all respects with (+)-chondocurine. Changes in structural assignments were therefore required, and the corrected structures shown in Scheme III were adopted: the phenolic function on ring A′ was kept at C-7′, the only difference being in the extent of *N*-methylation.[10]

NMR spectroscopy showed the presence in (+)-tubocurarine chloride of only three *N*-methyl groups instead of four. Upon addition of NaOD one of the *N*-methyl signals moved to higher field, giving conclusive proof that this nitrogen is originally in the form of a protonated tertiary amine which can revert to the free base in the presence of sodium hydroxide.

A sodium in liquid ammonia cleavage of (+)-tubocurarine chloride yielded the optically inactive Emde product **13** together with (+)-1-(4′-hydroxybenzyl)-2-methyl-6-methoxy-7-hydroxytetrahydroisoquinoline **(14)**, thus settling the identity of the protonated nitrogen in the alkaloid.

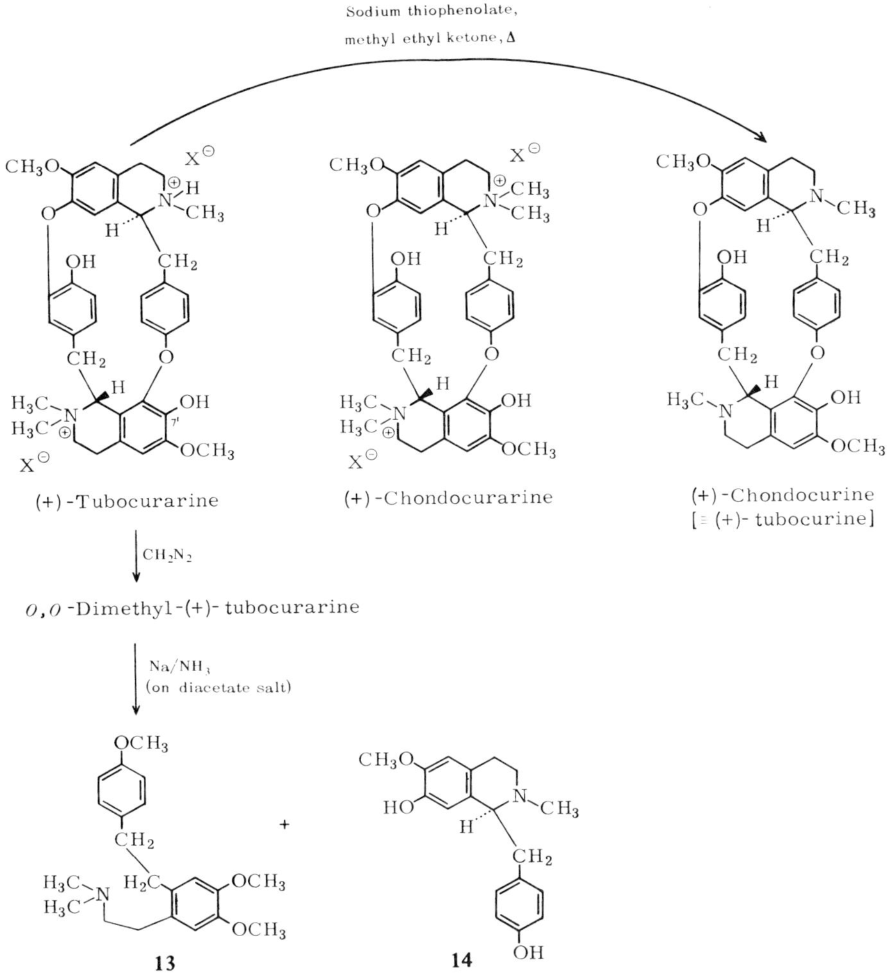

Scheme III

(+)-Chondocurarine chloride is the dimethochloride of (+)-chondocurine, differing from (+)-tubocurarine chloride in the degree of *N*-methylation.[10]

V. Repanduline: A Ketonic Bisbenzylisoquinoline Alkaloid

The yellow alkaloid (+)-repanduline, $C_{37}H_{36}O_7N_2$, investigated by Harley-Mason and co-workers, has been found in a variety of *Daphnandra* species (Monimiaceae). In *Daphnandra tenuipes* Perkins it is accompanied by the bisbenzylisoquinoline nortenuipine of established structure and absolute configuration.

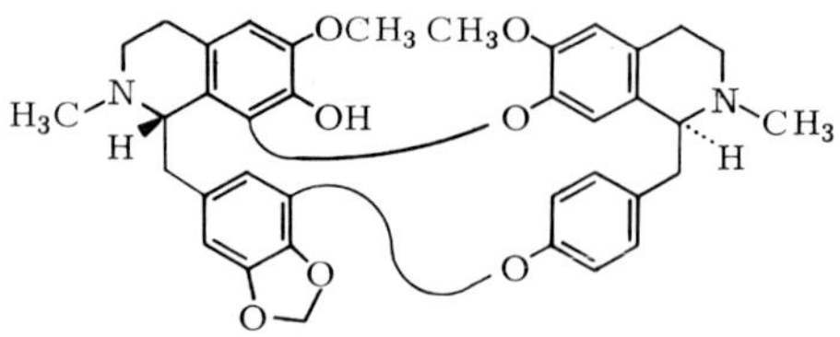

Nortenuipine

Repanduline contains two *N*-methyl groups, one methylenedioxy, and one methoxyl, but the most striking feature of the alkaloid is that it exhibits a sharp carbonyl band in the IR at $5.87\ \mu$ ($1703\ \text{cm}^{-1}$) which is replaced by a hydroxyl band at $2.81\ \mu$ ($3560\ \text{cm}^{-1}$) after reduction with sodium borohydride.

Potassium permanganate oxidation of the alkaloid furnished repandulinic acid. Potassium in liquid ammonia reduction of repanduline produced hemirepanduline, a benzylisoquinoline whose structure was confirmed by synthesis. The interesting point about hemirepanduline is that it includes two methoxyl groups, while the parent alkaloid possesses only one methoxyl.[10a]

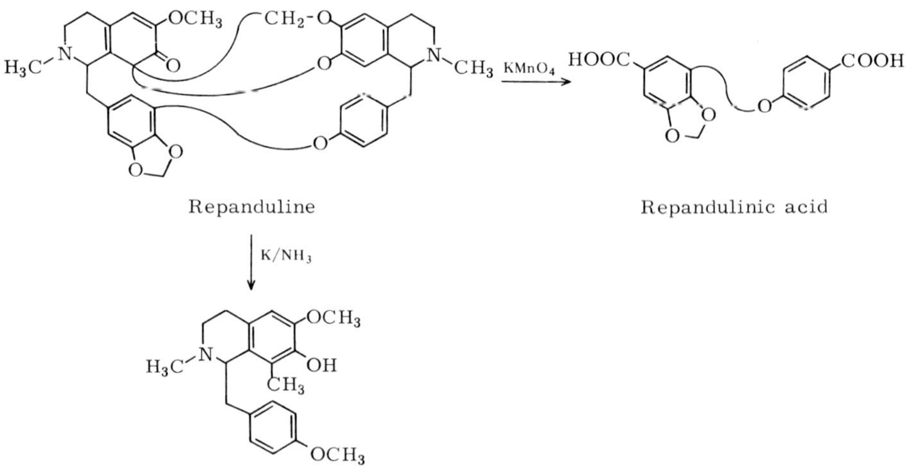

Repanduline

Repandulinic acid

Hemirepanduline

Reduction of repanduline with sodium borohydride gave repandulinol, which could be oxidized by the Oppenauer method back to repanduline. Catalytic reduction of repanduline led to dihydrorepanduline, and reduction of repandulinol gave dihydrorepandulinol, which could also be obtained from the sodium borohydride reduction of dihydrorepanduline.

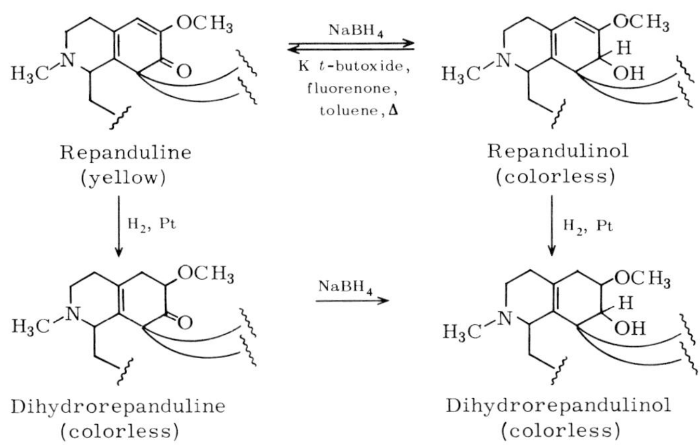

Acid-catalyzed hydrolysis of the enol ether function of repandulinol gave de-*O*-methylrepandulinol which exhibited an IR carbonyl band at 5.80 μ (1724 cm^{-1}). Sodium methoxide treatment of de-*O*-methylrepandulinol resulted in β-elimination and formation of isode-*O*-methylrepandulinol which gave an intense green coloration with ferric chloride characteristic of catechols. Reduction of de-*O*-methylrepandulinol with sodium borohydride afforded de-*O*-methyldihydrorepandulinol – a product which showed a positive reaction with the periodate-Schiff reagent for 1,2-diols.[10a]

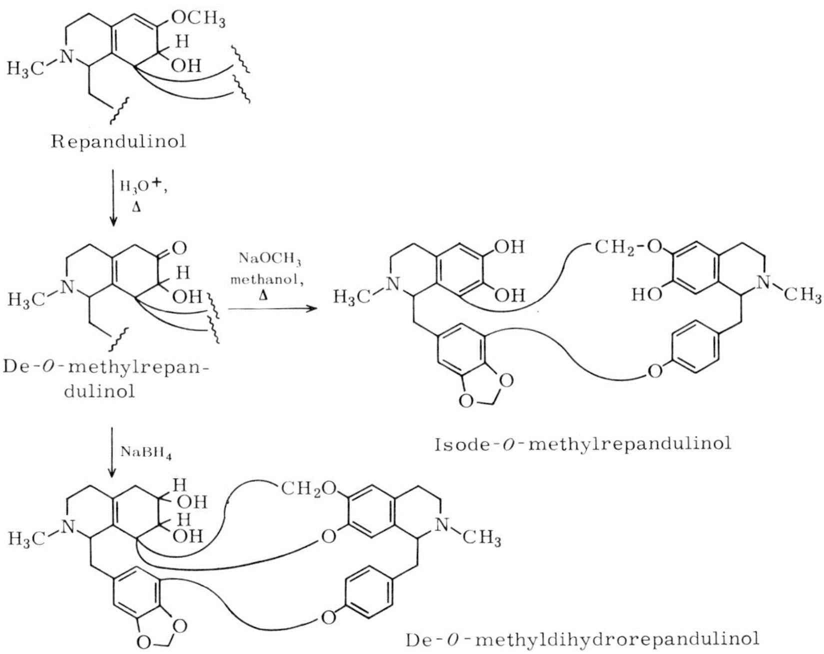

Although the structural assignment for repanduline is supported by NMR and mass spectral data, it is not possible to eliminate altogether the alternate structure drawn below. The structure given above to repanduline is favored on biogenetic grounds since it is conceivable that repanduline is formed by oxidation of one of the *O*-methyl groups of nortenuipine.[10a]

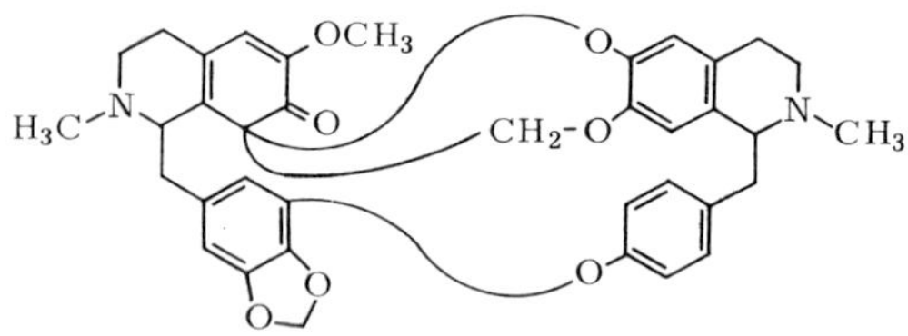

Alternate structure for repanduline

VI. Syntheses of Bisbenzylisoquinolines

To date, all but one of the syntheses that have led to naturally occurring dimeric benzylisoquinolines have proceeded through Ullmann condensations to form the diphenyl ether linkages. Several such Ullmann syntheses are available, but in this section only a few will be considered: that of liensinine which contains one diphenyl ether linkage; those of cepharanthine, isotetrandrine, and phaeanthine which incorporate two diphenyl ethers; and that of *N*-methyldihydromenisarine which has three such linkages. All of these schemes depend upon the judicious use of protective groups and blocking devices. The last synthesis which will be described is based on the electrolytic oxidation of a phenolic tetrahydrobenzylisoquinoline.

A. The Kametani Synthesis of a Diastereoisomeric Mixture of Liensinines

In 1962, a research group at the Chinese Academy of Sciences in Shanghai reported on its extensive degradative work on the new bisbenzylisoquinoline alkaloid liensinine obtained from the seed of the lotus, *Nelumbo nucifera* Gaertn. (Nymphaceae). As a result of these efforts, the alkaloid was shown to possess the structure shown in Scheme IV.[11]

The Kametani synthesis (Scheme IV) led to a mixture of diastereoisomeric liensinines which were not separated. It is worth noting the use made of the differences in reactivity between tosylates and benzyl ethers. Aromatic tosylates are generally base-sensitive, whereas benzyl ethers are not, so that specific phenolic functions could be unveiled as desired at the appropriate stages. Ullmann condensation of the aromatic bromide **15** with the phenol **16** followed by acid-catalyzed debenzylation, afforded a mixture of the diastereoisomeric liensinines, which could not be readily separated.[12,13]

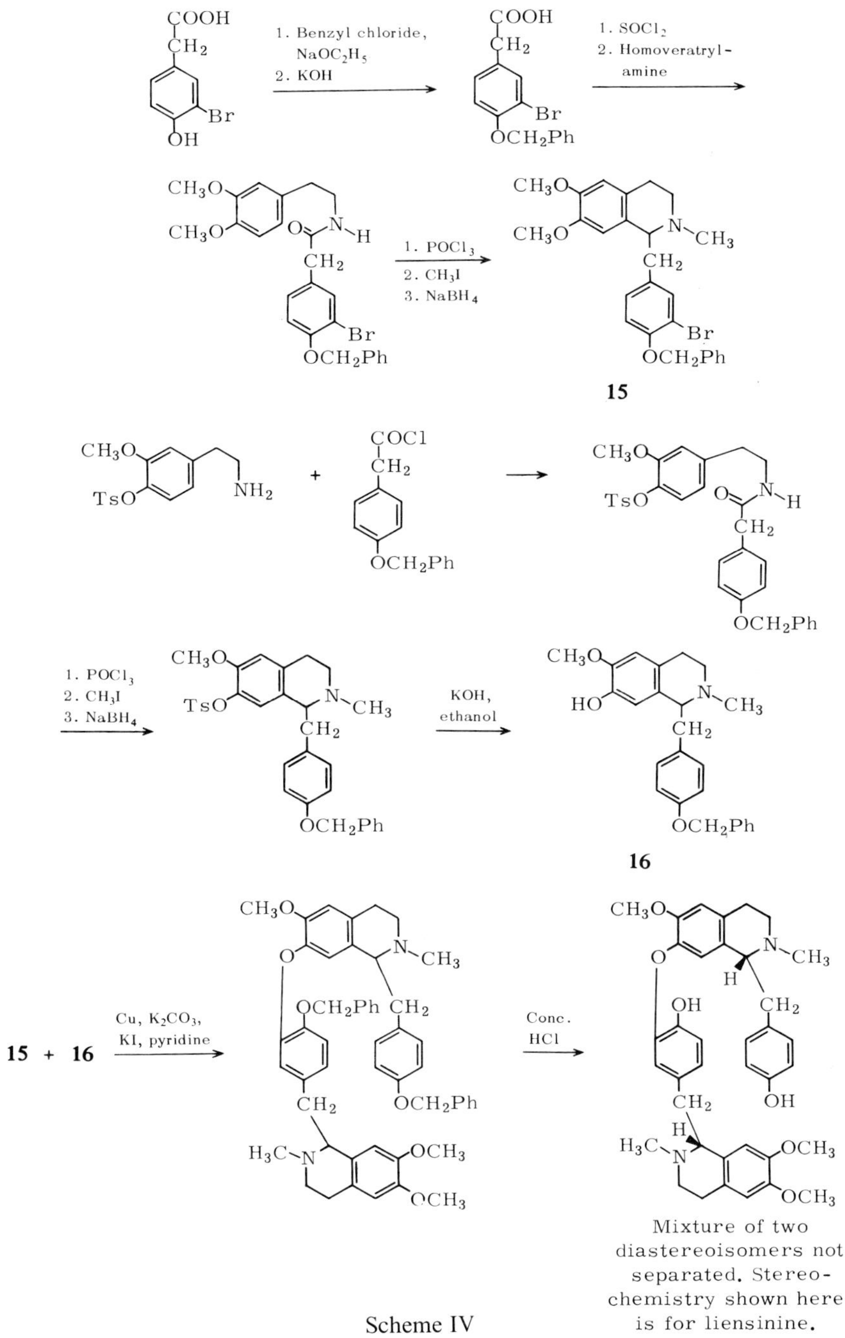
1. Benzyl chloride, $NaOC_2H_5$
2. KOH
1. $SOCl_2$
2. Homoveratrylamine
1. $POCl_3$
2. CH_3I
3. $NaBH_4$
15
1. $POCl_3$
2. CH_3I
3. $NaBH_4$
KOH, ethanol
16
15 + 16
Cu, K_2CO_3, KI, pyridine
Conc. HCl
Mixture of two diastereoisomers not separated. Stereochemistry shown here is for liensinine.

Scheme IV

B. The Tomita Synthesis of (±)*-Cepharanthine*

The synthesis of (±)-cepharanthine was achieved through the dual Bischler–Napieralski cyclization of the key cyclic bislactam **22**.[14] One precursor to this key intermediate was the substituted amino urethane **19**, which was prepared from species **17** and **18** as indicated in Scheme V.

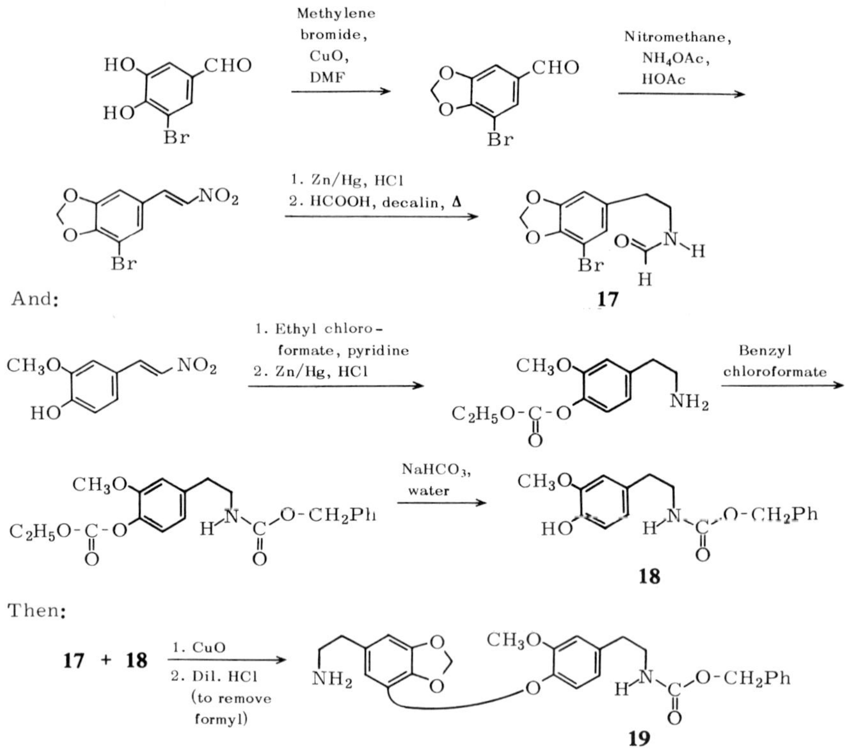

Scheme V

Another important precursor was the acid ester **20**, which was generated as shown.

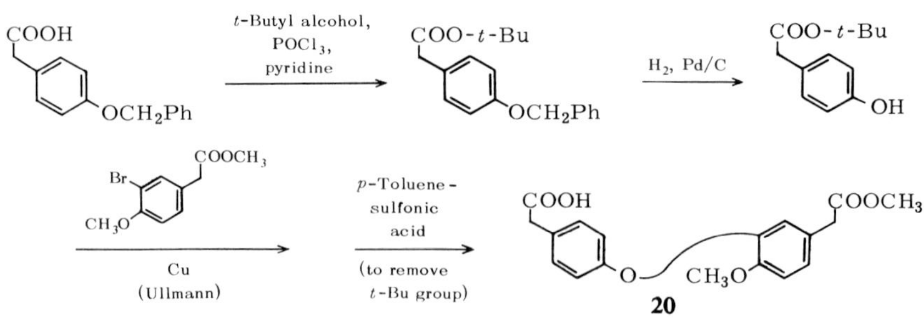

Condensation of the amino urethane **19** with the carboxylic acid **20** gave rise to the urethane **21** which was converted to the desired bislactam **22**. Bischler–Napieralski cyclization of **22** gave a bisimine which could be reduced to a bis secondary amine either with Adams catalyst or with sodium borohydride. Since the ratios of the two diastereoisomers obtained from each of these reductive procedures were different, the available mixtures of secondary amines were combined, *N*-methylated using formaldehyde and sodium borohydride, and then separated by chromatography. One of the products isolated proved to be (±)-cepharanthine (Scheme VI).

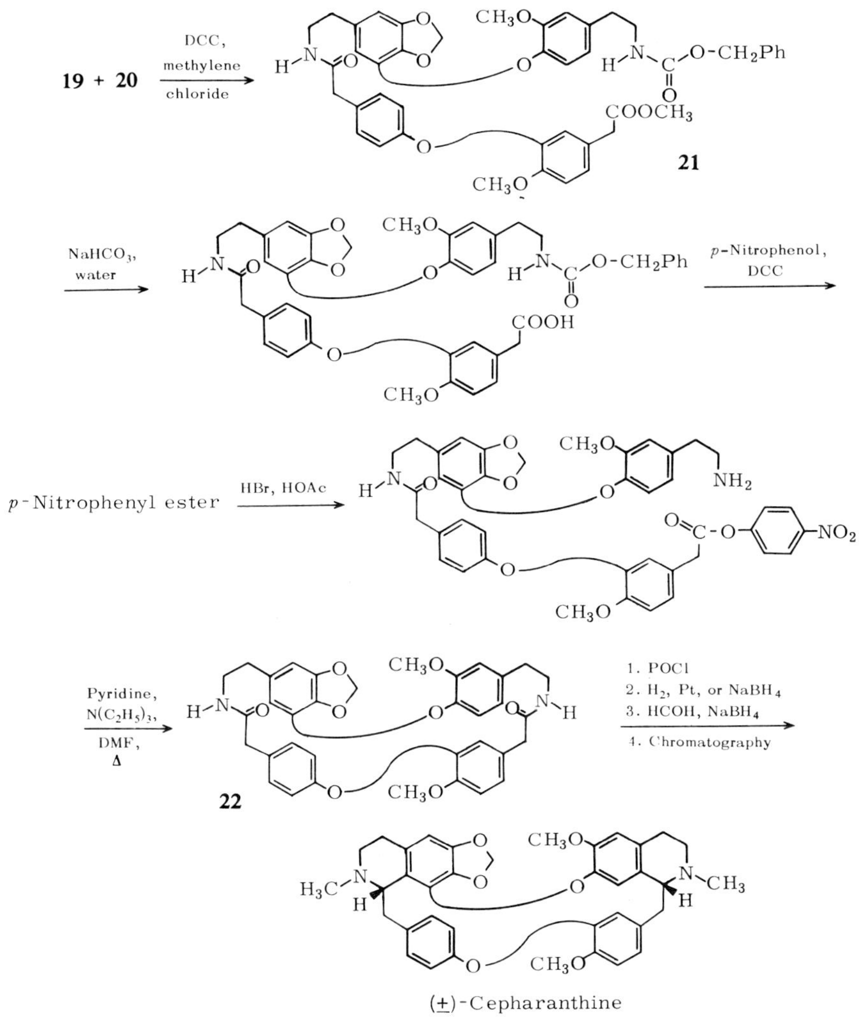

Scheme VI

The success of this synthesis rests upon the exploitation of specific O and N protective groups, and the following remarks are pertinent:

(a) A urethane function is more difficult to saponify than a carbonate ester, hence the transformation to **18**. But the benzyl urethane function is hydrolyzed in acid more readily than the *p*-nitrophenyl ester to afford the immediate precursor of the bislactam **22**.

(b) A *t*-butyl ester group is resistant to catalytic hydrogenolysis, but is easily hydrolyzed in acid, thus allowing the preparation of the acid ester **20**.

(c) The condensation of **19** with **20** involves the reaction of an acid with an amine, all the other functional groups being protected.

(d) The reaction of an amine with a carboxylic ester of *p*-nitrophenol is a superior method of peptide synthesis. The cyclization to the bislactam **22** with formation of a 26-membered ring was achieved by such a procedure.

C. The Inubushi Synthesis of Naturally Occurring (+)-Isotetrandrine and (−)-Phaeanthine

The first intermediate required, (−)-*O*-benzyl-8-bromolaudanidine (**23**) was synthesized from 3-benzyloxy-4-methoxyphenylacetic acid and 3-bromo-4,5-dimethoxyphenylethylamine as shown in Scheme VII.[15]

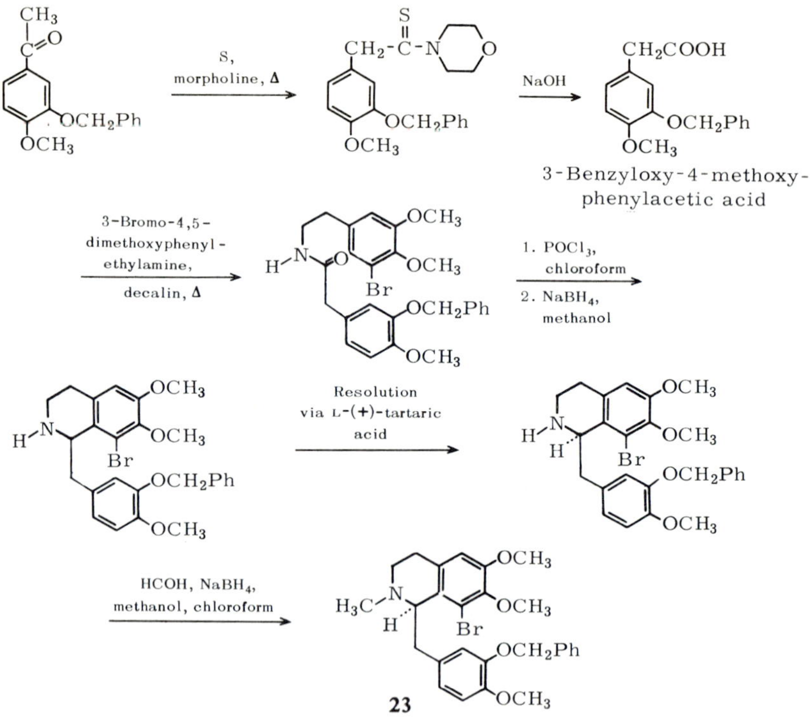

Scheme VII

The second intermediate that was needed was *N-t*-butoxycarbonyl-4-hydroxy-3-methoxyphenylethylamine (**24**), and its preparation is described below. The *t*-butoxycarbonyl group can be readily removed under mild acid conditions but is resistant to hydrogenolysis and cleavage by alkali under relatively mild conditions.

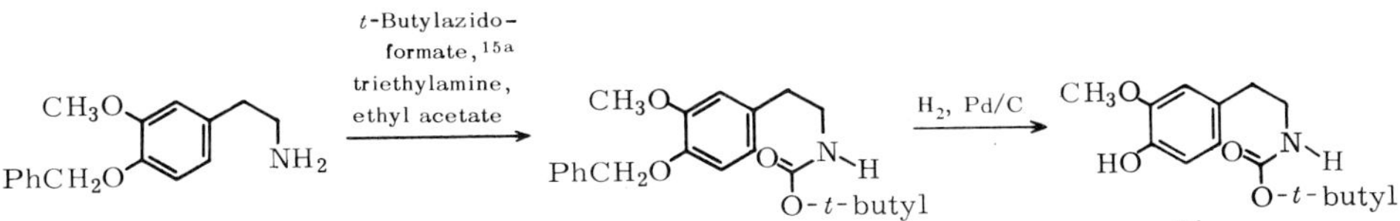

24

Ullmann condensation of **23** with **24** afforded the diphenyl ether **25** in 50% yield. Catalytic debenzylation was followed by another Ullmann condensation with the third required intermediate, methyl *p*-bromophenylacetate, to give the bisdiphenyl ether **26** in 48% yield.

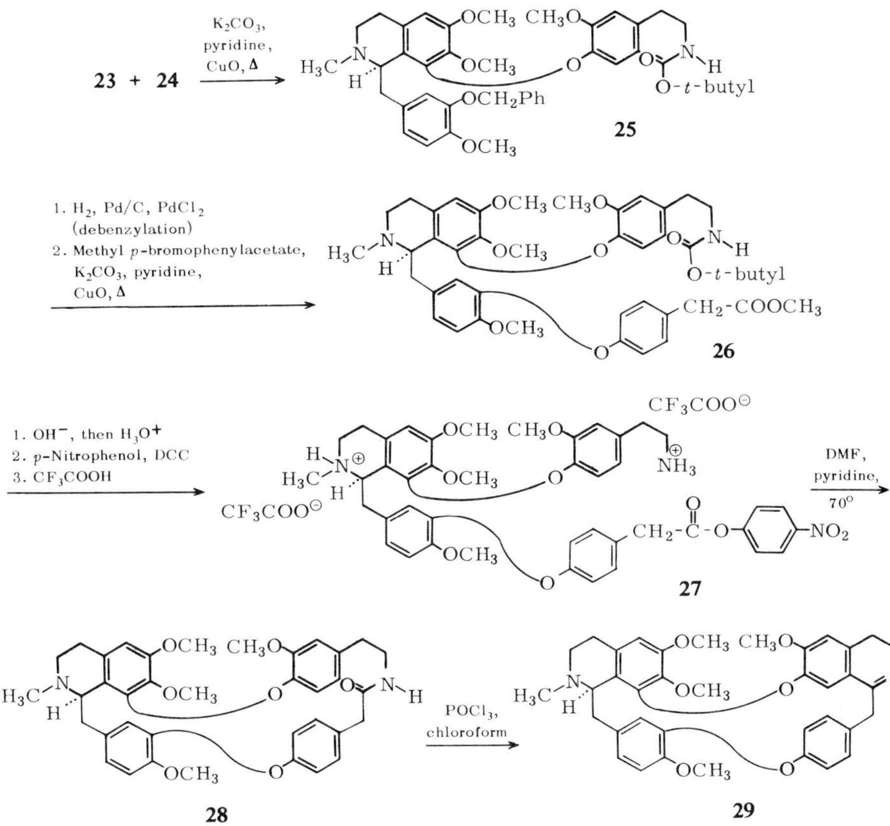

Scheme VIII

The amide linkage furnishing a large ring was formed using the *p*-nitrophenyl ester method. The carbomethoxy group of **26** was saponified preferentially, and the resulting acid converted into the *p*-nitrophenyl ester with *p*-nitrophenol and DCC. The *t*-butoxycarbonyl group was then removed with trifluoroacetic acid to give a trifluoroacetate salt (**27**) which on heating cyclized to the desired amide **28**. Bischler–Napieralski cyclization gave rise to the imine **29** (Scheme VIII).

The reduction of the imine **29** was studied using a variety of reducing agents. With sodium borohydride in methanol, a 3 : 2 ratio of bisbenzylisoquinolines **30** and **31** was obtained. When zinc in ethanol and sulfuric acid was utilized, only species **31** could be isolated. No stereospecificity could be observed using catalytic hydrogenation over Adams catalyst in ethanol containing a trace of concentrated hydrochloric acid. Finally, *N*-methylation of species **30** and **31** yielded (+)-isotetrandrine and (−)-phaeanthine, respectively.[15]

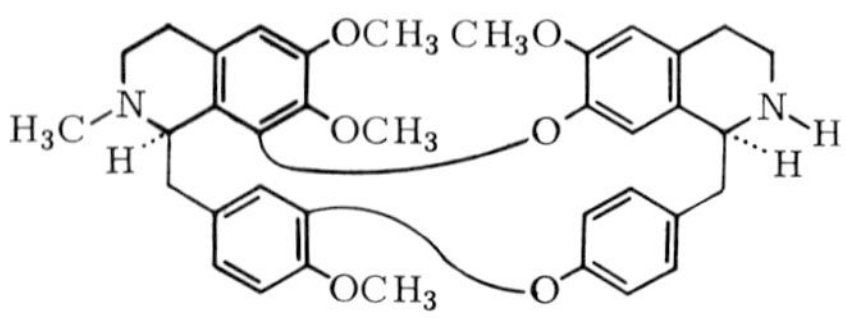

30

↓ HCOH, NaBH$_4$

(+)-Isotetrandrine

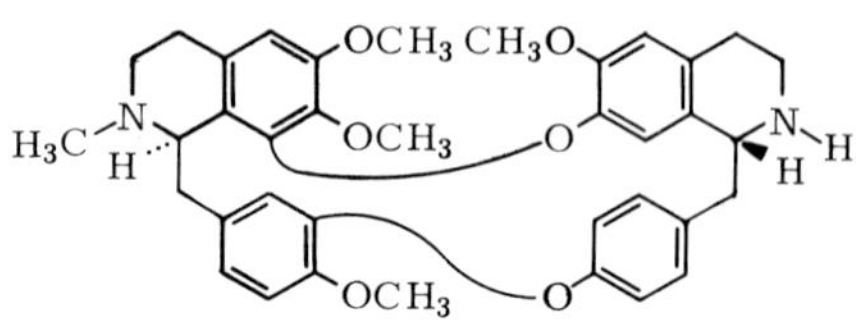

31

↓ HCOH, NaBH$_4$

(−)-Phaeanthine

Essentially the same approach as described above has been followed in the synthesis of naturally occurring (+)-*O*-methylthalicberine.[15b]

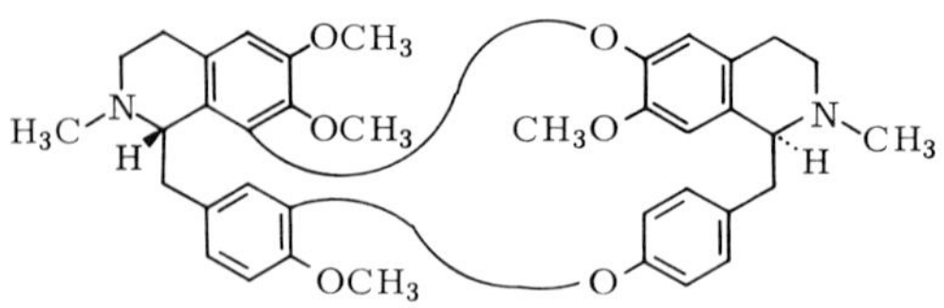

(+)-*O*-Methylthalicberine

D. The Tomita–Ueda Synthesis of (±)-*N-Methyldihydromenisarine*

In this preparation of a bisbenzylisoquinoline with three diphenyl ether linkages, an Ullmann reaction was used to build the important symmetrical diamine **32** which incorporates the diphenylene dioxide bridge (Scheme IX).

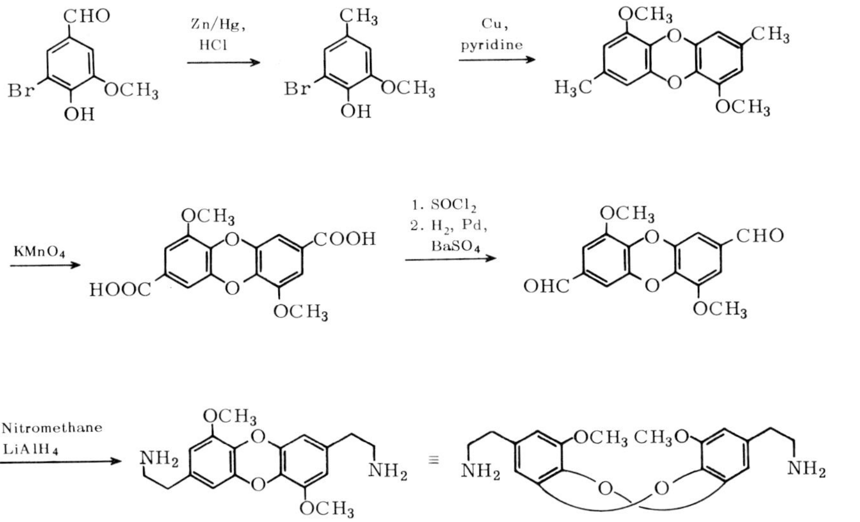

32

Scheme IX

Condensation with the diacid **33** followed by Bischler–Napieralski cyclization, reduction, and Eschweiler–Clarke *N*-methylation furnished (±)-*N*-methyldihydromenisarine (**32**).

32 +

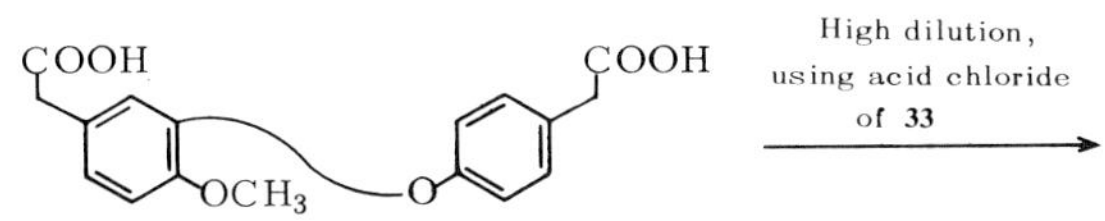

33

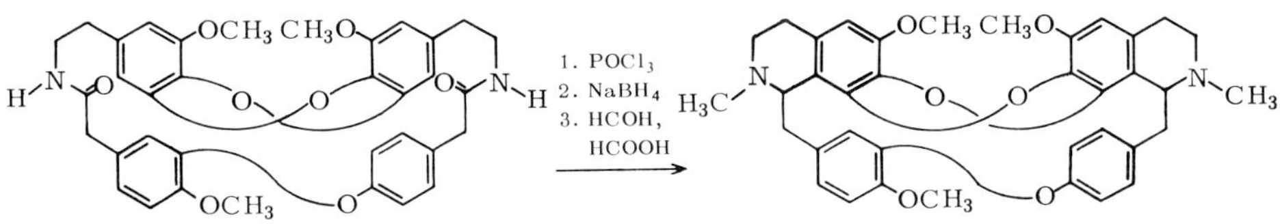

(±)-**34**

The dimer (±)-**34** was spectrally identical with the product derived from the reduction and *N*-methylation of naturally occurring (+)-menisarine.[16,17]

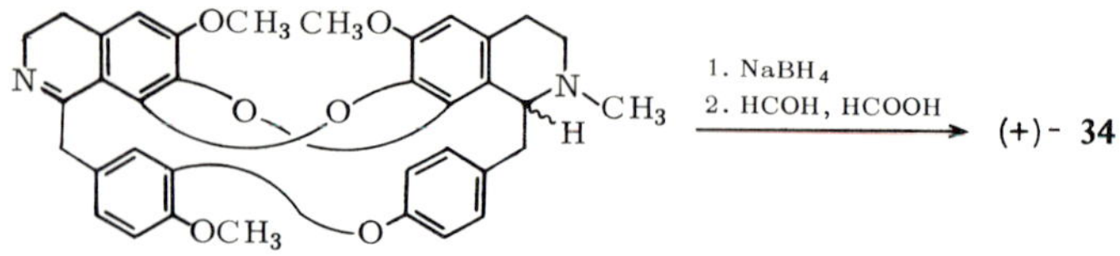

(+)-Menisarine

An interesting modification of the above sequence led to the preparation of (±)-*O*-methyltiliacorine.[17a]

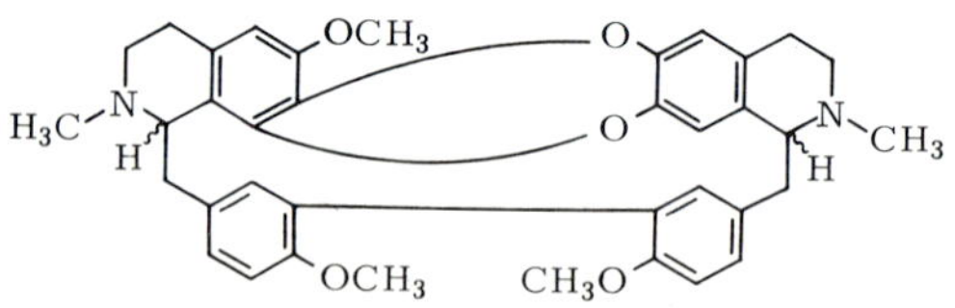

(±)-*O*-Methyltiliacorine

E. The Bobbitt–Hallcher Electrolytic Synthesis of an Isomeric Mixture of Dauricine Analogs

The electrolytic oxidation of the sodium salts of armepavine and *N*-norarmepavine led to the same sort of fragmentation of the C-1 to C-α bond as observed upon enzymic oxidation (Chapter 2, Section VII, A). However, the electrolytic oxidation of the sodium salt of (±)-*N*-carbethoxy-*N*-norarmepavine led to the dimers **34a** and **35**. The latter was converted, without isolation, into a mixture of dauricine analogs via *O*-benzylation, reduction with lithium aluminum hydride, and hydrogenolytic debenzylation. This transformation represents the first preparation of an analog of a natural bisbenzylisoquinoline by oxidation of a phenolic monomeric benzylisoquinoline (Scheme IX a).[17b]

Some other protective devices beside those mentioned above which could be used in benzylisoquinoline and bisbenzylisoquinoline synthesis are the methylsulfonyloxy group,[18] the phenacyl group for phenols,[19] and the phenacylsulfonyl group for amines.[20,21]

VII. Conversion of Oxyacanthine-Type Alkaloids to the Trilobine Series

One of the more interesting transformations in the bisbenzylisoquinoline series concerns the conversion of alkaloids of the oxyacanthine subgroup which have only two diphenyl ether linkages into compounds of the trilobine series, with three such linkages.[22,23] Thus the alkaloid (+)-epistephanine of the oxyacanthine subgroup was first reduced and *N*-methylated to *N*-methyldihydroepistephanine. The latter compound on treatment with hydrobromic acid at 100° gave the tetraphenolic base **35**. Dehydration of this base with hydrobromic acid at 150° followed by *O*-methylation with diazomethane supplied (−)-isotrilobine, antipodal with natural (+)-isotrilobine (Scheme X).

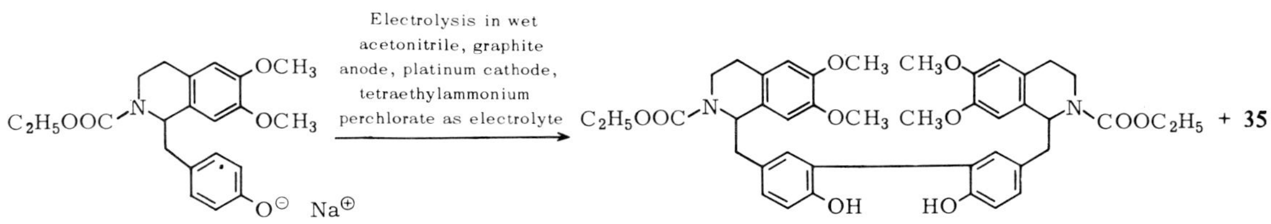

(±)-*N*-Carbethoxy-*N*-norarmepavine

34a
(Isomeric mixture)

Then using **35**:

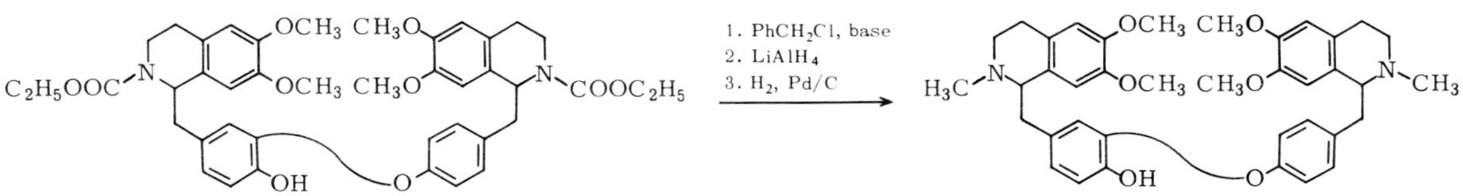

35
(Isomeric mixture)

(Mixture of isomeric dauricine analogs)

Scheme IXa

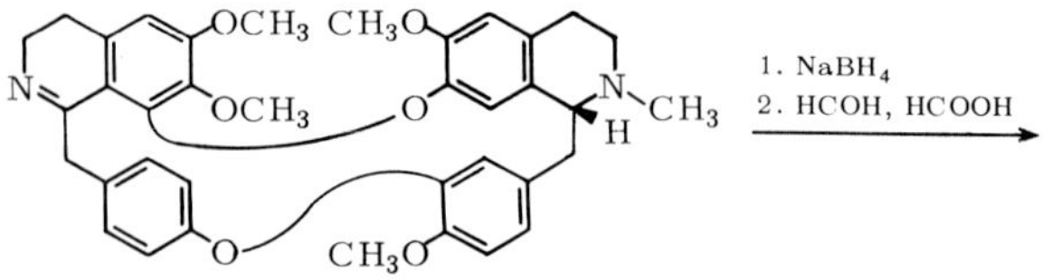

(+)-Epistephanine

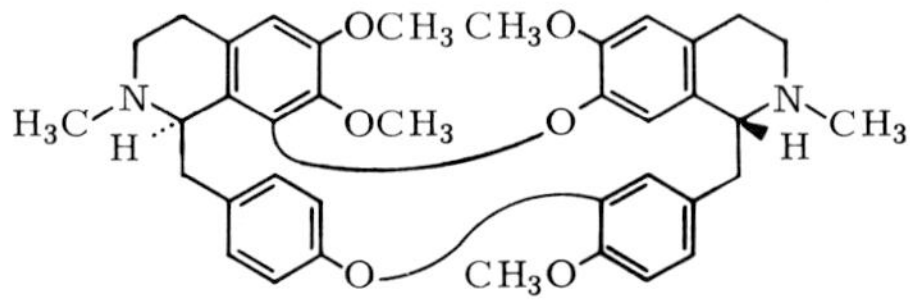

N-Methyldihydroepistephanine

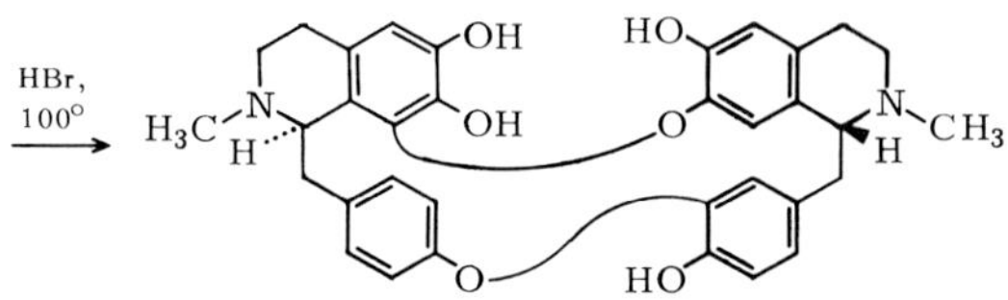

35

(obtained as dihydrobromide salt)

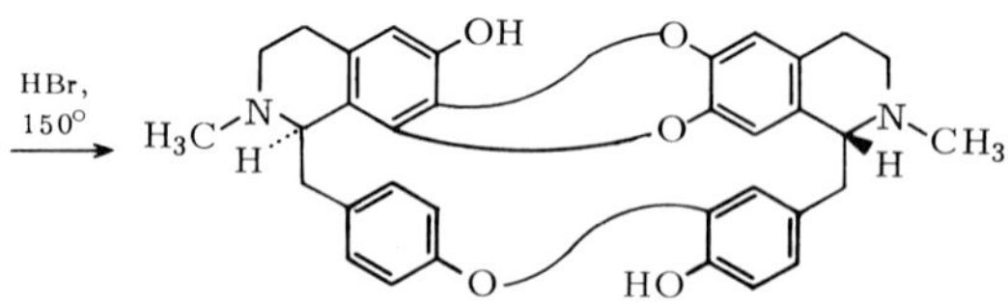

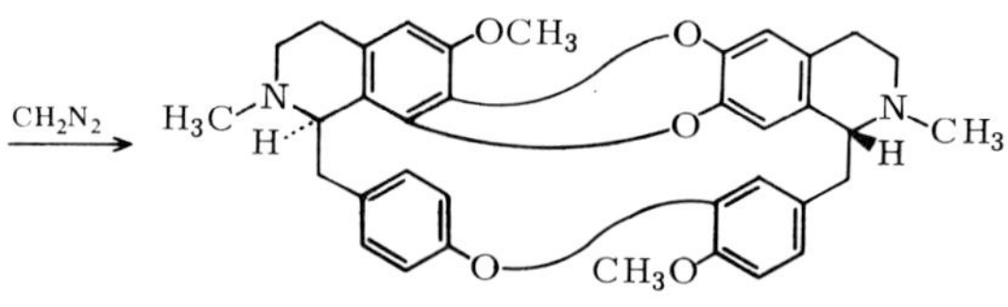

(−)-Isotrilobine

Scheme X

In the sequence presented above, the sodium borohydride reduction of (+)-epistephanine afforded one clean stereochemical product which was *N*-methylated and hydrolyzed to **35**. It is known that the alternate reduction of (+)-epistephanine with zinc in sulfuric acid affords a mixture of diastereoisomers.[24,25] The stereochemistry of reduction in this series is, therefore, significantly different from that in the isotetrandrine–phaeanthine system discussed in Section VI, C.[15]

VIII. Some Reagents for Interrelating Benzylisoquinolines, Bisbenzylisoquinolines, and Other Isoquinolines

A. N-Demethylation of a Tertiary Amine

An efficient method for effecting the *N*-demethylation of a tertiary amine involves treatment of the amine with phenyl chloroformate. A urethane is produced which can be hydrolyzed to the corresponding secondary amine.[26] Recently, the use of ethyl chloroformate in lieu of phenyl chloroformate has been reported.[26a] The Polonovski reaction using the amine oxide and acetic anhydride is an older and somewhat less satisfactory method.[27] The amine oxide can also be demethylated by reaction with liquid sulfur dioxide followed by hydrolysis with hydrochloric acid.[27a] It has been shown lately that treatment of a tertiary amine hydrochloride with platinum and oxygen at room temperature for several days leads to *N*-demethylation.[27b] Additionally, vinyl chloroformate in 1,2-dichloroethane at or near room temperature is an effective *N*-demethylating reagent.[27c] However, cleavage of ring B of the tetrahydrobenzylisoquinoline is a possibility with many of these reagents. (See also Chapter 1, Section IV, K.) An interesting photooxidative method for *N*-demethylation has recently appeared.[27d]

B. N-Demethylation of a Quaternary Salt

The thiophenolate[28] and selenophenolate[29] anions can achieve the transformation of a quaternary *N*-metho salt to the tertiary amine by way of an S_N2 process. Refluxing the quaternary acetate in an aprotic solvent such as chloroform or acetonitrile also results in *N*-demethylation.[30]

C. O-Demethylation of an Ether

The standard method of *O*-demethylation is through acid hydrolysis using hydrobromic or hydroiodic acids. Some other reagents that could be useful are lithium iodide in collidine,[31] the thioethoxide anion in dimethylformamide,[32] dry methylmagnesium iodide,[33] pyridine hydrochloride,[34] or even the seldom used sodium in hot pyridine.[35] A superior reagent for this purpose is boron tribromide in methylene chloride: no heating is required and diphenyl ethers are unaffected.[36] Diborane and iodine at room temperature or below are also known to effect *O*-demethylation.[36a]

D. Deoxygenation of a Phenol

The method of choice for going from a phenol to its benzene analog is that of Musliner and Gates using 1-phenyl-5-chlorotetrazole.[37]

E. N-Methylation of a Secondary Amine

The Eschweiler–Clarke procedure usually works satisfactorily. Another method also referred to in this chapter involves the use of formaldehyde and sodium borohydride. A third method which is known to work well with tetrahydrobenzylisoquinolines relies on the reaction of the secondary amine with ethyl chloroformate to furnish a urethane. Reduction with lithium aluminum hydride then supplies the *N*-methyltetrahydrobenzylisoquinoline.[38] A reagent of potential use is trimethyl phosphate, which so far has been utilized in the *N*-methylation of aromatic amines.[38a]

F. N-Oxidation of a Tertiary Amine

A common reagent for preparing an *N*-oxide is 30% hydrogen peroxide, but a safer method uses *m*-chloroperbenzoic acid in chloroform at 0°–25°.[38b]

G. Methylenedioxy Formation

A catechol can be smoothly converted to its methylenedioxy derivative through the use of methylene chloride in the aprotic solvent dimethyl sulfoxide, using sodium hydroxide as the base.[38c]

H. Cleavage of a Methylenedioxy Group

Numerous examples are available in the literature of the hydrolysis of a methylenedioxy group to the corresponding catechol using sulfuric acid. Dimedone is often added as a formaldehyde trap. Refluxing a methylenedioxy derivative with methylmagnesium iodide in benzene leads to fission of the methylenedioxy ring with formation of the corresponding *o*-ethoxy phenol.[38d] Alternatively, reaction with phosphorus pentachloride followed by hydrolysis also results in formation of the catechol derivative.

IX. Biogenesis and Phenolic Oxidative Coupling

Bisbenzylisoquinolines are formed in nature through the phenolic oxidative coupling of simple benzylisoquinolines such as *N*-methylcoclaurine. A logical, but so far hypothetical sequence that could lead to dauricine, oxyacanthine, and berbamine would be as described in Scheme XI.

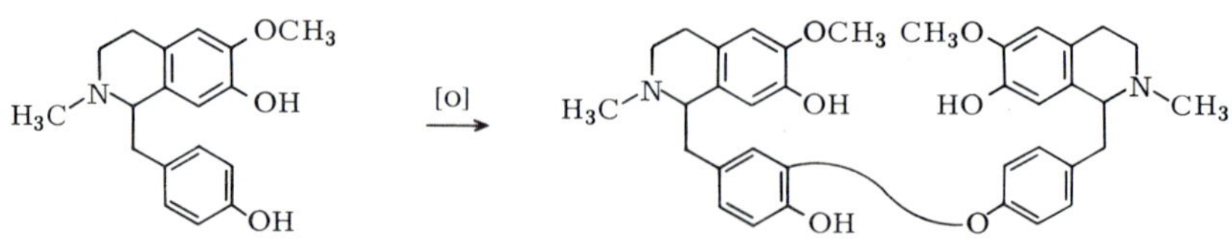

N-Methylcoclaurine Dimer

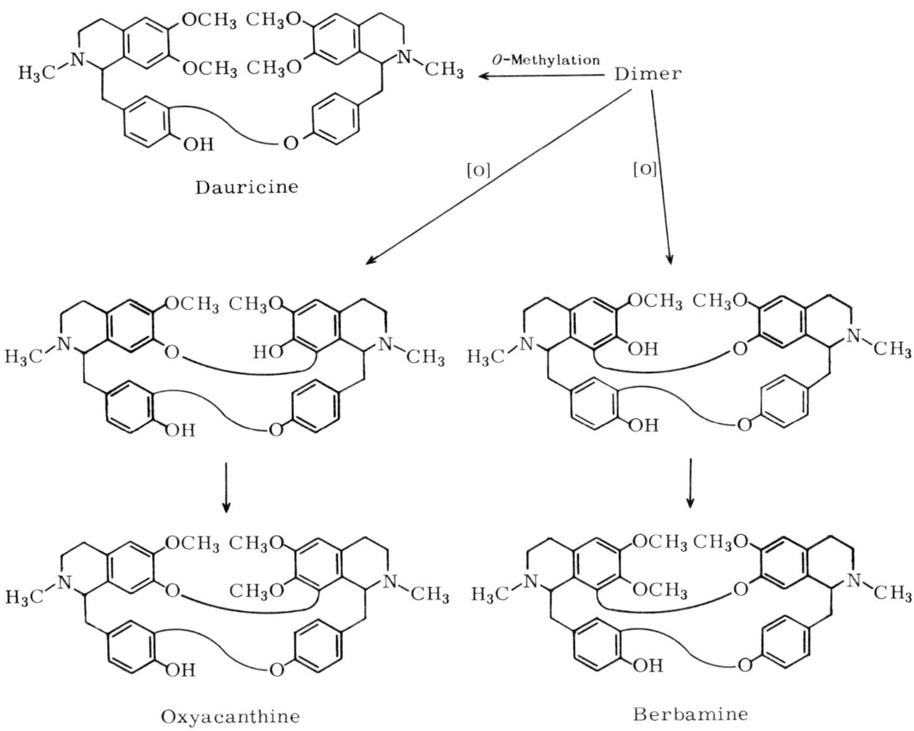

Scheme XI

Franck and his associates reported the first biogenetic-type synthesis of a benzylisoquinoline dimer. Oxidation of magnocurarine with potassium ferricyanide gave the dimer **36**.[39]

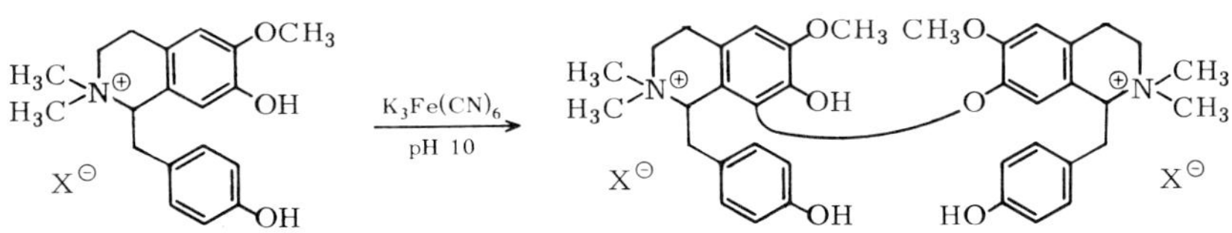

Magnocurarine **36**

A number of similar condensations have since been carried out. For example, oxidation of the (±)-isoquinoline **37** with ferric chloride afforded the product of intramolecular oxidation **38** as well as the dimer **39** as mixtures of diastereoisomers.[40,41] (See also Chapter 2, Section VII, A.)

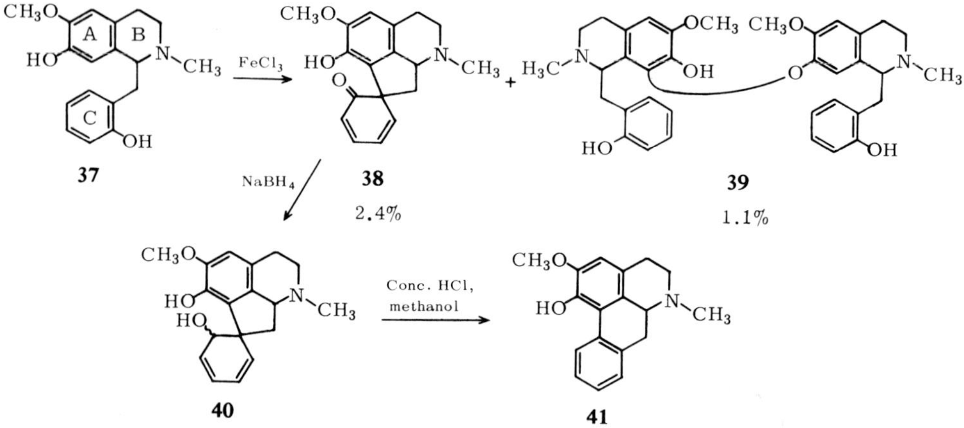

It is worth noting here in passing that the dienone **38** was reduced to the dienol **40**, which by a dienol–benzene rearrangement afforded the aporphine (±)-*N*-methylcaaverine (**41**).[40]

In the dimerization of benzylisoquinolines, it is usually easier for oxidative coupling to occur on ring A than on ring C of the molecule. A rare instance of a dimerization involving only ring C has been reported by Schofield and his group. Oxidation of (±)-armepavine methiodide with alkaline ferricyanide under conditions markedly more severe than those usually employed generated at least two new compounds, one of which was identified as a mixture of the racemates represented by expression **42**.[42]

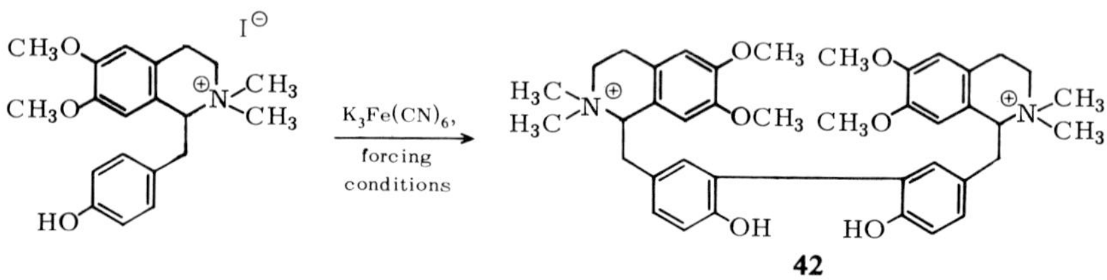

In spite of the foregoing, a biogenetic-type synthesis of a true analog of a bisbenzylisoquinoline alkaloid is an elusive aim which still remains to be achieved. The systems represented by the dimers **36**, **39**, and **42** have not been found in nature thus far. As reported earlier in this chapter, the electrolytic oxidation of a phenolic tetrahydrobenzylisoquinoline did lead to the preparation of analogs of dauricine.

Some interesting tracer studies have been carried out on the bisbenzylisoquinoline alkaloid (+)-epistephanine produced by *Stephania japonica* Miers (Menispermaceae). After labeled (±)-*N*-methylcoclaurine (**43**) had been fed to the plant, it was found that 98% of the activity of the (+)-epistephanine resided on the *N*-methyl group (○).[43]

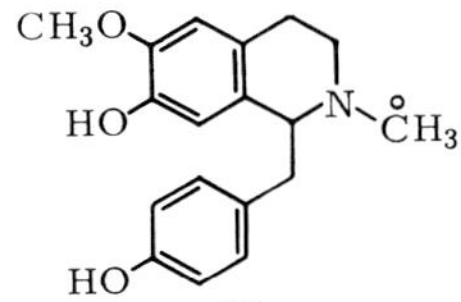

43

Labeled (±)-*N*-Methylcoclaurine

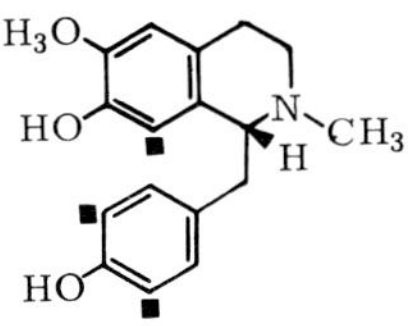

(−)-*N*-Methylcoclaurine

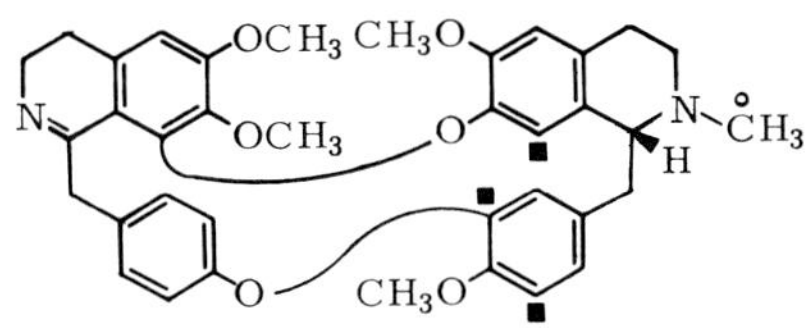

(+)-Epistephanine

Additionally, the (−)-enantiomer of *N*-methylcoclaurine was incorporated into (+)-epistephanine much more efficiently than its antipode, showing that racemization is unimportant in the plant. When (−)-*N*-methylcoclaurine was labeled on rings A and C as indicated, the labels appeared in only one of the two benzylisoquinoline moieties of epistephanine, indicating that *N*-methylcoclaurine provides only half the epistephanine molecule.[43]

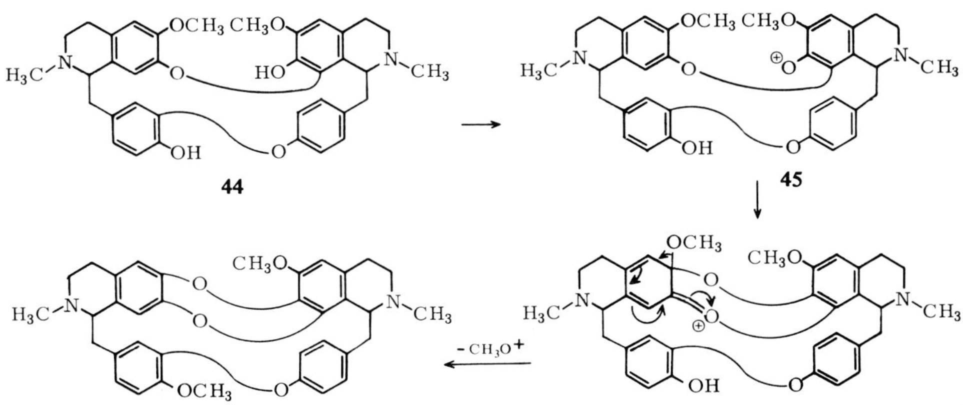

Isotrilobine

Scheme XII

Alkaloids with three diphenyl ether linkages present an interesting problem in genesis. The intermediate **44** mentioned earlier in this section may be converted in the plant to the oxonium ion (or free radical) **45**, which through loss of CH_3O^+ (or the corresponding free radical) can generate isotrilobine (Scheme XII).[44]

An alternate proposed pathway is based upon a 1,2 C → O shift as indicated below, leading to bases of the isotrilobine type.[45]

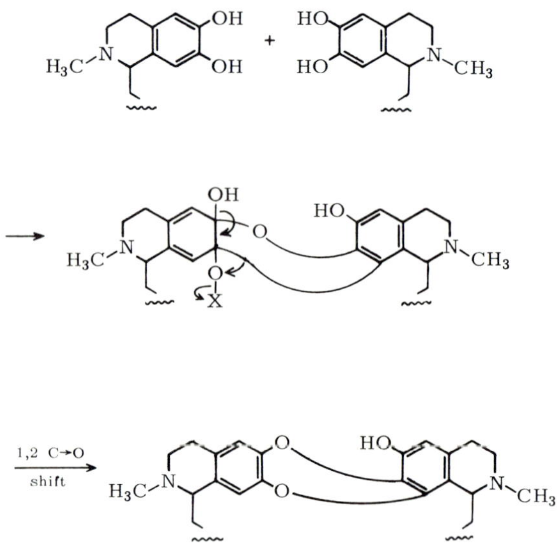

Isotrilobine type

In conclusion, it should be noted that head-to-head coupling, without accompanying tail-to-tail coupling, has never been observed in nature with the bisbenzylisoquinolines. Tail-to-tail coupling is quite common, as in the alkaloids belonging to the thalisopine, magnolamine, and dauricine subgroups. Head-to-tail coupling must be responsible for bases of the liensinine, isochondodendrine, curine, insularine, cissampareine, and hayatine types.

X. Pharmacology

(+)-Tubocurarine, one of the alkaloids of tube curare from Brazil and Peru, is often present in the arrow poisons used by South American Indians. The curare is derived from *Chondodendron* species (Menispermaceae), and (+)-tubocurarine is the most active constituent of the mixture. It is usually isolated as the crystalline dichloride salt. When injected into the bloodstream, it quickly blocks neuromuscular action so that respiration ceases and death results. Muscle paralysis is actually achieved by blocking the action of acetylcholine on the end-plate region of the muscle fiber. A single paralyzing dose in man (3–9 mg) lasts almost a half hour. The levorotatory enantiomer, also a natural product, is much less powerful in this respect.

Dr. Bernard Finch, in "Passport to Paradise," published by the Philosophical Library in 1960, gave the following account from the diary of a man struck by a poisoned arrow in the Peruvian jungle: "After the initial shock had worn off, I felt a gradual weakness and tiredness of my eyes. ... I was able to see only by separating the lids with my fingers, and even then I could not see properly as vision was blurred and the other members of the party appeared double. Gradually, I felt I could not close my mouth and my lower jaw fell open. My lips were wet and saliva poured onto my chest. I was lying in the bottom of the canoe in the hot sun, but could not raise my head or close my mouth owing to the extreme weakness of my neck muscles. I was bitten by innumerable insets and yet I found I could not brush them off my face as my arms and legs were weak and paralyzed. By now, I was unable to turn my head at all and the wound in my arm was throbbing. From my arms I felt the paralysis creeping down my chest so that it was impossible to breathe without effort." Unfortunately, the quotation does not indicate whether the man died, and if so, how he was able to manage this final entry in his diary.[45a]

(+)-Tubocurarine chloride is used in abdominal surgery in very small doses as a complement to anesthetics since it causes paralysis of the abdominal muscles without stopping the natural movement of the intestines.[46] It actually blocks the transmission of nerve impulses at the myoneural junction to skeletal muscle. The drug is ineffective when taken orally, probably because it does not penetrate the intestinal walls. This fact also explains why animals that have been killed with an injection of tube curare can be eaten with impunity. (+)-*O*,*O*-Dimethyltubocurarine iodide, obtained by *O*-methylation of the two phenolic groups of tubocurarine, is also used to relax skeletal muscles. It is about three to six times more potent than the parent alkaloid and does not cause respiratory depression. It is longer acting than tubocurarine chloride.[47]

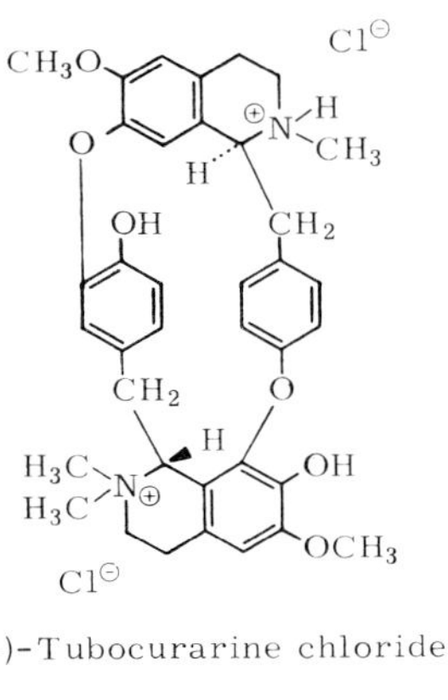

(+)-Tubocurarine chloride

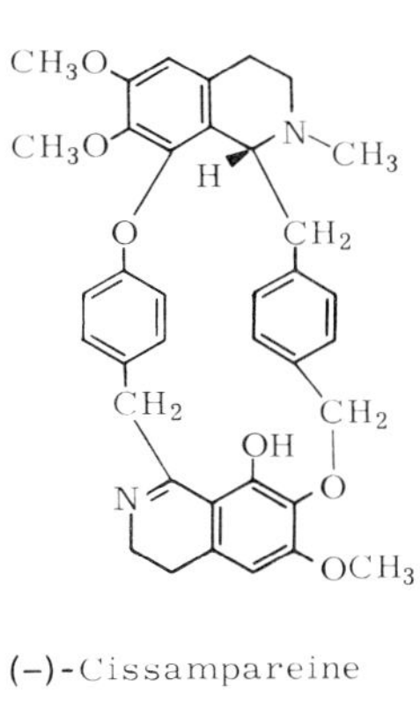

(−)-Cissampareine

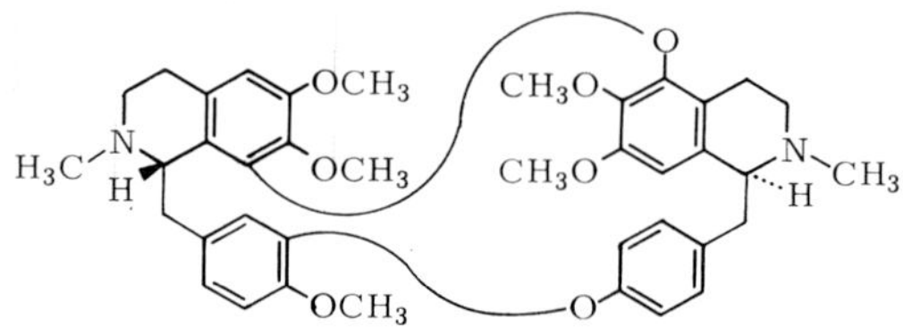

(–)-Thalidasine

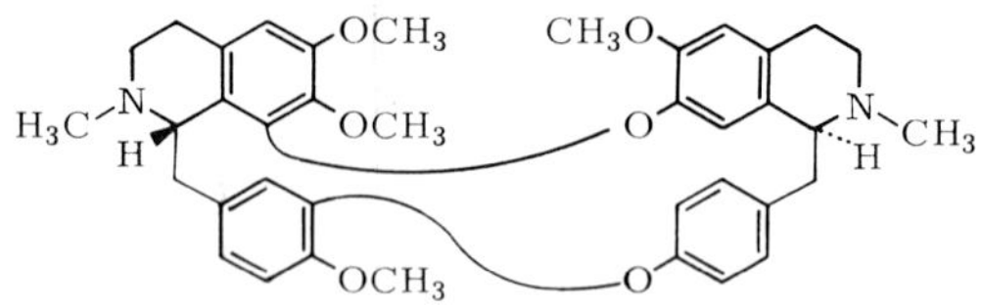

(+)-Tetrandrine

With the recent structural revision of (+)-tubocurarine, a reconsideration of the structure of (+)-*O*,*O*-dimethyltubocurarine is warranted. In particular, depending upon the mode of preparation of this material, it is possible that *O*-methylation was accompanied by *N*-methylation, leading to a bisdimethylammonium derivative which should be more active than (+)-tubocurarine since it cannot be readily dequaternized. Similarly, given that it is the bisonium structure that is responsible for the activity, it is not surprising to find that (+)-chondocurarine, which corresponds to the bisdimethylammonium derivative of (+)-tubocurarine, is from two to four times as potent.[47a]

The alkaloids cissampareine,[48] thalidasine,[49] and tetrandrine[50] have shown promising tumor-inhibiting properties in initial tests.

The oxyacanthine type alkaloid (+)-cepharanthine is claimed to be highly effective against human tuberculosis and leprosy.[50a]

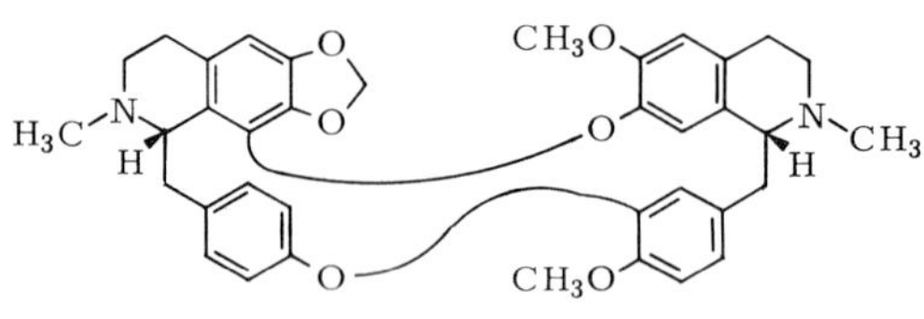

(+)-Cepharanthine

As an extension of the finding that (+)-cepharanthine may be useful against human tuberculosis, it was determined that a number of bisbenzylisoquinolines found in *Thalictrum rugosum* Ait. (*T. glaucum* Desf.) (Ranunculaceae) are active *in vitro* against *Mycobacterium smegmatis* ATCC 607, and thus possess potential antitubercular activity. These include the known alkaloids thalidasine and obamegine which are strongly active, and the new bases thalrugosine and thalrugosidine which are only weakly active.[50b]

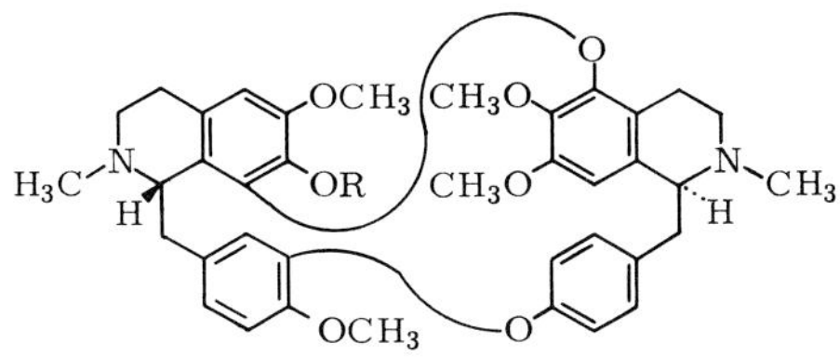

(−)-Thalidasine, R = CH_3
(−)-Thalrugosidine, R = H

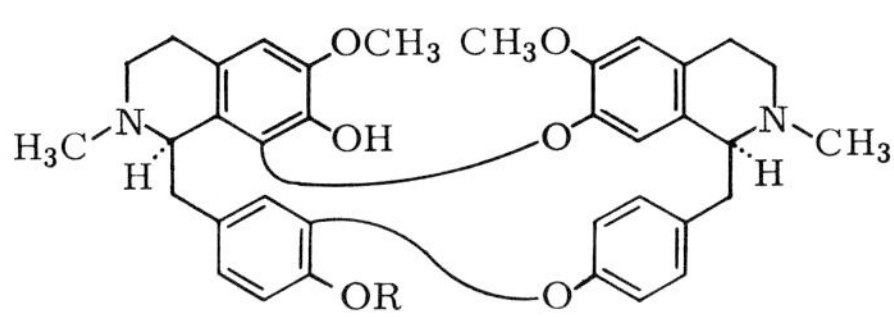

(+)-Obamegine, R = H
(+)-Thalrugosine, R = CH_3

The synthetic bisbenzylisoquinoline salt **46**, laudexium methyl sulfate, is a curarimimetic agent sometimes used as a skeletal muscle relaxant. Its side effects are respiratory depression and hypotension.[51] It is obtained by refluxing the tetrahydrobenzylisoquinoline laudanosine with decamethylene diiodide, followed by anion exchange.

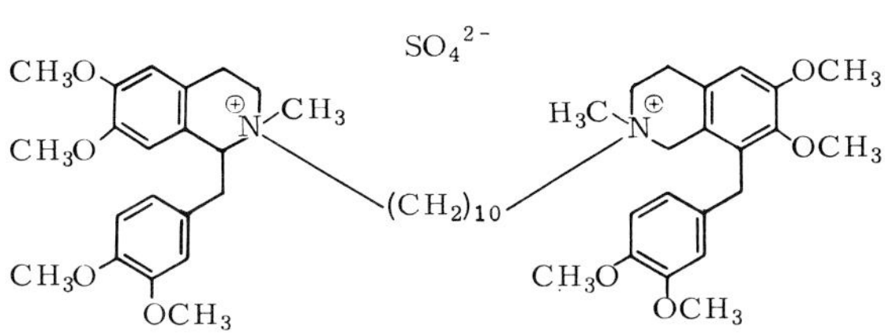

46

Laudexium methyl sulfate

XI. Absolute Configuration

The determination of the absolute configuration of bisbenzylisoquinolines is usually based upon cleavage of the dimer with sodium in liquid ammonia. The products are simple benzylisoquinolines whose absolute configurations can be readily established by comparison with authentic samples, or by ORD measurements. The absolute configuration of a bisbenzylisoquinoline is often designated in terms of a (+) or a (−) sign at each of the two asymmetric centers, whereby (+)=L=*S* and (−)=D=*R*.

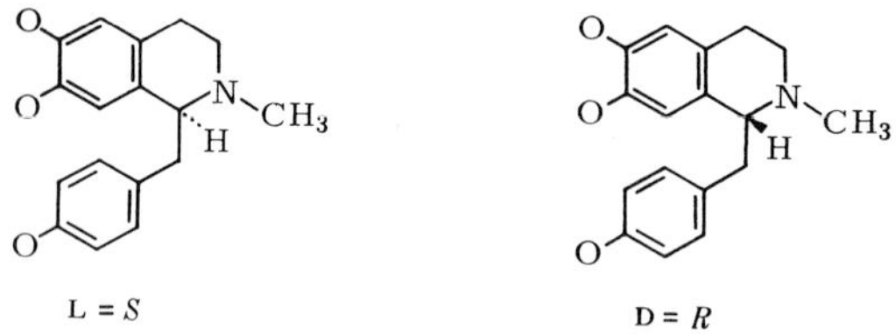

Correlations have also been drawn between the ORD curves of bisbenzylisoquinolines and their absolute configurations.[1a,52,53]

XII. The Isomerization of (+)-Oxyacanthine to (−)-Repandine

Heating oxyacanthine with one equivalent of hydrochloric acid results in racemization at C-1′ and in the formation of the diastereoisomeric base repandine; this rearrangement has been shown to be reversible. *O*-Methyloxyacanthine and *O*-methylrepandine, on the other hand, do not isomerize under the identical acid conditions. No mechanism has been offered for this interesting transformation,[54–56] but it is relevant to note that it is possible to racemize optically active monomeric benzylisoquinolines by treatment with 36% hydrobromic acid. (See Chapter 2, Section VI, F.)

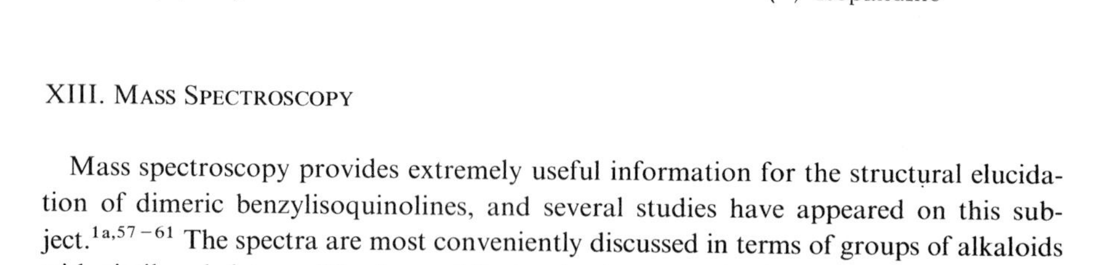

XIII. Mass Spectroscopy

Mass spectroscopy provides extremely useful information for the structural elucidation of dimeric benzylisoquinolines, and several studies have appeared on this subject.[1a,57–61] The spectra are most conveniently discussed in terms of groups of alkaloids with similar skeletons. The favored fission is always at the doubly benzylic positions at C-1 and C-1′.

The fragmentation pattern that will be discussed first is that for the alkaloid tetrandrine of the berbamine series, in which the doubly charged ion *m/e* 198 is the base peak. This ion can lose the elements of dimethyl ether to form the *m/e* 175 peak. All of the other ions indicated are singly charged species (Scheme XIII).

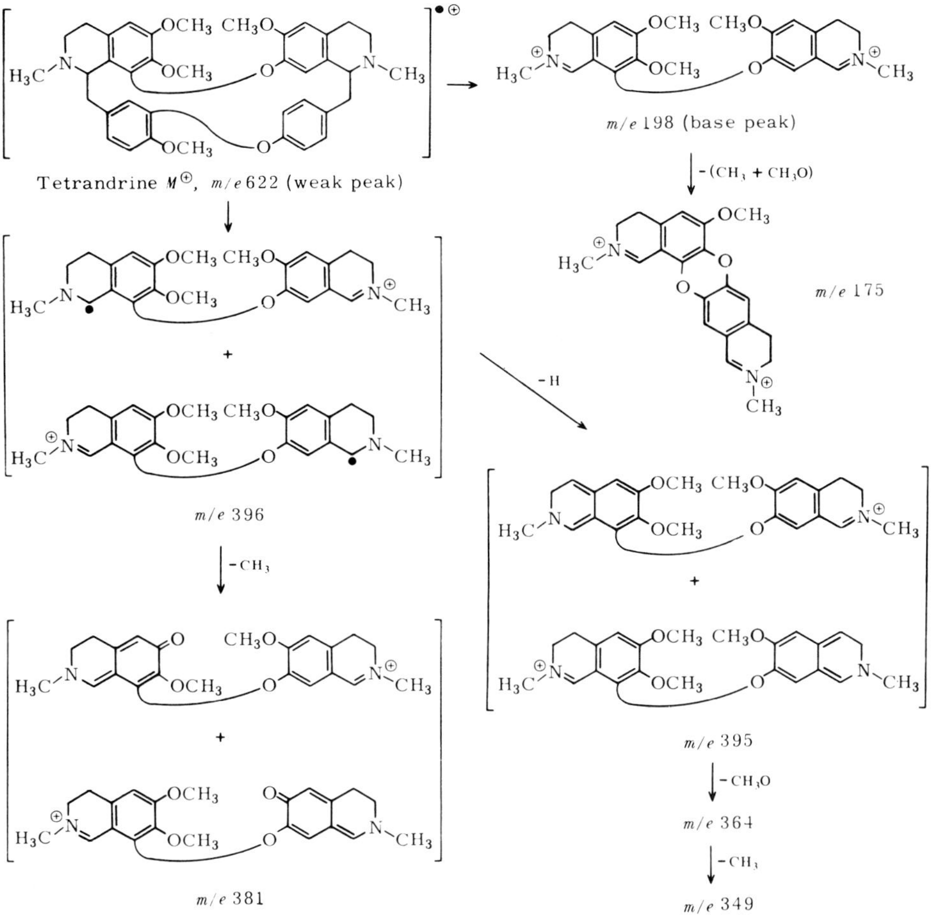

Scheme XIII

Tetrandrine also shows weak but characteristic peaks at m/e $(M-191)$ and $(M-137)$.

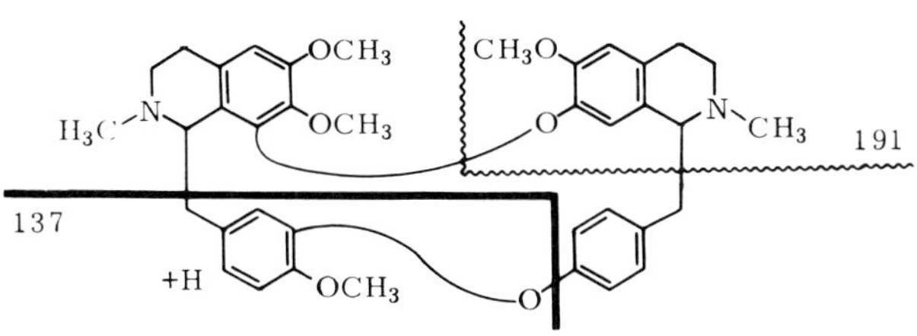

On the other hand, oxyacanthine-type bases will exhibit peaks at *m/e* $(M-191)$ and $(M-107)$.

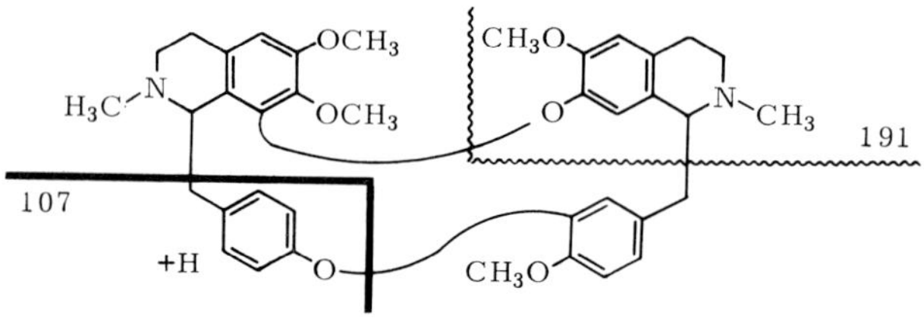

The $(M-137)$ peak is usually less intense than the $(M-191)$ peak in the berbamine series, while in the oxyacanthine series it is the $(M-107)$ peak which is more intense than the $(M-191)$ peak.[1a]

The most intense ion peak of cycleanine, a symmetrical isochondodendrine-type alkaloid, occurs at *m/e* 312. This fragment is formed by fission at the dotted lines with hydrogen transfer.[58]

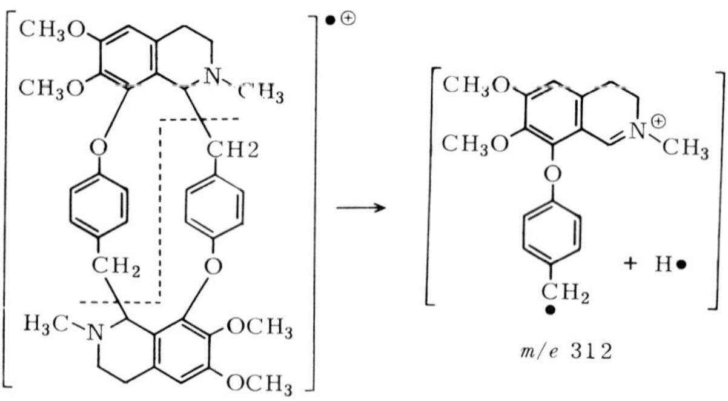

Cycleanine $M^{\oplus}$,

m/e 622

The structural revisions of tubocurarine, chondocurarine, and chondocurine led in turn to a change in the structural assignment for the alkaloid chondrofoline, a member of the curine subgroup. Chondrofoline had originally been formulated as analyzing for $C_{35}H_{36}O_6N_2$, with two secondary amino groups. But the mass spectrum of the alkaloid confirmed it as a higher homolog with the formula $C_{37}H_{40}O_6N_2$, while the NMR spectrum clearly showed two *N*-methyl groups. The main cleavage upon electron impact is similar to that for cycleanine and results in the formation of a fragment which shows a prominent *m/e* 297 peak.[58a]

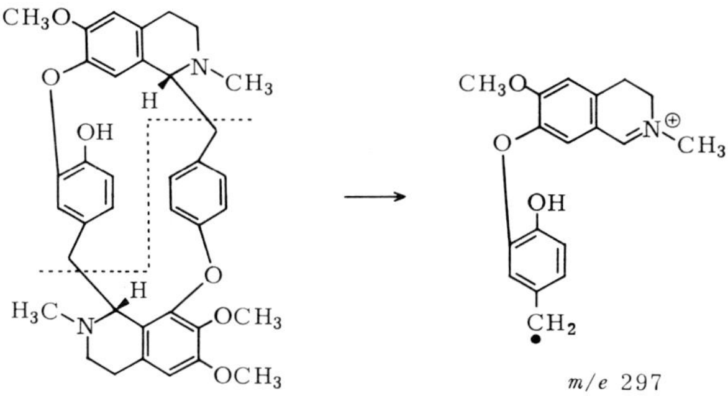

Chondrofoline
(new structure)

XIV. NMR Spectroscopy

The NMR spectra of bisbenzylisoquinolines is a complex subject. Some generalizations have been made relating chemical shifts of *O*-methyl and *N*-methyl groups to structure and stereochemistry.[62] In the berbamine–oxyacanthine series, a methoxyl group at C-12 is found between δ3.87 and 3.95, while one at C-7 is shielded and appears between δ3.02 and 3.20. A C-6 methoxyl is usually found in the vicinity of δ3.75. A C-6′ methoxyl shows up in the range δ3.3–3.65: when the asymmetric carbons are paired (+ −), the absorption is near δ3.6, but when the pairing is (+ +) or (− −), the value is around δ3.35.[62] These generalizations should be used with caution since exceptions are possible.[1a] One generalization that seems to hold throughout is that for alkaloids of the oxyacanthine type both *N*-methyl resonances occur near δ2.55, whereas alkaloids of the berbamine series give well-separated peaks near δ2.3 and 2.6.[1a,62] In the diastereoisomeric curine-chondocurine system, the 12′-*O*-methyl protons resonate at relatively high field, δ3.28–3.43, while the 7′-*O*-methyl protons are downfield, δ3.62–3.63. The two diastereoisomeric series can be clearly differentiated from one another particularly by the chemical shifts of the *N*-methyl protons which come around δ2.24 and 2.28 in the curine series where the asymmetric centers are paired (+ +) or (− −), and δ2.06 and 2.27 in the chondocurine series where the centers are paired (+ −).[58a]

It has also been reported that the chemical shift of an *N*-methyl group in some instances changes with variations in concentration,[1b] while the chemical shift of the C-8 proton of *O*-benzylated bisbenzylisoquinolines may be time-dependent because of conformational changes.[53]

XV. UV Spectroscopy

Ordinary bisbenzylisoquinoline alkaloids exhibit the usual benzylisoquinoline spectrum with a maximum near 283 mμ (3.70) and a minimum near 260 mμ (3.45).

There is also a strong absorption near 225 mμ. UV spectroscopy cannot readily differentiate between the different bisbenzylisoquinoline subgroups. The unusual bisbenzylisoquinoline repanduline shows λ_{max}^{EtOH} 283 and 326 mμ.[63]

REFERENCES

1. For a recent review on the bisbenzylisoquinolines, see M. F. Grundon, *Progr. Org. Chem.* **6**, 38 (1964).
1a. Also see a monograph of two-hundred and one pages, in Japanese, honoring Professor M. Tomita, published in 1967 by a Committee for the Retirement of Professor Tomita.
1b. M. Curcumelli-Rodostamo and M. Kulka, *in* "The Alkaloids" (R. H. F. Manske, ed.), Vol. 9, p. 133. Academic Press, New York, 1967.
1c. For a review on Thalictrum alkaloids which include several bisbenzylisoquinolines, see P. L. Schiff, Jr. and R. W. Doskotch, *Lloydia* **33**, 403 (1970), and N. M. Mollov, H. B. Dutschewska, and V. St. Georgiev, *in* "Recent Developments in the Chemistry of Natural Carbon Compounds" (R. Bognár, V. Bruckner, and Cs. Szántay, eds.), Vol. 4, p. 193. Publishing House of the Hungarian Academy of Sciences, Budapest, 1971.
1d. W. Döpke, *Arch. Pharm.* **295**, 605 (1962).
2. F. von Bruchhausen and P. H. Gericke, *Arch. Pharm.* (*Weinheim*) **269**, 115 (1931).
3. F. von Bruchhausen, H. Oberembt, and A. Feldhaus, *Justus Liebigs Ann. Chem.* **507**, 144 (1933).
4. I. R. C. Bick, P. S. Clezy, and W. D. Crow, *Aust. J. Chem.* **9**, 111 (1956).
5. P. A. Sartoretto and F. J. Sowa, *J. Amer. Chem. Soc.* **59**, 603 (1937).
6. Y. Sasaki and M. Suzuki, *Chem. Pharm. Bull.* **18**, 1 (1970).
7. M. Tomita, E. Fujita, and F. Murai, *J. Pharm. Soc. Jap.* **71**, 226 (1951).
8. Y. Inubushi, K. Nomura, and M. Miyawaki, *J. Pharm. Soc. Jap.* **83**, 282 (1963).
9. Z. F. Ismailov, A. U. Rakhmatkariev, and S. Y. Yunusov, *Uzb. Khim. Zh.* 56 (1961); *Chem. Abstr.* **58**, 3469 (1963).
9a. M. R. Falco, J. X. de Vries, Z. Macció, and I. R. C. Bick, *Chem. Commun.* p. 1056 (1971).
10. A. J. Everett, L. A. Lowe, and S. Wilkinson, *Chem. Commun.* p. 1020 (1970).
10a. J. Harley-Mason, A. S. Howard, W. I. Taylor, M. J. Vernengo, I. R. C. Bick, and P. S. Clezy, *J. Chem. Soc., C* p. 1948 (1967); I. R. C. Bick, J. H. Bowie, J. Harley-Mason, and D. H. Williams, *ibid.* p. 1951; K. Aoki and J. Harley-Mason, *ibid.* p. 1957.
11. P.-C. P'an, Y.-L. Chou, T.-C. Sun, and I.-S. Kao, *Sci. Sinica* **11**, 321 (1962); *Chem. Abstr.* **58**, 3467 (1963).
12. T. Kametani, S. Takano, K. Masuko, and F. Sasaki, *Chem. Pharm. Bull.* **14**, 67 (1966).
13. For the synthesis of the related isoliensinine, see T. Kametani, S. Takano, H. Iida, and M. Shinbo, *J. Chem. Soc., C* p. 298 (1969).
14. M. Tomita, K. Fujitani, and Y. Aoyagi, *Tetrahedron Lett.* p. 1201 (1967).
15. Y. Inubushi, Y. Masaki, S. Matsumoto, and F. Takami, *J. Chem. Soc., C* p. 1547 (1969).
15a. M. A. Insalaco and D. S. Tarbell, *Org. Syn.* **50**, 9 (1970).
15b. E. Fujita and A. Sumi, *Chem. Pharm. Bull.* **18**, 2591 (1970).
16. S. Ueda, *J. Pharm. Soc. Jap.* **82**, 714 (1962).
17. M. Tomita, S. Ueda, and A. Teraoka, *Tetrahedron Lett.* p. 635 (1962).
17a. B. Anjaneyulu, T. R. Govindachari, and N. Viswanathan, *Tetrahedron* **27**, 439 (1971).
17b. J. M. Bobbitt and R. C. Hallcher, *Chem. Commun.* p. 543 (1971).
18. T. Kametani, S. Takano, K. Masuko, and S. Kuribara, *J. Pharm. Soc. Jap.* **85**, 166 (1965).
19. J. B. Hendrickson and C. Kandall, *Tetrahedron Lett.* p. 343 (1970).
20. J. B. Hendrickson and R. Bergeron, *Tetrahedron Lett.* p. 345 (1970).
21. For a discussion of protective groups, see J. F. W. McOmie, *Advan. Org. Chem.* **3**, 191 (1963).

22. M. Tomita and H. Furukawa, *J. Pharm. Soc. Jap.* **83**, 676 (1963).
23. Y. Inubushi and M. Kozuba, *Pharm. Bull.* **2**, 215 (1954).
24. H. Furukawa, *J. Pharm. Soc. Jap.* **86**, 253 (1966).
25. Y. Watanabe, *J. Pharm. Soc. Jap.* **80**, 166 (1960).
26. J. D. Hobson and J. G. McCluskey, *J. Chem. Soc., C* p. 2015 (1967).
26a. G. Krauss and K. Nador, *Tetrahedron Lett.* p. 57 (1971).
27. Max Polonovski and Michel Polonovski, *Bull. Soc. Chim. Fr.* **41**, 1190 (1927).
27a. M. P. Cava and M. Srinivasan, *J. Org. Chem.* (in press) (1971).
27b. R. D. Birkenmeyer and L. A. Dolak, *Tetrahedron Lett.* p. 5049 (1970).
27c. R. A. Olofson and R. C. Schnurr, (in press) (1972).
27d. J. H. E. Lindner, H. J. Kuhn, and K. Gollnick, *Tetrahedron Lett.* p. 1705 (1972).
28. M. Shamma, N. C. Deno, and J. F. Remar, *Tetrahedron Lett.* p. 1375 (1966).
29. V. Šimánek and A. Klásek, *Tetrahedron Lett.* p. 3039 (1969).
30. N. D. V. Wilson and J. A. Joule, *Tetrahedron* **24**, 5493 (1968).
31. I. T. Harrison, *Chem. Commun.* p. 616 (1969).
32. G. I. Feutrill and R. N. Mirrington, *Tetrahedron Lett.* p. 1327 (1970).
33. A. L. Wilds and W. B. McCormack, *J. Amer. Chem. Soc.* **70**, 4127 (1948).
34. J. C. Sheehan, W. F. Erman, and P. A. Cruickshank, *J. Amer. Chem. Soc.* **79**, 147 (1957).
35. V. Prey, *Chem. Ber.* **76**, 156 (1943); H. G. Richey, Jr., Ph. D. Thesis, Harvard University (1958).
36. J. F. W. McOmie and D. E. West, *Org. Syn.* **49**, 50 (1969).
36a. L. H. Long and G. F. Freeguard, *Nature* **207**, 403 (1965).
37. W. J. Musliner and J. W. Gates, Jr., *J. Amer. Chem. Soc.* **88**, 4271 (1966).
38. M. P. Cava and K. T. Buck, *Tetrahedron* **25**, 2795 (1969).
38a. W. A. Sheppard, *Org. Syn.* **49**, 111 (1969).
38b. J. C. Craig and K. K. Purushothaman, *J. Org. Chem.* **35**, 1721 (1970).
38c. W. Bonthrone and J. W. Cornforth, *J. Chem. Soc., C* p. 1202 (1969).
38d. S. Cabiddu, A. Maccioni, and M. Secci, *Gazz. Chim. Ital.* **98**, 800 (1968).
39. B. Franck, G. Blaschke, and G. Schlingloff, *Angew. Chem., Int. Ed. Engl.* **3**, 192 (1964).
40. T. Kametani and I. Noguchi, *J. Chem. Soc., C* p. 502 (1969).
41. See also T. Kametani and I. Noguchi, *Chem. Pharm. Bull.* **17**, 1977 (1969).
42. A. M. Choudhury, I. G. C. Coutts, A. K. Durbin, K. Schofield, and D. H. Humphreys, *Chem. Commun.* p. 1341 (1968).
43. D. H. R. Barton, G. W. Kirby, and A. Wiechers, *Chem. Commun.* p. 266 (1966).
44. A. R. Battersby, *in* "Oxidative Coupling of Phenols" (W. I. Taylor and A. R. Battersby, eds.), p. 142. Marcel Dekker, New York 1967.
45. D. H. R. Barton and T. Cohen, *Festschr. Prof. Dr. Arthur Stoll siebzigsten Geburtstag 1957* p. 117 (1957).
45a. For an excellent historical account of the curare alkaloids, see M. G. Kreig, "Green Medicine," p. 223. Rand McNally, Chicago, Illinois, 1964.
46. J. D. Dutscher, *J. Amer. Chem. Soc.* **74**, 2221 (1952).
47. "The Dispensary of the U.S.A.", 25th ed., p. 472. Lippincott, Philadelphia, Pennsylvania, 1960.
47a. O. Carrier, Jr., *in* "Medicinal Chemistry," (A. Burger, ed.), 3rd ed., p. 1581. Wiley (Interscience), New York, 1970.
48. S. M. Kupchan, S. Kubota, E. Fujita, S. Kobayashi, T. H. Block, and S. A. Telang, *J. Amer. Chem. Soc.* **88**, 4212 (1966).
49. S. M. Kupchan, T.-H. Yang, G. S. Vasilikiotis, M. H. Barnes, and M. L. King, *J. Org. Chem.* **34**, 3884 (1969).
50. R. E. Perdue, Jr. and J. L. Hartwell, *Morris Arbor. Bull.* **20**, 35 (1969).
50a. A. Lewis and R. G. Shepherd, *in* "Medicinal Chemistry" (A. Burger, ed.), 3rd ed., p. 409. Wiley (Interscience), New York, 1970.

50b. L. A. Mitscher, W. Wu, R. W. Doskotch, and J. L. Beal, *Chem. Commun.* p. 589 (1971).
51. Allen and Hansburys Ltd., Brit. Pat. 695,298 (1953); E. P. Taylor, *J. Chem. Soc., London* p. 142 (1952).
52. A. R. Battersby, I. R. C. Bick, W. Klyne, J. P. Jennings, P. M. Scopes, and M. J. Vernengo, *J. Chem. Soc., London* p. 2239 (1965).
53. T. Kametani, H. Iida, K. Sakurai, S. Kano, and M. Ihara, *Chem. Pharm. Bull.* **17**, 2120 (1969).
54. F. von Bruchhausen and H. Schultze, *Arch. Pharm.* (*Weinheim*) **267**, 623 (1929).
55. I. R. C. Bick and A. R. Todd, *J. Chem. Soc., London* p. 2170 (1948).
56. J. Knabe and P. Horn, *Arch. Pharm.* (*Weinheim*) **300**, 726 (1967).
57. M. Shamma, B. S. Dudock, M. P. Cava, K. V. Rao, D. R. Dalton, D. C. DeJongh, and S. R. Shrader, *Chem. Commun.* p. 7 (1966).
58. M. Tomita, T. Kikuchi, K. Fujitani, A. Kato, H. Furukawa, Y. Aoyagi, M. Kitano, and T. Ibuka, *Tetrahedron Lett.* p. 857 (1966).
58a. J. Baldas, I. R. C. Bick, Q. N. Porter, and M. J. Vernengo, *Chem. Commun.* p. 132 (1971).
59. J. Baldas, Q. N. Porter, I. R. C. Bick, and M. J. Vernengo, *Tetrahedron Lett.* p. 2059 (1966).
60. D. C. DeJongh, S. R. Shrader, and M. P. Cava, *J. Amer. Chem. Soc.* **88**, 1052 (1966).
61. J. Baldas, I. R. C. Bick, G. K. Douglas, and Q. N. Porter, *Aust. J. Chem.* **21**, 2305 (1968).
62. I. R. C. Bick, J. Harley-Mason, N. Sheppard, and M. J. Vernengo, *J. Chem. Soc., London* p. 1896 (1961).
63. I. R. C. Bick, K. Doebel, W. I. Taylor, and A. R. Todd, *J. Chem. Soc., London* p. 692 (1953).

Chapter 6 / THE CULARINES

Occurrence: Fumariaceae
Number: 4
Structures:

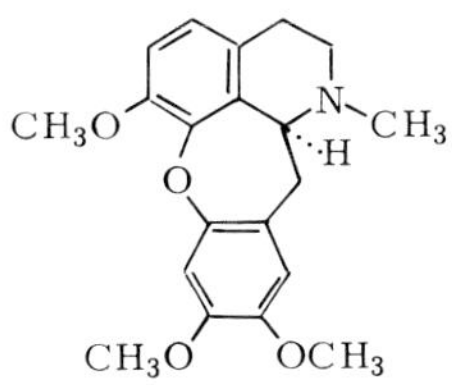

(+)-Cularine

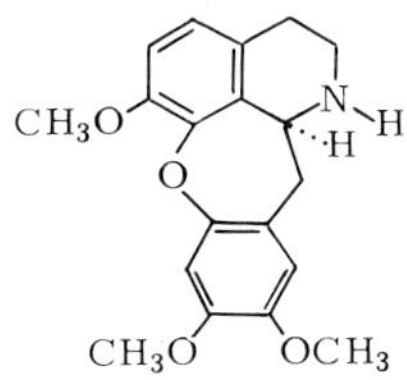

(+)-Cularimine

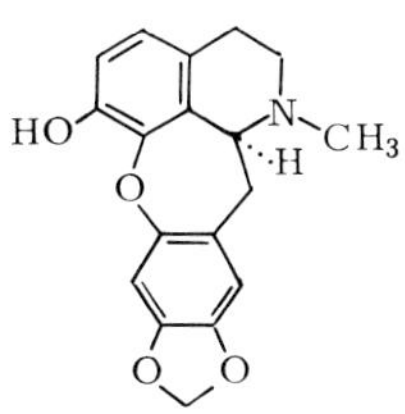

(+)-Cularicine

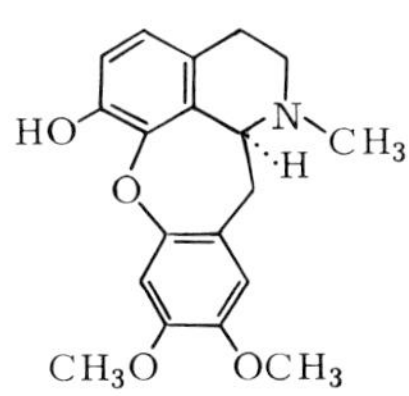

(+)-Cularidine

I. Introduction

The intramolecular oxidative coupling of benzylisoquinolines of type **1** in plants may lead to tetracyclic bases called the cularine alkaloids and represented by structure

2. These compounds incorporate a dihydrooxepine system, and their occurrence in nature seems to be limited to the genera Dicentra and Corydalis, both of which belong to the fumitory (Fumariaceae) family.

The four alkaloids of the cularine group that have been fully characterized are cularine, cularimine, cularidine, and cularicine.

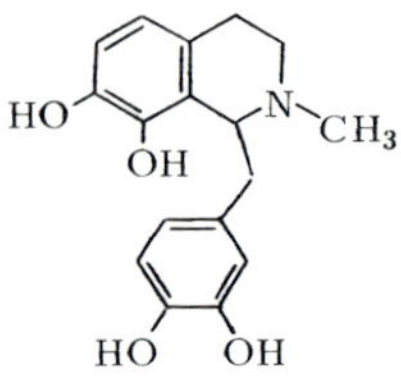

1

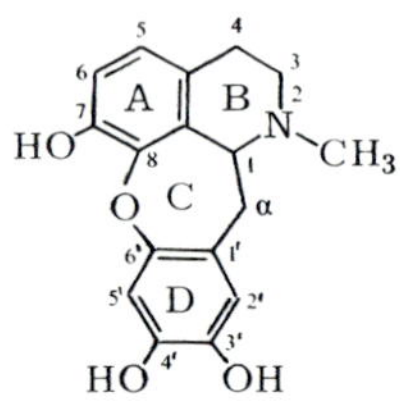

2

II. Cularine: Structural Determination

Cularine was first isolated by Manske in the 1930s,[1] and the structural elucidation was reported in 1950.[2] The alkaloid, $C_{20}H_{23}O_4N$, contains one *N*-methyl group, three methoxyls, and one additional oxygen atom as part of a diphenyl ether linkage. Two separate degradative sequences were carried out in order to obtain the structure.

In the first sequence, a series of Hofmann degradations supplied the substituted oxepine **3** which upon permanganate oxidation afforded the colorless triacid **4** and the yellow xanthone acid **5**. Both acids were characterized as their methyl esters. Additionally, catalytic reduction of **3** with Adams catalyst absorbed 2 moles of hydrogen and yielded the tricyclic ether **6** (Scheme I).

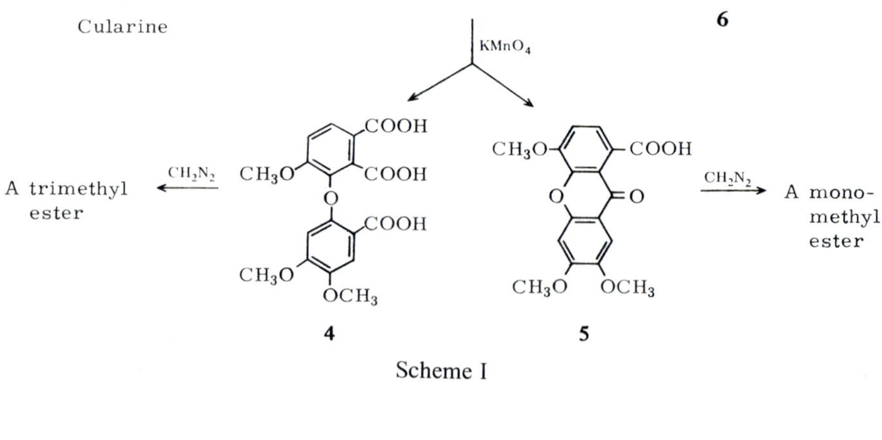

Scheme I

In the second sequence, sodium in liquid ammonia cleavage of cularine generated the phenolic benzylisoquinoline **7** which could be oxidized with permanganate to 4-methoxyphthalic acid. Compound **7** could not be *O*-methylated with diazomethane. It was therefore treated with excess methyl iodide and alkali. The resulting material, presumably the corresponding quaternary *O*-methyl ether, was subjected to excess dimethyl sulfate and base so as to carry out a double Hofmann degradation. The product from this procedure was the resin **8** which on oxidation yielded 4-methoxyphthalic acid and asaronic acid (Scheme II).

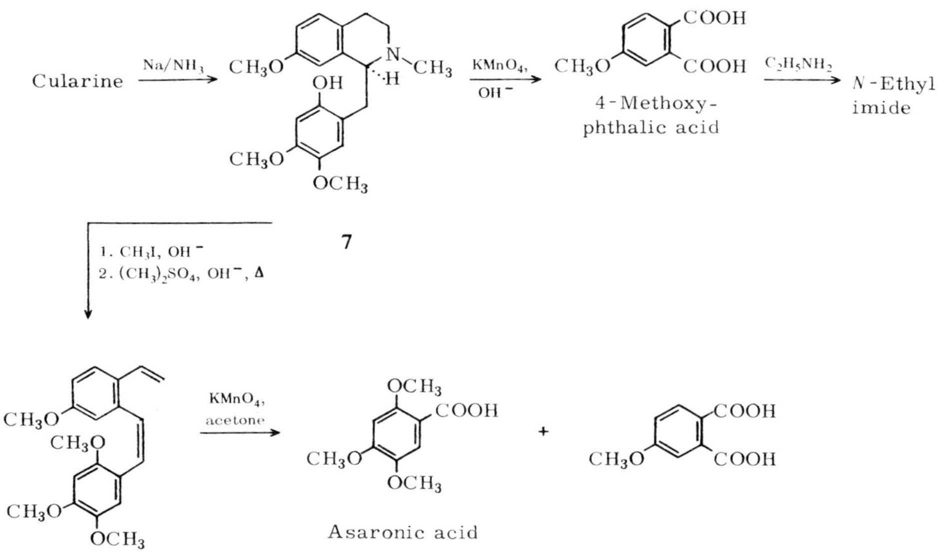

7

8

Scheme II

The isolation of 4-methoxyphthalic and asaronic acids helped establish the positions of the methoxyl substituents. The remaining structural features of cularine were then derived from a consideration of the nature of the degradation products.[2]

III. Syntheses of Cularine and Its Analogs

Several syntheses of the cularine system are presently available in the literature, all but one of which are due to Japanese efforts.

In the first synthesis of cularine to be carried out, the tricyclic ketone **9**, prepared earlier by Manske through an Ullmann reaction, was condensed with aminoacetaldehyde diethyl acetal. The resulting Schiff base **10** was treated with sulfuric acid to afford a small yield of the crystalline isoquinoline **11**. Sodium borohydride reduction of the isoquinoline methiodide then furnished racemic cularine (Scheme III).[3]

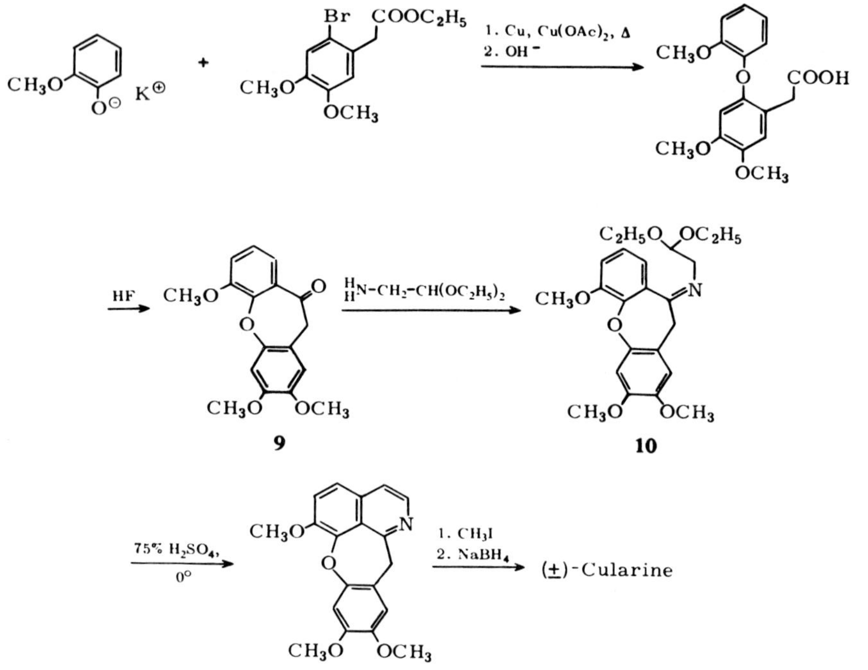

9 10 11

Scheme III

A second synthesis of cularine begins with an Ullmann condensation and is outlined in Scheme IV. Noteworthy is the use of the Erlenmeyer azlactone sequence for side-chain homologation.[4,4a]

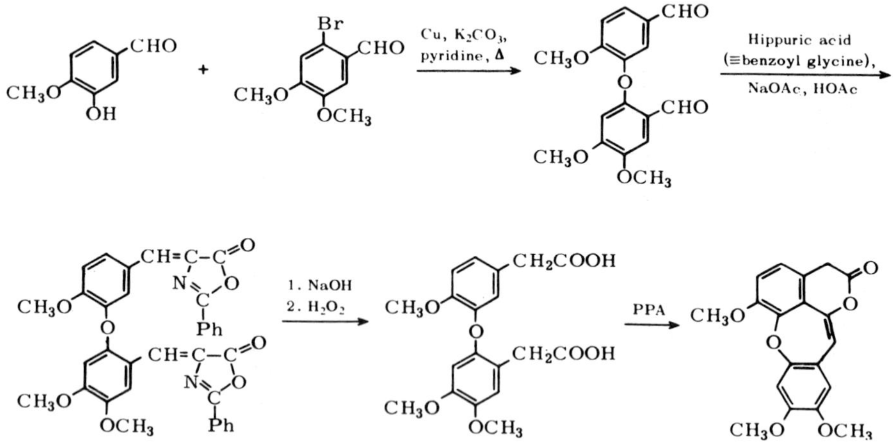

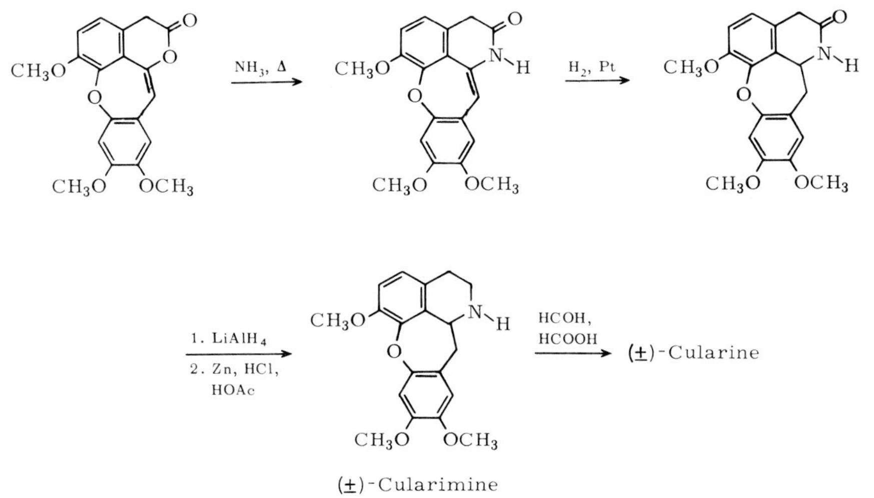

Scheme IV

The (±)-cularimine obtained was also resolved using (+)-tartaric acid, and the resulting (+)-cularimine was *N*-methylated to yield (+)-cularine, identical with the natural product.[5]

Iida and co-workers have carried out a preparation of a cularine analog (**12**) which differs from the two outlined above in that the Ullmann condensation was utilized in the final step, rather than in the early stages of the synthesis (Scheme V).[6]

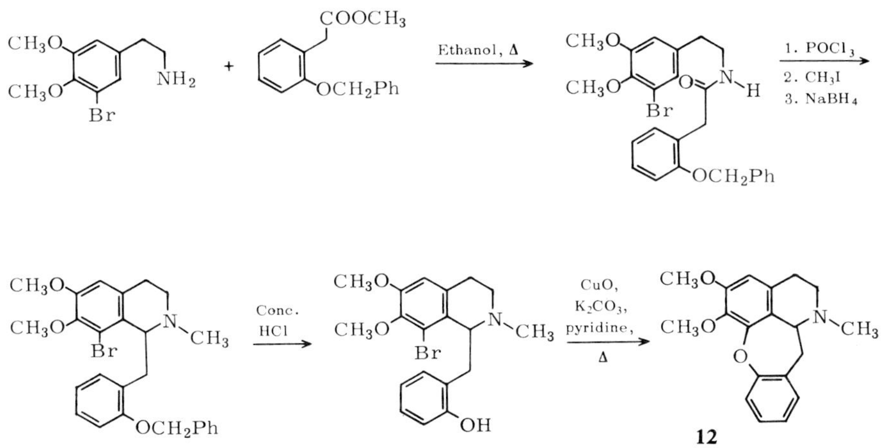

12

Scheme V

In a second Iida synthesis, the Ullmann condensation was also used in the final step. However, the aryl halide moiety involved in the condensation constituted the precursor of ring D of the cularine system, rather than of ring A as in the previous synthesis (Scheme VI).[7] Following the separation of the tetrahydrobenzylisoquinolines **13** and **14**, Ullmann cyclization of the former yielded the desired cularine analog **15**.

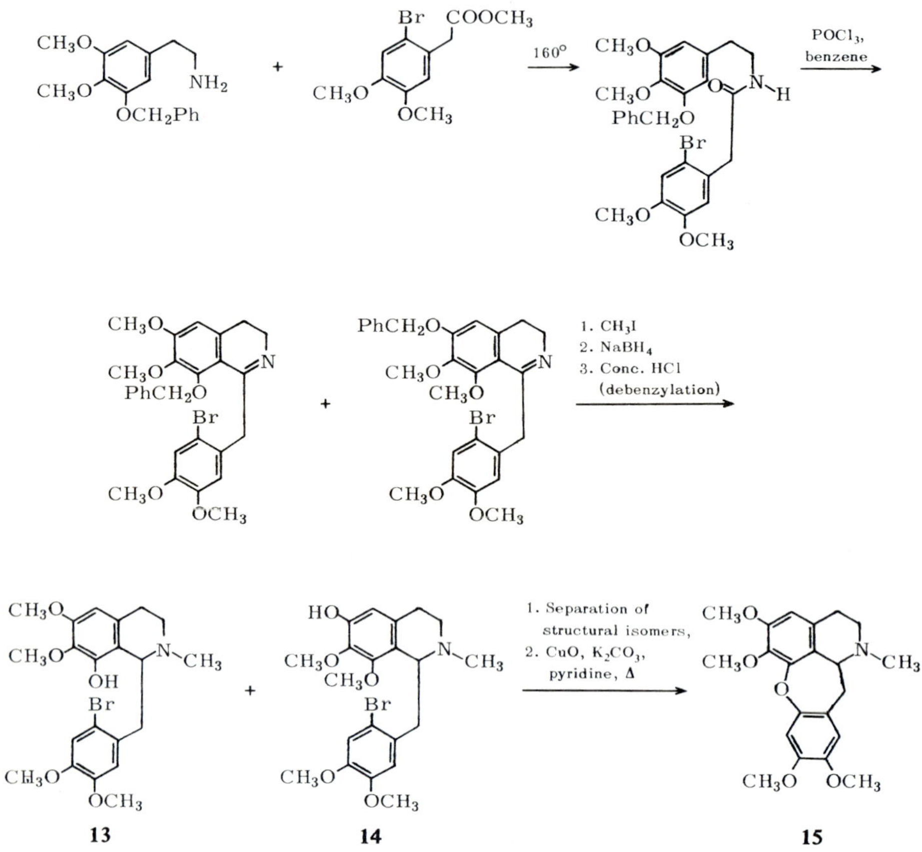

Scheme VI

Taking advantage of the fact that the ethoxycarbamido group is a good substituent for accelerating the Bischler–Napieralski reaction and is easily hydrolyzed to an amino group which can in turn be diazotized and reduced, Ishiwata and co-workers have achieved a synthesis of cularine which includes an Ullmann condensation in the final step, but is nevertheless different from any of the approaches described above. The starting material was 3-bromo-4-methoxyphenylethylamine (Scheme VII).[7a]

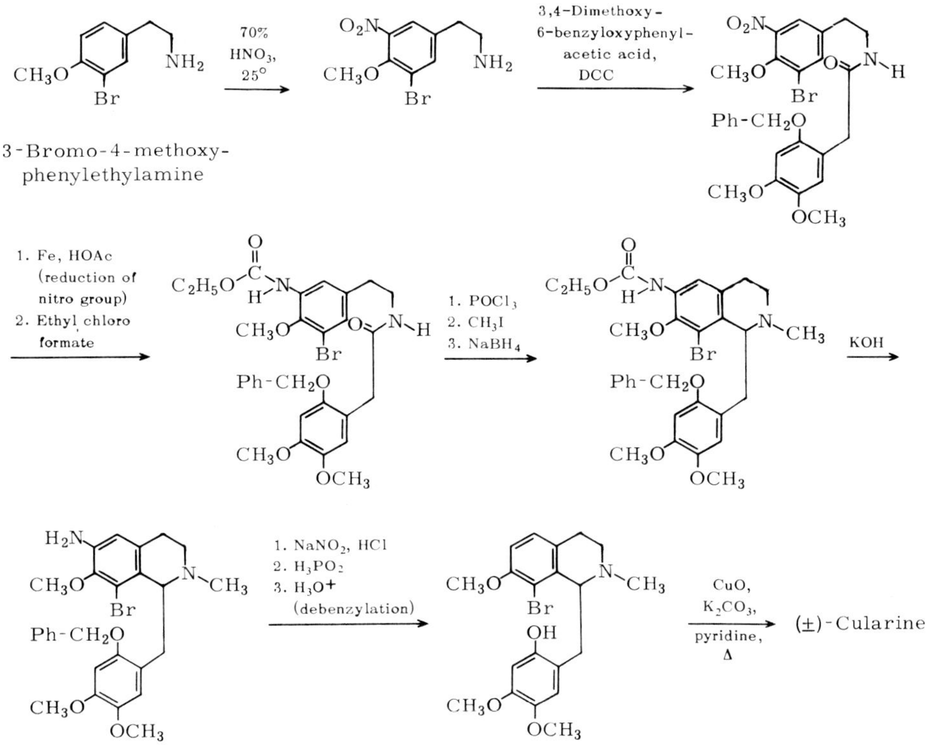

Scheme VII

A synthesis of a cularine system which relies upon phenolic oxidative coupling is that of Kametani, Kikuchi, and Fukumoto. Oxidation of the diphenolic tetrahydrobenzylisoquinoline **16** was achieved with potassium ferricyanide. The resulting diastereoisomeric mixture was assigned structure **17**. Acid-catalyzed rearrangement then gave the cularine analog **18** whose structure was confirmed by an alternate synthesis involving an Ullmann reaction.[8,9] It is noteworthy that it is the phenolate moiety that migrates in going from **17** to **18**, while in the acid-catalyzed rearrangement of *p*-toluquinol it is the methyl group that shifts preferentially (Chapter 1, Section VI, A).

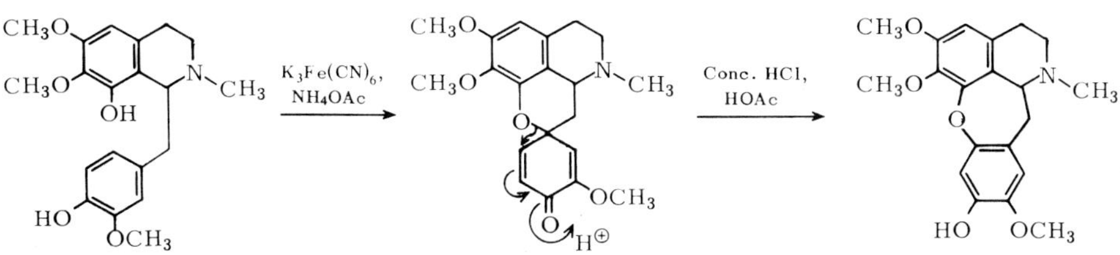

16 **17** **18**

Interestingly enough, an attempt to prepare (±)-cularine by this approach was not successful since the dienones obtained failed to rearrange to the cularine skeleton upon treatment with acid (Scheme VII a).[9a] (*cf.* Chapter 32.)

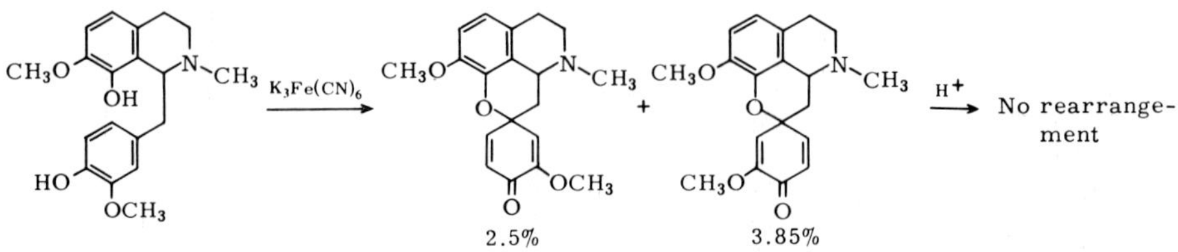

Scheme VII a

An alternate synthesis by Jackson and Stewart, also using phenolic oxidative coupling, did furnish (±)-cularine. In this synthesis, the formation of the isoquinoline nucleus involves a new and mild variant of the Pomeranz–Fritsch cyclization which is particularly suitable for the 7,8-oxygenated compounds. This variant also provides a direct route to the fully aromatic isoquinolines, in contrast to the Bobbitt modification which leads more readily to tetrahydroisoquinolines, Scheme VII b.[9b]

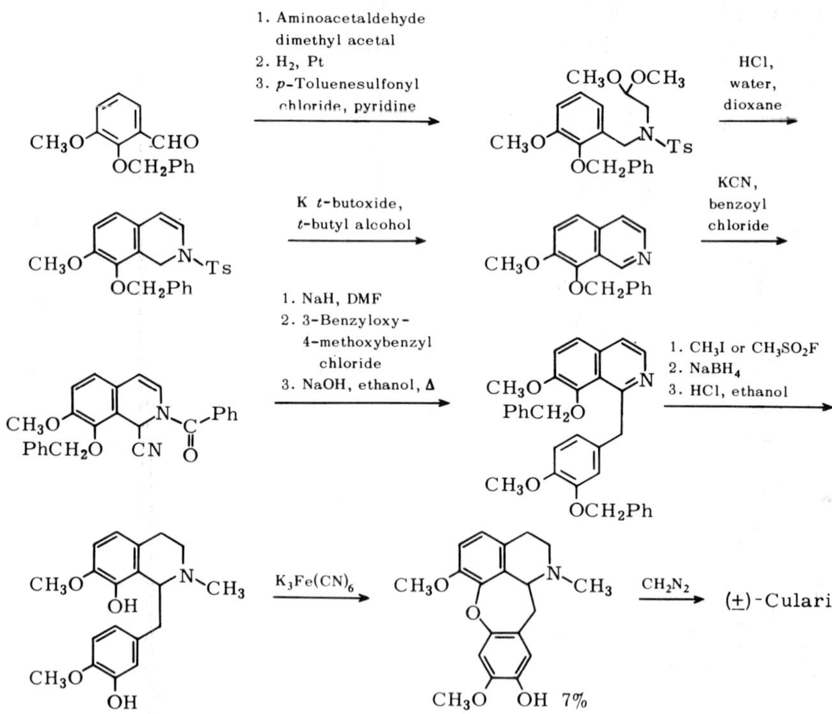

Scheme VII b

Another synthesis of (±)-cularine by phenolic oxidative coupling is available which relies upon the Pictet–Spengler reaction rather than the Pomeranz–Fritsch cyclization for the formation of the isoquinoline system (Scheme VIIc).[9a] Note the formation of the two positional isomers resulting from the ferricyanide oxidation.

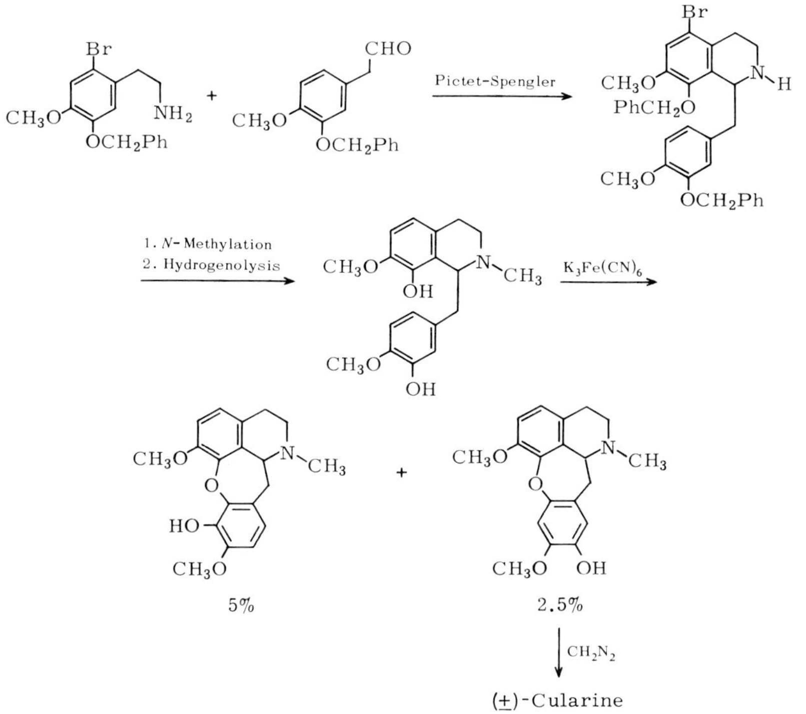

Scheme VIIc

No feeding experiments with labeled precursors have yet been conducted regarding the formation of cularine bases in nature. Any of three routes presently appear possible, i.e., through indirect phenolic coupling as in Scheme VIIa, through direct phenolic coupling as in Scheme VIIb, or yet through a sequence which includes cyclization to an 11-membered ring followed by formation of the tetrahydroisoquinoline system.[9b] This third possibility appears the least likely, but should nevertheless be considered (Scheme VIId).

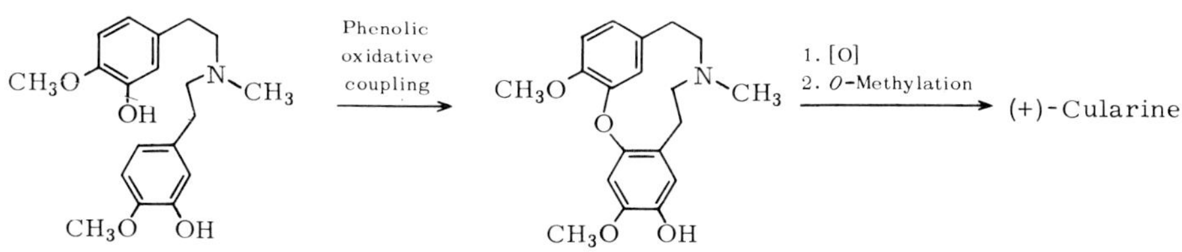

Scheme VIId

IV. Absolute Configuration

The absolute configuration of naturally occurring (+)-cularine was arrived at through its chemical correlation with L-romneine of established absolute configuration (Scheme VIIe). Bromination of L-romneine and Ullmann condensation with guaiacol led to a diphenyl ether which was cleaved with sodium in liquid ammonia. The two diphenols obtained were selectively mono-*O*-methylated with diazomethane, thus furnishing two pseudo phenolic products, one of which, **7**, corresponded in all respects to the sodium in liquid ammonia derivative from cularine.[9c] Since all the other cularine alkaloids have been related chemically to (+)-cularine, they too must possess the alpha-hydrogen configuration at C-1.

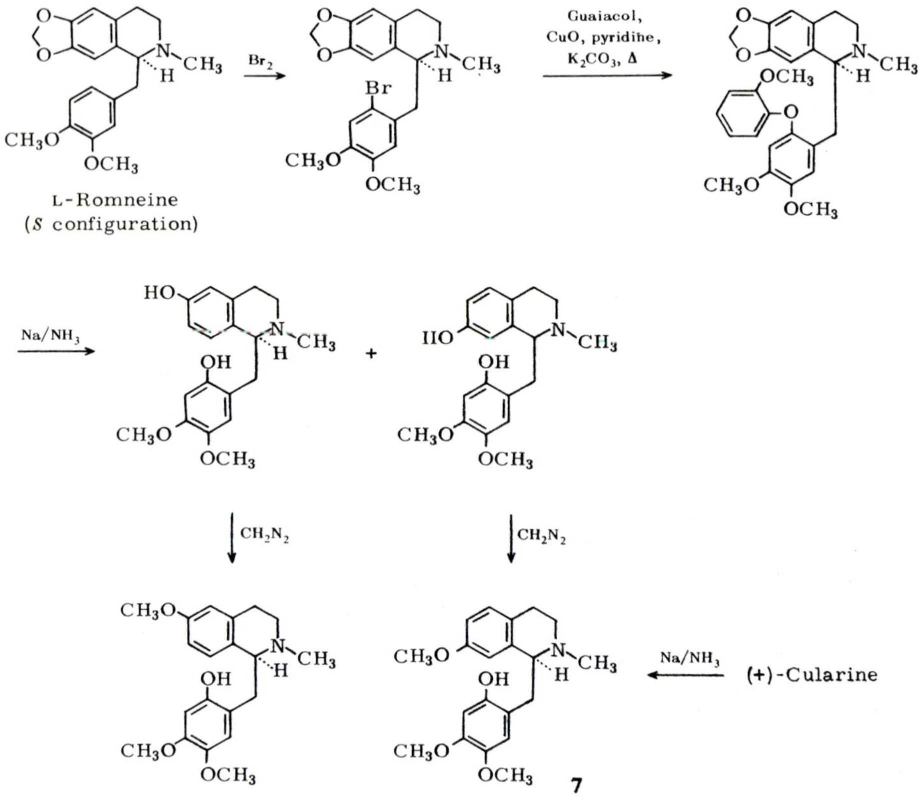

Scheme VIIe

V. NMR Spectroscopy and Conformation

A detailed study of the NMR spectrum of cularine at 100 MHz involving spin-decoupling experiments furnished the chemical shifts indicated in expression **19** below.[10]

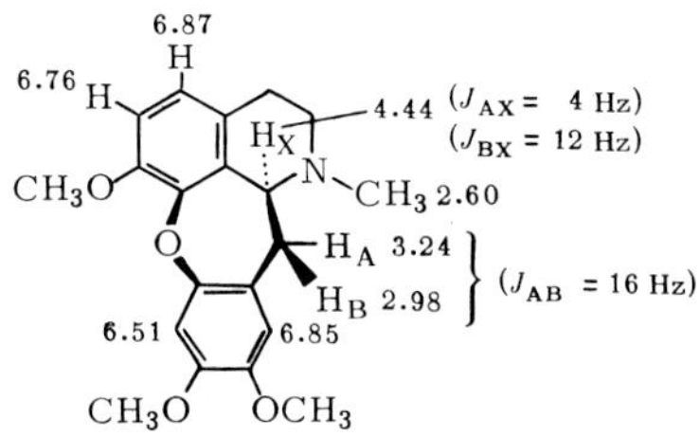

19

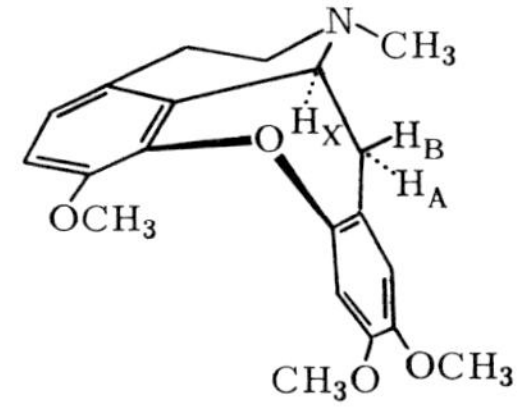

19a

H_X appeared as two doublets at δ4.44. Double irradiation of this absorption showed $J_{AB} = 16$ Hz, so that $J_{AX} = 4$ Hz and $J_{BX} = 12$ Hz. These values are in accord with the dihedral angles incorporated in conformation **19a**, and the relevant data found and calculated are tabulated below.

	J (Hz)	
	Calculated	Found
$H_A - H_X$ Dihedral angle about 60°	$J_{AX} = 3-4$	$J_{AX} = 4$
$H_B - H_X$ Dihedral angle about 180°	$J_{BX} = 8-14$	$J_{BX} = 12$

The fact that the H_X absorption is relatively downfield at δ4.44 as compared to the equivalent hydrogen in an aporphine or a tetrahydrobenzylisoquinoline points to the proximity of H_X to the oxygen of the diphenyl ether bridge. This requirement is indeed fulfilled by conformation **19a** for cularine, so that the molecule is essentially V-shaped.[10]

The absorptions of the *O*-methyl groups of cularine are at δ3.79, 3.84, and 3.89, and it is believed that the peak furthest downfield is due to the C-7 methoxyl. The NMR spectra of some of the other cularine alkaloids have also been reported.[11]

VI. Mass Spectroscopy

Although the cularine alkaloids show a strong molecular ion peak, the base peak is due to the $M-15$ ion, corresponding to the loss of a methyl group. In the case of cularine, loss of this group results in formation of the stable quinonoid cation **20**.[12]

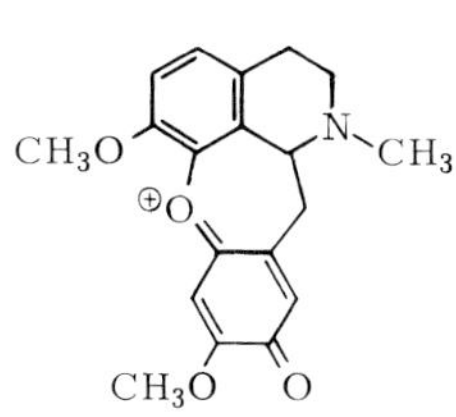

20

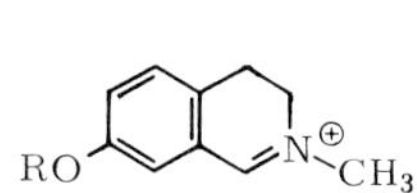

21

A useful but minor peak is due to the ion **21**, which represents the upper half of the cularine system. This ion can indicate whether the substituent R on ring A is hydroxyl or methoxyl and is thus of diagnostic value.[11]

VII. UV Spectroscopy

Cularine[10]: λ_{max}^{EtOH} 206 sh, 229 sh, 274, 283, and 295 sh mμ (5.21, 4.12, 3.63, 3.79, and 3.48)
λ_{min}^{EtOH} 256 mμ (3.13)

REFERENCES

1. R. H. F. Manske, *Can. J. Res., Sect. B* **16**, 81 (1938).
2. R. H. F. Manske, *J. Amer. Chem. Soc.* **72**, 55 (1950).
3. T. Kametani and K. Fukumoto, *J. Chem. Soc., London* p. 4289 (1963).
4. T. Kametani, S. Shibuya, S. Seino, and K. Fukumoto, *Tetrahedron Lett.* p. 25 (1964).
4a. T. Kametani, H. Iida, and C. Kibayashi, *J. Heterocycl. Chem.* **7**, 339 (1970).
5. T. Kametani and S. Shibuya, *Tetrahedron Lett.* p. 1897 (1965).
6. H. Iida, T. Kikuchi, K. Sakurai, and T. Watanabe, *J. Pharm. Soc. Jap.* **89**, 645 (1969).
7. H. Iida, T. Kikuchi, S. Tanaka, and M. Shinbo, *J. Pharm. Soc. Jap.* **89**, 1169 (1969).
7a. S. Ishiwata, T. Fujii, N. Miyaji, Y. Satoh, and K. Itakura, *Chem. Pharm. Bull.* **18**, 1850 (1970).
8. T. Kametani, T. Kikuchi, and K. Fukumoto, *Chem. Commun.* p. 546 (1967); *Chem. Pharm. Bull.* **16**, 1003 (1968).
9. T. Kametani, H. Iida, T. Kikuchi, M. Mizushima, and K. Fukumoto, *Chem. Pharm. Bull.* **17**, 709 (1969).
9a. T. Kametani, K. Fukumoto, and M. Fujihara, *Chem. Commun.* p. 352 (1971).
9b. A. H. Jackson and G. W. Stewart, *Chem. Commun.* p. 149 (1971). For related studies in oxidative coupling see T. Kametani, K. Fukumoto, and M. Fujihara, *Biorg. Chem.* **1**, 40 (1972), and *J. Chem. Soc., Perkin Trans.* **1**, 394 (1972).
9c. J. Kunitomo, K. Morimoto, K. Yamamoto, Y. Yoshikawa, K. Azuma, and K. Fujitani, *Chem. Pharm. Bull.* **19**, 2197 (1971).
10. N. S. Bhacca, J. C. Craig, R. H. F. Manske, S. K. Roy, M. Shamma, and W. A. Slusarchyk, *Tetrahedron* **22**, 1467 (1966).
11. T. Kametani, S. Shibuya, and C. Kibayashi, *Tetrahedron Lett.* p. 3215 (1966).
12. M. Ohashi, J. M. Wilson, H. Budzikiewicz, M. Shamma, W. A. Slusarchyk, and C. Djerassi, *J. Amer. Chem. Soc.* **85**, 2807 (1963).

Chapter 7 / THE CULARINE-MORPHINE DIMERS

Occurrence: Fumariaceae
Number: 3
Structures:

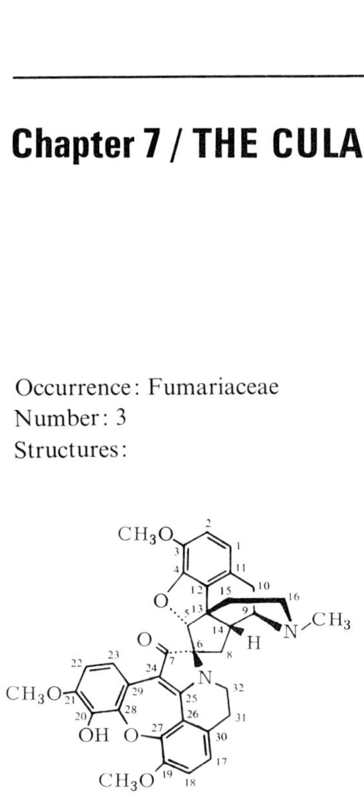

Cancentrine

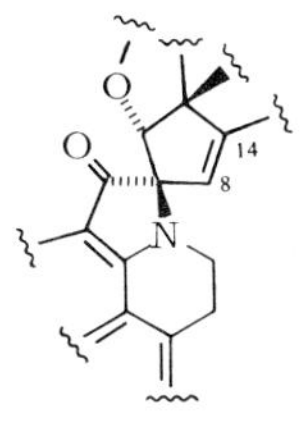

Dehydrocancentrine A

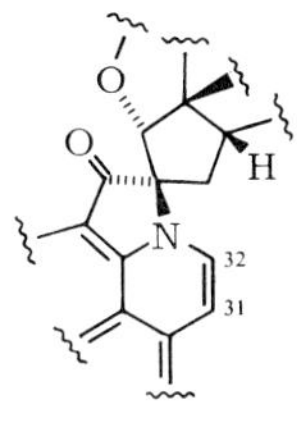

Dehydrocancentrine B

I. Structural Elucidation

The yellow alkaloid cancentrine, $C_{36}H_{34}O_7N_2$, was isolated by Manske in 1932 from *Dicentra canadensis* (Goldie) Walp,[1] but its structure was not elucidated until 1970.[2] Cancentrine possesses two nitrogen atoms pointing to its dimeric nature. The UV spectrum exhibits several maxima characteristic of a highly conjugated system, and the IR spectrum shows hydroxyl absorption at 2.90 μ (3450 cm^{-1}) and a conjugated carbonyl peak at 6.01 μ (1665 cm^{-1}). NMR spectroscopy shows three aromatic methoxyl groups and one *N*-methyl group. Additionally, cancentrine possesses one phenolic group which can be either acetylated or *O*-methylated by the Rodionov method.[3] Of the two nitrogen atoms, only one is basic since the alkaloid yields only a monohydrochloride salt.

Cancentrine methiodide undergoes Hofmann degradation to cancentrine methine which upon hydrogenation and *O*-methylation gives cancentrine dihydromethine-*O*-methyl ether. The yellow hydrobromide salt of this *O*-methyl ether was then used for an X-ray single-crystal structure analysis.

Following the determination of structure for cancentrine methine-*O*-methyl ether hydrobromide, the problem was to deduce the structure of cancentrine itself by first locating the point of attachment of the basic nitrogen and then establishing the site of the phenolic function in the natural base (Scheme I).

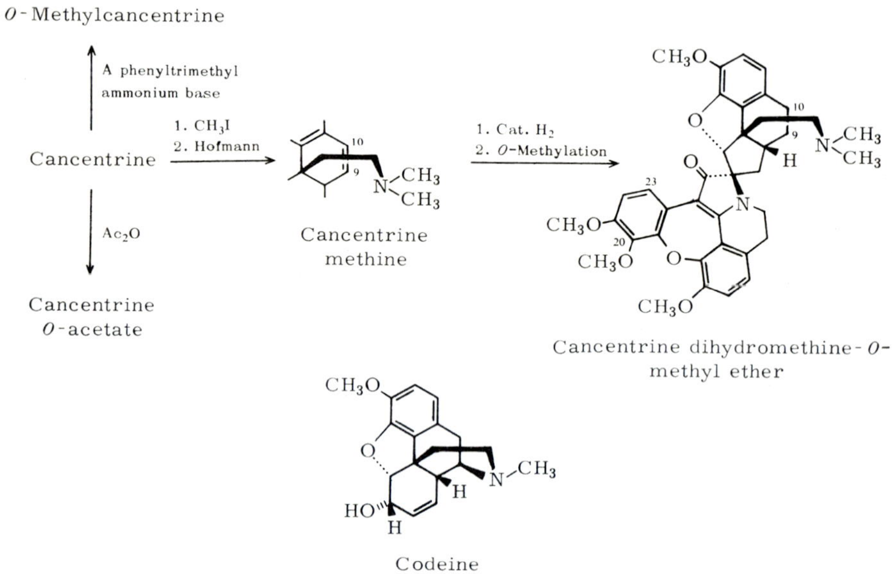

Scheme I

Cancentrine and all of its derivatives must possesses the same fundamental chromophoric system since they exhibit very similar UV spectra and also show an IR carbonyl band at or near 6.01 μ (1665 cm^{-1}).

The most conspicuous feature in the NMR spectrum of cancentrine methine is an AB quartet in the vinylic region due to the C-9, 10 double bond. It follows that in the alkaloid the basic nitrogen must be connected either at C-9 or C-10. The close similarity in the NMR spectra of codeine and cancentrine indicated that the nitrogen atom in the latter compound must be bonded at C-9, especially since joining at C-10 is biogenetically unlikely (Scheme I).

To establish the position of the phenolic group in cancentrine, the NMR spectra of the alkaloid and its *O*-acetate ester were compared. In each case, the aromatic protons at C-1 and 2, C-22 and 23, and C-17 and 18 appear as AB quartets. However, one of the aromatic protons which was centered at δ7.51 in the alkaloid is at δ7.88 in the acetate – a downfield shift of this order upon acetylation being characteristic of a proton para to a phenolic hydroxyl group. The phenolic function in cancentrine must, therefore, be at C-20, and the aromatic proton whose downfield shift is observed is at C-23. It was also found that double irradiation of any of the three methoxyl groups of cancentrine caused in each case an N.O.E. (nuclear Overhauser effect) of about 25% in the absorption of the adjacent aromatic proton. Such a result is possible only if the phenolic group is at C-20 where it is not vicinal to any aromatic hydrogen.

Cancentrine is thus the first member of a completely new class of dimeric benzylisoquinoline alkaloids involving cularine and morphine. An interesting biogenetic

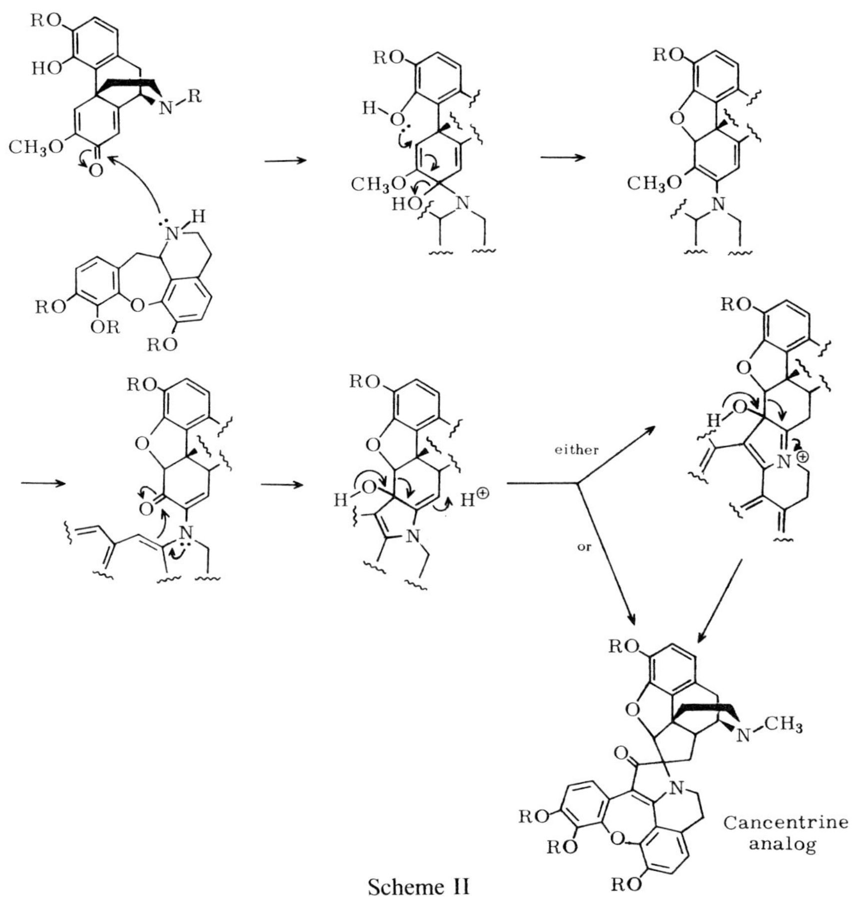

Scheme II

scheme has been put forth to explain the formation of cancentrine (Scheme II). A morphinandienone alkaloid, pallidine, has been found in *Corydalis pallida* var. *tenuis* Yatabe (Fumariaceae), a close relative of *D. canadensis,* from which the cularine–morphine dimers are obtained. The precursors of the cularine–morphine dimers could, therefore, be a morphinandienone analog and a cularine base. The initial condensation would be between the carbonyl of the dienone and the secondary amine of the cularine base.[4]

II. Dehydrocancentrine A and Dehydrocancentrine B

Dehydrocancentrine A and B are found in *D. canadensis* where they accompany cancentrine. Dehydrocancentrine A is a yellow compound which possesses two hydrogen atoms less than cancentrine. The DMSO-d_6 NMR spectrum showed a sharp singlet at δ5.26 (1 *H*) attributed to the C-8 vinylic hydrogen.

Dehydrocancentrine B, on the other hand, is a red alkaloid which also has the same elemental analysis as the A isomer. The NMR spectrum exhibits an AB system due to the C_{31}–C_{32} double bond which is partially masked by the aromatic hydrogens, but one-half of it is clearly seen at δ6.25 (1 *H*) with a coupling constant $J_{AB} = 7.0$ Hz.

Either the A or the B isomer could be reduced with Adams catalyst to cancentrine. Otherwise, dehydrogenation of cancentrine over 5% palladium on carbon in boiling naphthalene gave dehydrocancentrine A and B in low yield. The mass spectra of the two dehydro alkaloids were in accord with the structural assignments.[5]

III. UV Spectroscopy

Cancentrine[2]	λ_{max}^{EtOH} 213, 230 sh, 268, 291 sh, 330 sh, and 435 sh mμ (4.80, 4.63, 4.32, 4.22, 3.62, and 3.82)
Cancentrine methine[2]	λ_{max}^{EtOH} 227, 270, and 435 mμ (4.61, 4.38, and 3.75)
Cancentrine dihydro-methine-*O*-methyl ether[2]	λ_{max}^{EtOH} 208, 230 sh, 270, 330 sh, and 433 mμ (4.57, 4.37, 4.28, 3.61, and 3.77)
Dehydrocancentrine A[5]	λ_{max}^{EtOH} 216, 269, 296 sh, and 445 mμ (4.77, 4.36, 4.29, and 3.87)
Dehydrocancentrine B[5]	λ_{max}^{EtOH} 216 sh, 242, 270 sh, 310 sh, 370, 446, 492, and 525 sh mμ (4.86, 4.78, 4.23, 4.16, 3.90, 4.00, 3.95, and 3.85)

REFERENCES

1. R. H. F. Manske, *Can. J. Res., Sect. B* 7, 258 (1932); **16**, 81 (1938).
2. G. R. Clark, R. H. F. Manske, G. J. Palenik, R. Rodrigo, D. B. MacLean, L. Baczynskyj, D. E. F. Gracey, and J. K. Saunders, *J. Amer. Chem. Soc.* **92**, 4998 (1970).
3. W. Rodionov, *Bull. Soc. Chim. Fr.* [4] **39**, 305 (1926).
4. R. Rodrigo, R. H. F. Manske, D. B. MacLean, L. Baczynskyj, and J. K. Saunders, *Can. J. Chem.* (in press) (1971).
5. D. B. MacLean, L. Baczynskyj, R. Rodrigo, and R. H. F. Manske, *Can. J. Chem.* (in press) (1971).

Chapter 8 / THE DIBENZOPYRROCOLINES

Occurrence: Lauraceae
Number: 2
Structures:

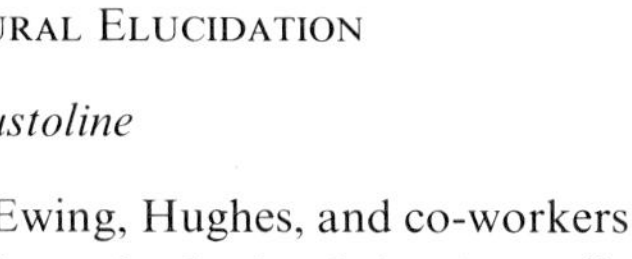

(−)-Cryptaustoline

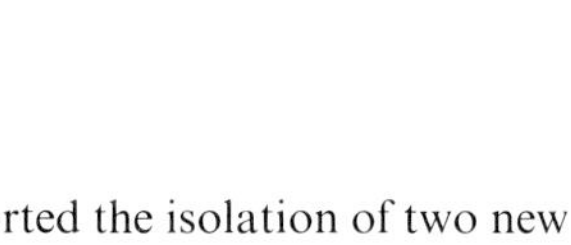

(−)-Cryptowoline

I. STRUCTURAL ELUCIDATION

A. Cryptaustoline

In 1952 Ewing, Hughes, and co-workers reported the isolation of two new quaternary alkaloids from the bark of the Australian shrub *Cryptocaria bowiei* (Hook.) Druce. The two alkaloids, cryptaustoline and cryptowoline, were isolated as the sparingly soluble iodides.[1]

Cryptaustoline, which incorporates one phenolic hydroxyl and three *O*-methyl groups, was first *O*-methylated, and then subjected to successive Hofmann degradations. It can be seen from degradative Scheme I that the isolation of formaldehyde and the *N*-ethyl imide of *m*-hemipinic acid helped to define rings A and B in *O*-methylcryptaustoline.

In the complementary reaction sequence delineated in Scheme II in which the methine **1** was further degraded, the final products obtained were the aldehyde **3** and the amino

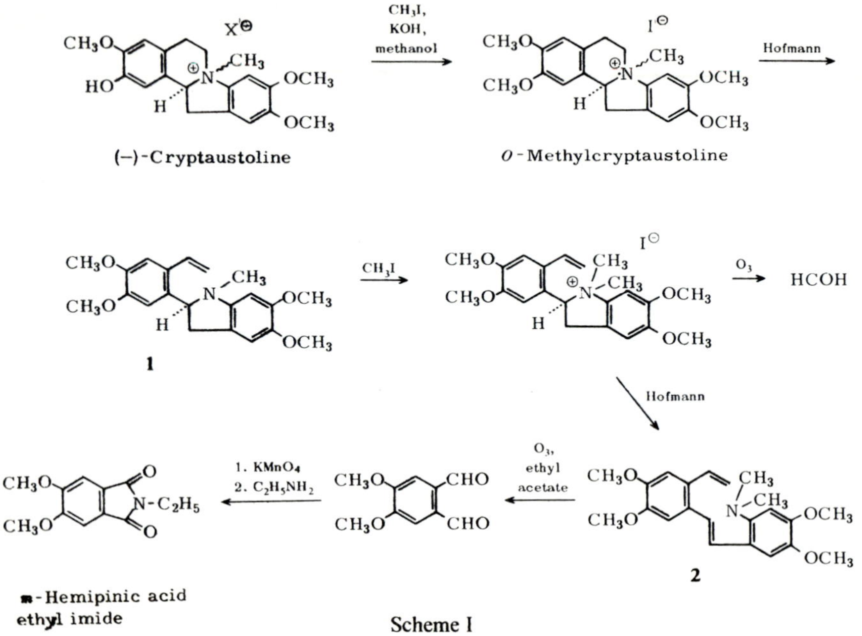

Scheme I

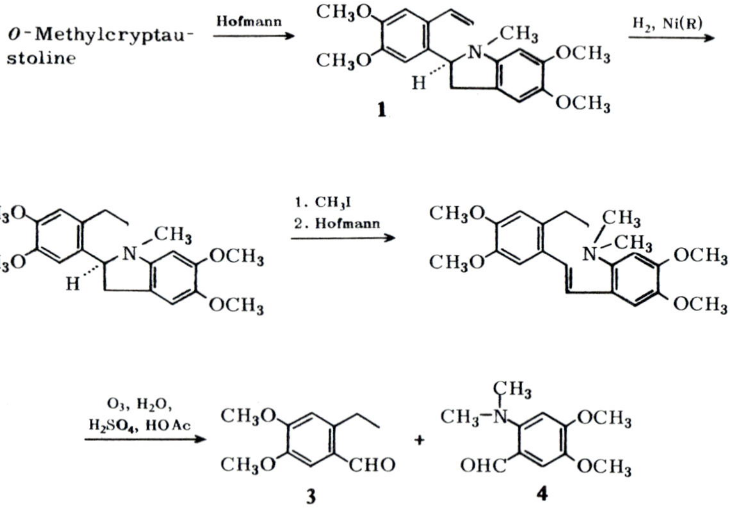

Scheme II

aldehyde **4**, both of which were known compounds. In particular, the identification of the amino aldehyde **4** completely described ring D in *O*-methylcryptaustoline.

Finally, the dihydrostylbene **5**, obtained from the bismethine **2** as shown in Scheme III, was characterized by comparison with an authentic sample.

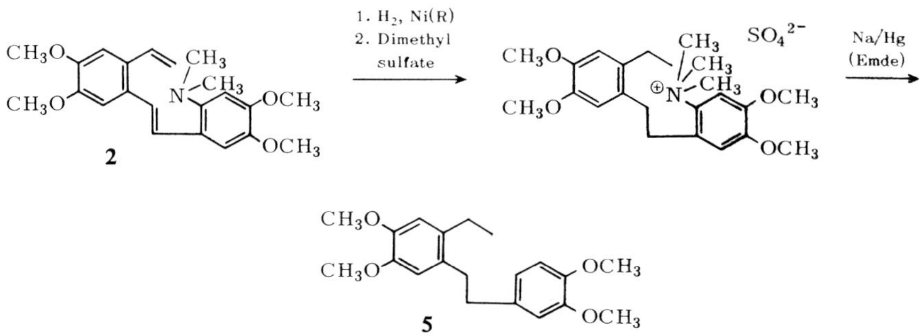

Scheme III

The position of the phenolic function in cryptaustoline was settled by degrading *O*-ethylcryptaustoline to the known 2-ethyl-4-methoxy-5-ethoxybenzaldehyde.[1]

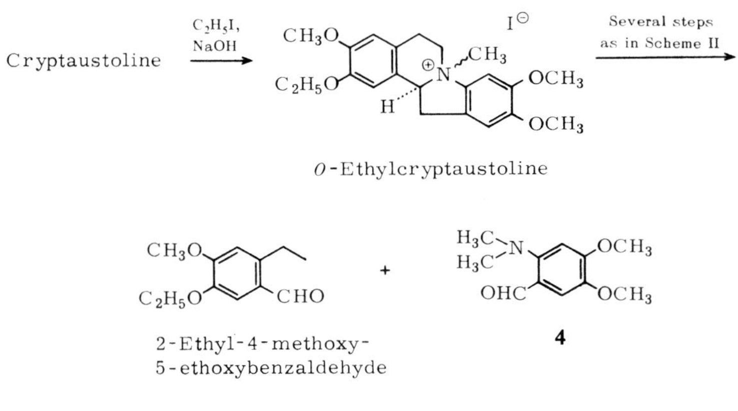

B. *Cryptowoline*

From the outset, it was believed that the only difference between cryptaustoline and cryptowoline was that the latter had a methylenedioxy group in place of two methoxyls in ring D. It was not surprising, therefore, that degradation of *O*-ethylcryptowoline by a route paralleling Scheme II afforded 2-ethyl-4-methoxy-5-ethoxybenzaldehyde and 6-dimethylaminopiperonal.[1]

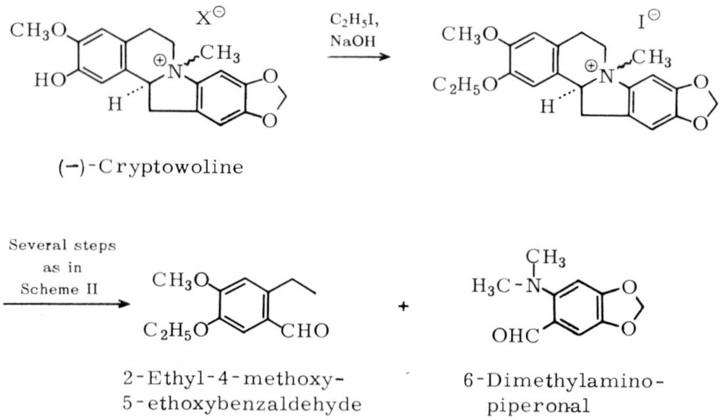

II. Syntheses of Dibenzopyrrocolines

A. *Using an Oxidative Approach*

In 1932, Robinson and Schöpf independently reported on some early efforts at phenolic oxidative coupling.[2,3] Their oxidation of the tetrahydrobenzylisoquinoline laudanosoline did not generate the expected aporphine, but gave instead the dibenzopyrrocoline salt **6**—a tetracyclic system which at that time had not yet been isolated from plant sources.

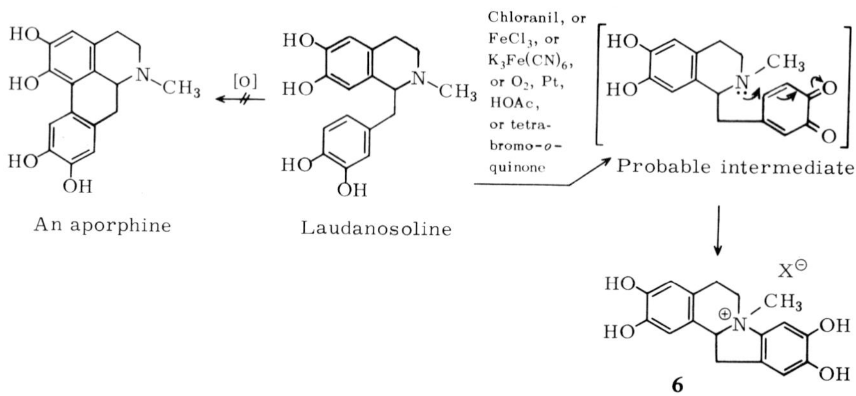

Following their characterization of cryptaustoline and cryptowoline, Hughes and co-workers therefore oxidized (+)-laudanosoline hydrobromide with chloranil to the optically active dibenzopyrrocoline salt **7**. Immediate *O*-methylation of this salt followed by addition of sodium iodide afforded (−)-*O*-methylcryptaustoline iodide, identical with material derived from the natural product. This transformation parallels closely nature's route to the dibenzopyrrocolines and testifies to Robinson's and Schöpf's early insight into alkaloid biogenesis.[4]

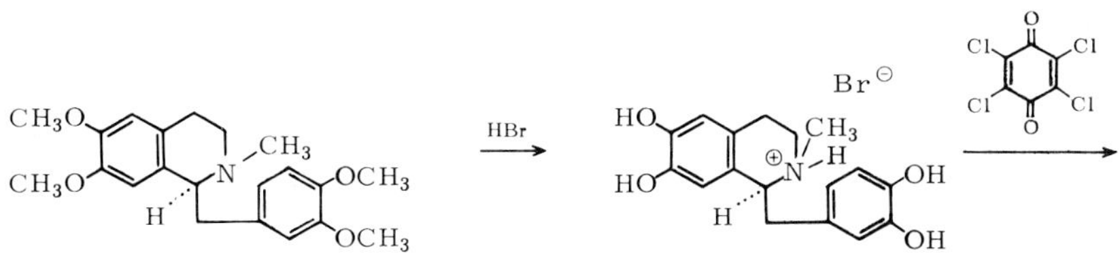

(+)-Laudanosine (+)-Laudanosoline hydrobromide

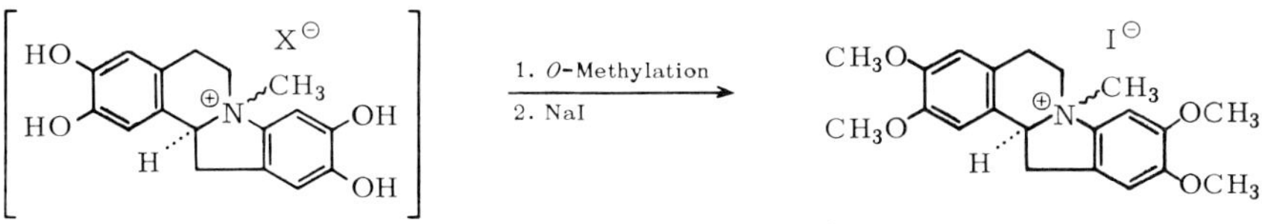

7 (−)-*O*-Methylcryptaustoline iodide

Although the above sequence determined the stereochemistry of (−)-cryptaustoline at C-13, the stereochemistry at N-7 remains unknown.

In a variation on the above synthetic scheme, it has been shown that if a benzylisoquinoline containing a secondary amino group such as norlaudanosoline is oxidized with potassium ferricyanide, the yellow pyrrocoline derivative **8** is formed. This substituted indole readily undergoes further oxidation to give the fluorescent species **9**. The facile complete aromatization of **8** is of interest since 3,4-dihydroisoquinolines are usually not readily dehydrogenated.[5]

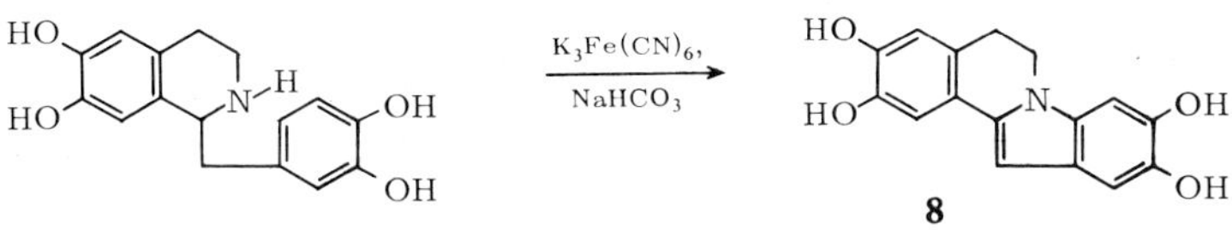

Norlaudanosoline **8**

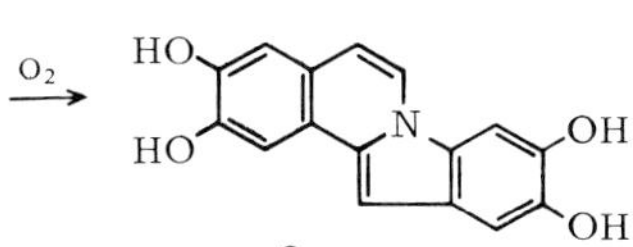

9

B. Using a Benzyne Intermediate

Cyclization of the brominated benzylisoquinoline **10** using sodium amide afforded the dibenzopyrrocoline free base **11**. (±)-Cryptaustoline iodide was then obtained by *N*-methylation followed by acid hydrolysis and treatment with potassium iodide.[6]

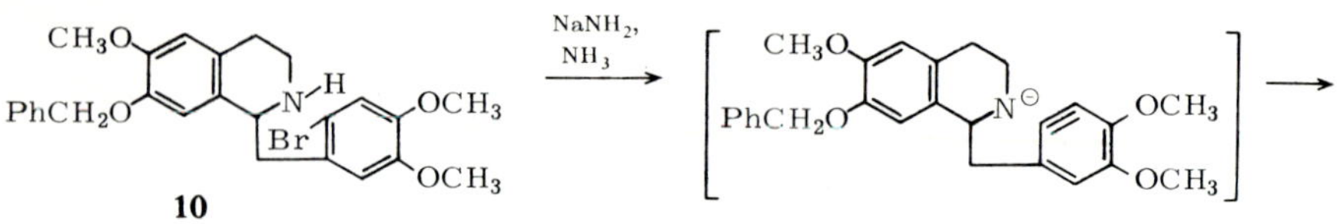

10

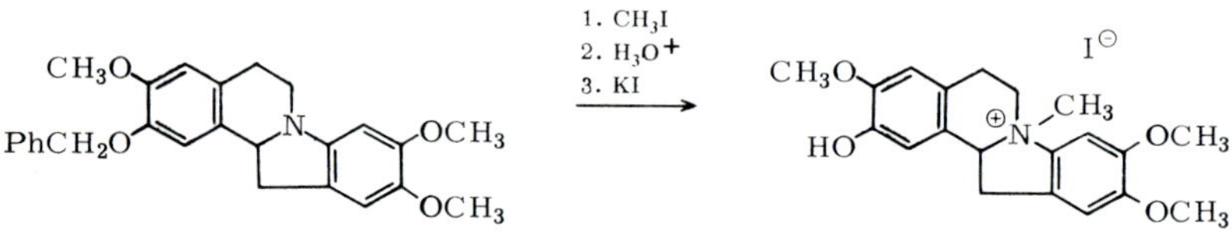

11

(±)-Cryptaustoline iodide

The same approach was used for the synthesis of ($\pm$)-cryptowoline except that the starting material was the chlorinated benzylisoquinoline **12**.[6]

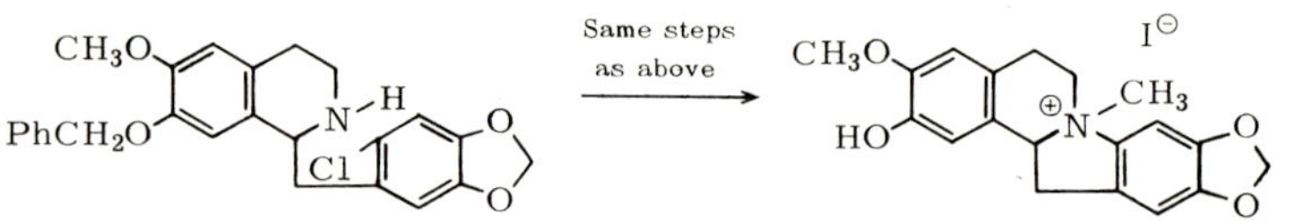

12

(±)-Cryptowoline iodide

In the same year in which the aforementioned work was published, the details of a very closely related but independent preparation of cryptowoline iodide were described. The optically active bromobenzylisoquinoline **13** was cyclized with potassium amide in liquid ammonia through the intermediacy of a benzyne. But, epimerization occurred, so that the final product was racemic cryptowoline.[7]

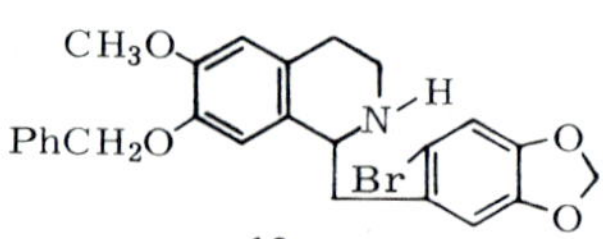

13

III. Mass Spectroscopy

The mass spectrum of cryptaustoline iodide has been recorded.[8] The ion of highest mass corresponds to the loss of methyl iodide from the salt and may be labeled the molecular ion. There is a strong peak at *m/e* 326 due to the loss of a hydrogen atom from the molecular ion, but the base peak is due to the loss of two hydrogens from the molecular ion (Scheme IV).

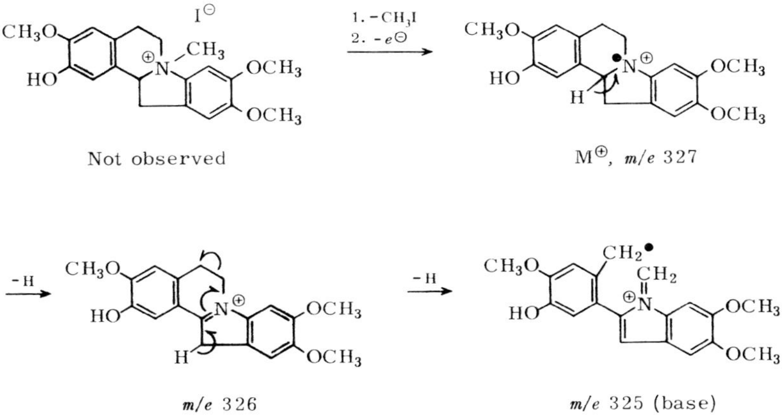

Scheme IV

Some further transformations that the *m/e* 325 ion can undergo are outlined in Scheme V.

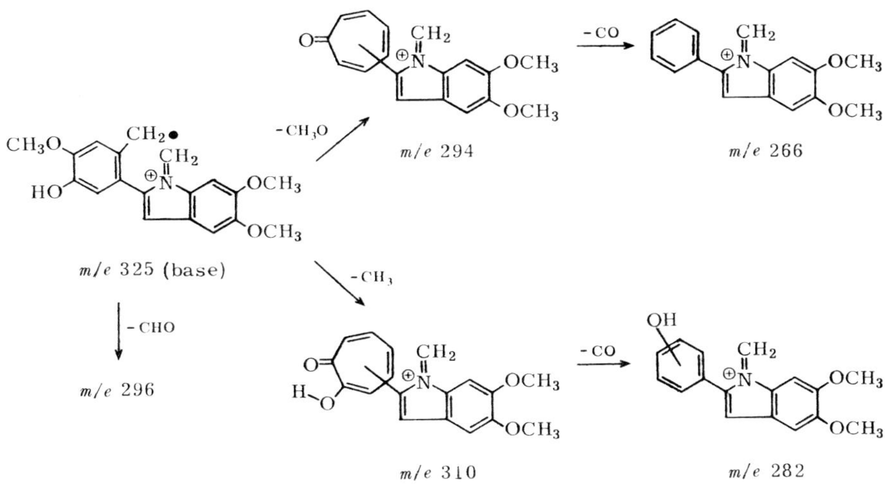

Scheme V

IV. NMR Spectroscopy

The following data were obtained in deuteriochloroform solution for the free base **14**: six methylene hydrogens as multiplets, $\delta 2-3.8$; a quartet at $\delta 4.79$ representing the C-13 hydrogen; four *O*-methyl singlets situated at δ3.74, 3.79, 3.87, and 3.87; and finally, four aromatic hydrogen singlets at δ6.31, 6.49, 6.68, and 6.71.[6]

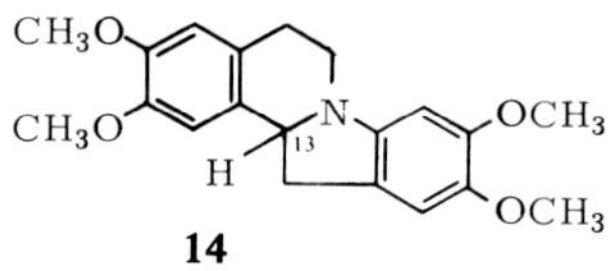

14

REFERENCES

1. J. Ewing, G. K. Hughes, E. Ritchie, and W. C. Taylor, *Nature* (*London*) **169**, 618 (1952); *Aust. J. Chem.* **6**, 78 (1953).
2. R. Robinson and S. Sugasawa, *J. Chem. Soc., London* p. 789 (1932).
3. C. Schöpf and K. Thierfelder, *Justus Liebigs Ann. Chem.* **497**, 22 (1932).
4. G. K. Hughes, E. Ritchie, and W. C. Taylor, *Aust. J. Chem.* **6**, 315 (1953).
5. J. Harley-Mason, *J. Chem. Soc., London* p. 1465 (1953).
6. T. Kametani and K. Ogasawara, *J. Chem. Soc., C* p. 2208 (1967).
7. F. Benington and R. D. Morin, *J. Org. Chem.* **32**, 1050 (1967).
8. T. Kametani and K. Ogasawara, *Chem. Pharm. Bull.* **16**, 1498 (1968).

Chapter 9 / THE PROAPORPHINES

Occurrence: Euphorbiaceae, Lauraceae, Menispermaceae, Monimiaceae, Nymphaceae, and Papaveraceae

Approximate Number: 23

Some Proaporphines of Interest:

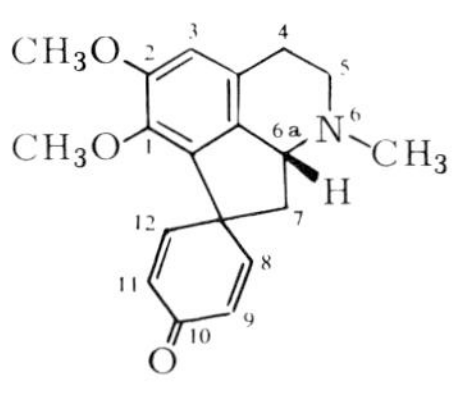

(+)-Pronuciferine

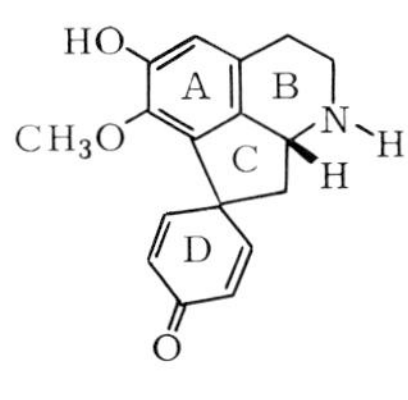

(+)-Crotonosine

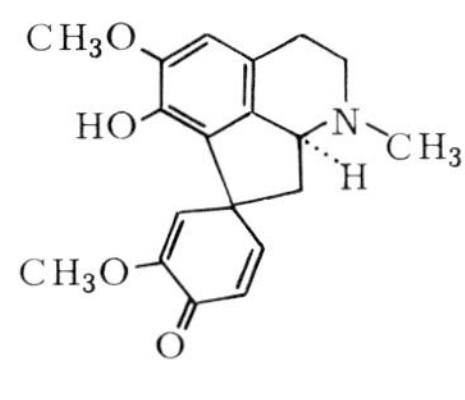

(−)-Orientalinone

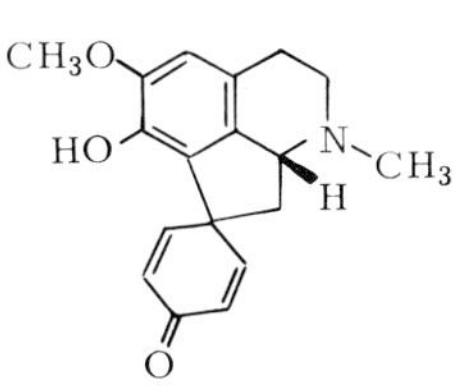

(+)-Glaziovine

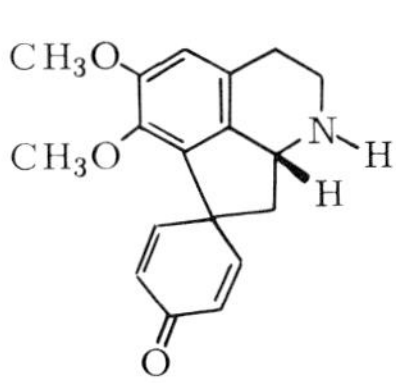

(+)-Stepharine

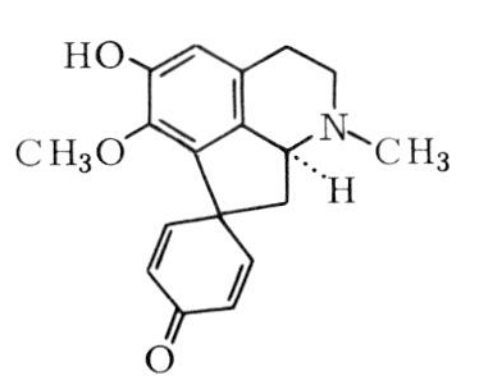

(−)-*N*-Methylcrotonosine

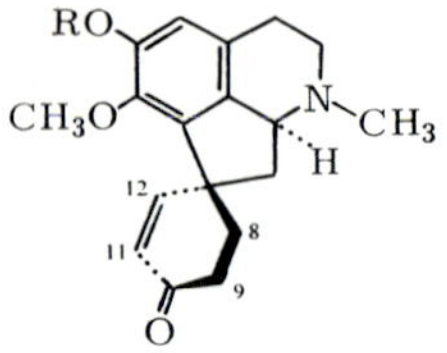

(+)-Linearisine, R = H
(+)-Amuronine, R = CH_3

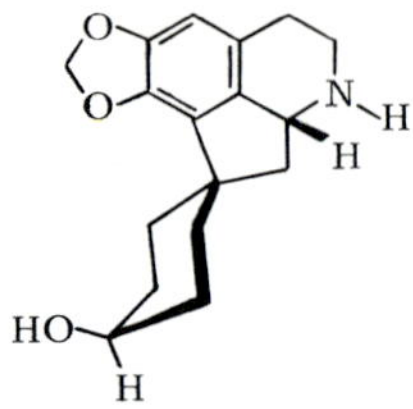

(+)-Litserisine

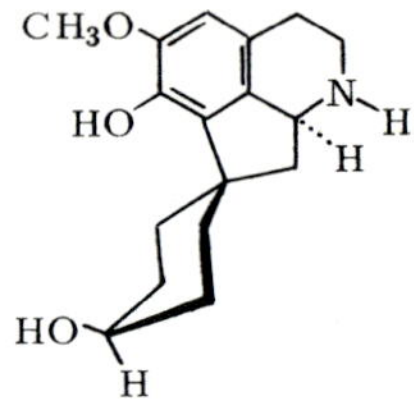

(−)-Oreoline

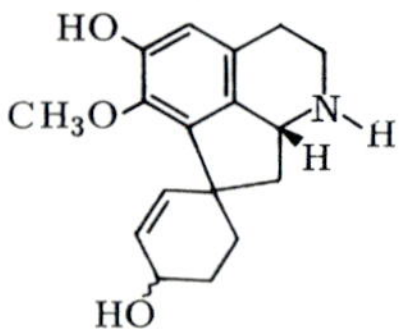

(−)-Jaculadine[1]

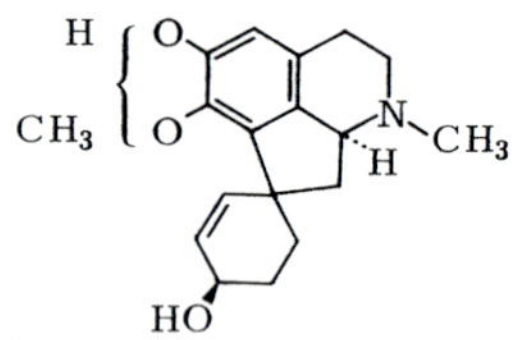

(+)-Discolorine[1]

I. Introduction

The proaporphine alkaloids occur in a variety of oxidation states.[2,2a] (+)-Pronuciferine possesses a cyclohexadienone system which is present in many other proaporphines such as (+)-crotonosine and (−)-orientalinone. There also exists a series of reduced proaporphines exemplified by (+)-linearisine, (+)-litserisine, and (+)-discolorine in which the double bonds of the original dienone system have been partially or completely reduced.

The proaporphine numbering system is as indicated for (+)-pronuciferine. Because of the asymmetry at C-6a one side of the dienone system is not equivalent to the other. The lower numbers (C-8 and C-9) are assigned to the side of the dienone above the mean plane of the molecule, and the higher numbers (C-11 and C-12) to the side below, as shown for (+)-linearisine.

II. Structural Elucidation

The first two proaporphines to have their structures elucidated were pronuciferine, studied by Bernauer,[3] and crotonosine, investigated by Barton, Stuart, and co-workers.[4] The structures of both alkaloids were revealed in 1963.

A. *Pronuciferine*

The alkaloid (+)-pronuciferine, $C_{19}H_{21}O_3N$, has been found in *Nelumbo nucifera* Gaertn. (Nymphaceae), *Croton linearis* Jacq. (Euphorbiaceae), *Stephania glabra* Miers (Menispermaceae), and in a variety of *Papaver* species.

Its chemistry is typical of that of any proaporphine possessing a dienone system, and the more important reactions are shown in Scheme I.[3]

(a) With mineral acid, (+)-pronuciferine underwent a dienone–phenol rearrangement to (−)-1,2-dimethoxy-10-hydroxyaporphine.

(b) Upon reduction of (+)-pronuciferine with sodium borohydride, two dienols,

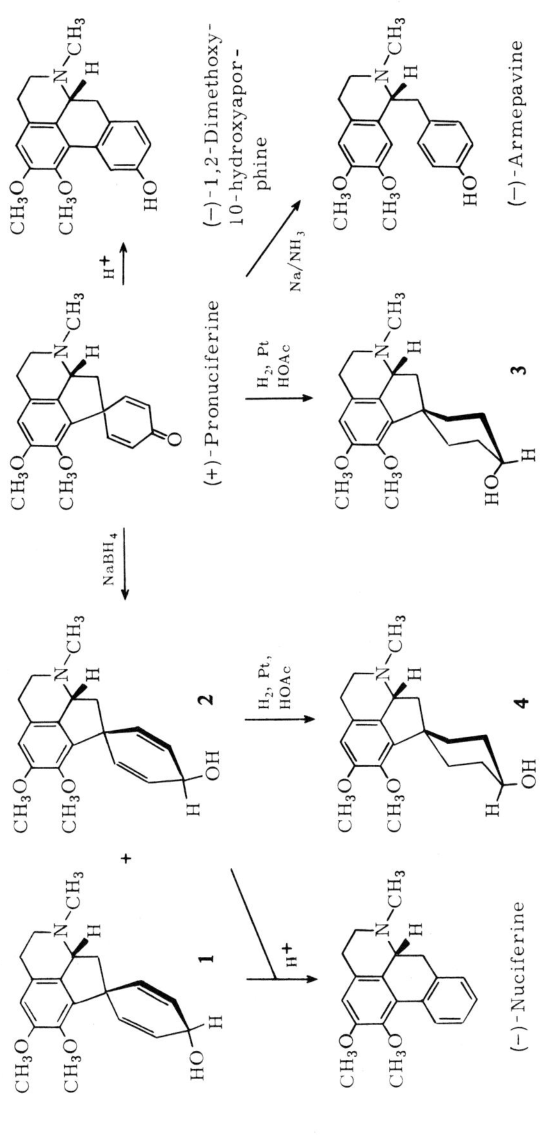

(+)-Pronuciferine
NaBH4
H+
(−)-1,2-Dimethoxy-10-hydroxyaporphine
H2, Pt HOAc
Na/NH3
(−)-Armepavine
1
2
3
4
(−)-Nuciferine

Scheme I

1 and **2**, were obtained. These in mineral acid underwent a dienol–benzene rearrangement to yield the aporphine (−)-nuciferine.

(c) (+)-Pronuciferine could be reduced with Adams catalyst to a hexahydro derivative, formulated as **3**, assuming that the catalyst approached from the less-hindered side of the molecule.

(d) Alternatively, dienols **1** and **2** could be separated chromatographically. They were assigned the stereochemistry indicated when it was found that upon catalytic reduction, alcohol **2** yielded a hexahydropronuciferine (**4**) different from its diastereoisomer **3** of established stereochemistry.[3]

(e) At a later date, it was determined that proaporphines with a dienone system may be hydrogenolyzed by means of sodium in liquid ammonia to yield benzylisoquinolines. In the present case, reduction of (+)-pronuciferine afforded (−)-armepavine.[5]

B. Crotonosine

As with pronuciferine, the structural elucidation of (+)-crotonosine, $C_{17}H_{17}O_3N$, originally obtained from *Croton linearis* Jacq., involves the study of the dienone–phenol and the dienol–benzene rearrangements. These parallel closely the pronuciferine case and will not be discussed in detail.

Noteworthy is the procedure for the determination of the position of the phenolic function in ring A of (+)-crotonosine. The noraporphine (−)-apocrotonosine, obtained from the dienone–phenol rearrangement of (+)-crotonosine, was subjected to base-catalyzed deuterium exchange. NMR spectroscopy then indicated that three aromatic hydrogens had been replaced by deuteriums. Since only those hydrogens ortho or para to phenolic functions undergo exchange, it follows that the ring A phenolic function must be at C-2 rather than at C-1.[4,6]

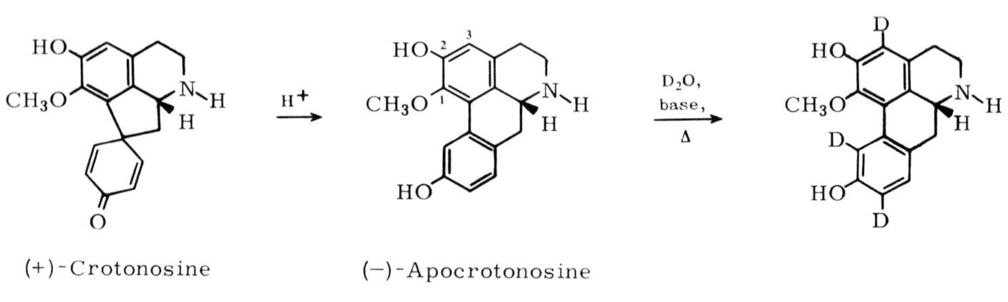

(+)-Crotonosine (−)-Apocrotonosine

C. Orientalinone

The *Papaver* alkaloid (−)-orientalinone, $C_{19}H_{21}O_4N$, is unusual in that it had actually been synthesized in the racemic form and had had its chemistry investigated prior to the isolation of the natural product. Reduction of (±)-orientalinone with sodium borohydride followed by acid-catalyzed dienol–benzene rearrangement produced the aporphine (±)-isothebaine.[7]

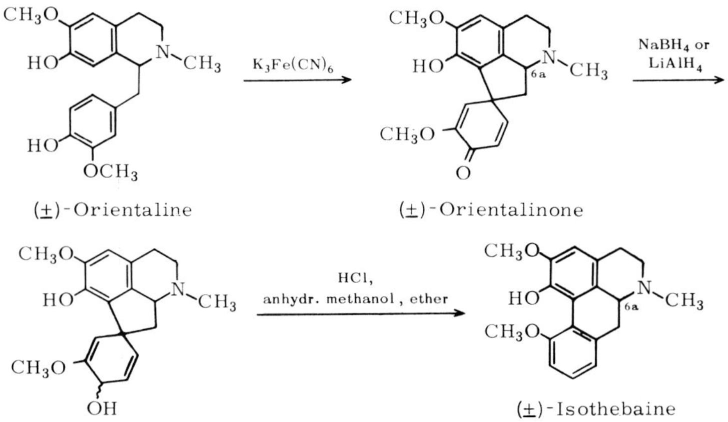

In a later experiment, synthetic (−)-orientalinone (α-H at C-6a), corresponding to the natural product and derived from (+)-orientaline, was reduced with sodium borohydride to a mixture of dienols. Without separation, these were rearranged in acid to (+)-isothebaine (α-H at C-6a), identical with the natural base.[8,8a]

D. The Reduced Proaporphines

The structural elucidation of the reduced proaporphines is based primarily upon their interrelationship with proaporphines possessing dienone systems. For example, the structure of (+)-linearisine was obtained through comparison of (−)-dihydrolinearisine with the known and enantiomeric (+)-*N*-methyltetrahydrocrotonosine (Scheme II).[9] The configuration of (+)-linearisine at C-7a was settled by the use of circular dichroism – a topic which will be covered in Section VI of this chapter.

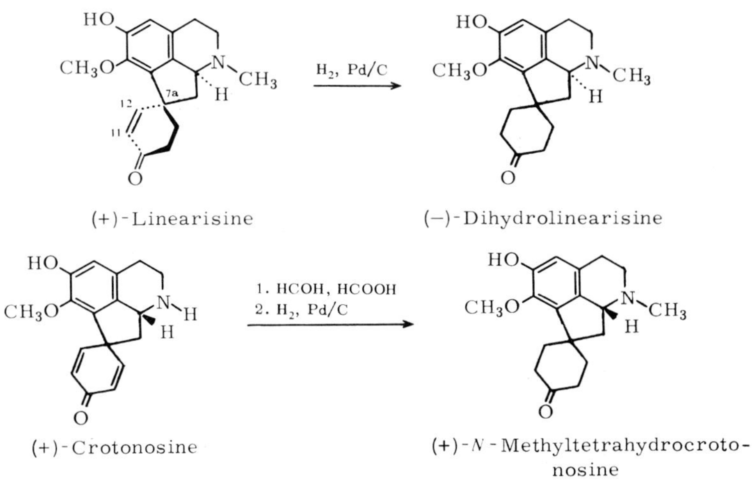

Scheme II

In the case of the cyclohexanolic proaporphine (−)-oreoline, also known as (−)-oridine, the alkaloid was converted to its *O,N*-dimethyl methiodide derivative **5**, and this salt was compared with racemic hexahydropronuciferine methiodide which is identical with the methiodide of racemic base **3**. Additionally, (−)-oreoline upon treatment with acid lost the elements of water and afforded the polycyclic nonphenolic base **6** which could readily be *N*-acetylated.[10]

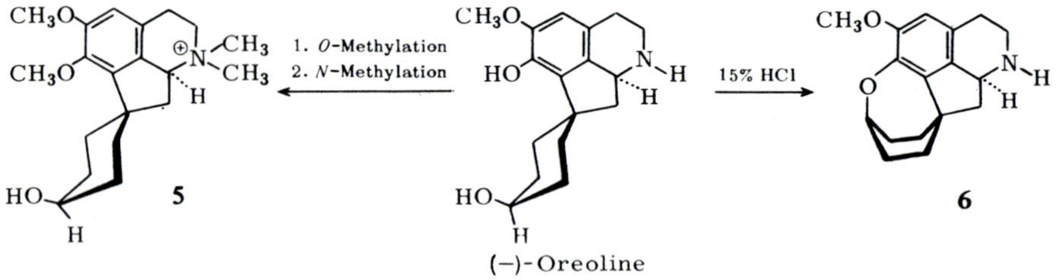

III. NONBIOGENETIC SYNTHESES

A. An Approach Using Methyl Vinyl Ketone

An early synthesis of a proaporphine is that of pronuciferine carried out by Bernauer. The known amino ester **7** was converted to the tricyclic ketone **8**. Homologation through a Darzens glycidic ester sequence provided the aldehyde **9**, which was condensed in the presence of potassium *t*-butoxide and methyl vinyl ketone to give (±)-amuronine, a naturally occurring base in the dextrorotatory base. Oxidation with DDQ then furnished (±)-pronuciferine (Scheme III).[11]

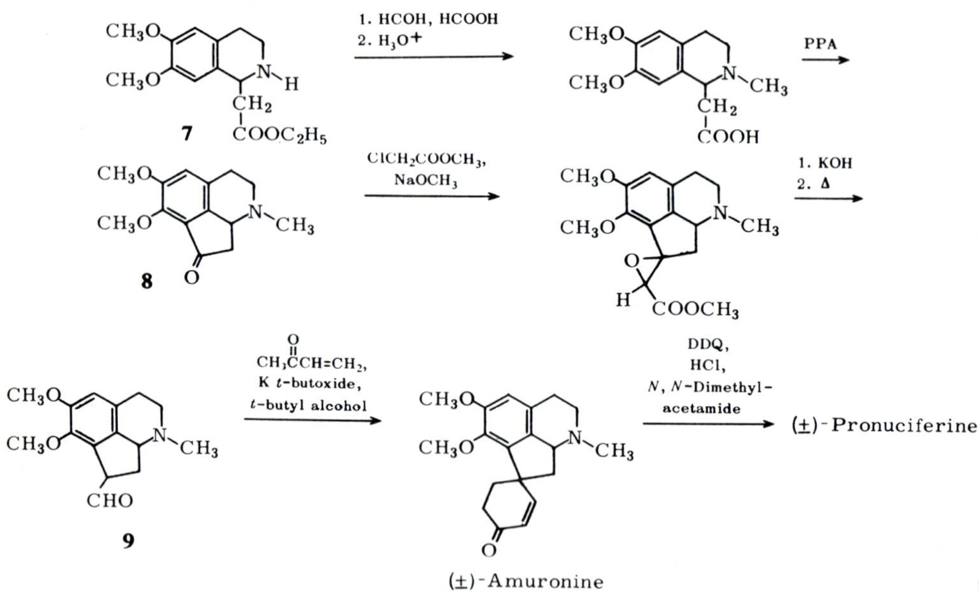

Scheme III

B. Friedel–Crafts Acylation and Pomeranz–Fritsch Cyclization

In this approach, developed by Huffman and Opliger, a Friedel–Crafts acylation succeeded by a Bobbitt modification of the Pomeranz–Fritsch cyclization sequence supplied hexahydrostepharine which was *N*-methylated to one of the two known hexahydropronuciferines (Scheme IV).[12]

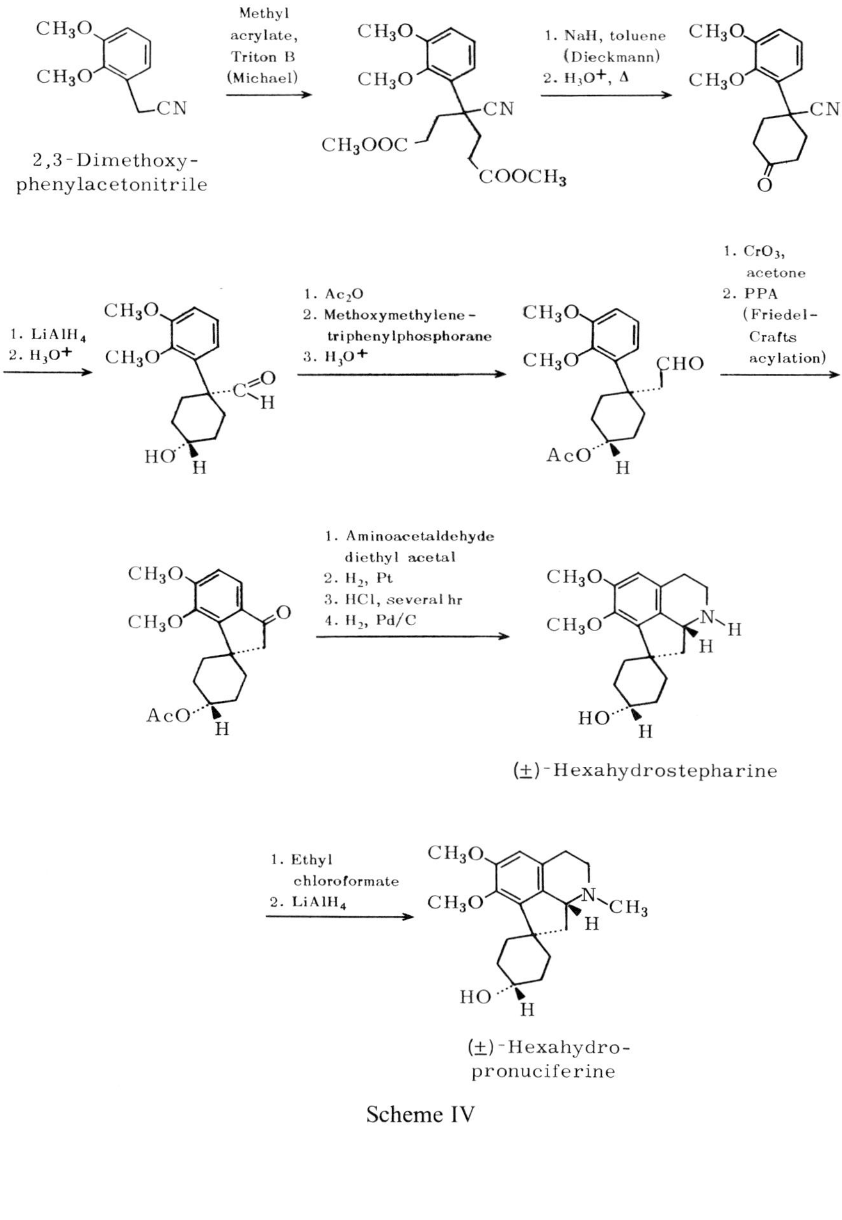

Scheme IV

C. Pschorr Cyclization

When the 8-aminobenzyltetrahydroisoquinoline **10** was diazotized and the diazonium salt decomposed with an excess of sodium acetate at room temperature over a period of 3 hours, two compounds were isolated, the expected (±)-2-hydroxy-1,10-dimethoxyaporphine in trace amounts and (±)-homolinearisine, the latter in 10% yield. The naturally occurring base (−)-homolinearisine corresponds to (−)-*N*-methylcrotonosine.[13–15]

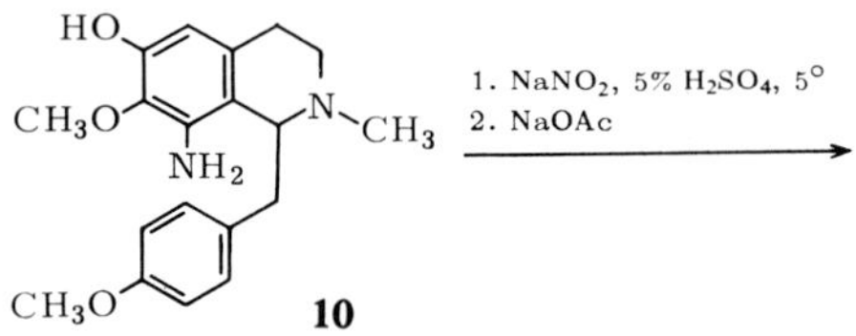

10

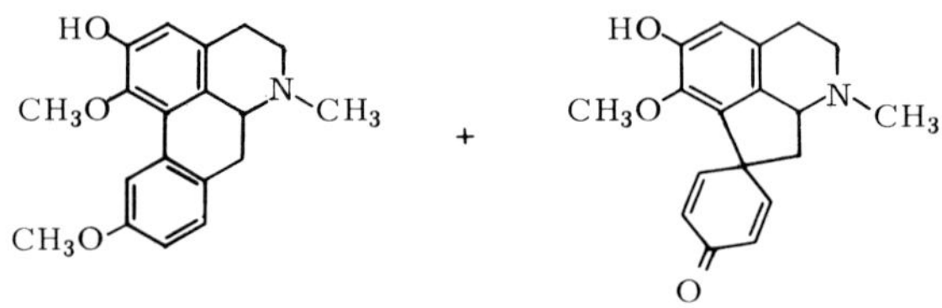

(±)-2-Hydroxy-1,10-dimethoxyaporphine (±)-Homolinearisine

D. A Photochemical Route

Irradiation of the brominated tetrahydrobenzylisoquinoline **10a** with a high pressure mercury lamp for two hours in the presence of sodium borohydride led to a dienol which furnished (±)-pronuciferine on oxidation with manganese dioxide.[14a]

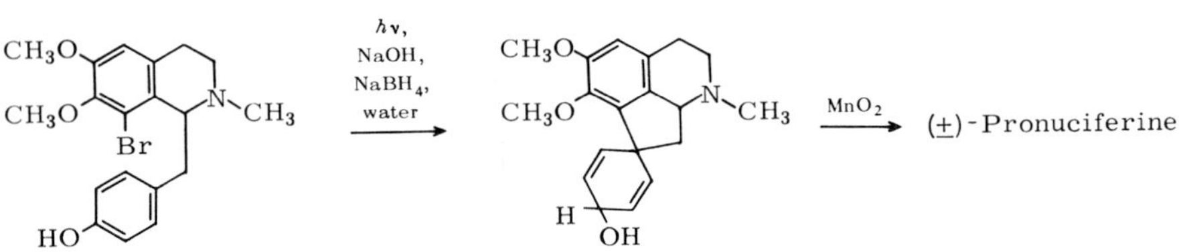

10a

Irradiation of the same material, **10a**, in the absence of sodium borohydride led directly to (±)-pronuciferine in 10% yield. When copper powder was added, and methanol replaced with ethanol, the yield increased to 17%.[14b]

Photolysis of the brominated tetrahydroisoquinoline **10b** supplied a mixture of diastereoisomeric *O*-methylorientalinones,[14b] and photolysis of **10c** gave (±)-glaziovine.[14c]

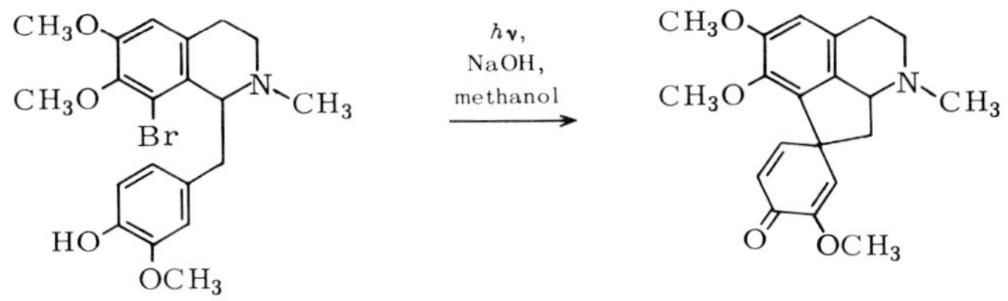

10b — *O*-Methylorientalinone diastereoisomers

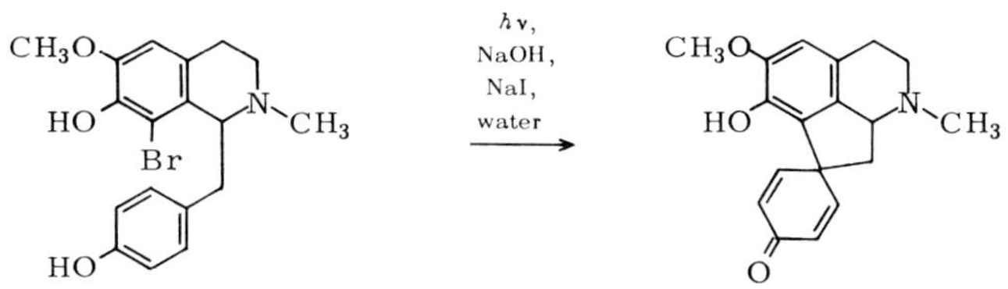

10c — (±)-Glaziovine (10%)

IV. Biogenetic Studies Using Phenolic Oxidative Coupling

Any adequate treatment of the proaporphines must cite the historically important paper published by Barton and Cohen in 1957 discussing the role of phenolic oxidative coupling in alkaloid biogenesis.[16] Noting that certain aporphines were either unsubstituted or only monosubstituted in ring D, it was postulated that the biogenesis of these compounds had to proceed through what were at that time hypothetical intermediates – the species presently known as proaporphines. Such dienones could rearrange in acid to aporphines or be reduced first to dienols, which could subsequently undergo rearrangement to aporphines lacking a ring D substituent.

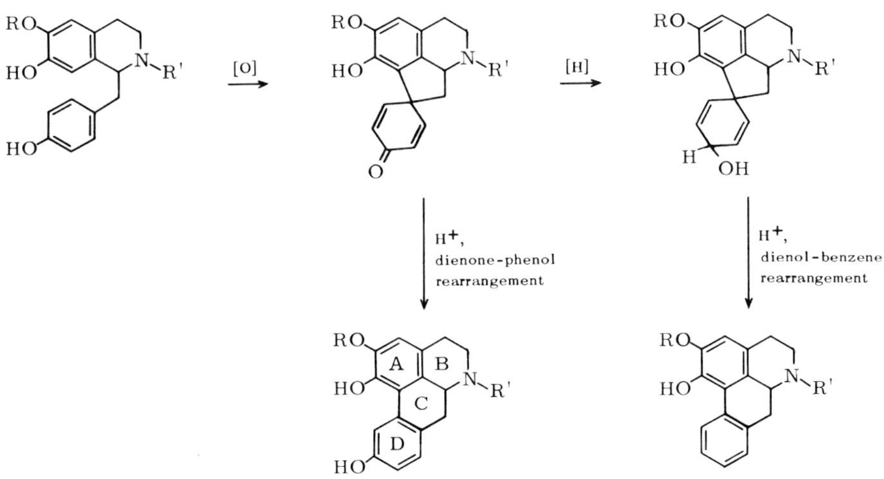

This suggestion received impressive support in 1963 when the structures of crotonosine and pronuciferine were elucidated. Since that time, additional experimental data have been forthcoming so that there can be no doubt of the existence in nature of the sequence benzylisoquinoline → proaporphine → aporphine.

In the first synthesis of a proaporphine through phenolic oxidative coupling, potassium ferricyanide treatment of (±)-orientaline afforded a mixture of the proaporphine (±)-orientalinone and its diastereoisomer.[7,17] Interestingly enough, two years later the isolation from *Papaver orientale* L. (Papaveraceae) of this same dienone in the levorotatory form was reported.[8a,18]

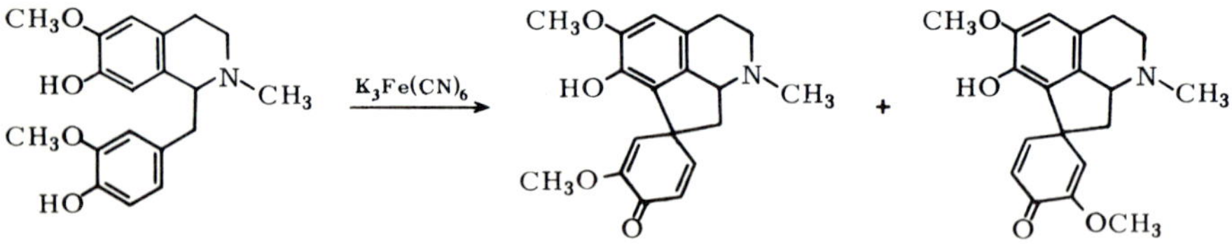

(±)-Orientaline (±)-Orientalinone Diastereoisomer of (±)-orientalinone

Another synthesis of (±)-orientalinone involved the oxidative coupling of the benzylisoquinoline **11** to yield the dienones **12** and **13**. Subsequent reduction of one of these two dienones, here represented by expression **13**, with sodium borohydride produced a mixture of dienols which was not separated but was converted to orientalinone by aqueous acid. Reaction of orientalinone with concentrated hydrochloric acid in acetic acid gave rise to the aporphine (±)-isocorytuberine, while under anhydrous acid conditions (±)-corydine was generated (Scheme V).[19] (See also this chapter, Section II, C and Chapter 10, Section VIII, C.)

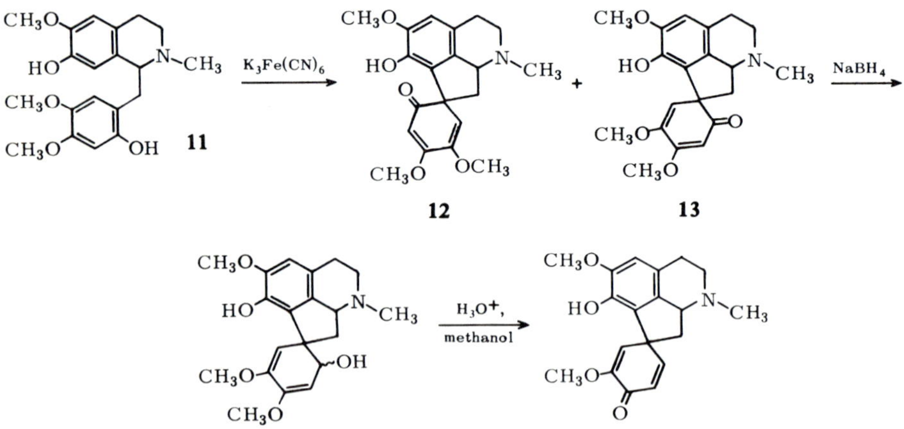

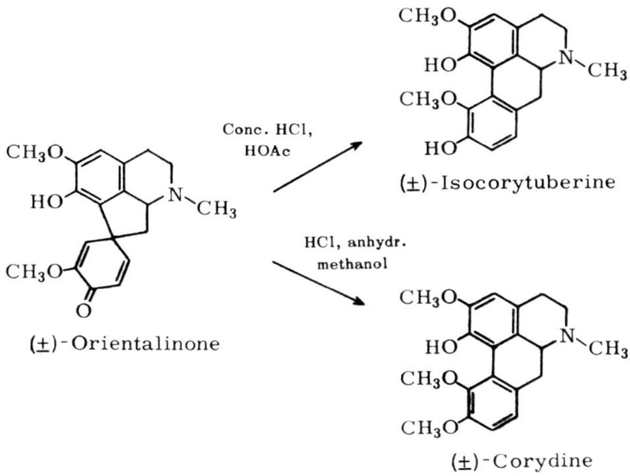

Scheme V

In a separate study, the mixture of diastereoisomeric dienones **12** and **13** was *O*-methylated with diazomethane and then reduced with sodium borohydride. Treatment of the resulting dienol mixture **14** with aqueous methanolic hydrochloric acid gave (±)-*O*-methylorientalinone and the aporphine (±)-pseudocorydine. Alternatively, reaction of the mixture **14** with hydrogen chloride in dry methanol led to the aporphines (±)-glaucine and (±)-*O*-methylcorydine (Scheme VI).[20]

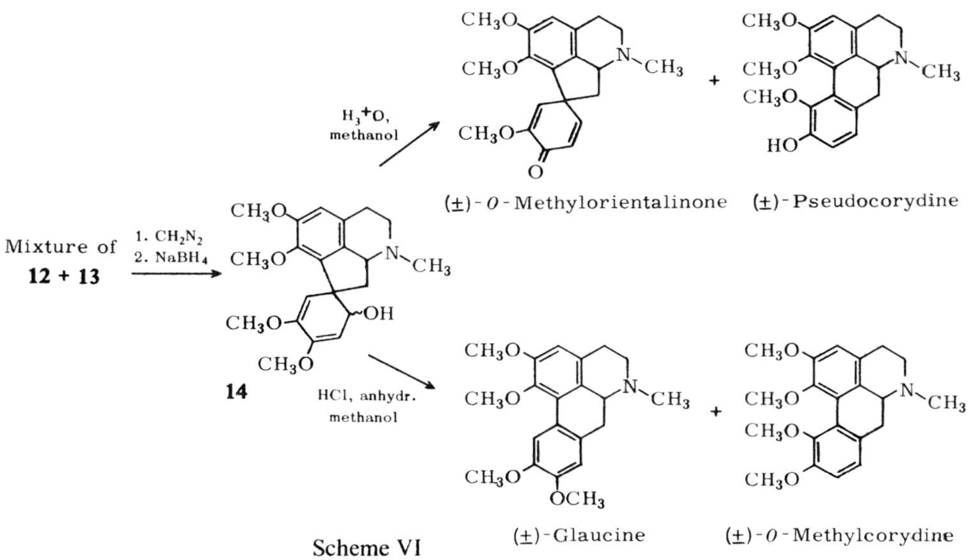

Scheme VI

When (±)-*N*-methylcoclaurine was oxidized with potassium ferricyanide, the known natural optically active proaporphine glaziovine was obtained in the racemic form and in low yield. *O*-Methylation then furnished (±)-pronuciferine.[21]

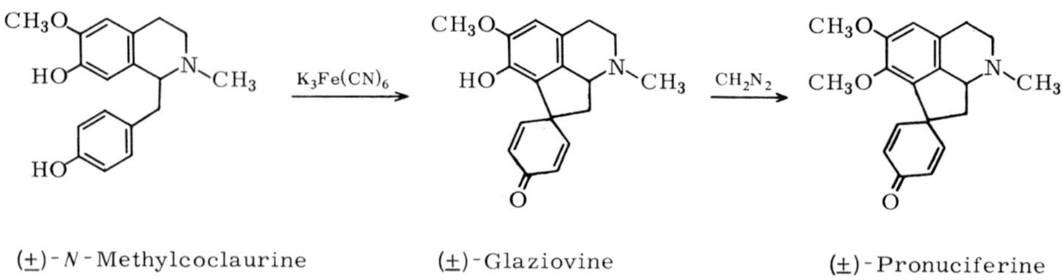

(For another example of the oxidative coupling of a benzylisoquinoline to a dienone see Chapter 5, Section IX.)

V. Biogenetic Studies with Labeled Compounds

Barton, Haynes, and co-workers have shown that when (+)-coclaurine, labeled with tritium as shown, is fed to *Croton linearis* Jacq., radioactive crotonosine in which the label is alpha to the carbonyl group is isolated. Furthermore, only the dextrorotatory isomer and not the levorotatory species was an efficient precursor for (+)-crotonosine.[22]

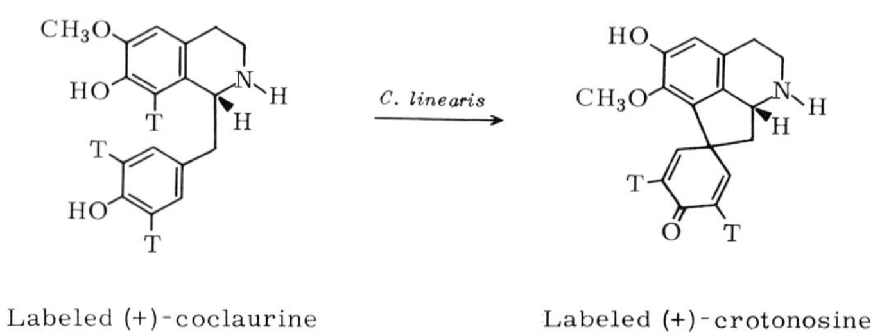

Labeled (+)-isococlaurine (**15**) was not incorporated in *C. linearis*. This negative result supports Barton's theory, which demands cyclization only ortho and para to the free phenolic hydroxyl groups during oxidative cyclization. It is apparent, therefore, that the transformation of the proaporphine **16** to (+)-crotonosine must occur in the plant and that it involves a demethylation–remethylation mechanism, the details of which are not presently clear.[23]

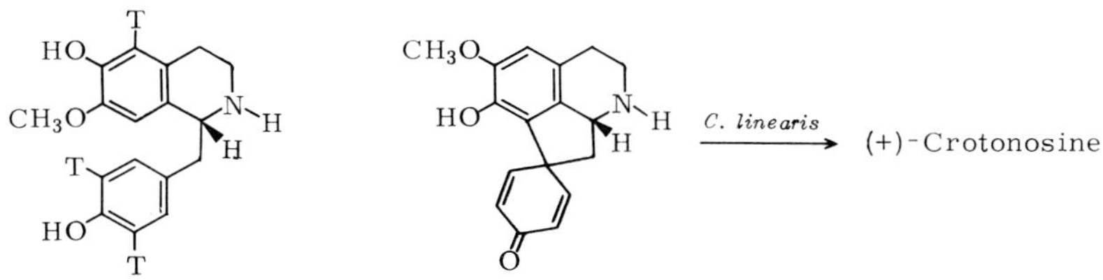

15 **16**

In parallel experiments with *Papaver dubium* L., tritium-labeled (+)-*N*-methylcoclaurine was efficiently incorporated into the aporphine alkaloid (+)-roemerine although the exact position of the label in the product was not specifically determined.[24]

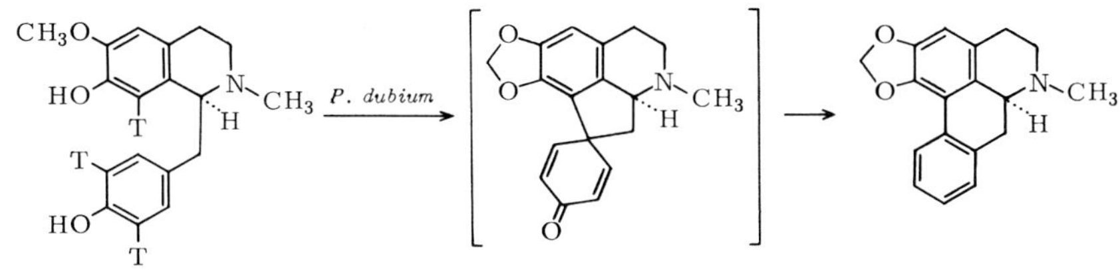

Labeled (+)-*N*-methyl coclaurine

Labeled (+)-roemerine

Labeled (+)-orientaline, when fed to *Papaver orientale* L., was incorporated 28 times more efficiently into the aporphine alkaloid (+)-isothebaine than the levo enantiomer; therefore, the expected steric relationship obtains between precursor and product.[8a,25] Noteworthy is the previously mentioned fact that (−)-orientalinone has been found among the alkaloids present in *P. orientale* (Scheme VII).[17,25]

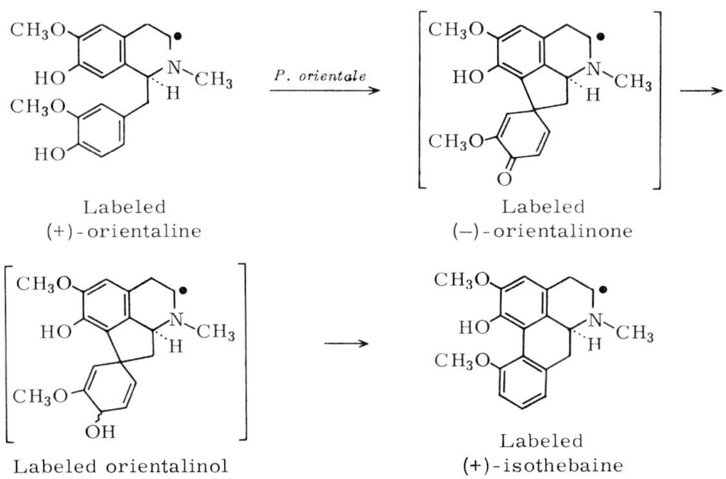

Scheme VII

It has also been observed that *C. linearis* Jacq. can achieve the conversion of labeled (+)-linearisine into (+)-crotonosine, although in poor yield (0.01–0.03%).[25a] This transformation requires isomerization of the C-6a asymmetric center as well as *N*-demethylation, and may not be one of the principal biogenetic pathways in the plant.

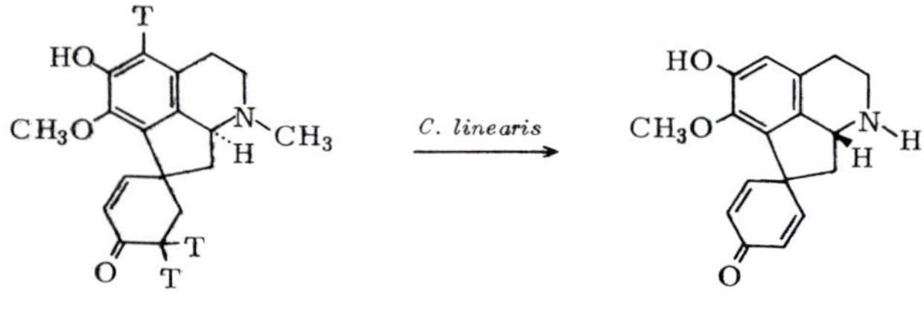

Labeled (+)-linearisine → Labeled (+)-crotonosine

(For a more complete discussion of aporphine biogenesis see Chapter 10, Section VIII, C.)

VI. Absolute Configuration and Circular Dichroism

The absolute configurations of simple proaporphines such as pronuciferine and crotonosine can be readily determined either through acid-catalyzed rearrangement to aporphines of known absolute configuration or through sodium in liquid ammonia reduction to benzylisoquinolines of established stereochemistry.

The proaporphines containing a cyclohexenone system, such as (+)-linearisine, present a more complex problem. As indicated in Section II, D, the absolute configuration at C-6a in (+)-linearisine was determined by comparing dihydrolinearisine with a derivative of (+)-crotonosine. The stereochemistry at C-7a was then obtained by Snatzke and Wollenberg from a detailed study of the CD curve of the alkaloid. Molecular models and NMR chemical shifts indicated that the most favored configurations for (+)-linearisine could be represented by expressions **17** and **18**. From comparisons with known enone systems in the steroid series, it was deduced that configuration **17** should give rise to a positive Cotton effect near 334 mμ, whereas the curve for **18** should be negative. The fact that the Cotton effect for linearisine was positive indicated the correctness of expression **17**.[26]

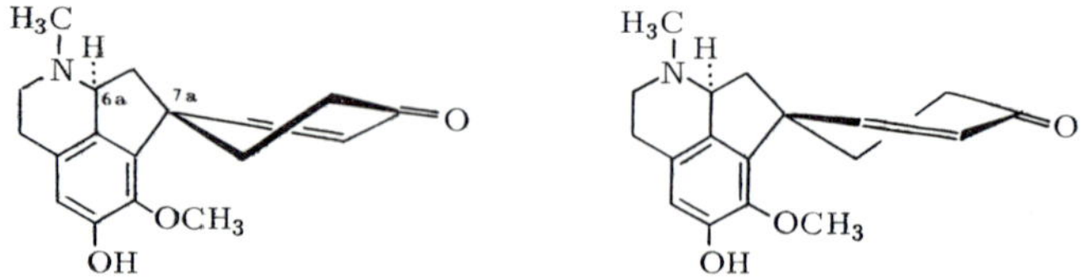

17 **18**

(+)-Linearisine

A study of the optical rotatory dispersion curves of the proaporphines has also been carried out, allowing the assignment of absolute configuration at C-6a to the alkaloid (−)-oreoline.[27]

Pronuciferine is now known to occur naturally both in the dextro- and the levorotatory forms.[28]

VII. Pharmacology

(+)-Pronuciferine and (+)-crotonosine may have some potential as local anesthetics,[29] while the alkaloid (+)-glaziovine has shown antidepressive activity.[30]

VIII. Mass Spectroscopy

The molecular ion peak is usually the base peak for the proaporphines. The main peaks shown in the mass spectrum of pronuciferine are rationalized in Scheme VIII.[31,32]

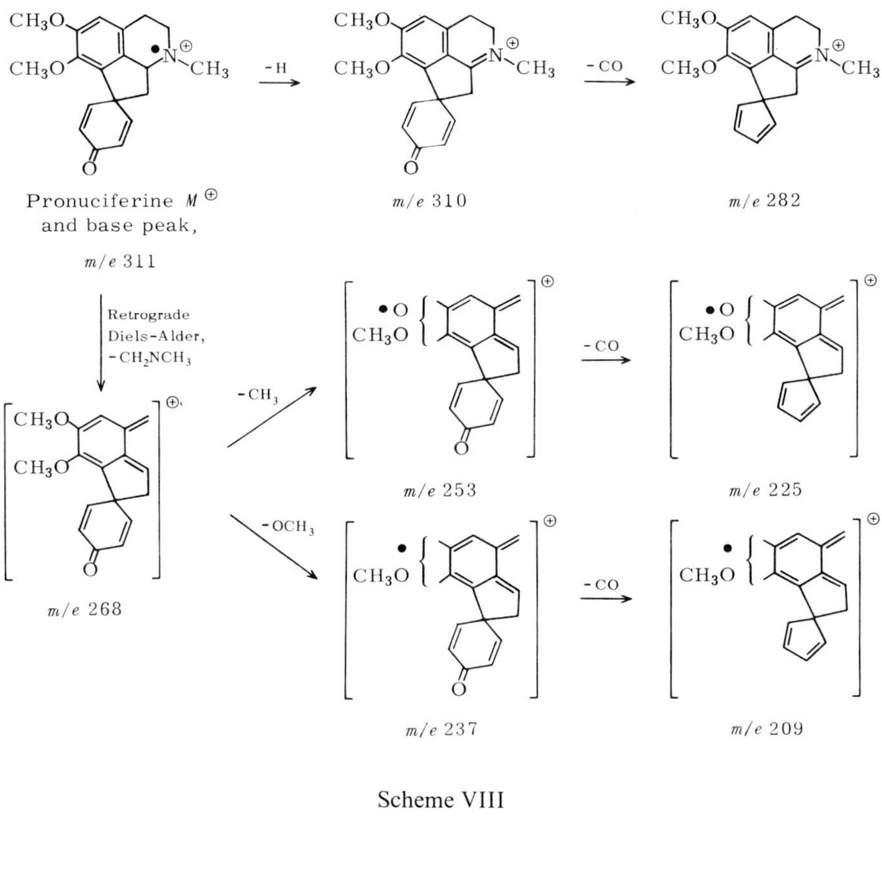

Scheme VIII

IX. NMR Spectroscopy

For the proaporphines of the dienone type, the C-9 and C-11 vinylic protons appear between δ6.1 and 6.6. The C-8 and C-12 hydrogens, being in an environment of lesser electron density, are situated further downfield between δ6.7 and 7.3. Some average values for the coupling constants are $J_{8,9} = J_{11,12} \approx 10$ Hz, $J_{9,11} \approx 1.5$ Hz, and $J_{8,12} \approx 2.5$ Hz.

A methoxyl group at C-1 is shielded by the cyclohexadienone system and can be readily detected since it appears relatively upfield near δ3.6 instead of at the more usual site of δ3.8. The *N*-methyl group is found at its expected position near δ2.4.[2,9]

An alternate method exists for differentiating between a 1-hydroxy-2-methoxy and the reverse 1-methoxy-2-hydroxy arrangement in ring A of a proaporphine. For this purpose, the alkaloid is *O*-acetylated, and the NMR spectrum of the acetate derivative recorded. The C-3 aromatic proton normally occurs between δ6.5 and 6.7. But, if an *O*-acetyl group is present at C-2, the C-3 proton will be shifted downfield between δ6.7 and 6.9 due to the deshielding effect of the acetoxyl group. This analytical procedure can also be of assistance in the determination of the position of a phenolic function in other types of isoquinoline alkaloids.[33] The use of NMR spectroscopy in connection with deuterium exchange studies has already been noted in Section II, B.

The 100 MHz NMR spectrum of pronuciferine has been discussed in detail with particular emphasis on long-range coupling constants.[34] Slavík, Sedmera, and Bláha have also shown that NMR spectroscopy can be useful in settling the stereochemistry of the C-10 hydroxyl group in partially reduced proaporphines.[35]

X. UV and IR Spectroscopy

The UV spectra of proaporphines possessing a dienone ring usually show maxima near 230 and 285 mμ (4.4 and 3.5). Pronuciferine itself exhibits λ_{max}^{EtOH} 230 and 282 mμ (4.41 and 3.49). The partially reduced proaporphine linearisine has λ_{max}^{EtOH} 228, 282, and 288 mμ (4.30, 3.19, and 3.22), while the cyclohexanolic proaporphine litserisine shows λ_{max}^{EtOH} 240 sh and 291 mμ (3.57 and 3.54).[2] The UV spectrum of orientalinone has been recorded, λ_{max}^{EtOH} 231, 242, and 284 mμ (4.30, 4.14, and 3.77).

For proaporphines with the dienone system, the C=O band appears in the IR between 5.98 and 6.04 μ (1656 and 1673 cm^{-1}), while the C=C band is between 6.06 and 6.23 μ (1605 and 1650 cm^{-1}).[2]

REFERENCES

1. K. L. Stuart, D. Byfield, C. Chambers, and G. E. M. Husbands, *J. Chem. Soc., C* p. 1228 (1970).
2. For a review on the proaporphines, see K. L. Stuart and M. P. Cava, *Chem. Rev.* **68**, 321 (1968).
2a. For another review see K. Bernauer and W. Hofheinz, *Fortschr. Chem. Org. Naturst.* **26**, 245–283 (1968).

3. K. Bernauer, *Helv. Chim. Acta* **46**, 1783 (1963); **47**, 2122 (1964).
4. L. J. Haynes, K. L. Stuart, D. H. R. Barton, and G. W. Kirby, *Proc. Chem. Soc., London* p. 280 (1963).
5. M. P. Cava, K. Nomura, R. H. Schlessinger, K. T. Buck, B. Douglas, R. F. Raffauf, and J. A. Weisbach, *Chem. Ind. (London)* p. 282 (1964).
6. L. J. Haynes, K. L. Stuart, D. H. R. Barton, and G. W. Kirby, *Proc. Chem. Soc., London* p. 261 (1964).
7. A. R. Battersby and T. H. Brown, *Proc. Chem. Soc., London* p. 85 (1964).
8. A. R. Battersby, T. H. Brown, and J. H. Clements, *J. Chem. Soc., London* p. 4550 (1965).
8a. A. R. Battersby, T. J. Brocksom, and R. Ramage, *Chem. Commun.* p. 464 (1969).
9. L. J. Haynes, K. L. Stuart, D. H. R. Barton, and G. W. Kirby, *J. Chem. Soc., C* p. 1676 (1966).
10. I. Mann and S. Pfeifer, *Pharmazie* **22**, 124 (1967).
11. K. Bernauer, *Experientia* **20**, 380 (1964); *Helv. Chim. Acta* **51**, 1119 (1968).
12. J. W. Huffman and C. E. Opliger, *Tetrahedron Lett.* p. 5243 (1969); *J. Org. Chem.* **36**, 111 (1971).
13. S. Ishiwata and K. Itakura, *Chem. Pharm. Bull.* **18**, 416 and 1841 (1970).
14. For related work, see S. Ishiwata, K. Itakura, and K. Misawa, *Chem. Pharm. Bull.* **18**, 1219 (1970).
14a. Z. Horii, Y. Nakashita, and C. Iwata, *Tetrahedron Lett.* p. 1167 (1971).
14b. T. Kametani, T. Sugahara, H. Sugi, S. Shibuya, and K. Fukumoto, *Chem. Commun.* p. 724 (1971).
14c. T. Kametani, H. Sugi, S. Shibuya, and K. Fukumoto, *Chem. Ind. (London)* p. 818 (1971).
15. Also see S. Ishiwata and K. Itakura, *Chem. Pharm. Bull.* **18**, 1224 (1970).
16. D. H. R. Barton and T. Cohen, *Festschr. Prof. Dr. Arthur Stoll Siebzigsten Geburtstag 1957* p. 117 (1957).
17. A. R. Battersby and T. H. Brown, *Chem. Commun.* p. 170 (1966).
18. K. Heydenreich and S. Pfeifer, *Pharmazie* **21**, 121 (1966).
19. A. H. Jackson and J. A. Martin, *J. Chem. Soc., C* p. 2222 (1966).
20. M. Shamma and W. A. Slusarchyk, *Chem. Commun.* p. 528 (1965).
21. T. Kametani and H. Yagi, *J. Chem. Soc., C* p. 2182 (1967).
22. L. J. Haynes, K. L. Stuart, D. H. R. Barton, D. S. Bhakuni, and G. W. Kirby, *Chem. Commun.* p. 141 (1965).
23. D. H. R. Barton, D. S. Bhakuni, G. M. Chapman, and G. W. Kirby, *J. Chem. Soc., C* p. 1295 (1967).
24. D. H. R. Barton, D. S. Bhakuni, G. M. Chapman, and G. W. Kirby, *Chem. Commun.* p. 259 (1966).
25. A. R. Battersby, R. T. Brown, J. H. Clements, and G. G. Iverach, *Chem. Commun.* p. 230 (1965).
25a. K. L. Stuart and L. Graham, *Chem. Commun.* p. 392 (1971).
26. G. Snatzke and G. Wollenberg, *J. Chem. Soc., C* p. 1681 (1966).
27. F. Šantavý and M. Maturová, *Planta Med.* **15**, 311 (1967); and F. Šantavý, P. Sedmera, G. Snatzke, and T. Reichstein, *Helv. Chim. Acta* **54**, 1084 (1971).
28. S. Pfeifer and L. Kuhn, *Pharmazie* **22**, 221 (1967).
29. R. St. C. Gaskin and P. C. Feng, *J. Pharm. Pharmacol.* **19**, 195 (1967).
30. S. A. Siphar, Belg. Pat. 688,814 (1966).
31. M. Tomita, A. Kato, T. Ibuka, and H. Furukawa, *Tetrahedron Lett.* p. 2825 (1965).
32. M. Baldwin, A. G. Loudon, A. Maccoll, L. J. Haynes, and K. L. Stuart, *J. Chem. Soc., C* p. 154 (1967).
33. L. J. Haynes, G. E. M. Husbands, and K. L. Stuart, *J. Chem. Soc., C* p. 1680 (1966).
34. W. Regel and W. von Philipsborn, *Helv. Chim. Acta* **52**, 1354 (1969).
35. J. Slavík, P. Sedmera, and K. Bláha, *Collect. Czech. Chem. Commun.* **35**, 1558 (1970).

Chapter 10 / THE APORPHINES

Occurrence: Anonaceae, Araceae, Aristolochiaceae, Berberidaceae, Euphorbiaceae, Fumariaceae, Hernandiaceae, Lauraceae, Magnoliaceae, Menispermaceae, Monimiaceae, Nymphaceae, Papaveraceae, Ranunculaceae, and Rhamnaceae

Approximate Number: 85

Some Aporphines of Interest:

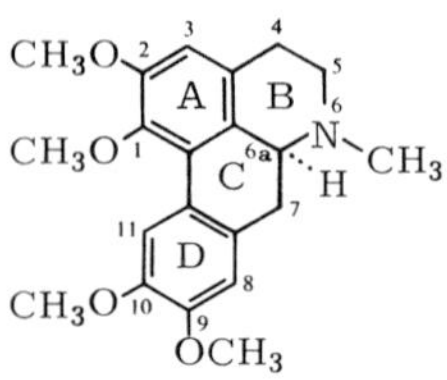

(+)-Glaucine

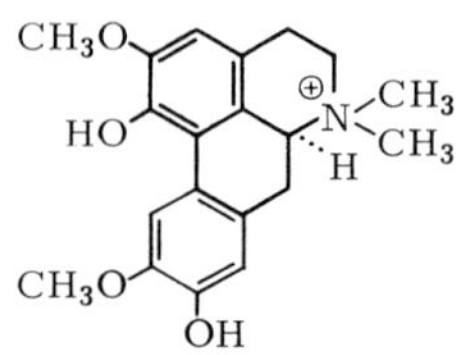

(+)-Laurifoline

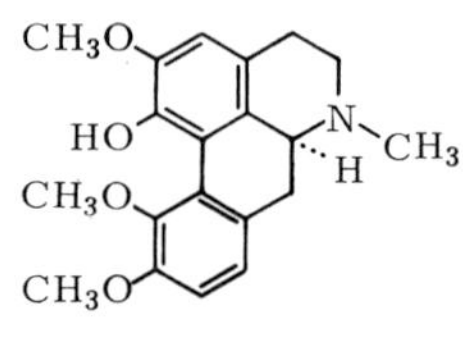

(+)-Corydine

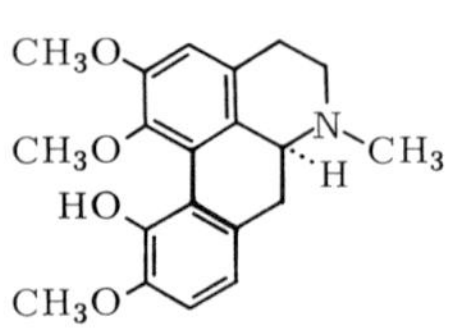

(+)-Isocorydine

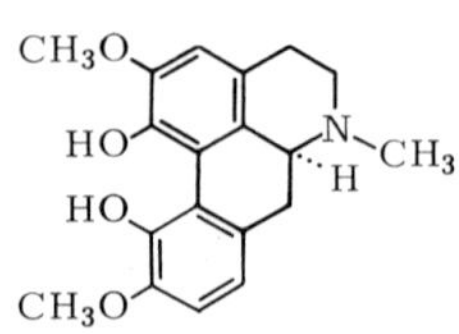

(+)-Corytuberine

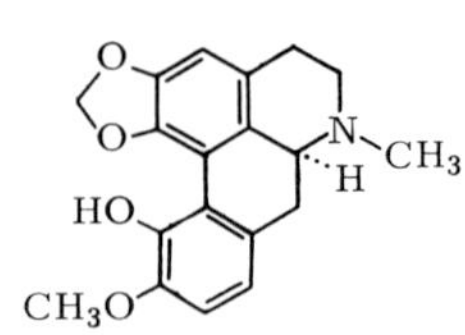

(+)-Bulbocapnine

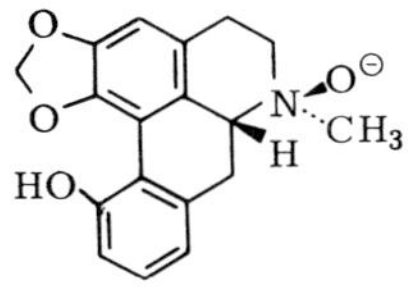

(−)-Laurepukine

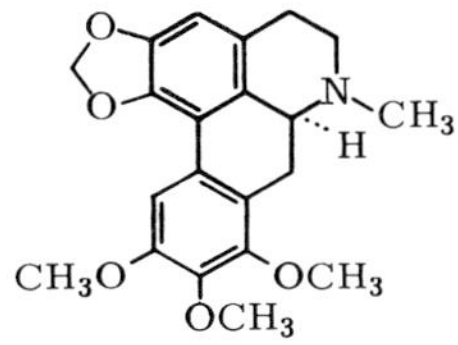

(+)-Ocopodine

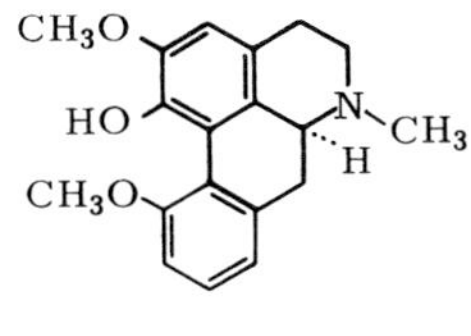

(+)-Isothebaine

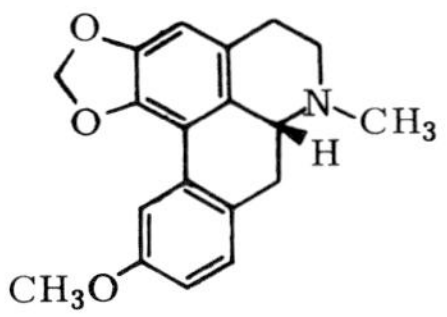

(−)-Laureline

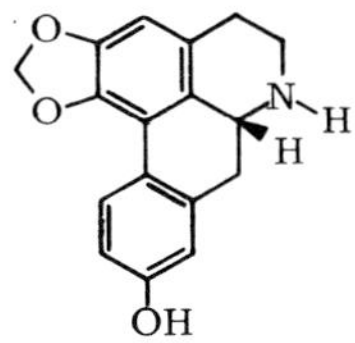

(−)-Anolobine

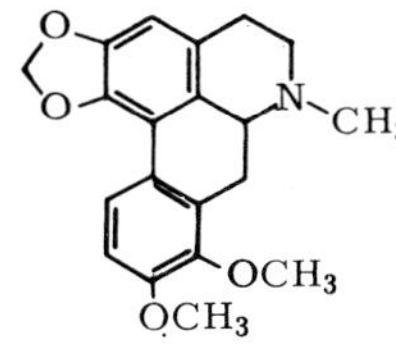

(−)-Crebanine

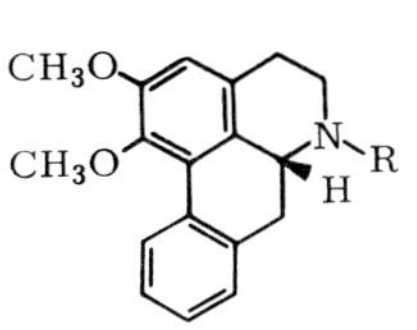

(−)-Nuciferine, R = CH_3
(−)-Nornuciferine, R = H

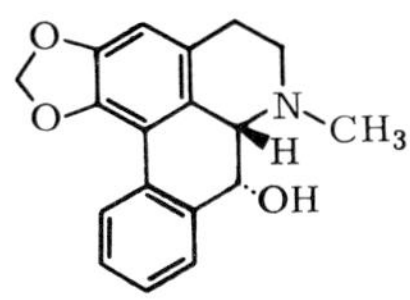

(−)-Ushinsunine

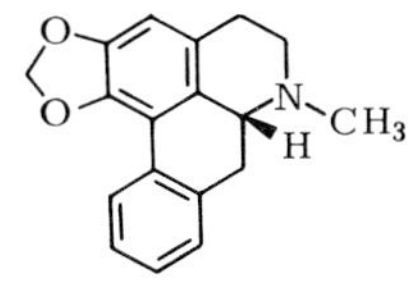

(−)-Roemerine

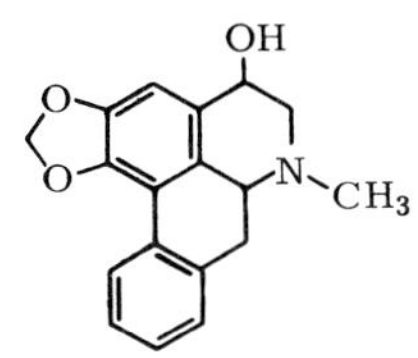

Steporphine

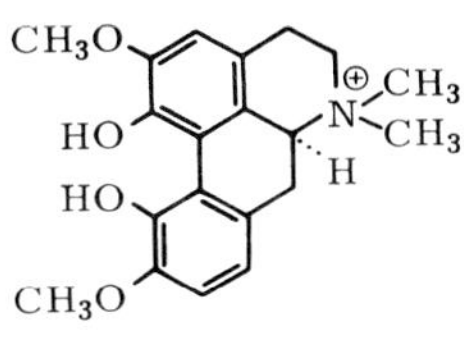

(+)-Magnoflorine

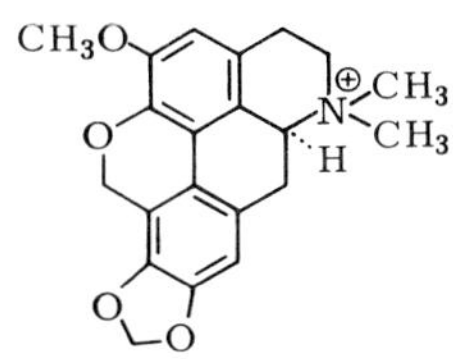

(+)-Thalphenine
(See Chapter 32)

I. Introduction

The aporphines constitute the next to the largest group of isoquinoline alkaloids, being second only to the bisbenzylisoquinolines.[1,1a,1b] A methyl group is usually attached to the nitrogen atom, making that nitrogen tertiary. If the nitrogen is secondary, the alkaloid is called a noraporphine. Several quaternary aporphine salts with two methyl groups attached to the nitrogen are also known.

Positions 1 and 2 are always substituted, the substituents being hydroxyl, methoxyl, or methylenedioxy. Other positions that may be found substituted are 9, 10, and 11, and less often 3 and 8. In a few cases, a hydroxyl function is located at C-7, while steporphine is the only aporphine alkaloid known to be oxygenated at C-4.

Glaucine, corytuberine, corydine, isocorydine, and bulbocapnine were among the first aporphines to have their structures elucidated.[2]

II. A Classical Degradation of an Aporphine

The alkaloid (+)-bulbocapnine, $C_{19}H_{19}O_4N$, was studied first by Gadamer and then by Späth, and their findings are summarized in Scheme I. The bulbocapnine nitrogen atom is tertiary, and the four oxygens are accounted for by a phenolic hydroxyl, a methoxyl, and a methylenedioxy group. The chief reactions used were successive Hofmann degradations and also oxidations with potassium permanganate and with nitric acid.[3,4]

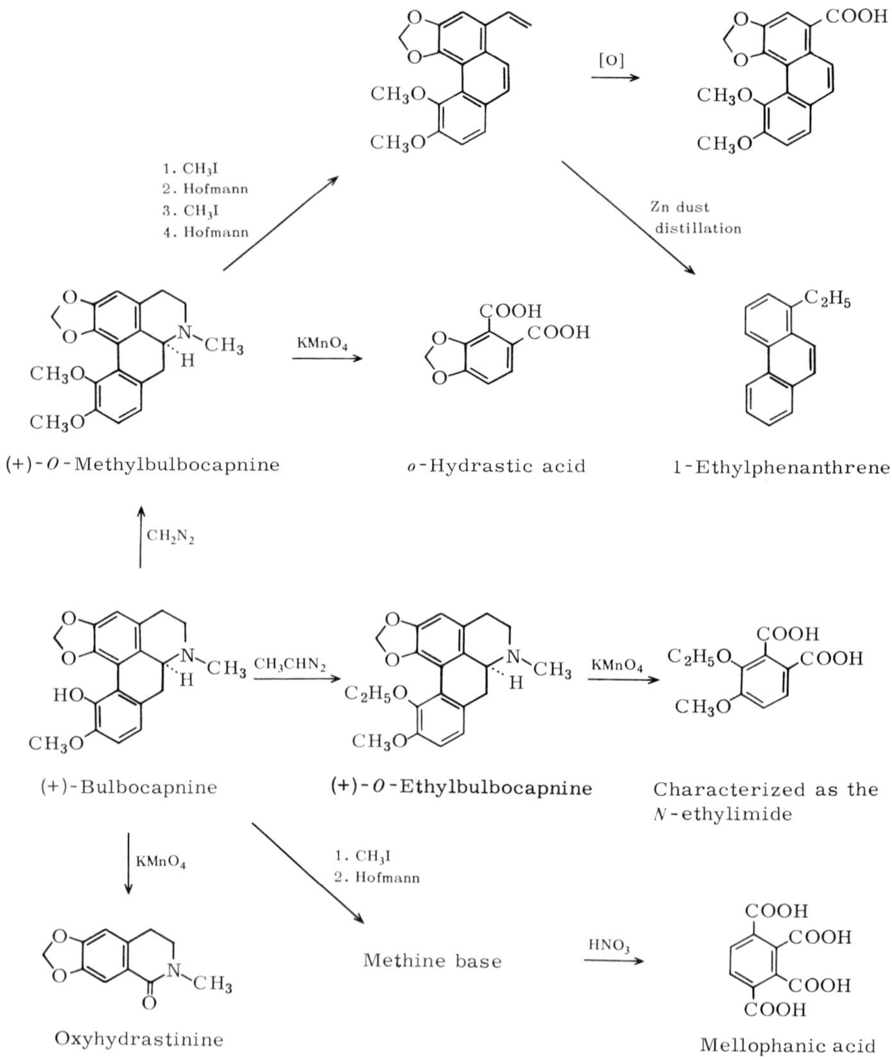

Scheme I

Nowadays, such extensive degradations are not usually carried out. Rather, use is made of spectral data coupled with the appropriate transformation of the new alkaloid into a known aporphine. As a final proof of structure a total synthesis may be achieved.

III. Apomorphine

As early as 1869 it was noted that acid-catalyzed rearrangement of morphine yields the base (−)-apomorphine. In the early 1900's the structure of apomorphine was determined through degradation, and then confirmed through synthesis. Apomorphine is not, therefore, a natural product and it lacks the usual substituents at C-1 and C-2.[5] Thebaine and codeine undergo similar acid-catalyzed rearrangements to yield morphothebaine and apocodeine, respectively.

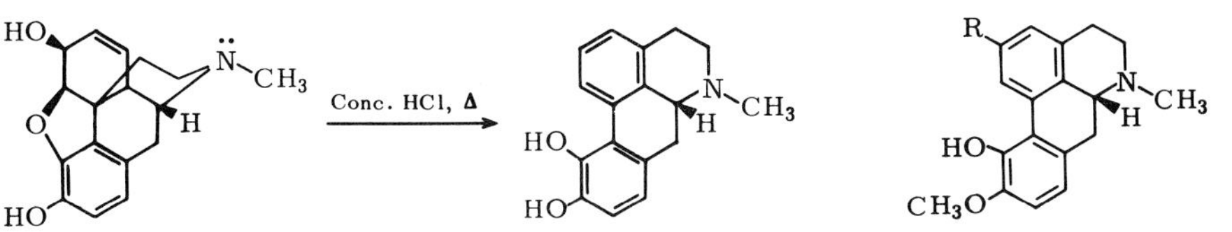

Morphine (−)-Apomorphine (−)-Apocodeine, R = H
(−)-Morphothebaine, R = OH

A related rearrangement is that of the morphinandienone base isosalutaridine to 2,10-dimethoxy-3,9-dihydroxyaporphine again under the influence of acid.[5a]

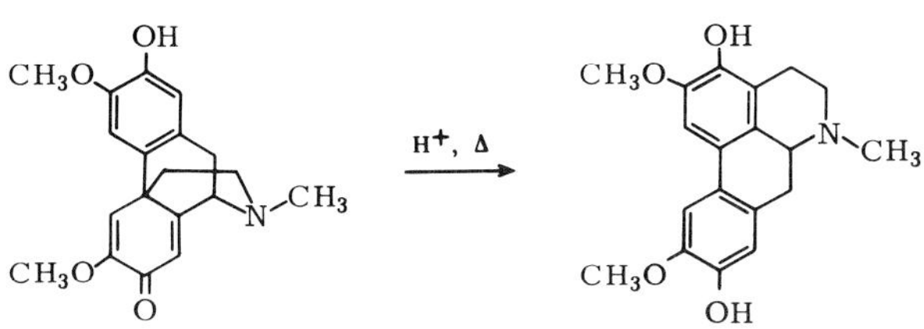

Isosalutaridine 2,10-Dimethoxy-3,9-dihydroxyaporphine

IV. Nonbiogenetic Syntheses of Aporphines

A. *Syntheses Using the Pschorr Cyclization*

A classical method of aporphine synthesis due to Pschorr and Gadamer involves nitration of a benzylisoquinoline followed by reduction, diazotization, and Pschorr cyclization. This procedure led to the first synthesis of an aporphine, that of glaucine (Scheme II).[2]

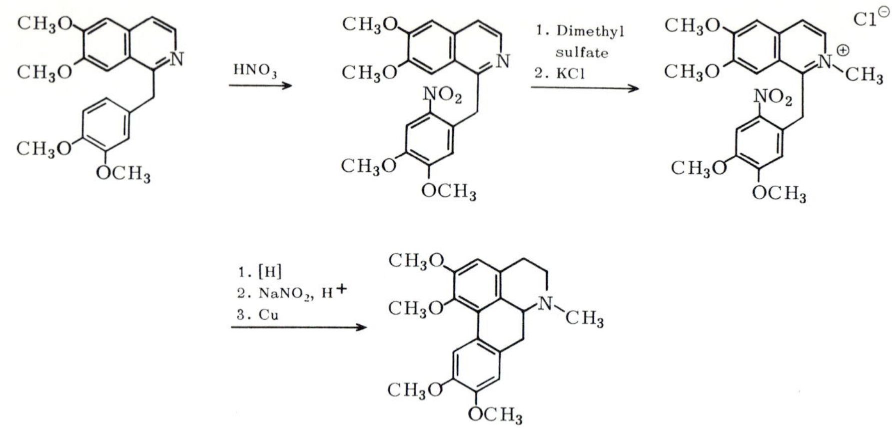

(±)-Glaucine

Scheme II

A much more common variant of the above approach involves condensation between a substituted phenethylamine and an *o*-nitrophenylacetyl chloride. The synthesis of the aporphine caaverine is a good example of such a scheme in which both the phenolic function and the nitrogen of ring B were protected by benzyl groups (Scheme III).[6]

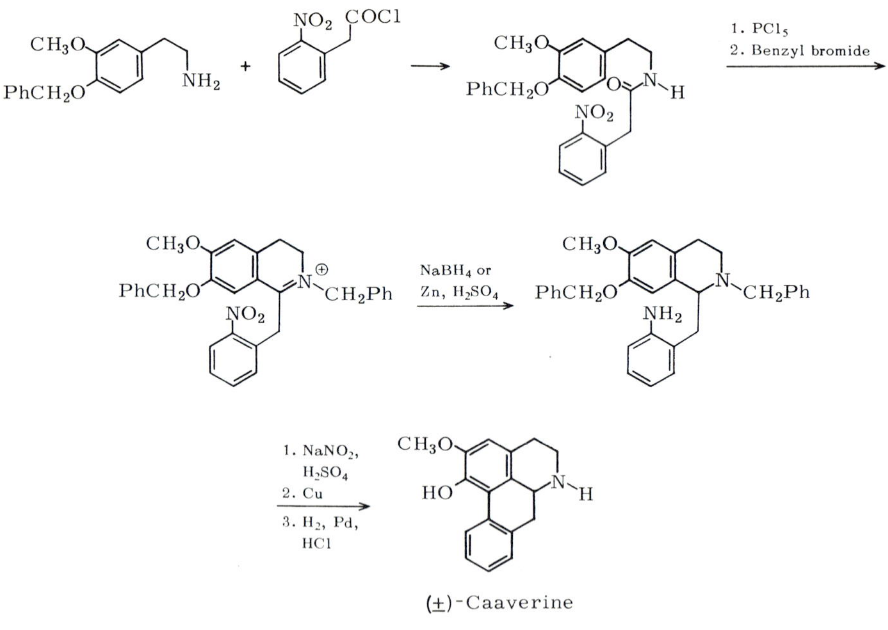

(±)-Caaverine

Scheme III

An Arndt–Eistert homologation can be applied to produce the amide required for the Bischler–Napieralski cyclization, as in the synthesis of the racemic form of the alkaloid pukateine (Scheme IV).[7]

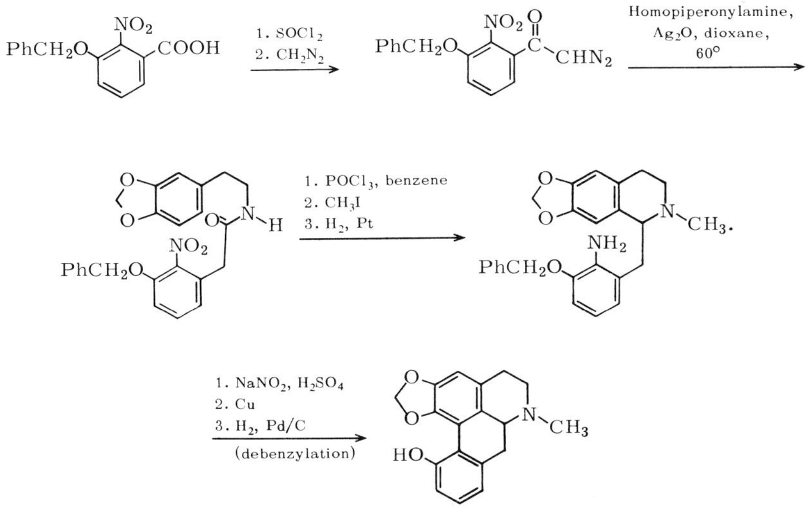

(±)-Pukateine

Scheme IV

The preferential reduction of an amide to an amine by use of diborane, as well as the hydrogenation of an aromatic nitro group to a primary amine using hydrazine and palladium, are good methods for obtaining the intermediate required for Pschorr cyclization as in the Cava synthesis of (±)-anonaine (Scheme V).[8]

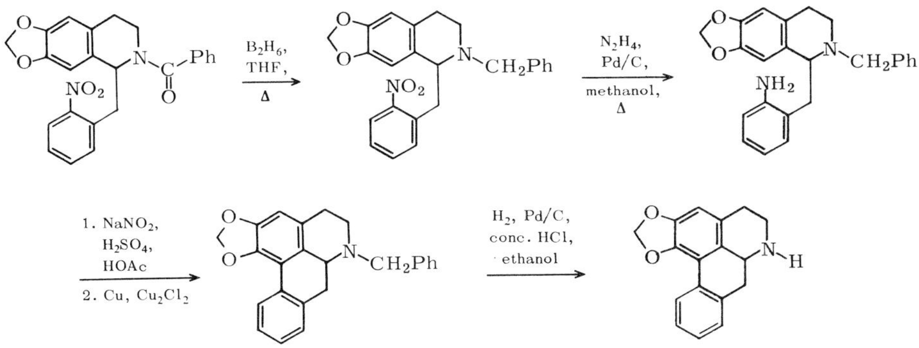

(±)-Anonaine

Scheme V

Another approach involving Pschorr cyclization begins with the condensation of an *o*-nitrotoluene with a quaternary isoquinoline salt. This synthetic sequence was first developed by Gadamer, and was improved by Weisbach and his group (Scheme VI).[9]

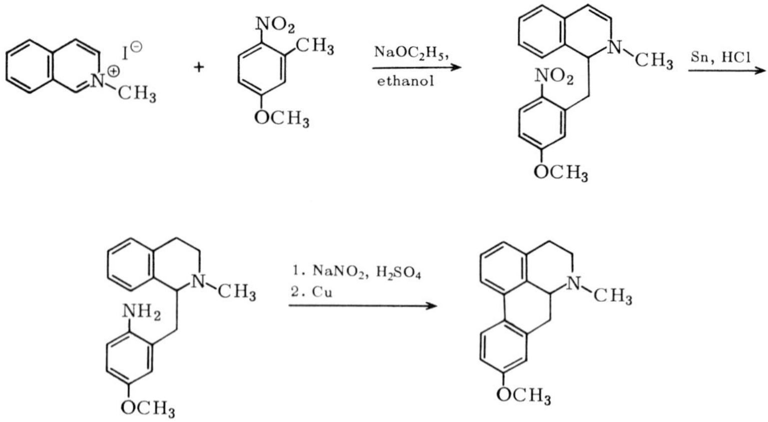

Scheme VI

The amino group required for Pschorr cyclization may also be situated in ring A of the benzylisoquinoline. Pschorr cyclization of the amino isoquinoline **1** yielded 1,2,10-trimethoxyaporphine, the proaporphine (±)-pronuciferine, and the benzylisoquinoline **2** (Scheme VII).[10,11]

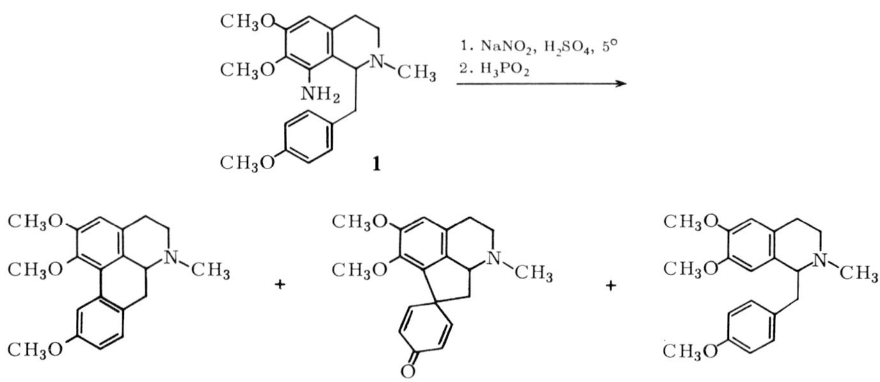

(±)-1,2,10-Trimethoxyaporphine (±)-Pronuciferine **2**

Scheme VII

Since an aromatic primary amine is required for the Pschorr cyclization, generation of a benzyne intermediate in the presence of potassium amide in liquid ammonia can also lead to an aporphine synthesis (Scheme VIII).[12,13]

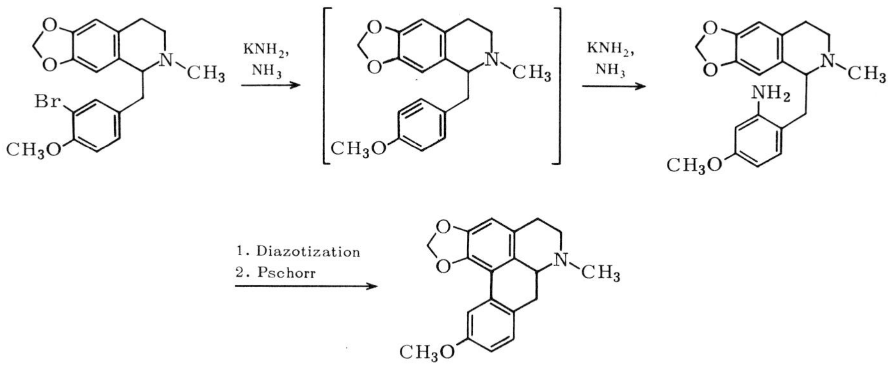

(±)-Laureline

Scheme VIII

The use of Reissert compounds in the construction of aporphines has been exploited to good avail by Neumeyer and co-workers. The key step involves the generation of a 1-(2′-nitrobenzyl)isoquinoline by the alkylation of a Reissert compound with *o*-nitrobenzyl chloride followed by reduction and Pschorr cyclization (Scheme IX).[14,15]

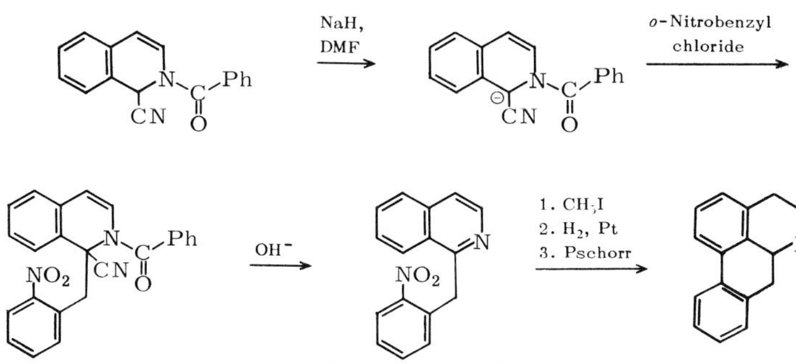

1-(2′-Nitrobenzyl)isoquinoline

(±)-Aporphine

Scheme IX

This specific approach was recently extended to the preparation of the aporphines oconovine, *N*-methylovigerine,[16,17] and apomorphine.[15]

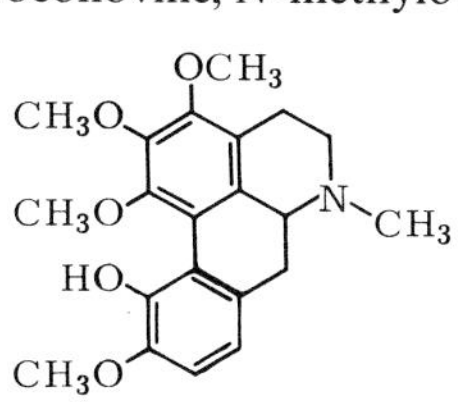

Oconovine

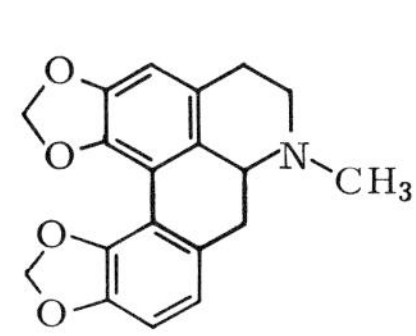

N-Methylovigerine

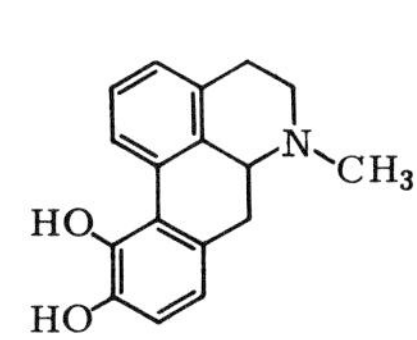

Apomorphine

An interesting synthetic method for the elaboration of 7-hydroxyaporphines and 6a,7-dehydroaporphines is known. Oxidation of 1-(2′-nitrobenzyl)isoquinoline with sodium dichromate in glacial acetic acid provided the corresponding ketone which was reduced with sodium borohydride to the carbinol. Phosgene reacted readily with the carbinol to yield the oxazolidone **2a**. Selective reduction of the nitro group was accomplished with hydrogen and palladium, and further catalytic reduction with platinum led to the *cis* derivative **2b**. Pschorr ring closure then produced the oxazolo-aporphine **2c**. Hydrolysis of **2c** with trifluoroacetic acid led exclusively to dehydronor-aporphine. Reduction of **2c** with lithium aluminum hydride gave *cis*-hydroxyaporphine. Finally, treatment of **2c** with methyllithium resulted in the isolation of *cis*-7-hydroxy-noraporphine (Scheme IXa).[17a]

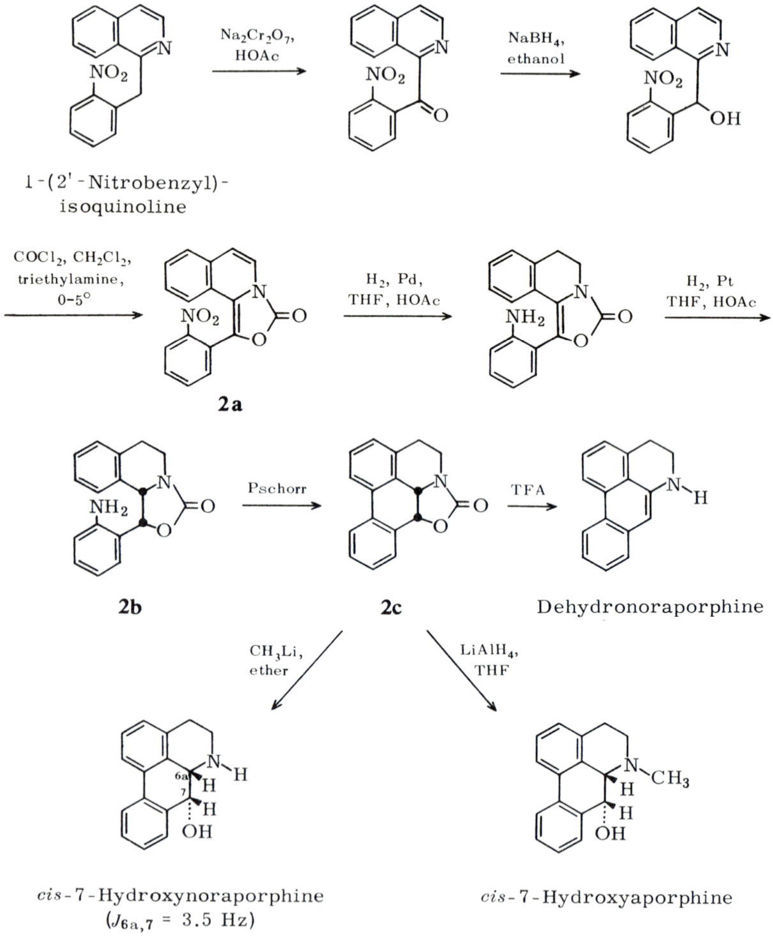

Schema IXa

B. Photochemical Routes

A variety of photochemical routes to the aporphines that do not involve Pschorr cyclization have been developed, starting in 1966.

Cava and co-workers found that the reaction of the benzylimine **3** with ethyl chloroformate gave the benzylidene urethane **4**. Irradiation of **4** in ethanol in the presence of iodine and cupric acetate produced the dehydroaporphine urethane **5**. Reduction of **5** could be best achieved with lithium aluminum hydride and aluminum chloride. The product, dehydronuciferine, was identical with the material obtained by oxidation of nuciferine with mild permanganate. Reduction of dehydronuciferine with zinc amalgam furnished (±)-nuciferine (Scheme X).[18,19] (See also Chapter 16, Section III, H.)

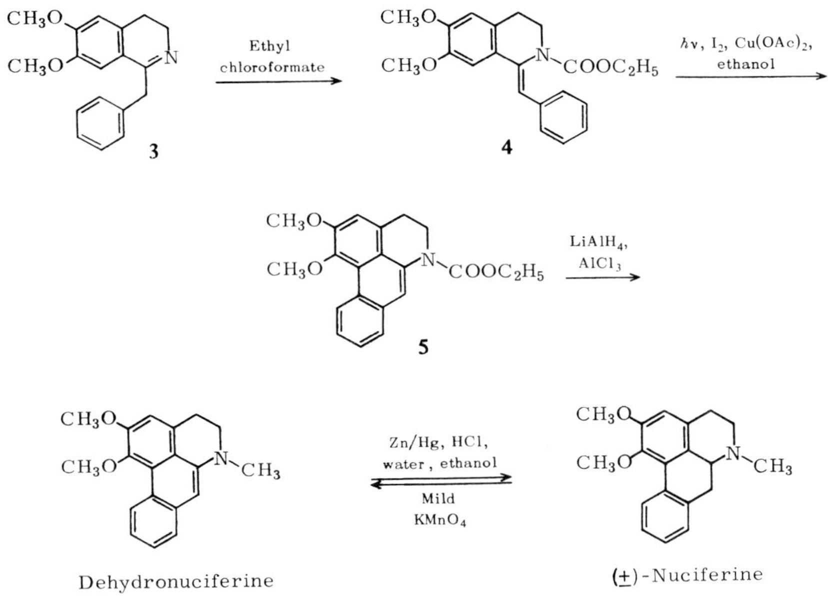

Scheme X

Yang and co-workers carried out simultaneously and independently a series of transformations somewhat similar to that described above. They noted that photocyclization does not occur if the nitrogen substituent is a methyl group instead of a carbethoxy. The availability of a free electron pair making the nitrogen basic precludes the photochemical process.[20]

Kupchan has offered an alternative photochemical pathway which involves photocyclization of 1-(2′-iodobenzyl)tetrahydroisoquinoline derivatives. Again it was noted that a basic nitrogen hinders the photochemical process so that the hydrochloride salt or the acetyl derivative of the iodotetrahydrobenzylisoquinoline **6** had to be utilized (Scheme XI).[21]

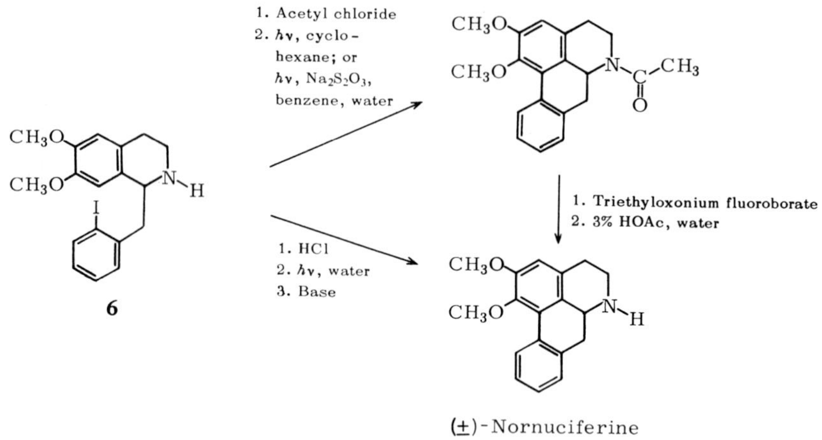

Scheme XI

In a recent development, the chloro urethane **7** was photolyzed under nitrogen in methanol solution in the presence of calcium carbonate as an acid scavenger. Hydrogen chloride was smoothly eliminated to give the dehydroaporphine urethane **5**. A probable mechanism for this nonoxidative process is shown in Scheme XII.[19]

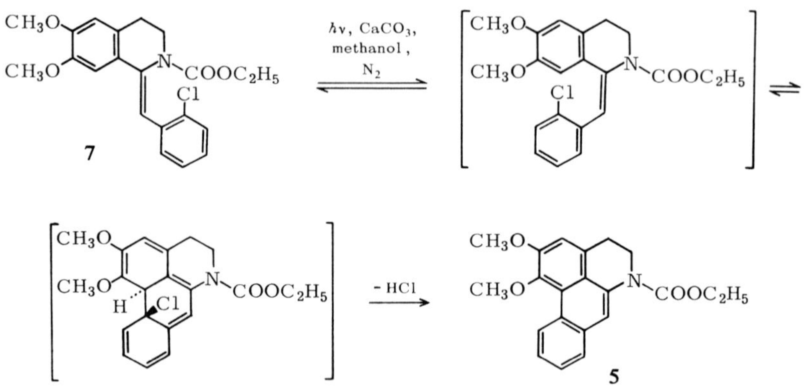

Scheme XII

A photo-Pschorr cyclization has been utilized to obtain the aporphine skeleton. Photolysis of the diazonium salt shown below gave rise to a morphinandienone derivative and an aporphine. Debenzylation of the latter produced 6,10-dimethylhernovine.[21a]

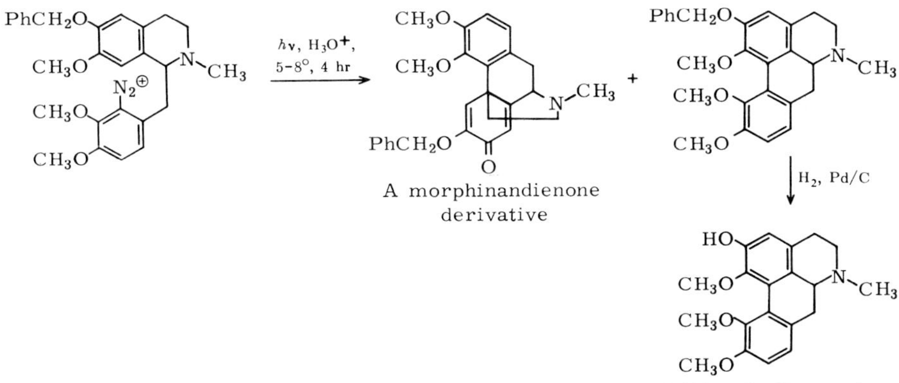

A morphinandienone derivative

6,10-Dimethylhernovine

Yet another approach to the aporphines involves phenolic oxidative coupling. Several such syntheses have been carried out, and these are discussed in Section VIII, Biosynthesis and Biogenetic Syntheses.

V. Some Reactions of Aporphines

A. Oxidation

The oxidation of an aporphine by mild permanganate to a dehydroaporphine has already been noted in Section IV, B,[22] and it has also been possible to racemize (+)-*O*-methylbulbocapnine by first oxidizing with iodine and subsequently reducing with zinc in dilute acid.[1] Stronger oxidation of a nonphenolic aporphine yields an oxoaporphine, and this subject is discussed in Chapter 13. Dehydroaporphines are best obtained through iodine oxidation of nonphenolic aporphines.[27a]

Aporphines carrying a phenolic function at C-1 or C-11 or two phenolic functions at C-10 and C-11 are readily oxidized by air, iodine, or mercuric acetate to colored

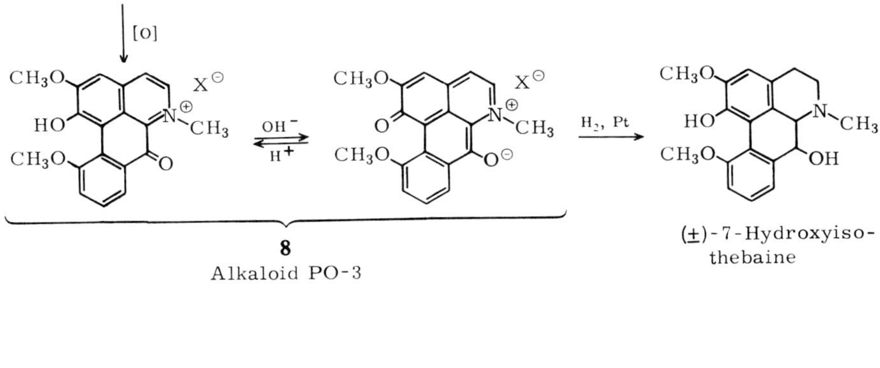

(+)-Isothebaine

8
Alkaloid PO-3

(±)-7-Hydroxyisothebaine

products. (+)-Isothebaine which is found in *Papaver orientale* L. (Papaveraceae) was thus oxidized to the highly aromatic green salt **8**, and this same product was also found as an alkaloid in the same plant. As part of the structural proof of the salt **8**, called alkaloid PO-3, the material was reduced with Adams catalyst to 7-hydroxyisothebaine.[23]

The oxidation of the unusually substituted aporphine apomorphine is pH-dependent. In acid or neutral solution the *ortho*-quinone **9** is formed, while in base species **10** and **11** are produced.[24]

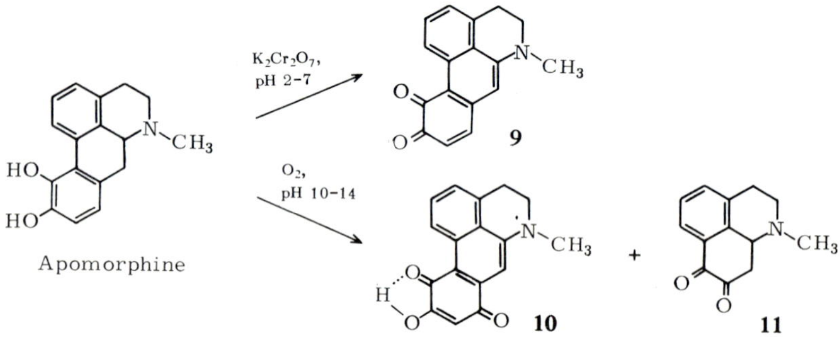

B. Cleavage of Ring B

When an aporphine is refluxed with an acid chloride or with an acid anhydride, an optically inactive *N*-acyl or *N*-aroyl derivative is obtained which is no longer basic. The ring cleavage involved, illustrated below for the case of boldine, has been used extensively in aporphine characterization and points to the fact that formation of an amide derivative under forcing conditions does not necessarily constitute proof that the initial amine is primary or secondary.[25]

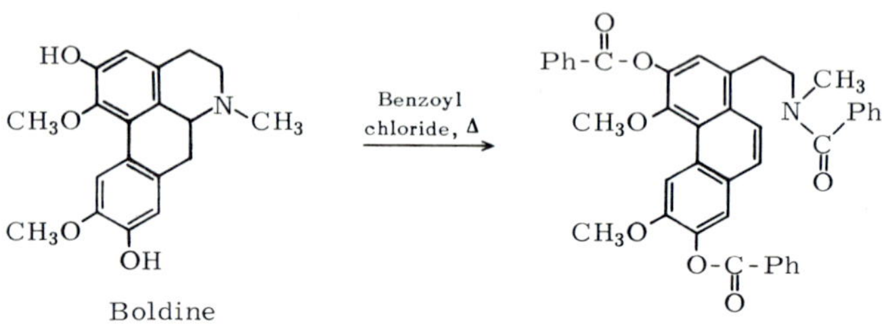

Ring cleavage can also be achieved with cyanogen bromide in chloroform, in which case the *N*-cyano derivative is obtained.[25a]

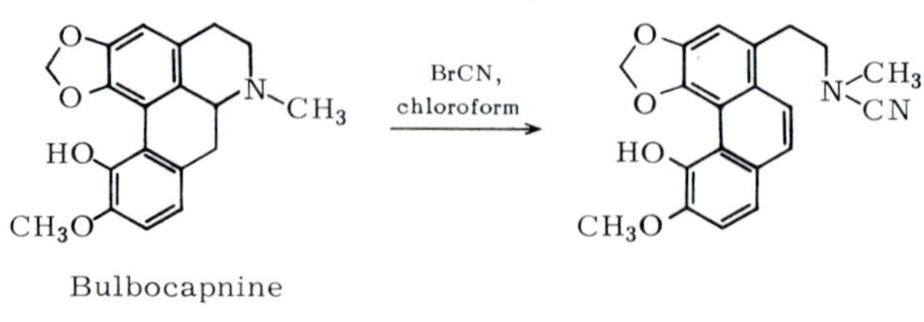

C. O-Methylation, O-Demethylation, and N-Demethylation

Treatment of either of the alkaloids boldine or isoboldine with dihydropyran and then with diazomethane, and acid hydrolysis of the protective dihydropyranyl grouping results in the formation of *N*-methyllaurotetanine. A phenolic function is therefore more reactive when at C-9 than when situated at either C-1 or C-2.[26] In the same vein, mono-*O*-methylation of boldine and isoboldine with diazomethane gave *O*-methylboldine and *O*-methylisoboldine, respectively, so that in both instances reaction occurred at the C-9 phenol (Scheme XIII).[6]

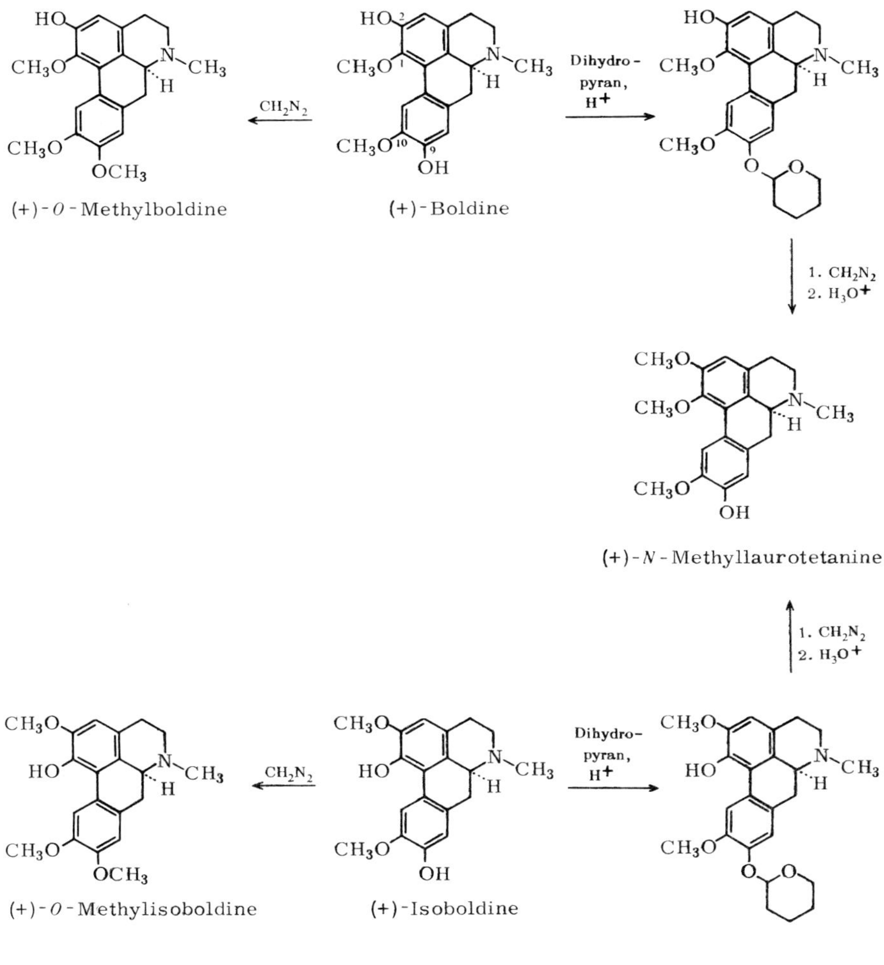

Scheme XIII

An example of the selective *O*-demethylation of an aporphine system has been recorded. Reaction of nornuciferine or any of its *N*-alkyl derivatives with hydriodic acid at 150° led to hydrolysis of the more hindered C-1 methoxyl.[27]

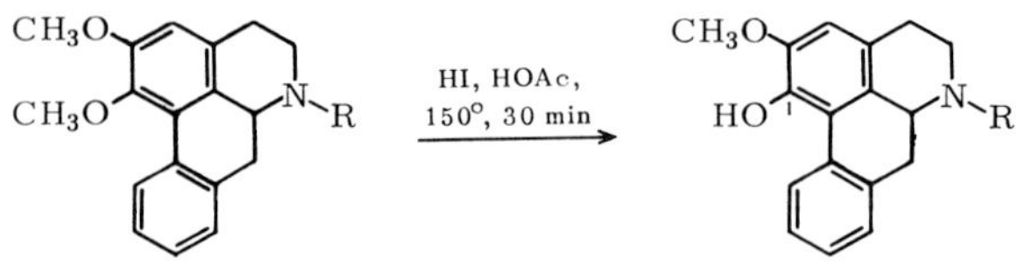

N-Demethylation of an aporphine may be achieved first by conversion to the *N*-oxide, and then by reaction with liquid sulfur dioxide followed by acid-catalyzed hydrolysis,[17] but cleavage of ring B to yield a phenanthrene is an ever present side reaction.

D. Sodium in Liquid Ammonia Reduction

Extensive studies of the sodium in liquid ammonia reduction of aporphines have been carried out. A methoxyl at C-1, being hindered, is hydrogenolized off by this method, while a methylenedioxy group, if present in the molecule, is converted into a hydroxyl function. The reduction of nantenine is illustrative of these changes.[1a] A C-9 methoxyl also undergoes facile hydrogenolysis.

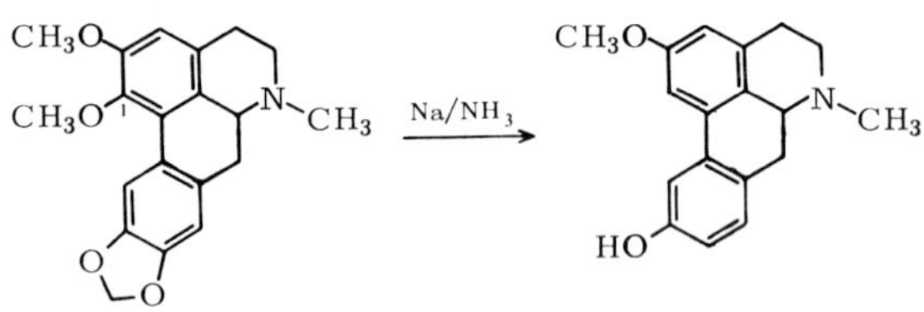

Nantenine

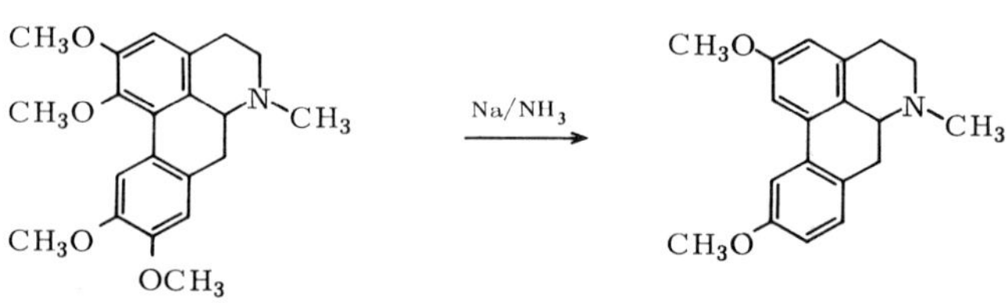

Glaucine 2,10-Dimethoxyaporphine

VI. Stereochemistry and Absolute Configuration

Shamma has pointed out that aporphines are not planar, but can exist in either of the two stereochemical arrangements indicated here.[28]

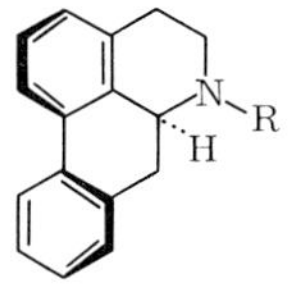

L-S Configuration

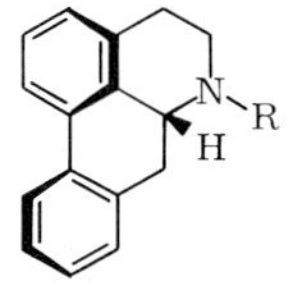

D-R Configuration

Bentley and Cardwell first observed that since naturally occurring (+)-glaucine and (−)-morphothebaine, a rearrangement product from codeine, are enantiomeric at C-6a and the absolute configuration of (−)-morphothebaine is known to be of the D series, (+) glaucine must belong to the L configuration. They also generalized that all aporphines that are appreciably dextrorotatory at the sodium D line belong to the L series, whereas the levorotatory aporphines are of the D configuration.[29]

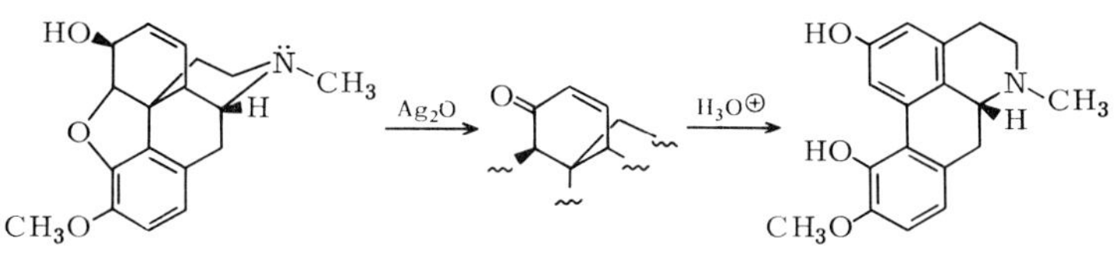

Codeine (known absolute configuration) — Codeinone — (−)-Morphothebaine (Related to codeine as shown)

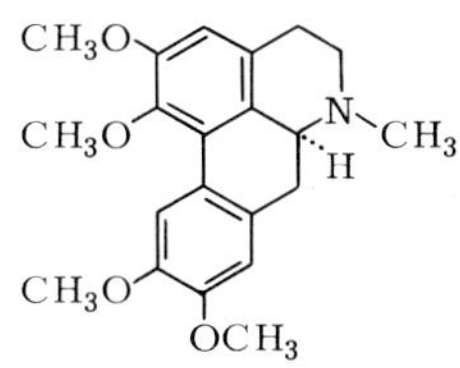

(+)-Glaucine

The absolute configuration of natural (+)-bulbocapnine is also known with certainty and fits the above rotation rule since Ayer and Taylor have converted the alkaloid by treatment with sodium in liquid ammonia into (+)-morphothebaine of known absolute configuration.[30] Additionally, a three-dimensional X-ray analysis of bulbocapnine methiodide has been carried out, the details of which are discussed in the section that follows.

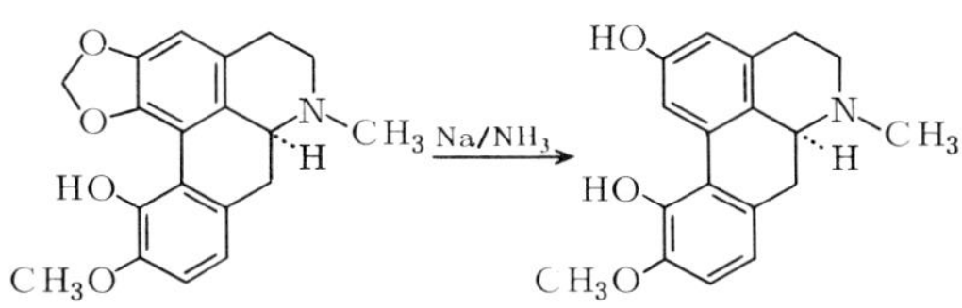

(+)-Bulbocapnine — (+)-Morphothebaine

Studies of the ORD curves of aporphines have been published.[31,32] Craig and Roy have noted that aporphines exhibit a Cotton effect of high amplitude centered between 235 and 245 mμ. This curve is independent of the substitution at the 1, 2, 3, 9, 10, and

11 positions and is diagnostic of the absolute configuration; if the Cotton effect is positive, the alkaloid belongs to the L series, and if the Cotton effect is negative, the compound must be of the D series.[33]

Simple specific rotations at the sodium D line, besides being fairly reliable indicators of the absolute configuration, offer a simple way of differentiating between C-1,2,9,10- and C-1,2,10,11-substituted aporphines. The 1,2,9,10-series exhibits rotations of +119° or less, whereas the 1,2,10,11-substituted aporphines show values of +139° or more. The occurrence of racemic aporphines in nature is rare.[34]

The circular dichroism of some of the aporphines has also been discussed.[35]

VII. X-Ray Crystallography

Besides confirming the absolute configuration, a three-dimensional X-ray analysis of bulbocapnine methiodide has shown that the angle of twist of the biphenyl system is 29.9°. The distance between O-1 and O-11 is 2.74 Å which is just under the sum of the van der Waals radii of the two oxygens, which is 2.80 Å. C-11b is 0.15 Å above the plane of ring D, the angle between the C-11a and C-11b bond and the plane of ring D being 4.5°. Furthermore, the O-1 atom lies 0.17 Å above the plane of ring A, while O-11 is 0.20 Å below plane D. The angle between the C-1 to O-1 bond and the plane of ring A is 6.4° and that between the C-11 to O-11 and the ring D plane is 6.6°. The angles O-1–C-1–C-11b, C-1–C-11b–C-11a, C-11b–C-11a–C-11, and C-11a–C-11–O-11 are all larger than 120°.[36]

All these data indicate that the biphenyl system in bulbocapnine methiodide is appreciably strained.

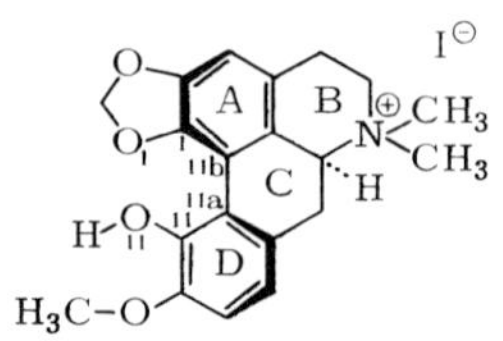

Bulbocapnine methiodide

VIII. Biosynthesis and Biogenetic Syntheses

Aporphines are formed in plants by intramolecular phenolic oxidative coupling of benzylisoquinolines. Mention was made in Chapter 8, Section II, A, of how Robinson and Schöpf independently reported in 1932 on their unsuccessful attempts to perform such an oxidation on *N*-methyllaudanosoline. While describing his results, Robinson predicted at that time that the desired coupling could probably be brought about if the basicity of the nitrogen atom were repressed either by acylation or by quaternary salt formation.

In spite of these pregnant ideas on the biosynthesis of the aporphines, a hiatus of some 25 years followed. In 1957, Barton presented the novel and important thesis that aporphines could be formed in nature through the intermediacy of proaporphines.[37]

In the discussion that follows the *in vitro* oxidations of benzylisoquinolines to aporphines and proaporphines will be considered first, to be succeeded by the description of *in vivo* experiments with labeled precursors.

A. Direct in Vitro Phenolic Oxidative Coupling of Tetrahydrobenzylisoquinolines to Aporphines

In 1962, Franck and co-workers reported the first synthesis of an aporphine by phenolic oxidative coupling using a quaternary salt. Treatment of *N*-methyllaudanosoline methiodide with aqueous ferric chloride gave a good yield of the aporphine salt **12**.[38]

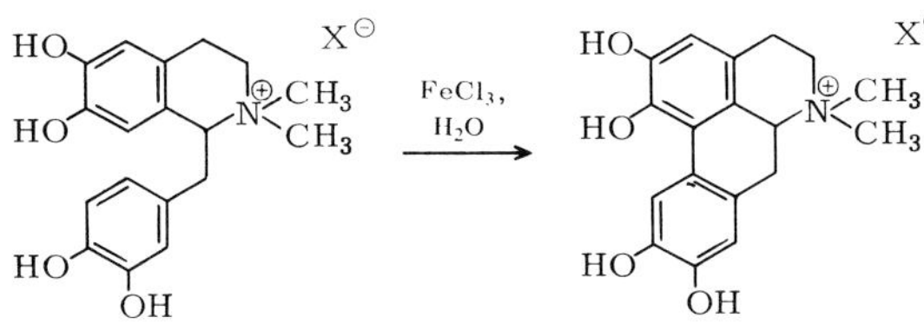

N-Methyllaudanosoline methiodide — **12**

Similarly, but at a later date, the triphenolic benzylisoquinoline salt **13** upon oxidation with ferric chloride furnished the aporphine salt **14**, isolated as the picrate.[39]

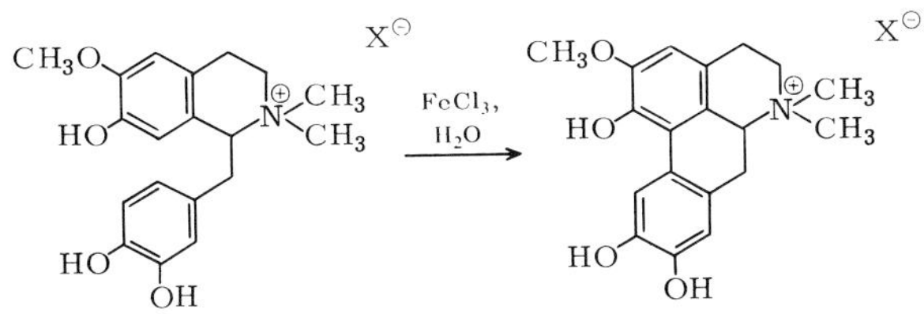

13 — **14**

Under nearly identical experimental conditions the quaternary benzylisoquinoline alkaloid (+)-tembetarine was found to furnish the aporphine (+)-laurifoline in low yield.[40]

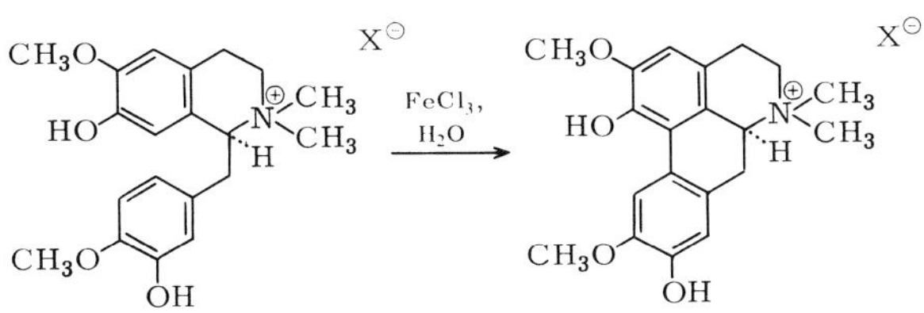

(+)-Tembetarine — (+)-Laurifoline

Similarly the quaternary salt **15** could be cyclized to the aporphine **16**.[41]

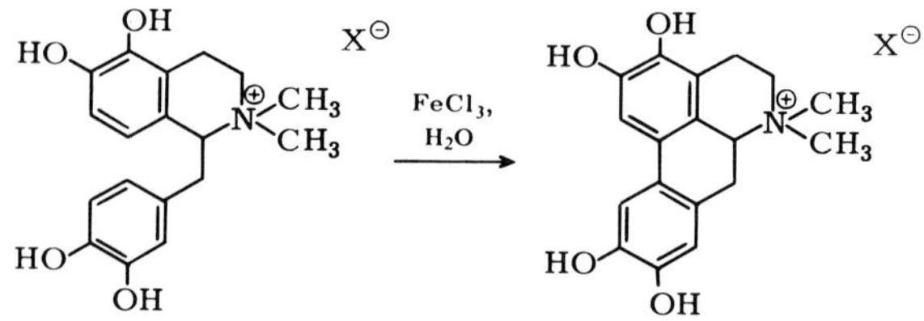

15 **16**

Interestingly enough, the pentahydroxylated tetrahydroisoquinoline salt **17** generated the aporphine **18** rather than a cularine derivative.[41]

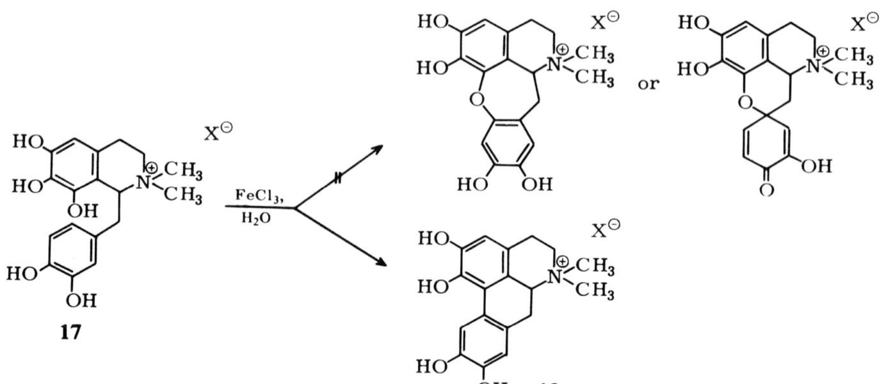

17 **18**

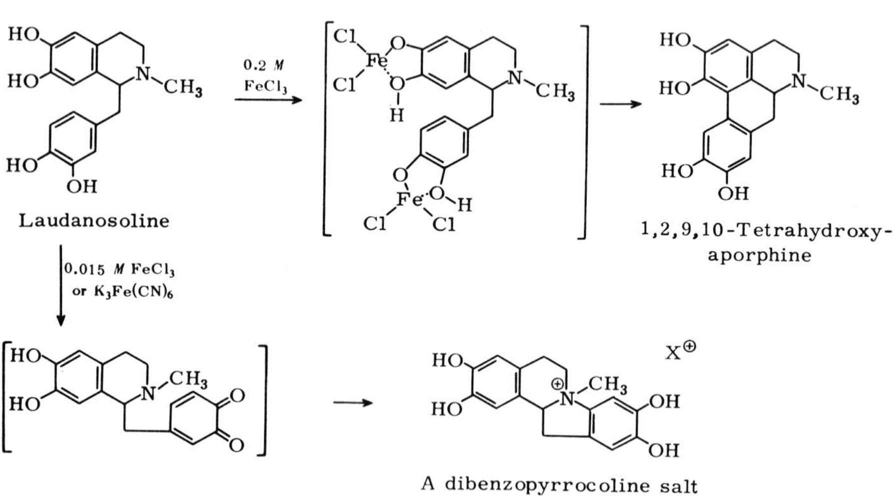

Scheme XIV

Franck and Tietze have shown that phenolic benzylisoquinolines may be converted to aporphines in good yields not only through initial quaternization of the nitrogen atom, but also by repressing the formation of *o*-quinone intermediates through complexation with ferric chloride. Thus laudanosoline is converted into 1,2,9,10-tetrahydroxyaporphine by a concentrated solution of ferric chloride which complexes with the phenolic functions of the benzylisoquinoline while simultaneously acting as the oxidizing agent (Scheme XIV).[42,43]

The benzylisoquinoline reticuline plays an important role in alkaloid biogenesis, and several studies of its oxidation, usually by means of potassium ferricyanide or ferric chloride, have been carried out.[43–47] The two products found are isoboldine and isosalutaridine, the latter species belonging to the morphinandienone series. (See also Chapter 2, Section VII, A.)

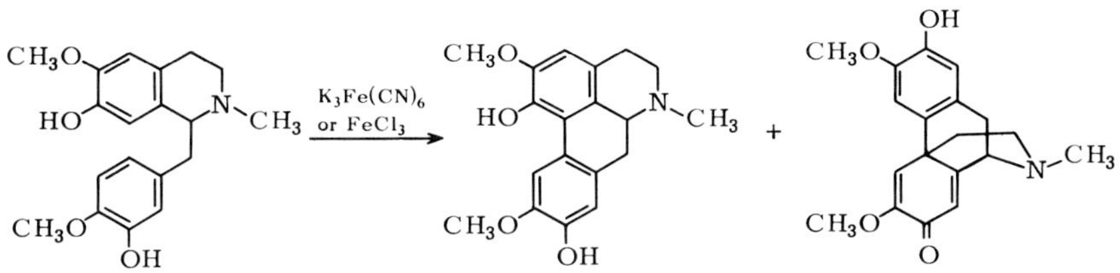

Reticuline Isoboldine Isosalutaridine

B. *In Vitro Phenolic Oxidative Coupling of Tetrahydrobenzylisoquinolines to Proaporphines and Rearrangement to Aporphines*

Several studies have appeared dealing with the oxidation of phenolic tetrahydrobenzylisoquinolines to proaporphines, which were then rearranged in acid to aporphines. (See Chapter 5, Section IX and Chapter 9, Sections II, C, and IV

As indicated in Chapter 9, Section II, C, phenolic oxidative coupling has allowed Battersby and his group to carry out the first synthesis of the aporphine alkaloid isothebaine. (±)-Orientaline upon oxidation with alkaline potassium ferricyanide gave a mixture of two dienones, one in 20% yield and the other in 1% yield. The major dienone was separated and reduced with lithium aluminum hydride to a mixture of two dienols which without separation was rearranged in acid to (±)-isothebaine in high yield. In subsequent work, (+)-isothebaine identical with the natural base was obtained after starting with (+)-orientaline.[48,49]

Some newer reagents for oxidative coupling are manganic tris(acetylacetonate),[50] vanadium oxytrichloride,[51] vanadium tetrachloride,[52,53] and the ferric chloride-dimethyl formamide complex.[53a]

C. *Experiments with Labeled Precursors*

The benzylisoquinoline orientaline is the precursor for isothebaine in *Papaver orientale* L. through the intermediacy of the proaporphines orientalinone and orientalinol.[49]

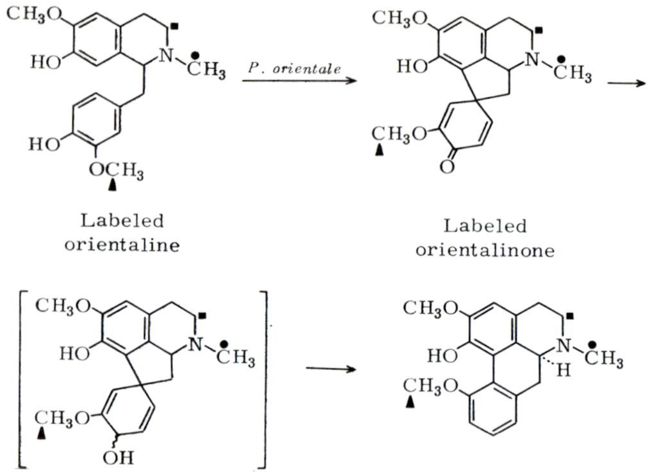

Labeled orientaline

Labeled orientalinone

Labeled orientalinol

Labeled (+)-isothebaine

Barton and co-workers have shown that tritium-labeled (+)-*N*-methylcoclaurine when fed into *Papaver dubium* L. is incorporated stereospecifically into (+)-roemerine. Neither (−)-*N*-methylcoclaurine nor isococlaurine, which lacks a properly positioned phenolic function for coupling, could be incorporated.[54,55]

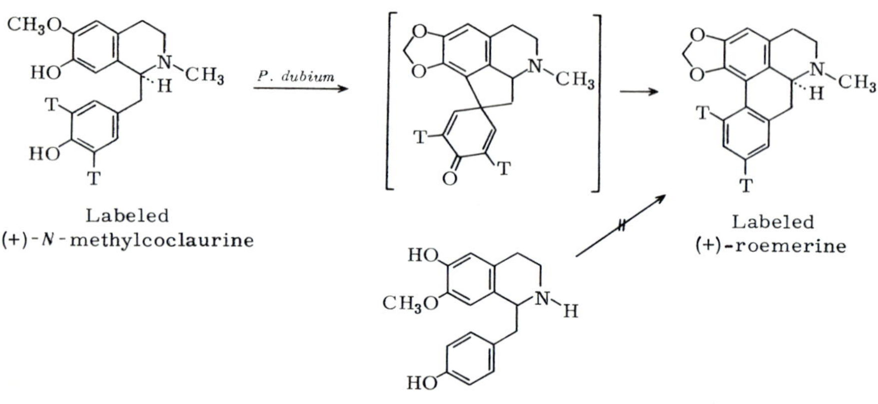

Labeled (+)-*N*-methylcoclaurine

Labeled (+)-roemerine

Labeled isococlaurine

An accompanying feeding experiment using (±)-*N*-methylcoclaurine labeled with ^{14}C on the *O*- and the *N*-methyl groups gave an unexpected result. The derived roemerine showed that a significant fraction of the methoxyl activity had been lost during biosynthesis. It was previously known that an *o*-methoxy phenol cyclizes to furnish the methylenedioxy group,[56,57] but this process had not previously been observed accompanied by demethylation–remethylation.[54]

In other experiments, it was found that (±)-coclaurine could be a precursor for anonaine in *Anona reticulata* Sieber, ex A. DC (Anonaceae), and for mecambrine in *Meconopsis cambrica* (L.) Vig. (Papaveraceae). Treatment of mecambrine with hot hydrochloric acid gave the aporphine mecambroline.[54] The additional finding that labeled mecambrine is well incorporated into mecambroline in *M. cambrica* indicates that formation of the methylenedioxy group from the *o*-methoxyphenol precursor has already been consummated by the time the proaporphine stage is reached (Scheme XV).[55]

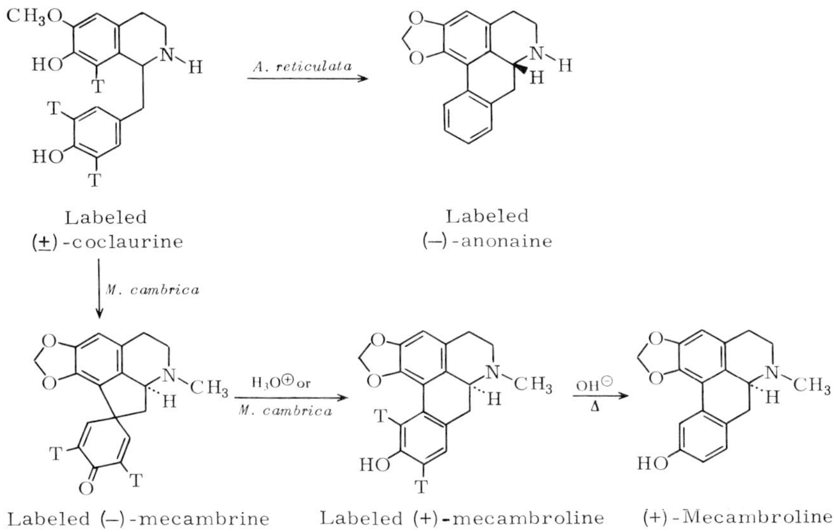

Scheme XV

The aforementioned data clearly support the dienol–benzene and the dienone–phenol biogenetic pathways to the aporphine alkaloids. The evidence in favor of direct coupling of a benzylisoquinoline to an aporphine is more ambiguous. Blaschke has found that feeding (±)-reticuline labeled on the *N*-methyl group to the larkspur, *Corydalis cava* Schweigg. et Korte (Fumariaceae), led to labeled bulbocapnine.[58]

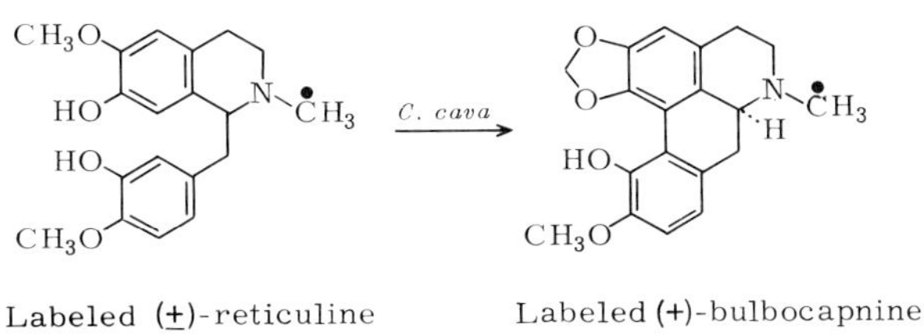

In a supplementary experiment, it was determined that (±)-reticuline labeled with tritium as shown and mixed with (±)-reticuline labeled on the *N*-methyl when supplied to *C. cava* resulted in the formation of tritium-labeled bulbocapnine. Bromination of the product resulted in loss of the tritium label.[59] It should be added that direct phenolic oxidative coupling would prevail only if an *O*-demethylation and *O*-remethylation mechanism were inoperative.

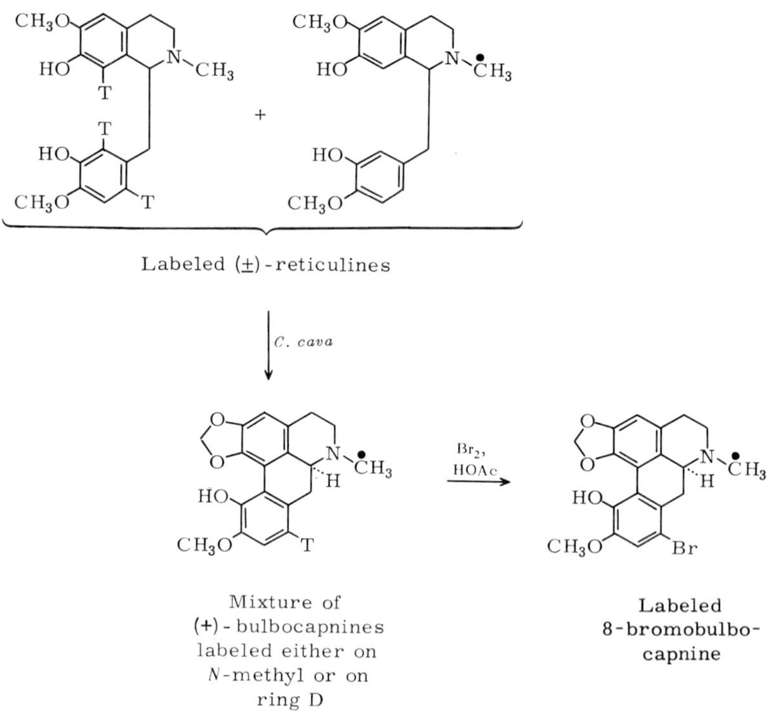

Significant results have been forthcoming lately from Battersby's laboratory regarding the biogenesis of aporphines dioxygenated in ring D. Working with *Dicentra eximia* (Ker) Torr. (Fumariaceae), it was firmly established after several experiments with differently labeled compounds that (±)-laudanosoline, (±)-reticuline, and (±)-orientaline were ineffective as precursors for the aporphines corydine, dicentrine, and glaucine, but a compound which was readily incorporated into all three alkaloids was 4′-*O*-methylnorlaudanosoline which originates from norlaudanosoline, also a good precursor. 4′-*O*-Methylnorlaudanosoline is converted into norprotosinomenine which must be incorporated into the three aporphine alkaloids by way of the two neoproaporphine intermediates indicated (Scheme XVa).[59a]

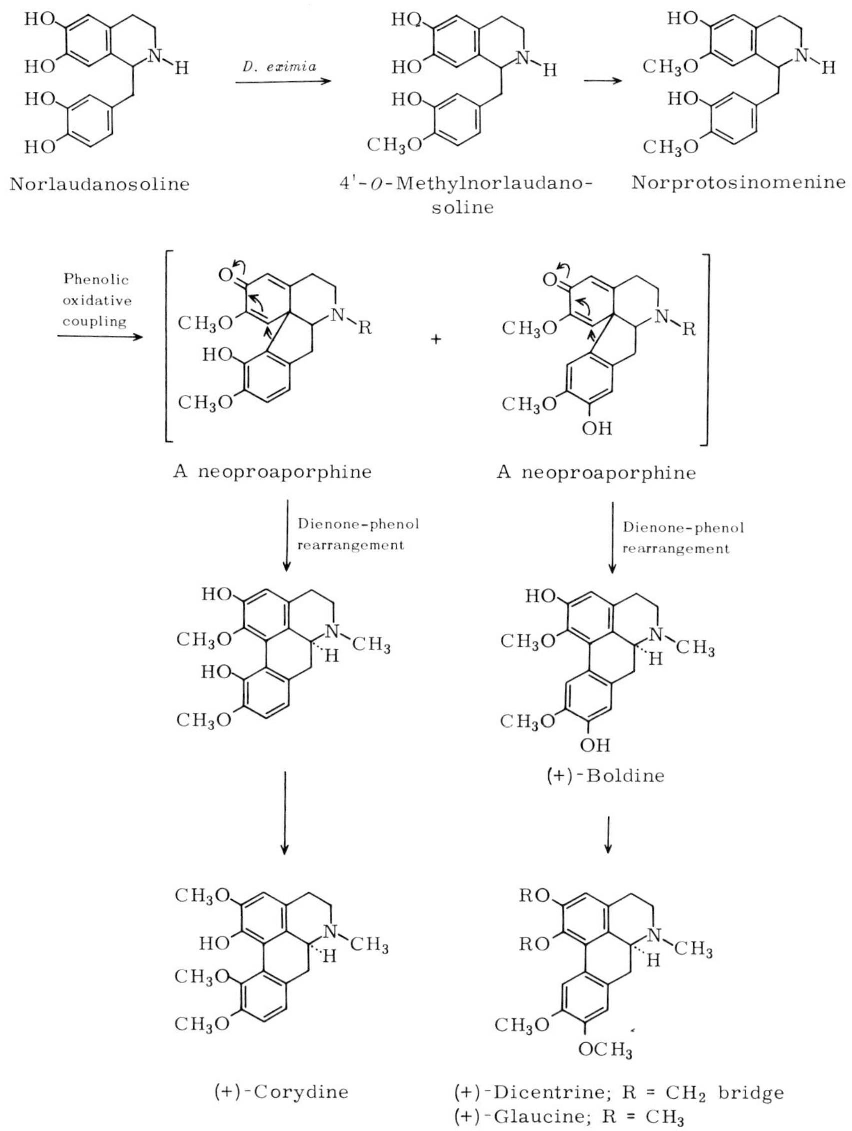

Scheme XVa

The abovementioned results dramatize the importance of experiments with labeled precursors, since they establish that direct phenolic coupling – which *a priori* would have been the most logical pathway for the formation of the 1,2,9,10- and the 1,2,10,11-tetrasubstituted aporphines in nature – is definitely not operative at least in this instance.

Neoproaporphines have not yet been isolated from plant sources, but it is safe to assume that they will be found some day as natural products.

The alkaloid (−)-crebanine has an unusual 1,2,8,9-substitution pattern and may be formed from orientalinone as shown below.[60]*

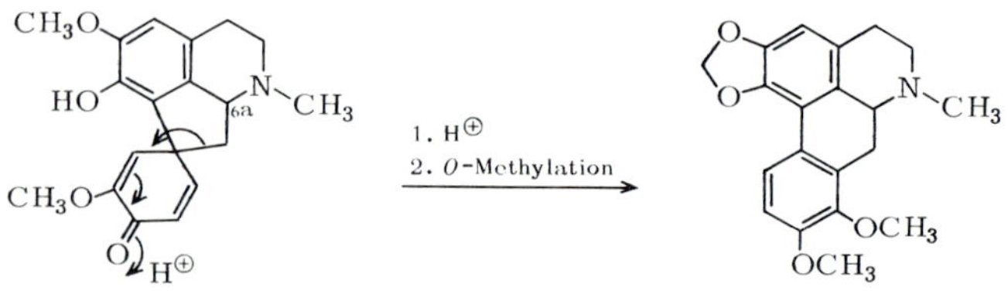

Orientalinone (−)-Crebanine

IX. The Relationship between the Ring D Substituents and the Absolute Configuration

Shamma has noted a correlation between the ring D substituents in the aporphines and the absolute configuration.[34,61] Naturally occurring aporphines are usually dextrorotatory and belong to the L configuration. But those aporphines monosubstituted or unsubstituted in ring D which do not originate from the family Papaveraceae generally are levorotatory and of the D configuration. For example, 9,10- and 10,11-substituted aporphines such as glaucine, corytuberine, bulbocapnine, and ocopodine are all dextrorotatory and possess a C-6a alpha hydrogen. By contrast, laureline, which is monosubstituted, and nuciferine, which is unsubstituted in ring D, are levorotatory and have a beta hydrogen at C-6a.[61] Aporphines originating in the Papaveraceae are exceptional in that they may belong to either series regardless of their substitution pattern.[61,62]

X. Pharmacology

The aporphines have a very wide range of physiological activity.

Apocodeine may have useful emetic activity.[63] Bulbocapnine affects the central nervous system and causes catatonia. Boldine is only slightly toxic and does not cause addiction. It has a mild sedative, diuretic, and antiparasitic action, and also increases the secretions of the liver and salivary glands.[64] Isothebaine depresses the central nervous system.[65] Glaucine and dicentrine cause narcosis in animals, and with larger doses convulsions.[66] Glaucine has also shown antitussive properties.[67] Laurifoline chloride has some hypotensive activity, while corytuberine accelerates respiration

* The term orientalinone refers to any of four isomers differing in stereochemistry at C-6a and at the spiro center. Obviously a specific orientalinone isomer is involved in each biogenetic transformation.

and slows the pulse.[66] Finally, one of the insect feeding inhibitory substances in the leaves of *Cocculus trilobus* DC. is the aporphine alkaloid isoboldine.[68]

Apomorphine hydrochloride is a prompt and efficient emetic, producing results in 10 to 15 minutes. It induces emesis through central action only. In larger doses it can have a hypnotic effect. At times it causes euphoria, restlessness, and tremors. There have been instances of serious depression and even death of feeble patients from the use of therapeutic doses, but apomorphine hydrochloride is useful in cases of poisoning where it is desired promptly to empty the stomach. It is best used hypodermically.[69] Subcutaneous injection of apormorphine temporarily controls involuntary movement – the major side effect induced by L-dopa in Parkinsonism patients – for about one hour. Tremor in Parkinsonism patients not receiving L-dopa was also eliminated by apomorphine for a maximum of three hours.[69a]

Xylopine is supposed to possess sedative and analgesic activity (Simes, Milan).

XI. Mass Spectroscopy

The main fragmentation pathways for the aporphines are outlined in Scheme XVI.[70,71]

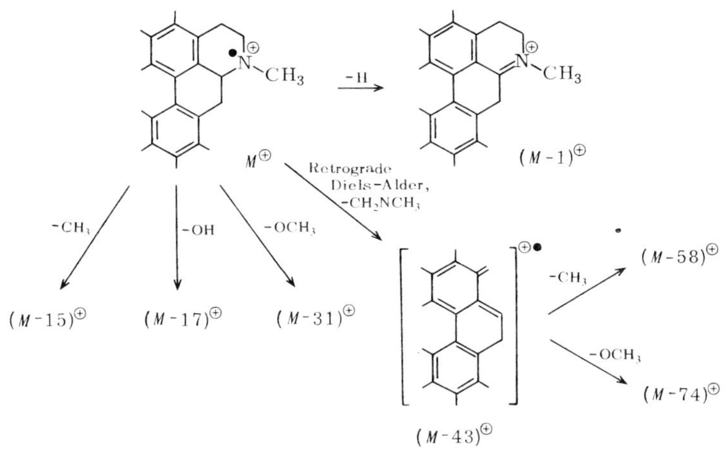

Scheme XVI

The mass spectra of 1,2,9,10-substituted aporphines differ in some important respects from the 1,2,10,11-series. The $(M-1)^+$ is the base peak in the 1,2,9,10-series. In the 1,2,10,11-series, the base peak is the molecular ion, while the $(M-1)^+$ peak has 50% or less of the intensity of the base peak. The $(M-15)^+$, $(M-17)^+$, and $(M-31)^+$ peaks are usually more intense in the 1,2,10,11-series.[71] Confirmation of the identity of the molecular ion is best obtained through chemical ionization mass spectrometry.[71a]

XII. NMR Spectroscopy

The NMR spectra of aporphines have been adequately summarized in papers by Bick[72] and by Baarschers[73] and their co-workers, and in other papers.[44,73a]

A methoxyl group at C-1 will consistently appear further upfield ($\delta 3.4-3.7$) than if at C-2, C-9, or C-10 ($\delta 3.8-3.9$). A methoxyl at C-11 will have an intermediate chemical shift ($\delta 3.6-3.8$).

Methoxyl bands may be further separated by using benzene as a solvent. Methoxyl groups which are not hindered will undergo an appreciable upfield shift. In the case of glaucine, the C-1 methoxyl is the most hindered and is therefore the one that will show the least change in chemical shift.[74]

A methylenedioxy group at C-1,2 will be split into two doublets because of the asymmetry of the twisted biphenyl system, $\delta 5.9$ and 6.0 ($J \approx 10$ Hz). But a methylenedioxy group at C-9,10 will appear as a singlet.

The aromatic hydrogen that is often but not always the furthest upfield is at C-3 ($\delta 6.5-6.7$), and the one furthest downfield is at C-11 ($\delta 7.6-8.2$). The other aromatic hydrogens are usually between $\delta 6.7$ and 7.4.

An interesting item concerns the splitting of the C-8,9 protons. An AB quartet for these ortho hydrogens is found if ring D contains methoxyl groups at C-10 and C-11. When the methoxyl group at C-11 is replaced by hydroxyl, no observable splitting is usually found, and the C-8 and C-9 proton absorptions overlap as a singlet.

The use of NMR spectroscopy in conjunction with deuterium exchange experiments to locate the position of a phenolic function has already been noted (Chapter 9, Section II, B).[75]

The influence of alkali on the NMR spectra of phenolic aporphines has been studied. In the NaOH–DMSO system, the aromatic protons experience a characteristic upfield shift when anion formation occurs. The usual upfield shifts are as follows: ortho, $0.47-0.54$ ppm; meta, $0.25-0.42$ ppm, and para, 0.86 ppm. Meta shifts of 0.57 ppm have, however, been observed.[76] Protons attached to a nonphenolic aromatic ring undergo an upfield shift of $0.11-0.26$ ppm. The only downfield shift observed when an aporphine is converted into its anion takes place when a phenolic hydroxyl is present at C-1 or at C-11. It is the neighboring proton at C-11 or C-1, respectively, that undergoes a downfield shift of about 1 ppm.[76]

Following are actual examples of spectral data for some aporphines.

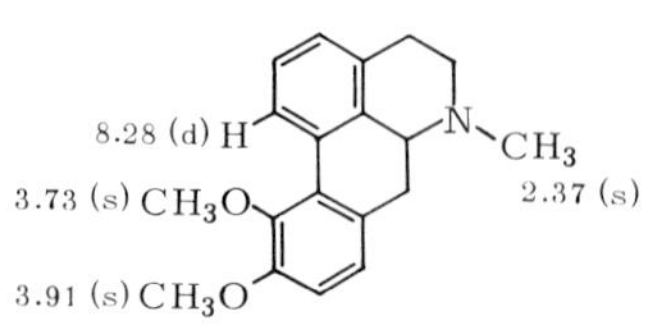

Apomorphine dimethyl ether[15]

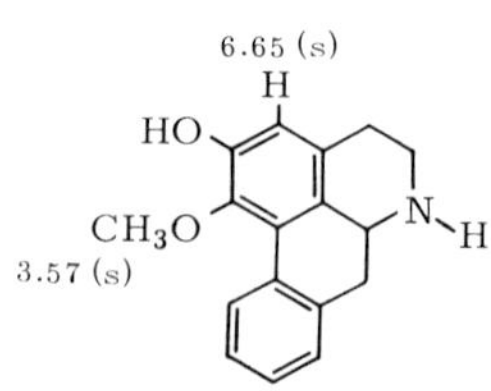

Asimilobine[76a]

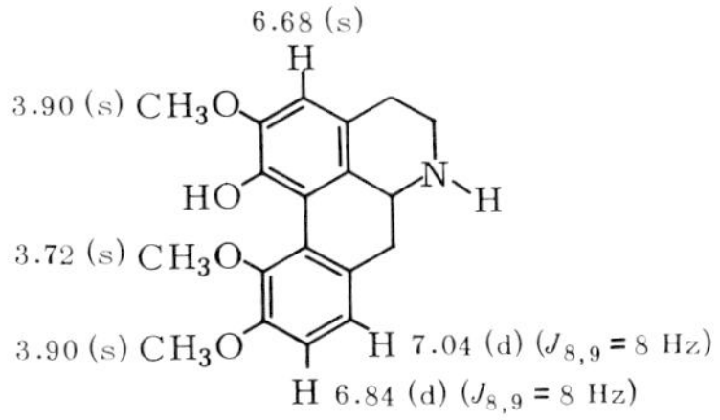

Norcorydine[76a]

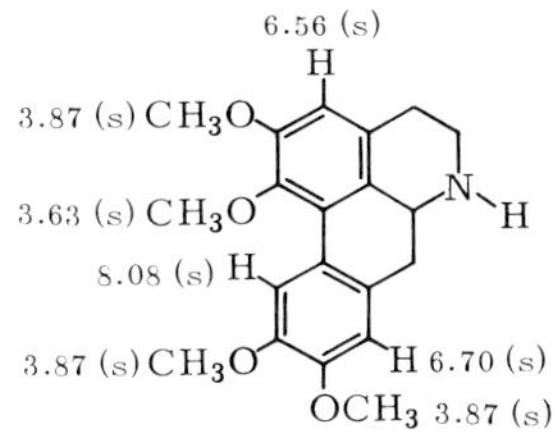

Norglaucine[76b]

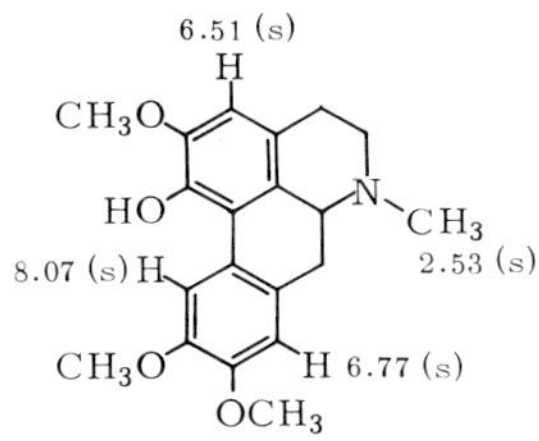

Three methoxyl singlets at δ 3.83, 3.89, and 3.89

Thaliporphine[76c]

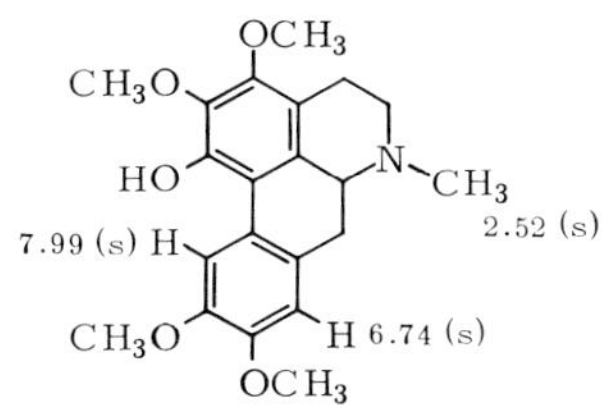

Four methoxyl singlets at δ 3.88, 3.88, 3.92, and 3.92

Preocoteine[76c]

XIII. UV SPECTROSCOPY

UV spectroscopy can be a valuable tool in the structural elucidation of aporphines (Table I). Those aporphines substituted at C-1,2,9,10 show maxima near 220, 282, and 305 mμ, while C-1,2,10,11-aporphines exhibit maxima around 220, 270, and 305 mμ.[28] Aporphines with 1,2,9-substitution have peaks close to 233 sh, 280, and 310 sh mμ; and for the 1,2,10-series, the peaks are around 226, 266, 275, and 305 mμ. For the 1,2-disubstituted aporphines, the maxima lie in the vicinity of 234, 273, and 312 mμ.[1,1a]

The UV spectra of monophenolic aporphines in basic solution are also instructive. The presence of a phenolic function at C-9 results in a strong absorption between 315 and 330 mμ upon the addition of base.[78]

XIV. USHINSUNINE: A C-7 HYDROXYLATED APORPHINE

(−)-Ushinsunine, $C_{18}H_{17}O_3N$, has been found in a variety of *Michelia* species (Magnoliaceae). The alkaloid has one methylenedioxy function, one *N*-methyl group, and a readily acetylatable alcoholic function. Ushinsunine undergoes facile dehydration on treatment with phosphorus oxychloride to yield the optically inactive dehydro-

TABLE I

UV ABSORPTION MAXIMA OF APORPHINES

Substitution pattern	Aporphine	UV absorption λ_{max}	References
1,2,9,10	Glaucine	218, 280 – 282, and 302 mμ (4.58, 4.18, and 4.16)	1
	Domesticine (1-Hydroxy-2-methoxy-9,10-methylenedioxyaporphine)	221, 283, and 310 mμ (4.56, 4.01, and 4.17)	76d
1,2,10,11	Isocorydine	220, 268, and 302 mμ (4.6, 4.2, and 3.8)	1
1,2,8	1,2,8-Trimethoxyaporphine	271 mμ (4.26)	1
1,2,9	Xylopine (1,2-Methylenedioxy-9-methoxynoraporphine)	217 and 281 mμ (4.49 and 4.45)	1a
	Anolobine	239 and 280 mμ (4.16 and 4.32)	77
1,2,10	1,10-Dihydroxy-2-methoxyaporphine	218, 266, 275, and 307 mμ (4.58, 4.01, 4.12, and 3.96)	1a
1,2,11	1,2,11-Trimethoxyaporphine	230, 270, and 300 mμ (4.1, 4.1, and 3.9)	1
1,2	Nuciferine	230, 272, and 310 mμ (4.26, 4.19, and 3.32)	1
1,2,3	Guatterine (1,2-Methylenedioxy-3-methoxy-7-hydroxyaporphine)	242 and 281 mμ (4.27 and 4.26)	1a
10,11	*O,O*-Dimethylapomorphine	216, 269, and 306 mμ (4.66, 4.28, and 3.36)	15
1,2,3,9,10	Purpureine (1,2,3,9,10-Pentamethoxyaporphine)	273 sh, 282, 303, and 312 sh mμ (4.26, 4.36, 4.33, and 4.29)	76e

ushinsunine. Under Clemmensen reductive conditions, the well-characterized (±)-roemerine was obtained. Oxidation of ushinsunine with manganese dioxide gave the yellow oxoaporphine alkaloid liriodenine whose structure had already been elucidated (Scheme XVII).[78a]

The NMR spectrum of (−)-ushinsunine showed $J_{6a,7} = 2.5$ Hz indicating that the two protons are in a *cis* relationship to one another. The absolute configuration was established through catalytic hydrogenolysis of the C-7 hydroxyl group to yield the known alkaloid (−)-roemerine.[78b]

XV. STEPORPHINE: A C-4 HYDROXYLATED APORPHINE

The structural elucidation of the unusual aporphine alkaloid steporphine is based almost entirely on spectroscopic methods.[79] Steporphine, $[\alpha]_D -90.6°$ (methanol),

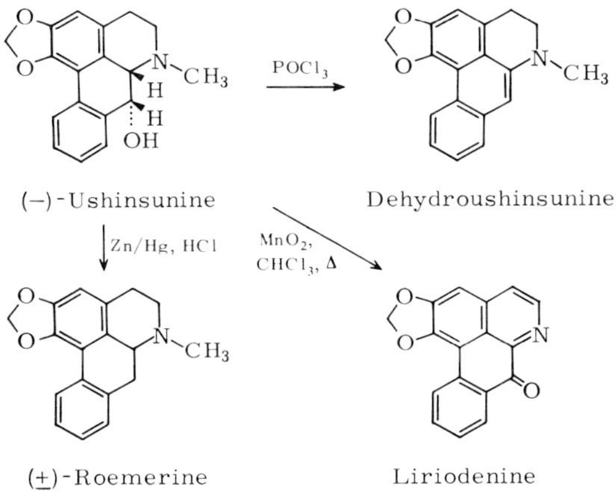

Scheme XVII

shows a molecular ion peak at m/e 295 for $C_{18}H_{17}O_3N$. The UV spectrum exhibited λ_{max}^{EtOH} 238, 273, 293, and 312 mμ (4.25, 4.25, 3.87, and 3.55), somewhat characteristic of ring D unsubstituted aporphines, and showed no bathochromic shift upon the addition of base, denoting the absence of a phenolic group.

The IR spectrum showed hydroxyl absorption at 2.86 – 2.98 μ (3360 – 3500 cm^{-1}) and methylenedioxy absorption at 7.17, 7.43, 9.52, and 10.64 μ (1395, 1345, 1050, and 940 cm^{-1}).

The NMR spectrum is summarized in the following diagram. Of particular importance is the downfield absorption of the C-3 aromatic proton at δ6.84. A C-3 proton in an aporphine usually comes at δ6.7, so that the present deshielding must be assigned to the proximity of the C-4 oxygen atom. Additionally, the C-4 proton appears as a clean triplet centered at δ4.47 and split by the C-5 methylene hydrogens.

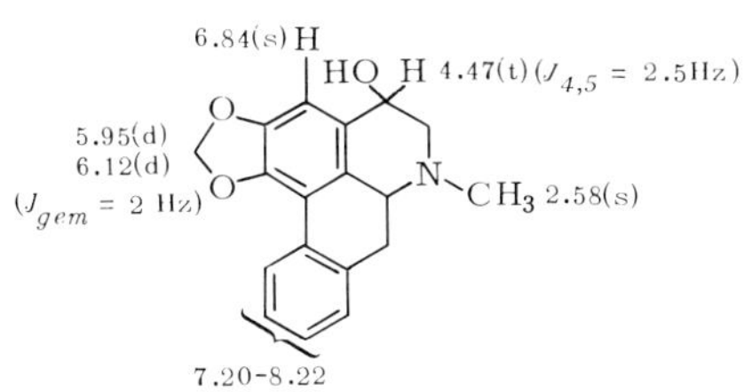

NMR spectral values for steporphine

The alkaloid gave a monoacetate derivative upon treatment with acetic anhydride and pyridine, which showed an acetyl absorption in the IR at 5.78 μ (1730 cm^{-1}).

The mass spectrum of steporphine was also very informative. The base peak was at *m/e* 252 and can be depicted as derived from the molecular ion by a retrograde Diels–Alder fragmentation. Other strong peaks were at *m/e* 294 and 276.

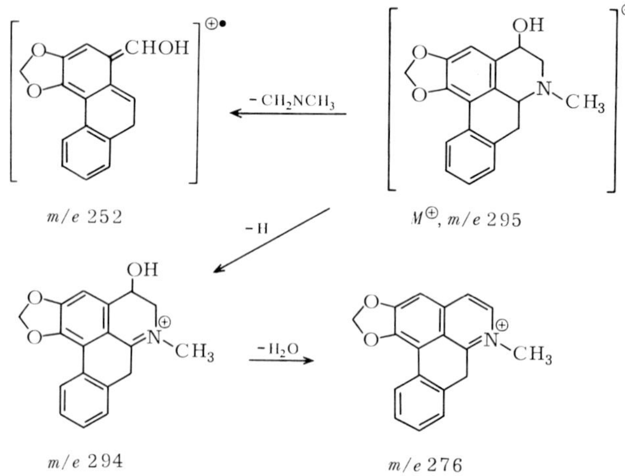

XVI. The Dehydroaporphines

The dehydroaporphines are a recently discovered, optically inactive, subgroup of the aporphine alkaloids with unsaturation at C-6a(7). The first member of this subgroup to be isolated from natural sources was dehydrodicentrine.[80,81] For purposes of characterization, oxidation of the known aporphine dicentrine with mild permanganate yielded dehydrodicentrine in 50% yield.[22,80]

Dehydroocopodine[1b] and dehydroglaucine[82] have also been obtained from plants. Dehydroaporphines are sometimes found as dehydroaporphine–benzylisoquinoline dimers (see Chapter 12).

The NMR spectrum of dehydrodicentrine is of interest and is summarized here[80]:

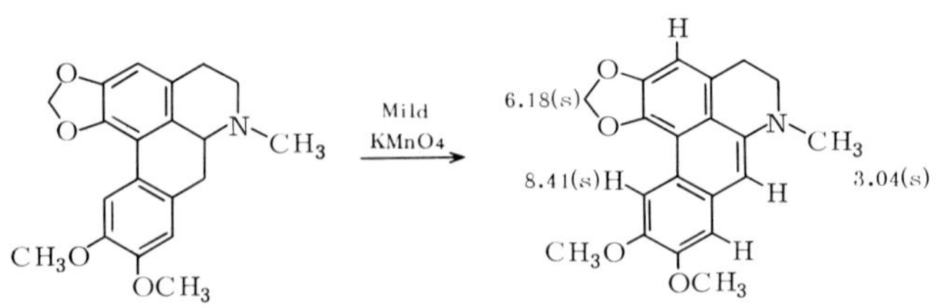

Dicentrine

Dehydrodicentrine
(Three unsplit aromatic protons at δ6.51, 6.87, and 7.02)

The low-field C-11 aromatic proton and *N*-methyl protons are typical of dehydroaporphines.

The UV spectra of the dehydroaporphines indicate a highly conjugated system, and three examples follow. The presence of an absorption maximum or shoulder between 252 and 265 mμ is diagnostic of a C-6a to C-7 double bond.

Dehydrodicentrine[80]	λ_{max}^{EtOH} 263, 302, and 340 mμ (4.74, 3.85, and 4.10)
Dehydronuciferine[18]	λ_{max}^{EtOH} 253, 264, 293, and 327 mμ (4.97, 4.98, 4.26, and 4.49)
Dehydronoraporphine hydroiodide[17a] (not a natural product)	λ_{max}^{EtOH} 255 and 323 mμ (4.62 and 3.94)
Dehydroocopodine[83]	λ_{max}^{EtOH} 220, 262 sh, 267, and 340 mμ (4.37, 4.65, 4.66, and 4.50)

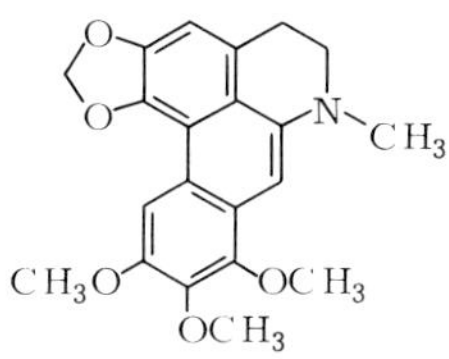

Dehydroocopodine[1b]

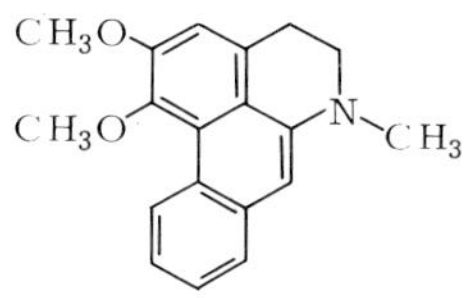

Dehydronuciferine[18]

REFERENCES

1. For reviews on the aporphines, see M. Shamma and W. A. Slusarchyk, *Chem. Rev.* **64**, 59 (1964); and R. H. F. Manske in "The Alkaloids" (R. H. F. Manske, ed.), Vol. **4**, p. 119. Academic Press, New York, 1954.

1a. For a more recent review see M. Shamma in "The Alkaloids" (R. H. F. Manske, ed.), Vol. **9**, p. 1. Academic Press, New York, 1967.

1b. Also see M. P. Cava and A. Venkateswarlu, *Annu. Pep. Med. Chem.* p. 331 (1969).

2. R. H. F. Manske, *in* "The Alkaloids" (R. H. F. Manske, ed.), Vol. **4**, p. 119. Academic Press, New York, 1954.

3. J. Gadamer and F. Kuntze, *Arch. Pharm.* (*Weinheim*) **249**, 598 (1911).

4. E. Späth, H. Holter, and R. Posega, *Chem. Ber.* **61**, 322 (1928).

5. For a discussion of apomorphine, see K. W. Bentley, *in* "The Chemistry of the Morphine Alkaloids," p. 302. Oxford Univ. Press, London and New York, 1954; G. Stork, *Alkaloids* **2**, 195 (1952).

5a. B. Franck, H. I. Lubs, and G. Dunkelmann, *Angew. Chem. Int. Ed. Engl.* **6**, 969 (1967).

6. R. Tschesche, P. Welzel, R. Moll, and G. Legler, *Tetrahedron* **20**, 1435 (1964).

7. F. Zymalkowski and K. H. Happel, *Chem. Ber.* **102**, 2959 (1969).

8. M. P. Cava and D. R. Dalton, *J. Org. Chem.* **31**, 1281 (1966).

9. J. A. Weisbach, C. Burns, E. Macko, and B. Douglas, *J. Med. Chem.* **6**, 91 (1963).

10. S. Ishiwata and K. Itakura, *Chem. Pharm. Bull.* **17**, 1298 and 2216 (1969).

11. See also A. R. Battersby, A. K. Bhatnagar, P. Hackett, C. W. Thornber, and J. Staunton, *Chem. Commun.* p. 1214 (1968).

12. M. S. Gibson and J. M. Walthew, *Chem. Ind. (London)* p. 185 (1965); M. S. Gibson, G. W. Prenton, and J. W. Walthew, *J. Chem. Soc., C* p. 2234 (1970).
13. For the synthesis of (±)-domesticine by the benzyne approach, see S. V. Kessar, S. Batra, and S. S. Gandhi, *Indian. J. Chem.* **8**, 468 (1970).
14. J. L. Neumeyer, B. R. Neustadt, and J. W. Weintraub, *Tetrahedron Lett.* p. 3107 (1967).
15. J. L. Neumeyer, K. H. Oh, K. K. Weinhardt, and B. R. Neustadt, *J. Org. Chem.* **34**, 3786 (1969); J. L. Neumeyer, B. R. Neustadt, and K. K. Weinhardt, *J. Pharm. Sci.* **59**, 1850 (1970).
16. M. P. Cava and M. V. Lakshmikantham, *J. Org. Chem.* **35**, 1867 (1970).
17. M. P. Cava and M. Srinivasan, (in press).
17a. J. L. Neumeyer and F. E. Granchelli, *Tetrahedron Lett.* p. 5261 (1970).
18. M. P. Cava, S. C. Havlicek, A. Lindert, and R. J. Spangler, *Tetrahedron Lett.* p. 2937 (1966).
19. M. P. Cava, M. J. Mitchell, S. C. Havlicek, A. Lindert, and R. J. Spangler, *J. Org. Chem.* **35**, 175 (1970).
20. N. C. Yang, G. R. Lenz, and A. Shani, *Tetrahedron Lett.* p. 2941 (1966); also N. C. Yang, A. Shani, and G. R. Lenz, *J. Amer. Chem. Soc.* **88**, 5369 (1966).
21. S. M. Kupchan and R. M. Kanojia, *Tetrahedron Lett.* p. 5353 (1966); also S. M. Kupchan, J. L. Moniot, R. M. Kanojia, and J. B. O'Brien, *J. Org. Chem.* **36**, 2413 (1971).
21a. T. Kametani, M. Koizumi, K. Shishido, and K. Fukumoto, *J. Chem. Soc., C* p. 1923 (1971).
22. N. M. Mollov and H. B. Dutschewska, *Tetrahedron Lett.* p. 853 (1966).
23. V. Preininger, J. Hrbek, Jr., Z. Samek, and F. Šantavý, *Arch. Pharm.* (*Weinheim*) **302**, 808 (1969); see also *Chem. Abstr.* **76**, 25454h (1972).
24. K. Rehse, *Arch. Pharm.* (*Weinheim*) **302**, 487 (1969).
25. S. R. Johns, J. A. Lamberton, and A. A. Sioumis, *Aust. J. Chem.* **19**, 2339 (1966).
25a. E. S. Smissman, A. C. Makriyannis, and E. J. Walaszek, *J. Med. Chem.* **13**, 640 (1970).
26. R. Tschesche, P. Welzel, and G. Legler, *Tetrahedron Lett.* p. 445 (1965).
27. R. J. Vavrek, J. G. Cannon, and R. V. Smith, *J. Pharm. Sci.* **59**, 823 (1970).
27a. M. P. Cava, A. Venkateswarlu, M. Srinivasan, and D. L. Edie, (in press).
28. M. Shamma, *Experientia* **16**, 484 (1960).
29. K. W. Bentley and H. M. E. Cardwell, *J. Chem. Soc., London* p. 3252 (1955).
30. W. A. Ayer and W. I. Taylor, *J. Chem. Soc., London* p. 472 (1956).
31. C. Djerassi, K. Mislow, and M. Shamma, *Experientia* **18**, 53 (1962).
32. M. J. Vernengo, *Experientia* **17**, 420 (1961).
33. J. C. Craig and S. K. Roy, *Tetrahedron* **21**, 395 (1965).
34. M. Shamma and M. J. Hillman, *Experientia* **25**, 544 (1969).
35. K. Kotera, Y. Hamada, and R. Mitsui, *Tetrahedron* **24**, 2463 (1968).
36. T. Ashida, R. Pepinsky, and Y. Okaya, see reference 1.
37. D. H. R. Barton and T. Cohen, *Festschr. Prof. Dr. Arthur Stoll siebzigsten Geburtstag 1957* p. 117 (1957).
38. B. Franck, G. Blaschke, and G. Schlingloff, *Tetrahedron Lett.* p. 439 (1962); B. Franck and G. Blaschke, *Justus Liebigs Ann. Chem.* **695**, 144 (1966).
39. T. Kametani and I. Noguchi, *J. Chem. Soc., C* p. 1440 (1967).
40. S. M. Albonico, A. M. Kuck, and V. Deulofeu, *Chem. Ind.* (*London*) p. 1580 (1964).
41. B. Franck, G. Blaschke, and G. Schlingloff, *Angew. Chem., Int. Ed. Engl.* **3**, 192 (1964).
42. B. Franck and L.-F. Tietze, *Angew. Chem., Int. Ed. Engl.* **6**, 799 (1967).
43. See also B. Franck, G. Dunkelmann, and H. J. Lubs, *Angew. Chem., Int. Ed. Engl.* **6**, 1075 (1967).
44. A. H. Jackson and J. A. Martin, *Chem. Commun.* p. 420 (1965); *J. Chem. Soc., C* pp. 2061 and 2222 (1966).
45. W. W.-C. Chan and P. Maitland, *J. Chem. Soc., C* p. 753 (1966).
46. T. Kametani, K. Fukumoto, A. Kozuka, H. Yagi, and M. Koizumi, *J. Chem. Soc., C* p. 2034 (1969).
47. T. Kametani, T. Sugahara, H. Yagi, and K. Fukumoto, *Tetrahedron* **25**, 3667 (1969).

48. A. R. Battersby, T. H. Brown, and J. H. Clements, *J. Chem. Soc., London* p. 4550 (1965).
49. A. R. Battersby, R. T. Brown, J. H. Clements, and G. G. Iverach, *Chem. Commun.* p. 230 (1965); also A. R. Battersby, T. J. Brocksom, and R. Ramage, *ibid.* p. 464 (1969).
50. M. J. S. Dewar and T. Nakaya, *J. Amer. Chem. Soc.* **90**, 7134 (1968).
51. M. A. Schwartz, R. A. Holton, and S. W. Scott, *J. Amer. Chem. Soc.* **91**, 2800 (1969).
52. W. L. Carrick, G. L. Karapinka, and G. T. Kwiatkowski, *J. Org. Chem.* **34**, 2388 (1969).
53. For a discussion of experimental conditions for phenolic oxidative coupling, see H. Musso, *in* "Oxidative Coupling of Phenols" (W. I. Taylor and A. R. Battersby, eds.), p. 1. Marcel Dekker, New York, 1967.
53a. S. Tobinaga and E. Kotani, *J. Amer. Chem. Soc.* **94**, 309 (1972).
54. D. H. R. Barton, D. S. Bhakuni, G. M. Chapman, and G. W. Kirby, *Chem. Commun.* p. 259 (1966).
55. D. H. R. Barton, D. S. Bhakuni, G. M. Chapman, and G. W. Kirby, *J. Chem. Soc., London* p. 2134 (1967).
56. D. H. R. Barton, R. H. Hesse, and G. W. Kirby, *J. Chem. Soc., London* p. 6379 (1965).
57. A. R. Battersby, R. J. Francis, E. A. Ruveda, and J. Staunton, *Chem. Commun.* p. 89 (1965).
58. G. Blaschke, *Arch. Pharm. (Weinheim)* **301**, 432 (1968).
59. G. Blaschke, *Arch. Pharm. (Weinheim)* **303**, 358 (1970).
59a. A. R. Battersby, J. L. McHugh, J. Staunton, and M. Todd, *Chem. Commun.* p. 985 (1971).
60. A. R. Battersby, *in* "Oxidative Coupling of Phenols" (W. I. Taylor and A. R. Battersby, eds.), p. 124. Marcel Dekker, New York, 1967.
61. M. Shamma, *Experientia* **18**, 64 (1962).
62. J. Slavík, L. Slavíková, and L. Dolejš, *Collect. Czech. Chem. Commun.* **33**, 4066 (1968).
63. M. V. Koch, J. G. Cannon, and A. M. Burkman, *J. Med. Chem.* **11**, 977 (1968).
64. H. Kreitmair, *Pharmazie* **7**, 507 (1952).
65. B. L. Konson and P. P. Saksonov, *Farmakol. Toksikol. (Moscow)* **9**, 14 (1946).
66. A. K. Reynolds and L. O. Randall, "Morphine and Related Drugs," p. 223. Univ. of Toronto Press, Toronto, 1957.
67. V. V. Berezhinskaya, E. E. Aleshinskaya, and Y. A. Aleshina, *Farmakol. Toksikol. (Moscow)* **31**, 44 (1968); *Chem. Abstr.* **68**, 94521 g (1968).
68. K. Wada and K. Munakata, *J. Agr. Food Chem.* **16**, 471 (1968).
69. "The Dispensary of the U.S.A.," 25th ed., p. 102. Lippincott, Philadelphia, Pennsylvania, 1955.
69a. S. Düby, G. C. Cotzias, A. Steck, and P. S. Papavasiliou, as quoted in *Chem. Eng. News*, April 19, 1971, p. 49; G. C. Cotzias, P. S. Papavasiliou, C. Fehling, B. Kaufman, and E. Mena, *New Engl. J. Med.* **282**, 31 (1970) and references cited therein.
70. M. Ohashi, J. M. Wilson, H. Budzikiewicz, M. Shamma, W. A. Slusarchyk, and C. Djerassi, *J. Amer. Chem. Soc.* **85**, 2807 (1963).
71. A. H. Jackson and J. A. Martin, *J. Chem. Soc., C* p. 2181 (1966).
71a. H. M. Fales, H. A. Lloyd, and G. W. A. Milne, *J. Amer. Chem. Soc.* **92**, 1590 (1970).
72. I. R. C. Bick, J. Harley-Mason, N. Sheppard, and M. J. Vernengo, *J. Chem. Soc., London* p. 1896 (1961).
73. W. H. Baarschers, R. R. Arndt, K. Pachler, J. A. Weisbach, and B. Douglas, *J. Chem. Soc., London* p. 4778 (1964).
73a. T.-H. Yang, *J. Pharm. Soc. Jap.* **82**, 804 (1962).
74. H. M. Fales and K. S. Warren, *J. Org. Chem.* **32**, 501 (1967).
75. See also M. Tomita, S.-T. Lu, and S.-J. Wang, *J. Pharm. Soc. Jap.* **85**, 827 (1965).
76. K. G. R. Pachler, R. R. Arndt, and W. H. Baarschers, *Tetrahedron* **21**, 2159 (1965).
76a. S. R. Johns, J. A. Lamberton, C. S. Li, and A. A. Sioumis, *Aust. J. Chem.* **23**, 363 (1970).
76b. S. R. Johns, J. A. Lamberton, C. S. Li, and A. A. Sioumis, *Aust. J. Chem.* **23**, 423 (1970).
76c. M. Shamma, R. J. Shine, and B. S. Dudock, *Tetrahedron* **23**, 2887 (1967); M. Shamma, M. J. Hillman, R. Charubala, and B. R. Pai, *Indian J. Chem.* **7**, 1056 (1969).

76d. T. R. Govindachari, N. Viswanathan, R. Charubala, and B. R. Pai, *Indian J. Chem.* **7**, 841 (1969).
76e. P. E. Sonnet and M. Jacobson, *J. Pharm. Sci.* **60**, 1254 (1971).
77. Holubek and Štrouf, Spectrum No. 22.
78. M. Shamma, S. Y. Yao, B. R. Pai, and R. Charubala, *J. Org. Chem.* **36**, 3253 (1971).
78a. T.-H. Yang, *J. Pharm. Soc. Japan* **82**, 798 and 804 (1962).
78b. J. Kunitomo, M. Miyoshi, E. Yuge, T.-S. Yang, and C.-M. Chen, *Chem. Pharm. Bull.* **19**, 1502 (1971).
79. J. Kunitomo, Y. Okamoto, E. Yuge, and Y. Nagai, *Tetrahedron Lett.* p. 3287 (1969).
80. M. P. Cava, Y. Watanabe, K. Bessho, M. J. Mitchell, A. I. daRocha, B. Hwang, B. Douglas, and J. A. Weisbach, *Tetrahedron Lett.* p. 2437 (1968).
81. H. G. Kiryakov and P. Panov, *Dokl. Bolg. Akad. Nauk* **22**, 1019 (1969).
82. H. G. Kiryakov, *Chem. Ind.* (*London*) p. 1807 (1968).
83. M. P. Cava and A. Venkateswarlu, *Tetrahedron* **27**, 2639 (1971).

Chapter 11/PAKISTANAMINE: A PROAPORPHINE-BENZYLISOQUINOLINE DIMER

Occurrence: Berberidaceae
Structure:

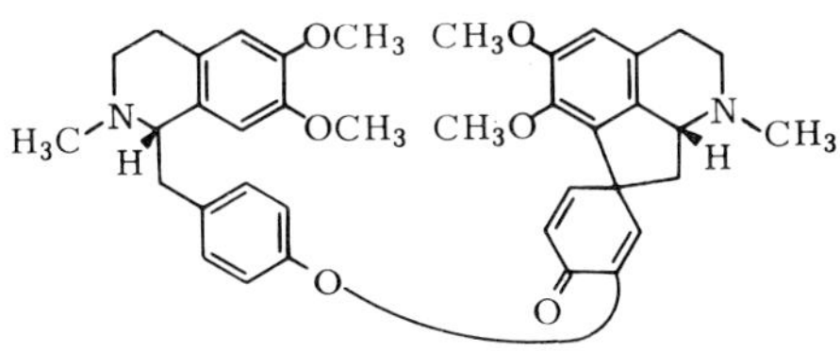

(+)-Pakistanamine

I. Structural Elucidation and Chemical Conversions

Shamma, Miana, and co-workers have reported the isolation and characterization of pakistanamine, the first known proaporphine–benzylisoquinoline dimer. The alkaloid was found in *Berberis baluchistanica* Ahrendt (Berberidaceae), where it is accompanied by the aporphine–benzylisoquinoline dimer pakistanine (**1a**)[1] (see Chapter 12).

Acid-catalyzed dienone–phenol rearrangement of (+)-pakistanamine gave 1-*O*-methylpakistanine (**1b**), and further *O*-methylation with diazomethane produced the known *O*,*O*-dimethylpakistanine (**1c**) of established absolute configuration (Scheme I).

Sodium borohydride reduction of pakistanamine yielded pakistanaminol which was subjected to an acid-catalyzed dienol–benzene rearrangement. The product was 1,2-

dimethoxy-9-[(+)-armepavinyloxy]aporphine (**2**). The most significant feature of the NMR spectrum of **2** was the C-11 one proton doublet at $\delta 8.30$ ($J_{10,11} = 10$ Hz).

The presence of pakistanine and pakistanamine in the same plant lends substantial support for the biogenetic scheme involving the sequence: bisbenzylisoquinoline → proaporphine–benzylisoquinoline dimer → aporphine–benzylisoquinoline dimer.[1]

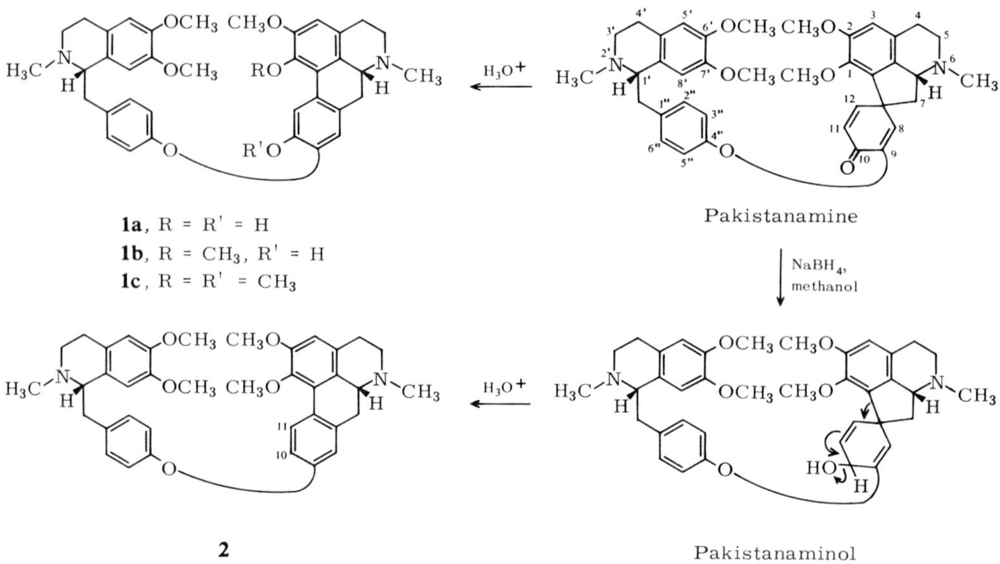

Scheme I

II. Mass Spectroscopy

The parent peak in the mass spectrum of pakistanamine corresponded to the molecular ion and was at *m/e* 622, while the base peak, at *m/e* 206, was due to rings A and B of the benzylisoquinoline portion of the alkaloid. Other important fragments, in decreasing order of intensity, were at *m/e* 416, 310, 326, and 281.

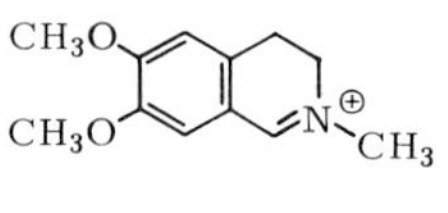

m/e 206
($C_{12}H_{16}O_2N$)

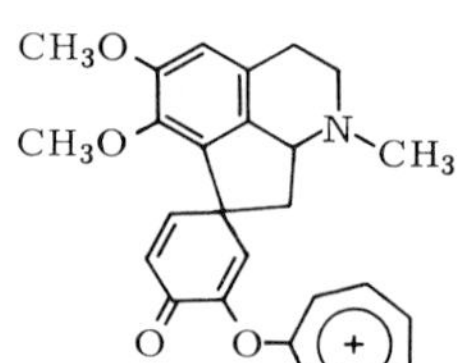

m/e 416
($C_{22}H_{26}O_4N$)

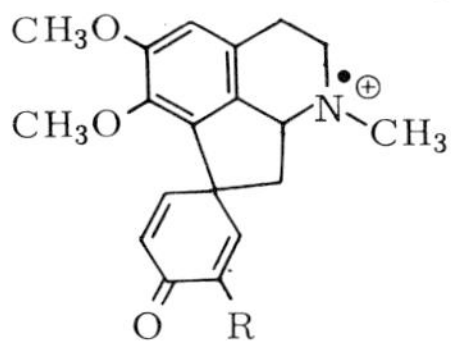

m/e 310, R = •
m/e 326, R = O•
($C_{19}H_{20}O_3N$ and $C_{19}H_{20}O_4N$)

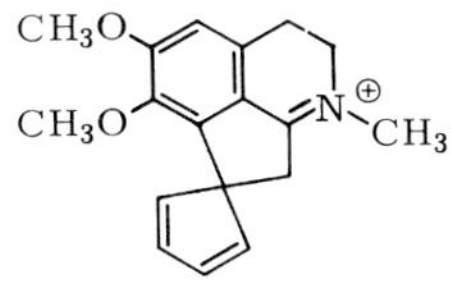

m/e 281
($C_{18}H_{19}O_2N$)

III. NMR Spectroscopy

The three vinylic hydrogens in ring D of the proaporphine half of pakistanamine could be clearly observed as noted below.

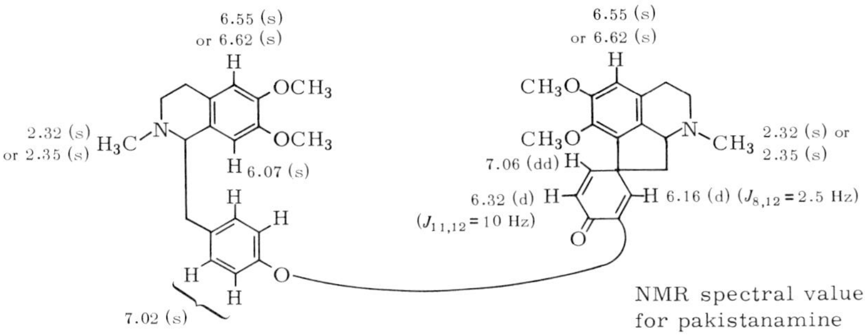

NMR spectral values for pakistanamine

Four methyl singlets at δ3.57, 3.64, 3.80, and 3.82.

IV. UV and IR Spectroscopy

Pakistanamine[1]: λ_{max}^{EtOH} 225 sh, 280, and 310 sh mμ (4.63, 4.12, and 3.61)

λ^{CHCl_3} 5.99 μ (1670 cm^{-1}) and 6.10 μ (1640 cm^{-1})

REFERENCE

1. M. Shamma, J. L. Moniot, S. Y. Yao, G. A. Miana, and M. Ikram, *J. Amer. Chem. Soc.* **95**, 1381 (1972).

Chapter 12 / THE APORPHINE-BENZYLISOQUINOLINE DIMERS

Occurrence: Berberidaceae, Hernandiaceae, and Ranunculaceae
Number: 8
Structures:

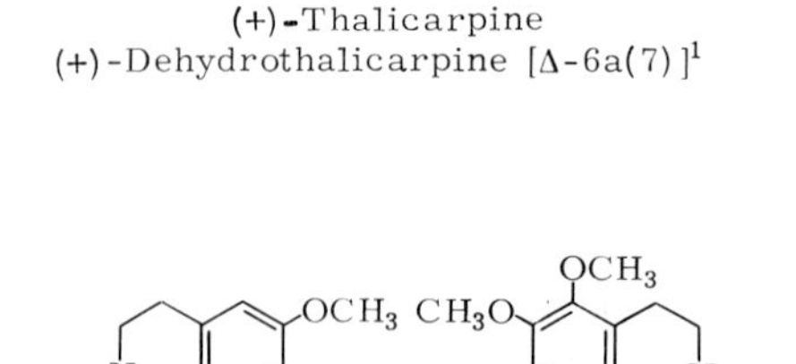

(+)-Thalicarpine
(+)-Dehydrothalicarpine [Δ-6a(7)][1]

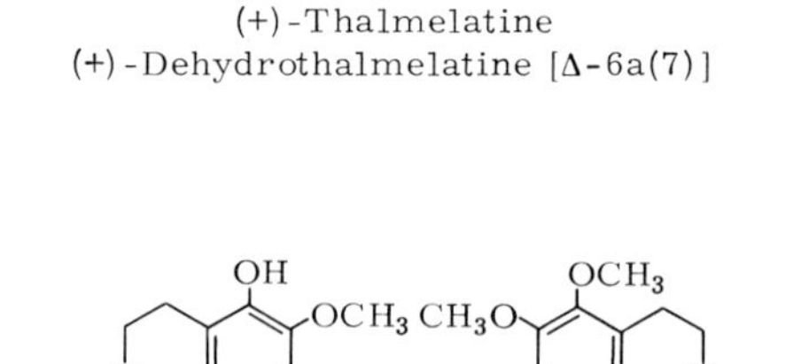

(+)-Thalmelatine
(+)-Dehydrothalmelatine [Δ-6a(7)]

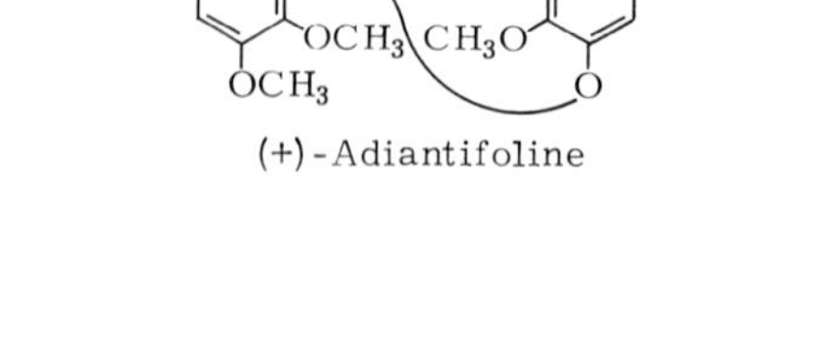

(+)-Adiantifoline

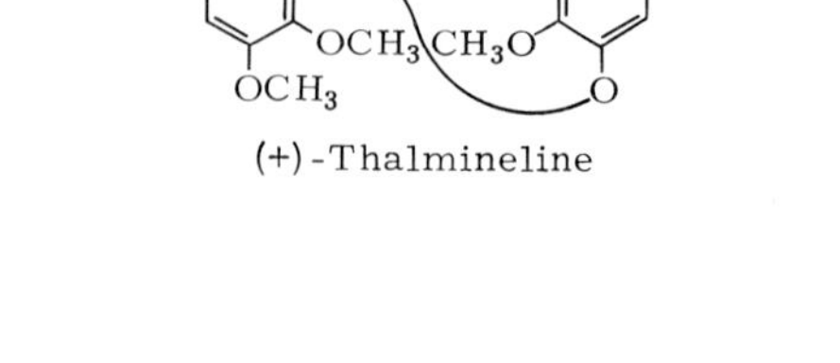

(+)-Thalmineline

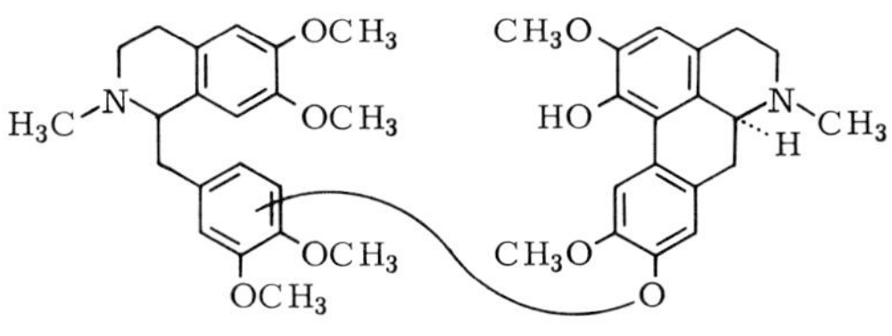

(+)-Fetidine[1a](?)

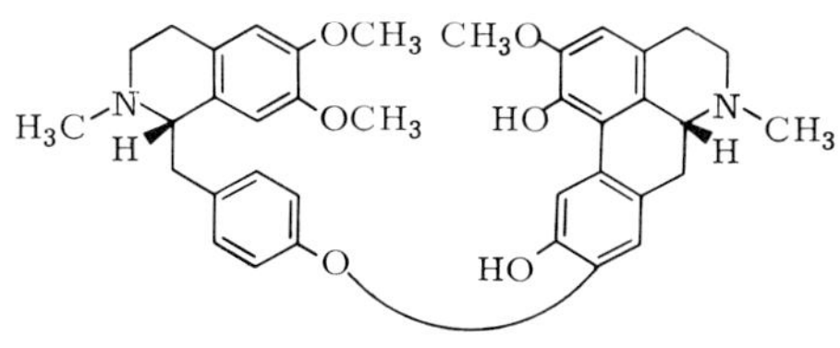

(+)-Pakistanine

I. Introduction

The dimeric aporphine–benzylisoquinoline alkaloids have been found in the genus *Thalictrum*, family Ranunculaceae, except for thalicarpine, which has been found in *Thalictrum* species and also in *Hernandia ovigera* L. (Hernandiaceae). Pakistanine, which is present in a member of the Berberidaceae, is an unusual aporphine–benzylisoquinoline very recently characterized.

From a historical standpoint, the first aporphine–benzylisoquinoline alkaloid to be fully characterized was thalicarpine, with which dehydrothalicarpine and thalmelatine were later correlated chemically.

II. Structural Elucidation and Synthesis

A. Thalicarpine

The elemental analysis of thalicarpine supported the formula $C_{41}H_{48}O_8N_2$. The alkaloid has seven *O*-methyl and two *N*-methyl groups with the remaining oxygen atom existing as part of a diphenyl ether linkage.

Sodium in liquid ammonia cleavage afforded (−)-6′-hydroxylaudanosine and (+)-2,10-dimethoxyaporphine, together with the tetracyclic amino ketone **1** formed through further reduction of the aforementioned aporphine (Scheme I).

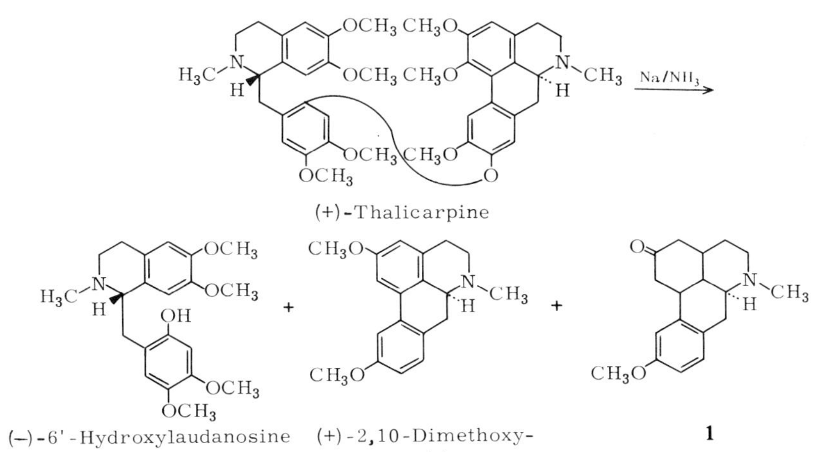

(+)-Thalicarpine

(−)-6′-Hydroxylaudanosine (+)-2,10-Dimethoxyaporphine **1**

Scheme I

The sodium in liquid ammonia cleavage of thalicarpine did not settle conclusively the site of attachment of the benzylisoquinoline moiety to the aporphine skeleton. Final proof of structure came from synthesis. Ullmann condensation of (+)-6′-bromolaudanosine with (+)-*N*-methyllaurotetanine yielded (+)-thalicarpine, identical with the natural base.[2]

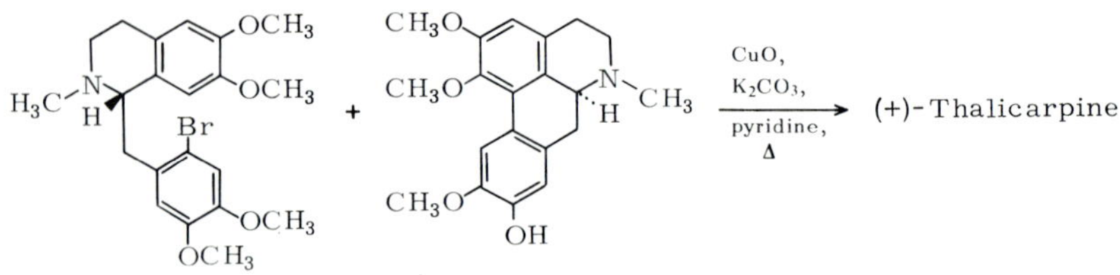

(+)-6′-Bromolaudanosine (+)-*N*-Methyllaurotetanine

The NMR and UV spectra of thalicarpine were also of significant assistance in the structural determination. The alkaloid shows one aromatic proton singlet at δ8.19, indicating that the C-11 position in thalicarpine must be unsubstituted, and the UV spectrum exhibits maxima at 282 and 301 mμ, pointing to a 1,2,9,10-substitution pattern for the aporphine moiety.

B. Dehydrothalicarpine

Dehydrothalicarpine was assigned the molecular formula $C_{41}H_{46}O_8N_2$, two hydrogen atoms less than thalicarpine, on the basis of elemental analyses. The NMR spectrum showed the presence of two *N*-methyl and seven *O*-methyl groups, but one of the *N*-methyl groups appeared at the unusually low field of δ2.99. The UV spectrum showed maxima at 265 and 331 mμ, pointing to a more highly conjugated chromophore than in thalicarpine.[3,4]

Sodium in liquid ammonia cleavage yielded 6′-hydroxylaudanosine, 2,10-dimethoxydehydroaporphine, and the oxoaporphines **2** and **3**, the last two compounds being oxidation products from the reaction work-up. These data led to the postulation that the alkaloid should be formulated as a dehydrothalicarpine, and indeed catalytic reduction of dehydrothalicarpine led to thalicarpine. Alternatively, oxidation of thalicarpine with 2,3-dichloro-5,6-dicyano-1,4-benzoquinone gave dehydrothalicarpine. The downfield δ2.99 *N*-methyl absorption in dehydrothalicarpine was then assigned to the aporphine moiety where the nitrogen atom is part of a highly conjugated enamine

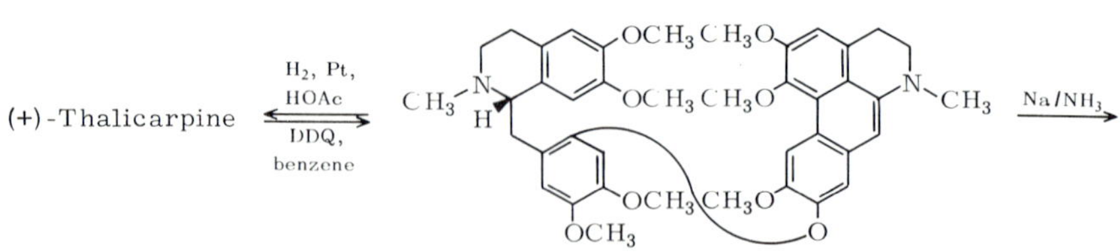

(+)-Dehydrothalicarpine

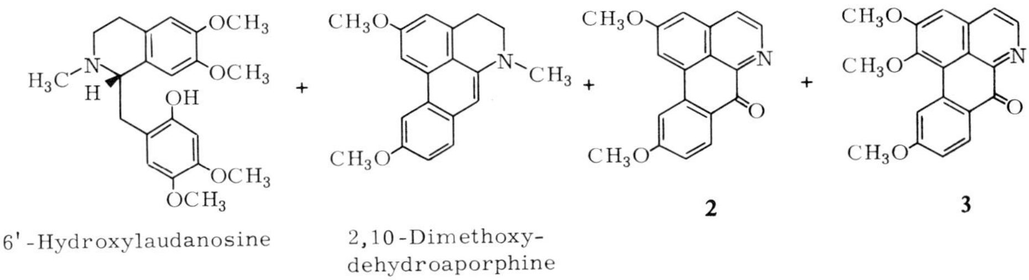

Scheme II

system (Scheme II). The *N*-methyl groups in simple dehydroaporphines are also known to appear relatively downfield. (See Chapter 10, Section XVI.)

C. *Thalmelatine*

Thalmelatine, $C_{40}H_{46}O_8N_2$, was investigated by the Bulgarian team of Mollov and Dutschewska. A phenolic function is present in the alkaloid, together with six methoxyls and two *N*-methyl groups. It was readily ascertained that this alkaloid must correspond to de-*O*-methylthalicarpine since thalicarpine was obtained upon *O*-methylation.

In order to settle the position of the phenolic function, *O*-ethylthalmelatine was oxidized with potassium permanganate and the product shown to be the known lactam **4** (Scheme III).[5]

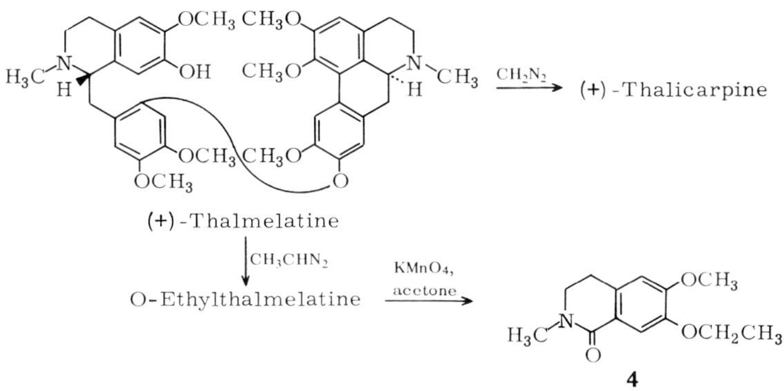

Scheme III

D. *Adiantifoline*

The structural work on adiantifoline parallels that for thalicarpine and consists mainly of a sodium in liquid ammonia cleavage coupled with spectral analysis. The alkaloid was synthesized by the Ullmann condensation of (+)-6′-bromolaudanosine with (+)-1,2,3,10-tetramethoxy-9-hydroxyaporphine.[6]

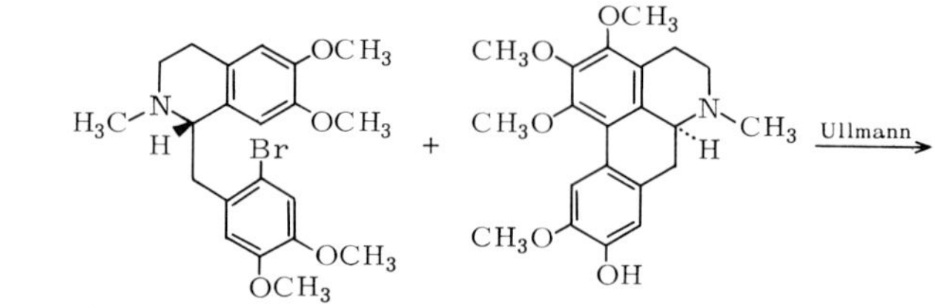

(+)-6'-Bromolaudanosine (+)-1,2,3,10-Tetramethoxy-9-hydroxyaporphine

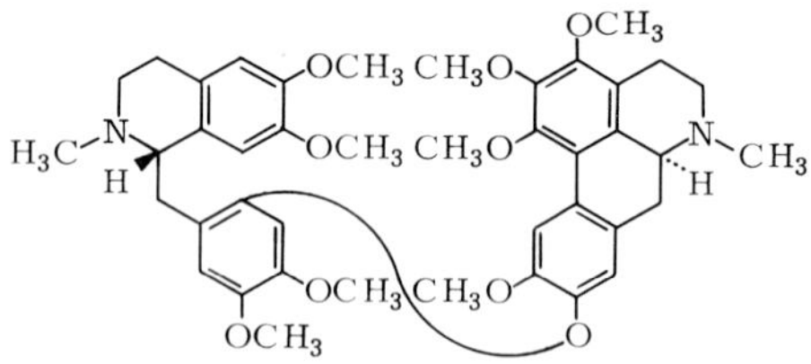

(+)-Adiantifoline

E. *Thalmineline*

The data on the phenolic alkaloid thalmineline are almost exclusively spectral in nature. A small molecular ion could be observed at *m/e* 742, $C_{42}H_{50}O_{10}N_2$. Facile cleavage of the C-1′ to C-α bond resulted in strong peaks at *m/e* 222 and 206, while the remainder of the thalmineline molecule gave rise to peaks at *m/e* 519, 520, and 521.[7]

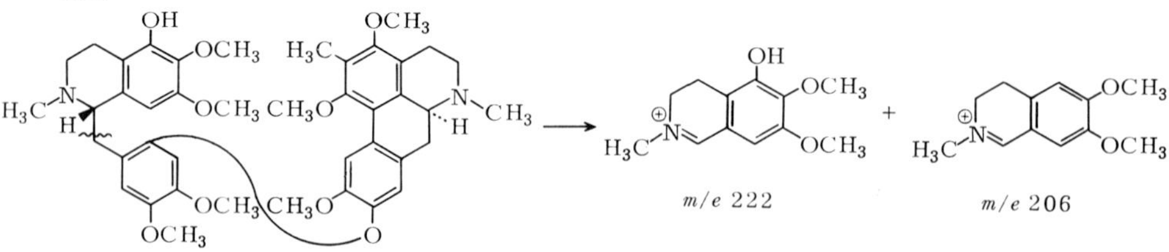

(+)-Thalmineline *m/e* 222 *m/e* 206

The NMR spectrum for the alkaloid is summarized and interpreted in Section VI.

III. Biosynthesis

No studies with labeled precursors have yet been reported, but a sequence involving proaporphines has been suggested for the aporphine–benzylisoquinoline dimers (Scheme IV).[8,16]

IV. Pharmacology

Thalicarpine has shown weak hypotensive activity in the cat. Respiratory toxicity and weak adrenolytic activity accompanied this action.[9]

Thalicarpine, thalmelatine, and hernandaline (Section VIII) have shown cytotoxic activity towards monolayer KB cell cultures, indicating they may have potential as

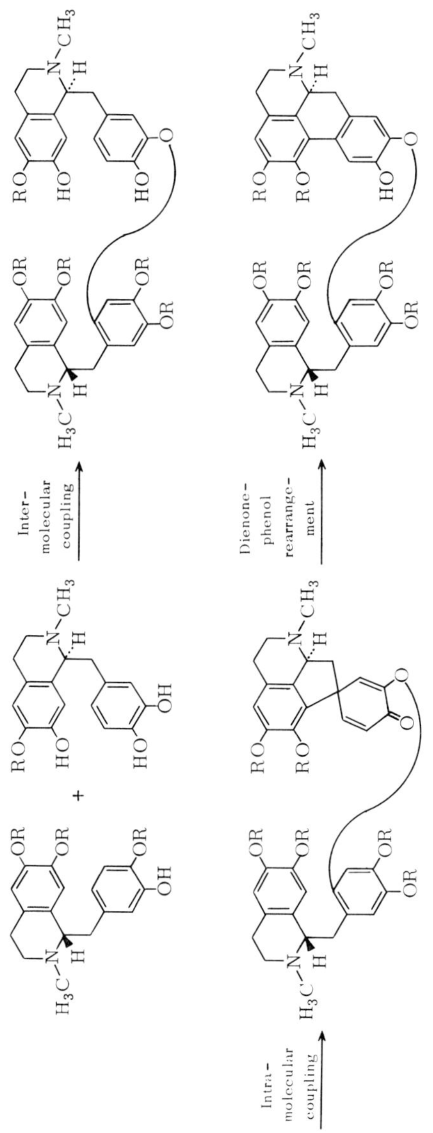
Intra-
molecular
coupling
Inter-
molecular
coupling
Dienone-
phenol
rearrange-
ment

Scheme IV

anticancer drugs. Furthermore, thalicarpine has significant inhibitory activity against the Walker intramuscular carcinosarconia 256 in rats,[9a] and has been selected for clinical trials.

V. Circular Dichroism

For thalicarpine and adiantifoline three clearly defined Cotton effect curves, two negative and one positive, have been observed.[10]

Thalicarpine: $[\Theta]_{306}$ $-16{,}300$, $[\Theta]_{275}$ $-18{,}000$, $[\Theta]_{239}$ $+242{,}000$
Adiantifoline: $[\Theta]_{305}$ $-33{,}800$, $[\Theta]_{275}$ $-31{,}200$, $[\Theta]_{241}$ $+234{,}000$

VI. NMR Spectroscopy

Proton magnetic resonance is extremely useful in the structural elucidation of the aporphine–benzylisoquinoline dimers. The values of the chemical shifts for these alkaloids are given in Table I.

TABLE I

Chemical Shifts (δ) of Aporphine–Benzylisoquinoline Dimers

	$N-CH_3$	$O-CH_3$	ArH
Thalicarpine[11,12]	δ2.45 (3H)	δ3.60 (C-1) (3H)	δ6.21 (1H)
	2.48 (3H)	3.71 (3H)	6.53 (1H)
		3.79 (3H)	6.60 (1H)
		3.81 (3H)	6.63 (1H)
		3.83 (3H)	6.68 (1H)
		3.91 (3H)	8.23 (C-11) (1H)
		3.95 (3H)	
Dehydrothalicarpine[4]	2.42 (3H)	3.26 (C-1) } (21H)	6.23 (1H)
	2.99 (3H),	3.56	6.47 (1H)
	aporphine	3.78	6.52 (1H)
	moiety.	3.95	6.69 (1H)
		4.01	6.82 (1H)
		4.08	6.99 (1H)
			9.25 (C-11) (1H)
Adiantifoline[10]	2.44 (3H)	3.59 (C-1) (3H)	6.24 (1H)
	2.47 (3H)	3.78 (9H)	6.55 (2H)
		3.82 (3H)	6.60 (1H)
		3.89 (3H)	8.08 (C-11) (1H)
		3.94 (3H)	
		3.96 (3H)	
Thalmineline[7]	2.45 (3H)	3.47 (C-1) } (24H)	5.71 (C-8′) (1H)
	2.50 (3H)	3.70	6.43 (1H)
		3.73	6.48 (1H)
		3.76	6.66 (1H)
		3.83	7.98 (C-11) (1H)
		3.89	

VII. UV Spectroscopy

Thalicarpine[2]	λ_{max}^{EtOH} 282 and 301 mμ (4.26 and 4.02)
Dehydrothalicarpine[4]	λ_{max}^{EtOH} 265 and 331 mμ (4.70 and 4.13)
Adiantifoline[10]	λ_{max}^{EtOH} 283, 302, and 312 mμ (4.51, 4.39, and 4.34)
Thalmineline[7]	λ_{max}^{EtOH} 283 mμ (5.46 ?)

VIII. (+)-Hernandaline, an Elaborated Aporphine Alkaloid

Cava and co-workers have reported the isolation of the novel alkaloid hernandaline from *Hernandia ovigera* L. (Hernandiaceae). This base represents the first example of a naturally occurring elaborated aporphine, intermediate to the aporphine–benzylisoquinoline dimers.[13]

Hernandaline, $C_{29}H_{31}O_7N$, showed a conjugated carbonyl group at 5.98 μ (1672 cm^{-1}). The NMR spectrum revealed the presence of an *N*-methyl group (δ2.49), five methoxyls (δ3.90, 3.90, 3.90, 3.81, and 3.72), five aromatic protons (δ6.48, 6.65, 6.79, 7.40, and 8.20), and one aldehydic proton at δ10.41.

The downfield aromatic proton at δ8.20 could be assigned to the C-11 hydrogen of an aporphine system, and indeed the UV spectrum of the alkaloid, λ_{max}^{EtOH} 216, 278, and 304 mμ (4.36, 4.40, and 4.20) pointed to the presence of a C-1,2,9,10-tetrasubstituted aporphine.

The expression below, therefore, appeared as a likely formulation for hernandaline, and this postulate was confirmed by synthesis. Ullmann condensation of (+)-*N*-methyllaurotetanine with excess 6-bromoveratraldehyde afforded crystalline (+)-hernandaline, identical in all respects with the natural material.[13]

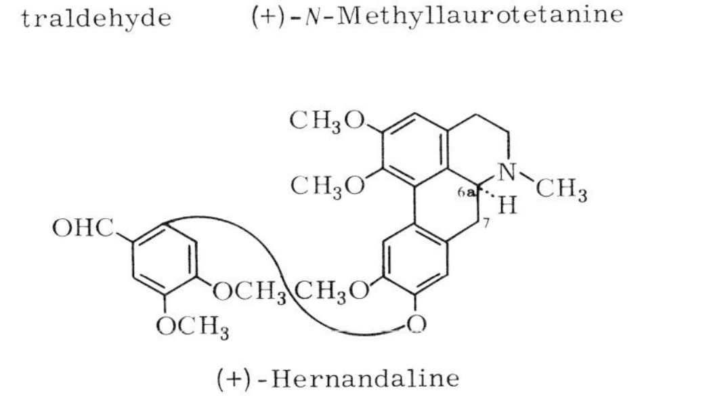

(+)-Hernandaline

Structures of the dehydrohernandaline type can be generated in the laboratory by the mild permanganate oxidation of aporphine–benzylisoquinoline dimers. For example, oxidation of thalicarpine furnished 6a,7-dehydrohernandaline.[10]

IX. The Kupchan–Liepa Total Synthesis of (+)-Thalicarpine via Synthetic (+)-Hernandaline

The initial step in this synthesis was the critical and high yield Ullmann condensation leading to the diaryl ether **5** (Scheme V).[14] In the next step the ether **5** was condensed with 2-methyl-6,7-dimethoxy-3,4-dihydroisoquinolinium iodide in sodium hydride and *N,N*-dimethylacetamide to produce the tetrahydroisoquinoline **6**. Reduction of the nitro function followed by diazotization and ring closure generated the aporphine derivative **7**. Vilsmeier–Haack formylation then gave (±)-hernandaline. This racemate proved to be spectrally identical with a sample of (+)-hernandaline obtained by oxidation of (+)-thalicarpine,[15] and was resolved using the (−)-α-bromocamphor-π-sulfonate salt. Condensation of synthetic (+)-hernandaline with the Reissert compound derived from 6,7-dimethoxyisoquinoline yielded the benzoate ester **8** as a mixture of diastereoisomers. Reduction and hydrogenolysis with zinc in acetic acid led to a mixture of two diastereoisomeric amines **9** which in turn gave two products upon Eschweiler–Clarke *N*-methylation. Prolonged stirring of a solution of these diastereoisomers in 60% aqueous ethanol led to crystallization of (+)-thalicarpine, identical in all respects with the natural product.[14]

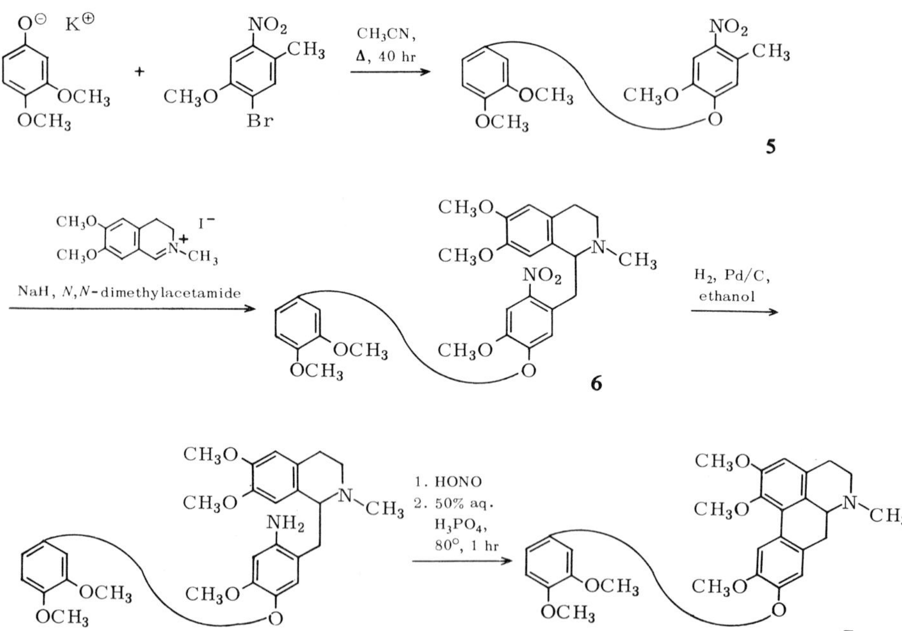

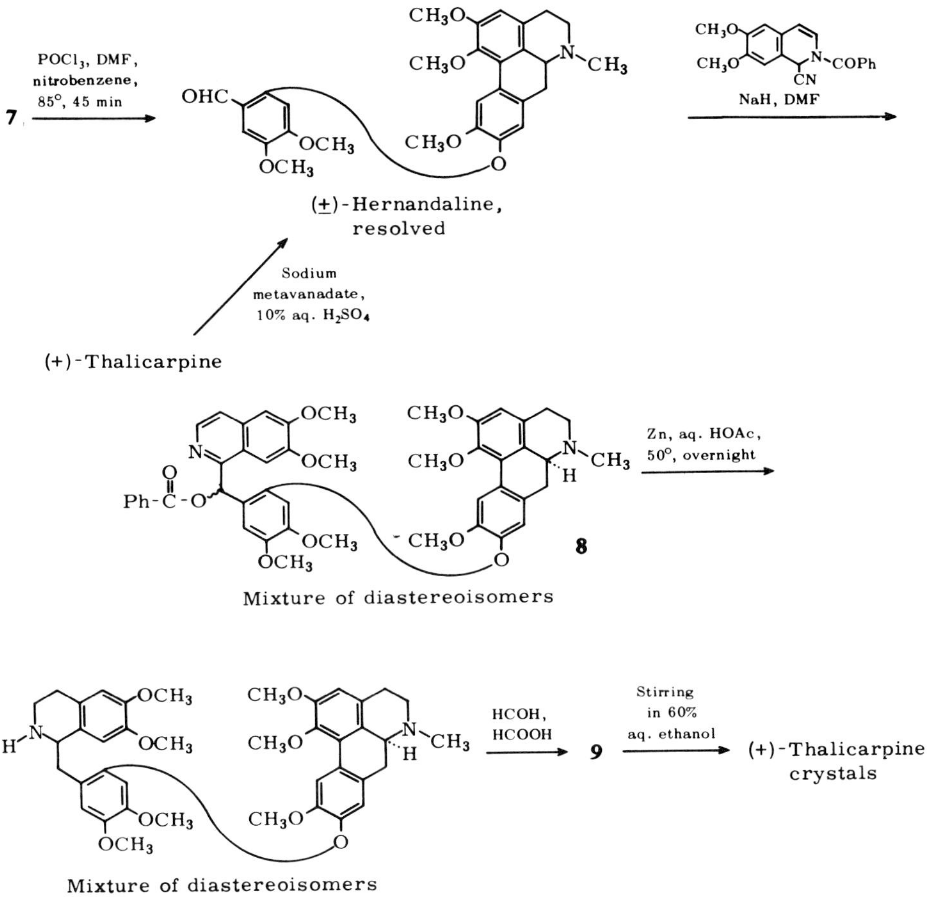

Scheme V

X. Pakistanine, A Novel Type Aporphine–Benzylisoquinoline Dimer

(+)-Pakistanine, $C_{37}H_{40}O_6N_2$, is the first and so far sole aporphine–benzylisoquinoline dimer found in a member of the Berberidaceae, *Berberis baluchistanica* Ahrendt. The alkaloid could be *O*-methylated with diazomethane to the nonphenolic *O*,*O*-dimethylpakistanine, $C_{39}H_{44}O_6N_2$. Comparison of the mass spectrum of pakistanine, *m/e* 608 (M^+), 402 ($M - x$), 312 ($M - y$), 296 ($M - z$), and 206 (x, base), with that of *O*,*O*-dimethylpakistanine, *m/e* 636 (M^+), 430 ($M - x$), 340 ($M - y$), 324 ($M - z$), and 206 (x, base), established that pakistanine possessed a diphenolic aporphine moiety bonded through a diphenyl ether linkage to an armepavine-like residue.[16]

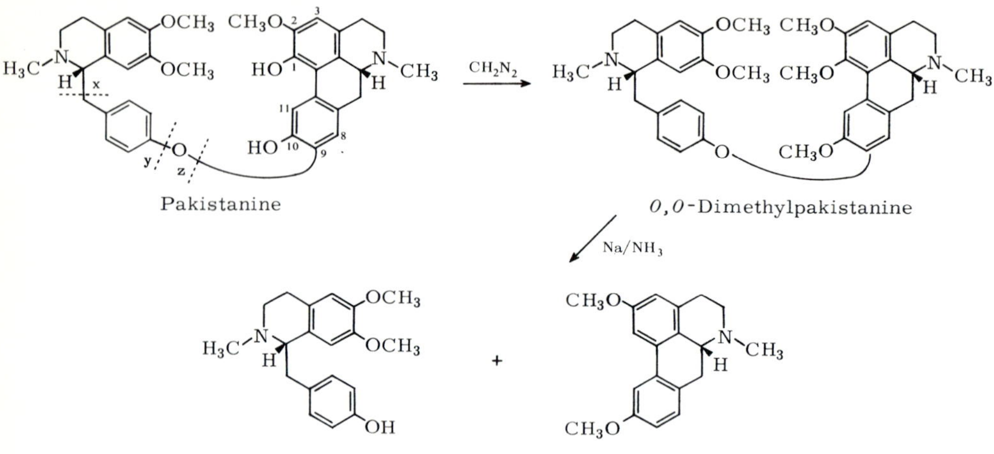

Pakistanine → O,O-Dimethylpakistanine → L-(+)-Armepavine + (−)-2,10-Dimethoxyaporphine

The NMR spectrum of pakistanine is summarized below. The presence of the one proton singlet at δ8.11 should be noted, indicating a C-11 proton. The spectrum of O,O-dimethylpakistanine shows two additional methoxyl resonances, one at δ3.72 and the other at δ3.90, the former being characteristic of a C-1 methoxyl. Since the phloroglucinol test for an *ortho*-diphenol function was negative for pakistanine, a methoxyl group could be assigned to the C-2 position.

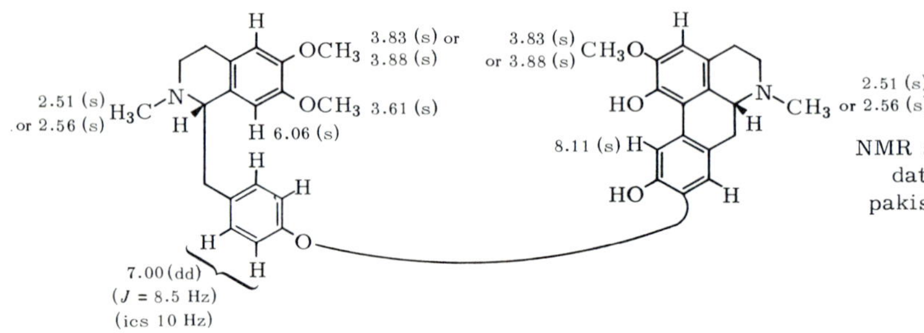

NMR spectral data for pakistanine

Three aromatic singlets at δ6.54, 6.56, and 6.71

Sodium in liquid ammonia cleavage of O,O-dimethylpakistanine yielded L-(+)-armepavine and (−)-2,10-dimethoxyaporphine. The terminus of the diphenyl ether bridge must be at C-9 rather than at C-3 or C-8 to account for the singlet aromatic proton signals in the NMR spectrum of pakistanine.

The UV spectrum of pakistanine, λ_{max}^{EtOH} 218, 270 sh, 277, and 307 mμ (4.61, 4.13, 4.21, and 4.07), while different from that of thalicarpine and its analogs, was found to be very close to that of 9-phenylboldine, λ_{max}^{EtOH} 218, 270 sh, 276, and 303 mμ (log e 4.30, 4.23, 4.25, and 3.82).[16]

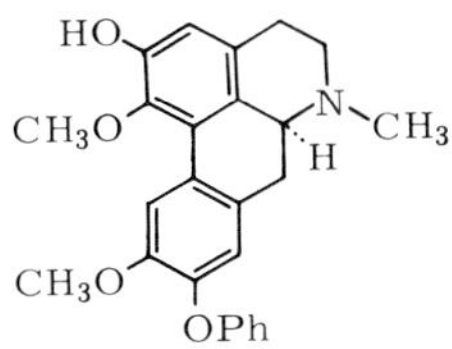

9-Phenylboldine

The biogenesis of pakistanine can be best understood in terms of the known bisbenzylisoquinoline alkaloid (+)-berbamunine found in *Berberis amurensis* Rupr. *var. japonica*, a close relative of *B. baluchistanica*. Starting with berbamunine, intramolecular phenolic oxidative coupling can afford the proaporphine–benzylisoquinoline **10**. Dienone–phenol rearrangement and selective *O*-methylation can then generate pakistanine.

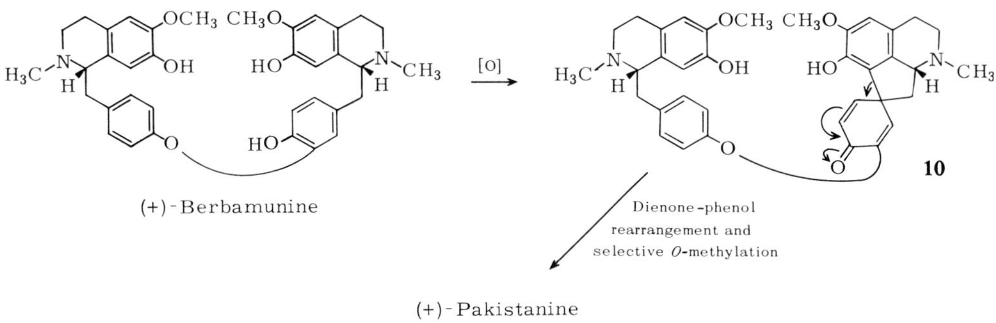

REFERENCES

1. Isolated from a Thalictrum species by N. M. Mollov and H. B. Dutschewska, *Chem. Ind. (London)*, p. 770 (1966).
1a. Z. F. Ismailov and S. Yu. Yunusov, *Khim. Prir. Soedin.* **6**, 142 (1970); *Chem. Abstr.* **73**, 35590b (1970).
2. M. Tomita, H. Furukawa, S.-T. Lu, and S. M. Kupchan, *Chem. Pharm. Bull.* **15**, 959 (1967).
3. H. B. Dutschewska and N. M. Mollov, *Chem. Ind. (London)* p. 770 (1966).
4. S. M. Kupchan, T.-H. Yang, M. L. King, and R. T. Borchardt, *J. Org. Chem.* **33**, 1052 (1968).
5. N. M. Mollov and H. B. Dutschewska, *Tetrahedron Lett.* p. 2219 (1964).
6. R. W. Doskotch, J. D. Phillipson, A. B. Ray, and J. L. Beal, *Chem. Commun.* p. 1083 (1969); and *J. Org. Chem.* **36**, 2409 (1971).
7. J. Reisch, H. Alfes, T. Kaniewska, and B. Borkowski, *Tetrahedron Lett.* p. 2113 (1970).
8. G. E. M. Husbands, Ph. D. Thesis, University of the West Indies (1967), as suggested by K. L. Stuart.

9. S. M. Kupchan, K. K. Chakravarti, and N. Yokoyama, *J. Pharm. Sci.* **52**, 985 (1963).
9a. J. L. Hartwell and B. J. Abbott, *Adv. Pharm. Chemotherapy* **7**, 117 (1969).
10. R. W. Doskotch, P. L. Schiff, Jr., and J. L. Beal, *Tetrahedron Lett.* p. 4999 (1968). See also N. M. Mollov and H. B. Dutschewska, *Tetrahedron Lett.* p. 853 (1966).
11. S. M. Kupchan and N. Yokoyama, *J. Amer. Chem. Soc.* **85**, 1361 (1963).
12. See also M. Tomita, S.-T. Lu, and Y.-Y. Chen, *J. Pharm. Soc. Jap.* **86**, 763 (1966).
13. M. P. Cava, K. Bessho, B. Douglas, S. Markey, and J. A. Weisbach, *Tetrahedron Lett.* p. 4279 (1966).
14. S. M. Kupchan and A. J. Liepa, *Chem. Commun.* p. 599 (1971).
15. *cf.* H. B. Dutschewska and N. M. Mollov, *Chem. Ber.* **100**, 3135 (1967), where potassium permanganate in acetone was used for the oxidation.
16. M. Shamma, J. L. Moniot, S. Y. Yao, G. A. Miana, and M. Ikram, *J. Amer. Chem. Soc.* **95**, 1381 (1972).

Chapter 13 / THE OXOAPORPHINES

Occurrence: Anonaceae, Araceae, Hernandiaceae, Lauraceae, Magnoliaceae, Menispermaceae, Monimiaceae, Papaveraceae, and Ranunculaceae[1]

Number: 17

Structures:

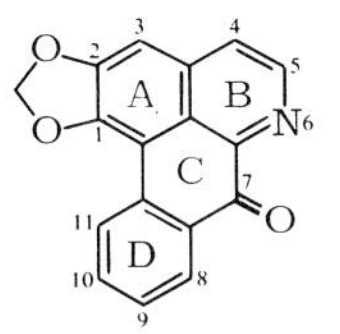

Liriodenine

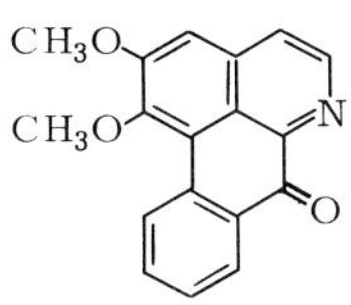

Lysicamine

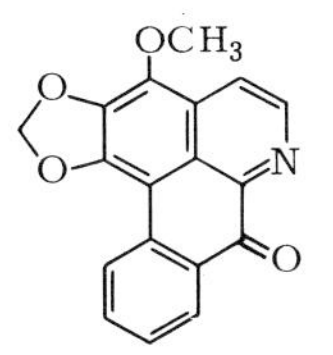

Atherospermidine

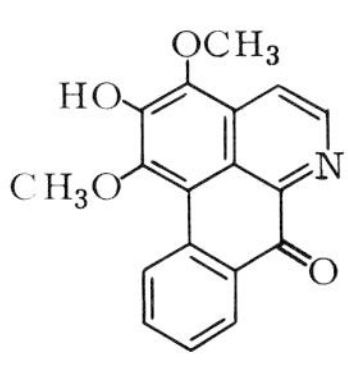

Moschatoline

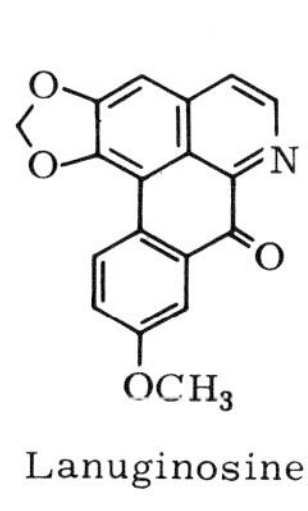

Lanuginosine

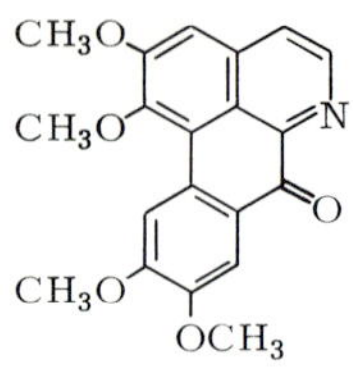

1,2,9,10-Tetramethoxy-oxoaporphine

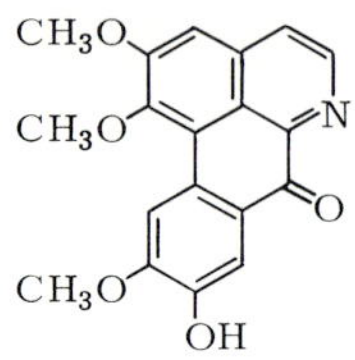

Atheroline

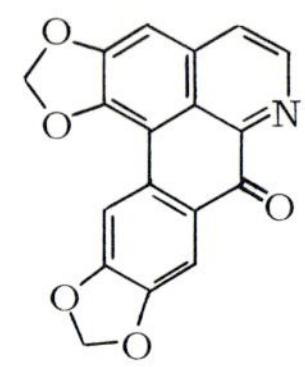

Cassameridine

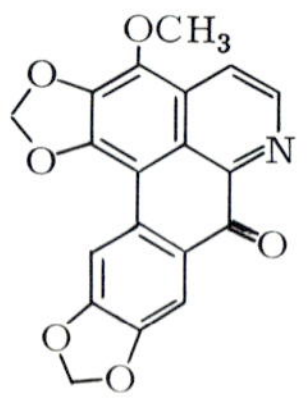

Cassamedine

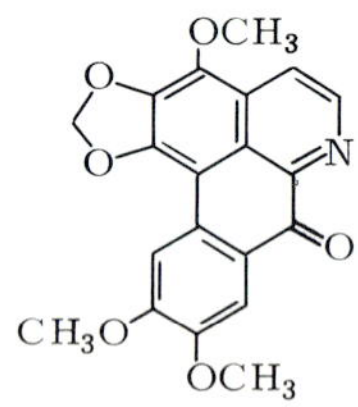

Thalicminine

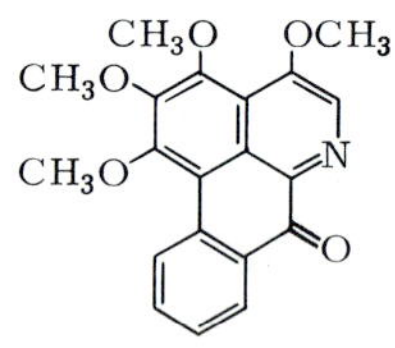

Imenine

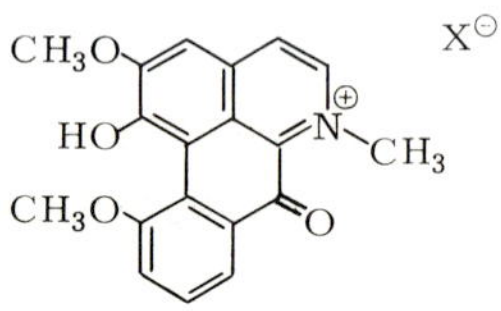

Alkaloid PO-3 (See Chapter 10, Section V, A)

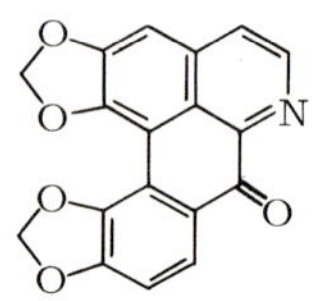

Hernandonine (ovigerinone)

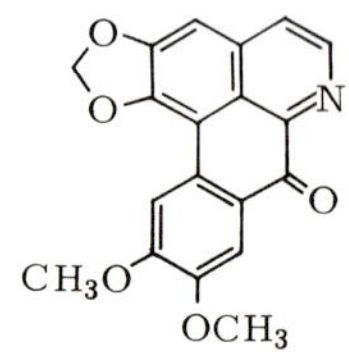

Dicentrinone

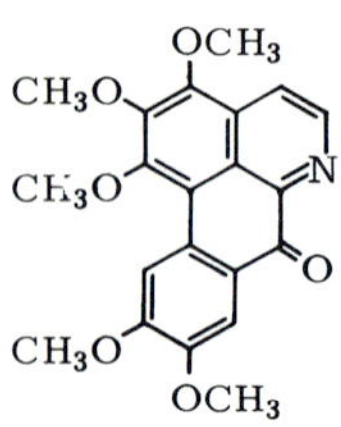

Oxopurpureine

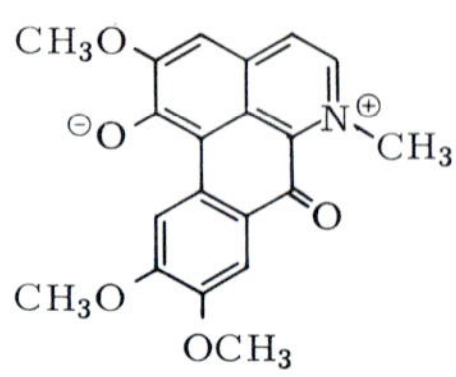

Corunnine (zwitterionic form)

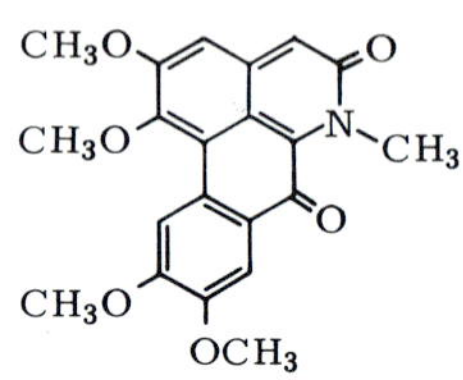

Pontevedrine

I. Introduction

The oxoaporphines are most probably derived in plants by oxidation of the corresponding aporphines. The free bases possess a bright yellow or orange color which turns pink or red upon the addition of mineral acid. Imenine, whose structure was established by X-ray analysis, is the first oxoaporphine found to be oxygenated at C-4, while the alkaloid PO-3, first mentioned in Chapter 10, Section V, A, is of special interest since it is the first known example of a naturally occurring quaternary oxoaporphine. Oxoaporphines are high melting and show a decomposition point rather than a melting point.

II. Liriodenine: Structural Elucidation and Synthesis

W. I. Taylor described in 1961 the structural elucidation and synthesis of liriodenine, the first oxoaporphine to be fully characterized.[1a]

The alkaloid is present in *Liriodendron tulipifera* L. (Magnoliaceae), commonly known as the tulip tree, which is widely distributed throughout the eastern United States. The yellow color of the heartwood is partially due to this optically inactive base which analyzes for $C_{17}H_9O_3N$.

The presence of a conjugated ketone was indicated by a sharp peak at 6.1 μ (1639 cm^{-1}) and by the formation of an oxime derivative. Liriodenine also incorporates one methylene-dioxy group with IR absorption at 6.71, 7.04, 7.35, 8.93, 9.52, and 10.42 μ (1490, 1420, 1360, 1120, 1050, and 960 cm^{-1}) but contains no hydroxyl, *N*-methyl or *O*-methyl groups. The UV spectrum presented a complex picture indicative of a highly conjugated aromatic system.

Chromic acid oxidation of liriodenine afforded 1-azaanthraquinone-4-carboxylic acid (**1**) which could be decarboxylated to the azaanthraquinone **2** of established structure. Considering that members of the Magnoliaceae family were known to produce aporphines and other isoquinoline alkaloids, the structure given in Scheme I was proposed for liriodenine.[1a]

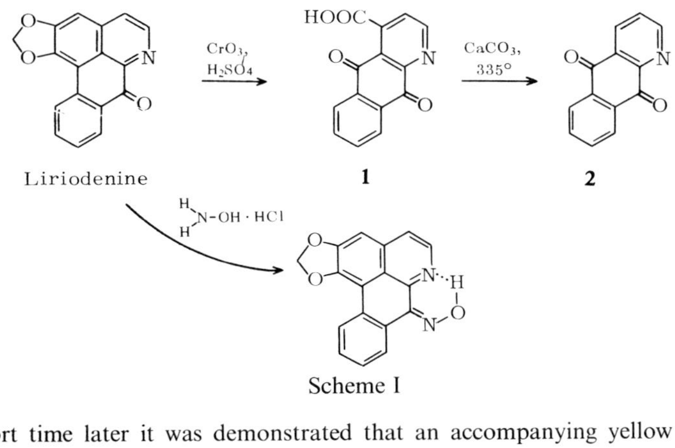

Scheme I

A short time later it was demonstrated that an accompanying yellow alkaloid in the tulip tree is 1,2,9,10-tetramethoxyoxoaporphine, and significantly a colorless alkaloid also shown to be present is the aporphine (+)-glaucine.

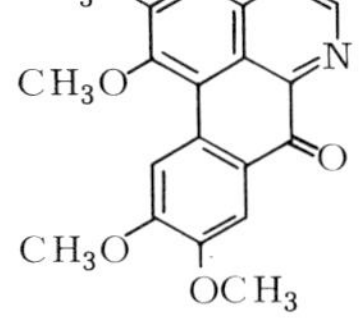

1,2,9,10-Tetramethoxy-oxoaporphine

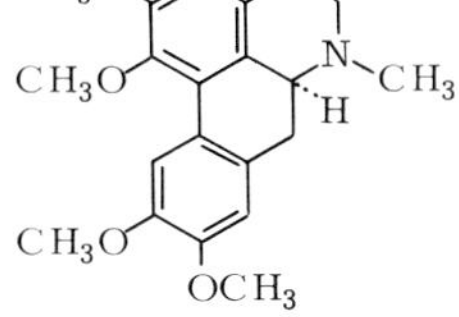

(+)-Glaucine

A straightforward synthesis of liriodenine using the Pschorr cyclization was carried out as confirmatory evidence for the structural assignment (Scheme II).[1a]

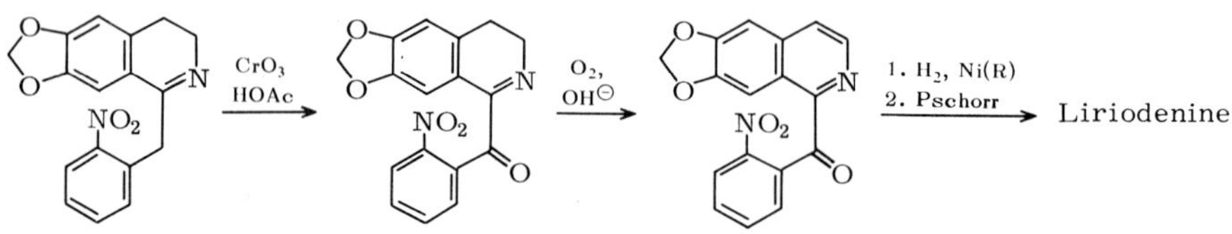

Scheme II

III. The Oxidation of Aporphines and the Reduction of Oxoaporphines

The presumed oxidation of an aporphine into an oxoaporphine in nature can be duplicated in the laboratory, usually using the chromium trioxide in pyridine complex[2,3] or manganese dioxide[4] as oxidizing agent. Thus liriodenine was obtained by oxidation of either ushinsunine or roemerine, while 1,2,9,10-tetramethoxyoxoaporphine was produced from glaucine. Oxoaporphines can therefore be prepared *in vitro* either by synthesis via Pschorr cyclization or by oxidation of the corresponding aporphine base.

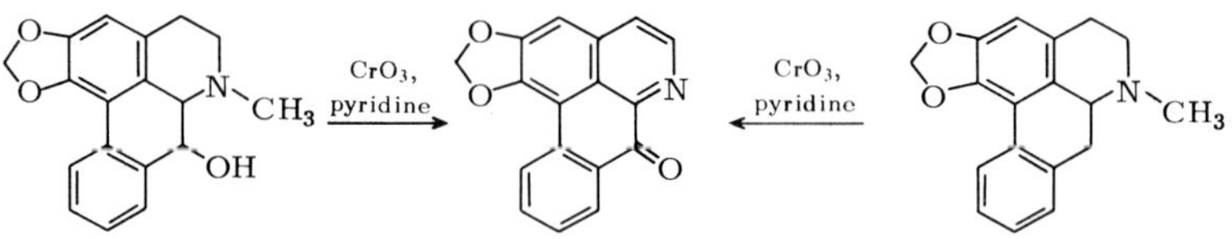

Ushinsunine   Liriodenine   Roemerine

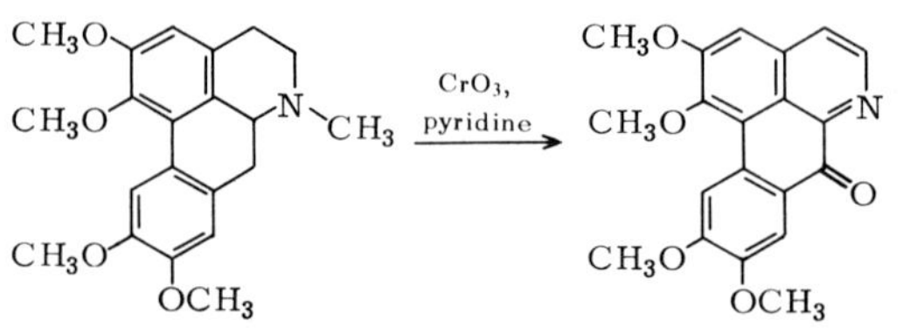

Glaucine   1,2,9,10-Tetramethoxyoxoaporphine

It has also been found that simple passage of air through a solution of *t*-butyl alcohol containing anonaine results in a 30% yield of liriodenine.[5] Oxidation of nonphenolic aporphines with iodine in dioxane affords the corresponding dehydroaporphines; but iodine in ethanol oxidation of nonphenolic noraporphines proceeds all the way to the oxoaporphine stage. Dehydroaporphines may also be efficiently oxidized to oxoaporphines by peracetic acid or by benzoyl peroxide.[5a]

Oxoaporphines can be reduced with zinc in hydrochloric acid or under Clemmensen conditions to noraporphines; lanuginosine was reduced to xylopine[6] and liriodenine to anonaine.[7]

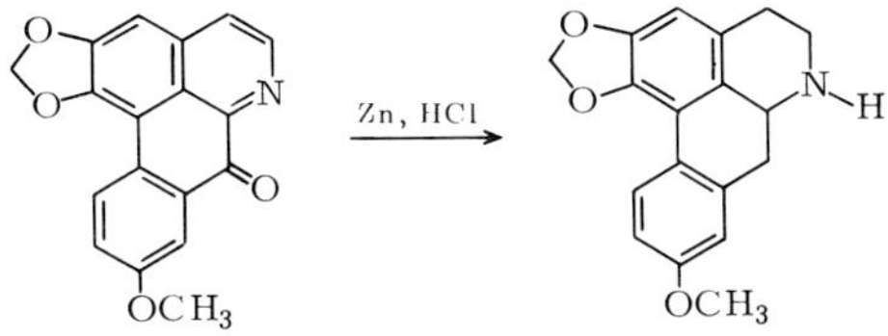

Lanuginosine (±)-Xylopine

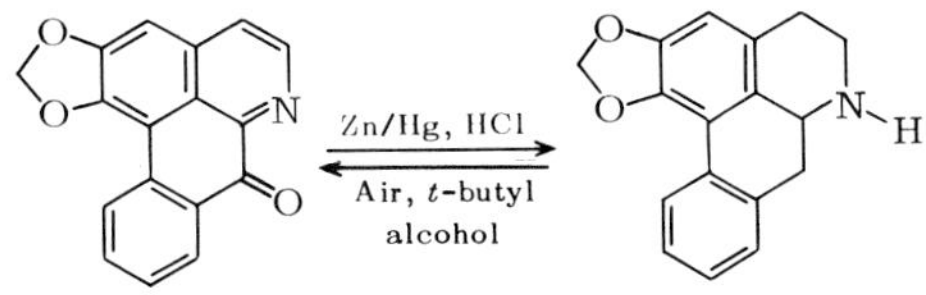

Liriodenine (±)-Anonaine

If it is the methiodide salt of an oxoaporphine which is reduced, an aporphine is produced[7]:

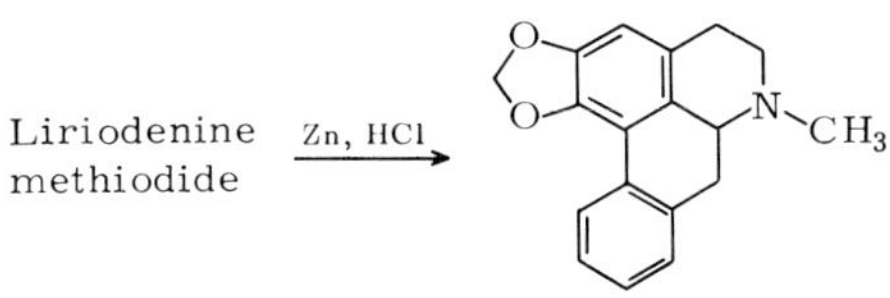

(±)-Roemerine

IV. An Unusual Demethoxylation Reaction

It has been observed that catalytic hydrogenation of 1,2,9,10-tetramethoxyoxoaporphine in acetic acid using Adams catalyst affords an excellent yield of the C-1 demethoxylated derivative **3**.[8] Removal of an aromatic methoxyl group through hydrogenolysis is unusual, and this cleavage may be characteristic of oxoaporphines. The C-1 methoxyl is clearly the most hindered of the four methoxyl groups present.

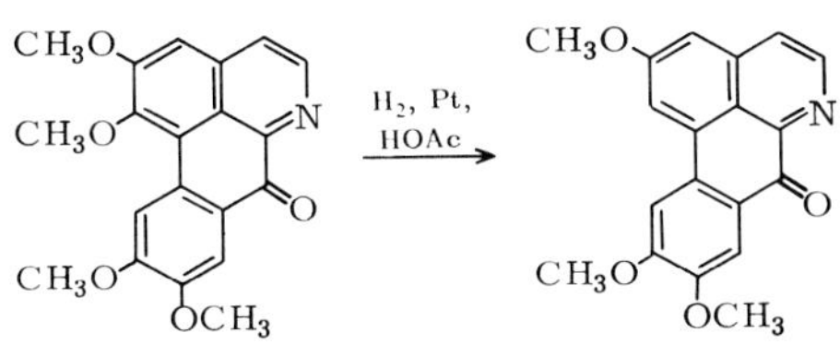

3

V. Pharmacology

Liriodenine has shown significant cytotoxic inhibitory activity *in vitro* against cells derived from human carcinoma of the nasopharynx.[9]

VI. NMR Spectroscopy

The NMR spectral data available for the oxoaporphines have been summarized in the diagrams that follow. Trifluoroacetic acid was the usual solvent. It will be noted that by analogy with the aporphines a C-3 hydrogen appears at higher field than the remaining aromatic hydrogens, and a C-11 hydrogen is always found appreciably downfield.

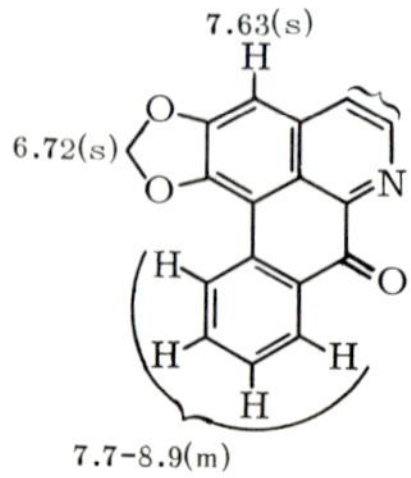

Liriodenine[6,9,10]
in trifluoroacetic acid

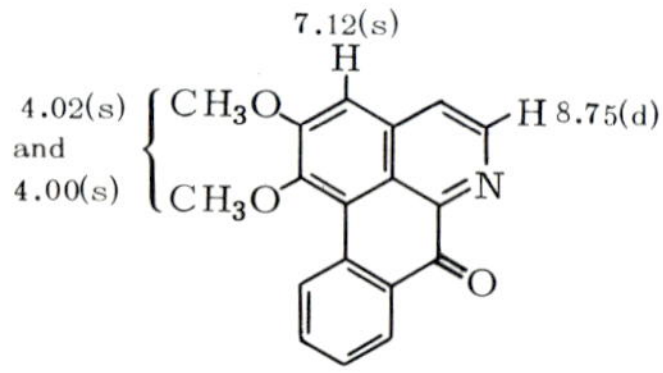

Lysicamine[11]

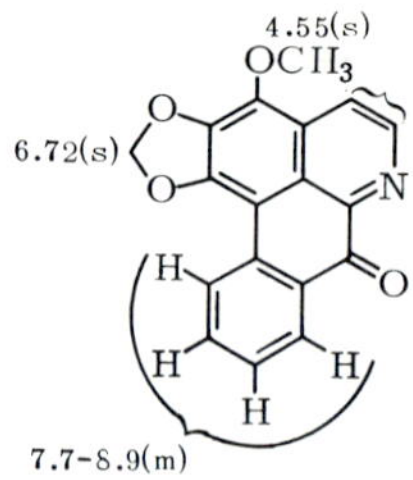

Atherospermidine[6,10,12]
in trifluoroacetic acid

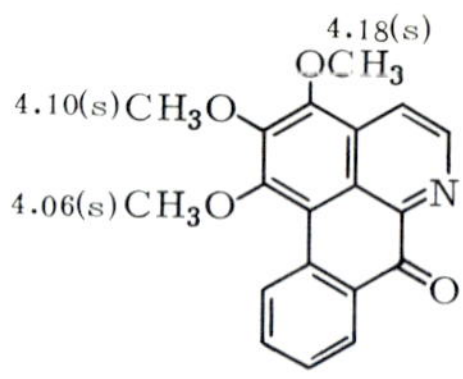

O-Methylmoschatoline[13]

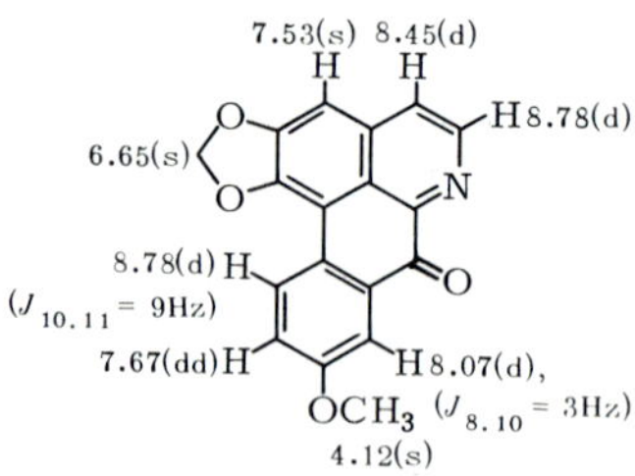

Lanuginosine[6]
in trifluoroacetic acid

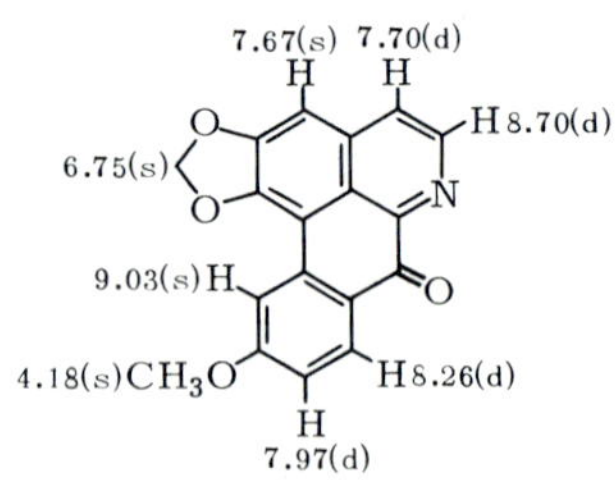

1,2-Methylenedioxy-10-methoxyoxoaporphine[14]
in trifluoroacetic acid
(synthetic)

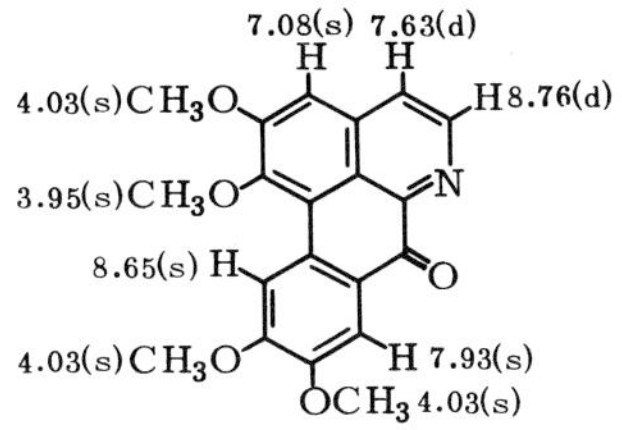

O-Methylatheroline[13]

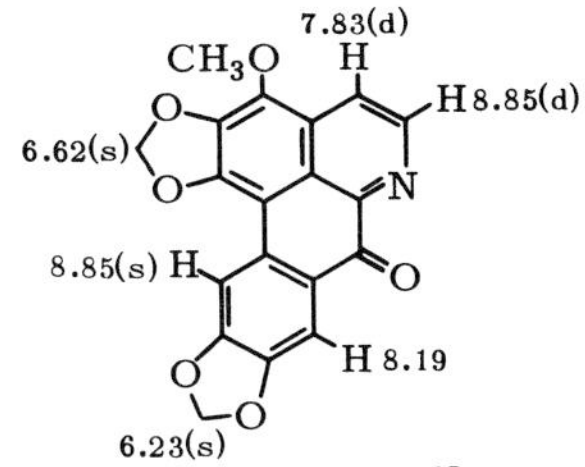

Cassamedine[15]
in trifluoroacetic acid

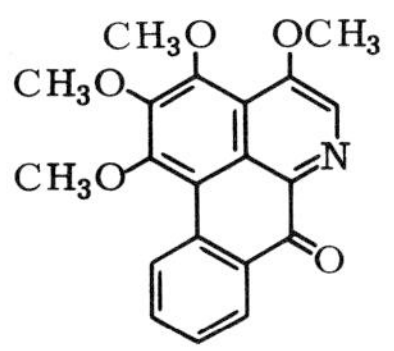

Imenine[16]
Four methoxyl singlets at
δ4.05, 4.10, 4.15, and 4.25

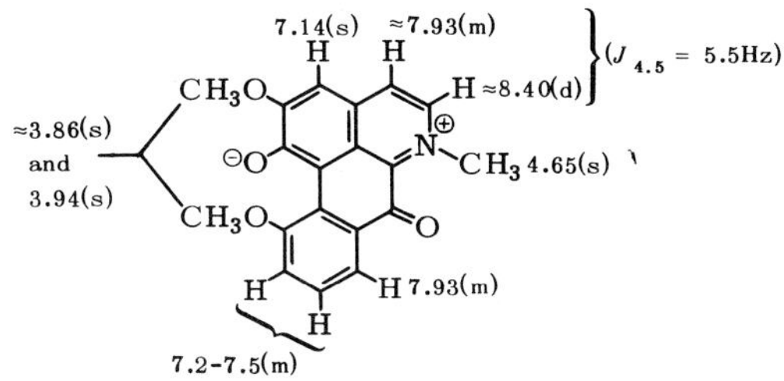

Alkaloid PO-3 in
DMSO-d_6[17,18]

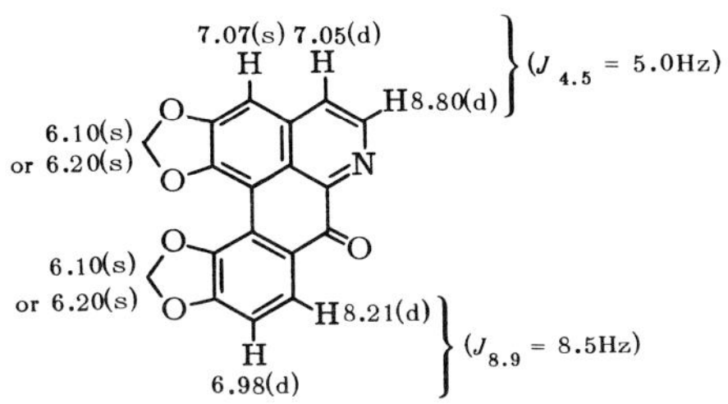

Hernandonine
in deuteriochloroform[19,19a]

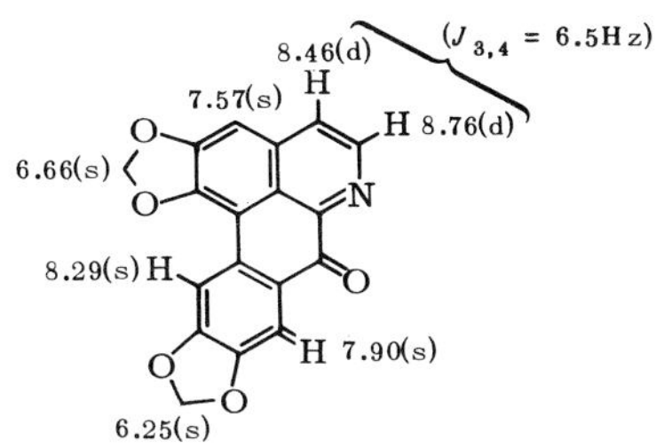

Cassameridine[15a]
in trifluoroacetic acid

VII. UV Spectroscopy

The UV spectra of oxoaporphines are complex. Most of the data available have been culled from the literature and are presented below. A peak at 281 – 282 mμ seems to be indicative of a 1,2-methylenedioxy-3-methoxyoxoaporphine system unsubstituted at C-11.

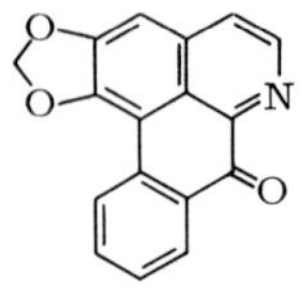

Liriodenine[20]

λ_{max}^{EtOH} 247, 268, 309, and 413 mμ
(4.22, 4.13, 3.62, and 3.82)
λ_{min}^{EtOH} 258, 292, and 340 mμ
(4.08, 3.51, and 3.16)
$\lambda_{max}^{0.1\ N\ HCl\ in\ EtOH}$ 257, 277, 329, 392, and 455 mμ
(4.33, 4.26, 3.67, 3.69, and 3.58)
$\lambda_{min}^{0.1\ N\ HCl\ in\ EtOH}$ 269, 307, 362, and 426 mμ
(4.20, 3.53, 3.55, and 3.52)

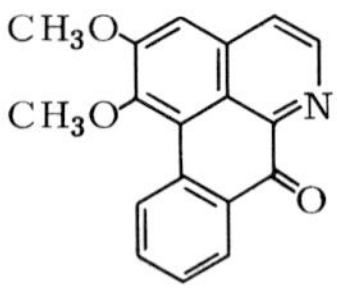

Lysicamine

λ_{max}^{EtOH} 237, 272, 317, and 405 mμ[3]
(4.48, 4.41, 3.81, and 3.91)
λ_{min}^{EtOH} 227, 253, 297, and 344 mμ[3]
(4.42, 4.24, 3.72, and 3.57)
λ_{max}^{EtOH} 235, 270, 307, and 400 mμ[11]
(4.47, 4.41, 376, and 3.94)
$\lambda_{max}^{0.1\ N\ HCl\ in\ EtOH}$ 249, 276, 306, and 453 mμ[11]
(4.33, 4.44, 3.82, and 3.58)

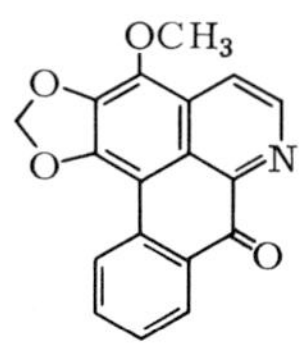

Atherospermidine[10, 21]

λ_{max}^{EtOH} 247, 281, and 312 sh mμ
(4.38, 4.52, and 3.95)
$\lambda_{max}^{0.1\ N\ HCl}$ 262 and 283 mμ
(4.24 and 4.16)

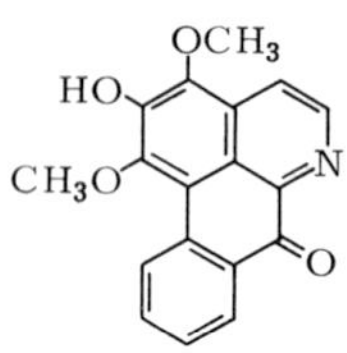

Moschatoline[10, 13]

λ_{max}^{EtOH} 237, 272, 315 sh, 374, and 440 mμ
(4.47, 4.41, 4.10, 3.55, and 3.67)
$\lambda_{max}^{0.05\ N\ HCl,\ EtOH,\ H_2O}$ 246, 281, 390, and 496 mμ
(4.37, 4.40, 3.63, and 3.36)
$\lambda_{max}^{0.05\ N\ NaOH,\ EtOH,\ H_2O}$ 247, 283, 310, 407, and 517 mμ
(4.42, 4.31, 4.25, 3.99, and 3.33)

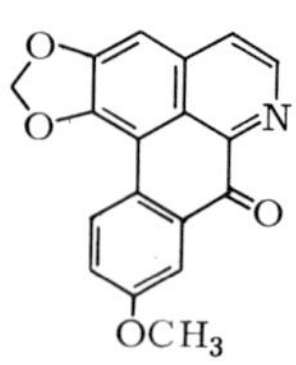

Lanuginosine[6]

λ_{max}^{EtOH} 246, 271, and 314 mμ
(4.46, 4.34, and 3.78)
$\lambda_{max}^{0.1\ N\ HCl}$ 257 and 284 mμ
(4.31 and 4.19)

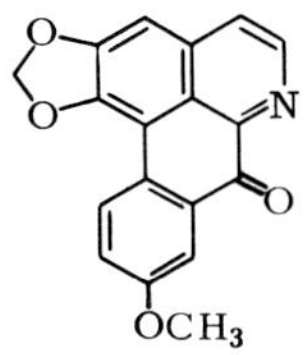

Lanuginosine[14]

λ_{max}^{EtOH} 246, 271, and 315 mμ
(4.54, 4.44, and 3.89)
$\lambda_{max}^{HCl,\ EtOH}$ 258, 283, and 334 mμ
(4.57, 4.47, and 3.83)

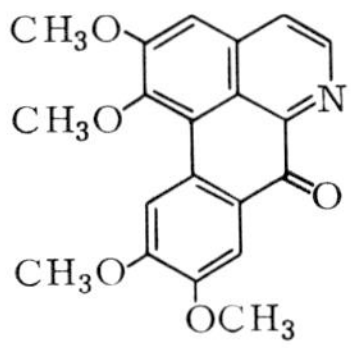

1,2,9,10-Tetramethoxy-oxoaporphine[3]

λ_{max}^{EtOH} 244, 273, 356, and 423–433 mμ
(4.46, 4.47, 4.04, and 3.87)
λ_{min}^{EtOH} 230, 258, and 323 mμ
(4.33, 4.31, and 3.82)

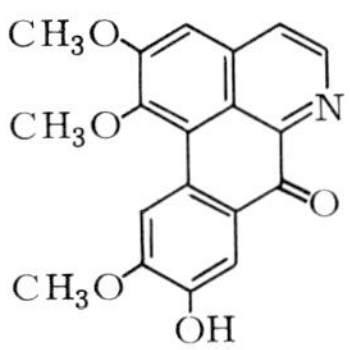

Atheroline[10,13]

λ_{max}^{EtOH} 244, 273, 292 sh, 355, 380 sh, and 435 mμ
(4.09, 4.17, 3.96, 3.90, 3.83, and 3.62)
$\lambda_{max}^{0.05\ N\ HCl,\ EtOH,\ H_2O}$ 257, 282, 385, and 500 mμ
(4.12, 4.12, 4.05, and 3.38)
$\lambda_{max}^{0.05\ N\ NaOH,\ EtOH,\ H_2O}$ 252, 294, 320, 390, and 535 mμ
(4.04, 3.99, 3.98, 3.74, and 3.46)

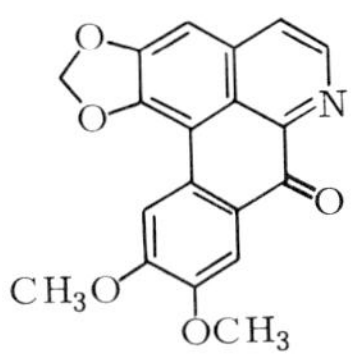

1,2-Methylenedioxy-9,10-dimethoxyoxoaporphine
≡ Dicentrinone[21a]

λ_{max}^{EtOH} 250, 272, 313 sh, 351, 392, and 438 mμ
(4.69, 4.62, 4.17, 4.22, 4.39, and 4.29)
$\lambda_{max}^{HCl,\ EtOH}$ 260, 292, 382, and 506 mμ
(4.69, 4.62, 4.30, and 3.64)
λ_{max}^{EtOH} 213, 250, 272, 310 sh, 352, 396, and 433 mμ
(4.57, 4.54, 4.45, 4.05, 4.07, 3.62, and 3.60)

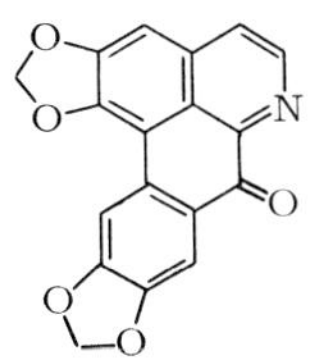

Cassameridine[15]

λ_{max}^{EtOH} 251, 274, 323, 353, 388, and 440 mμ
(4.46, 4.40, 4.08, 3.91, 3.85, and 3.73)
$\lambda_{max}^{HCl,\ EtOH}$ 261, 290, 385, and 500 mμ
(4.62, 4.59, 4.31, and 3.62)

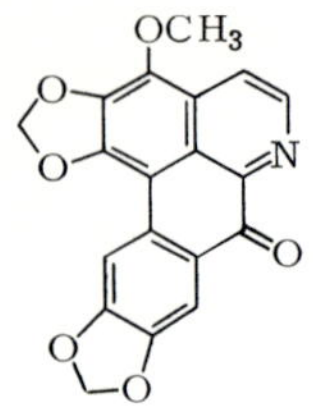

Cassamedine[15]

λ_{max}^{EtOH} 252, 281, 324, 364, and 460 mμ
(4.47, 4.53, 4.12, 3.97, and 3.76)
$\lambda_{max}^{HCl, EtOH}$ 272, 286, 408, and 534 mμ
(4.49, 4.50, 4.10, and 3.40)

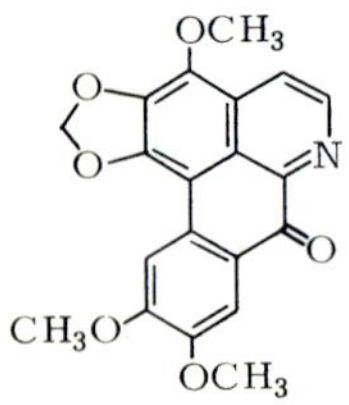

Thalicminine[22]

$\lambda_{max}^{EtOH-CHCl_3}$ 252, 282, 364, and 456 mμ
(4.29, 4.43, 3.91, and 3.72)

Oxopurpureine[23]

λ_{max}^{EtOH} 251, 282, 354, 392, and 456 mμ
(4.37, 4.54, 3.86, 3.94, and 3.78)

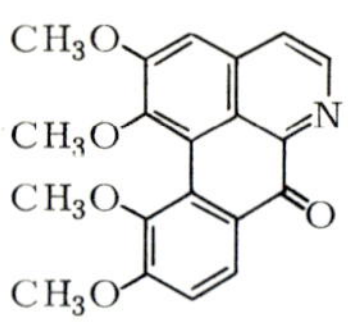

1,2,10,11-Tetramethoxy-oxoaporphine
(not a natural product)

λ_{max}^{EtOH} 222, 275, 360, and 405 mμ
(4.42, 4.23, 3.82, and 3.79)
λ_{min}^{EtOH} 263, 325, and 382 mμ
(4.15, 3.52, and 3.74)

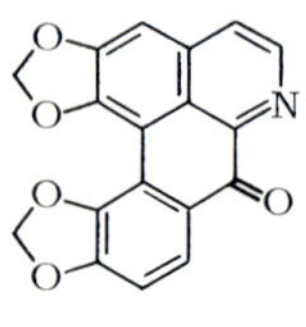

Hernandonine[19]

λ_{max}^{EtOH} 222, 265, 364, and 426 mμ
(4.55, 4.37, 4.03, and 3.99)

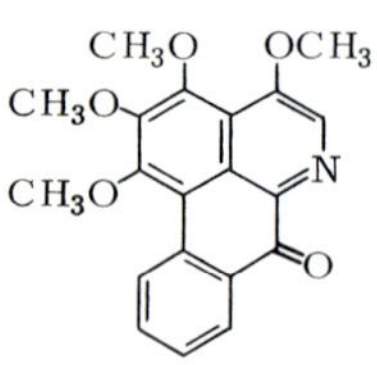

Imenine[16]

λ_{max}^{EtOH} 240, 275, 345, and 438 mμ
(4.15, 4.38, 3.58, and 3.42)

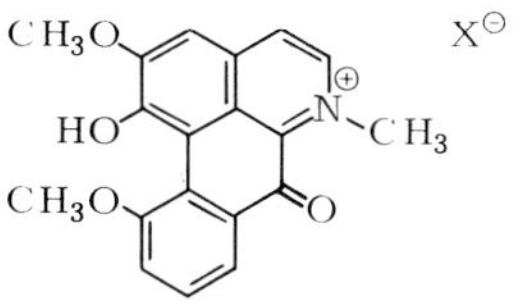

Alkaloid PO-3[18]

λ_{max}^{EtOH} 310, 430, and 645 mμ
(3.7, 3.7, and 4.5)
(values approximated from chart)

The UV spectra of phenolic oxoaporphines show significant bathochromic shifts upon addition of base. These shifts may be of diagnostic value in establishing the position of the phenolic function. Moschatoline and atheroline show acidic properties and readily dissolve in aqueous sodium carbonate, and atheroline has been noted to produce a blue salt. They both show strong bathochromic shifts in their UV spectra upon addition of base. On the other hand, the synthetic 1-hydroxy-2,9,10-trimethoxyoxoaporphine shows an even stronger bathochromic shift. These results may be rationalized by considering the resonance of the corresponding anions.[10,13]

It is possible to draw canonical resonating forms **4** and **5** for the anions derived from moschatoline and atheroline, respectively, in which the net charge is on the nitrogen atom. However, the charge is on the oxygen carbonyl atom in anion **6**, obtained from the synthetic oxoaporphine.*

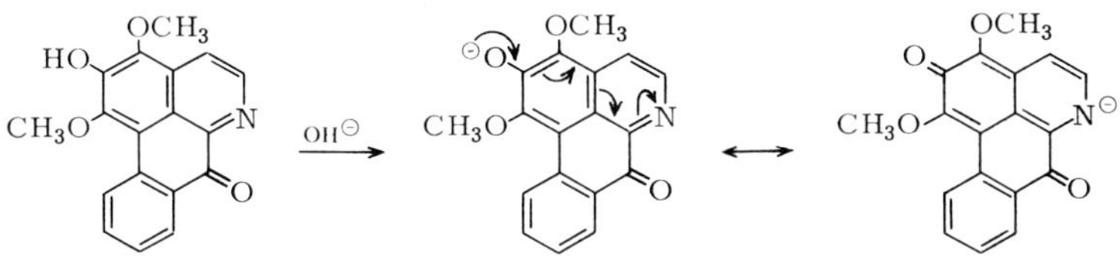

Moschatoline **4**

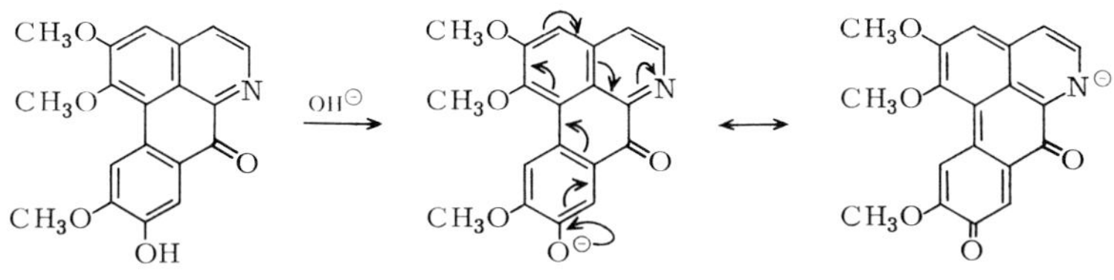

Atheroline **5**

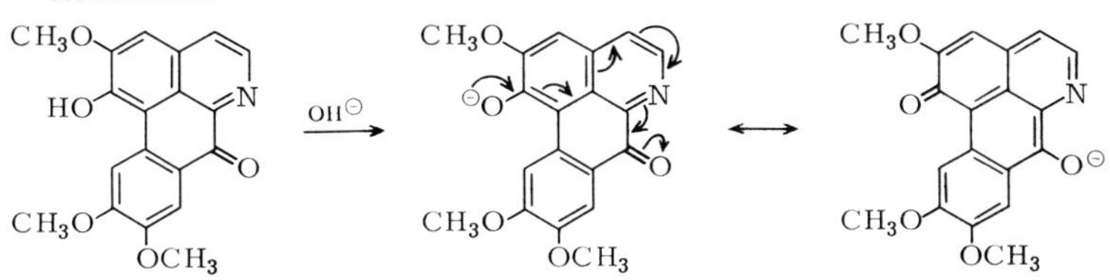

1-Hydroxy-2,9,10-trimethoxyoxoaporphine (synthetic) **6**

* The explanation offered here is somewhat different from that presented by the authors.

VIII. Corunnine and Pontevedrine, Two New Oxoaporphines

The Spanish team of Ribas, Sueiras, and Castedo has reported on the isolation and characterization of two highly colored oxoaporphines, corunnine and pontevedrine, which were found in *Glaucium flavum* Cr. var. *Vestitum* (Papaveraceae) together with the known yellow oxoaporphine 1,2,9,10-tetramethoxyoxoaporphine.[24]

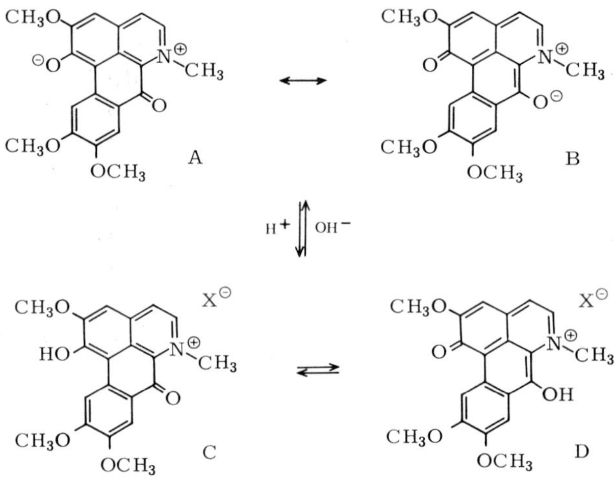

Corunnine

Corunnine, $C_{20}H_{17}O_5N$, crystallizes as violet needles and appears to give a positive ferric chloride test. The alkaloid is green in neutral or basic solution and red in acid solution. The IR spectrum showed λ^{KBr} 6.17 μ (1620 cm^{-1}) weak and 6.38 μ (1567 cm^{-1}) medium.

The NMR spectrum taken in trifluoroacetic acid revealed the presence of three methoxyl singlets at δ4.55 (6H) and 4.80 (3H), one N–CH_3 group at δ5.36, doublets at δ8.75 and 8.95 due to the C-4 and C-5 hydrogens ($J_{4,5}$ = 6 Hz), and finally three aromatic one proton singlets at δ7.93, 8.33, and 9.30 (C-11H).

Since there appears to be some evidence that oxoaporphine salts carrying a phenolic function at C-1 of C-11 are green,[18] the C-1 position was assigned a hydroxyl group, so that corunnine can be represented by forms A, B, C, or D, with the latter two prevailing in acid solution.

Pontevedrine, on the other hand, is a red, crystalline compound which analyzes for $C_{21}H_{19}O_6N$. It is insoluble in base in spite of the fact that it gives a positive ferric chloride test; and interestingly enough its UV spectrum is not affected by acid or base. The IR spectrum shows peaks at 6.02 μ (1660 cm^{-1}) strong, 6.19 μ (1615 cm^{-1}) weak, and 6.29 μ (1590 cm^{-1}).

The NMR spectrum of pontevedrine in deuteriochloroform showed four methoxyl singlets at δ3.96 (3H), 4.00 (3H), and 4.10 (6H), an *N*-methyl singlet at δ3.50, and four aromatic one proton singlets at δ6.96, 7.00, 7.70, and 8.80 (C-11H).

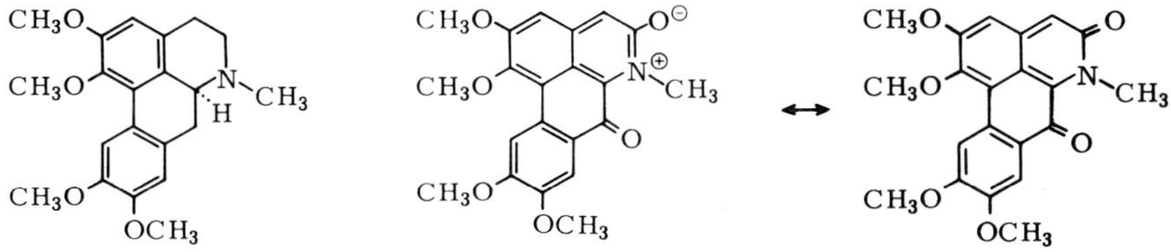

(+)-Glaucine Pontevedrine

Treatment of glaucine with a large excess of chromium trioxide in pyridine complex was then found to give in low yield a mixture of dehydroglaucine, corunnine, pontevedrine, and 1,2,9,10-tetramethoxyoxoaporphine.

In an attempt to quaternize 1,2,9,10-tetramethoxyoxoaporphine with methyl iodide in commercial acetone, it was found that instead corunnine and a small amount of pontevedrine were unexpectedly produced. Shorter reaction times led mostly to the desired methiodide salt, and subsequent refluxing of an acetone solution of this salt afforded corunnine as the only reaction product (Scheme III).[24]

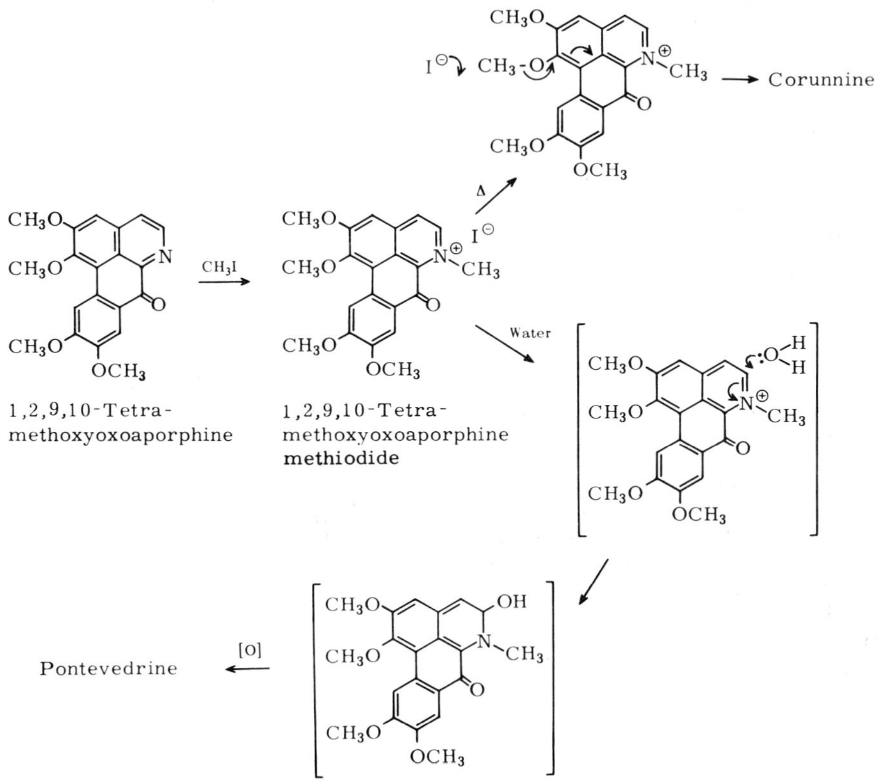

Scheme III

Corunnine[24]	$\lambda_{max}^{EtOH\ or\ OH^{\ominus}}$ 258, 325, 400, 440 sh, and 630 mμ (4.13, 4.32, 3.54, 3.42, and 3.35)
	$\lambda_{max}^{H^{\oplus}}$ 256, 295, and 385 mμ (4.23, 4.14, and 3.75)
Pontevedrine[24]	λ_{max}^{EtOH} 245, 312, 325, and 470 mμ (4.59, 4.28, 4.39, and 4.01)

REFERENCES

1. For a review on the oxoaporphines see M. Shamma and R. L. Castenson, *in* "The Alkaloids" (R. H. F. Manske, ed.), Vol. **14**. Academic Press, New York (in press).

1a. W. I. Taylor, *Tetrahedron* **14**, 42 (1961).

2. T.-H. Yang, *J. Pharm. Soc. Jap.* **82**, 794 (1962).

3. M. Tomita, T.-H. Yang, H. Furukawa, and H.-M. Yang, *J. Pharm. Soc. Jap.* **82**, 1574 (1962).

4. S.-S. Yang, W.-Y. Huang, L. C. Lin, and P.-Y. Yeh, *Chemistry* (*Taipei*) p. 144 (1961).

5. M. P. Cava and D. R. Dalton, *J. Org. Chem.* **31**, 1281 (1966).

5a. M. P. Cava, A. Venkateswarlu, M. Srinivasan, and D. L. Edie, (in press).

6. S. M. Kupchan, M. I. Suffness, and E. M. Gordon, *J. Org. Chem.* **35**, 1682 (1970).

7. T.-H. Yang, *J. Pharm. Soc. Jap.* **82**, 804 (1962).

8. J. Cohen and W. I. Taylor, *J. Org. Chem.* **28**, 3567 (1963).

9. D. Warthen, E. L. Gooden, and M. Jacobson, *J. Pharm. Soc.* **58**, 637 (1969).

10. I. R. C. Bick and G. K. Douglas, *Tetrahedron Lett.* p. 1629 (1964).

11. N. Katsui, K. Sato, S. Tobinaga, and N. Takeuchi, *Tetrahedron Lett.* p. 6257 (1966).

12. W. M. Harris and T. A. Geissman, *J. Org. Chem.* **30**, 432 (1965).

13. I. R. C. Bick and G. K. Douglas, *Tetrahedron Lett.* p. 4655 (1965).

14. S. K. Talapatra, A. Patra, and B. Talapatra, *Chem. Ind.* (*London*) p. 1056 (1969); for the synthesis of lanuginosine, see T. Govindachari, N. Viswanathan, S. Narayanaswami, and B. R. Pai, *Indian J. Chem.* **8**, 475 (1970).

15. M. P. Cava, K. V. Rao, B. Douglas, and J. A. Weisbach, *J. Org. Chem.* **33**, 2443 (1968).

15a. F. N. Lahey and K. F. Mak, *Tetrahedron Lett.* p. 4311 (1970).

16. M. D. Glick, R. E. Cook, M. P. Cava, M. Srinivasan, J. Kunitomo, and A. I. daRocha, *Chem. Commun.* p. 1217 (1969).

17. V. Preininger and F. Šantavý, *Acta Univ. Palacki. Olomuc., Fac. Med.* **43**, 5 (1966).

18. V. Preininger, J. Hrbek, Jr., Z. Samek, and F. Šantavý, *Arch. Pharm.* (*Weinheim*) **302**, 808 (1969).

19. K. Ito and H. Furukawa, *Tetrahedron Lett.* p. 3023 (1970).

19a. F. N. Lahey and K. F. Mak, *Aust. J. Chem.* **24**, 671 (1971).

20. M. A. Buchanan and E. E. Dickey, *J. Org. Chem.* **25**, 1389 (1960).

21. B. R. Pai and G. Shanmugasundaram, *Tetrahedron* **21**, 2579 (1965).

21a. M. P. Cava and A. Venkateswarlu, *Tetrahedron* **27**, 2639 (1971).

22. Kh. G. Pulatova, Z. F. Ismailov, and S. Yu. Yunusov, *Khim. Prir. Soedin.* **2**, 426 (1966); *Chem. Natur. Prod.* **2**, 349 (1966); *Chem. Abstr.* **68**, 13226g (1968).

23. P. E. Sonnet and M. Jacobson, *J. Pharm. Sci.* **60**, 1254 (1971).

24. I. Ribas, J. Sueiras, and L. Castedo, *Tetrahedron Lett.* p. 3093 (1971).

Chapter 14 / THE PHENANTHRENE ALKALOIDS

Occurrence: Anonaceae, Aristolochiaceae, Lauraceae, Monimiaceae and Ranunculaceae

Number: 7

Structures:

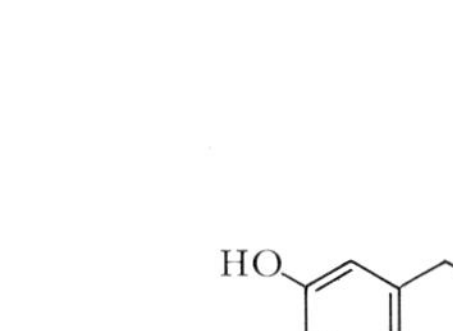

Atherosperminine

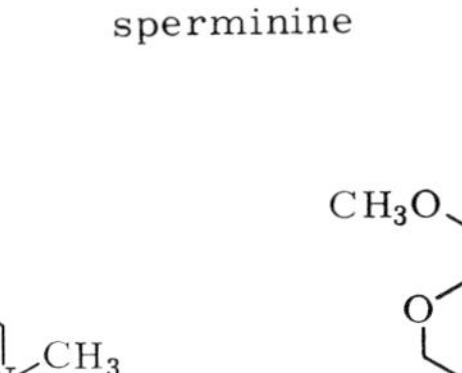

Methoxyatherosperminine

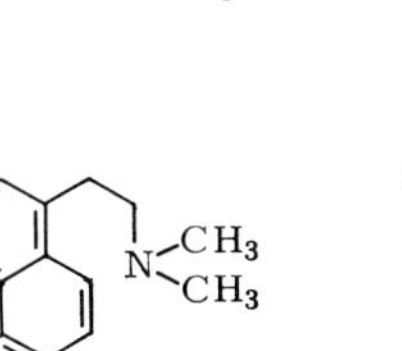

Uvariopsine

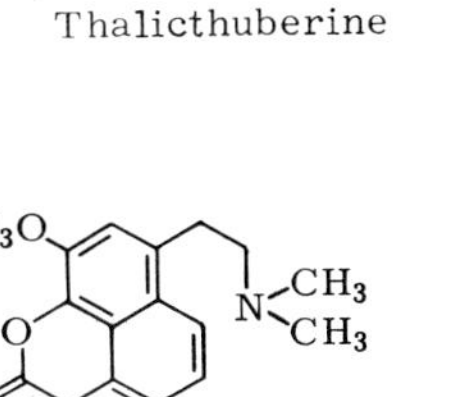

Thalicthuberine

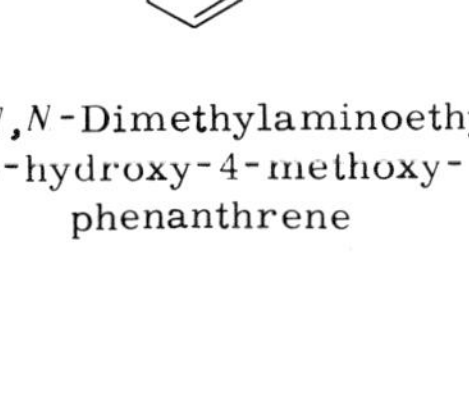

1-*N*,*N*-Dimethylaminoethyl-3-hydroxy-4-methoxyphenanthrene

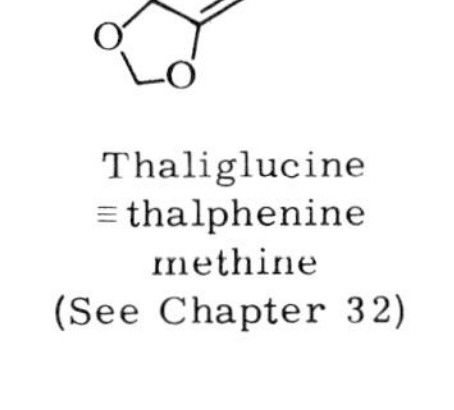

Thaliglucine
≡ thalphenine
methine
(See Chapter 32)

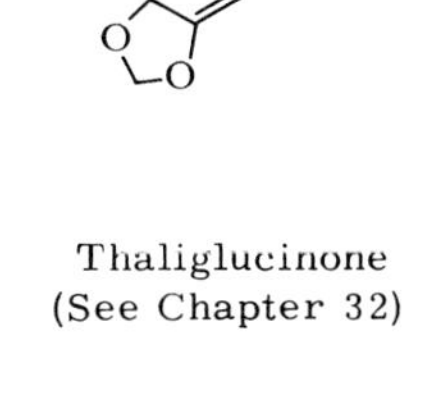

Thaliglucinone
(See Chapter 32)

I. STRUCTURAL ELUCIDATION AND SYNTHESIS

The phenanthrene alkaloids are a small group of optically inactive tertiary bases probably derived biogenetically from the Hofmann elimination of quaternary aporphine salts. They can, therefore, be included among the isoquinoline alkaloids.

The structural elucidation of the phenanthrene bases rests on spectral data, especially NMR, as well as upon chemical correlation with the corresponding aporphine *N*-metho salts. As an example of a chemical interrelationship, Hofmann degradation of the methiodide salt of synthetic 1,2,3-trimethoxyaporphine gave 1-*N*,*N*-dimethylaminoethyl-2,3,4-trimethoxyphenanthrene, identical in all respects with the alkaloid methoxyatherosperminine isolated from *Atherosperma moschatum* Labill. (Monimiaceae).[1]

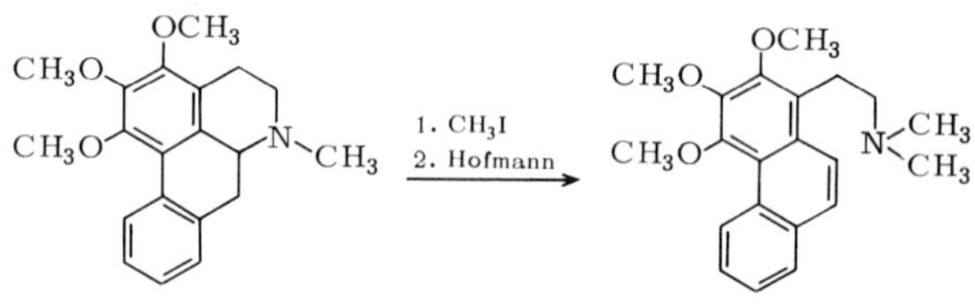

1,2,3-Trimethoxyaporphine Methoxyatherosperminine

Similarly, the naturally occurring base 1-*N*,*N*-dimethylaminoethyl-3-hydroxy-4-methoxyphenanthrene was first *O*-ethylated. The methiodide salt of the *O*-ethyl ether was then found to correspond to the methiodide of 1-methoxy-2-ethoxyaporphine methine prepared from 1-methoxy-2-ethoxyaporphine.[2,3]

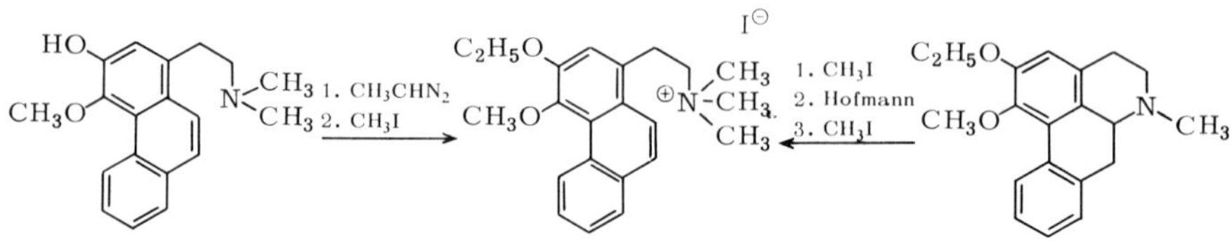

1-*N*,*N*-Dimethylaminoethyl-3-hydroxy-4-methoxyphenanthrene Methiodide salt 1-Methoxy-2-ethoxyaporphine

The phenanthrene alkaloids are always substituted at C-3,4 since their precursors, the aporphines, are found with substituents at these two positions corresponding to C-1,2 of the aporphine skeleton.

II. PHARMACOLOGY

The phenanthrene **1** derived from the cyanogen bromide treatment of the aporphine alkaloid bulbocapnine produces sedation in rats.[4]

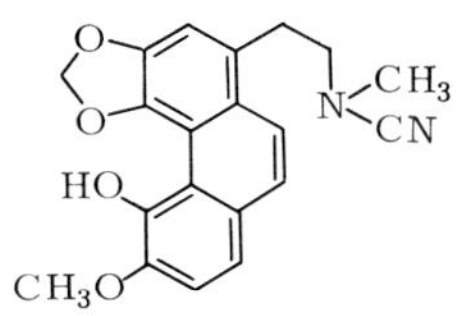

1

III. Mass Spectroscopy

The mass spectrum of thalicthuberine shows a base peak at *m/e* 58 for the dimethylaminomethylene moiety, $CH_2\overset{\oplus}{=}N(CH_3)_2$. The next most intense peak is the molecular ion at *m/e* 353 for $C_{21}H_{23}O_4N$. The third most intense peak is at *m/e* 295 and corresponds to $(M-58)^+$ (Scheme I).[5]

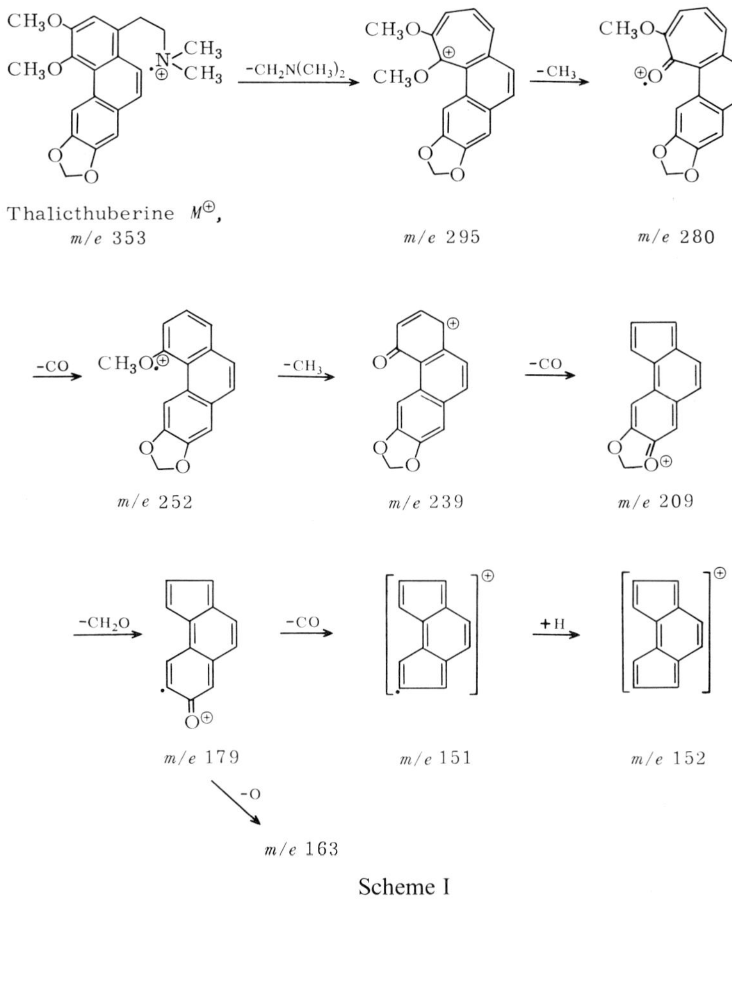

Scheme I

IV. NMR Spectroscopy

As with the aporphines, a C-4 methoxy in a phenanthrene alkaloid is found further upfield than the other methoxy groups. Similarly, the C-5 proton falls appreciably downfield and apart from the other aromatic hydrogens. In some cases the C-9,10-hydrogens cannot be differentiated easily from the hydrogens at C-6, 7, or 8. The C-2 proton is often found further upfield from the other aromatic hydrogens.

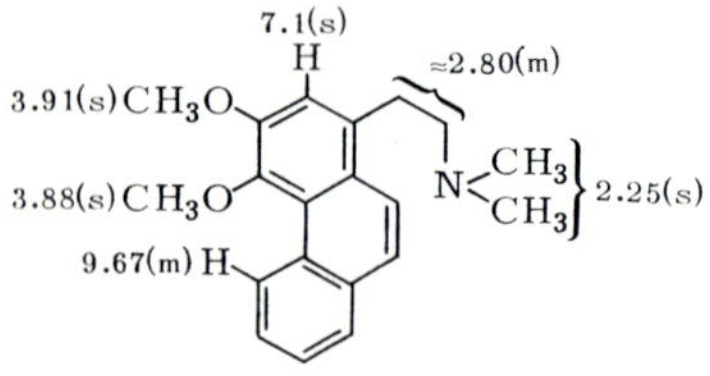

Atherosperminine[1]

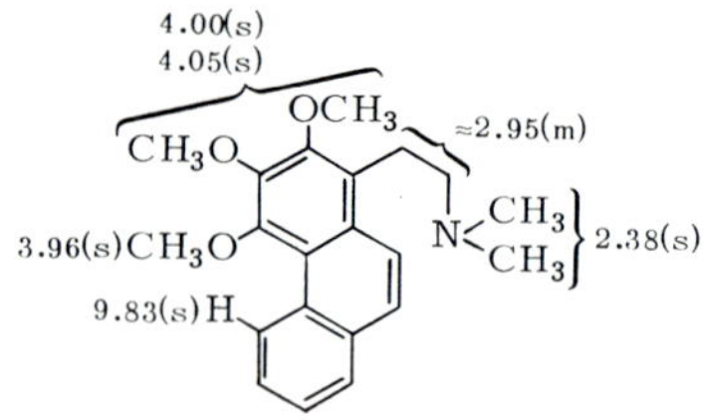

Methoxyatherosperminine[1]

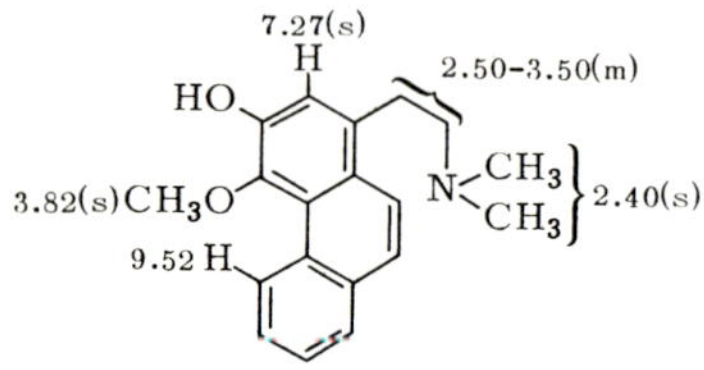

1-*N*,*N*-Dimethylaminoethyl-3-hydroxy-4-methoxyphenanthrene[2]

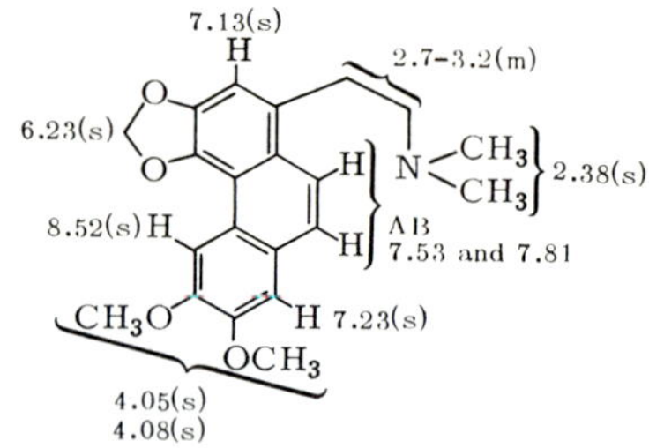

Dicentrine methine[6] (not a natural product)

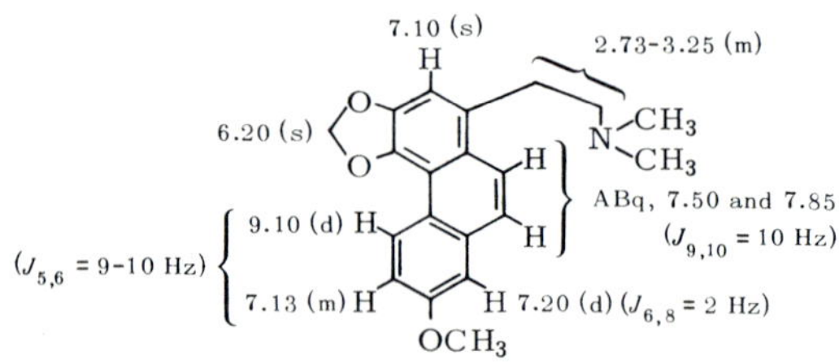

Uvariopsine[6a]

V. UV Spectroscopy

The phenanthrene alkaloids exhibit at least four peaks in the UV range, the most diagnostic of which falls between 255 and 266 mμ.

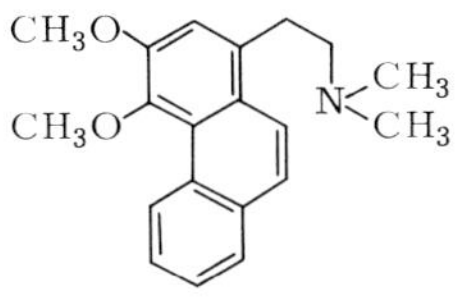

Atherosperminine[1]

Methiodide salt:
λ_{max}^{EtOH} 251 sh, 257, 279, 302, and 312 mμ
(4.69, 4.73, 4.08, 4.08, and 4.08)

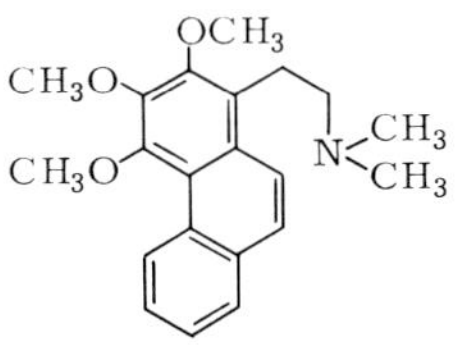

Methoxyatherosperminine[1]

Methiodide salt:
λ_{max}^{EtOH} 217, 259, 284, 295.5, and 308 mμ
(4.52, 4.80, 4.03, 4.05, and 4.11)

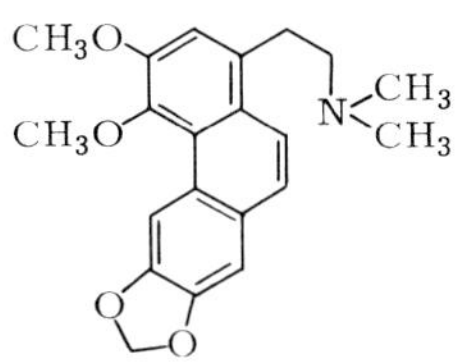

Thalicthuberine[7]

Free base:
λ_{max}^{EtOH} 261, 285, 310, and 345 mμ
(4.84, 4.50, 4.32, and 3.50)
λ_{max}^{EtOH} 230, 278, 300, and 340 mμ
(4.32, 4.49, 4.23, and 3.39)

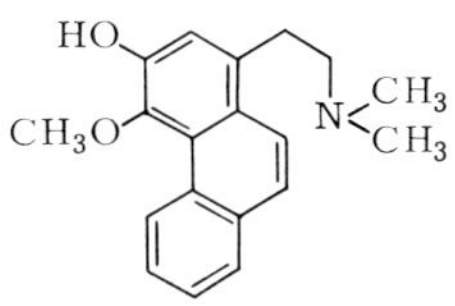

1-*N*,*N*-Dimethylaminoethyl-3-hydroxy-4-methoxyphenanthrene[2]

Oxalate salt:
λ_{max}^{EtOH} 210, 232, 255, and 310 mμ
(4.26, 4.32, 4.64, and 4.00)

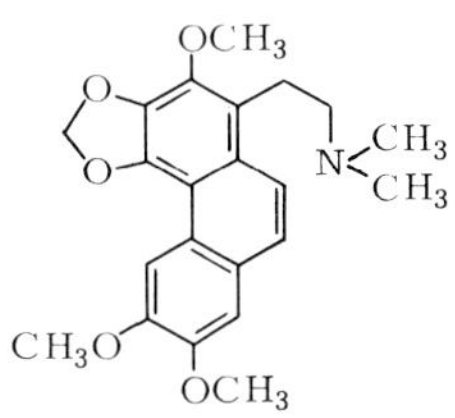

1-*N*,*N*-Dimethylaminoethyl-2,6,7-trimethoxy-3,4-methylenedioxyphenanthrene[8]
(not a natural product)

Free base:
λ_{max} 218, 266, 316, 326, 345, and 370 mμ
(4.22, 4.91, 4.08, 4.09, 3.50, and 3.60)

REFERENCES

1. I. R. C. Bick and G. K. Douglas, *Aust. J. Chem.* **18**, 1997 (1965).
2. H. A. Priestap, E. E. Ruveda, S. M. Albónico, and V. Deulofeu, *Chem. Commun.* p. 754 (1967).
3. M. Tomita, Y. Watanabe, and H. Furukawa, *J. Pharm. Soc. Jap.* **81**, 942 (1961).
4. E. S. Smissman, A. C. Makriyannis, and E. J. Walaszek, *J. Med. Chem.* **13**, 640 (1970).
5. M. Shamma and S. Y. Yao, unpublished results (1970). See also S. K. Akramov and S. Y. Yunusov, *Chem. Nat. Compds.* **4**, 332 (1968).
6. "Varian NMR Spectra Catalog," Spectrum No. 348. Varian Assoc., 1962.

6a. A. Bouquet, A. Cave, A. Cace, and R.-R. Paris, *C. R. Acad. Sci. (Paris), Ser. C* **271**, 1100 (1970).

7. E. Fujita and T. Tomimatsu, *J. Pharm. Soc. Jap.* **79**, 1252 (1959).
8. Kh. G. Pulatova, Z. F. Ismailov, and S. Yu. Yunusov, *Khim. Prir. Soedin* **2**, 426 (1966); *Chem. Natur. Prod.* **2**, 349 (1966); *Chem. Abstr.* **68**, 13226g (1968).

Chapter 15 / TASPINE

Occurrence: Berberidaceae and Euphorbiaceae
Number: 1
Structure:

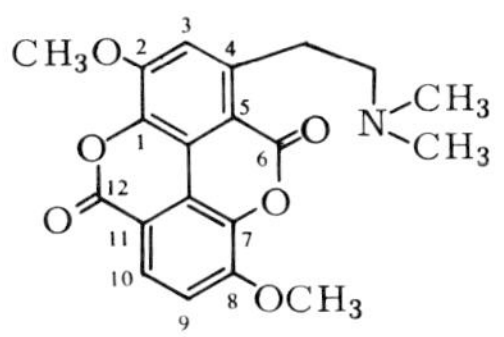

Taspine

I. Structural Elucidation

The alkaloid taspine, $C_{20}H_{19}O_6N$, isolated from *Leontice eversmannii* Bunge (Berberidaceae), *Caulophyllum robustum* Maxim (Berberidaceae), and other plants, is an optically inactive tertiary base with one *N*,*N*-dimethylamino function, two methoxyls, and two lactone groups.[1,2] Alkaline hydrolysis furnished a diacid, taspinic acid, $C_{20}H_{23}O_8N$, which could be recyclized to taspine in mineral acid. Oxidation of taspinic acid with chromic acid afforded β-dimethylaminopropionic acid. Potassium hydroxide fusion of taspine gave 2,3,2′,3′-tetrahydroxybiphenyl which upon permanganate oxidation yielded 2,3-dihydroxybenzoic acid (Scheme I).[1]

Hofmann degradation of taspine methiodide was accompanied by hydrolysis of the lactone groups so that the diphenolic diacid **1** was obtained. *O*-Methylation of this product followed by permanganate oxidation produced 5,6,5′,6′-tetramethoxybiphenyl-2,3,2′-tricarboxylic acid (**2**), also generated by oxidation of the aporphine salt *O*-methylcorydine methiodide (Scheme I).[1]

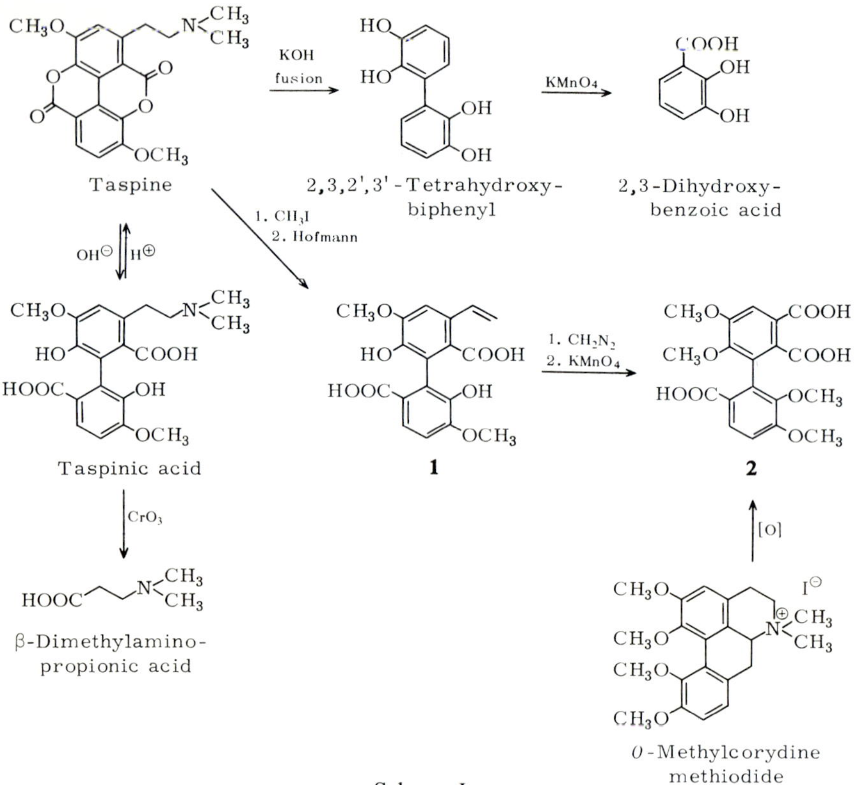

Scheme I

II. Biogenesis and Conversion of Magnoflorine into Taspine

Taspine occupies a unique position in the realm of alkaloid chemistry, with no close relative among other alkaloids. It is probably derived biogenetically from the widely occurring quaternary aporphine magnoflorine. Oxidation of its phenanthrene derivative (methine) would give rise to taspinic acid which could readily undergo cyclization to taspine:

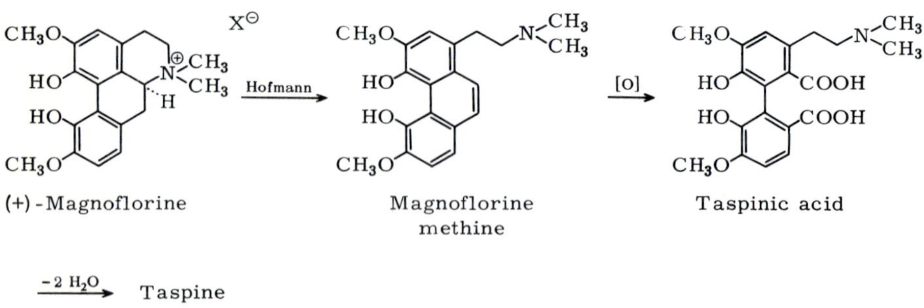

In an effort to emulate this sequence *in vitro*, Shamma and Moniot carried out the following series of reactions. Magnoflorine methine derived through Hofmann degradation of naturally occurring (+)-magnoflorine chloride was first acetylated with acetic anhydride in pyridine to afford diacetylmagnoflorine methine. Ozonolysis of this material gave the dialdehyde diacetate **3** as a yellow foam. An aqueous solution of **3** was oxidized with basic silver oxide, and the filtrate made strongly acidic with hydrochloric acid and refluxed for a half hour. Work-up provided white crystals of taspine hydrochloride from which taspine could be liberated (Scheme II).[3]

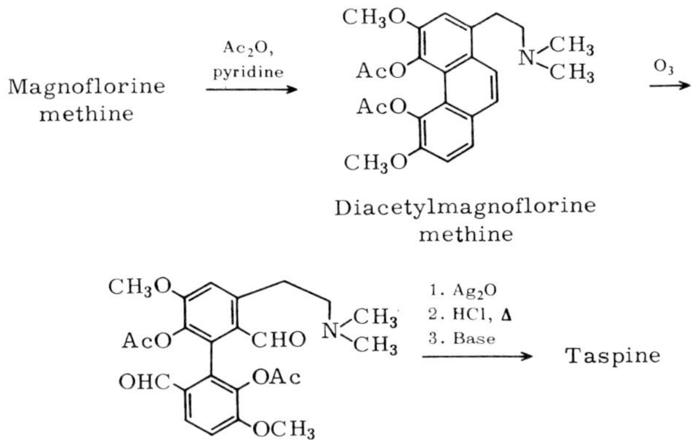

3

Scheme II

III. NMR and UV Spectroscopy

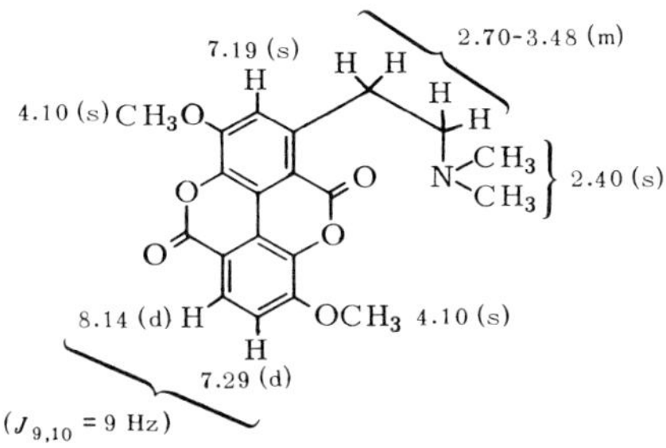

NMR spectral data for taspine[3]

Taspine shows λ_{max}^{MeOH} 246, 285, 330, and 345 mμ (4.83, 4.01, 3.89, and 3.97).[3,4]

REFERENCES

1. T. F. Platonova, A. D. Kusowkow, and Yu. N. Scheinker, *Zh. Obshch. Khim.* **26**, 2651 (1956).
2. L. N. Safronich, *Chem. Abstr.* **55**, 18892 (1961).
3. M. Shamma and J. L. Moniot, *Chem. Commun.* p. 1065 (1971).
4. Holubek and Štrouf, Spectrum No. 393.

Chapter 16 / THE PROTOBERBERINES AND RETROPROTOBERBERINES

Occurrence: Anonaceae, Berberidaceae, Convolvulaceae, Fumariaceae, Lauraceae, Menispermaceae, Papaveraceae, Ranunculaceae, and Rutaceae

Approximate Number: 70, including 4 retroprotoberberines

Some Protoberberines and Retroprotoberberines of Interest:

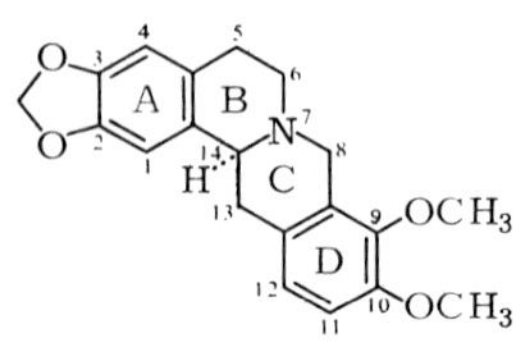

(–)-Canadine ≡ (–)-Tetrahydroberberine

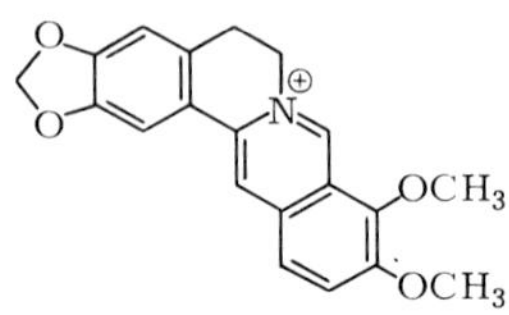

Berberine

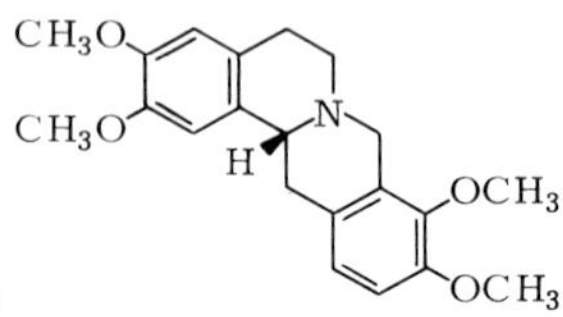

(+)-Tetrahydropalmatine (Enantiomer also a natural product)

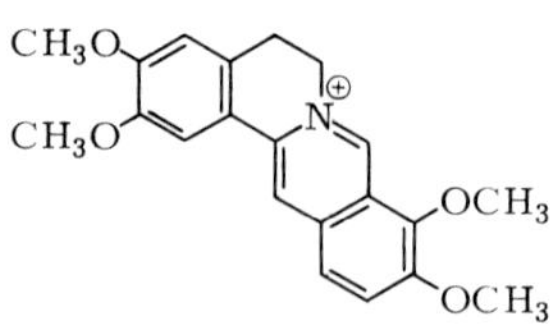

Palmatine

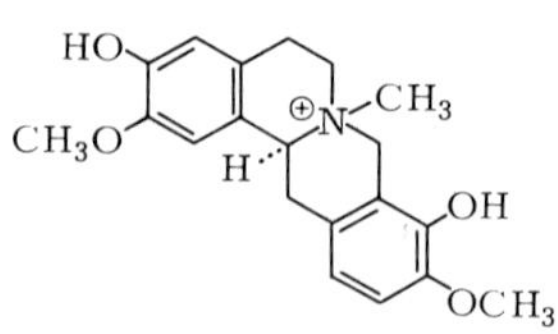

(–)-Steponine

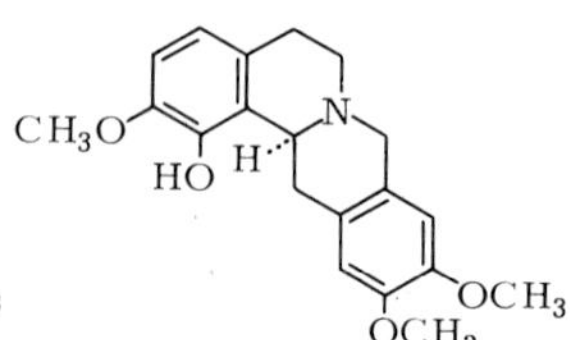

(–)-Caseadine

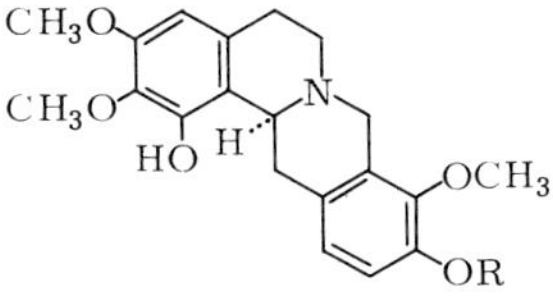

(−)-Capaurine, R = OCH_3
(−)-Capaurimine, R = H

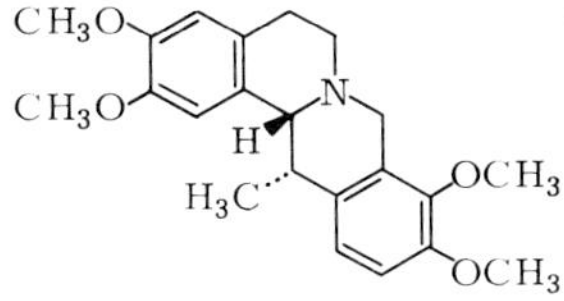

(+)-Corydaline

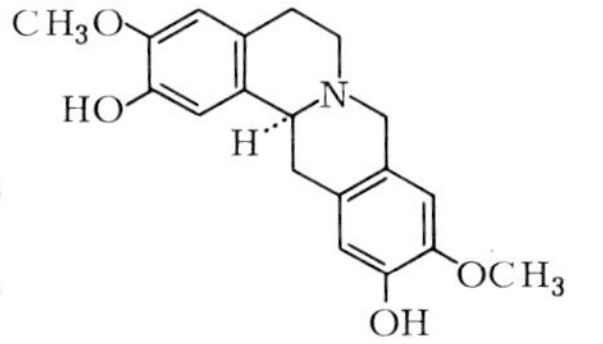

(−)-Coreximine

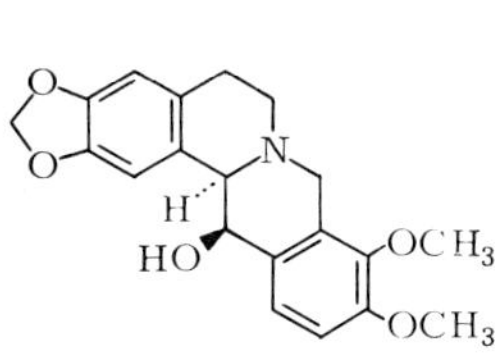

(−)-Ophiocarpine

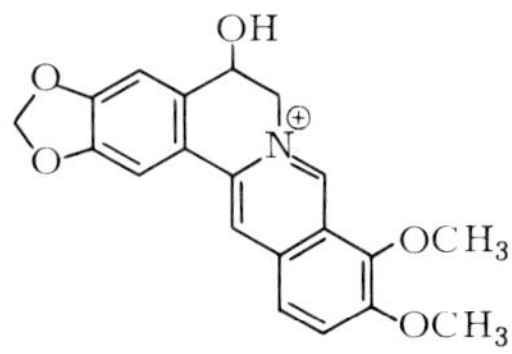

Berberastine

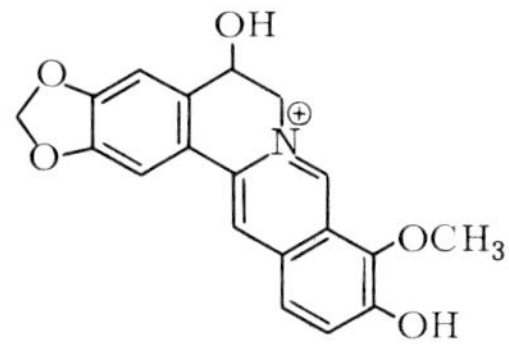

Thalidastine

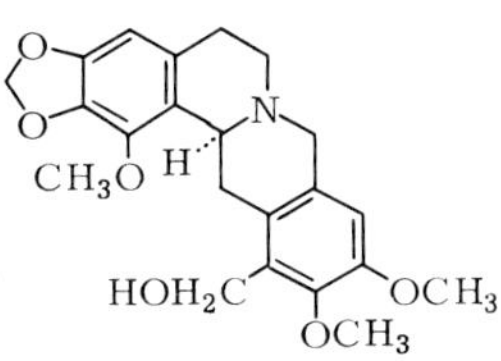

(−)-Mecambridine

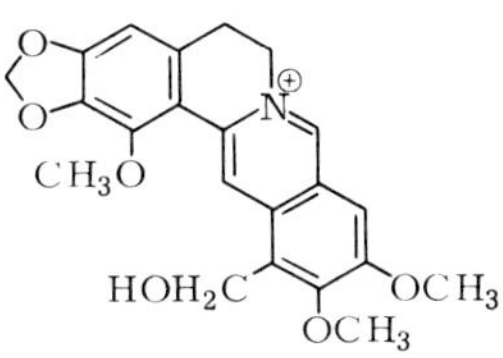

Alkaloid PO-5

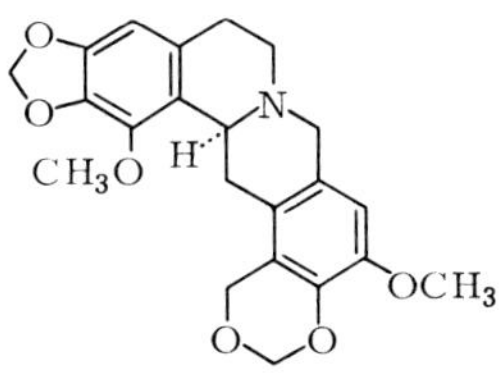

(−)-Orientalidine

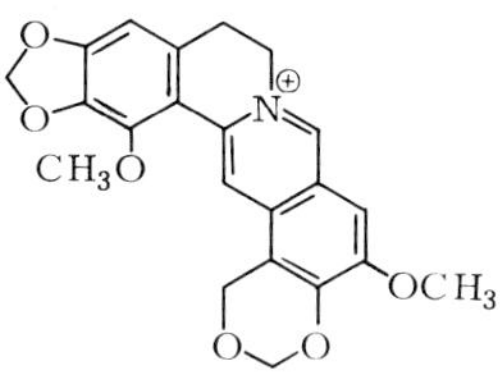

Alkaloid PO-4

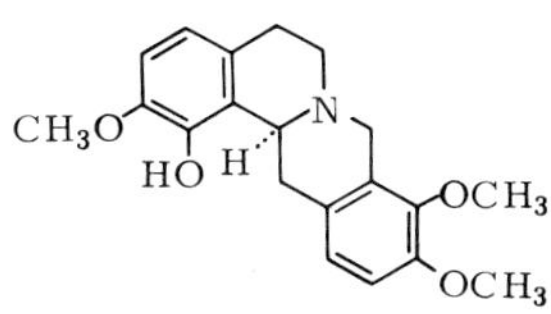

(−)-Caseanadine

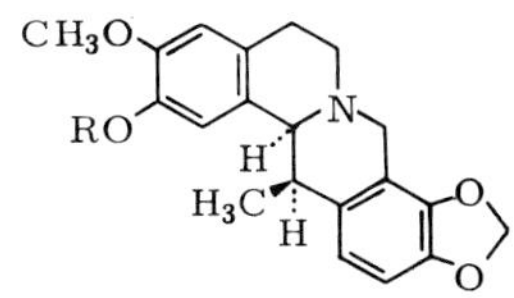

(±)-Cavidine, R = CH_3
(±)-Apocavidine, R = H

I. Introduction

Most protoberberine alkaloids exist in nature either as tetrahydroprotoberberines or as quaternary protoberberine salts, but some dihydroprotoberberines are also known.[1,2] Substituents are usually present at C-2 and C-3 and either at C-9 and 10 or at C-10 and 11. In some instances a hydroxyl or methoxyl substituent may be present at C-1. A methyl group is sometimes found at C-13, while in a few cases an alcoholic hydroxyl is located at C-13 or at C-5.

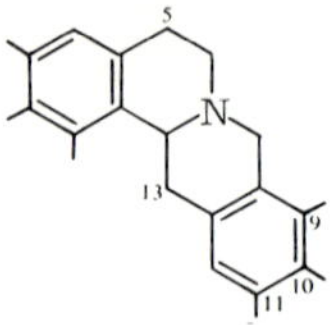

A tetrahydroproto-berberine

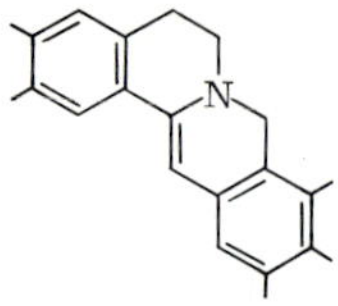

A dihydroprotober-berine

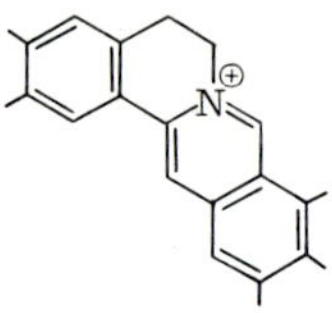

A quaternary protoberberine salt

Four retroprotoberberines are known; these are (−)-mecambridine, alkaloid PO-5, (−)-orientalidine, and alkaloid PO-4. They are characterized by the presence of one extra carbon atom as a side chain bonded to ring D. Retroprotoberberines may be formed from the conversion of a protoberberine to a benzylisoquinoline which then reverts back to the protoberberine stage, having acquired one extra carbon atom in the process. This theme will be developed further in Sections VII and IX of this chapter.

II. Chemical Degradation

Tetrahydroprotoberberines are easily oxidized with potassium permanganate to mixtures of phthalic acids and bicyclic lactams.* The characterization of such shards often permits the elucidation of the structure of the parent molecule. As a representative example, the degradation of (−)-steponine will be considered (Scheme I).

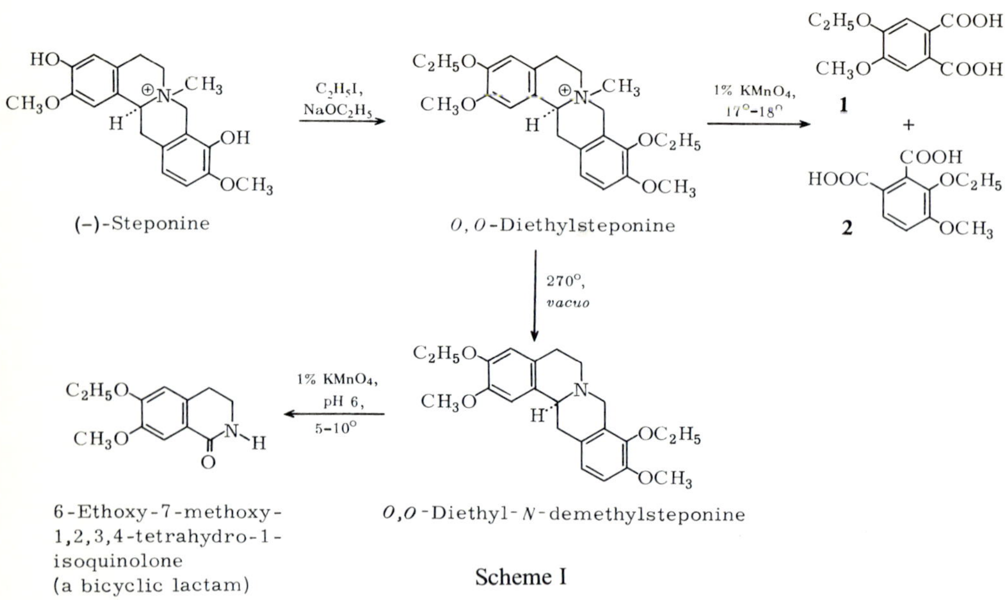

Scheme I

* Tetrahydroprotoberberines substituted at C-10 and 11, and with the C-11 substituent a methoxyl, apparently differ in their behavior from those of other tetrahydroprotoberberines and on oxidation do not yield bicyclic lactams.[1a]

Steponine occurs in *Stephania japonica* Miers (Menispermaceae) as a salt whose cation has the formulation $C_{20}H_{24}O_4N^+$. Two methoxyls, two phenolic hydroxyls, and one *N*-methyl group are incorporated in the alkaloid.[3,4]

O,O-Diethylsteponine upon mild oxidation with potassium permanganate yielded two known phthalic acid derivatives, **1** and **2**, which were characterized as their respective *N*-ethylphthalimides. *N*-Demethylation of *O,O*-diethylsteponine could be achieved by heating the iodide salt *in vacuo* at 270°. The resulting *O,O*-diethyl-*N*-demethylsteponine was oxidized with permanganate in acid to give the known lactam 6-ethoxy-7-methoxy-1,2,3,4-tetrahydro-1-isoquinolone. The only expression for steponine consistent with all the degradation products obtained is that indicated.

When a C-13 methyl group is present in a tetrahydroprotoberberine, a good degradative procedure involves Hofmann elimination followed by Emde reduction and ozonization of the reduction product. This sequence is described as it was applied to the alkaloid (+)-thalictricavine (Scheme II).[5] It should be noted, however, that if the C-13 and 14 hydrogens are trans to each other, the direction of Hofmann elimination takes a different course from that indicated in Scheme II (see Section V, B).

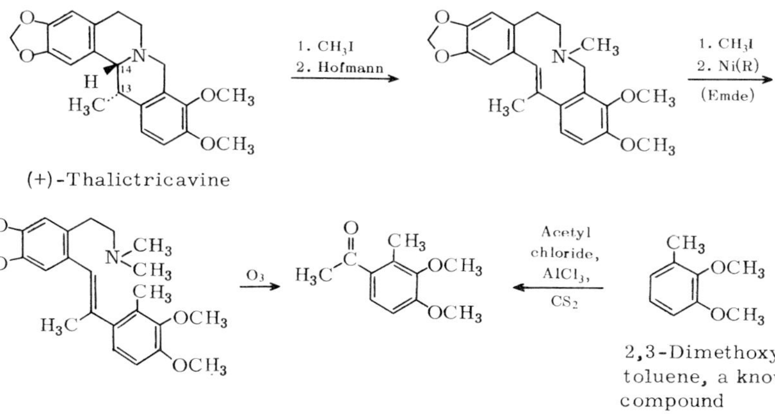

Scheme II

Since tetrahydroprotoberberines lend themselves readily to degradative studies, new quaternary protoberberines are usually reduced to their tetrahydro derivatives for characterization purposes.

III. Syntheses of Protoberberines

A. The Mannich Condensation

A classical preparation of tetrahydroprotoberberines involves Mannich condensation of a benzylisoquinoline such as **3** with formaldehyde in the presence of acid. The main product in this case is tetrahydropseudoberberine,* while a very minor product is the isomeric canadine. This approach, even though more than 50 years old, is still the method of choice for the construction of protoberberine systems.[6,7]

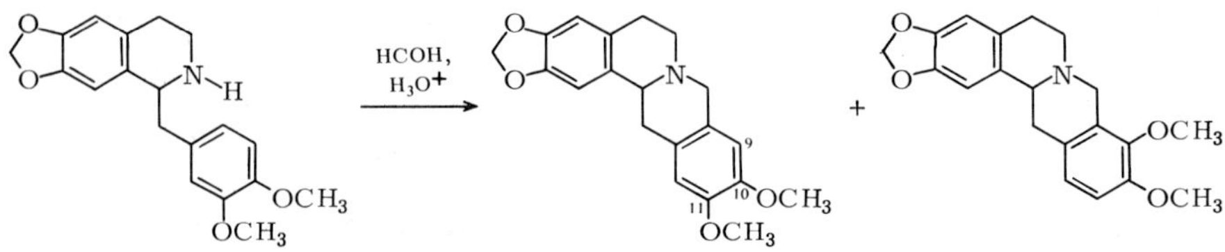

3 Tetrahydropseudo-berberine Canadine

When the two substituents in the bottom ring are hydroxyl, a mixture of 9,10- and 10,11-substituted products is obtained.[8,9] With norreticuline, which has one hydroxyl and one methoxyl in the bottom ring, two compounds can be obtained, scoulerine and coreximine. The pH of the reaction medium apparently determines the relative quantities of the two products. At pH 6.3 the ratio of scoulerine to coreximine is 2 : 1,[10] while at pH 7 apparently only or mostly coreximine is isolated.[11]

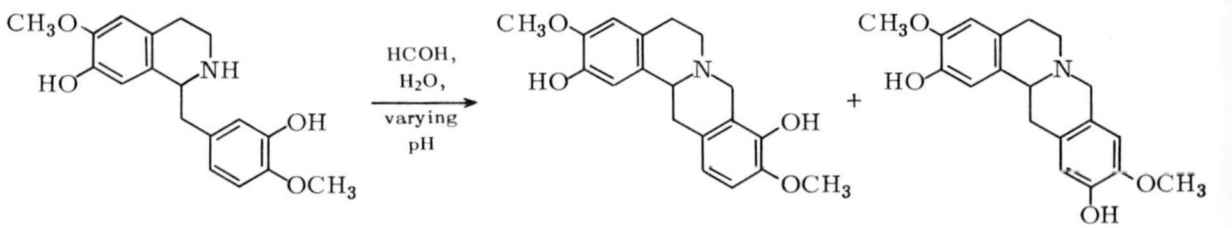

Norreticuline Scoulerine Coreximine

The synthesis of the racemic form of the alkaloid (−)-kikemanine involved a separation of 9,10- and 10,11-isomers and is the theme of Scheme III.[12]

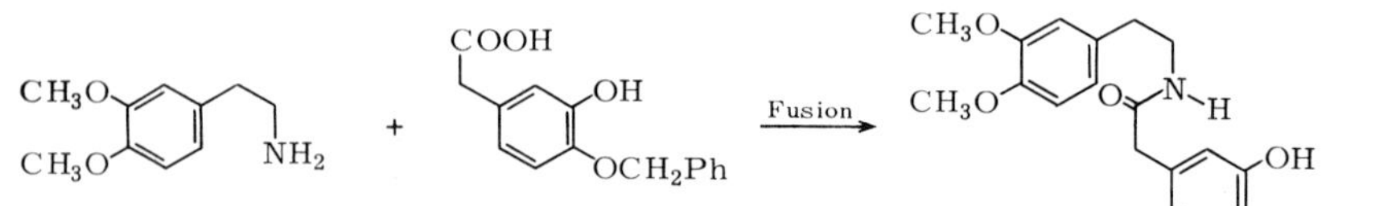

Homoveratrylamine 3-Hydroxy-4-benzyloxyphenylacetic acid

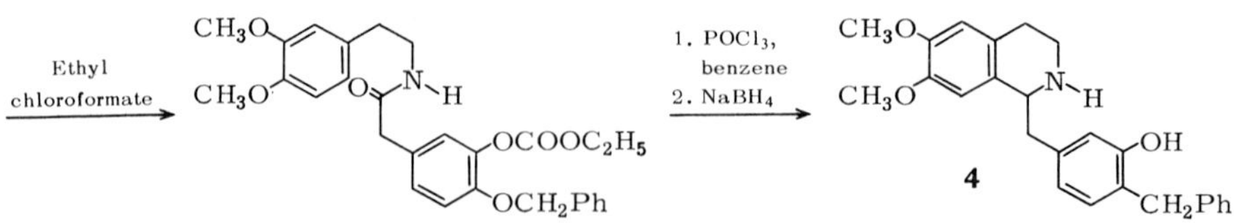

4

* The term "pseudo" in the protoberberine series refers to 10,11-substitution.

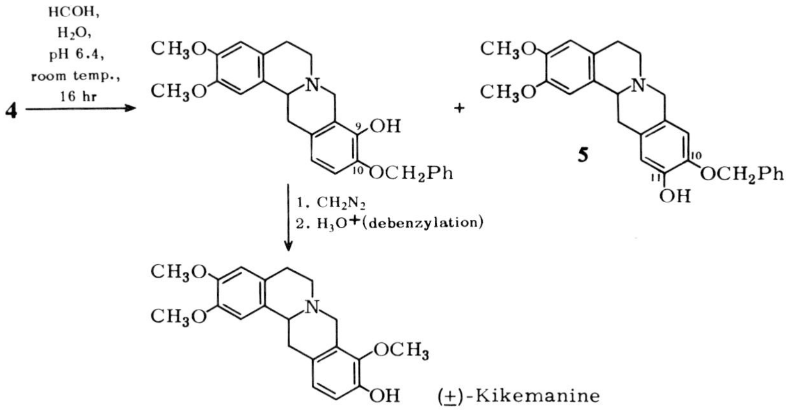

Scheme III

On the other hand, if the hydrochloride of **4** was heated with formalin in ethanol, condensation occurred para to the hydroxyl groups to give solely the tetrahydroprotoberberine **5**.[12] (For a further discussion of the effect of pH see Chapter 32.)

Alkaloids with 9,10-substitution can be obtained cleanly through the Mannich cyclization provided bromine is used as a protective group. The dibenzyloxytetrahydrobenzylisoquinoline **6** was selectively brominated in ring C. Subsequent debenzylation, cyclization, and reductive cleavage of the halogen afforded a preparation of scoulerine (Scheme IV).[13]

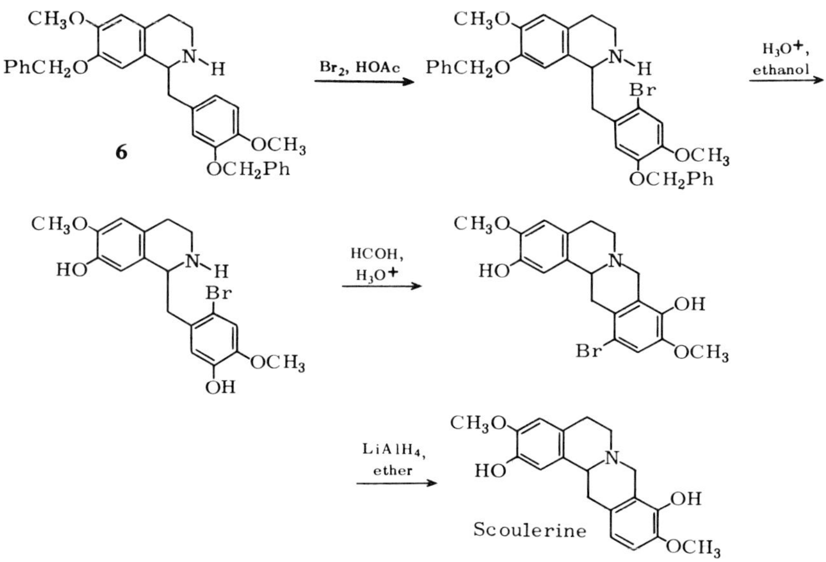

Scheme IV

Since the ethoxycarbamido substituent activates an aromatic ring, it has been possible to synthesize the tetrahydroprotoberberine *O*-methylcaseadine, again through a Mannich reaction. *O*-Methylcaseadine had originally been obtained via *O*-methylation of the unusually substituted alkaloid (−)-caseadine (Scheme V).[14]

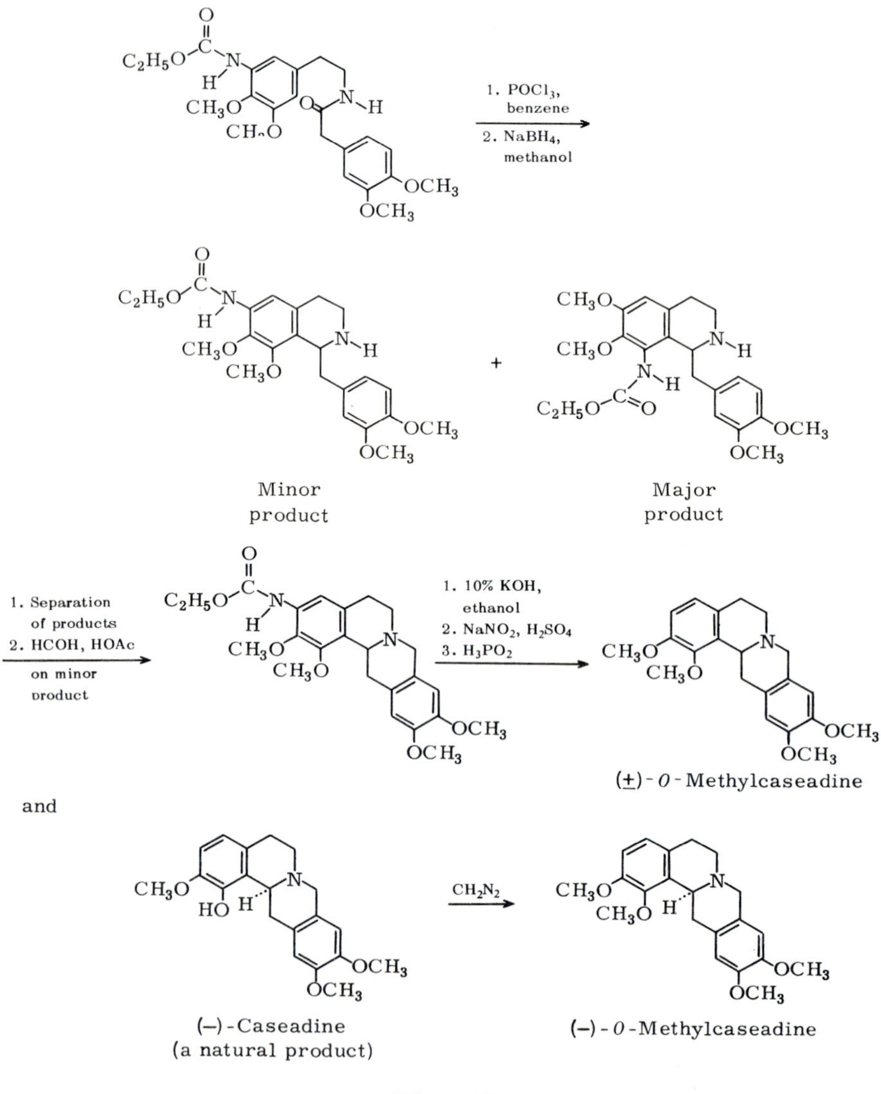

Scheme V

Metalation of laudanosine with butyllithium followed by condensation with formaldehyde and cyclization using methanesulfonyl chloride apparently yields the *N*-metho salt of tetrahydropalmatine in 10% overall yield. The occurrence of a 10,11-substituted tetrahydroprotoberberine *N*-metho salt as a by-product was not mentioned.[15]

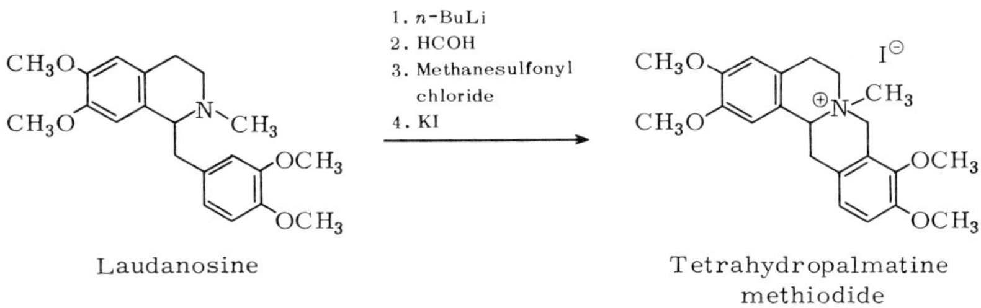

B. Variants of the Bischler–Napieralski Cyclization

An approach has been worked out which utilizes the Reimer–Tiemann reaction to obtain the important aldehydo acid **7**. In subsequent steps, **7** was converted to the lactam **8** which through a Bischler–Napieralski cyclization followed by reduction and hydrolysis furnished scoulerine (Scheme VI).[10]

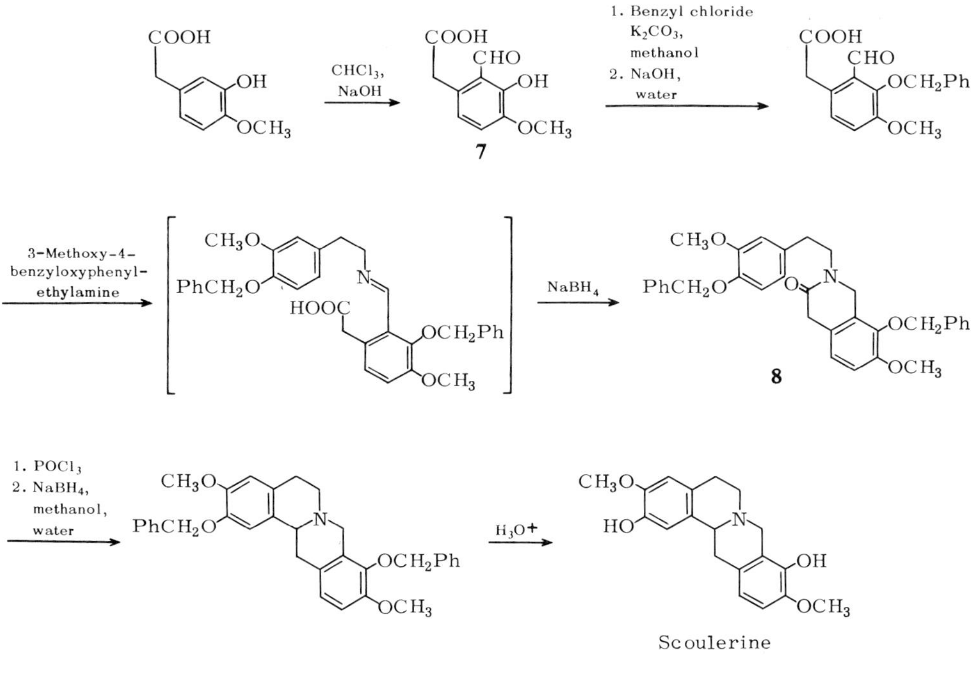

Scheme VI

The lactone **9** derived from the aldehydo acid **7** may also be employed (Scheme VII).[10,16]

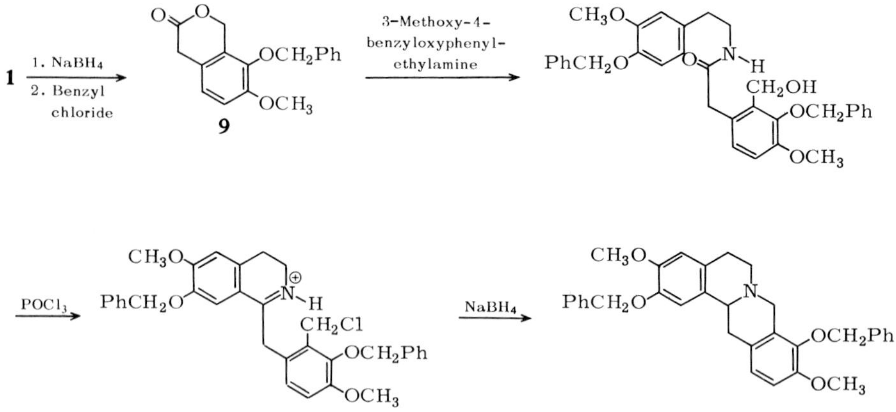

O,O-Dibenzylscoulerine

Scheme VII

One of the starting compounds can even be an anhydride, as in the condensation of homopiperonylamine with 4,5-dimethoxyhomophthalic anhydride. The resulting imide was then hydrolyzed to the corresponding acid amide, which in turn led to oxypseudoberberine (Scheme VIII).[17]

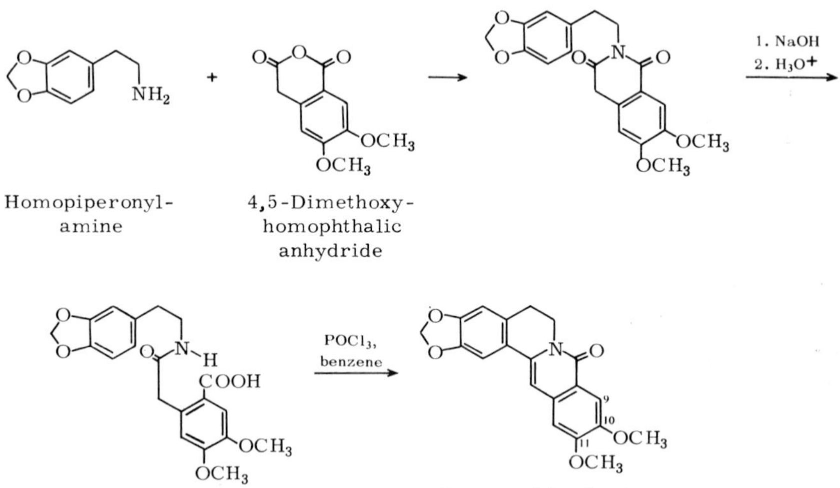

Oxypseudoberberine

Scheme VIII

9,10-Substituted protoberberines have also been obtained through the anhydride approach; 3,4-methylenedioxyhomophthalic anhydride has been prepared by the route shown in Scheme IX.[18] 3,4-Dimethoxyhomophthalic anhydride is also a known compound and has been prepared by an analogous sequence.

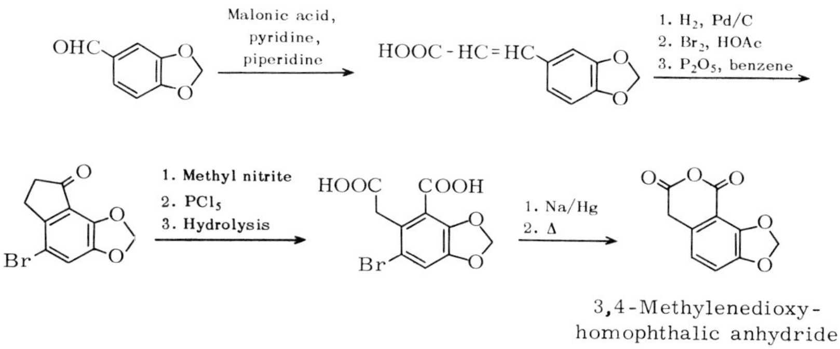

Scheme IX

The above schemes involve Bischler-Napieralski cyclization in the formation of ring B, but this same reaction has also been employed in the formation of ring C of a protoberberine[19]:

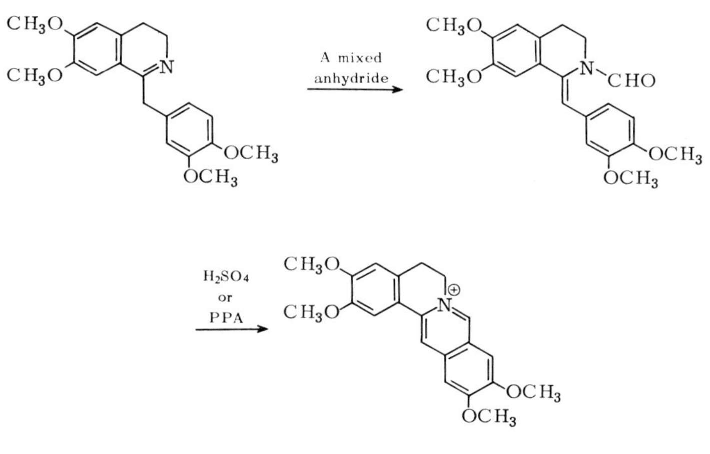

A modification of the above approach uses the Vilsmeier–Haack reagent to form ring C as in the two examples cited in Scheme X.[20]

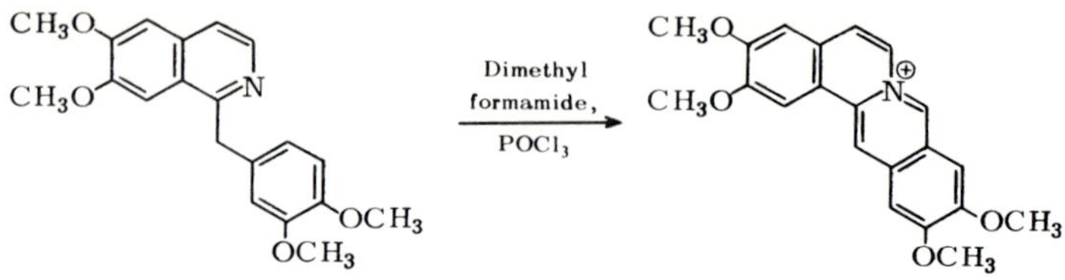

Papaverine

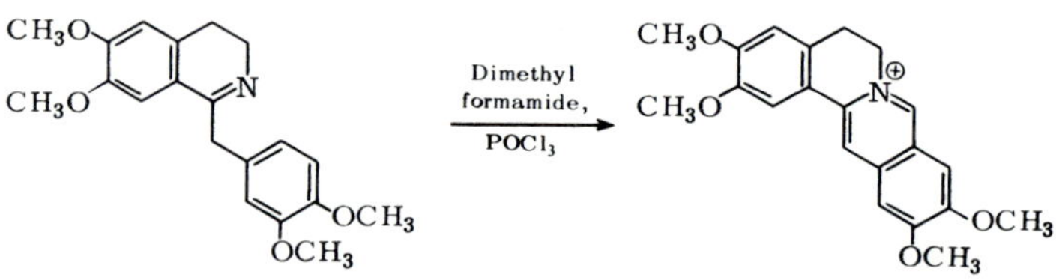

3,4-Dihydropapaverine

Scheme X

The benzylisoquinoline **10** will also undergo cyclization and *C*-alkylation, not necessarily in this order, in a reaction medium containing formaldehyde, an acid, and diethylamine. The product is the C-13-methylated dehydroprotoberberine salt **11**.[21]

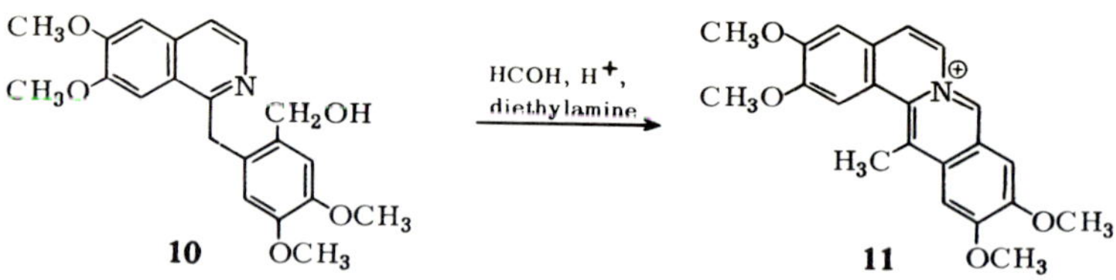

10 **11**

C. Syntheses through Phthalideisoquinolines

It was demonstrated in 1925 that tetrahydroprotoberberines can be prepared using the acid chloride of meconine α-carboxylic acid as an intermediate, but meconine α-carboxylic acid itself is not very readily prepared, making this approach impractical (Scheme XI).[22]

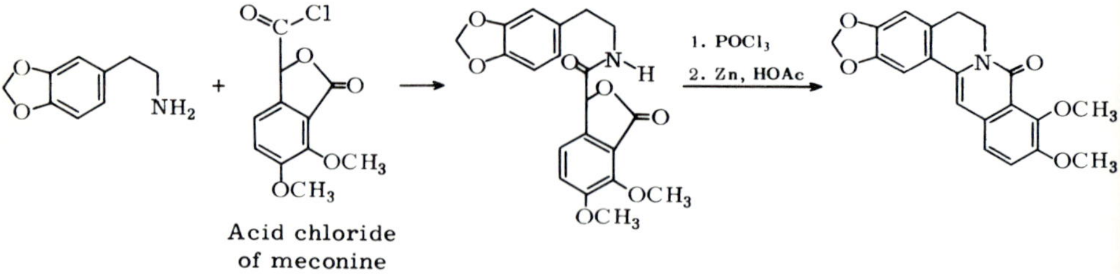

Acid chloride of meconine α-carboxylic acid

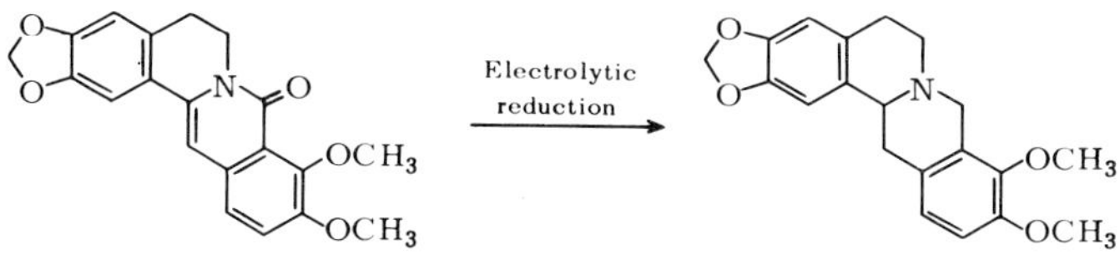

Canadine

Scheme XI

The C-13-hydroxylated alkaloid ophiocarpine was synthesized through the phthalide-isoquinoline **12**. Reduction with Adams catalyst yielded a mixture of diastereoisomers which upon further reduction with lithium aluminum hydride produced a diastereoisomeric mixture from which ophiocarpine could be isolated (Scheme XII).[23]

12

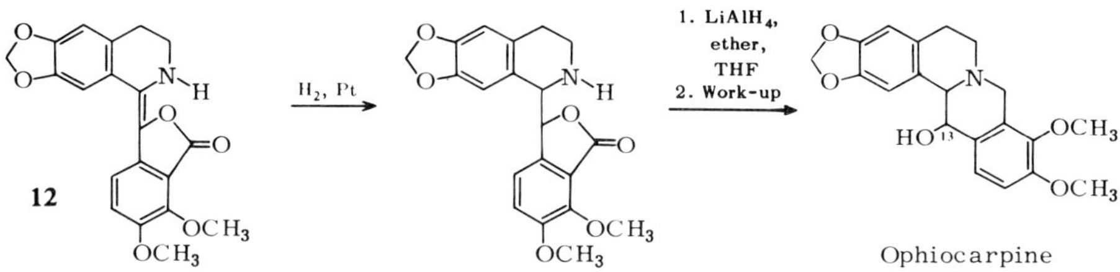

Ophiocarpine
and epiophiocarpine

Scheme XII

An alternative route to the C-13-hydroxylated protoberberines involves reduction of berberine chloride to dihydroberberine with subsequent hydration using Brown's hydroboration method. The yield of epiophiocarpine to ophiocarpine was in the ratio 4 : 1 since hydration occurs by cis addition.[24]

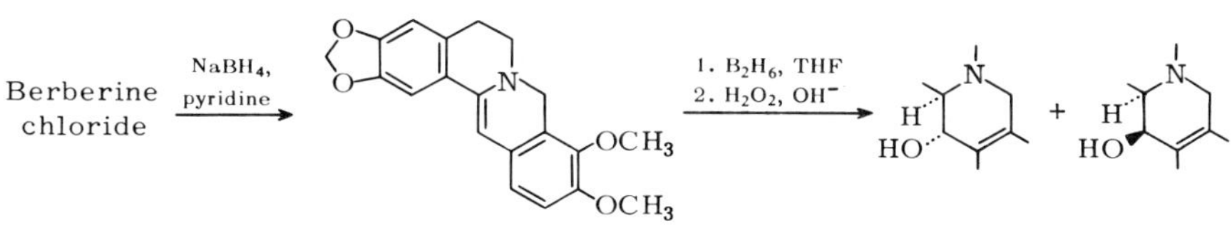

Dihydroberberine Epiophiocarpine Ophiocarpine

D. Syntheses Using a Dihydroisoquinoline Intermediate

Tetrahydroprotoberberines can be obtained from the acid-catalyzed ring closure of dihydroisoquinolines, as exemplified by the synthesis of norcoralydine (Scheme XIII).[25]

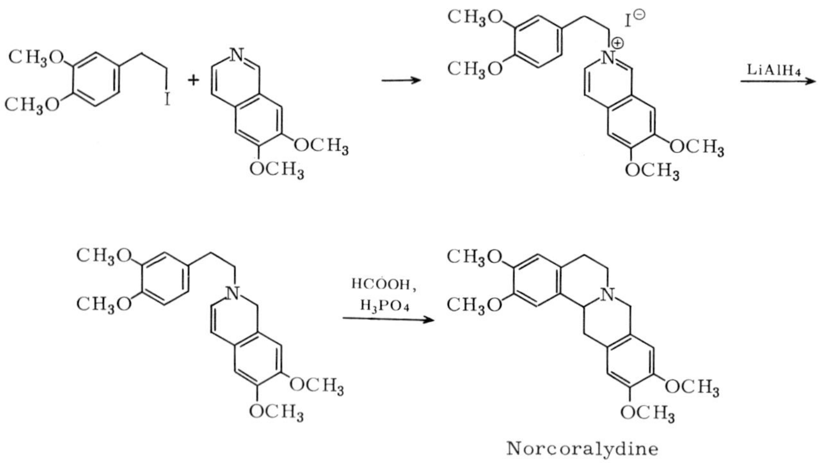

Scheme XIII

Dihydroisoquinolines such as **15** can also be generated through the Cannizzaro-type oxidation–reduction of an isoquinoline salt. It is interesting to note in this context that the isocarbostyril **13** is rapidly converted to the lactam **14** by concentrated hydrochloric acid at room temperature (Scheme XIV).[26]

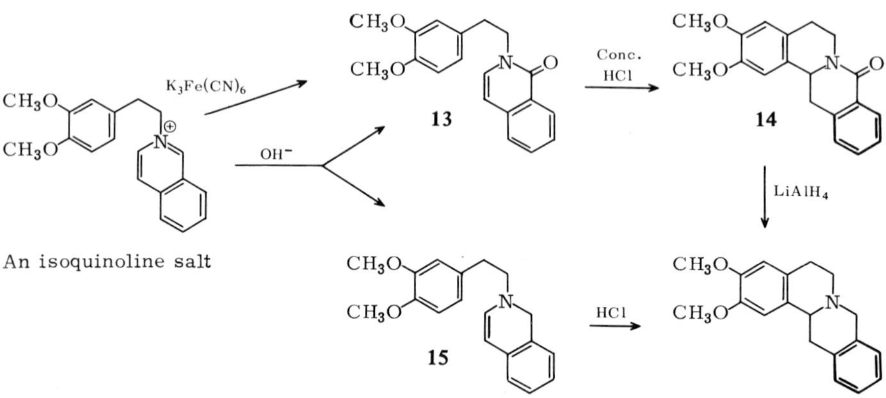

Scheme XIV

E. The Use of Methyl Vinyl Ketone

The enolic β-keto ester **16** which had been obtained earlier as part of a synthetic effort in the emetine series was treated with methyl vinyl ketone to afford the diketo ester **17**. Internal aldol condensation of **17** gave the unsaturated ketone **18**. Aromatization of ring D in **18** was achieved by heating the hydrobromide salt with cupric bromide in benzyl alcohol. Hydrolysis of the resulting benzyl ether was succeeded by a Musliner–Gates dehydroxylation so that the final product was the tetrahydroprotoberberine **19** (Scheme XV).[27]

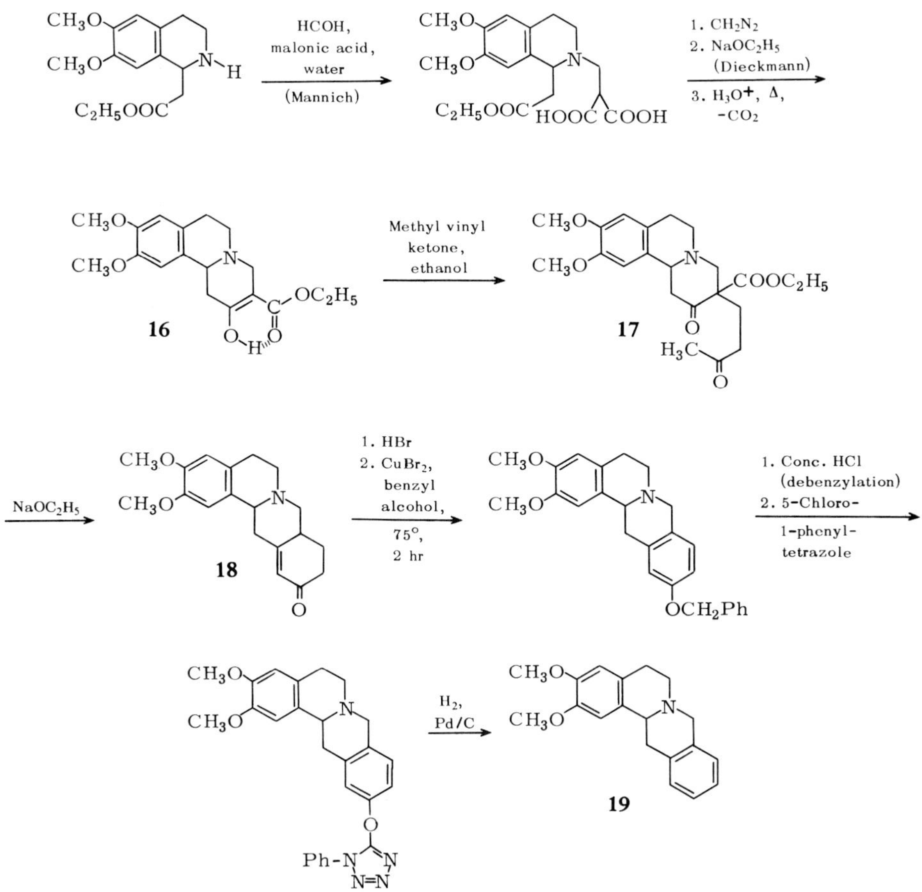

Scheme XV

F. The Pomeranz–Fritsch Cyclization

Two natural protoberberine salts are known with hydroxyl groups at C-5, berberastine and thalidastine. The following sequence was designed to afford a route to these compounds and their analogs.[28]

The known tetrahydroisoquinoline **20** was treated with glycidol. Without isolation, the amino glycol **21** was oxidized with periodic acid to provide the aldehyde **22**. When the latter compound was left in contact with 6 *N* hydrochloric acid at room temperature, cyclization occurred and the 5-hydroxytetrahydroprotoberberine **23** was isolated in 70% yield. (For another synthesis of a 5-hydroxylated tetrahydroprotoberberine see Chapter 32.)

Dehydrogenation of **23** with iodine gave the protoberberine salt **24** which readily dehydrated to the dehydroprotoberberine salt **25**. Finally, reduction with sodium borohydride gave rise to norcoralydine (Scheme XVI).

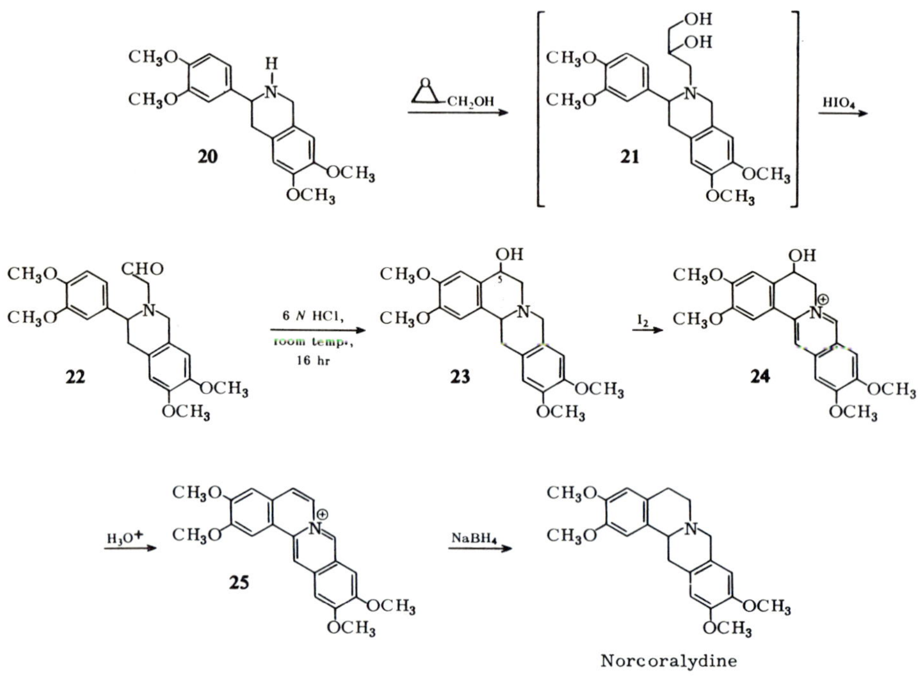

Scheme XVI

A related sequence worked out independently allowed for a three-step synthesis of norcoralydine starting from a 3-arylisoquinoline. The steps were quaternization with bromoacetaldehyde, treatment with dry hydrogen chloride to yield a cyclized product, and hydrogenation of the cyclized product to norcoralydine (Scheme XVII).[29]

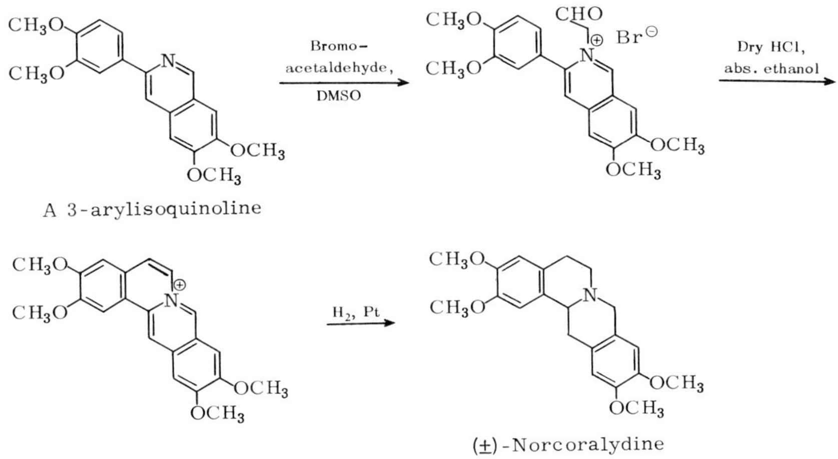

Scheme XVII

G. *Isoquinoline-1-carboxaldehydes as Intermediates*

Selenium dioxide oxidation of the isoquinoline **26** gave the isoquinoline-1-carboxaldehyde **27**. Condensation of the oxime of **27** with 2,3-dimethoxybenzyl bromide afforded the quaternary salt **28**. Acid-catalyzed cyclization then furnished dehydropalmatine bromide, which could be reduced catalytically to tetrahydropalmatine (Scheme XVIII). This sequence for the construction of the protoberberine nucleus has been applied in a wide number of instances.[30]

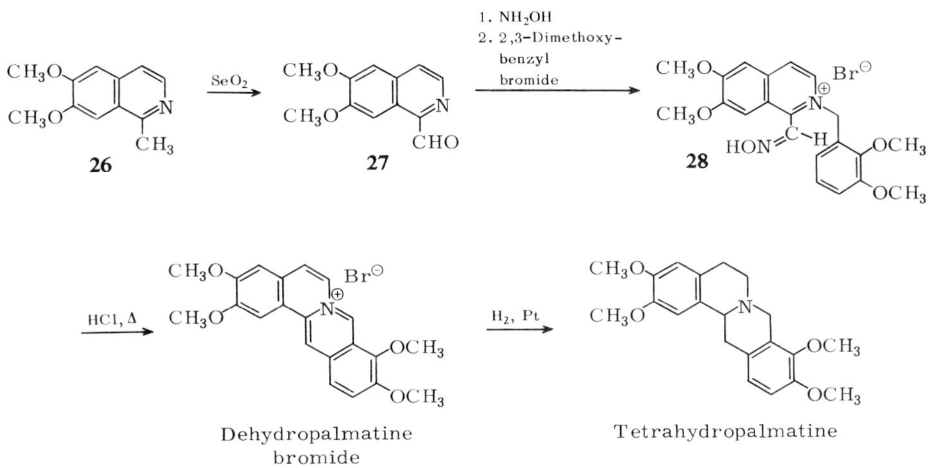

Scheme XVIII

The isoquinoline-1-carboxaldehyde approach has been extended to the construction of C-13-methylated tetrahydroprotoberberines as demonstrated in Scheme XIX.[31]

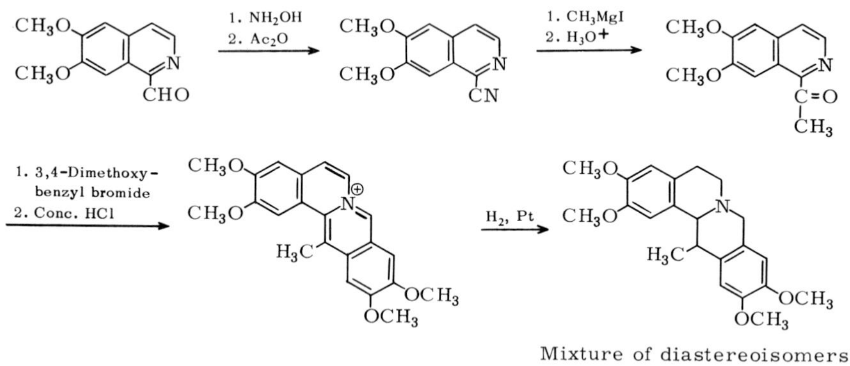

Scheme XIX

H. A Photochemical Route to an Oxoprotoberberine

Irradiation of 1-benzylidene-2-carbethoxytetrahydroisoquinoline (**29**) produced the dehydroaporphine **30** in 65% yield and the oxoprotoberberine **31** in 10–21% yield, depending upon reaction conditions.[32] (See also Chapter 10, Section IV, B.)

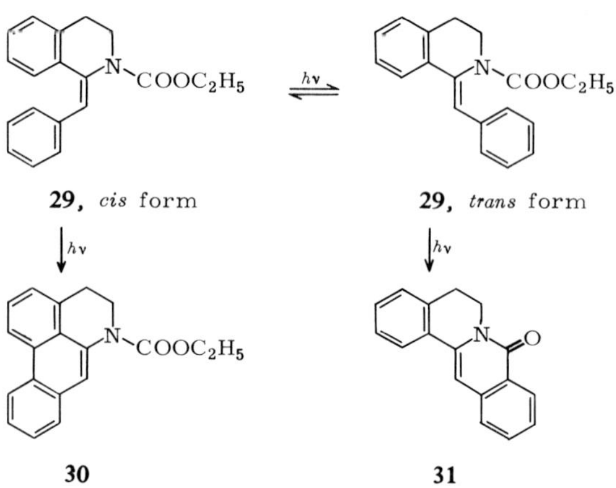

IV. Reactions of Protoberberines

A. Oxidation and Reduction

Quaternary protoberberine salts may be reduced to the corresponding tetrahydroprotoberberines with a variety of reducing agents. If, however, the reduction is carried out with a mixed metal hydride in a dry aprotic solvent, the reaction stops at the dihydroprotoberberine stage.[24,33,34] Reoxidation of the dihydro- or tetrahydroprotoberberine to the quaternary salt can be accomplished with iodine,[35] mercuric acetate, or simply by standing in air (Scheme XX).[35a]

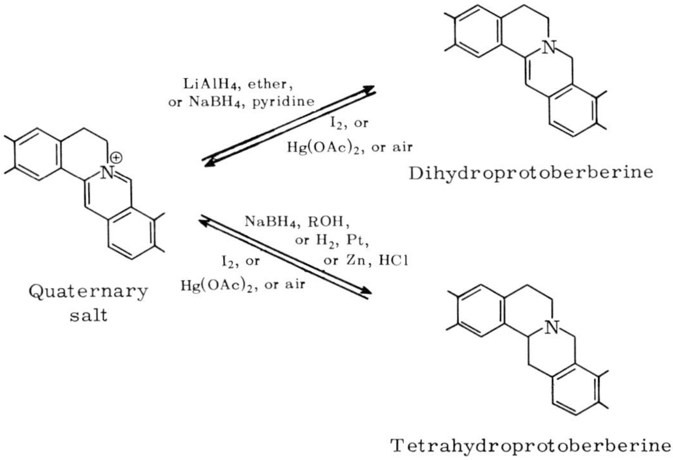

Scheme XX

Quaternary protoberberine salts are unstable in the presence of concentrated alkali. They undergo hydride transfer with formation of the oxy and dihydro derivatives.[36] If only the oxy derivative is desired, it can be obtained from the quaternary salt by oxidation with potassium ferricyanide (Scheme XXI).

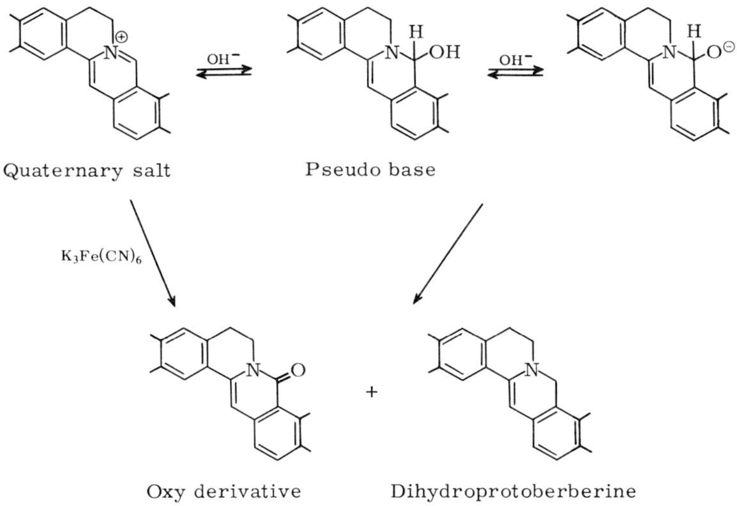

Scheme XXI

Isooxyberberine, formed by hot acid treatment of oxyberberine, has been shown to possess a phenolic function at C-9.[36a] Pyrolysis of berberine chloride also results in *O*-demethylation at C-9 with formation of berberrubine.

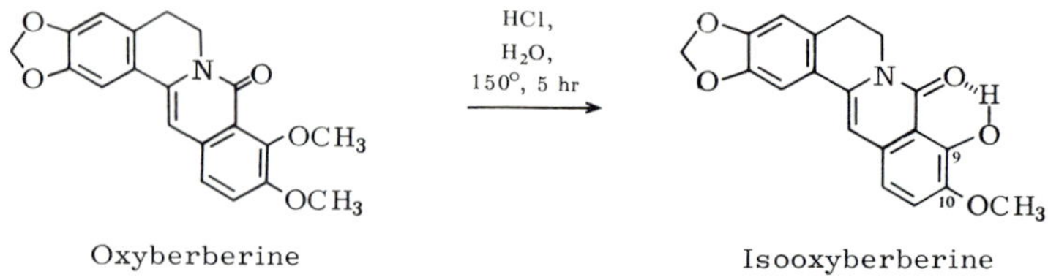

In a protic medium, a dihydroprotoberberine derivative will be in equilibrium with its immonium form. The latter species is unstable and undergoes rapid disproportionation to a mixture of the quaternary protoberberine salt and the tetrahydroprotoberberine.[33]

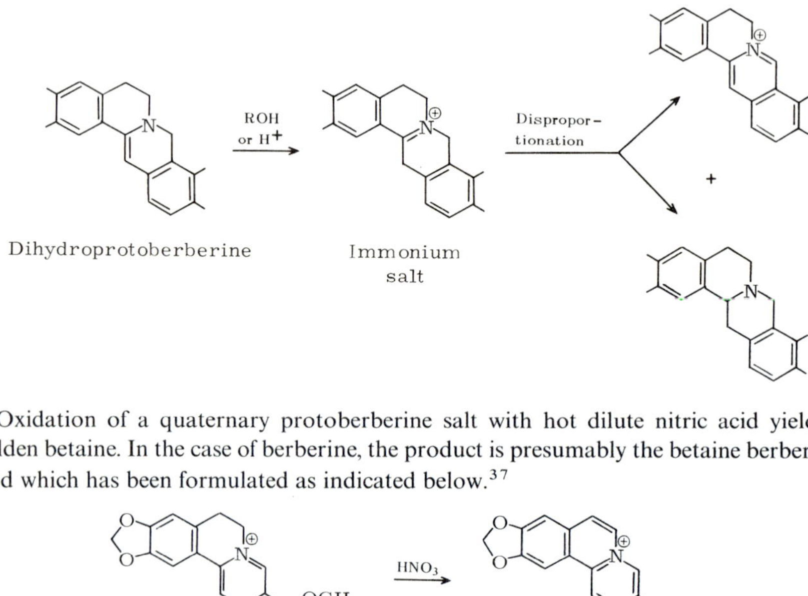

Oxidation of a quaternary protoberberine salt with hot dilute nitric acid yields a golden betaine. In the case of berberine, the product is presumably the betaine berberidic acid which has been formulated as indicated below.[37]

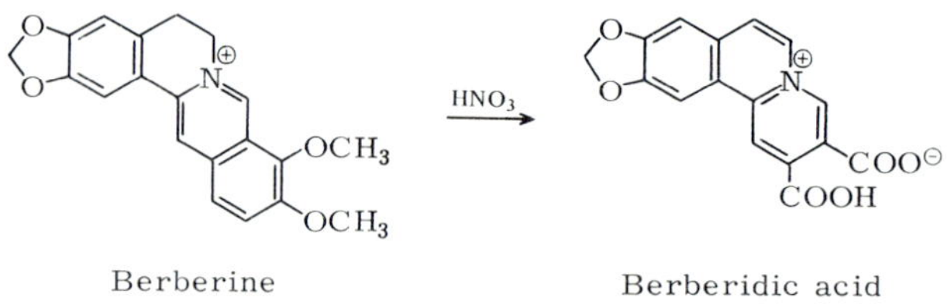

B. Berberine–Acetone and Its Derivatives

When berberine is treated with the nucleophilic anion of acetone, crystalline berberine–acetone is obtained. The chemistry of this adduct, sometimes called 8-acetonyldihydroberberine, and its derivatives has been studied in some detail.[38]

Permanganate oxidation of berberine–acetone yields two products, neoxyberberine–acetone and the yellow-orange berberine–phenol betaine. The latter compound can

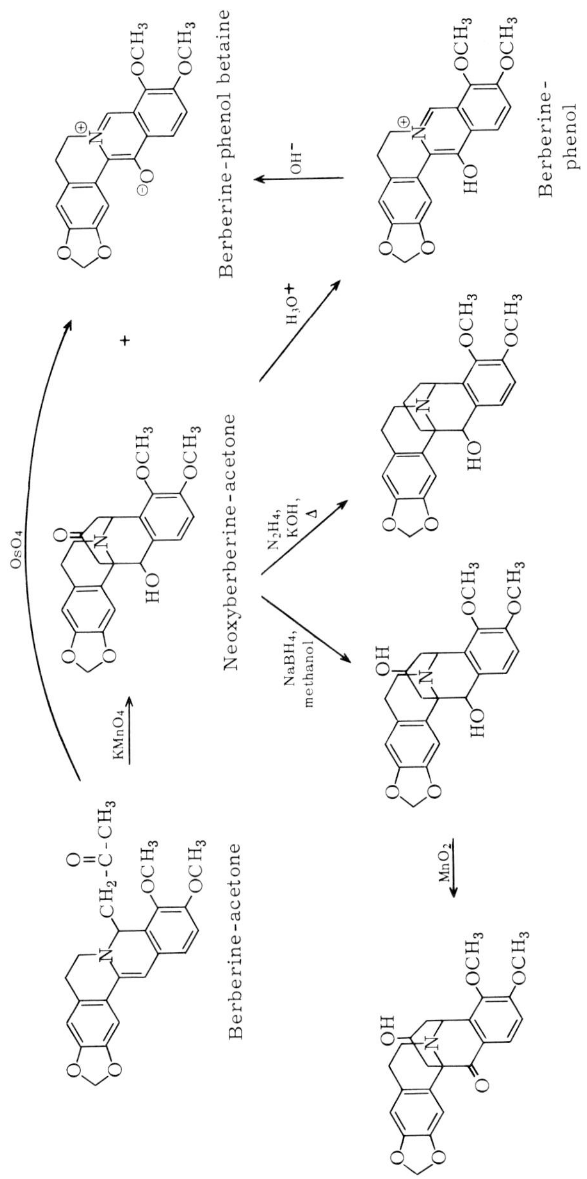
Berberine-acetone
KMnO4
OsO4
Neoxyberberine-acetone
Berberine-phenol betaine
+
H3O+
OH−
Berberine-phenol
N2H4, KOH, Δ
NaBH4, methanol
MnO2

Scheme XXII

also be obtained in pure form by reaction of berberine–acetone with osmium tetroxide. The reactions of neoxyberberine–acetone are summarized in Scheme XXII. It should be pointed out that acid treatment of neoxyberberine–acetone produces berberine–phenol, which is the protonated form of berberine–phenol betaine.

Borohydride reduction of berberine–phenol led to (±)-ophiocarpine. 13-Hydroxylated tetrahydroprotoberberines can therefore be obtained from quaternary protoberberines through the intermediacy of the acetone adducts.[38] The reduction of berberine–phenol takes a completely different course if carried out with zinc in acetic acid, the two products formed being tetrahydroberberine and the rearranged polycyclic base **32**.[39,39a] The transformation of berberine–phenol to **32** has an analogy in the Clemmensen reduction of the α-amino ketone **33** to the rearranged base **34** (Scheme XXIII).[40]

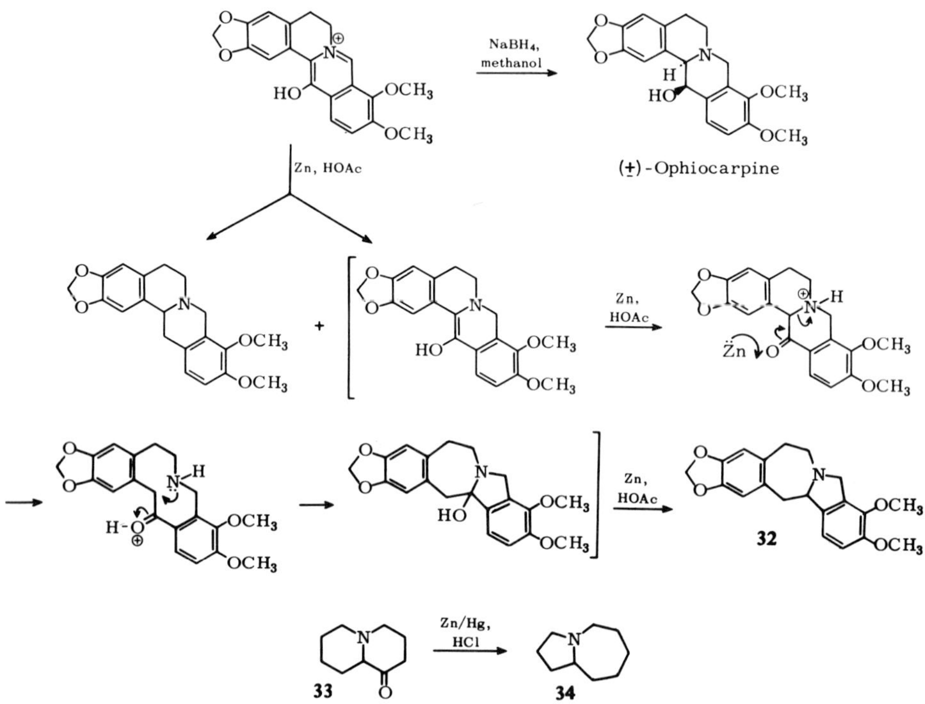

Scheme XXIII

The acetone derivative of a quaternary protoberberine salt permits clean enamine alkylation at C-13. The two quaternary alkaloids coptisine and corysamine were interrelated through this adduct.[41]

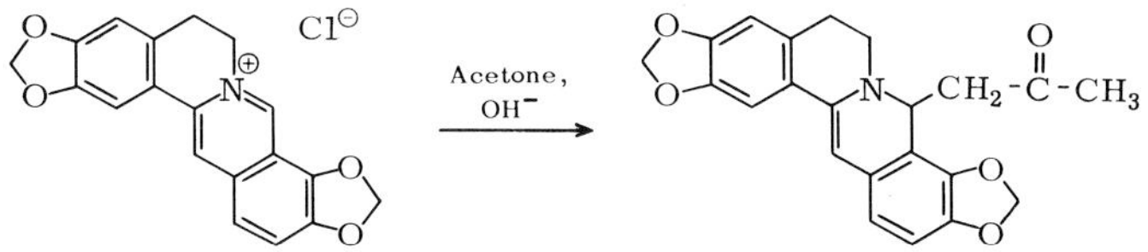

Coptisine   Coptisine-acetone

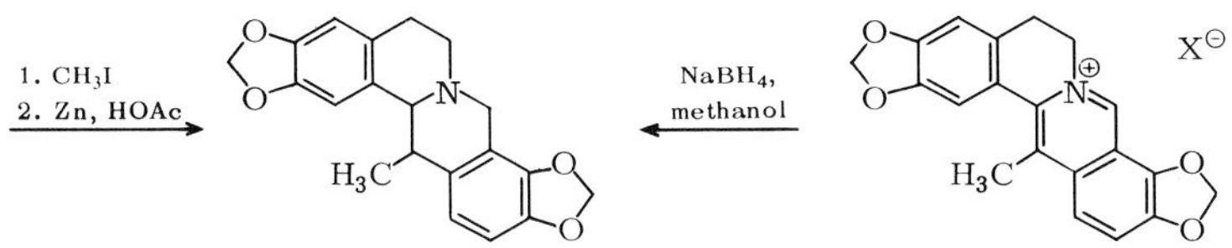

Tetrahydrocorysamine   Corysamine

Alkylation at C-13 has also been achieved directly by the reaction of the enamine system of 7(8)-dihydroberberine with methyl iodide. Reduction of the resulting salt afforded 13-methylcanadine.[1,33]

C. Cleavage with Cyanogen Bromide

Tetrahydroprotoberberines are readily cleaved with cyanogen bromide in ethanol–chloroform at the benzylic C-14 to N-7 bond. The solvolytic conditions of the reaction result in the introduction of an ethoxyl group at C-14. With canadine, the product was shown to be the tricyclic species **35**.[42]

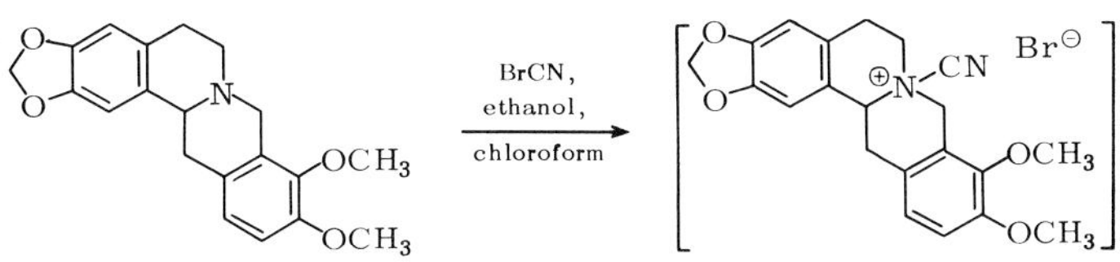

Canadine

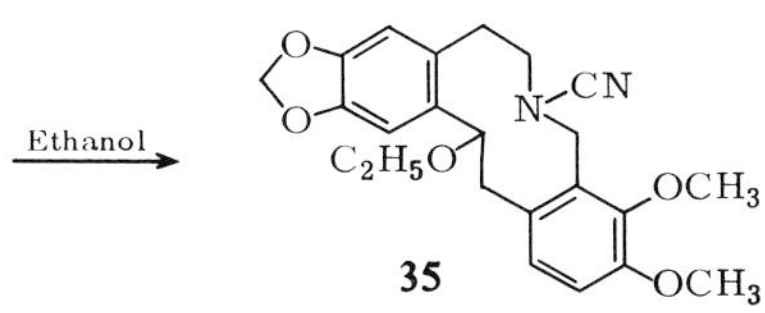

35

Under nonsolvolytic conditions, two products were isolated in the ratio of 1 : 2, namely **36** and **37**.[1,43]

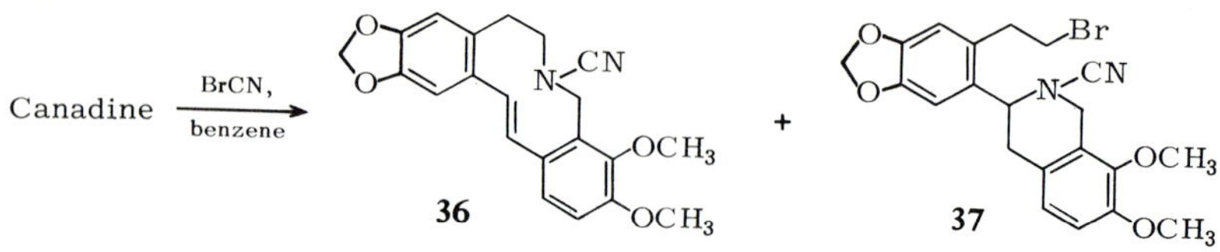

36 **37**

D. Racemization under Catalytic Conditions

Optically active tetrahydroprotoberberines may be racemized at C-14 when treated with hydrogen in the presence of Adams catalyst. It is probable that the alkaloid is first adsorbed on the catalytic surface and that the C-14—H bond is then ruptured homolytically to afford a free radical which is hydrogenated to the racemate. Palladium and Raney nickel are ineffective in this respect. Phenolic groups appear to slow down the racemization process.[44]

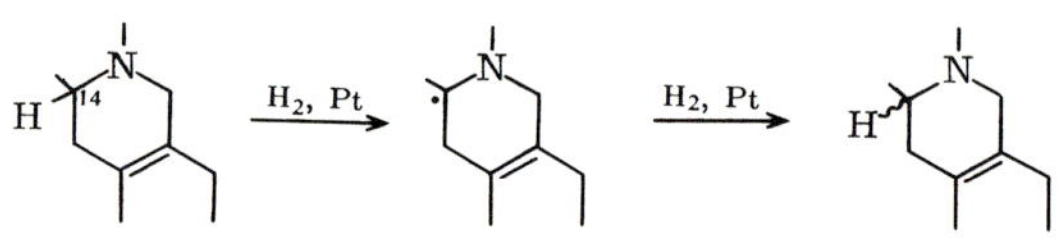

E. Pyrolysis of Tetrahydroberberine Methine Hydrochloride

Upon pyrolysis of the hydrochloride salt of (±)-tetrahydroberberine methine—the Hofmann degradation product from (±)-tetrahydroberberine – two bases were obtained, labeled Base A and Base B, which showed IR Bohlmann bands and were assigned the trans fused pyrrocoline structures designated below.[45]

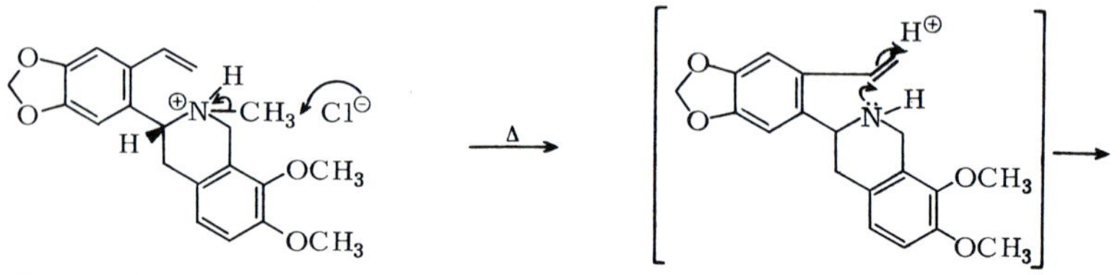

(±)-Tetrahydroberberine methine hydrochloride

(±)-Base A
(C-Methyl doublet at δ1.48; $J \approx 10$ Hz)

(±)-Base B
(C-Methyl doublet at δ1.32; $J \approx 10$ Hz)

F. Reaction of Quaternary Protoberberines with Grignard Reagents

The fact that the N-7 to C-8 bond in quaternary protoberberines is carbonyl-like has allowed the condensation of berberine with phenylmagnesium bromide. The derived C-8 phenyl base can be oxidized to yield benzoic acid.[46,47]

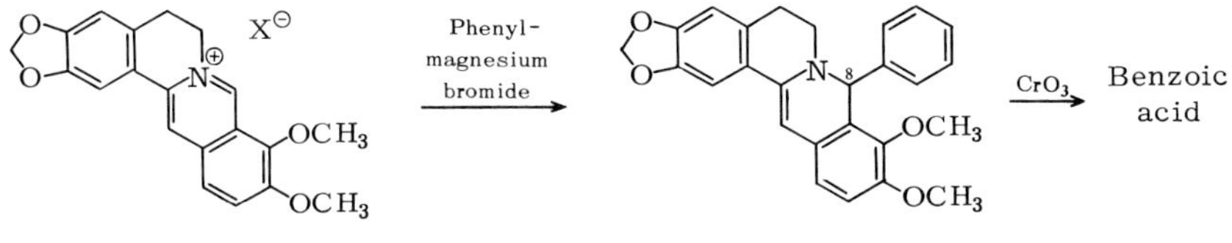

Berberine

Berberine also reacts with benzylmagnesium bromide to produce an 8-benzyldihydroprotoberberine. Reduction with sodium borohydride yields 8-benzylcanadine which can undergo cleavage of ring C through a Hofmann degradation.[48]

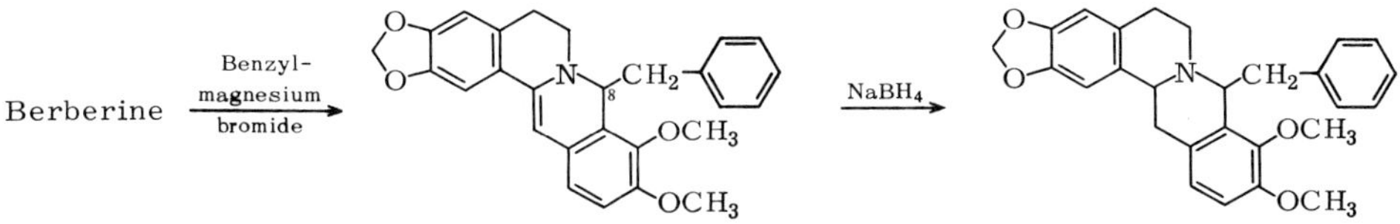

8-Benzylcanadine

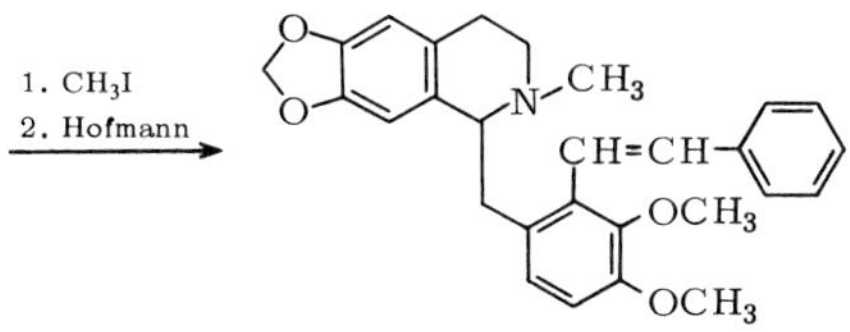

Both of the above reaction sequences have found use in the determination of the radioactive sites in berberine derived from the feeding of labeled precursors.

G. Cleavage of Tetrahydroprotoberberine Substituents with Sodium in Liquid Ammonia

A 2,3-methylenedioxy group is converted to a 2-hydroxyl function by sodium in liquid ammonia, and a C-9 methoxyl group is demethylated to a hydroxyl function. Reduction of canadine yielded the two products shown, indicating that fission of the C-9 methoxyl is slower than cleavage of the 2,3-methylenedioxy group.[49]

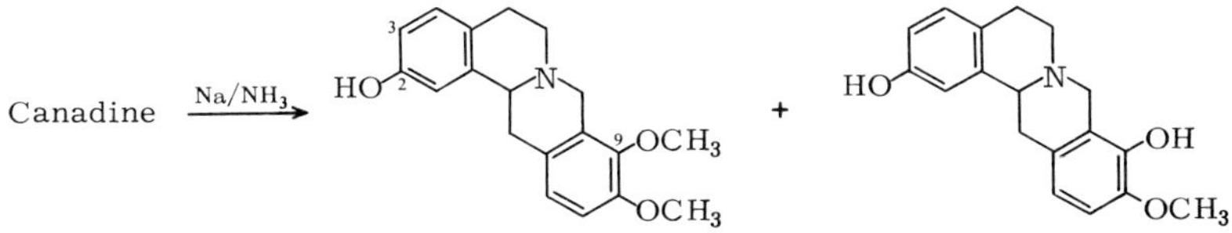

H. 2+4 Cycloaddition of Quaternary Berberine Salts

Berberine in acetonitrile solution reacts with ethyl vinyl ether in a manner typical of isoquinolinium salts. The Diels–Alder adduct is the bridged species **37a**.[49a]

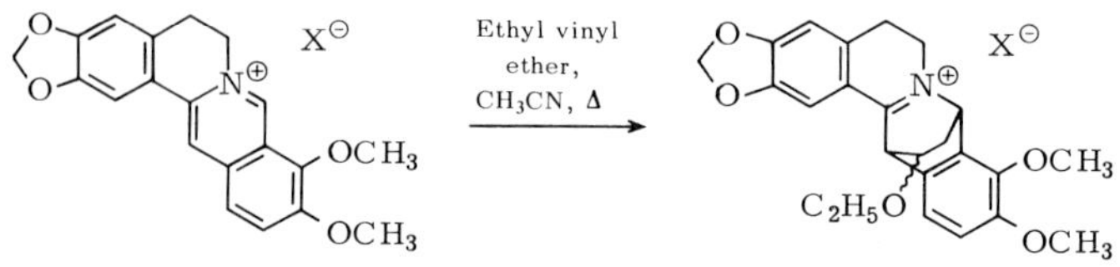

Berberine **37a**

V. Stereochemistry

A. Absolute Configuration

It will be recalled that Corrodi and Hardegger carried out the key experiment in the determination of the absolute configuration of the benzylisoquinolines. Their oxidation of (−)-*N*-norlaudanosine yielded *N*-*β*-carboxyethyl-L-aspartic acid of known absolute configuration. Treatment of (−)-*N*-norlaudanosine with formaldehyde in acid solution then gave (−)-norcoralydine.[50] The absolute configuration of (−)-norcoralydine is thus established since, as expected, racemization about the C-1 position of a tetrahydrobenzylisoquinoline does not occur during Mannich condensation.[51]

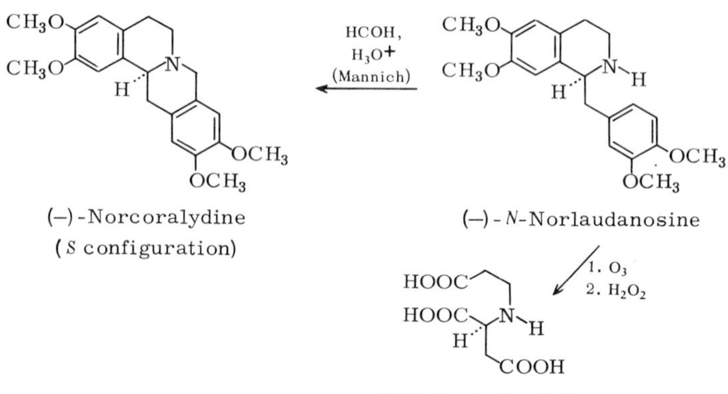

(−)-Norcoralydine
(*S* configuration)

(−)-*N*-Norlaudanosine

N-β-Carboxyethyl-L-aspartic acid

Changes in the nature of the substituents in rings A and D of a tetrahydroprotoberberine generally have only a minor effect on the specific rotation. It follows that in general those molecules that are levorotatory have the same absolute configuration as (−)-norcoralydine, while those that are dextrorotatory correspond to the mirror image.

ORD measurements can also be of substantial assistance in the assignment of absolute configuration.[51,52] Levorotatory tetrahydroprotoberberines exhibit a negative ORD curve with a trough near 240 mμ.

The CD curves of a few tetrahydroprotoberberines have also been recorded.[53] A useful nonempirical rule has been enunciated by the team of Snatzke, Hrbek, Hruban, Horeau, and Šantavý, which allows the determination of the absolute configuration of tetrahydroprotoberberines from the ^1L_b and ^1L_a bands of the CD.[53a]

B. Stereochemistry of the C-1- and C-13-Substituted Tetrahydroprotoberberines. The Application of Rates of Methiodide Formation to Alkaloid Structural Determination

The conformation of the tetrahydroprotoberberines is governed by the degree of steric interaction between a C-13 substituent and the C-1 hydrogen or, alternatively, between a C-1 substituent and the C-13 hydrogens.

To reach an understanding of the stereochemistry of the C-1- or C-13-substituted tetrahydroprotoberberines, it is best to proceed in a nonhistorical fashion and to consider first the results of kinetic studies performed on these compounds. The utility of rates of methiodide formation in the determination of alkaloid stereochemistry was first demonstrated by Shamma and Moss in 1961 in connection with the yohimbine and heteroyohimbine alkaloids.[54,55] The pseudo-first-order rates are usually obtained on 3 mg of the alkaloid in acetonitrile solution by adding 1 ml of methyl iodide and measuring the increase in the conductivity of the solution at 25°. As a preliminary consideration, the following data are relevant. Yohimbine which possesses the *normal* configuration shows a moderate rate of *N*-methylation of $48 \times 10^{-4}\ sec^{-1}$. α-Yohimbine, which is *allo* and possesses a hindered nitrogen, methylates very slowly, the value being $1.2 \times 10^{-4}\ sec^{-1}$. Finally, pseudoyohimbine has a very unhindered nitrogen, and the quaternization rate is extremely fast, $7 \times 10^{-2}\ sec^{-1}$.

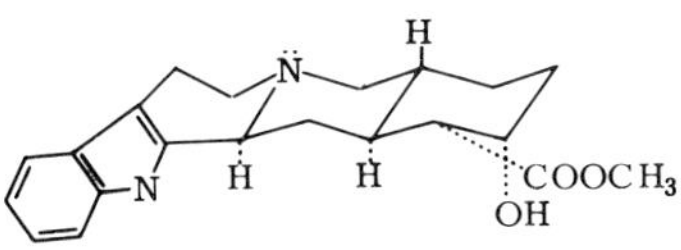

Yohimbine
$48 \times 10^{-4}\ sec^{-1}$

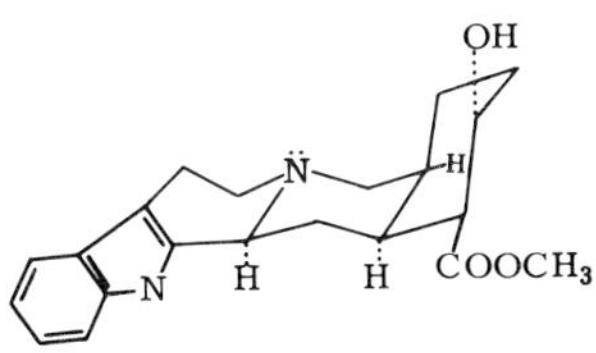

α-Yohimbine
$1.2 \times 10^{-4}\ sec^{-1}$

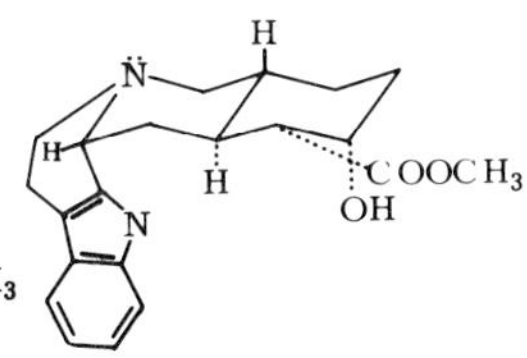

Pseudoyohimbine
$7 \times 10^{-2}\ sec^{-1}$

Those tetrahydroprotoberberines which are unsubstituted at both C-1 and C-13 exist mainly in the thermodynamically more stable *trans*-quinolizidine conformation **38**. This is indicated by the medium rate of *N*-methylation for canadine, $35 \times 10^{-4}\ sec^{-1}$. Conformation **38** is somewhat reminiscent of yohimbine's conformation.[56] Such trans-quinolizidine systems show Bohlmann bands in the IR near 3.57 μ (2800 cm^{-1})

and are also characterized by the absence of a C-14 proton signal below $\delta 3.8$ in the NMR spectrum.[27,56,57]

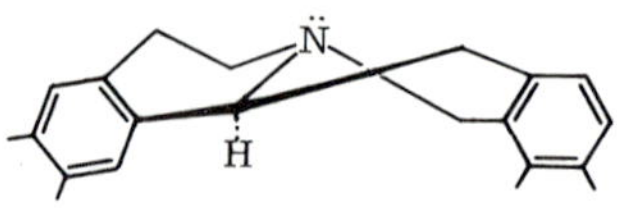

38

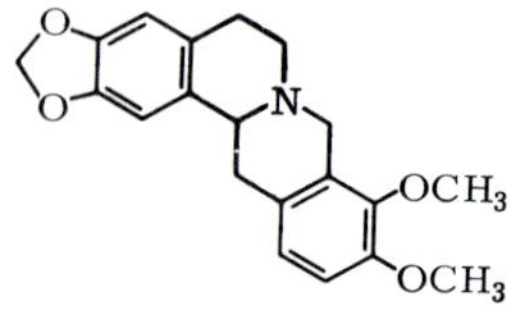

Canadine
35×10^{-4} sec^{-1}

In the case of the racemic 13-β-methyl derivative **39**, the rate of 1.1×10^{-4} sec^{-1} is substantially slower than that for the parent C-13-unsubstituted compound canadine. This result is consistent with a trans B/C fusion for **39** as shown, where approach of the methyl iodide is hindered by the axial C-13 methyl group.[56]

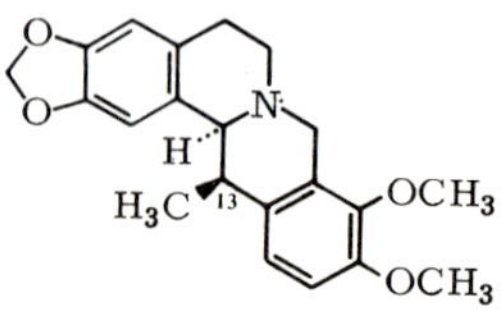

≡

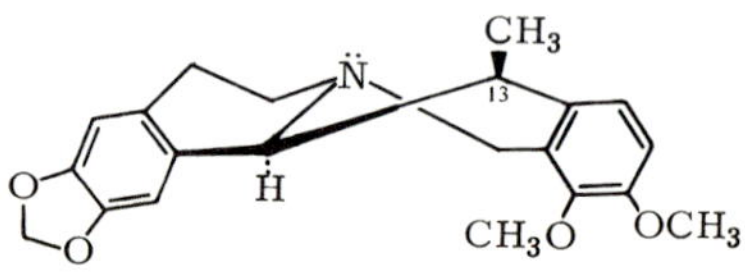

39
1.1×10^{-4} sec^{-1}

Methylation of the racemic α-isomer **40**, on the other hand, gave a very fast rate of 2.7×10^{-2} sec^{-1}. This value demonstrates the unhindered surroundings of the nitrogen atom in the molecule which must, therefore, exist mainly in the cis B/C conformation shown. Such an arrangement is somewhat similar to that for pseudoyohimbine, and in both cases the nitrogen atom is readily available for *N*-methylation.[56]

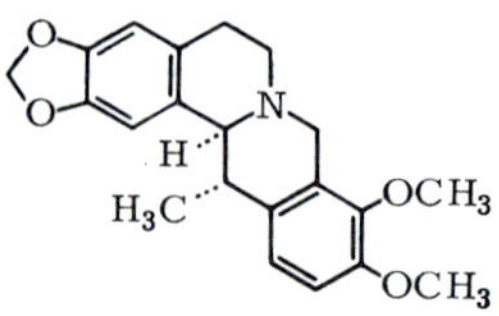

≡

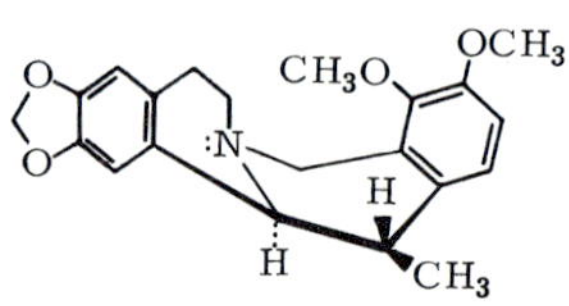

40
2.7×10^{-2} sec^{-1}

The alkaloid capaurine, found in *Corydalis* species, presents an interesting situation. The X-ray of capaurine hydrobromide revealed that the salt possesses a cis B/C system involving half-chair to half-chair fusion.[58] The pseudo-first-order rate for capaurine is 78×10^{-4} sec^{-1}. This relatively fast rate is in agreement with the X-ray data and points to the prevalence even in solution of a cis B/C juncture with two half-chair rings. The nitrogen atom is less hindered than in **38** but more hindered than in **40** and the rate reflects this intermediate degree of steric hindrance.[56] The accompanying alkaloid capaurimine is also cis B/C fused as shown by an X-ray analysis of capaurimine mono-*p*-bromobenzoate.[59]

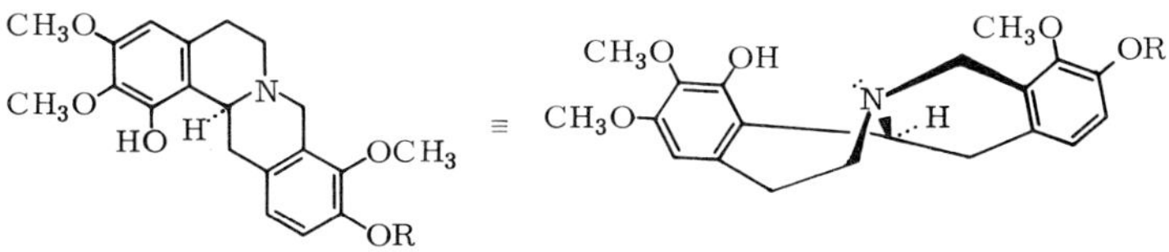

(−)-Capaurine, R = CH_3, 78 x 10^{-4} sec^{-1}
(−)-Capaurimine, R = H
(−)-Capaurimine mono-*p*-bromobenzoate, R = $CO\text{-}C_6H_4\text{-}Br$

To recapitulate, at least three discrete conformations for the tetrahydroprotoberberines can be recognized, and it is the degree of steric interaction between substituents at C-1 and C-13 that determines the conformation of a given molecule.*

Bersch[60] and Jeffs[61] had earlier correctly recognized the influence of the C-13 substituent on the chemistry and stereochemistry of the tetrahydroprotoberberines. Species **39** gave stilbene **41** upon Hofmann degradation, whereas the alternate styrene **42** was obtained from the tetrahydroprotoberberine **40**. These results are explicable on the grounds of a concerted trans elimination for the Hofmann degradation.

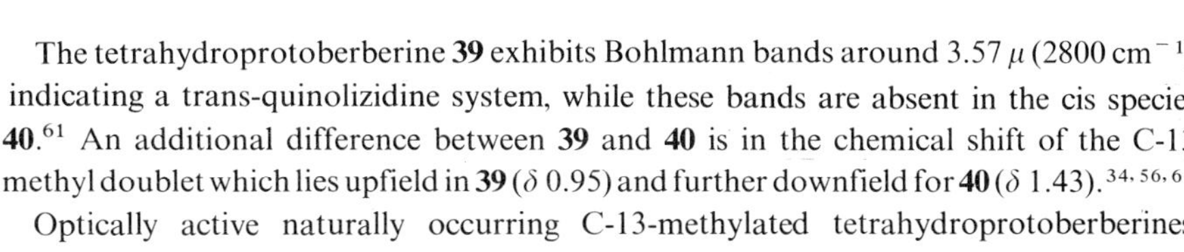

The tetrahydroprotoberberine **39** exhibits Bohlmann bands around 3.57 μ (2800 cm^{-1}) indicating a trans-quinolizidine system, while these bands are absent in the cis species **40**.[61] An additional difference between **39** and **40** is in the chemical shift of the C-13 methyl doublet which lies upfield in **39** (δ 0.95) and further downfield for **40** (δ 1.43).[34,56,62]

Optically active naturally occurring C-13-methylated tetrahydroprotoberberines have shown generally positive rotations and positive Cotton effects so that the C-14 hydrogen is beta.[1]

* For a fourth possible conformation of a tetrahydroprotoberberine see Chapter 19, Section III, A.

C. The Preparation of Diastereoisomeric C-13-Methylated Tetrahydroprotoberberines

The principal problem in the synthesis of C-13-methylated tetrahydroprotoberberines concerns the generation of a single diastereoisomer instead of a mixture of isomers. When a C-13-methylated protoberberine salt is reduced with sodium borohydride in ethanol at 0°, essentially one pure isomer is obtained, e.g., **43**, which possesses a B/C trans ring juncture. This stereochemistry results because the mixed hydride reduces the immonium ion by approaching from the less-hindered side.[34]

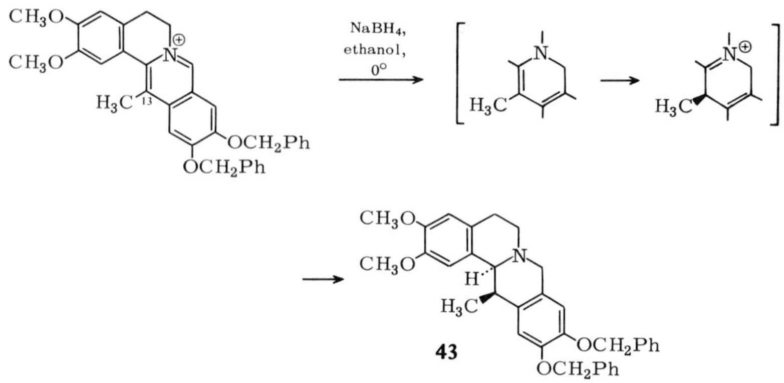

43

The alternate diastereoisomer **46** is available through the Mannich-type condensation of the appropriate tetrahydrobenzylisoquinoline **45**. The latter compound can be obtained cleanly by reduction of the imine **44** under the same conditions as described above. The single tetrahydrobenzylisoquinoline isomer **45** obtained is that predicted upon consideration of Cram's rule.[34]

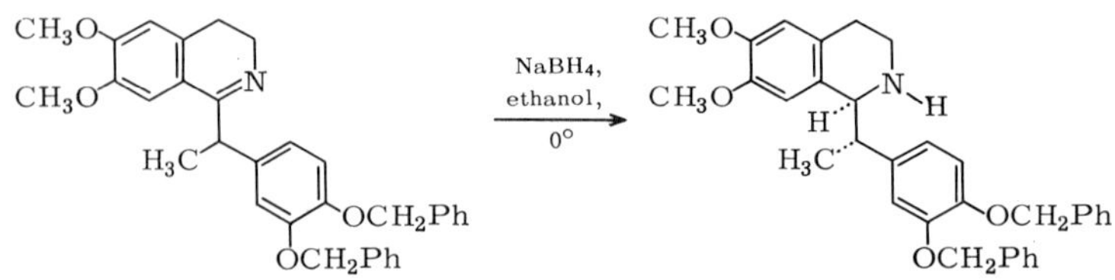

44 **45**

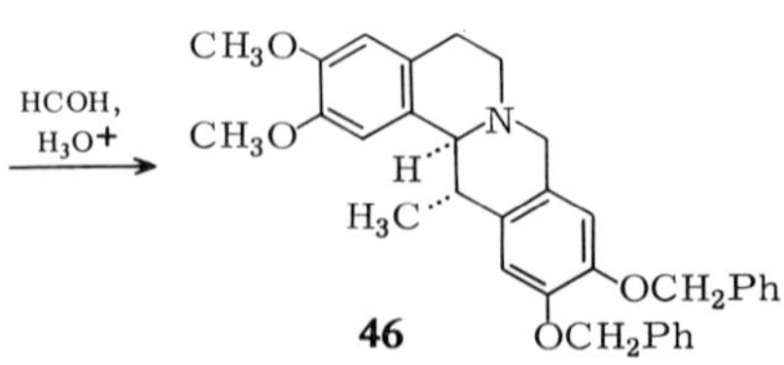

46

D. Ophiocarpine: A C-13-Hydroxylated Tetrahydroprotoberberine

The absolute configuration of the alkaloid (−)-ophiocarpine was deduced through the catalytic hydrogenolysis of (−)-ophiocarpine acetate, which yielded (−)-canadine and its racemate.[63]

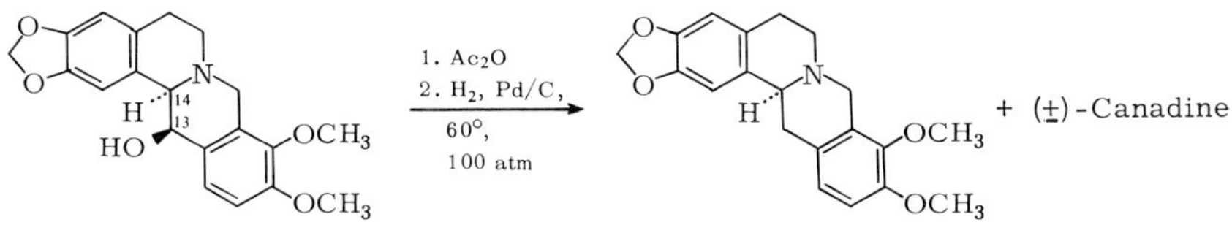

(−)-Ophiocarpine (−)-Canadine

The relative configuration of (−)-ophiocarpine at C-13 and C-14 can be deduced from physical measurements. The alkaloid has a wide IR absorption band near 2.84 μ (3526 cm^{-1}) and is thus strongly hydrogen-bonded, whereas the synthetic 13-epiophiocarpine has only a free hydroxyl peak at 2.78 μ (3597 cm^{-1}). Both compounds exhibit strong Bohlmann bands at 3.57 μ (2800 cm^{-1}), indicating that rings B and C are trans fused.

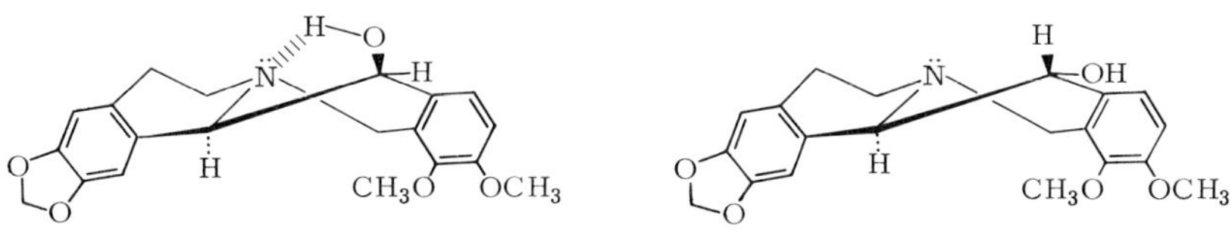

(−)-Ophiocarpine 13-Epiophiocarpine

NMR spectroscopy can be very useful in stereochemical assignments in the ophiocarpine and 13-epiophiocarpine series since $J_{13,14}$ is small ($\approx$1.5 Hz) when the C-13 and C-14 hydrogens are cis and large ($\approx$9 Hz) when they are trans.[64]

VI. Berberastine and Thalidastine: Two Quaternary Protoberberines Hydroxylated at C-5

Berberastine[65,66] and thalidastine[67] are two quaternary protoberberine alkaloids which possess an alcoholic hydroxyl at C-5. Berberastine, found in *Hydrastis canadensis* L., undergoes ready dehydration in acid to yield the known dehydroberberine ion.[65,66] With zinc in sulfuric acid one obtains the stable tetrahydroberberastine, which is a minor alkaloid in *H. canadensis*.[68] Tetrahydroberberastine does not exhibit the properties of a pseudo-base, so that the alcohol function must be at C-5 rather than at C-6.[65,66] The absolute configuration of berberastine and thalidastine at C-5 is unknown.

Thalidastine differs from berberastine by having a hydroxyl rather than a methoxyl group at C-10; its chemistry generally parallels that of berberastine.[67]

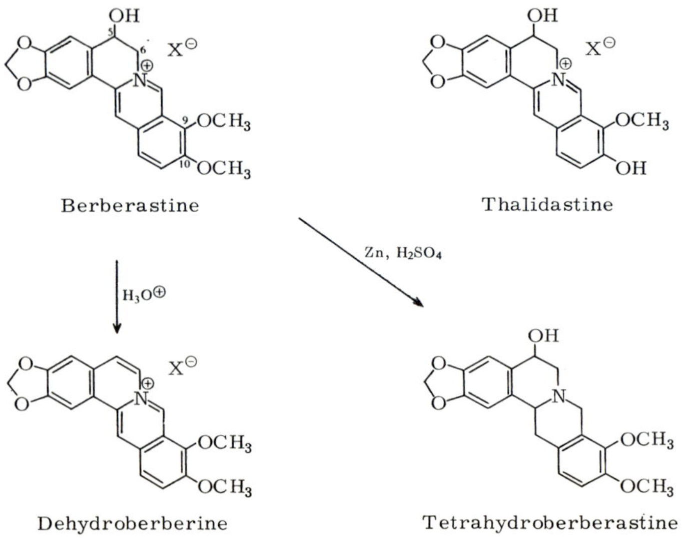

VII. Mecambridine and Alkaloid PO-5, Orientalidine and Alkaloid PO-4: Four Retroprotoberberine Alkaloids

The four alkaloids mecambridine, alkaloid PO-5, orientalidine, and alkaloid PO-4 were isolated from a number of Papaver species. The structures initially considered for mecambridine and orientalidine on the basis of spectral analysis combined with some chemical degradations are shown below.

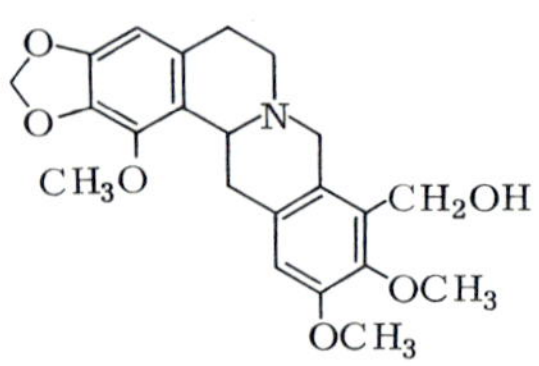

Original structure considered for mecambridine[69]

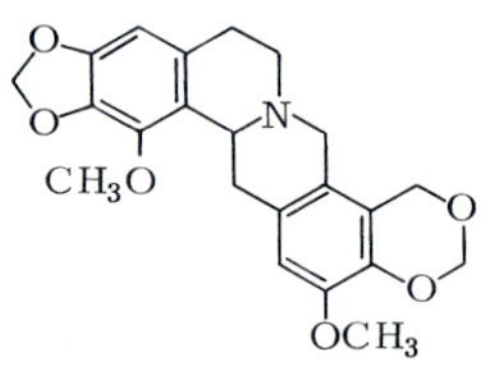

Original structure considered for orientalidine[70,71]

However, a comparison of the UV spectra of a series of tetrahydroprotoberberine bismethine derivatives showed that the structural assignments had to be modified as indicated below, so that mecambridine and orientalidine are 10,11,12- rather than 9,10,11-substituted.[72]

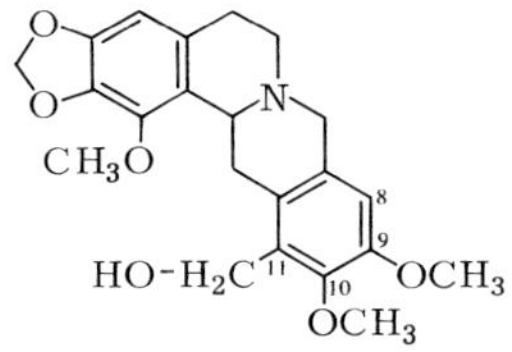

Mecambridine

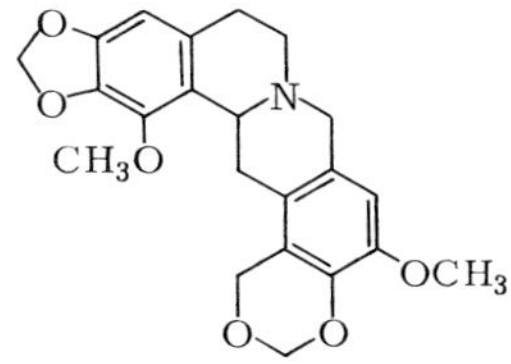

Orientalidine

It was noted that the position of the UV band of longest wavelength for tetrahydroprotoberberine bismethines is affected by aromatic substituents ortho to the trans ethylenic bridge which connects the two aromatic nuclei, resulting in a hypsochromic shift whenever such a substituent is present; for example,

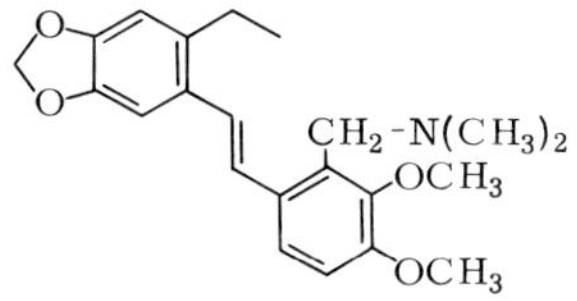

Dihydrocanadine bismethine
λ_{max} 333 mμ (4.35)

vs.

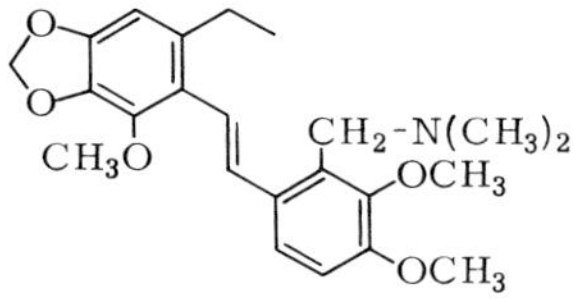

Dihydro-1-methoxycanadine bismethine
λ_{max} 308 mμ (4.17)

and

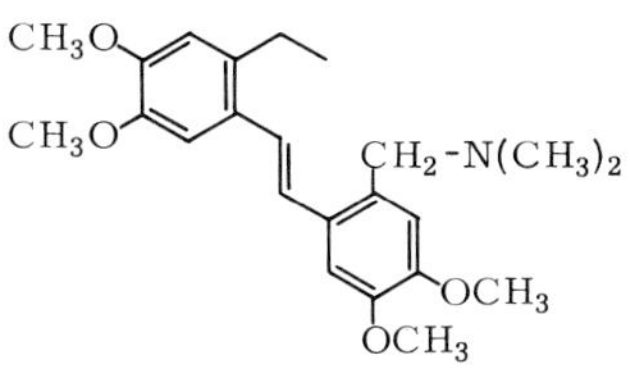

Dihydronorcoralydine bismethine
λ_{max} 326 mμ (4.26)

vs.

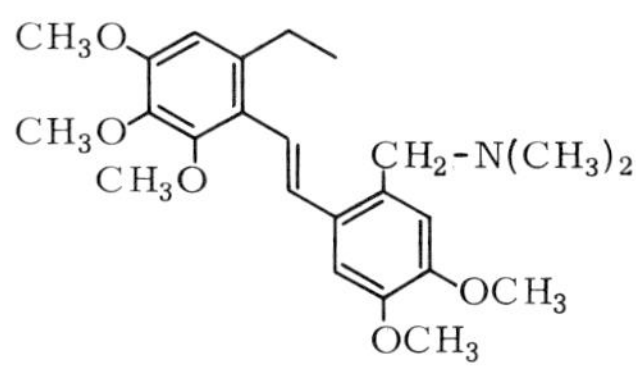

Dihydro-1-methoxynorcoralydine bismethine
λ_{max} 305 mμ (4.25)

Comparison of the UV spectrum of dihydro-1-methoxycanadine bismethine with that of dihydroorientalidine bismethine and of 1-methoxycanadine bismethine with that of mecambridine bismethine made it clear that the hypsochromic shifts observed in each of these pairs could be accounted for only by the presence of an additional ortho substituent, so that orientalidine and mecambridine are 10,11,12- and not 9,10,11-substituted.[72]

The quaternary protoberberines PO-5 and PO-4 were isolated at about the same time and from the same plants as mecambridine and orientalidine. Dehydrogenation of mecambridine and orientalidine with mercuric acetate, potassium permanganate, or chromium trioxide gave alkaloids PO-5 and PO-4, respectively, so that PO-5 is the quaternary analog of mecambridine and PO-4 the analog of orientalidine.[73,74]

VIII. Color Changes of Phenolic Quaternary Protoberberines

The alkaloid jatrorrhizine, which is yellow in neutral solution, turns to deep red in the presence of bicarbonate or hydroxylic base. On the other hand, columbamine, which is also yellow in neutral solution, shows no color change in bicarbonate and becomes only tan in hydroxide solution. The difference in the color behavior of these two alkaloids may be rationalized in terms of the resonance structures for the resulting anions. Jatrorrhizine can give rise to a favorable quinonoid structure with no charge separation, but such is not the case with columbamine, whose anion must exist as a betaine.[75]

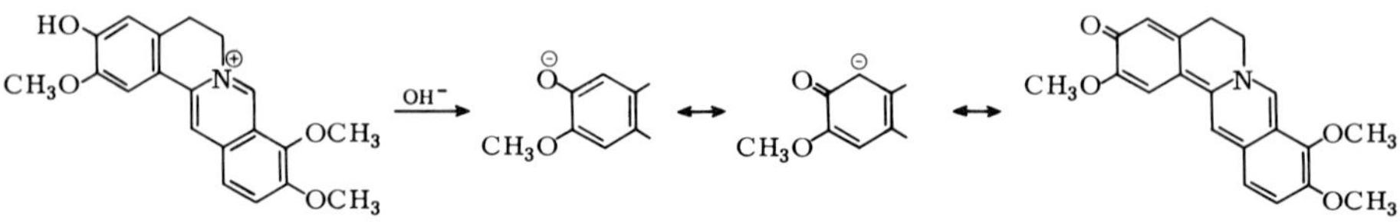

Jatrorrhizine Quinonoid form

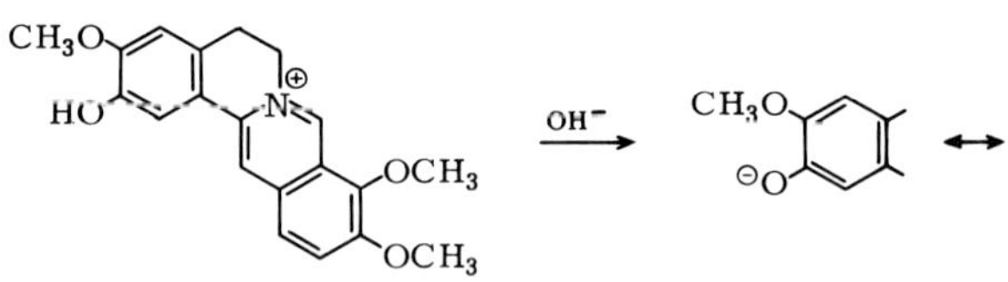

Columbamine

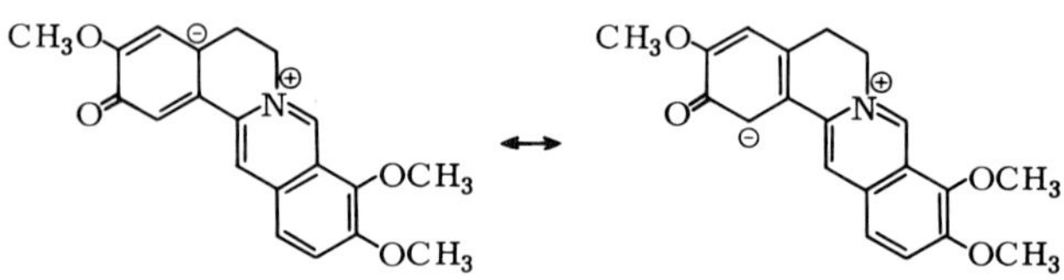

It should be mentioned in passing that a good method for separating quaternary protoberberine salts is by column chromatography of their chlorides on acid-washed alumina. Elution is carried out with chloroform gradually enriched with methanol.[75]

IX. Biosynthesis

Working with *Hydrastis canadensis* L. (Ranunculaceae), Spenser and co-workers have shown that tyrosine is a very efficient precursor for berberine and is incorporated into both the top and bottom parts of the alkaloid.[48] However, if labeled dopamine is fed, only one molecule of this species is incorporated into the alkaloid. Tyrosine must, therefore, give rise to two different intermediates during the biosynthetic process, one of them being dopamine.[68]

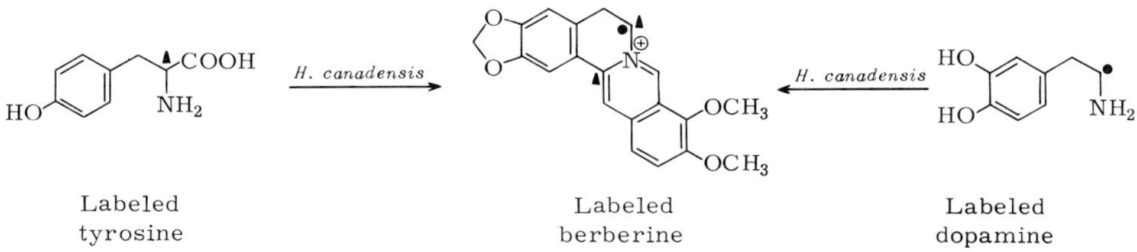

Barton[47] and Battersby[46] have independently demonstrated that an *N*-methyltetrahydrobenzylisoquinoline is converted to a protoberberine by oxidative cyclization of the *N*-methyl function rather than by a Mannich-type condensation with formaldehyde. Labeled reticuline gave rise to radioactive berberine – a transformation which also shows that the methylenedioxy group in nature is derived from an *o*-methoxyphenol[47] and that reticuline is indeed an important intermediate in protoberberine biogenesis. Furthermore, (+)-reticuline is incorporated much more efficiently than the levo enantiomer.[47] Noteworthy also is the finding that protosinomenine is not incorporated into berberine in *H. canadensis.*

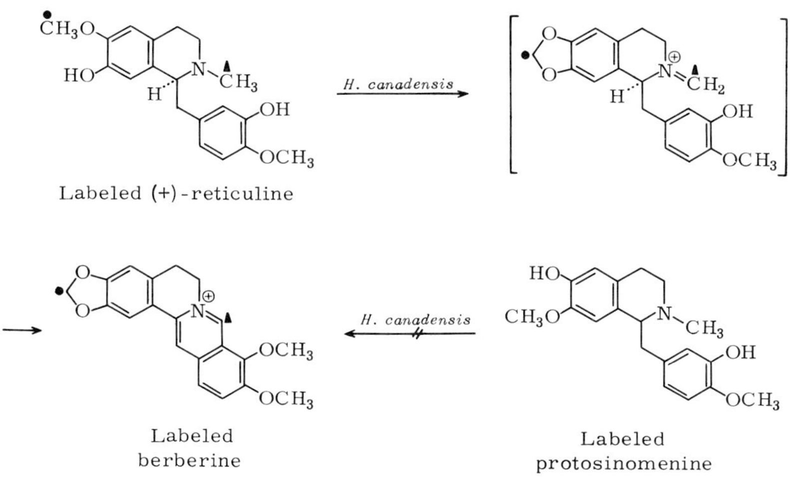

Feeding labeled (+)-laudanosoline to *Berberis japonica* Lindl. (Berberidaceae) also gave rise to labeled berberine.[46]

Labeled (+)-laudanosoline   Labeled berberine

Labeled (+)-reticuline is significantly incorporated into (−)-stylopine in *Chelidonium majus* L. (Papaveraceae), while incorporation of (−)-reticuline is a much less efficient process. Some of the tritium label at C-14 in (−)-stylopine was lost, indicating C-1 oxidation–reduction at the (+)-reticuline stage.[76]

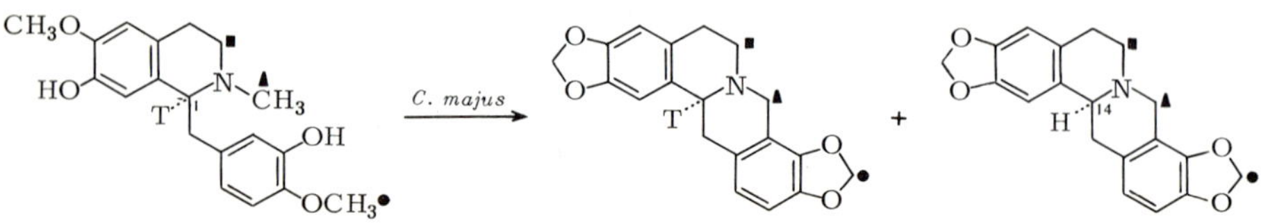

Labeled (+)-reticuline   Labeled (−)-stylopine   Labeled but detritiated (−)-stylopine

Coreximine, a 10,11-substituted tetrahydroprotoberberine, is also derived from reticuline since administration of labeled (±)-reticuline to intact opium poppies led to a relatively high incorporation into coreximine.[76a]

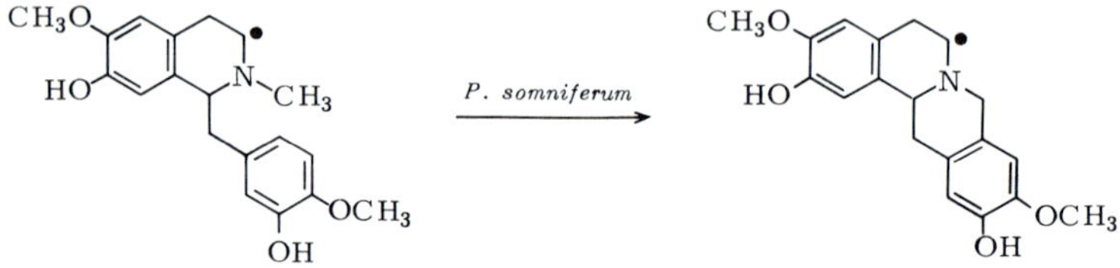

Labeled (±)-reticuline   Labeled coreximine

Both the C-8 carbon of a protoberberine and the methylene of the methylenedioxy group originate from the *S*-methyl function of methionine.[77]

Although not all of the individual steps are yet completely established, the following general sequence seems to prevail in the biosynthesis of berberine (Scheme XXIV).

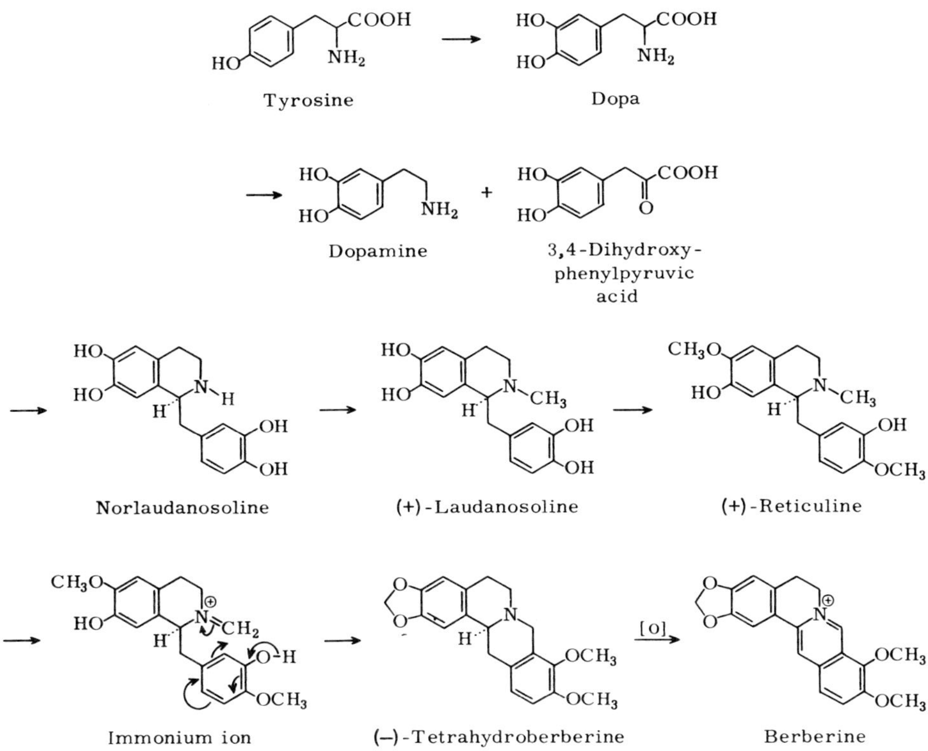

Scheme XXIV

The formation of the methylenedioxy group from an *o*-methoxyphenol can be visualized either in terms of an ionic or a free-radical mechanism, and the former is depicted here.

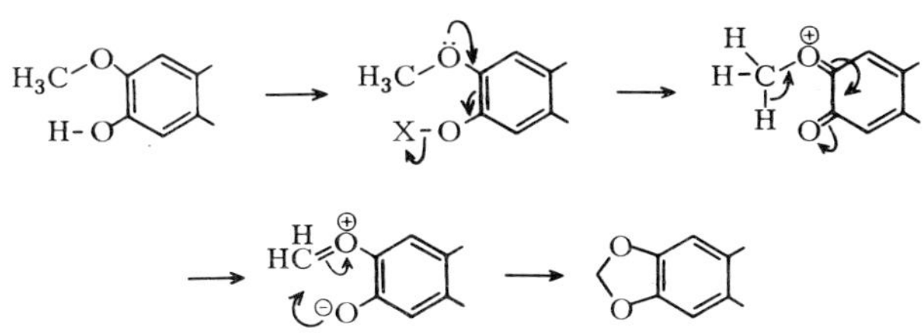

Turning to the 5-hydroxylated protoberberine alkaloids, berberastine, thalidastine, and tetrahydroberberastine, it has been shown that dopamine and noradrenaline are incorporated into berberastine much more efficiently than into canadine or berberine. The last two alkaloids, therefore, cannot be precursors for berberastine. Rather, the

C-5 hydroxyl must be introduced at some early stage which preceeds the formation of the necessary tetrahydrobenzylisoquinoline intermediate.[68]

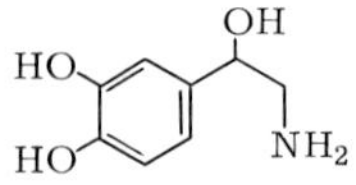

Noradrenaline

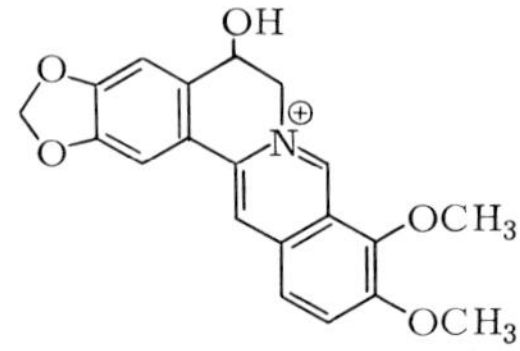

Berberastine

Labeled reticuline is also converted specifically into corydaline in *Corydalis cava* Schweigg et Korte (Fumariaceae). It follows that C-13-methylated tetrahydroprotoberberines are derived, like the non-C-13-methylated species, from the same tetrahydrobenzylisoquinoline precursor.[78]

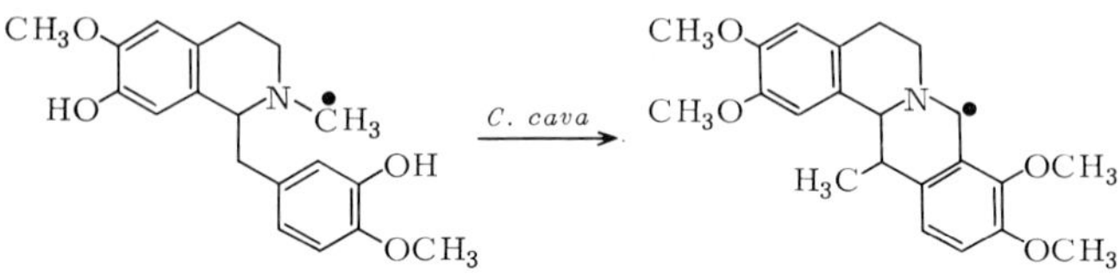

Labeled reticuline

Labeled corydaline

The special substitution pattern in mecambridine, PO-5, orientalidine, and PO-4 can be rationalized along biogenetic pathways. Oxidative cleavage of the N-7 to C-8 bond of a tetrahydroprotoberberine such as **47** yields a retrodihydrobenzylisoquinoline which can eventually undergo reduction, *N*-methylation, and recyclization to lead to these four retroprotoberberine alkaloids.[72]

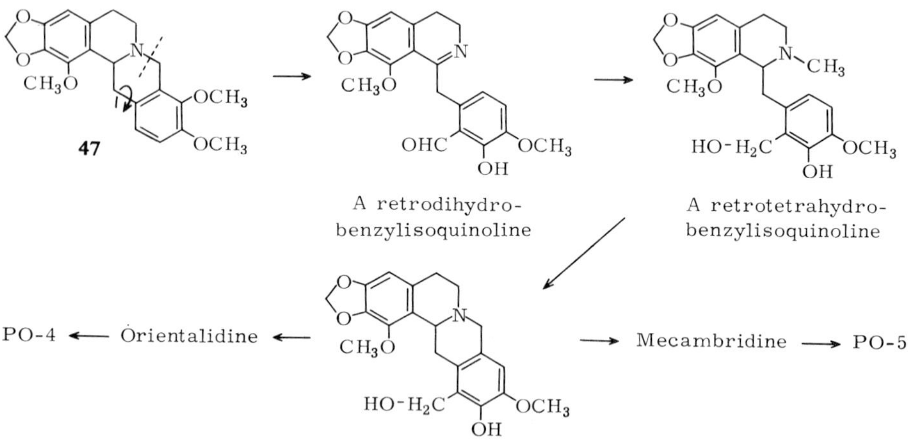

Another attractive possibility not previously considered would be through the phenolic ammonium salt **48**.

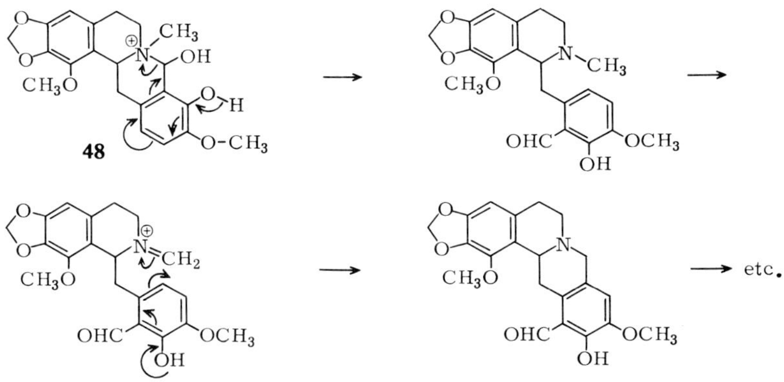

48

In fact, a phenolic group in **48** is not even required since this species is the salt of a pseudo-base which can readily open up to the required amino aldehyde form.

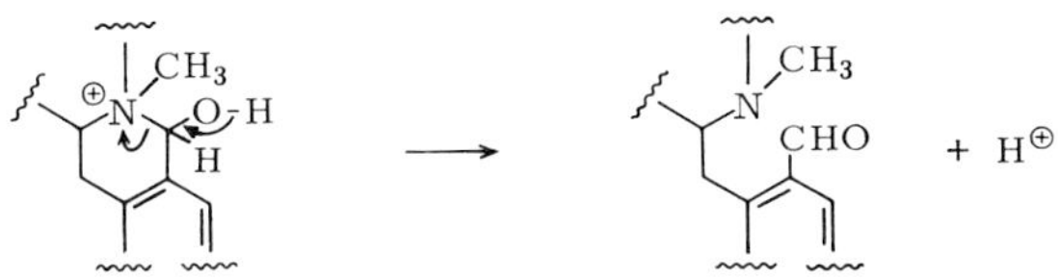

48 An amino aldehyde

Formal proof that retroprotoberberines originate from protoberberines reverting to retrotetrahydrobenzylisoquinolines is still lacking.

X. Pharmacology

A large number of tetrahydroprotoberberines substituted in rings A and D have shown promising tranquilizing properties in initial tests.[79]

Berberine which has some antibacterial and antiprotozoal activity has been shown to form a complex with DNA. The alkaloid is probably intercalated into supercoiled mitochondrial DNA to produce configurational changes in the DNA.[80] Furthermore, berberine has transient hypotensive activity in rats which can be prolonged by substituting the C-9 *O*-methyl group with *O*-*n*-C_4H_9 or *O*-*n*-C_5H_{11}.[81] It is also active *in vitro* against *Mycobacterium smegmaris* ATCC 607 and is thus of potential importance as an antitubercular drug.[81b]

The long-lasting and relatively strong adrenergic alpha-blocking effect of the alkaloid xylopinine on the blood pressure of rabbits, cats, and dogs has been established.[82] At least in guinea pigs this action can be related to the depression of the atrial beat.[83]

When berberine–acetone is heated to 100–110° with allyl bromide and chloroform in a sealed tube, 13-allylberberine bromide, which may have potential as an antiulcer drug,[84] is obtained.

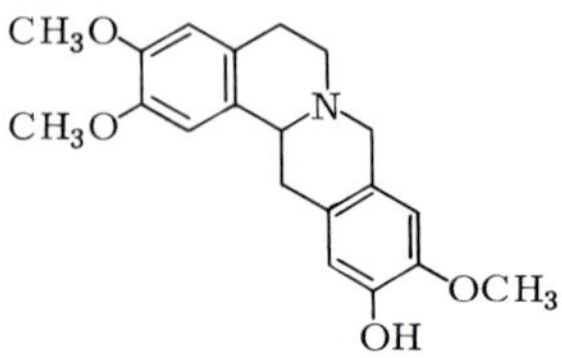

Xylopinine

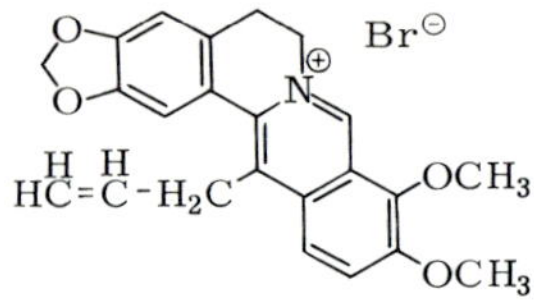

13-Allylberberine bromide

(−)-Canadine methochloride also has shown some hypotensive activity in anesthetized cats and dogs.[81a]

XI. Mass Spectroscopy

Tetrahydroprotoberberines undergo facile fission at the two benzylic bonds as shown in Scheme XXV in what amounts to a retrograde Diels–Alder condensation. Tetrahydroberberine itself gives intense peaks at *m/e* 174 and 176 and a lesser peak at 164 corresponding to the shards resulting from such cleavage. Mass spectrometry can, therefore, differentiate between substituents on the upper and the lower aromatic rings.[85]

9-Hydroxy-10-methoxy compounds may be differentiated from their 9-methoxy-10-hydroxy analogs. Those with the former substitution pattern preferentially expel a methyl group from the equivalent of ion C, yielding an ion at *m/e* 135, whereas the latter lose a hydrogen atom from their ion C equivalent, giving rise to a new ion at *m/e* 149 (Scheme XXV). Additionally, by careful analysis of peak intensities, it may even be possible to differentiate between C-9,10 and C-10,11 substitution.[86]

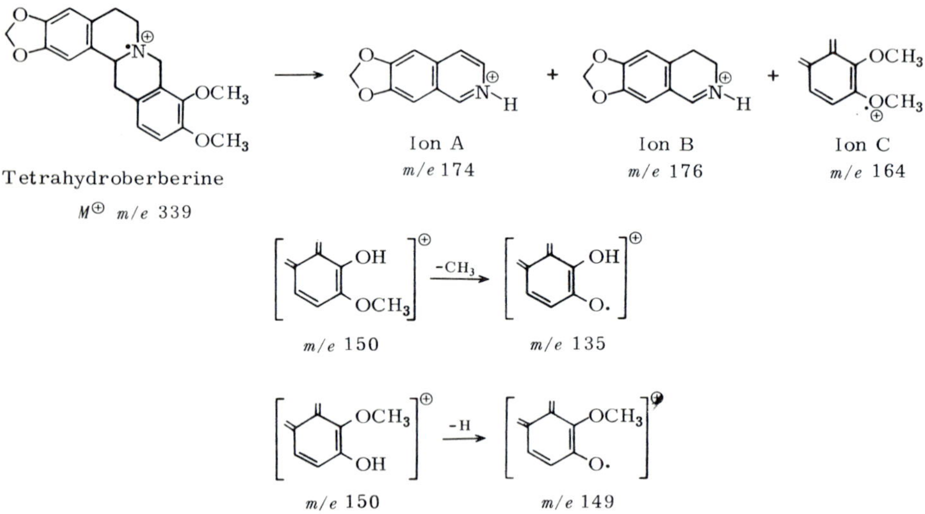

Scheme XXV

The mass spectrum of the pseudobase from berberine does not show a molecular ion due to thermal disproportionation to dihydroberberine and oxyberberine.[86a]

XII. NMR Spectroscopy

The oxygenation pattern of ring D for a tetrahydroprotoberberine may be deduced from an examination of the methylene protons at C-8. In the case of 9,10-substitution an AB quartet, or half an AB quartet when the high-field half is obscured by the methoxyl signals, is invariably observed with one doublet near $\delta 3.65$ and the other near $\delta 4.35$ ($J_{AB} \approx 16$ Hz). In alkaloids with 10,11-substitution, on the other hand, the C-8 protons appear as a broad singlet centered at $\delta 4.05$.[62,86]

Proton exchange coupled with NMR analysis can be a useful technique in structural elucidation. When caseadine was dissolved in about 12 *N* DCl and heated briefly, one of the aromatic protons was exchanged. Since it had been established that under these conditions only the hydrogens ortho or para to a phenolic function are exchanged, this result was useful in locating the hydroxyl group in the alkaloid. The proton exchanged in the present case is H_B. The AB quartet due to protons A and B collapsed into a broad singlet (1 H) after the deuterium exchange.[87]

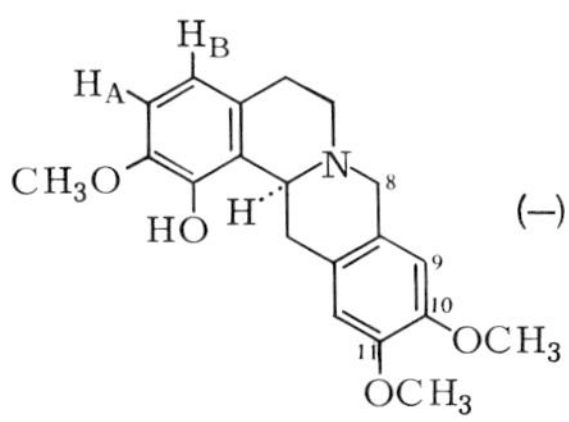

(—)-Caseadine

As a supplement to NMR chemical shifts and coupling constants, nuclear Overhauser effects (NOE) can be a powerful tool in organic structural and conformational analysis. If protons A and B are close to each other, then induced saturation of the B nuclei by double irradiation will result in enhancement of the intensity (NOE) of the absorption for the A nuclei. In the case of the alkaloid (±)-apocavidine, isolated from *Corydalis cava* Schweigg. et Korte (Fumariaceae), irradiation of the C-14 proton at $\delta 3.71$ caused a sharpening of the low-field aromatic proton at $\delta 6.78$, while irradiation of one of the C-5 protons at $\delta 2.68$ sharpened the absorption of the C-4 aromatic proton at $\delta 6.58$.

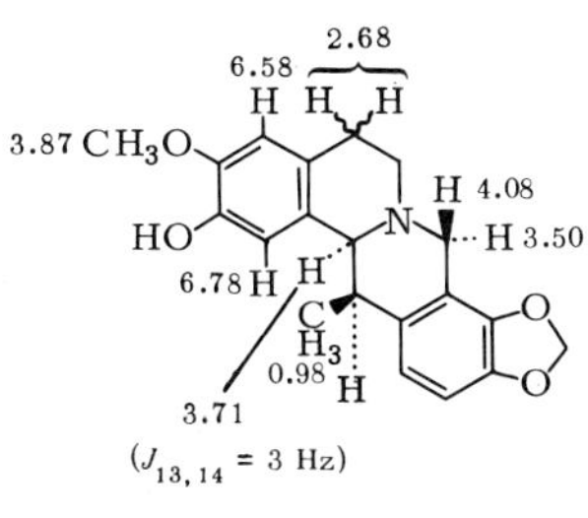

(±)-Apocavidine

Furthermore, irradiation of the methoxyl signal at $\delta 3.87$ caused a 23% increase in the C-4 aromatic hydrogen absorption at $\delta 6.58$, leaving the $\delta 6.78$ peak unaffected. The methoxyl group must therefore be assigned to C-3 and the hydroxyl to C-2.[62]

Some typical alkyl hydrogen chemical shifts are given below. In the trans-fused system of corydaline, the C-8 protons have a large difference in chemical shifts, $\delta 3.49$ and 4.19; but in the cis-fused system incorporated in mesocorydaline, this difference is quite small, namely $\delta 3.97$ vs. 4.14.[56,62] Note also the difference in chemical shifts between the C-13 methyl groups in corydaline and mesocorydaline.

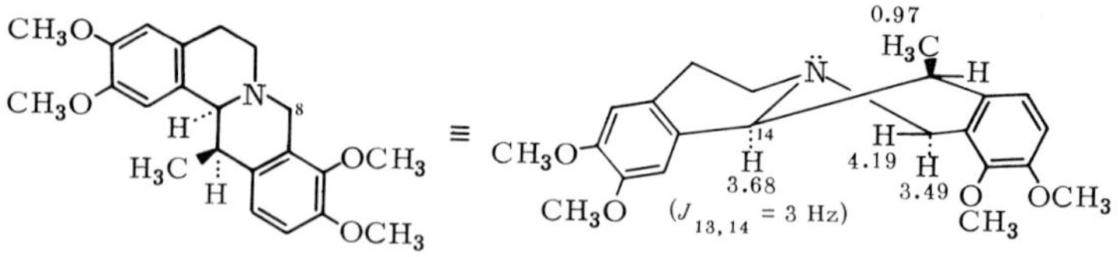

(±)-Corydaline

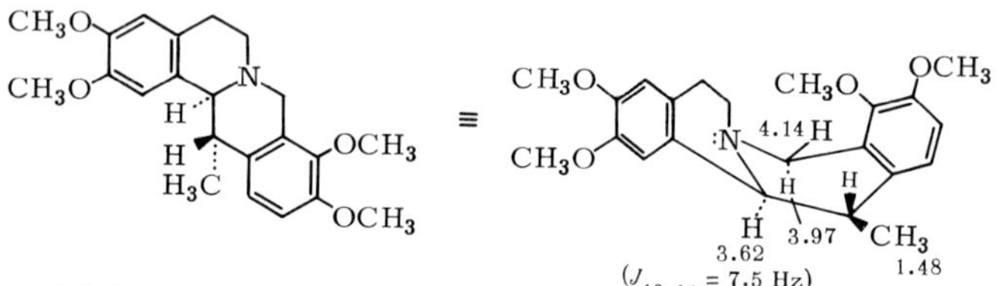

(±)-Mesocorydaline

The chemical shifts for a few quaternary protoberberine salts obtained in DMSO-d_6 have been listed.[73]

XIII. UV Spectroscopy

Tetrahydroprotoberberines absorb in the 282 – 289 mμ region with occasionally a shoulder near 230–240 mμ. Substantial absorption also occurs around 210 mμ, but generally this band has not been reliably recorded or has gone unmentioned. Tetrahydroprotoberberines substituted at C-2,3,10,11 cannot readily be differentiated from their C-2,3,9,10 analogs.[88] Introduction of a methylenedioxy instead of two methoxyl groups gives rise to a bathochromic shift of the UV absorption bands at 282–289 mμ and at 230–240 mμ.[88a]

A drastic alteration of the UV spectrum occurs, however, when changing from 9,10 to 10,11 substitution for the protoberberine salts. The 9,10-substituted salts show a minimum at 301 – 310 mμ, while their 10,11 counterparts show strong absorption in this region in the form of a peak or a shoulder. Since tetrahydroprotoberberines can be easily oxidized to their corresponding quaternary salts, UV spectroscopy can assist in the establishment of the ring D substitution pattern for the tetrahydroprotoberberine bases.[88]

A few spectral examples are cited here[88]:

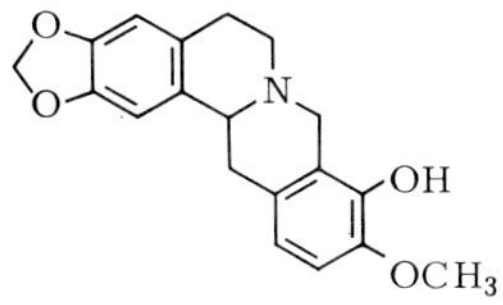

Nandinine

λ_{max}^{EtOH} 230 sh and 286 mμ (4.1 and 3.80)
λ_{min}^{EtOH} 252 mμ (2.3)

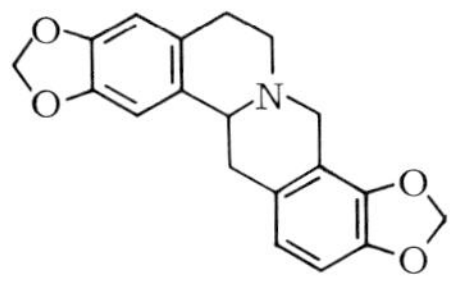

Tetrahydrocoptisine
(≡ stylopine)

λ_{max}^{EtOH} 237 and 289 mμ (3.85 and 3.89)
λ_{min}^{EtOH} 252 mμ (2.70)

Canadine
(≡ Tetrahydroberberine)

λ_{max}^{EtOH} 284 mμ (3.71)
λ_{min}^{EtOH} 252 mμ (2.76)

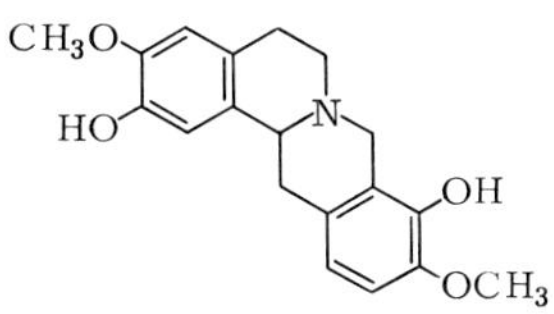

Scoulerine

λ_{max}^{EtOH} 230 and 283 mμ (4.20 and 3.85)
λ_{min}^{EtOH} 252 mμ (3.15)

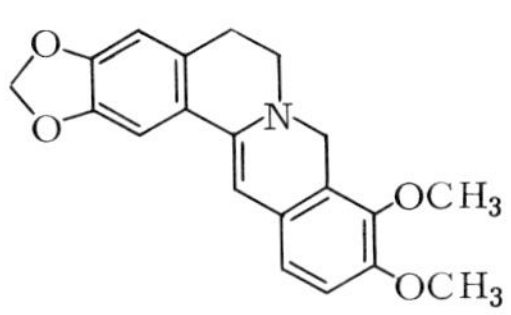

Dihydroberberine
(See also Ref. 35a)

λ_{max}^{EtOH} 280 and 368 mμ (4.10 and 4.25)
λ_{min}^{EtOH} 305 mμ (3.70)

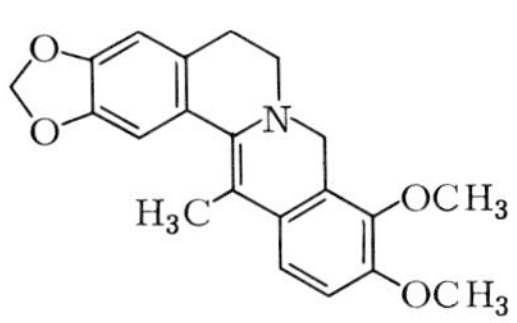

13-Methyldihydroberberine

λ_{max}^{EtOH} 280 and 360 mμ (4.0 and 4.5)
λ_{min}^{EtOH} 305 mμ (3.8)

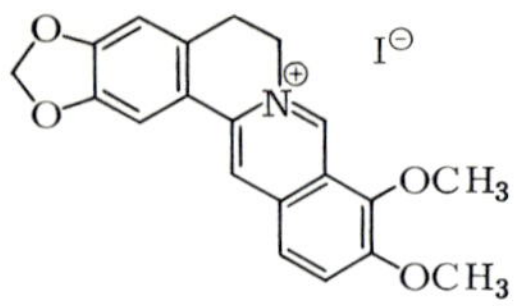

Berberine iodide

λ_{max}^{EtOH} 263, 345, and 423 mμ (4.4, 4.4, and 3.7)
λ_{min}^{EtOH} 250, 305, and 370 mμ (4.1, 3.8, and 3.7)

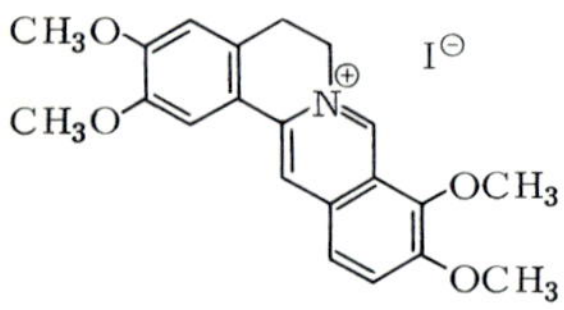

Palmatine iodide

λ_{max}^{EtOH} 265, 355, and 425 mμ (4.4, 4.5, and 4.0)
λ_{min}^{EtOH} 250, 305, and 380 mμ (4.3, 4.4, and 3.0)

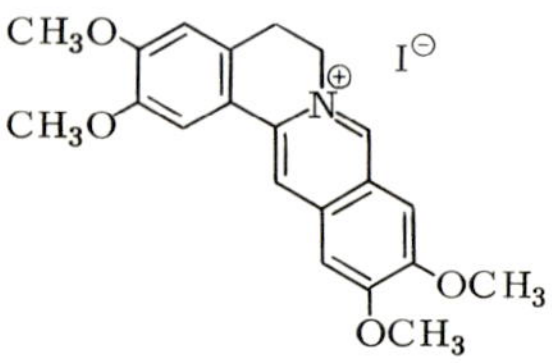

Pseudopalmatine iodide

λ_{max}^{EtOH} 265, 287, 310 sh, 345, and 375 mμ (4.3, 4.7, 4.5, 4.3, and 4.0)
λ_{min}^{EtOH} 250, 270, 335, and 365 mμ (4.1, 4.3, 4.3, and 3.9)

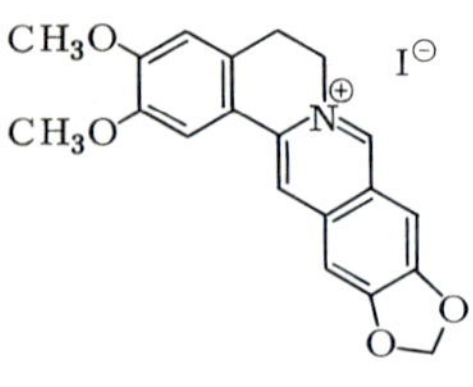

Pseudoepiberberine iodide

λ_{max}^{EtOH} 238, 262, 287, 310 sh, 338, and 365 mμ (4.3, 4.3, 4.4, 4.4, 4.2, and 3.8)
λ_{min}^{EtOH} 247, 270, and 332 mμ (4.2, 4.2, and 4.1)

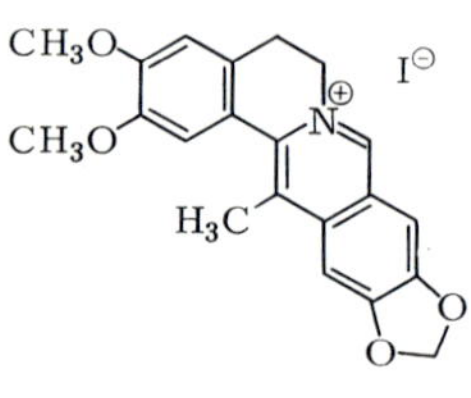

13-Methylpseudoepiberberine iodide

λ_{max}^{EtOH} 218, 260, 287, 305 sh, 338 sh, and 365 mμ (4.4, 4.4, 4.5, 4.4, 4.1, and 3.8)
λ_{min}^{EtOH} 247 and 270 mμ (4.2 and 4.3)

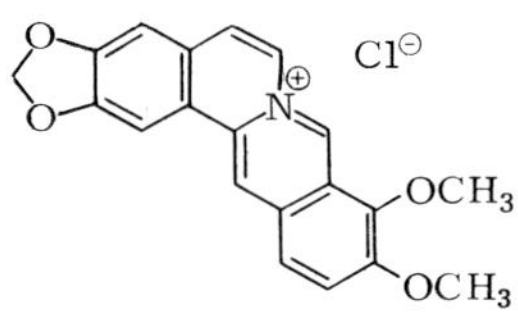

Dehydroberberine chloride
(dehydroprotoberberines are not natural products)

λ_{max}^{EtOH} 246, 278, 310, 348, and 460 mμ
λ_{min}^{EtOH} 257, 290.5, 332, and 405 mμ

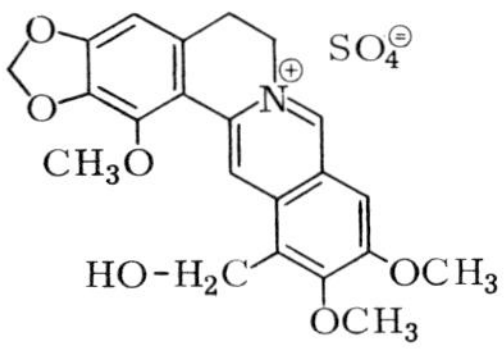

Alkaloid PO-5[73]

λ_{max}^{EtOH} 234 sh, 246, 261, 292, 337 sh, and 391 mμ (4.36, 4.40, 4.38, 4.55, 4.31, and 3.78)
λ_{min}^{EtOH} 253, 272, and 362 mμ (4.34, 4.30, and 3.65)

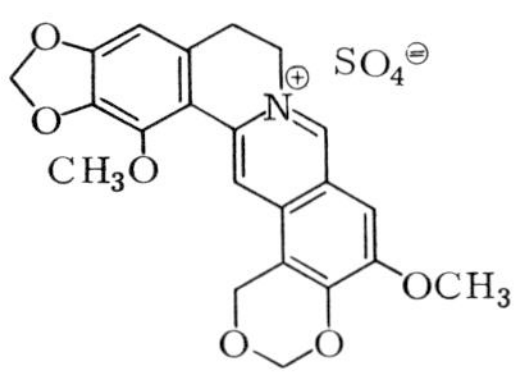

Alkaloid PO-4[73]

λ_{max}^{EtOH} 233 sh, 245, 262, 295, 341 sh, and 384 mμ (4.30, 4.32, 4.36, 4.65, 4.23, and 3.83)
λ_{min}^{EtOH} 272 and 366 mμ (4.27 and 3.75)

The effect of changes in solvent polarity and nature of the anion on the UV spectra of protoberberine salts has been discussed,[73] and some fluorescence spectra have been recorded.[89]

REFERENCES

1. For a review on the protoberberines, see P. W. Jeffs, *in* "The Alkaloids" (R. H. F. Manske, ed.), Vol. 9, p. 41. Academic Press, New York, 1967.

1a. Also see F. Santavý, *in* "The Alkaloids" (R. H. F. Manske, ed.), Vol. 12, p. 383. Academic Press, New York, 1970.

2. C. Tani, I. Imanishi, and J. Nishijo, *J. Pharm Soc. Jap.* **90**, 903 (1970).
3. M. Tomita, Y. Watanabe, and M. Fuse, *J. Pharm. Soc. Jap.* **77**, 274 (1957).
4. Y. Watanabe, *J. Pharm. Soc. Jap.* **77**, 278 (1957).
5. C. Tani, N. Takao, S. Takao, and K. Tagahara, *J. Pharm. Soc. Jap.* **82**, 751 (1962).
6. A. Pictet and A. Gams, *Chem. Ber.* **44**, 2480 (1911).
7. R. D. Haworth, W. H. Perkin, Jr., and J. Rankin, *J. Chem. Soc., London* **125**, 1686 (1924).
8. C. Schöpf, *Angew. Chem.* **50**, 797 (1937).
9. E. Späth and E. Mosetig, *Chem. Ber.* **59**, 1496 (1926).
10. A. R. Battersby, R. Southgate, J. Staunton, and M. Hirst, *J. Chem. Soc., C* p. 1052 (1966).
11. T. Kametani, K. Fukumoto, H. Agui, H. Yagi, K. Kigasawa, H. Sugahara, M. Hiiragi, T. Hayasaka, and H. Ishimaru, *J. Chem. Soc., C* p. 112 (1968).

12. T. Kametani, T. Honda, and M. Ihara, *Chem. Commun.* p. 1253 (1970); *J. Chem. Soc., C* p. 3318 (1971).
13. T. Kametani and M. Ihara, *J. Chem. Soc., C* p. 530 (1967).
14. S. Ishiwata and K. Itakura, *Chem. Pharm. Bull.* **18**, 1846 (1970). For an alternate synthesis of *O*-methylcaseadine using a Reissert compound, see M. P. Cava, M. V. Lakshmikantham, and M. J. Mitchell, *J. Org. Chem.* **34**, 2665 (1969).
15. N. S. Narasimhan and B. H. Bhide, *Chem. Ind. (London)* p. 621 (1969).
16. W. Meise and F. Zymalkowski, *Tetrahedron Lett.* p. 1475 (1969); W. Meise and F. Zymalkowski, *Arch. Pharm. (Weinheim)* **304**, 175 (1971). See also T. S. Stevens, *J. Chem. Soc., London* p. 663 (1935); as well as *Chem. Abstr.* **64**, 2135g (1966).
17. R. D. Haworth, W. H. Perkin, Jr., and H. S. Pink, *J. Chem. Soc., London* **127**, 1709 (1925).
18. R. D. Haworth, W. H. Perkin, Jr., and T. S. Stevens, *J. Chem. Soc., London* p. 1764 (1926).
19. N. T. Le Quang Thuan and J. Gardent, *Bull. Soc. Chim. Fr.* [5] p. 2401 (1966).
20. N. T. Le Quang Thuan and J. Gardent, *C. R. Acad. Sci., Ser. C* **267**, 1340 (1968).
21. P. Mathieu and J. Gardent, *C. R. Acad. Sci., Ser. C* **270**, 835 (1970).
22. W. H. Perkin, Jr., J. N. Rây, and R. Robinson, *J. Chem. Soc., London* **127**, 740 (1925).
23. T. R. Govindachari and S. Rajadurai, *J. Chem. Soc., London* p. 557 (1957).
24. I. W. Elliott, Jr., *J. Heterocycl. Chem.* **4**, 639 (1967).
25. A. R. Battersby, R. Binks, and P. S. Uzzell, *Chem. Ind. (London)* p. 1039 (1955); A. R. Battersby, D. J. LeCount, S. Garratt, and R. I. Thrift, *Tetrahedron* **14**, 46 (1961); J. W. Huffman and E. G. Miller, *J. Org. Chem.* **25**, 90 (1960).
26. D. W. Brown and S. F. Dyke, *Tetrahedron Lett.* p. 3587 (1964); S. F. Dyke and D. W. Brown, *Tetrahedron* **22**, 2429 (1966); T. R. Govindachari, K. Nagarajan, R. Charubala, and B. R. Pai, *Indian J. Chem.* **8**, 763 (1970).
27. A. Brossi, H. Bruderer, A. I. Rachlin, and S. Teitel, *Tetrahedron* **24**, 4277 (1968).
28. D. W. Brown, S. F. Dyke, G. Hardy, and M. Sainsbury, *Tetrahedron Lett.* p. 5177 (1968); and S. F. Dyke, D. W. Brown, M. Sainsbury, and G. Hardy, *Tetrahedron* **27**, 3495 (1971).
29. N. L. Dutta, M. S. Wadia, and A. A. Bindra, *Indian J. Chem.* **7**, 527 (1969).
30. C. K. Bradsher and N. L. Dutta, *J. Org. Chem.* **26**, 2231 (1961); **27**, 2213 (1962); also W. Augstein and C. K. Bradsher, *ibid.* **34**, 1349 (1969); H. F. Andrew and C. K. Bradsher, *Tetrahedron Lett.* p. 3069 (1966).
31. A. A. Bindra, M. S. Wadia, and N. L. Dutta, *Indian J. Chem.* **7**, 744 (1969).
32. N. C. Yang, A. Shani, and G. R. Lenz, *J. Amer. Chem. Soc.* **88**, 5369 (1966).
33. T. Takemoto and Y. Kondo, *J. Pharm. Soc. Jap.* **82**, 1408 (1962).
34. M. Shamma and C. D. Jones, *J. Amer. Chem. Soc.* **92**, 4943 (1970).
35. M. P. Cava, K. Nomura, S. K. Talapatra, M. J. Mitchell, R. H. Schlessinger, K. T. Buck, J. L. Beal, B. Douglas, R. F. Raffauf, and J. A. Weisbach, *J. Org. Chem.* **33**, 2785 (1968).
35a. W. Meise, *Liebigs Ann. Chem.* **748**, 55 (1971).
36. W. H. Perkin, Jr., *J. Chem. Soc., London* p. 722 (1918).
36a. T. R. Govindachari, K. Nagarajan, R. Charubala, and B. R. Pai, *Indian J. Chem.* **8**, 766 (1970).
37. A. Resplandy, *C. R. Acad. Sci., Ser. C* **247**, 2nd Sem., 2428 (1958).
38. J. Iwasa and S. Naruto, *J. Pharm. Soc. Jap.* **86**, 534 (1966); *Chem. Abstr.* **65**, 12247b (1966).
39. C. Schöpf and M. Schweikert, *Chem. Ber.* **98**, 2566 (1965).
39a. For new syntheses of compound **32** see H. O. Bernhard and V. Snieckus, *Tetrahedron Lett.* p. 4867 (1971); and S. Teitel, W. Klötzer, J. Borgese and A. Brossi, (in press).
40. N. J. Leonard, G. Fuller, and H. L. Dryden, Jr., *J. Amer. Chem. Soc.* **75**, 3727 (1953).
41. C. Tani, N. Takao, and S. Takao, *J. Pharm. Soc. Jap.* **82**, 748 (1962).
42. J. D. Albright and L. Goldman, *J. Amer. Chem. Soc.* **91**, 4317 (1969).
43. I. Sallay and R. H. Ayers, *Tetrahedron* **19**, 1397 (1963).
44. T. Kametani and M. Ihara, *J. Chem. Soc., C* p. 191 (1968).
45. C. Tani, S. Takao, and M. Sugiara, *J. Pharm. Soc. Jap.* **90**, 1012 (1970).

46. A. R. Battersby, *Proc. Chem. Soc., London* p. 189 (1963); A. R. Battersby, R. J. Francis, M. Hirst, and J. Staunton, *ibid.* p. 268.
47. D. H. R. Barton, *Proc. Chem. Soc., London* p. 293 (1963); D. H. R. Barton, R. H. Hesse, and G. W. Kirby, *ibid.* p. 267.
48. J. R. Gear and I. D. Spenser, *Can. J. Chem.* **41**, 783 (1963).
49. M. Tomita, M. Kozuka, H. Ohyabu, and K. Fujitani, *J. Pharm. Soc. Jap.* **90**, 82 (1970), and references cited therein.
49a. C. K. Bradsher and F. H. Day, *Tetrahedron Lett.* p. 409 (1971).
50. H. Corrodi and E. Hardegger, *Helv. Chim. Acta* **39**, 889 (1956).
51. T. Kametani and M. Ihara, *J. Chem. Soc., C* p. 1305 (1968).
52. J. C. Craig and S. K. Roy, *Tetrahedron* **21**, 401 (1965).
53. M. Legrand and R. Viennet, *Bull. Soc. Chim. Fr.* [5] p. 2798 (1966).
53a. G. Snatzke, J. Hrbek, Jr., L. Hruban, A. Horeau, and F. Šantavý, *Tetrahedron* **26**, 5013 (1970); G. Snatzke, G. Wollenberg, J. Hrbek, Jr., F. Šantavý, K. Blaha, W. Klyne, and R. J. Swan, *Tetrahedron* **25**, 5059 (1969).
54. M. Shamma and J. B. Moss, *J. Amer. Chem. Soc.* **83**, 5038 (1961).
55. M. Shamma and J. M. Richey, *J. Amer. Chem. Soc.* **84**, 1739 (1962); **85**, 2507 (1963).
56. M. Shamma, C. D. Jones, and J. A. Weiss, *Tetrahedron* **25**, 4347 (1969).
57. W. F. Trager, C. M. Lee, and A. H. Beckett, *Tetrahedron* **23**, 365 (1967).
58. T. Kametani, M. Ihara, K. Fukumoto, H. Yagi, H. Shimanouchi, and Y. Sasada, *Tetrahedron Lett.* p. 4251 (1968).
59. T. Kametani, M. Ihara, Y. Kitahara, C. Kabuto, H. Shimanouchi, and Y. Sasada, *Chem. Commun.* p. 1241 (1970); see also *J. Chem. Soc., C* p. 2541 (1971).
60. H. W. Bersch, *Arch. Pharm. (Weinheim)* **291**, 595 (1958).
61. P. W. Jeffs, *Experientia* **21**, 690 (1965).
62. C. K. Yu, D. B. MacLean, R. G. A. Rodrigo, and R. H. F. Manske, *Can. J. Chem.* **48**, 3673 (1970); T. R. Govindachari, K. Nagarajan, R. Charubala, B. R. Pai, and P. S. Subramanian, *Indian J. Chem.* **8**, 769 (1970).
63. M. Ohta, H. Tani, and S. Morozumi, *Tetrahedron Lett.* p. 859 (1963).
64. A. R. Battersby and H. Spenser, *J. Chem. Soc., London* p. 1087 (1965).
65. M. M. Nijland, *Pharm. Weekbl.* **96**, 640 (1961); *Chem. Abstr.* **55**, 2626371 h (1961).
66. M. M. Nijland, *Pharm. Weekbl.* **98**, 301 (1963); *Chem. Abstr.* **59**, 10142 (1963).
67. M. Shamma and B. S. Dudock, *Tetrahedron Lett.* p. 3825 (1965).
68. I. Monkovic and I. D. Spenser, *J. Amer. Chem. Soc.* **87**, 1137 (1965); *Can. J. Chem.* **43**, 2017 (1965).
69. S. Pfeifer, I. Mann, L. Dolejš, V. Hanuš, and A. D. Cross, *Tetrahedron Lett.* p. 83 (1967).
70. V. Preininger, A. D. Cross, J. W. Murphy, F. Šantavý, and T. Toube, *Collect. Czech. Chem. Commun.* **34**, 875 (1969).
71. See also K. Dėlenk-Heydenreich and S. Pfeiffer, *Pharmazie* **24**, 635 (1969).
72. V. Preininger, V. Šimánek, and F. Šantavý, *Tetrahedron Lett.* p. 2109 (1969).
73. V. Preininger, L. Hruban, V. Šimánek, and F. Šantavý, *Collect. Czech. Chem. Commun.* **35**, 124 (1970).
74. V. Šimánek, V. Preininger, P. Sedmera, and F. Šantavý, *Collect. Czech. Chem. Commun.* **35**, 1440 (1970).
75. M. P. Cava, T. A. Reed, and J. L. Beal, *Lloydia* **28**, 73 (1965).
76. A. R. Battersby, R. J. Francis, E. A. Ruveda, and J. Staunton, *Chem. Commun.* p. 89 (1965).
76a. E. Brochmann–Hanssen, C.-C. Fu, and G. Zanati, *J. Pharm. Sci.* **60**, 873 (1971).
77. R. N. Gupta and I. D. Spenser, *Can. J. Chem.* **43**, 133 (1965).
78. G. Blaschke, *Arch. Pharm. (Weinheim)* **301**, 439 (1968).
79. Roussel-UCLAF Patents, *Chem. Abstr.* **69**, 59475m (1968); **71**, 70785z, 70787b, 70788c, 91735y (1969); **72**, 32103p (1970); **73**, 15063d (1970).
80. A. K. Krey and F. E. Hahn, *Science* **166**, 755 (1969).

81. H. Fukuda, K. Watanabe, and Y. Kudo, *Chem. Pharm. Bull.* **18**, 1299 (1970).
81a. D. G. Patel, A. Tye, P. N. Patil, A. M. Burkman, and J. L. Beal, *Lloydia* **33**, 36 (1970).
81b. L. A. Mitscher, W. Wu, R. W. Doskotch, and J. L. Beal, *Chem. Commun.* p. 589 (1971).
82. N. Nakanishi, *Jap. J. Pharmacol.* **12**, 208 (1962).
83. S. Sasagawa, K. Kametani, and S. Kiyofugi, *Chem. Pharm. Bull.* **17**, 1 (1969).
84. J. Iwasa, S. Narito, K. Nakamura, and Y. Shoji, *Chem. Abstr.* **72**, 111667d (1970).
85. M. Ohashi, J. M. Wilson, H. Budzikiewicz, M. Shamma, W. A. Slusarchyk, and C. Djerassi, *J. Amer. Chem. Soc.* **85**, 2807 (1963).
86. C.-Y. Chen and D. B. MacLean, *Can. J. Chem.* **46**, 2501 (1968).
86a. G. Habermehl, J. Schunk, and G. Schaden, *Liebigs Ann. Chem.* **742**, 138 (1970).
87. C.-Y. Chen, D. B. MacLean, and R. H. F. Manske, *Tetrahedron Lett.* p. 349 (1968).
88. M. Shamma, M. J. Hillman, and C. D. Jones, *Chem. Rev.* **69**, 779 (1969), and references cited therein.
88a. L. Hruban and F. Šantavý, *Collect. Czech. Chem. Commun.* **32**, 3414 (1967); F. Šantavý, L. Hruban, V. Šimánek, and D. Walterová, *Collect. Czech. Chem. Commun.* **35**, 2418 (1970); L. Hruban, F. Šantavý, and S. Hegerova, *Collect. Czech. Chem. Commun.* **35**, 3420 (1970).
89. E. Sebe, S. Abe, N. Murase, and H. Sugaya, *J. Chin. Chem. Soc.* (*Taipei*) **15**, 146 (1968); *Chem. Abstr.* **72**, 43942n (1970).

Chapter 17 / THE BENZOPHENANTHRIDINES

Occurrence: Fumariaceae, Papaveraceae, and Rutaceae
Approximate Number: 30
Some Benzophenanthridines of Interest:

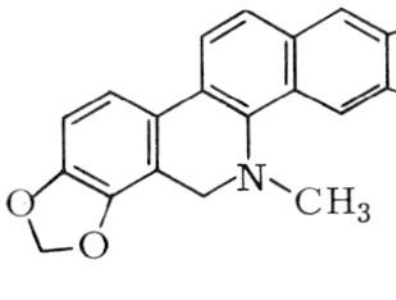

Sanguinarine

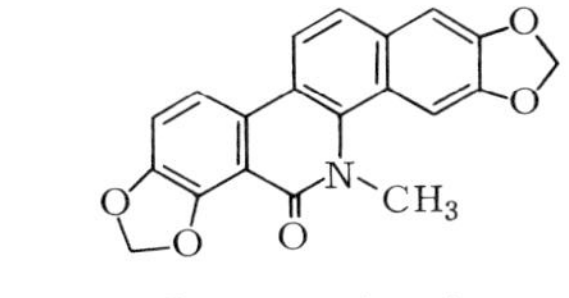

Chelerythrine

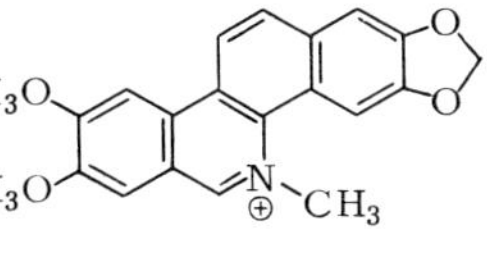

Dihydrosanguinarine

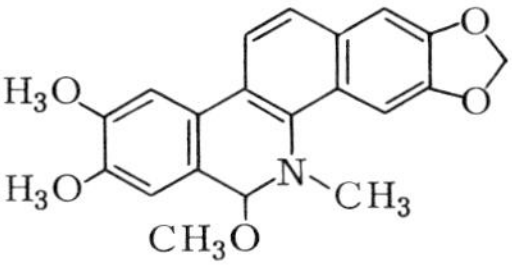

Oxysanguinarine

Nitidine

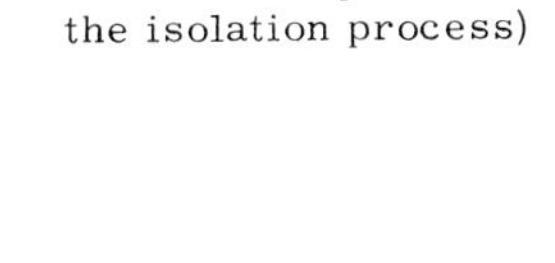

8-Methoxydihydronitidine
(may be an artefact produced during the isolation process)

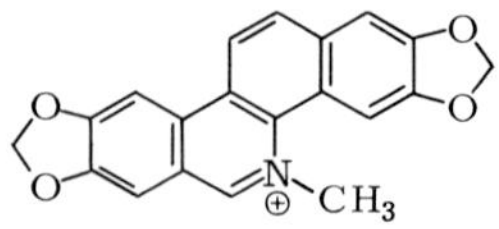

Avicine

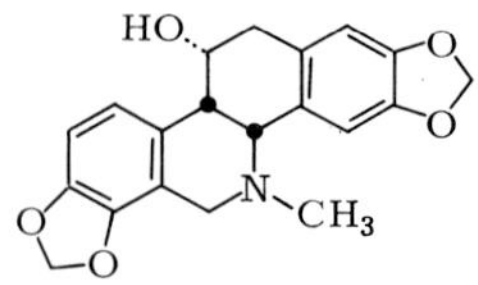

(+)-Chelidonine
[(−)and (±)-chelidonine are also natural products]

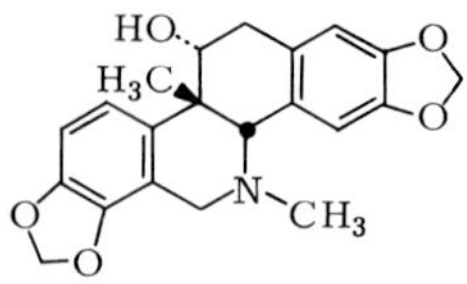

Corynoline

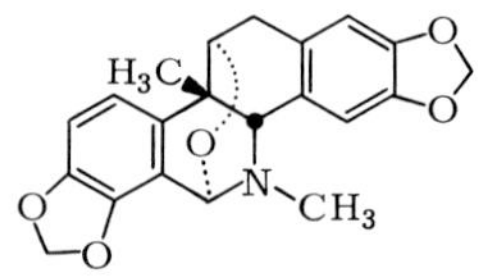

Corynoloxine

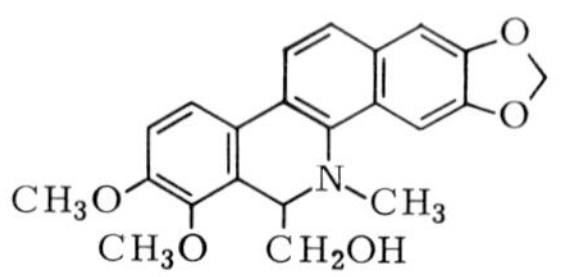

Bocconoline

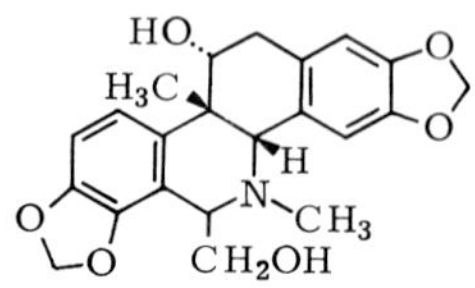

Corynolamine

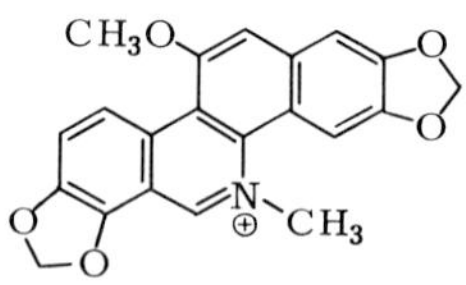

Chelirubine

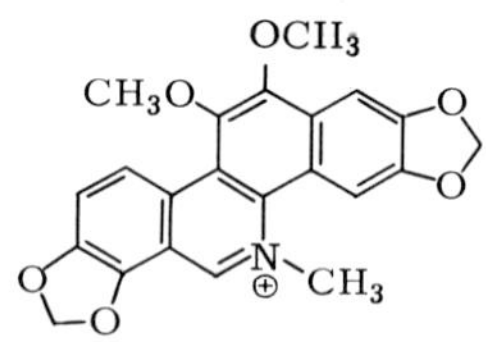

Macarpine

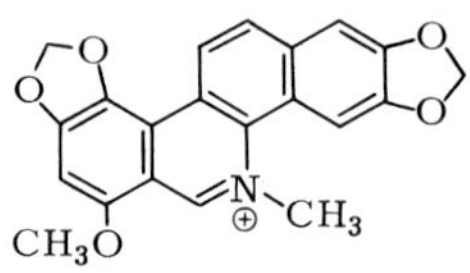

Bocconine

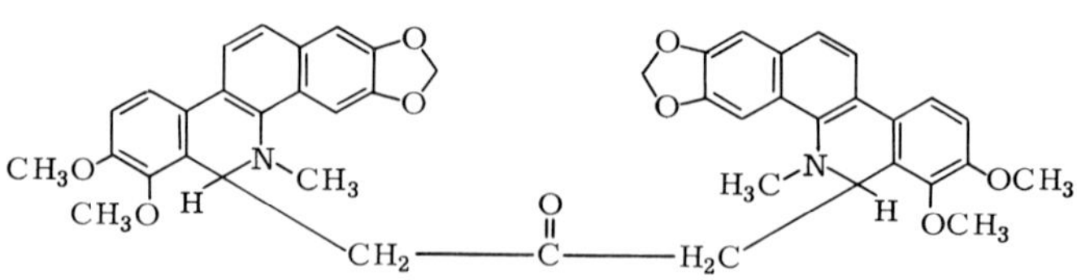

1,3-Bis(8-hydrochelerythrinyl)acetone

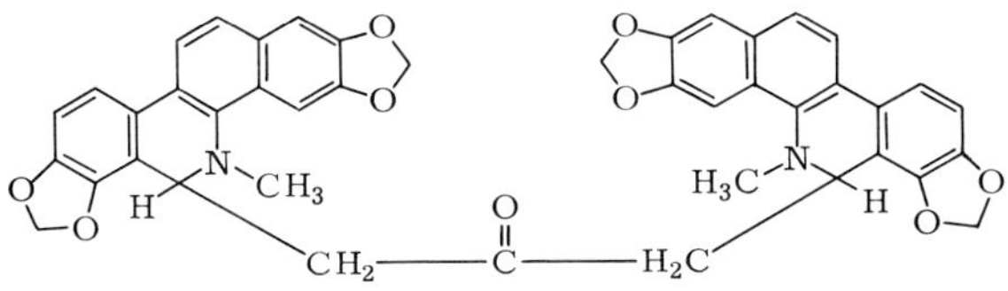

meso-1,3-Bis(8-hydrosanguinarinyl)acetone[1]
(+)-1,3-Bis(8-hydrosanguinarinyl)acetone[1a]

I. Introduction*

The benzophenanthridine alkaloids may be conveniently divided into three groups. The first and largest group, in which ring C is usually completely aromatic, includes as typical alkaloids sanguinarine and chelerythrine. Ring C is sometimes partially reduced or oxidized as in dihydrosanguinarine and oxysanguinarine, respectively. The substituents are usually located at C-2, 3, 9, and 10. But the ring D substituents may also appear at alternate positions C-10 and 11, as in nitidine and avicine.

The second group, which can be designated the alcoholic group, may be represented by the alkaloids chelidonine, corynoline, and corynoloxine. The distinguishing feature is an alcohol or a masked alcohol group at C-6.

The third grouping of benzophenanthridine bases was reported only in 1968 and includes such compounds as chelirubine and macarpine. These are completely aromatic species which invariably incorporate a methoxyl at C-6.

Alkaloids of the first group have been found in a few instances as acetone dimers, and it is conceivable that members of the third group may some day also be isolated in that state.

II. The Chemistry and Stereochemistry of Chelidonine

(+)-Chelidonine is the first benzophenanthridine alkaloid to have had its structure established. The degradative work was performed mainly by Gadamer, Späth, and von Bruchhausen, and the correct structure was proposed in 1930.[2]

A carefully purified sample of chelidonine was found to have the empirical formula $C_{20}H_{19}NO_5 \cdot H_2O$ and to possess one hydroxyl and two methylenedioxy groups. Zinc dust distillation of chelidonine gave benzo[*c*]phenanthridine. Oxidation of the alkaloid with hot permanganate yielded 3,4-methylenedioxyphthalic acid and hydrastic acid, providing an insight into the positions of the two methylenedioxy substituents.

Hofmann degradation of the methiodide salt of chelidonine produced the methine base **1** which upon permanganate oxidation afforded hydrastic acid and, more importantly, 2-dimethylaminomethylpiperonylic acid. The latter product established the posi-

* The numbering system for the benzophenanthridines presents a problem since there is no generally accepted scheme. The system presented here is new and has a twofold advantage. It is based on the proven biogenetic relationship between the protoberberines and the benzophenanthridines, and it assigns numbers to the bridgehead positions at the points of fusion of rings B and C.

tion of the methylenedioxy group in ring D of chelidonine relative to the nitrogen atom. When the methiodide salt of the methine **1** was treated with base, a tetracyclic ether best formulated as **2** was obtained which threw some light upon the position of the alcoholic group in chelidonine (Scheme I).

Further degradations of chelidonine involved its treatment with hot acetic anhydride to yield the optically inactive amide **3** which led to trimellitic acid upon nitric acid oxidation. With acetic anhydride at room temperature, on the other hand, optically active chelidonine acetate was generated which when treated repeatedly with mercuric acetate led to the aromatic benzophenanthridine alkaloid sanguinarine. Chelidonine could also be converted to deoxychelidonine by dehydration with thionyl chloride (Scheme I).

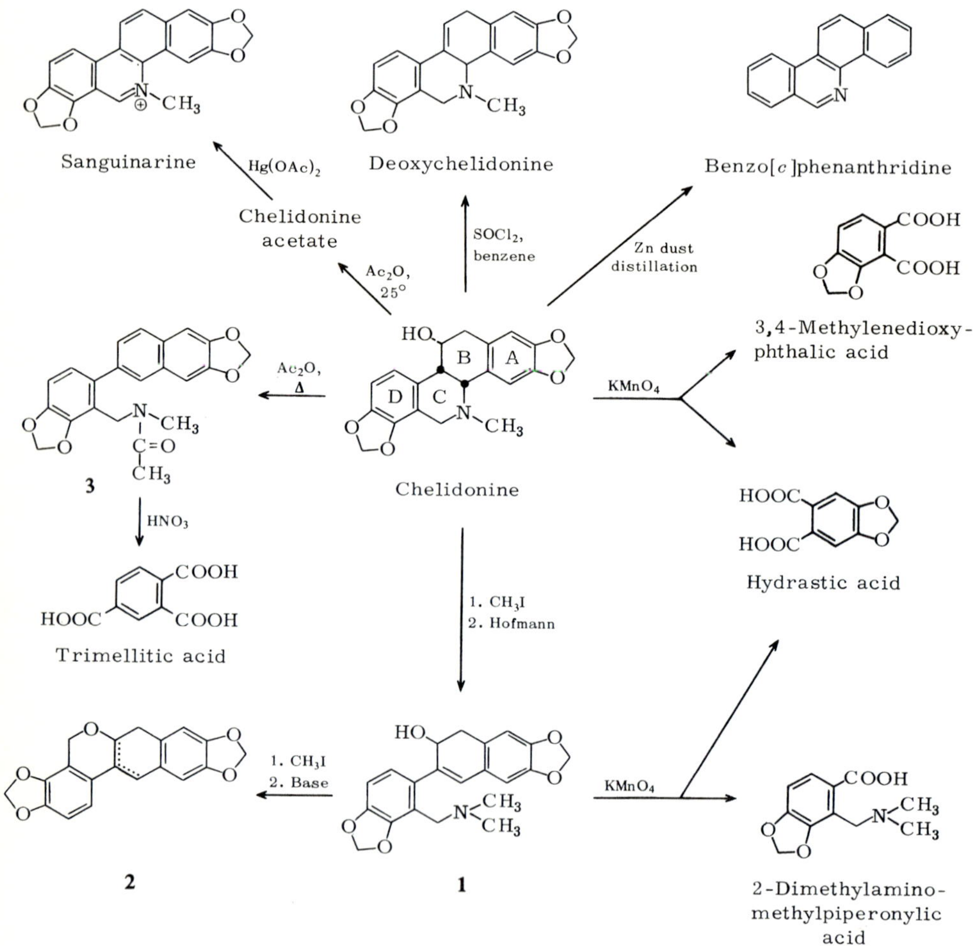

Scheme I

Attempts to oxidize the hydroxyl function of chelidonine were not successful in forming the corresponding ketone.[3] Oxidation under Oppenauer conditions gave rise to the yellow compound hydroxydihydrosanguinarine and the red betaine hydroxysanguinarine. Oxidation with mercuric acetate, chromic acid, or permanganate in acetone yielded mixtures of sanguinarine, dihydrosanguinarine, and oxysanguinarine (Scheme II).

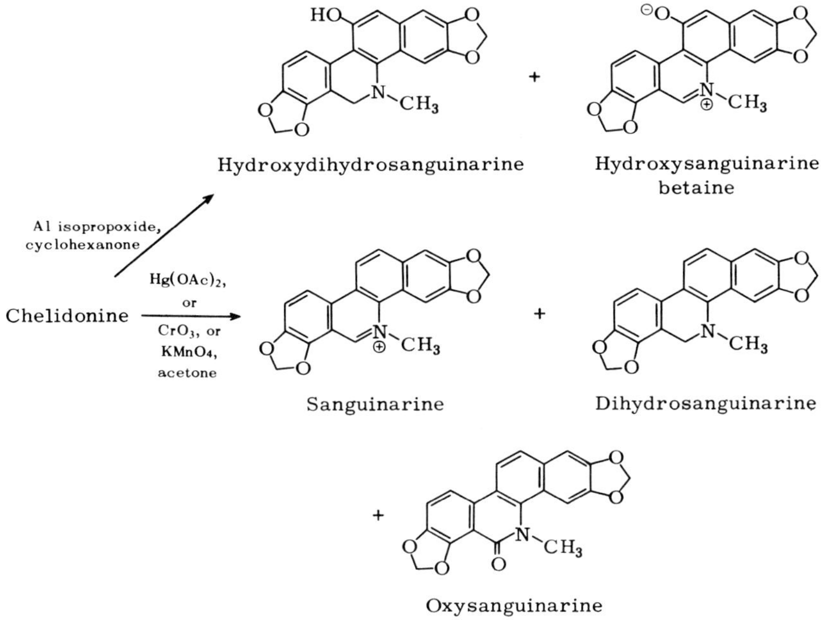

Scheme II

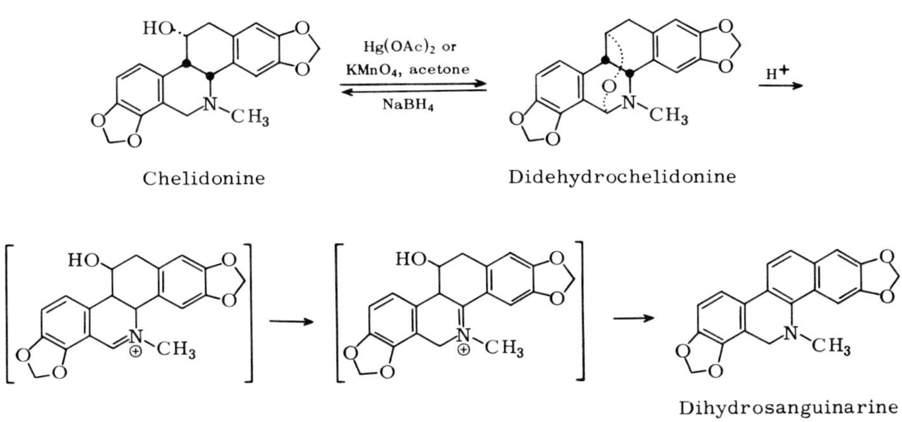

Scheme III

Yet another product formed from the mercuric acetate or mild permanganate oxidation of chelidonine is didehydrochelidonine. Gadamer was the first to obtain this compound more than 50 years ago, but its structure has only recently been elucidated. Didehydrochelidonine contains two hydrogen atoms less than chelidonine, and sodium borohydride reduction of the former gives the latter. Didehydrochelidonine does not show any alcohol absorption in the IR and undergoes facile transformation to dihydrosanguinarine. It has therefore been formulated as the bridged amino acetal shown in Scheme III.[4]

The relative stereochemistry of chelidonine was first considered by Bersch, who postulated a cis B/C fusion together with an axial hydroxyl at C-6 to explain the large amount of intramolecular hydrogen bonding showing up in the IR spectrum of the alkaloid between 3.3 and 3.4 μ (3030 and 2941 cm^{-1}).[5,6] The ready formation of didehydrochelidonine upon mild oxidation can be readily understood in terms of this steric arrangement. The alkaloid usually occurs in the dextrorotatory form, but it has also been isolated in the levo as well as the racemic forms.[7]

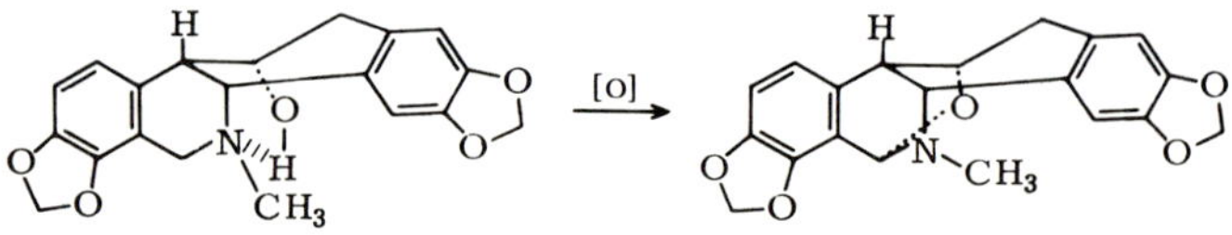

(+)-Chelidonine Didehydrochelidonine

It should be pointed out that the position of the alcoholic hydroxyl in chelidonine was originally set at C-6 because this is the most logical position for that function. The formation of the cyclic ether **2** by itself does not constitute conclusive chemical proof that the hydroxyl is at C-6 since the structure of this ether has never been confirmed by synthesis. The alternate possibility for chelidonine with the hydroxyl at C-5 does not appear plausible from a biogenetic viewpoint, and as will be seen in Section VII, is not supported by NMR data. The easy formation of didehydrochelidonine from chelidonine is a strong indication that the alcohol function is indeed at C-6.

Horeau has developed a useful method for the correlation of absolute configuration of secondary alcohols based upon a kinetic partial resolution during esterification with α-phenylbutyric anhydride, and this method has allowed the determination of the absolute configuration of (+)-chelidonine. Excess ($\pm$)-α-phenylbutyric anhydride was added to (+)-chelidonine. The optical rotation of the α-phenylbutyric acid recovered from the hydrolysis of the unreacted anhydride was found to be positive. Now it had been demonstrated empirically that when the configuration of the optically active alcohol is as shown in **3a**, the recovered α-phenylbutyric acid has the *S*-(+)-enantiomer in excess. It follows that (+)-chelidonine must be as drawn in stereo formula **3b**.[8]

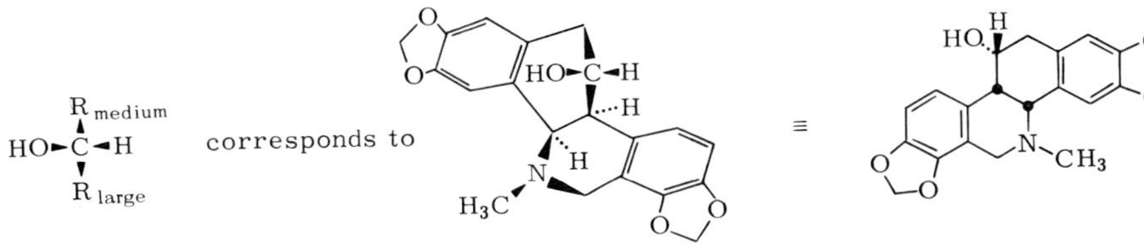

3a **3b** (+)-Chelidonine .

In the same paper in which Horeau gave his conclusions on (+)-chelidonine, Snatzke, Šantavý, and co-workers showed that analysis of the CD spectrum of (+)-chelidonine by a nonempirical method based on a sector rule leads to the identical stereochemical conclusions.[8]

III. The Chemistry of Sanguinarine and Chelerythrine

The structural elucidation of the completely aromatic alkaloids sanguinarine and chelerythrine was carried out following the establishment of structure for chelidonine.

Sanguinarine salts are red or orange and undergo oxidation–reduction in the presence of hydroxide ions to give oxysanguinarine and dihydrosanguinarine. Dihydrosanguinarine can also be conveniently obtained through sodium borohydride reduction of sanguinarine, while oxysanguinarine is available by the potassium ferricyanide oxidation of sanguinarine.

As with chelidonine, zinc dust distillation of sanguinarine gives benzo[*c*]phenanthridine, proving that the two alkaloids possess a common nucleus (Scheme IV),

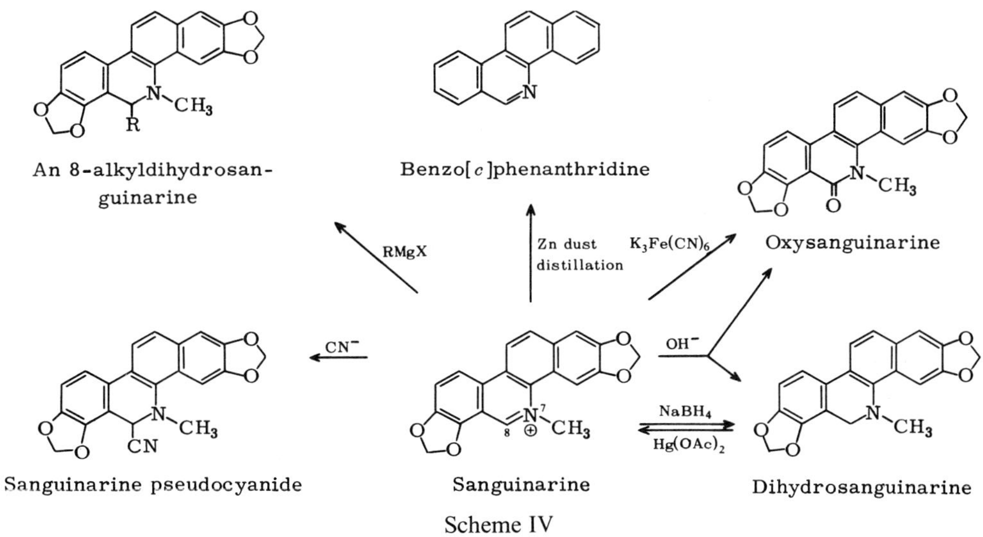

Scheme IV

and it has already been mentioned that chelidonine acetate can be transformed into sanguinarine through repeated treatment with mercuric acetate.[9]

The N-7 to C-8 double bond in sanguinarine undergoes definite carbonyl-like transformations. 8-Alkyldihydrosanguinarine derivatives can be produced by addition of Grignard reagents. With sodium or potassium cyanide, a stable, crystalline, and colorless pseudocyanide is generated. Recrystallization of the alkaloid from methanol or ethanol gives rise to 8-alkoxydihydrosanguinarines, and treatment with the anion of acetone yields 8-acetonyldihydrosanguinarine.[10]

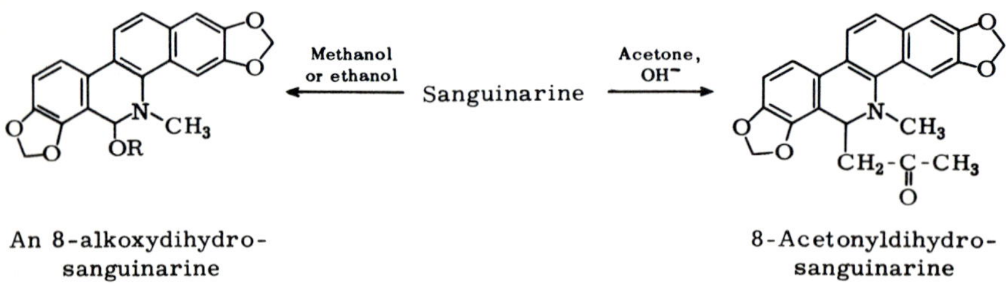

The chemistry of chelerythrine is similar to that of sanguinarine. The positions of the substituents in chelerythrine were determined by oxidation of the alkaloid with potassium permanganate. The products obtained were hydrastic acid and hemipinic acid *N*-methylimide.

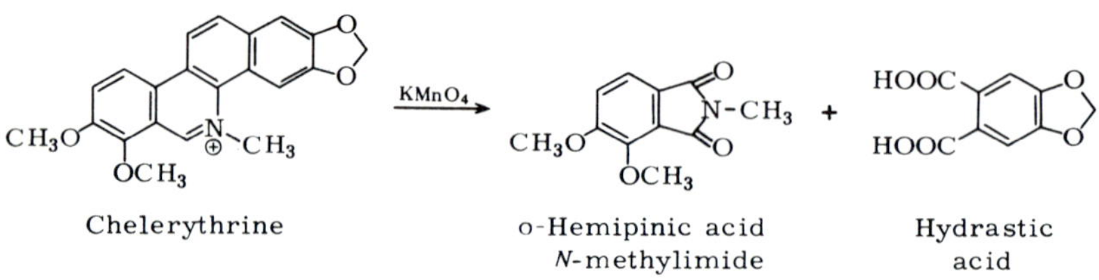

When dihydrosanguinarine and dihydrochelerythrine were each treated separately with hot sulfuric acid and phloroglucinol, the methylenedioxy groups were hydrolyzed. The corresponding phenols were *O*-methylated to afford the common product **4**, so that sanguinarine and chelerythrine differ only in the nature of their aromatic substituents (Scheme V).

Early work on sanguinarine was handicapped by the difficulty in separating this alkaloid from chelerythrine and protopine which are also present in its main natural source, the bloodroot plant, *Sanguinaria canadensis* L. (Papaveraceae). Nowadays, crude sanguinarine nitrate is commercially available (Aldrich Chemical Co.) and can be purified by sodium borohydride reduction followed by chromatography on alumina. Dihydrosanguinarine, dihydrochelerythrine, and protopine can be isolated from this

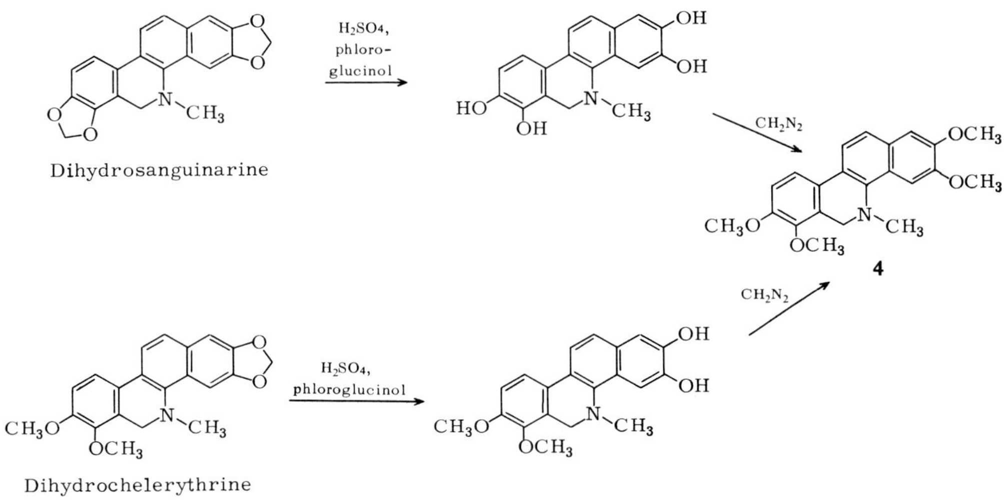

Scheme V

separation, and the dihydrosanguinarine can then be reoxidized to sanguinarine with mercuric acetate.[11]

Sanguinarine and chelerythrine may also be separated by paper chromatography using a butanol–acetic acid–water system.[12,13]

IV. The Structures of Corynoline and Corynoloxine: Two Unusual Benzophenanthridine Alkaloids

Two interesting variants of the alcoholic benzophenanthridine alkaloids have been obtained from *Corydalis incisa* Pers. (Fumariaceae).[14]

Corynoline analyzes for $C_{21}H_{21}O_5N$ and includes one alcoholic hydroxyl, two methylenedioxys, one *N*-methyl, and one *C*-methyl group. The alkaloid dehydrates readily with thionyl chloride to deoxycorynoline.

Oxidation of deoxycorynoline with potassium permanganate at 0° gave hydrastic acid characterized as its *N*-ethylimide. Heating deoxycorynoline at 300° *in vacuo* produced the totally aromatic species **5** which could easily be converted into dihydrosanguinarine. This transformation demonstrates the benzophenanthridine character of corynoline and proves the positions of the two methylenedioxy groups to be the same as in sanguinarine (Scheme VI).

Two successive Emde degradations were run on dihydrodeoxycorynoline derived from the catalytic reduction of deoxycorynoline. The product was the polycyclic compound **6** which was converted into 2-methyl-6,7-methylenedioxynaphthalene, thus settling the location of the *C*-methyl group in corynoline. Placement of the hydroxyl group at C-6 is in analogy with chelidonine (Scheme VI).

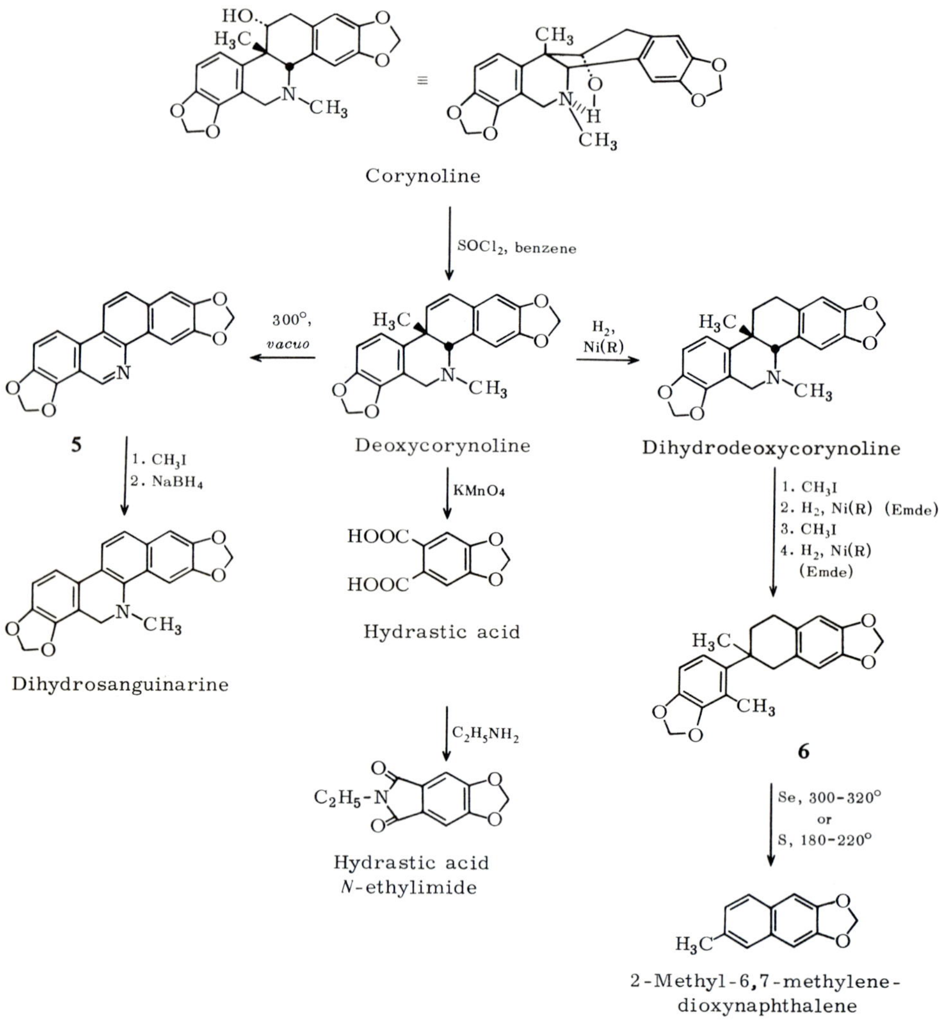

Scheme VI

When corynoline is treated with aqueous potassium permanganate at room temperature, one of the oxidation products is a compound termed corynoloxine also found with corynoline in *C. incisa*.[14,15] Formation of corynoloxine from corynoline proves the cis B/C ring fusion in both alkaloids, as also indicated by the strong intramolecular hydrogen absorption bands between 3 and 4 μ (3333 and 2500 cm^{-1}) in the IR spectrum of corynoline. The characterization of corynoloxine actually preceded that of its close relative didehydrochelidonine, and since corynoloxine is a natural product it would not be surprising if didehydrochelidonine is someday isolated from plant sources.

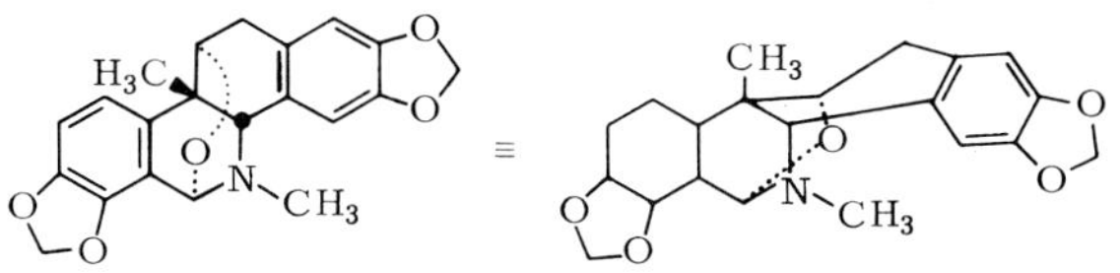

Corynoloxine

Lithium aluminum hydride reduction of corynoloxine gives corynoline, while treatment with hydrogen chloride yields the immonium salt **6a** which can also be obtained from the mercuric acetate oxidation of corynoline (Scheme VI a).[15a]

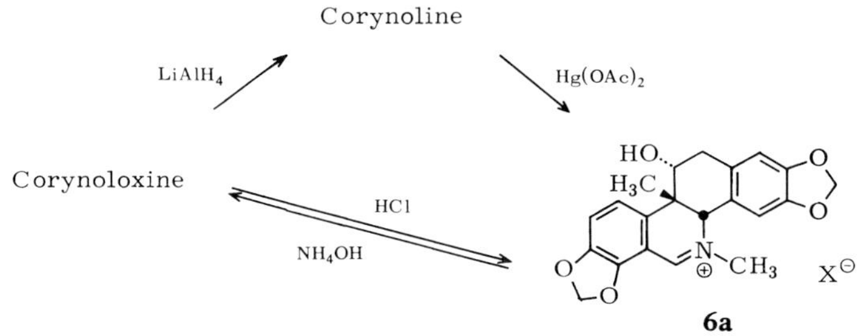

Scheme VI a

Emde degradation of corynoloxine methochloride led to the amino ethers **6b** and **6c** which upon catalytic hydrogenolysis gave rise to the ether **6d** (Scheme VI b).[15a]

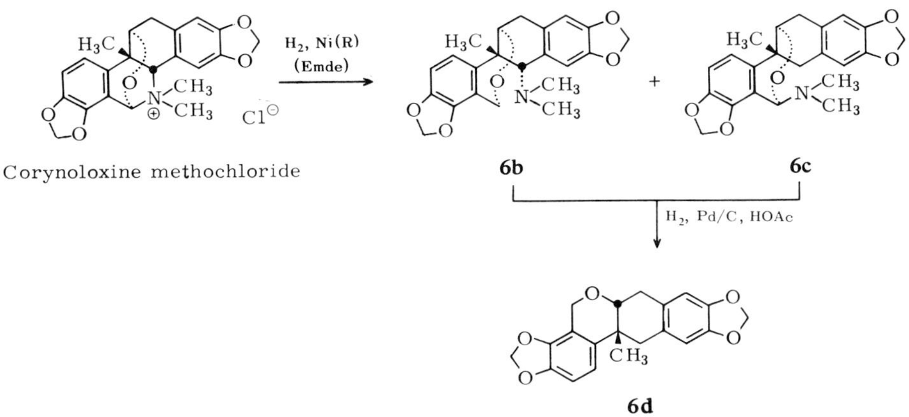

Scheme VI b

The same ether **6d** was also obtained through the sequence depicted in Scheme VI c. Emde degradation of corynoline acetate methochloride generated a hydromethine base, the methochloride salt of which upon treatment with base gave **6d**.[15b]

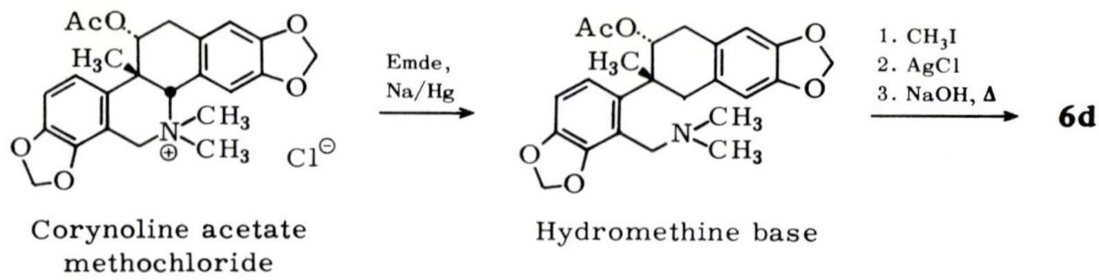

Scheme VIc

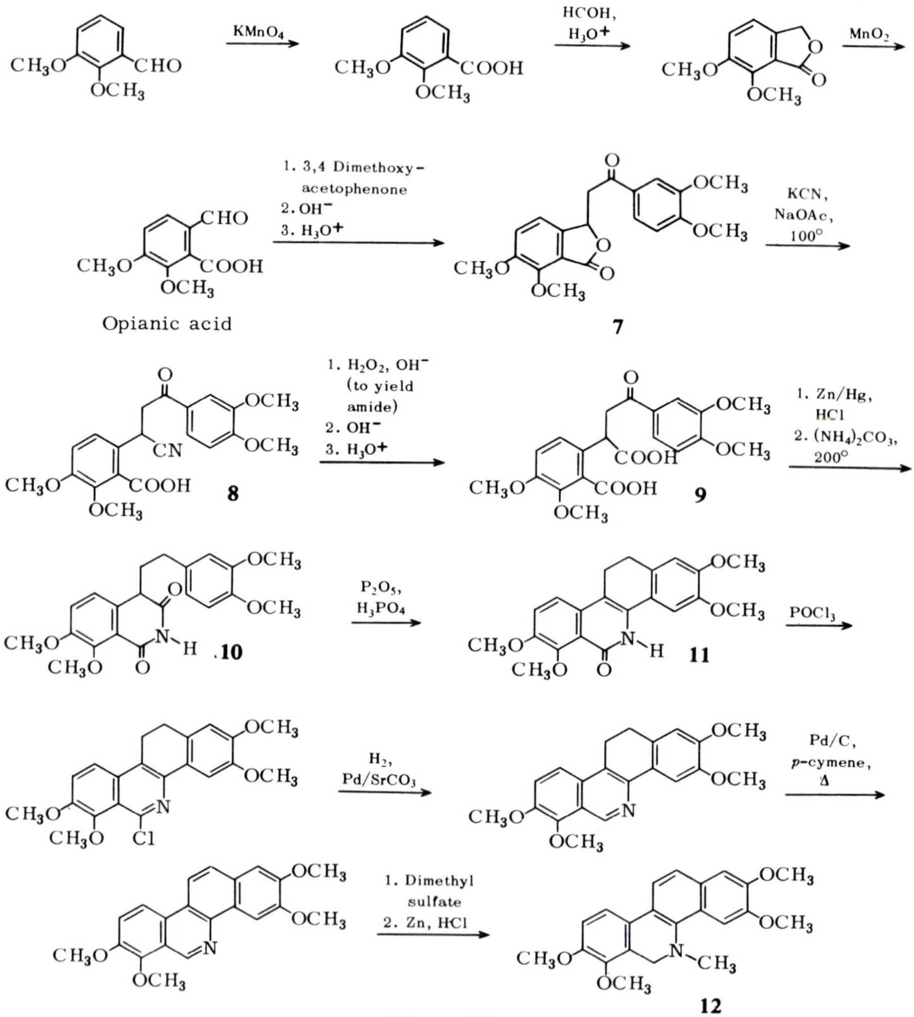

Scheme VII

V. Syntheses of Benzophenanthridines

A. The Bailey–Robinson Synthesis

Bailey and Robinson performed the first synthesis of a sanguinarine analog.[16,17] 2-Formyl-5,6-dimethoxybenzoic acid, commonly known as opianic acid, was condensed in base with 3,4-dimethoxyacetophenone to the γ-lactone **7**. Conversion to the cyanide **8** and then to the diacid **9** was followed by a Clemmensen reduction to remove the ketone function. The resulting diacid was heated with ammonium carbonate to afford the imide **10**. Cyclization to the tetracyclic lactam **11** and subsequent reductive and oxidative procedures led to the desired benzophenanthridine **12** (Scheme VII).

B. The Bailey–Worthing Synthesis of Chelerythrine

The Bailey–Robinson approach was extended to the synthesis of the naturally occurring alkaloid chelerythrine.[12] The diacid **13** was obtained by the modified procedure outlined in Scheme VIII. Since the imide derived from **13** could not be cyclized, it was necessary to utilize the diacid **13** directly to afford the pyrone **14**. The remainder of the synthesis followed in the expected manner.

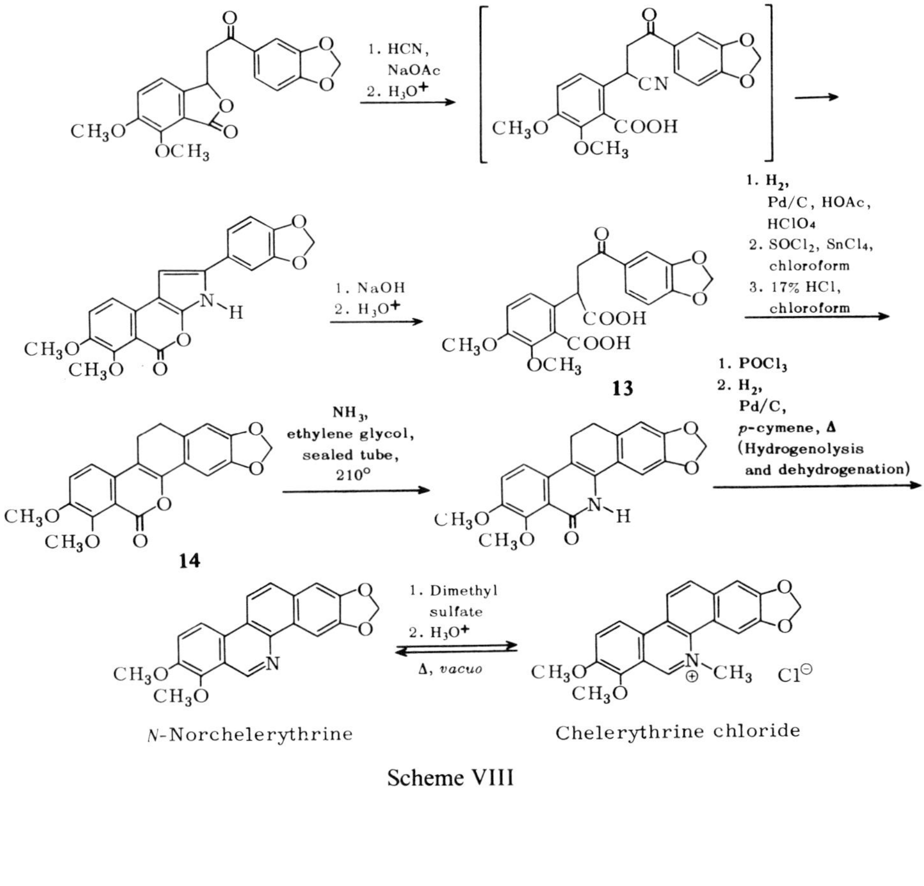

Scheme VIII

At a later date *N*-norchelerythrine was isolated as a natural product from *Toddalia aculeata* Pers. (Rutaceae).[18]

C. Another Modification of the Robinson Approach

The main steps in this synthesis were carried out as early as 1937, but the final transformations leading to the tetracyclic species **16** were not reported until 1950 (Scheme IX).[17,19]

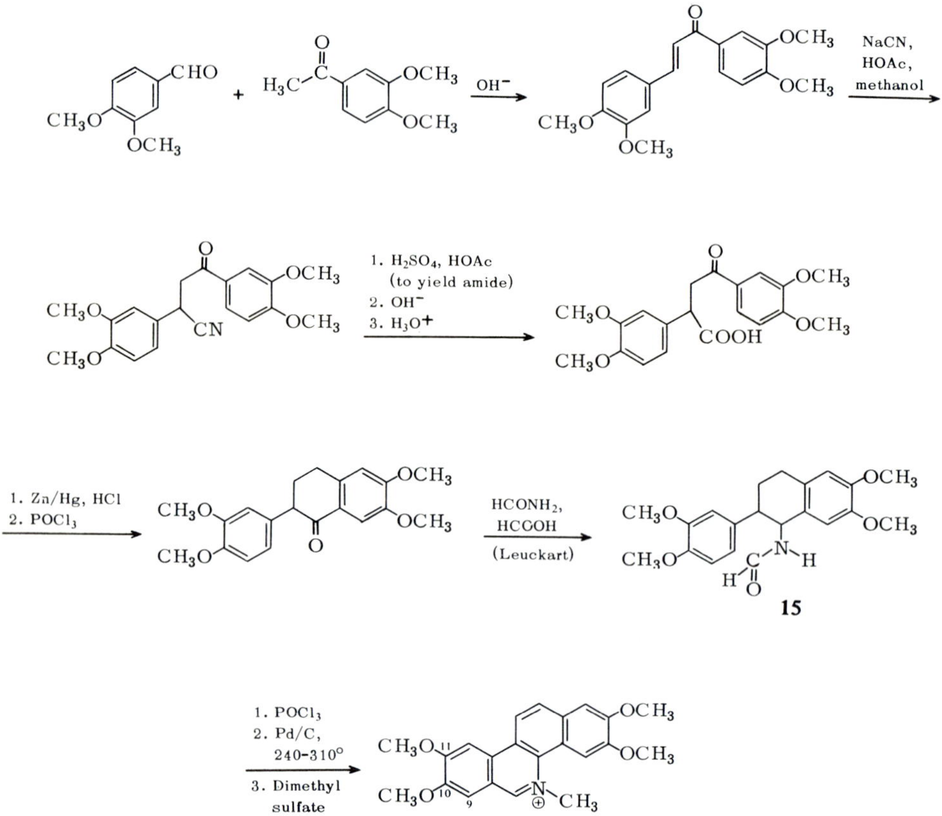

15

16

Scheme IX

The sequence in Scheme IX has been extended to the preparation of avicine[20] and nitidine.[21,22] But the approach cannot be used for the synthesis of C-9,10-substituted alkaloids without proper protection since Bischler–Napieralski cyclization of the amide **15** always proceeds on the less-hindered side of the aromatic ring.

D. The Synthesis of Sanguinarinc

In the first recorded preparation of sanguinarine the pathway followed is conceptually different from any of the Robinson syntheses.[23] Bobbitt had initially found that 1,2-dihydroisoquinolines will condense with aromatic aldehydes under acid conditions to produce 4-substituted isoquinolines, e.g.[24]:

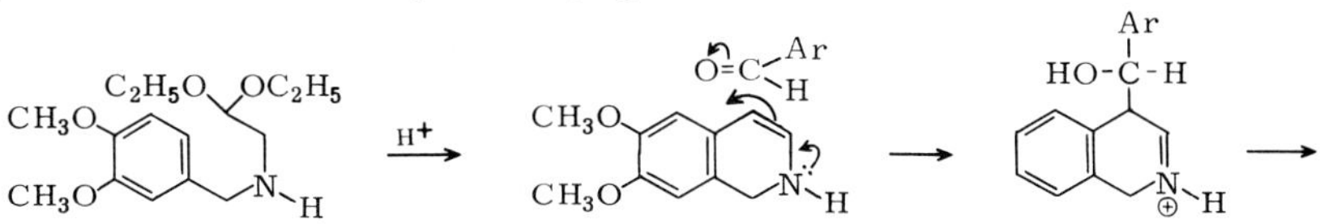

A 1,2-dihydroisoquinoline

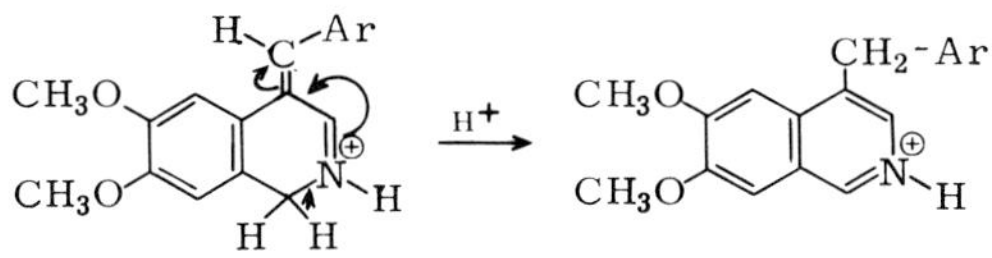

In a variation on this theme, Dyke and co-workers established that the condensation of 2,3-methylenedioxybenzaldehyde with aminoacetaldehyde dimethyl acetal followed by catalytic reduction afforded the amino acetal **17**. The 4-isoquinolylacetic acid **18** was then generated from **17** by treatment with glyoxylic acid. Subsequent condensation with 6-nitropiperonal gave the carboxylic acid **19**, and the remainder of the synthesis which includes a Pschorr cyclization follows in a straightforward manner (Scheme X).[23] Analogs of the alkaloids nitidine and avicine were synthesized by a similar procedure, again involving Pschorr cyclization.[25,26]

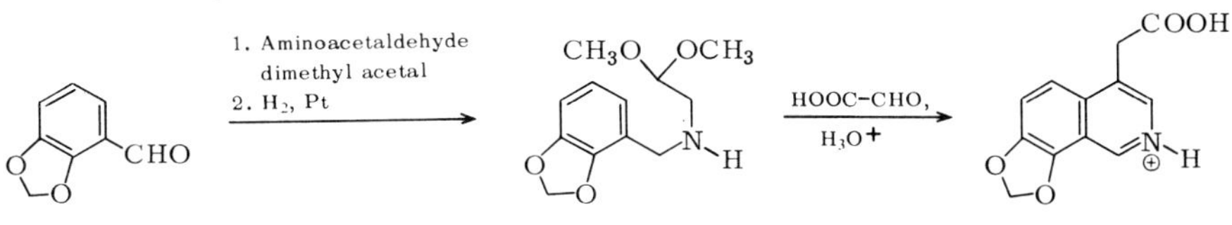

17 **18**

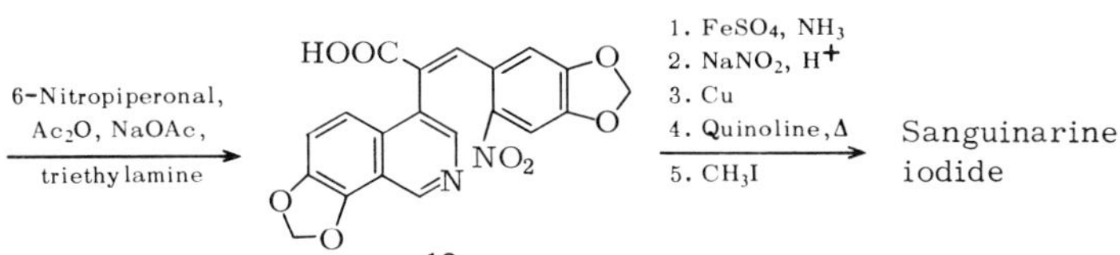

19

Scheme X

E. Cyclization through a Benzyne Intermediate

Treatment of an *o*-chloroanil with potassium amide in liquid ammonia lcads to the formation of a phenanthridine or a benzophenanthridine, as indicated below.

The reaction probably proceeds through a benzyne intermediate. This scheme has not yet been applied to the preparation of substituted benzophenanthridines.[27]

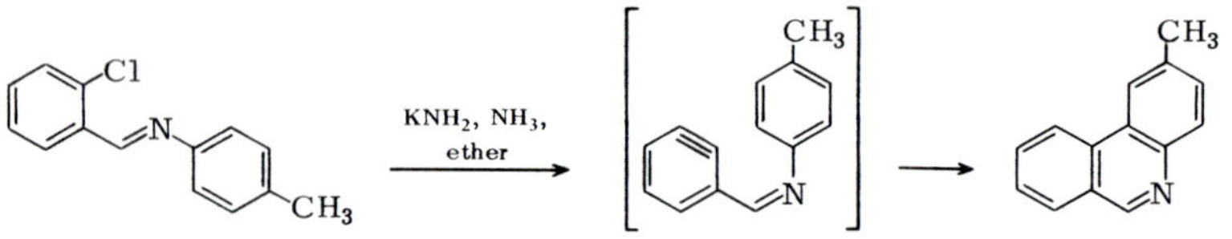

and

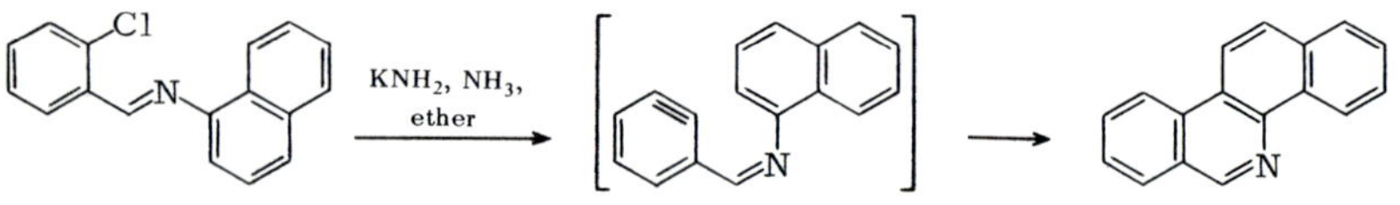

Benzo[c]phenanthridine

F. The Dyke–Sainsbury Photochemical Synthesis of the Aromatic Benzophenanthridine System

The importance of this benzophenanthridine synthesis lies in the fact that the final cyclization to the tetracyclic structure **20** was achieved photochemically rather than by a Pschorr reaction. The approach was then extended to the preparation of the analog **21**, but in this case the photocyclization step proceeded in lower yield (Scheme XI).[28]

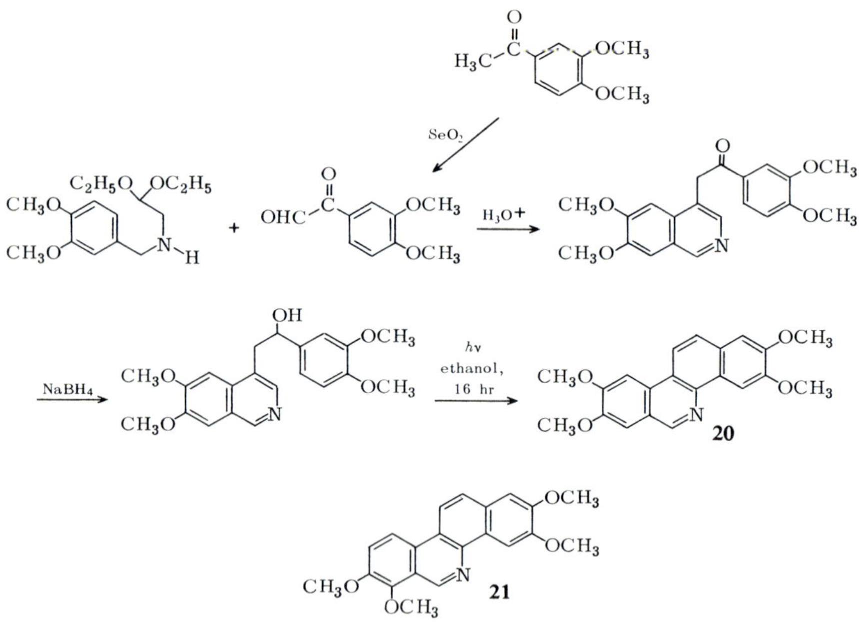

Scheme XI

(For the photochemical transformation of anhydroprotopine, a protopine derivative, into sanguinarine see Chapter 18, Section IV.)

G. A Photochemical Route to Alkaloids of the Chelidonine Type

When the benzoyl enamine **22** was irradiated, the trans-B/C-fused oxyphenanthridine **23** was obtained in 51% yield, and this product was reduced with lithium aluminum hydride to the trans-fused tertiary base **24**.[29]

Treatment of the photoproduct **23** with selenium afforded two compounds, the cis-fused *N*-methyllactam **25** and the aromatized tetracyclic lactam **26**. Reduction of **25** with lithium aluminum hydride afforded the cis amine **27**. This last product is a simple analog of chelidonine since the alkaloid is also cis-B/C-fused (Scheme XII).

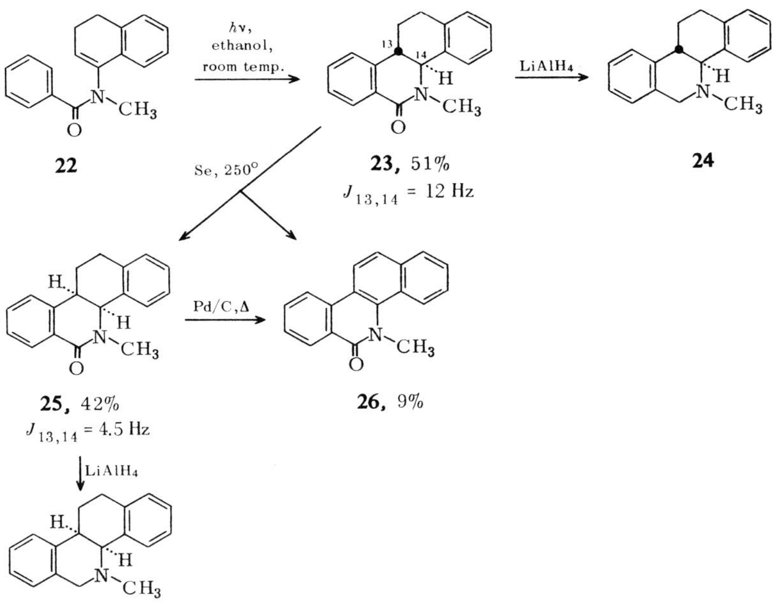

Scheme XII

H. The Sandoz Synthesis of Chelidonine

Oppolzer and Keller have evolved an interesting first synthesis of (±)-chelidonine based on the pyrolysis of a benzocyclobutene and intramolecular cycloaddition of the resulting *o*-quinodimethide (Scheme XIII). The known nitrile **28** was first converted to the anion **29** as depicted. Condensation with the benzylic bromide **30** furnished the urethane olefin **31** which through bromination and dehydrobromination provided the acetylenic urethane **32**. Pyrolysis of **32** presumably went through the *o*-quinodimethide **33** and led to the desired benzophenanthridene **34**. Hydroboration then gave in a

1 : 1 ratio the secondary alcohols **35** and **36**. Since chelidonine is cis B/C fused, the synthesis was continued using the alcohol **35**. The cis ketone **37** obtained through Jones oxidation of compound **35** was reduced with sodium borohydride in a stereospecific step which gave rise to the desired alcohol urethane as the sole product. Hydrogenolysis of the benzyloxycarbonyl group then produced (±)-norchelidonine which upon *N*-methylation furnished (±)-chelidonine, itself a natural product.[29a]

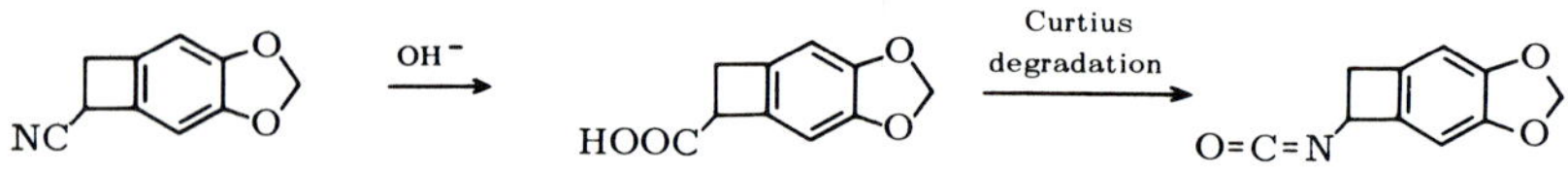

28

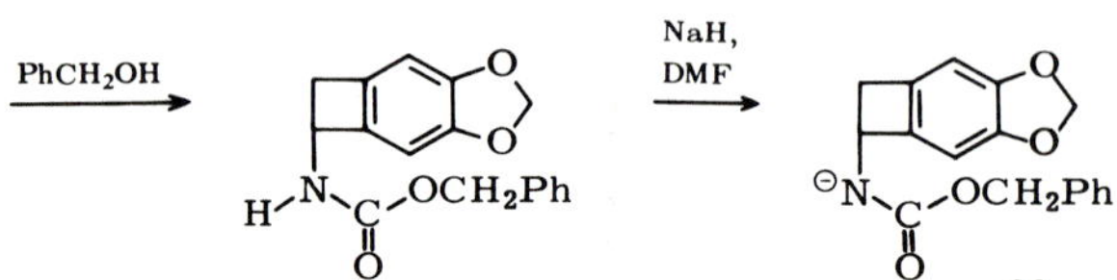

29

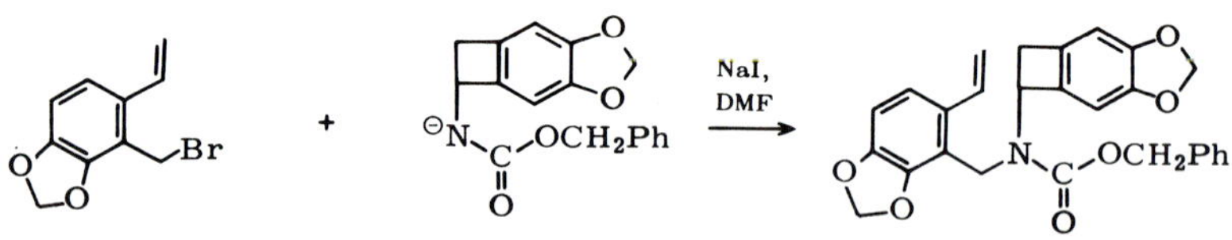

30 **29** **31**

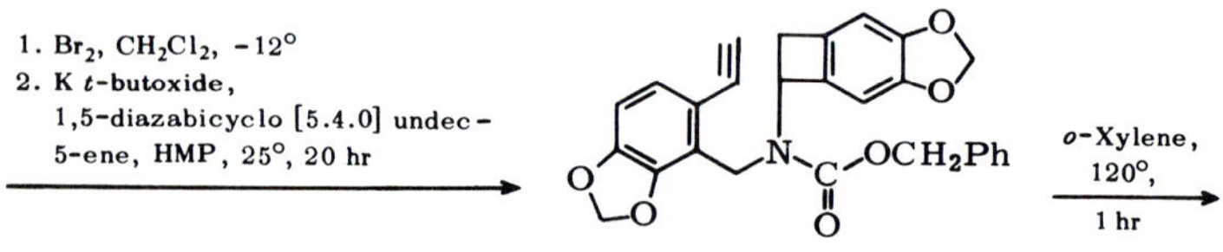

32

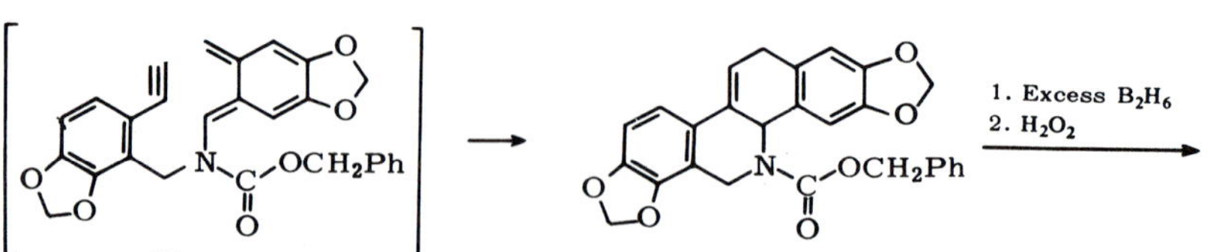

35

33 **34**

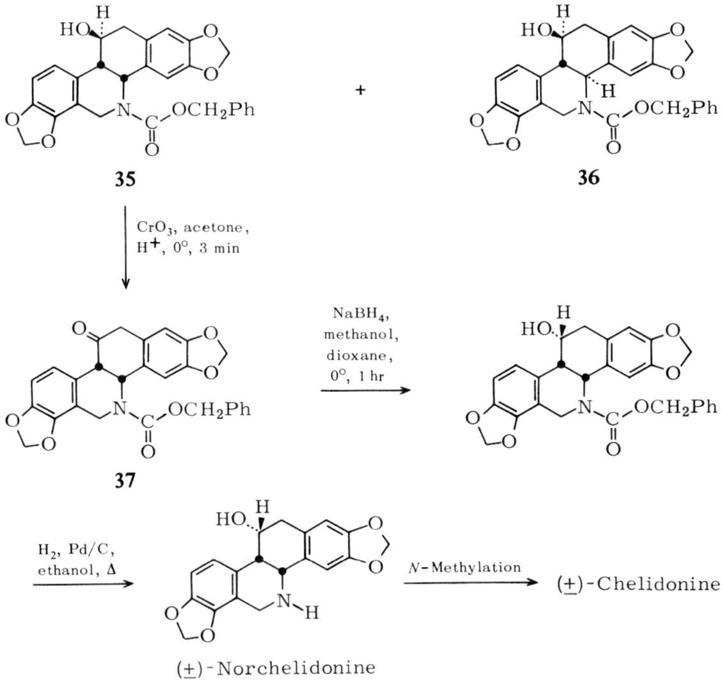

Scheme XIII

VI. Mass Spectroscopy

Since quaternary benzophenanthridine salts such as sanguinarine and chelirubine are nonvolatile, spectral results are best obtained on the corresponding pseudocyanides or acetonyl or dihydro derivatives.

The pseudocyanides show strong molecular ions which are also the base peaks. Another strong peak shown by the pseudocyanides is that arising by elimination of the cyano group.[30] If an acetonyl group is attached at C-8 in lieu of a cyano function, the base peak corresponds to the loss of $CH_3-CO-CH_2$ from the molecule, but the molecular ion can still be detected easily.[31]

In the cases of chelerythrine pseudocyanide and 8-acetonyldihydrochelerythrine, the *m*/*e* 348 ion produced can fragment by the initial loss of either CH_3 or CH_2O. Further fragmentation can be by loss of CO, CHO, or CH_2O.[31]

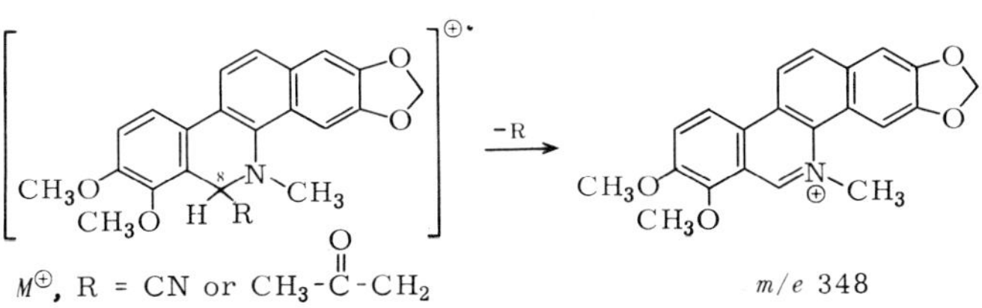

VII. NMR Spectroscopy

The NMR spectra of the aromatic benzophenanthridines can provide appreciable assistance in structural elucidation and are best obtained on the pseudocyanide, alcoholate, or dihydro derivatives.[30,31]

The spectrum of sanguinarine pseudocyanide in DMSO-d_6 shows a series of well-defined singlets and doublets. The 9,10-methylenedioxy group is split into two doublets because of the asymmetry at C-8, but the C-2,3-methylenedioxy resonance is a singlet at δ6.20. The aromatic hydrogens at C-11 and C-12, and at C-5 and C-6, each appear as doublets. The protons at C-6 and C-12 are situated relatively downfield as compared to those at C-5 and C-11, due to the peri nature of the former two. The C-1 and C-4 hydrogens show up as singlets at δ7.58 and 7.39, respectively.[30]

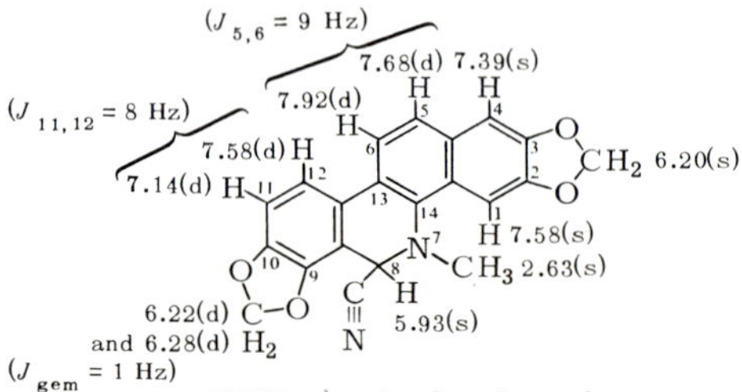

NMR spectral values for sanguinarine pseudocyanide (in DMSO-d_6)

The following assignments for dihydrosanguinarine and 8-methoxydihydrochelerythrine were confirmed by accompanying NOE experiments.[31] The solvent was deuteriochloroform.

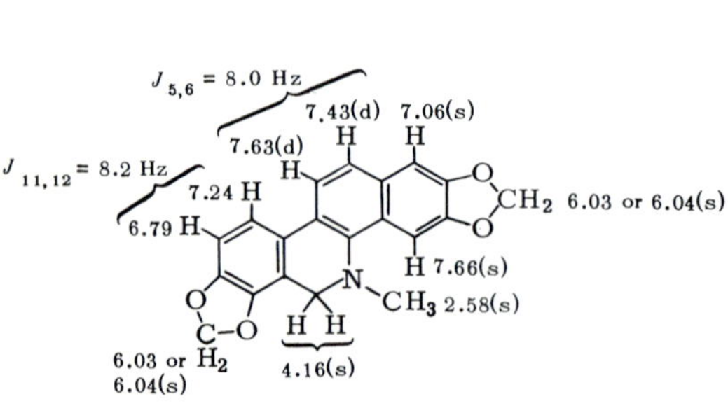

NMR spectral values for dihydrosanguinarine

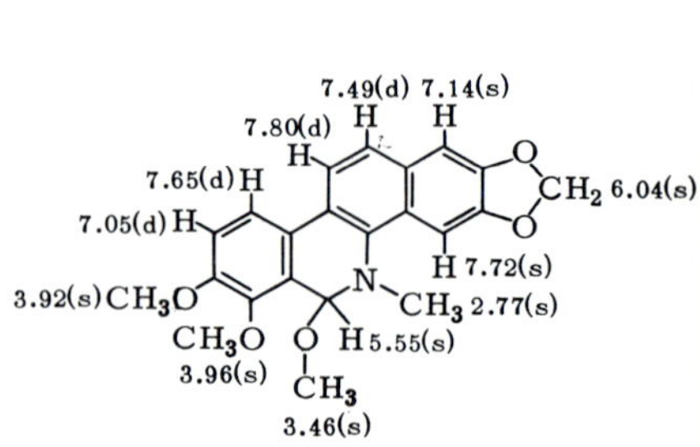

NMR spectral values for 8-methoxydihydrochelerythrine

Turning now to the alcoholic benzophenanthridines, careful 60-MHz NMR analysis of chelidonine has shown that the preferred conformation is that in which the cis-fused rings B and C both exist in half-chair conformations. Some of the chemical shifts and coupling constants are summarized below.[3,6,15,32]

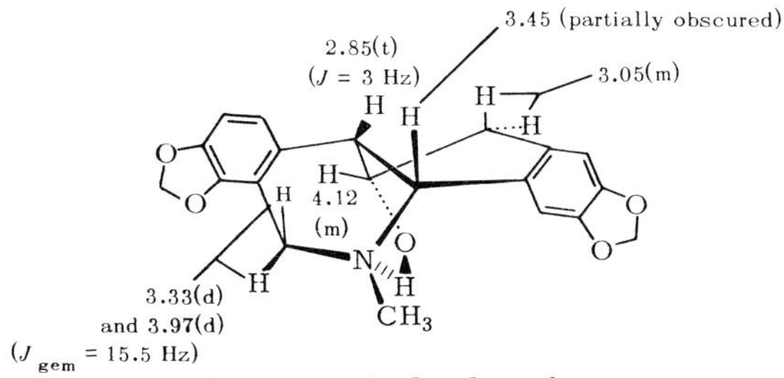

NMR spectral values for chelidonine

The NMR spectrum of chelidonine acetate indicated that the conformation of the molecule is slightly different from that of the parent alkaloid. More importantly, the small splitting of the C-14 proton, $J = 5$ Hz, is another proof for the cis B/C fusion in this series.[3,32] Double irradiation of the spectrum also provided clear-cut evidence that the acetoxy group is indeed at C-6. When the proton at $\delta 3.66$ was irradiated, the octet at $\delta 5.39$ collapsed to a quartet, and irradiation of the multiplet at $\delta 2.82$ caused the octet to collapse to an ill-defined doublet.[32]

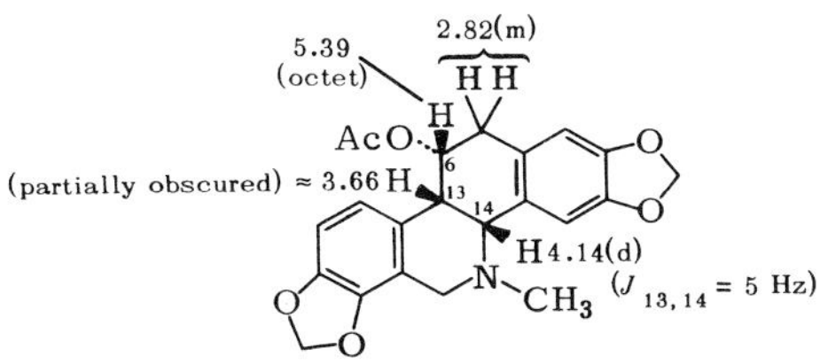

NMR spectral values for chelidonine acetate

The most salient feature in the NMR spectrum of corynoline is the *C*-methyl singlet at $\delta 1.13$.[15]

VIII. The C-6-Methoxylated Aromatic Benzophenanthridines

The five alkaloids of this group are chelirubine, chelilutine, sanguirubine, sanguilutine,[30] and macarpine.[33] These highly colored C-6-methoxylated alkaloids have been found in members of the Papaveraceae family, most notably in *Chelidonium majus* L. and *Sanguinaria canadensis* L. The structures were assigned on the basis of spectral data coupled with biogenetic analogy.

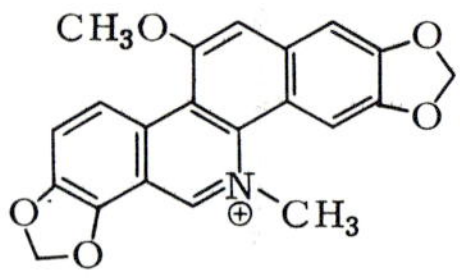

Chelirubine

Chelilutine

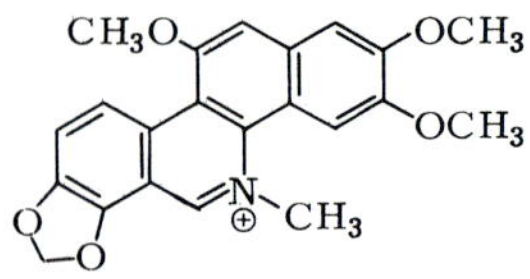

Sanguirubine

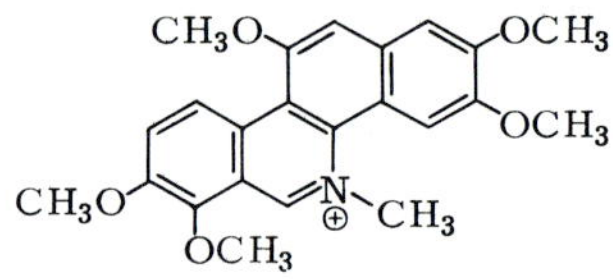

Sanguilutine

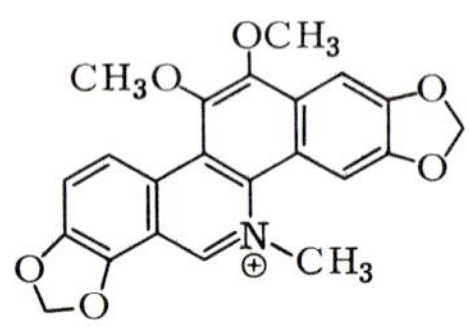

Macarpine

The UV spectra for this benzophenanthridine group resemble the corresponding spectra for sanguinarine and chelerythrine. The mass spectra of the corresponding pseudocyanides provided accurate values for the molecular weights and the elemental compositions.

The carefully recorded NMR spectra of the pseudocyanides in DMSO-d_6 were compared with that for sanguinarine pseudocyanide in the same solvent. Taking chelirubine as an example, the data indicated the presence of a methoxyl group, which was placed at C-6 by biogenetic analogy with chelidonine. The C-9,10-methylenedioxy group, being split, can be readily differentiated from the C-2,3-methylenedioxy absorption, which is a singlet.[30]

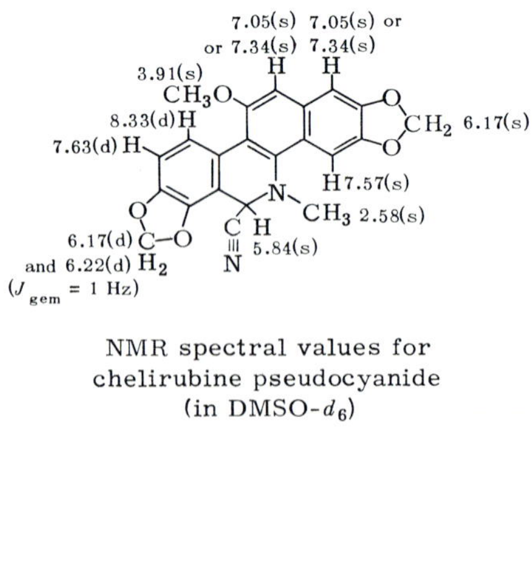

NMR spectral values for chelirubine pseudocyanide (in DMSO-d_6)

IX. Bocconine

The aromatic benzophenanthridine bocconine, $C_{21}H_{16}O_5N^+X^-$, was obtained from *Bocconia cordata* Willd. (Papaveraceae) where it was found together with sanguinarine and chelerythrine. The alkaloid possesses one *N*-methyl, one methoxyl, and two methylenedioxy groups, and reduction with sodium borohydride readily gave dihydrobocconine. Oxidation of the alkaloid with potassium ferricyanide afforded oxybocconine.[34]

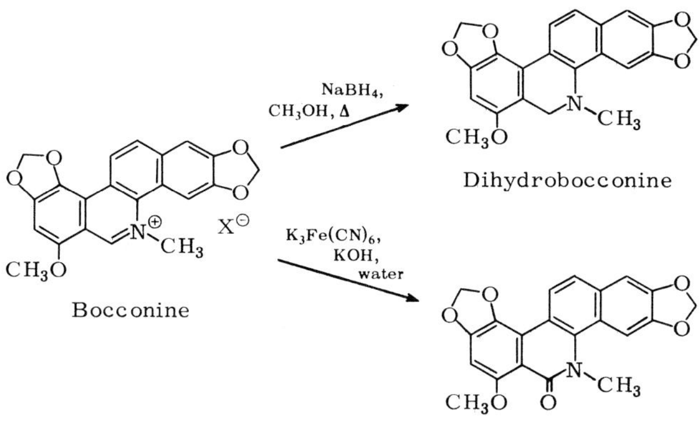

The NMR spectral values for dihydrobocconine and oxybocconine are listed below. They have been compared with those for the corresponding derivatives from sanguinarine and chelerythrine as well as with the values for tetralin and α-tetralone. The important points to note are:

(a) The downfield shift of the C-6 proton in dihydrobocconine due to the presence of an oxygenated substituent at C-12.

(b) Irradiation of the δ3.81 methoxyl absorption of dihydrobocconine produced a 47% NOE for the C-10 proton absorption at δ6.59.

(c) Irradiation of the δ4.07 C-8 methylene absorption of dihydrobocconine gave no NOE, so that a substituent must be present at C-9.[34]

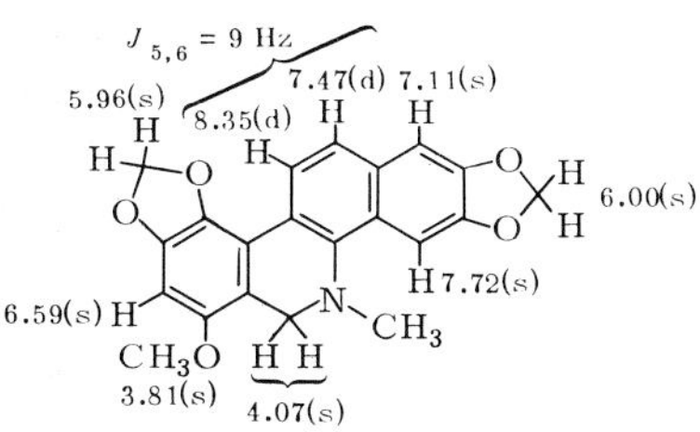

NMR spectral values for dihydrobocconine

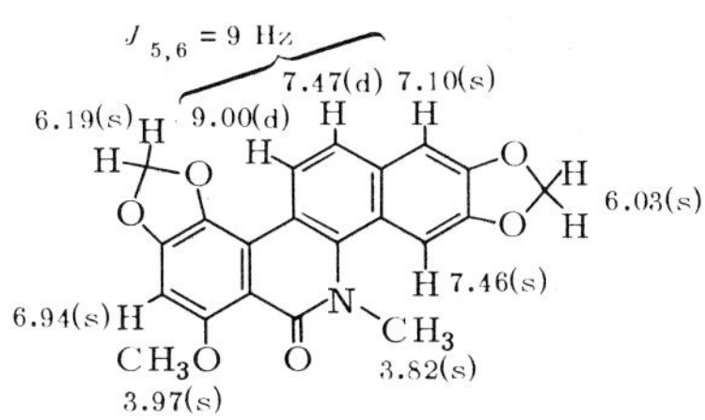

NMR spectral values for oxybocconine

It must be added that the substitution pattern in ring D of bocconine is so unusual that an X-ray analysis or a total synthesis is required before the structure assigned can be accepted as final.

X. Bocconoline and Corynolamine

Bocconia cordata Willd. (Papaveraceae) also contains the minor alkaloid bocconoline, $C_{22}H_{21}O_5N$, which has a hydroxymethyl substituent at C-8. The UV spectrum of bocconoline is superimposable on that of dihydrochelerythrine, underlining a similar substitution pattern at the benzenoid rings; and the IR spectrum shows the presence of a hydroxyl band. Acetylation of bocconoline moved downfield the absorption of two aliphatic protons, so that the alcohol function must be primary.[34a]

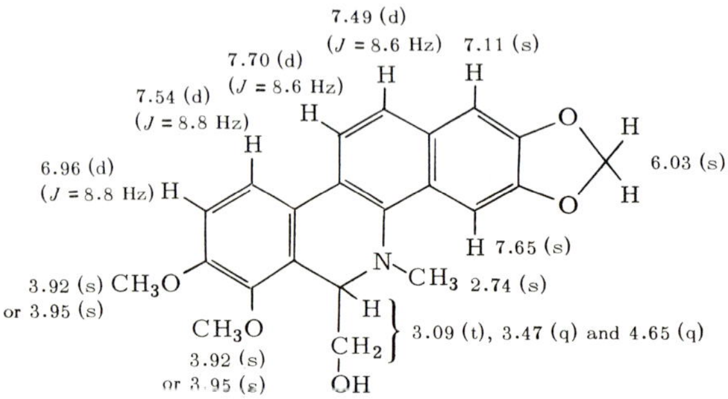

NMR spectral values for bocconoline

Corynolamine, isolated from *Corydalis incisa* Pers. (Fumariaceae) together with corynoline and corynoloxine, is structurally related to bocconoline since it too possesses a C-8 hydroxymethyl function.[34a]

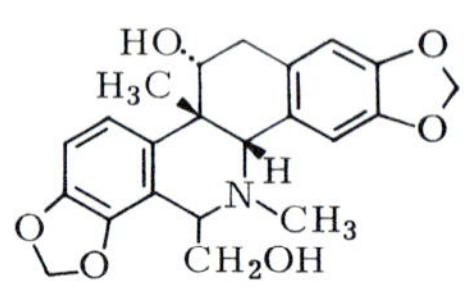

Corynolamine

XI. The Characterization of a Dimeric Benzophenanthridine Alkaloid

1,3-Bis(8-hydrochelerythrinyl)acetone, found in *Bocconia arborea* S. Wats (Papaveraceae), shows a weak molecular ion in the mass spectrum at *m/e* 752 ($C_{45}H_{40}O_9N_2$). Upon electron impact, hydrogen transfer accompanies the fragmentation process, yielding a neutral species presumably of mass 347 and an ion of *m/e* 405. This ion fragments with loss of 57 mass units to give the base peak at *m/e* 348 (Scheme XIV). Further fragmentation is along the lines indicated in Section VI.[31]

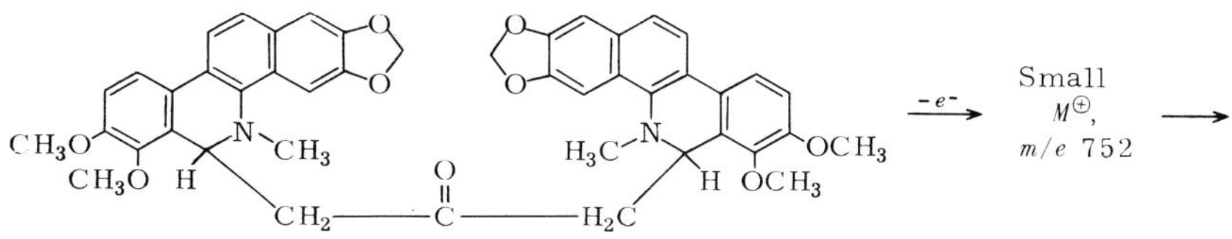

1,3-Bis(8-hydrochelerythrinyl) acetone

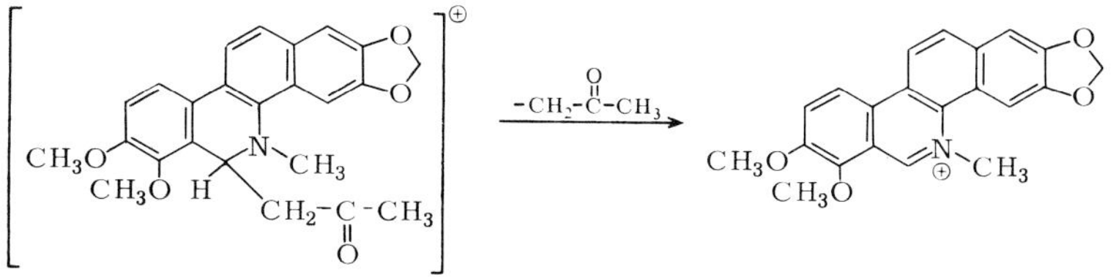

m/e 405 — Base peak, m/e 348

Scheme XIV

The NMR spectrum of the dimer resembles that for 8-acetonyldihydrochelerythrine, and indeed base-catalyzed condensation of chelerythrine with acetone dicarboxylic acid gave material identical with the natural base. It is worth noting that attempts to synthesize the dimer by mild base-catalyzed condensation of chelerythrine with 8-acetonyldihydrochelerythrine were unsuccessful, so that the dimer is probably a naturally occurring alkaloid rather than an artefact produced during the isolation process.[31]

XII. Biosynthesis

The suggestion[35] that the benzophenanthridines are formed in plants through cleavage of the 6,7 bond of protoberberines followed by joining of C-6 to C-13 has been substantiated by tracer experiments.

Feeding labeled tyrosine or dopamine to *Chelidonium majus* L. (Papaveraceae) resulted in formation of chelidonine labeled at C-6.[36]

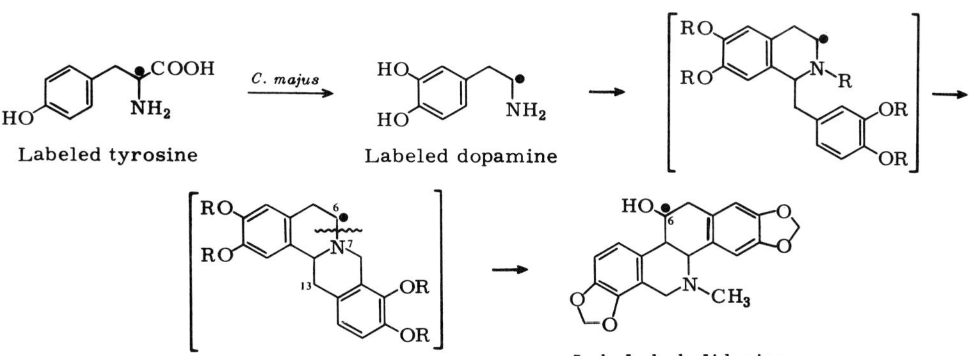

Labeled tyrosine — Labeled dopamine — Labeled chelidonine

(+)-Reticuline with multiple labels as indicated and with tritium at the asymmetric center was then fed to *C. majus*. Degradation of the radioactive (+)-chelidonine obtained showed that the labeling pattern corresponds exactly with that required to fit the benzylisoquinoline → protoberberine → benzophenanthridine hypothesis. Furthermore, the radioactive chelidonine contained no tritium, in keeping with the suggested dihydroisoquinoline intermediate **38**.[37] It was also determined that labeled (−)-scoulerine and (−)-stylopine were incorporated into (+)-chelidonine and that labeled (−)-scoulerine was a precursor for (−)-stylopine (Scheme XV).[38]

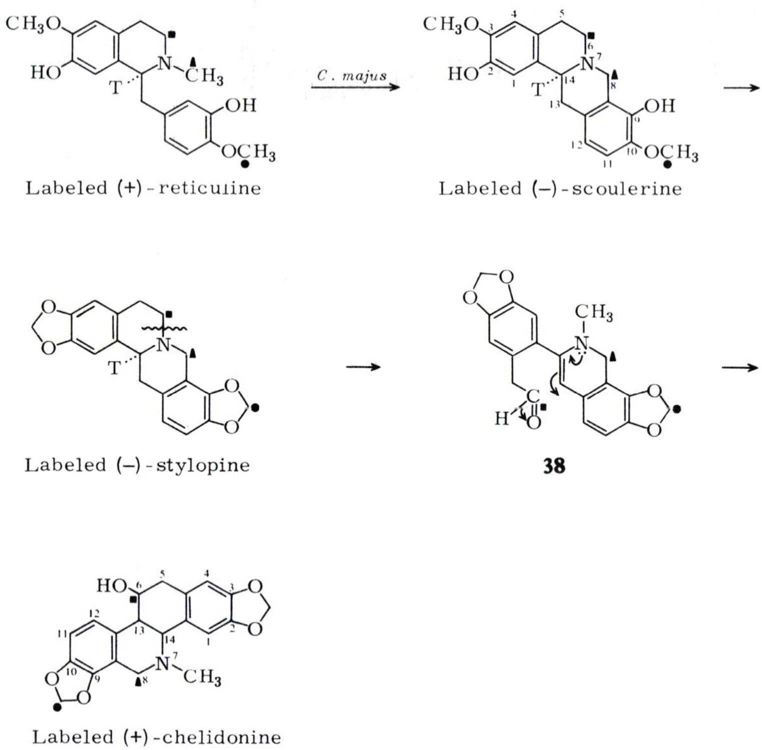

Scheme XV

Interestingly enough, it has not yet been established whether chelidonine is a precursor for sanguinarine or whether these two alkaloids are formed from stylopine or a similar tetrahydroprotoberberine by separate biogenetic pathways.

A C-13-methylated protoberberine such as tetrahydrocorysamine seems to be a probable biogenetic precursor for corynoline, but experimental results to support this statement are lacking at the present time.

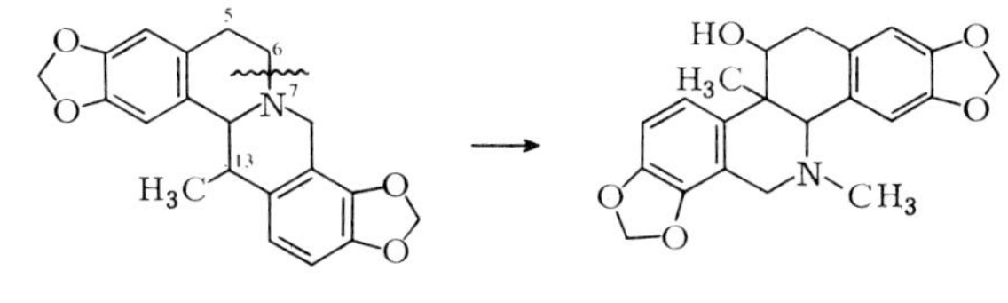

Tetrahydrocorysamine   Corynoline

It is also conceivable that a C-5-hydroxylated protoberberine may undergo fission at the C-6 to N-7 bond to lead to macarpine.

The chemically induced transformation of the tetrahydroprotoberberine salt canadine methosulfate into chelerythrine has been claimed but not substantiated.[39]

XIII. Pharmacology

8-Methoxydihydronitidine has shown some promise in the P-388 test as an anticancer drug.

Sanguinarine, chelerythrine, and bocconine possess nematocidal activity.[40]

Sanguinarine increases intraocular tension, and milk obtained from cattle grazing on plants which contain this alkaloid may be responsible for endemic primary glaucoma in man. When sanguinarine chloride was injected into rabbits, analysis of the urine revealed the presence of a green fluorescing metabolic product believed to be 3,4-benzacridine. The formation of this compound from sanguinarine is difficult to rationalize. If the fluorescing material has been correctly identified, its formation must be due to a profound metabolic alteration.[41]

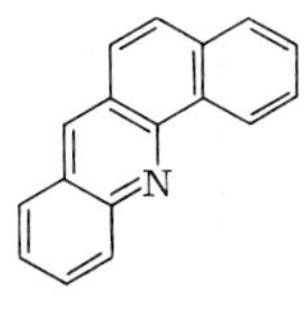

3,4-Benzacridine

Chelidonine produces moderate depression of the central nervous system and narcosis, while chelerythrine, which is less poisonous, has a depressant effect on the involuntary muscles.

XIV. UV Spectroscopy

Chelidonine[41a]	λ_{max}^{MeOH} 239 and 289 mμ (3.87 and 3.92)
Sanguinarine chloride[42]	λ_{max}^{MeOH} 234, 283, and 325 mμ (4.50, 4.52, and 4.18)
Dihydronitidine[43]	λ_{max}^{EtOH} 228, 278, and 311 mμ (4.61, 4.54, and 4.29)
Oxynitidine[43]	λ_{max}^{EtOH} 251, 277, 288, 320, 333, and 367 mμ (4.59, 4.72, 4.81, 4.20, 4.18, and 3.63)

Dihydroavicine[44]	λ_{max}^{EtOH} 232, 278, and 322 mμ (4.60, 4.50, and 4.33)
Oxyavicine[45]	λ_{max}^{EtOH} 248, 278, 289, 322, and 332 mμ (4.50, 4.70, 4.76, 4.21, and 4.19)
N-Norchelerythrine[12]	λ_{max}^{EtOH} 215, 243, 256, 277, 324, and 384 mμ (4.24, 4.58, 4.57, 4.71, 4.16, and 3.47)
Didehydrochelidonine[4]	λ_{max}^{EtOH} 235 sh and 293 mμ (3.94 and 3.78)
	λ_{min}^{EtOH} 257 mμ (3.04)
Macarpine chloride[33]	λ_{max}^{MeOH} 285 and 342 mμ (4.4 and 4.2)
	λ_{min}^{MeOH} 254 and 307 mμ (4.1 and 4.0)

The spectra of chelerythrine, dihydrochelerythrine, dihydrosanguinarine, oxysanguinarine, and oxychelerythrine have been published.[46]

The aromatic benzophenanthridine alkaloids exhibit characteristic fluorescence under UV light, and the following colors have been recorded: chelirubine, purple; sanguinarine, orange; sanguirubine, purple; macarpine, carmine red; chelilutine, orange; chelerythrine, yellow; and sanguilutine, orange.[13]

REFERENCES

1. H. K. Kim, N. R. Farnsworth, R. N. Blomster, and H. H. S. Fong, *J. Pharm. Sci.* **58**, 372 (1969).

1a. M. Tin-Wa, N. R. Farnsworth, H. H. S. Fong, and J. Trojanek, *Lloydia* **33**, 267 (1970); N. R. Farnsworth, private communication (1971).

2. F. von Bruchhausen and H. W. Bersch, *Chem. Ber.* **63**, 2520 (1930).

3. E. Seoane, *An. Real. Soc. Espan. Fis. Quim., Ser. B* **61**, 755 (1965).

4. M. H. Benn and R. E. Mitchell, *Can. J. Chem.* **47**, 3701 (1969).

5. H.-W. Bersch, *Arch. Pharm.* (*Weinheim*) **291**, 491 (1958).

6. F. Šantávý, M. Horák, M. Maturová, and J. Brabenec, *Collect. Czech. Chem. Commun.* **25**, 1344 (1960).

7. F. Šantavy, *in* "The Alkaloids" (R. H. F. Manske, ed.), Vol. 12, p. 417. Academic Press, New York, 1970.

8. G. Snatzke, J. Hrbek, Jr., L. Hruban, A. Horeau, and F. Šantavý, *Tetrahedron* **26**, 5013 (1970).

9. E. Späth and F. Kuffner, *Chem. Ber.* **64**, 370 (1931).

10. For examples of the oxidation, reduction and pseudocyanide formation of benzophenanthridines, see P. J. Scheuer, M. Y. Chang, and C. E. Swanholm, *J. Org. Chem.* **27**, 1472 (1962).

11. A. Brossi and R. Borer, *Lloydia* **28**, 199 (1965).

12. A. S. Bailey and C. R. Worthing, *J. Chem. Soc., London* p. 4535 (1956).

13. J. Slavík and L. Slaviková, *Collect. Czech. Chem. Commun.* **25**, 1667 (1960).

14. N. Takao, *Chem. Pharm. Bull.* **11**, 1306 and 1312 (1963).

15. S. Naruto, S. Arakawa, and H. Kaneko, *Tetrahedron Lett.* p. 1705 (1968).

15a. N. Takao, *Chem. Pharm. Bull.* **19**, 247 (1971).

15b. N. Takao, H.-W. Bersch, and S. Takao, *Chem. Pharm. Bull.* **19**, 259 (1971).

16. A. S. Bailey and R. Robinson, *J. Chem. Soc., London* p. 1375 (1950).

17. A. S. Bailey, R. Robinson, and R. S. Staunton, *J. Chem. Soc., London* p. 2277 (1950).

18. T. R. Govindachari and N. Viswanathan, *Indian J. Chem.* **5**, 280 (1967).

19. T. Richardson, R. Robinson, and E. Seijo, *J. Chem. Soc., London* p. 835 (1937).

20. K. W. Gopinath, T. R. Govindachari, and N. Viswanathan, *Tetrahedron* **14**, 322 (1961).

21. H. R. Arthur and Y. L. Ng, *J. Chem. Soc., London* p. 4010 (1959).
22. K. W. Gopinath, T. R. Govindachari, P. C. Parthasarathy, and N. Viswanathan, *J. Chem. Soc., London* p. 4012 (1959).
23. S. F. Dyke, B. J. Moon, and M. Sainsbury, *Tetrahedron Lett.* p. 3933 (1968); M. Sainsbury, S. F. Dyke, and B. J. Moon, *J. Chem. Soc., C* p. 1797 (1970).
24. J. M. Bobbitt, J. M. Kiely, K. L. Khanna, and R. Ebermann, *J. Org. Chem.* **30**, 2247 (1965); J. M. Bobbitt, D. P. Winter, and J. M. Kiely, *ibid.* p. 2459.
25. S. F. Dyke, M. Sainsbury, and B. J. Moon, *Tetrahedron* **24**, 1467 (1968).
26. For related work, see W. J. Gensler, K. T. Shamasundar, and S. Marburg, *J. Org. Chem.* **33**, 2861 (1968); D. W. Brown, S. F. Dyke, M. Palfreyman, and M. Sainsbury, *Tetrahedron Lett.* p. 5615 (1968).
27. S. V. Kessar and M. Singh, *Tetrahedron Lett.* p. 1155 (1969).
28. S. F. Dyke and M. Sainsbury, *Tetrahedron* **23**, 3161 (1967).
29. I. Ninomiya, T. Naito, and T. Mori, *Tetrahedron Lett.* p. 3643 (1969).
29a. W. Oppolzer and K. Keller, *J. Amer. Chem. Soc.* **93**, 3836 (1971).
30. J. Slavik, L. Dolejš, V. Hanuš, and A. D. Cross, *Collect. Czech. Chem. Commun.* **33**, 1619 (1968).
31. D. B. MacLean, D. E. F, Gracey, J. K. Saunders, R. Rodrigo, and R. H. F. Manske, *Can. J., Chem.* **47**, 1951 (1969).
32. C.-Y. Chen and D. B. MacLean, *Can. J. Chem.* **45**, 3001 (1967).
33. J. Slavík, L. Slavíkova, and K. Haisova, *Collect. Czech. Chem. Commun.* **32**, 4420 (1967).
34. M. Onda, K. Abe, K. Yonezawa, N. Esumi, and T. Suzuki, *Chem. Pharm. Bull.* **18**, 1435 (1970).
34a. H. Ishii, K. Hosoya, and N. Takao, *Tetrahedron Lett.* p. 2429 (1971).
35. R. Robinson, "The Structural Relations of Natural Products". Oxford Univ. Press (Clarendon), London and New York, 1955; R. B. Turner and R. B. Woodward, *in* "The Alkaloids" (R. H. F. Manske and H. L. Holmes, eds.), Vol. 3, p. 54. Academic Press, New York, 1953.
36. E. Leete and Sister J. B. Murrill, *Tetrahedron Lett.* p. 147 (1964).
37. A. R. Battersby, R. J. Francis, E. A. Ruveda, and J. Staunton, *Chem. Commun.* p. 89 (1965).
38. A. R. Battersby, R. J. Francis, M. Hirst, R. Southgate, and J. Staunton, *Chem. Commun.* p. 602 (1967).
39. S. Prasad, A. K. Wahi, and S. Ghosal, *J. Indian Chem. Soc.* **46**, 371 (1969).
40. M. Onda, K. Abe, and K. Yonezawa, *Chem. Pharm. Bull.* **16**, 2005 (1968).
41. S. A. E. Hakim, V. Mijović, and J. Walker, *Nature (London)* **189**, 201 (1961).
41a. Holubek and Štrouf, Spectrum No. 52.
42. Holubek and Štrouf, Spectrum No. 241.
43. H. R. Arthur, W. H. Hui, and Y. L. Ng, *J. Chem. Soc., London* p. 1840 (1959).
44. G. E. Foster and J. MacDonald, *J. Pharm. Pharmacol.* **3**, 127 (1951).
45. H. R. Arthur, W. H. Hui, and Y. L. Ng, *J. Chem. Soc., London* p. 4007 (1959).
46. C. Tani and N. Takao, *J. Pharm. Soc. Jap.* **82**, 755 (1962). See also L. Hruban, F. Šantavý, and S. Hegerova, *Collect. Czech. Chem. Commun.* **35**, 3420 (1970).

Chapter 18 / THE PROTOPINES

Occurrence: Berberidaceae, Fumariaceae, Papaveraceae, Ranunculaceae, and Rutaceae
Number: 17
Structures:

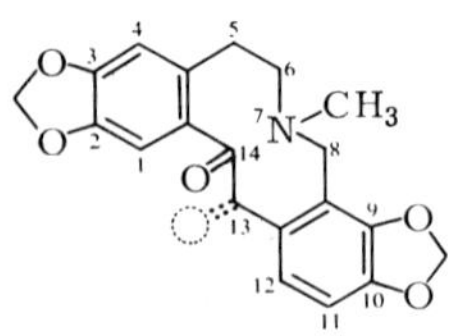

Protopine and 13-oxoprotopine[1]

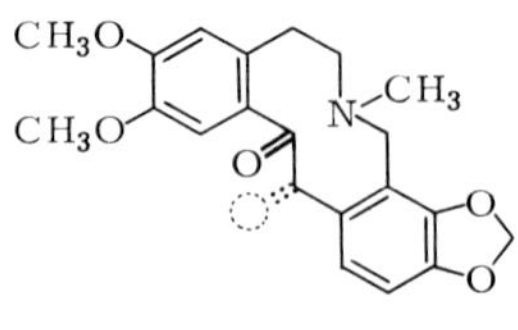

Cryptopine and 13-oxocryptopine[1]

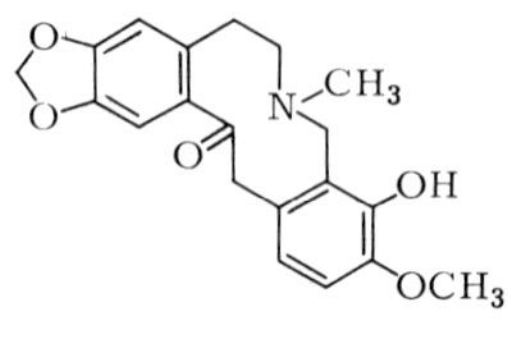

Hunnemanine

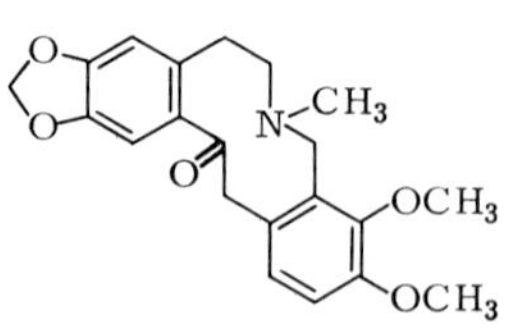

Allocryptopine

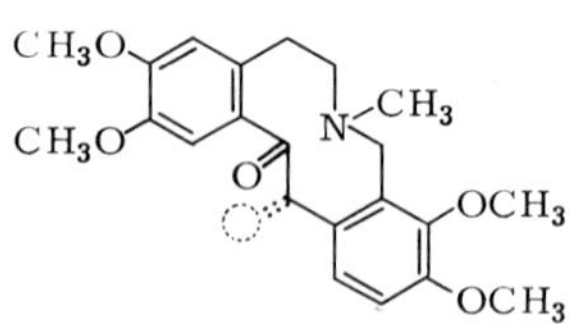

Muramine and 13-oxomuramine[1]

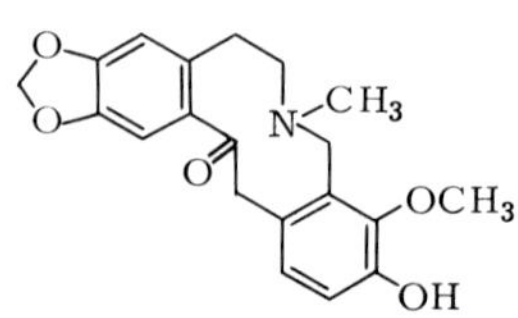

Thalictrisine[2]

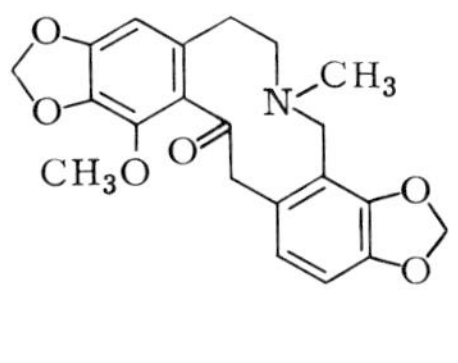

Coulteropine

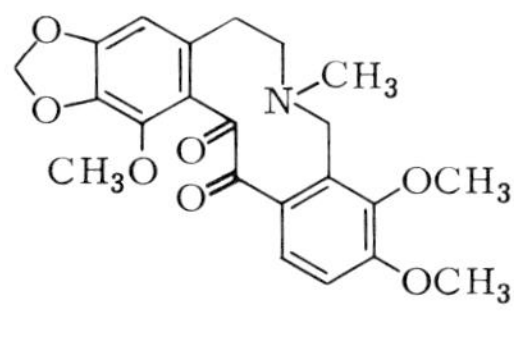

1-Methoxy-13-oxoallocryptopine

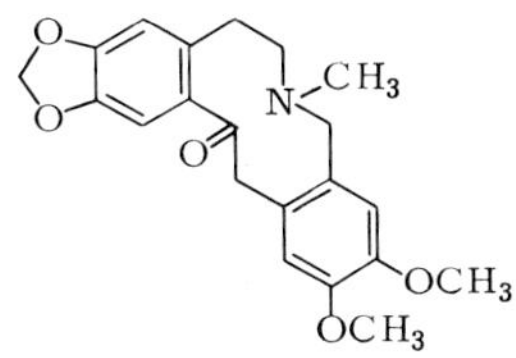

Fagarine II

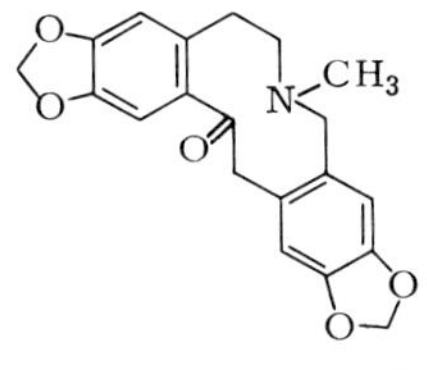

Pseudoprotopine[3]

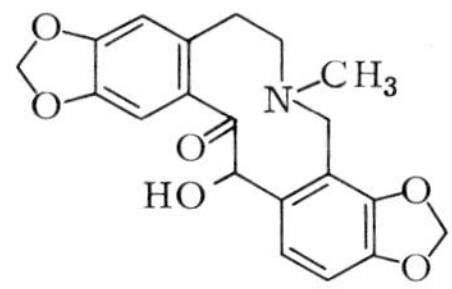

(+)-Ochrobirine*

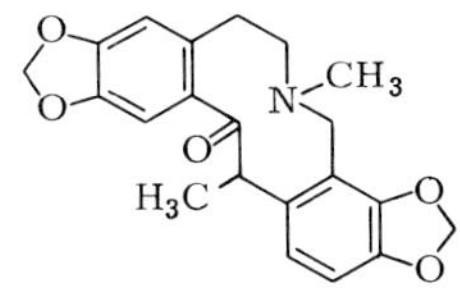

(+)-and (±)-Corycavamine

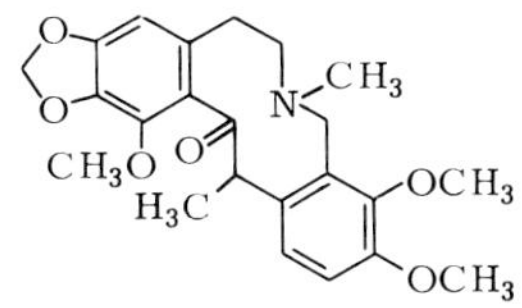

(+)-and (±)-Corycavidine

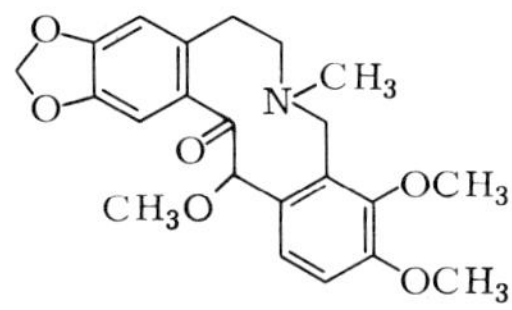

Oreophiline (?)[3a]

I. Introduction

The protopine alkaloids do not incorporate an isoquinoline system. They are characterized rather by the presence of a ten-membered ring which contains a tertiary nitrogen atom and a C-14 ketonic group. Since they are clearly derived biogenetically from tetrahydroprotoberberine precursors, they are always included among the isoquinoline alkaloids.[1]

(+)-Ochrobirine possesses a hydroxyl function at C-13, and corycavamine and corycavidine, which are known in the dextrorotatory as well as the racemic forms, have C-13 methyl groups. Four 13-oxoprotopine type alkaloids are known, and these are shown in the diagrams above.

Protopine and the aporphine magnoflorine are two of the most widely distributed isoquinoline alkaloids.

II. Structural Elucidation of Cryptopine and Protopine

The structural elucidation of cryptopine and protopine, the first protopine-type alkaloids to be chemically investigated, is due almost entirely to W. H. Perkin, Jr., who in two monumental papers published more than 50 years ago described the chemistry of these two alkaloids.[4]

* There is also a spirobenzylisoquinoline alkaloid of the same name.

Cryptopine occurs in opium and was first isolated in 1867. The base analyzes for $C_{21}H_{23}O_5N$ and is optically inactive. When oxidized with permanganate, the alkaloid yielded *m*-hemipinic acid, indicating the presence of two methoxyl groups in the molecule.

Perkin investigated several degradative routes for cryptopine. The one which is probably the most significant involves initial treatment of the alkaloid with dimethyl sulfate. The methosulfate salt upon reduction with sodium amalgam underwent an Emde degradation together with reduction of the carbonyl group to yield the amino alcohol **1**. Alcohol **1** could be dehydrated with acetyl chloride, and permanganate oxidation of the resulting stilbene yielded aldehydes **2** and **3**, which were further oxidized to acids **4** and **5** (Scheme I).

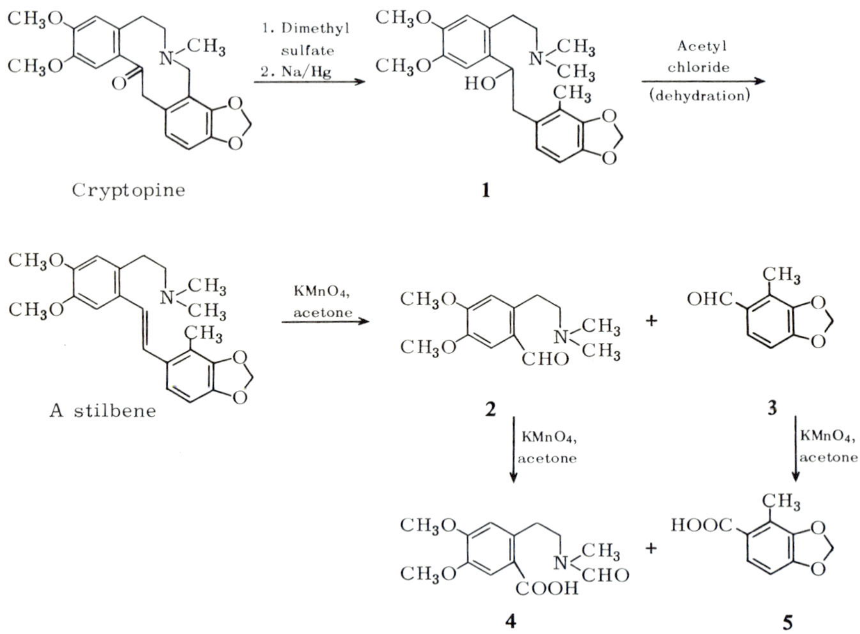

Scheme I

The degradative sequence of Scheme I is so explicit in the data which it yields that it later became the standard chemical method for the structural elucidation of the other protopine alkaloids. Its one weakness, however, is that it did not settle with certainty the position of the ketonic function in cryptopine, since the stilbene could have been formed regardless of whether the ketone was originally at C-13 or C-14.

However, when cryptopine was reduced directly with sodium amalgam, the dihydrocryptopine so obtained could be converted to a pair of diastereoisomeric tetrahydroprotoberberine *N*-metho salts, **6** and **7**, through treatment with acetyl chloride. The ketonic function in cryptopine therefore had to be located at C-14 (Scheme II).

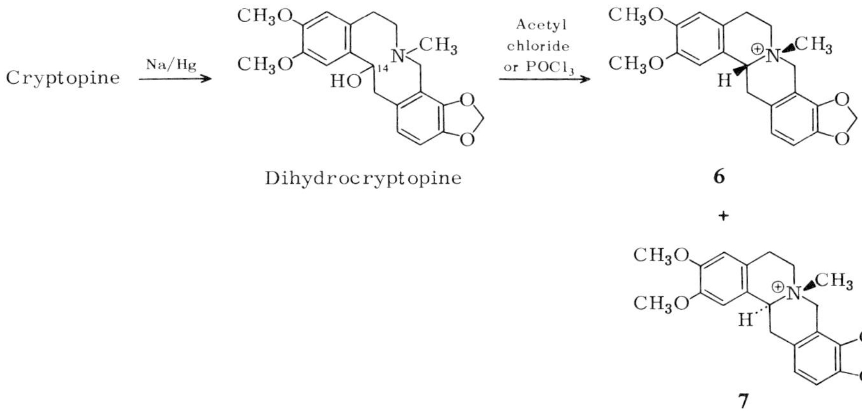

Scheme II

One of the more interesting aspects of cryptopine chemistry is that the alkaloid does not show ketonic properties. The carbonyl absorption in the free base appears at 5.97 μ (1675 cm^{-1}), indicating the presence of a transannular ground-state interaction between the carbonyl and the basic nitrogen, so that the carbonyl group is appreciably of the amide type: $O{=}C \leftarrow :N{-}CH_3$.[5] In cryptopine methiodide, however, the carbonyl absorption is at 5.93 μ (1686 cm^{-1}), and no amide interaction is present.[5] The carbonyl band is altogether absent in cryptopine perchlorate, and this salt must possess the polycyclic structure indicated.[5-7]

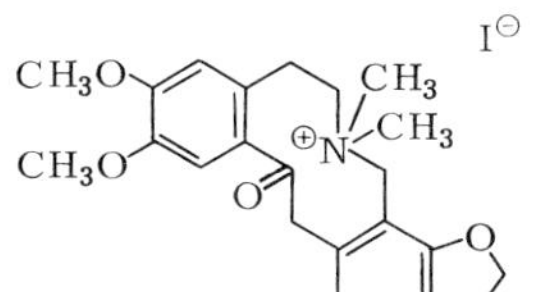

Cryptopine methiodide

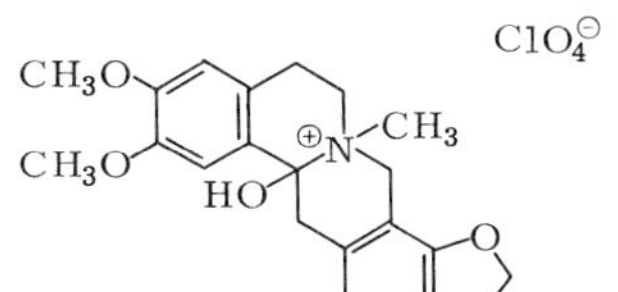

Cryptopine perchlorate

The chemistry of protopine closely parallels that of cryptopine, and some of the same degradative procedures were used for its structural elucidation.

III. Formation of Protoberberines from Protopines

The transformation of dihydrocryptopine to the tetracyclic salts **6** and **7** represents one of the early conversions of a protopine system to a protoberberine.

Another example of such a transformation was again obtained by Perkin.[8] Treatment of cryptopine with phosphorous oxychloride gave isocryptopine chloride, which is a dihydroprotoberberine salt; *N*-demethylation to the enamine **8** could be achieved by heating.

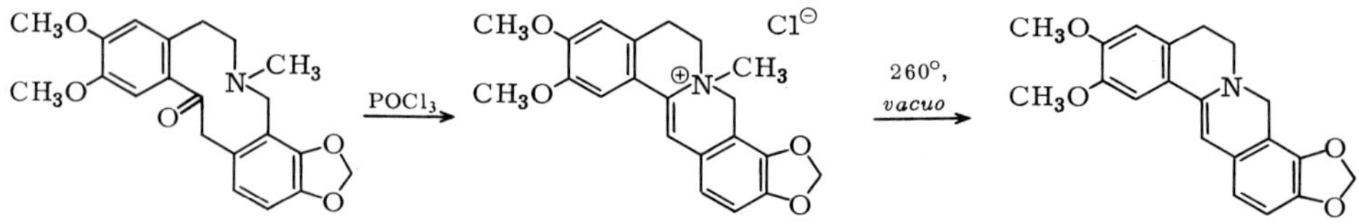

Cryptopine Isocryptopine chloride Dihydroepiberberine (**8**)

Several years later, compound **8** was reduced to yield (±)-sinactine, an alkaloid which is found in nature both in the racemic and the levorotatory forms.[9,10]

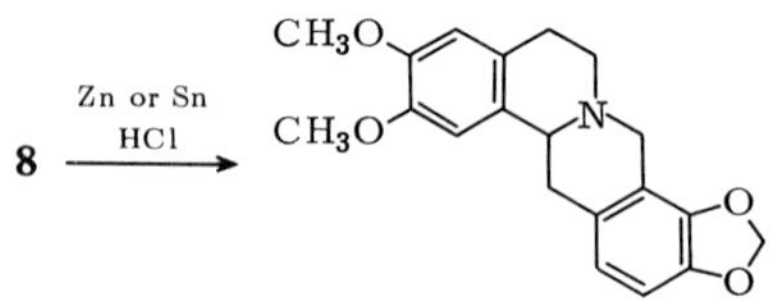

(±)-Sinactine

In the characterization of fagarine II, the alkaloid was first reduced with sodium amalgam to the corresponding alcohol. When the alcohol hydrochloride was heated on a water bath at atmospheric pressure, dehydration and cyclization occurred to give the quaternary tetrahydroprotoberberine **9**, isolated as the iodide salt. Heating this product *in vacuo* provided the known base tetrahydropseudoberberine (Scheme III).[11]

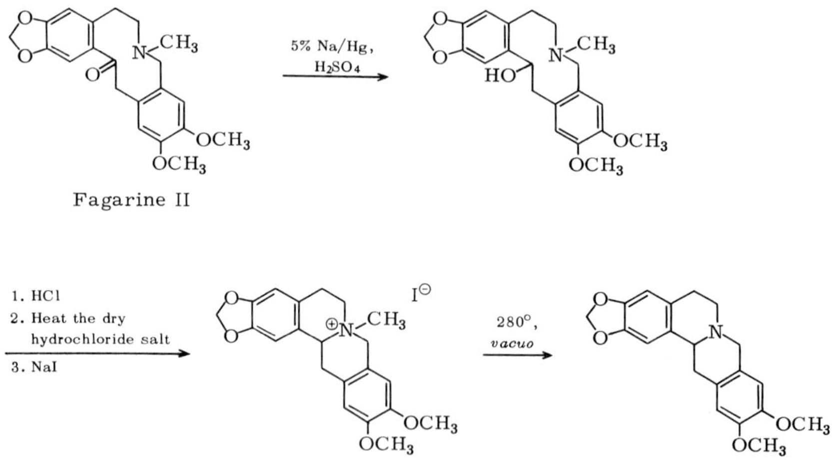

Fagarine II

9

(±)-Tetrahydropseudo-berberine

Scheme III

Transannular reaction can also be brought about by photochemical means. Irradiation of a solution of cryptopine in ethanol produced epiberberine in 34% yield. The process was even more efficient when chloroform was used as solvent.[12]

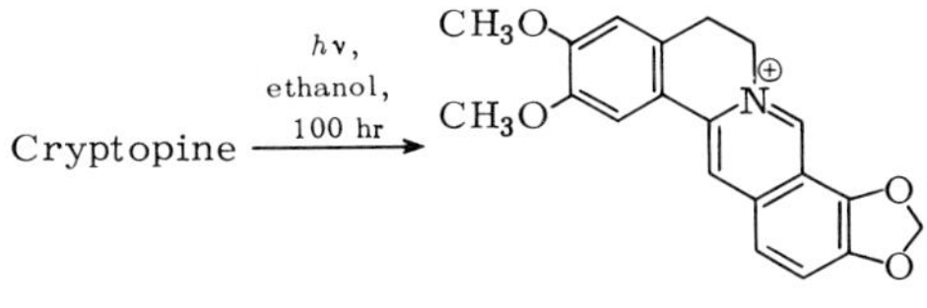

IV. The Chemistry of Anhydrocryptopine and the Conversion of Protopine to Sanguinarine

In his classical work on cryptopine, Perkin found that isocryptopine chloride when treated with hot alkali produced the olefinic base anhydrocryptopine. He than proceeded

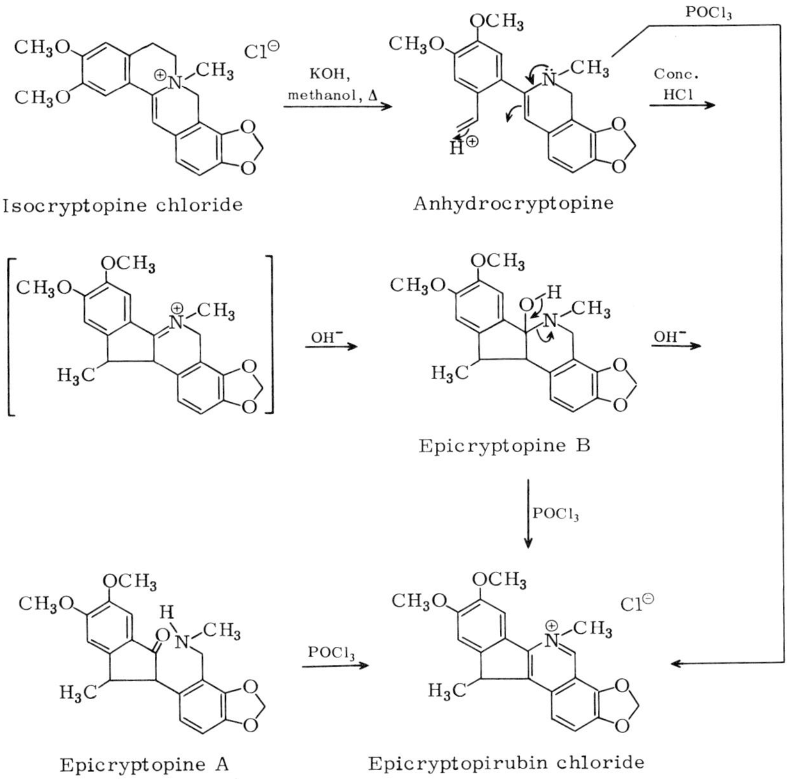

Scheme IV

to investigate the products of the reaction of anhydrocryptopine with both concentrated and dilute hydrochloric acid. Lacking the modern physical tools of spectroscopy, erroneous structures were assigned to these products. A reconsideration of the chemistry of anhydrocryptopine has, therefore, been undertaken during the past few years. Only the corrected structures will be indicated here.

With concentrated hydrochloric acid and after proper work-up, anhydrocryptopine yields epicryptopine B and epicryptopine A. Either of these compounds upon treatment with phosphorus oxychloride readily affords a red salt, epicryptopirubin chloride. Alternatively, epicryptopirubin chloride can be obtained directly from anhydrocryptopine by treatment with phosphorus oxychloride (Scheme IV).[13]

If the methosulfate salt of anhydrocryptopine is first formed and then treated with concentrated hydrochloric acid, pseudocryptopine chloride is obtained[13a]:

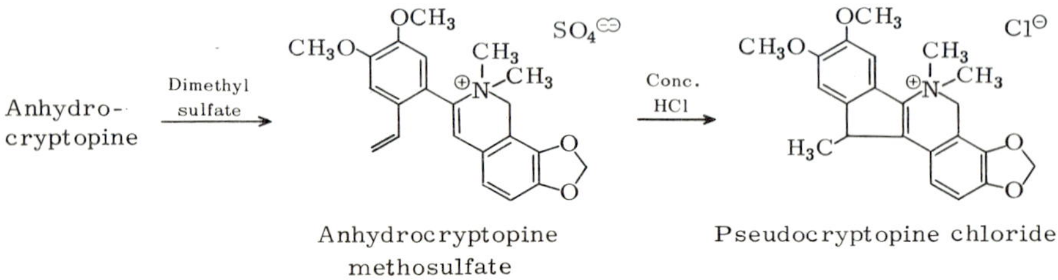

Treatment of anhydrocryptopine with dilute hydrochloric acid takes a different course. Two products are initially obtained, namely, the diastereoisomeric mixture of polycyclic alcohols **10** which readily undergoes dehydration to the olefin **11** (Scheme V).[13-15]

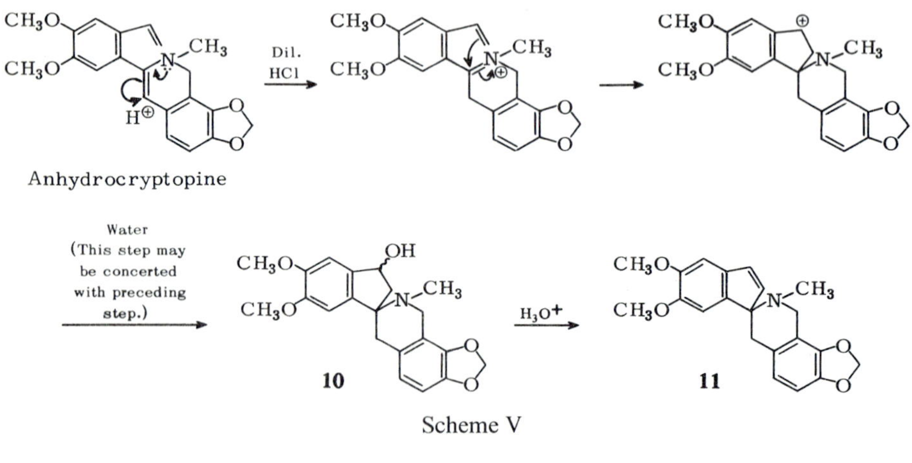

Scheme V

The most interesting reaction of anhydroprotopine, the analog of anhydrocryptopine, involves its transformation into the benzophenanthridine alkaloid sanguinarine. Onda and co-workers found that irradiation of anhydroprotopine dissolved in benzene produced an unstable material which when immediately reduced afforded dihydrosanguinarine. Oxidation with dichlorodicyanoquinone then furnished sanguinarine (Scheme VI).[16]

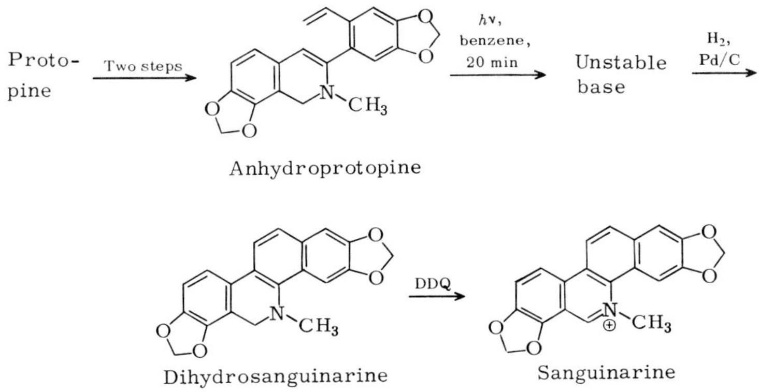

Scheme VI

V. The Conversion of Tetrahydroprotoberberines into Protopine Bases

Haworth and Perkin published in 1926 the first laboratory preparation of cryptopine and protopine. In the cryptopine synthesis, which closely parallels that for protopine, the key intermediate was the polycyclic **12**, which could be obtained from the Hofmann degradation of tetrahydroepiberberine. The by-product was the methine base **13**, which was separated from **12** by fractional crystallization. Reaction of base **12** with perbenzoic acid to give the *N*-oxide **14**, followed by acid-induced rearrangement, led to cryptopine (Scheme VII).[17]

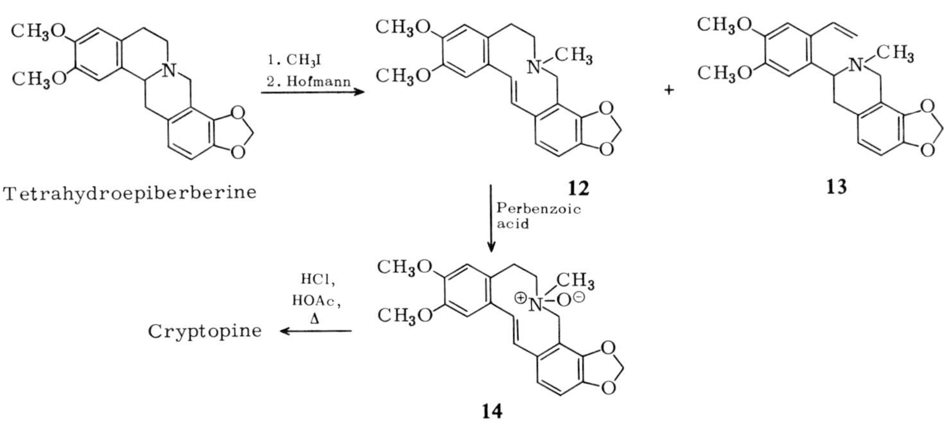

Scheme VII

The mechanism of the unusual transformation of the *N*-oxide **14** to cryptopine may be rationalized on the basis of an acid-catalyzed transannular interaction between the oxygen of the *N*-oxide and the olefinic linkage.[18]

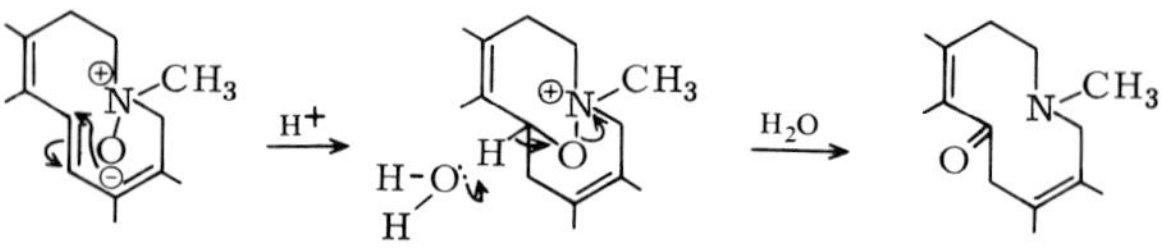

14 Cryptopine

Bentley and Murray have presented an alternate synthesis of cryptopine in which the *N*-oxide of tetrahydroepiberberine was treated with potassium chromate to afford the carbinolamine **15**. Quaternization with methyl iodide then supplied cryptopine hydroiodide.[19]

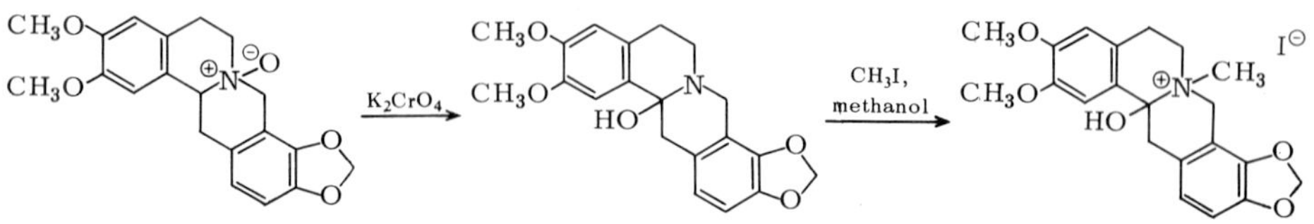

Tetrahydroepiberberine *N*-oxide **15** Cryptopine hydroiodide

VI. Mass Spectroscopy

The protopine alkaloids exhibit characteristic mass spectra due to the presence of the ten-membered heterocyclic ring. The base peak in the mass spectrum of the non-phenolic species cryptopine is due to ion **16**, whose formation can be visualized as a fragmentation of the molecular ion by a cyclic bond shift. The positive charge remains almost exclusively at the quinonoidal fragment, to the detriment of the accompanying lactam moiety.[20]

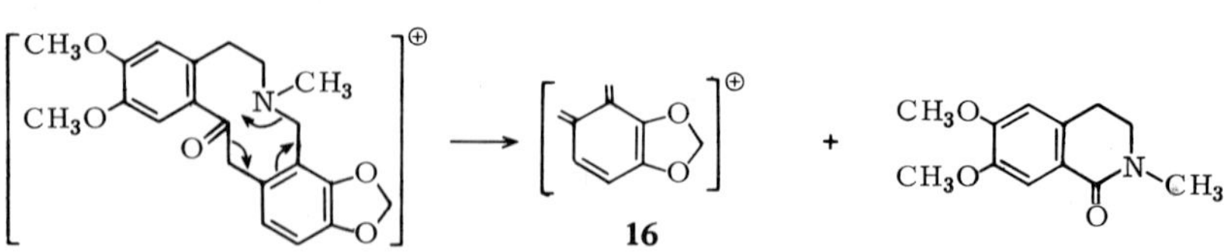

Cryptopine $M^{\oplus}$, *m/e* 369 **16** *m/e* 148

In the case of the phenolic alkaloid hunnemanine, in which a proton source is available from within the molecule, the base peak is at *m/e* 206 and is assigned to the protonated form of the lactam fragment. The second most intense peak is that due to the quinonoidal ion **17**.[20]

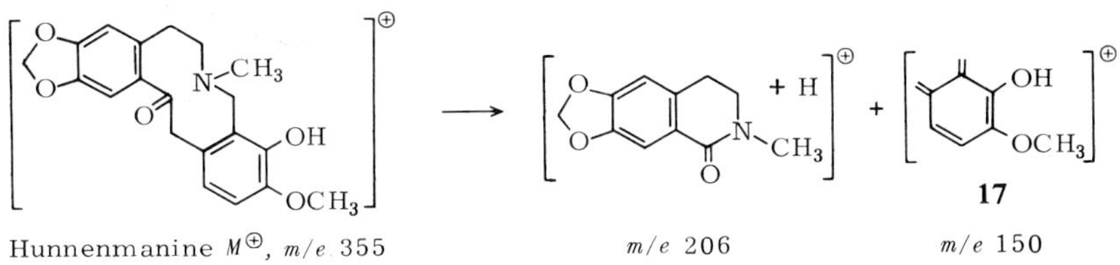

VII. 13-Oxomuramine, a Protopine Base Oxygenated at C-13

13-Oxomuramine, a light yellow alkaloid obtained from *Papaver nudicaule* var. *croceum* (Papaveraceae), is the first protopine base found to be oxygenated at C-13.[21,22] It analyzes for $C_{22}H_{25}O_6N$ and possesses one *N*-methyl group. Four of the oxygen atoms are incorporated as methoxyls and the remaining two must be parts of carbonyls since the alkaloid exhibits IR bands between 5.97 and 6.02 μ (1675 and 1660 cm^{-1}).

Mercuric acetate oxidation[23] of the accompanying alkaloid muramine led to 13-oxomuramine, identical with the natural product. Additionally, a mass spectral analysis of 13-oxomuramine showed a base peak at *m/e* 178. The next most prevalent ion was at *m/e* 150 due to the loss of carbon monoxide from the base ion (Scheme VIII).[24]

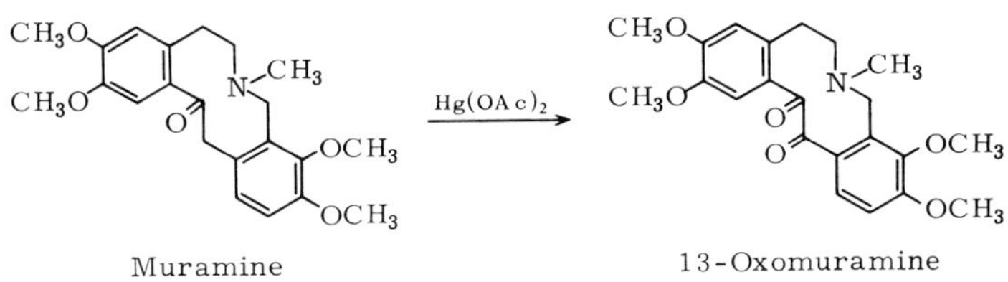

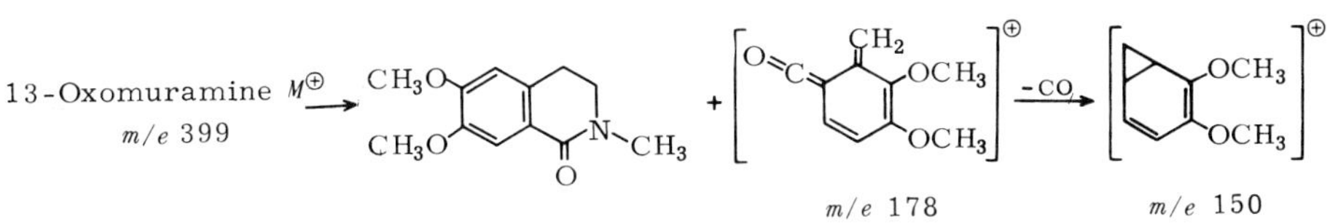

Scheme VIII

13-Oxoprotopine alkaloids are generally less basic than their simple protopine parents.

VIII. Fumaridine and Fumaramine, Two New Protopine Alkaloids?

Yunusov and co-workers have recently isolated two new polycyclic bases from *Fumaria parviflora* and *F. vaillantii* (Fumariaceae), fumaridine and fumaramine, to which the structures depicted here were assigned. Further investigations are warranted on these two compounds.[24a]

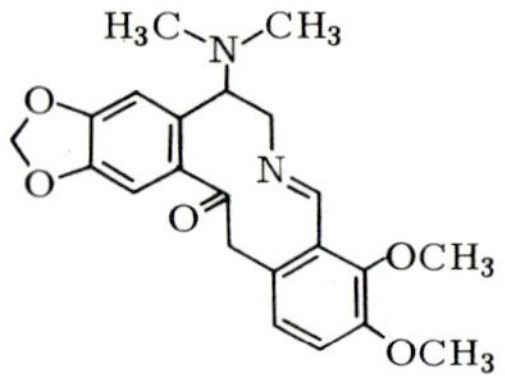

Fumaridine (?)

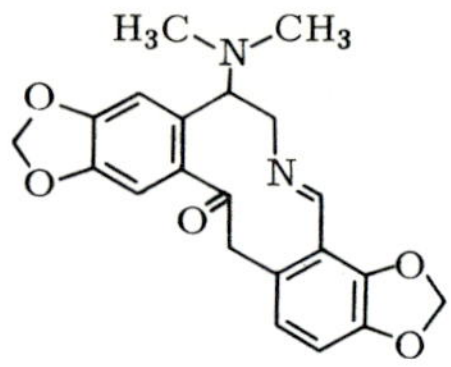

Fumaramine (?)

IX. Biosynthesis

Protopines are formed in nature by oxidation of protoberberines. Labeled (+)-reticuline hydrochloride when fed to *Dicentra spectabilis* Lem. (Fumariaceae) gave rise to labeled protopine.[25]

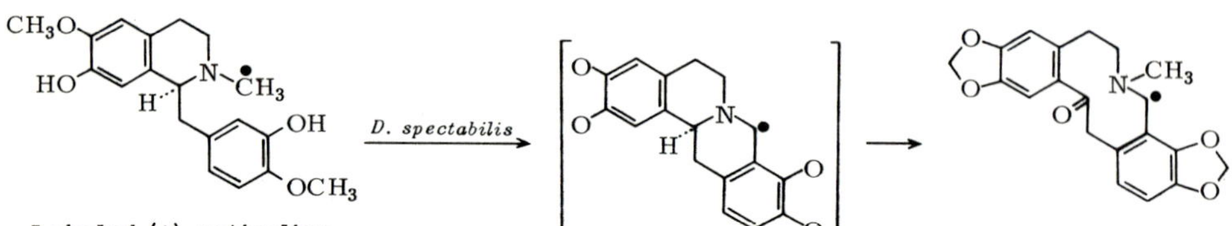

Labeled (+)-reticuline

Labeled protopine

The same pattern emerged from a study with *Chelidonium majus* L. (Papaveraceae). (+)-Reticuline tritiated at C-1 and labeled with carbon-14 was significantly incorporated into protopine as well as into the benzophenanthridine chelidonine and the protoberberine (−)-stylopine. The incorporation of (−)-reticuline on the other hand was very sluggish.[26]

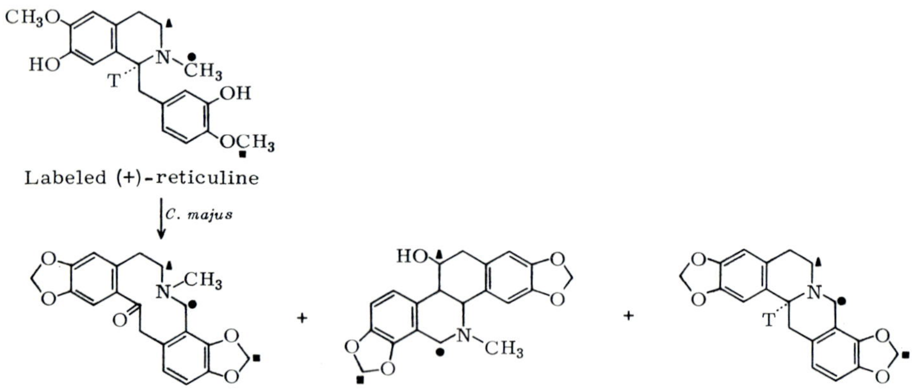

Labeled (+)-reticuline

Labeled protopine

Labeled chelidonine

Labeled (−)-stylopine (some of the tritium label lost)

The labeled tetrahydrobenzylisoquinoline alkaloidal salt (+)-tembetarine chloride, which corresponds to the methochloride salt of (+)-reticuline, when fed to *D. spectabilis* was incorporated into protopine only to a very small extent. The *N*-metho salts of reticuline are, therefore, poor precursors for protopine.[27]

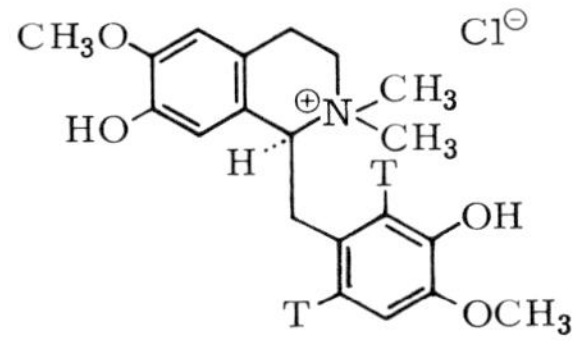

Labeled (+)-tembetarine chloride

X. Pharmacology

Protopine-type alkaloids slow the heart beat and have antifibrillatory properties. They also increase the coronary artery flow. Protopine and cryptopine stimulate the uterus, but the effect is of brief duration.[28]

XI. NMR Spectroscopy

The NMR spectrum of protopine at room temperature exhibits the chemical shifts indicated below.[29] The bands for the methylene protons in the ten-membered ring were somewhat broader than expected, due in part to unresolved long-range coupling as well as to incomplete averaging caused by ring inversion.

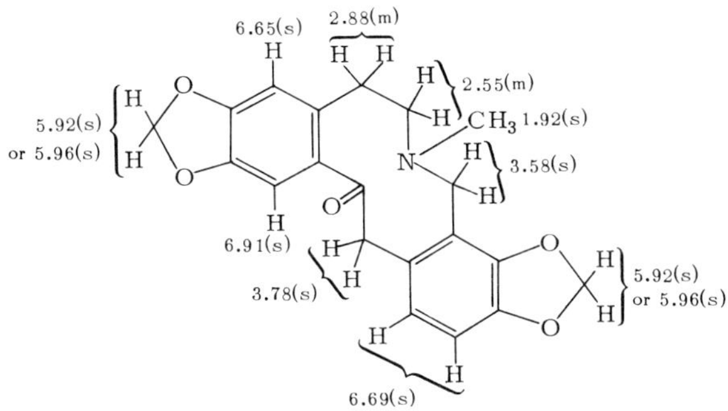

NMR spectral values for protopine at room temperature

When the spectrum was run at $-45°$, a sharp but complex spectrum appeared. The *N*-methyl, methylenedioxy, and aromatic bands were unaffected, but the methylene singlet peaks which were originally situated at $\delta 3.58$ and 3.78 as broad singlets now appeared as AB quartets partly superimposed on each other. This phenomenon is due to the slowing down of the ten-membered ring inversion on the NMR time scale. It can be calculated that the first-order rate constant for ring inversion at $-10°$ is about 100 sec^{-1}, and the free energy of activation for the inversion process, $\Delta F^{\ddagger}$, is about 13 kcal/mole. Ring inversion is a mechanism for racemization, so that protopine cannot be resolved at or near room temperature.[30]

The NMR spectrum of cryptopine closely resembles that of protopine and is described in the following diagram.[31]

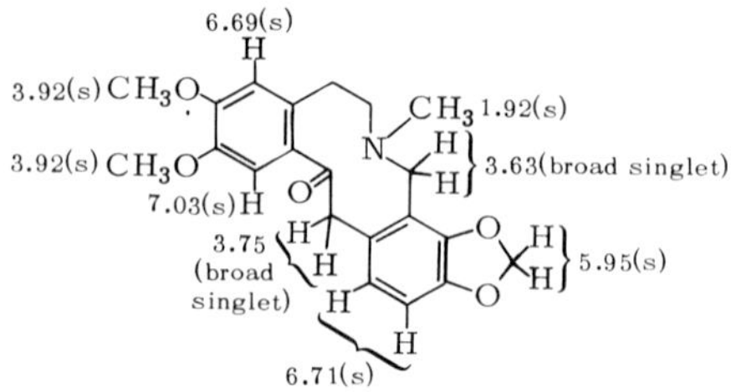

NMR spectral values for cryptopine at room temperature

An interesting series of changes was observed when the NMR spectrum of coulteropine was taken at varying concentrations of trifluoroacetic acid (TFA) in deuteriochloroform. With 5% TFA the *N*-methyl singlet, which in the free base is at $\delta 2.05$, is shifted downfield to $\delta 2.98$, and the spectrum in general fits expression **18**, in which coulteropine has simply been protonated.

When, however, coulteropine was dissolved in pure TFA and the spectrum recorded immediately, the data indicated the presence of the polycyclic salt **19**. The same spectrum was obtained when coulteropine hydrobromide was dissolved in 9% TFA in deuteriochloroform, and the spectrum rerun immediately. The *N*-methyl peak for species **19** also falls at $\delta 2.98$, but the other features of the spectrum of **19** differ from those for the tricyclic salt **18**. NMR data also showed that as time elapsed, species **19** in acid solution changed to a mixture of **18** and **19**. Coulteropine free base could be recovered unchanged after basification from either salt.[32] It should be noted that the salt **18** was observed spectrally but was not isolated in crystalline form.

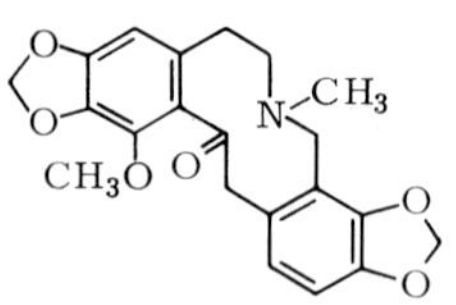

Coulteropine

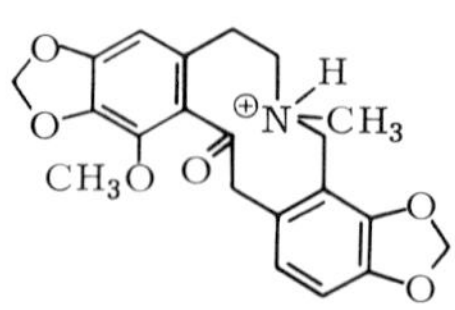

18

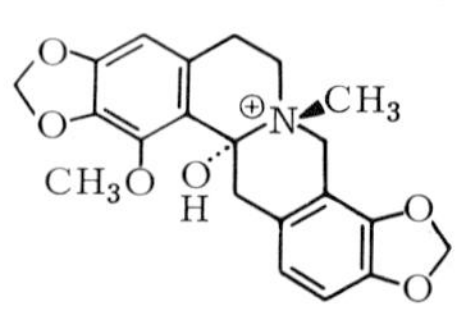

19

The NMR spectra for a number of other protopine bases have been recorded and discussed in the literature.[1,33,34]

XII. X-Ray Analysis

As previously mentioned, IR and NMR evidence had made it clear that acid salts of the protopines cannot be represented by a simple *N*-protonated structure, the absence of any carbonyl absorption indicating that closure of the ten-membered ring had occurred. For the alkaloid coulteropine, the hydrobromide salt can have either the trans structure **19** or the alternate cis B/C structure. An X-ray study of the salt showed that the trans species **19** had been formed.[32]

The stereochemistry at C-13 for the C-13 methylated or hydroxylated protopines is unknown.

XIII. UV Spectra

The protopine alkaloids show a maximum between 285 and 293 mμ and a peak or shoulder between 232 and 240 mμ which has sometimes gone unrecorded.

Protopine	λ_{max}^{EtOH} 293 mμ (3.93)[23]
	λ_{max}^{EtOH} 290 mμ (4.00)[35]
	$\lambda_{max}^{1\ N\ HCl,\ EtOH}$ 240 and 290 mμ (4.00 and 3.96)[35]
Coulteropine[36]	λ_{max}^{EtOH} 286 mμ (3.85)
	λ_{min}^{EtOH} 265 mμ
13-Oxoprotopine[23]	λ_{max}^{EtOH} 288 and 317 mμ (3.97 and 3.93)
13-Oxomuramine[21-23]	λ_{max}^{MeOH} 231, 287, and 306 sh mμ (4.35, 4.30, and 4.10)
	λ_{min}^{MeOH} 221 and 252 mμ (4.22 and 3.66)

REFERENCES

1. For a concise summary of recent advances in protopine chemistry, see F. Šantavý, *in* "The Alkaloids" (R. H. F. Manske, ed.), Vol. 12, p. 390. Academic Press, New York, 1970.
2. Kh. S. Umarov, Z. F. Ismailov, and S. Yu. Yunusov, *Khim. Prir. Soedin.* **4**, 329 (1968); *Chem. Abstr.* **70**, 88017q (1969).
3. S. R. Johns, J. A. Lamberton, H. J. Tweeddale, and R. I. Willing, *Aust. J. Chem.* **22**, 2233 (1969), and R. M. Sotelo and D. Giacopello, *Austral. J. Chem.* **25**, 385 (1972).

3a. S. Pfeifer and I. Mann, *Abh. Deut. Akad. Wiss. Berlin, Kl. Chem., Geol. Biol.* p. 315 (1966); *Chem. Abstr.* **67**, 32853e (1967).

4. W. H. Perkin, Jr., *J. Chem. Soc., London* **109**, 815 (1916); **115**, 713 (1919).
5. F. A. L. Anet, A. S. Bailey, and R. Robinson, *Chem. Ind. (London)* p. 944 (1953).
6. E. H. Mottus, H. Schwartz, and L. Marion, *Can. J. Chem.* **31**, 1144 (1953).
7. N. J. Leonard, T. W. Milligan, and T. L. Brown, *J. Amer. Chem. Soc.* **82**, 4075 (1960).
8. W. H. Perkin, Jr., *J. Chem. Soc., London* **113**, 492 (1918).
9. K. Goto and Z. Kitasato, *J. Chem. Soc., London* p. 1234 (1930).
10. E. Späth and E. Mosettig, *Chem. Ber.* **64**, 2048 (1931).
11. J. Comín and V. Deulofeu, *Tetrahedron* **6**, 63 (1959).

12. X. A. Dominguez, J. G. Delgado, W. P. Reeves, and P. D. Garner, *Tetrahedron Lett.* p. 2493 (1967).
13. S. F. Dyke and D. W. Brown, *Tetrahedron* **24**, 1455 (1968).
13a. S. F. Dyke and D. W. Brown, *Tetrahedron* **25**, 5375 (1969).
14. M. Onda, K. Abe, and K. Yonezawa, *Chem. Pharm. Bull.* **16**, 2005 (1968).
15. See also M. Onda, K. Yonezawa, and K. Abe, *Chem. Pharm. Bull.* **17**, 2565 (1969).
16. M. Onda, K. Yonezawa, and K. Abe, *Chem. Pharm. Bull.* **17**, 404 (1969).
17. R. D. Haworth and W. H. Perkin, Jr., *J. Chem. Soc., London* p. 1769 (1926).
18. P. B. Russell, *J. Amer. Chem. Soc.* **78**, 3115 (1956).
19. K. W. Bentley and A. W. Murray, *J. Chem. Soc., London* p. 2491 (1963).
20. L. Dolejš, V. Hanuš, and J. Slavík, *Collect. Czech. Chem. Commun.* **29**, 2479 (1964).
21. H. Flentje and W. Döpke, *Pharmazie* **21**, 321 (1966).
22. M. Maturová, D. Pávláskova, and F. Šantavý, *Planta Med.* **10**, 345 (1962).
23. N. J. Leonard and R. R. Sauers, *J. Org. Chem.* **22**, 63 (1957).
24. V. Hanuš, L. Dolejš, B. Müller, and W. Döpke, *Collect. Czech. Chem. Commun.* **32**, 1759 (1967).
24a. I. A. Israilov, M. S. Yunusov, and S. Yu. Yunusov, *Khim. Prir. Soedin.* **6**, 588 (1970); *Chem. Abstr.* **74**, 42528m (1971).
25. D. H. R. Barton, R. H. Hesse, and G. W. Kirby, *Proc. Chem. Soc., London* p. 267 (1963).
26. A. R. Battersby, R. J. Francis, E. A. Ruveda, and J. Staunton, *Chem. Commun.* p. 89 (1965).
27. D. H. R. Barton, R. B. Boar, D. A. Widdowson, V. Deulofeu, and S. M. Albonico, *J. Chem. Soc., C* p. 807 (1969).
28. E. L. McCauley and A. K. Reynolds, *in* "The Alkaloids" (R. H. F. Manske, ed.), Vol. 5, pp. 79 and 164. Academic Press, New York, 1955.
29. "Varian NMR Spectra Catalog," Vol. 1, Spectrum No. 339. Varian Assoc., Palo Alto, California, 1962.
30. F. A. L. Anet and M. A. Brown, *Tetrahedron Lett.* p. 4881 (1967).
31. M. Shamma and L. R. Bennet, unpublished results (1971).
32. F. R. Stermitz, R. M. Coomes, and D. R. Harris, *Tetrahedron Lett.* p. 3915 (1968).
33. A. D. Cross, L. Dolejš, V. Hanuš, M. Maturová, and F. Šantavý, *Collect. Czech. Chem. Commun.* **30**, 1335 (1965).
34. D. Giacopello and V. Deulofeu, *Tetrahedron Lett.* p. 2859 (1966).
35. O. E. Edwards and K. L. Handa, *Can. J. Chem.* **39**, 1801 (1961).
36. F. R. Stermitz, L. Chen, and J. I. White, *Tetrahedron* **22**, 1095 (1966).

Chapter 19 / THE PHTHALIDEISOQUINOLINES

Occurrence: Berberidaceae, Fumariaceae, Papaveraceae, and Ranunculaceae
Number: 11
Structures:

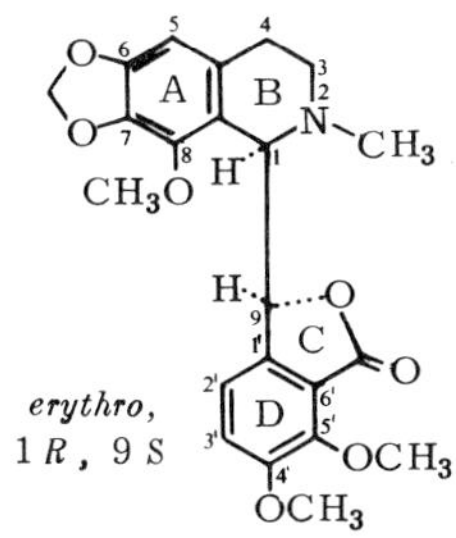

(−)-α-Narcotine
(racemate, called (±)-gnoscopine, also naturally occurring)

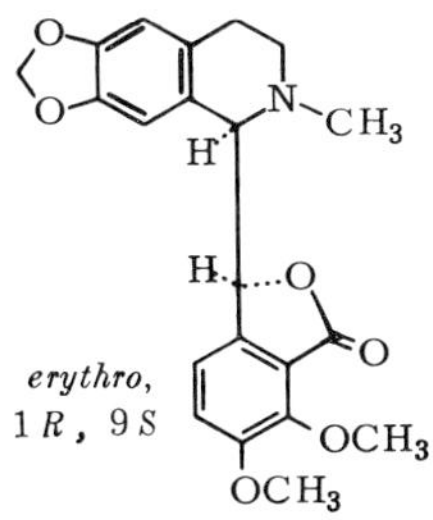

(−)-β-Hydrastine

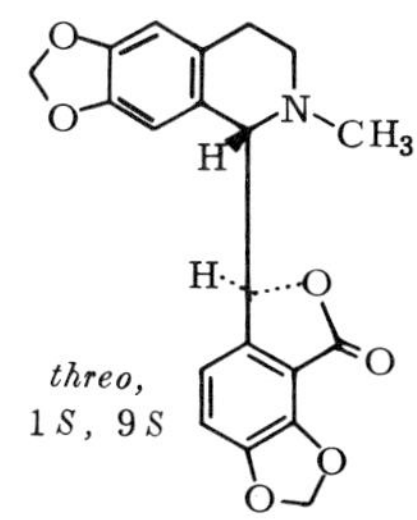

(+)-Adlumidine
(enantiomer, called (−)-capnoidine, also naturally occurring)

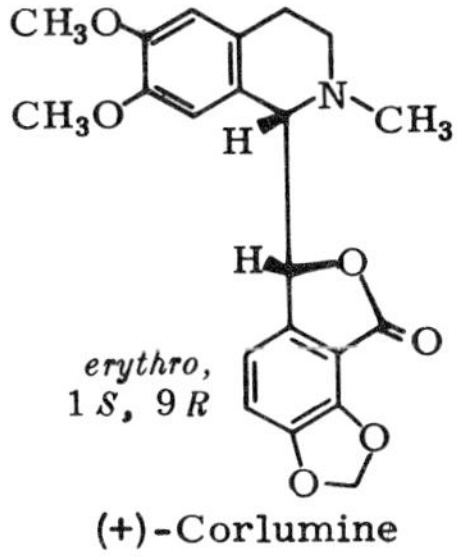

(+)-Corlumine

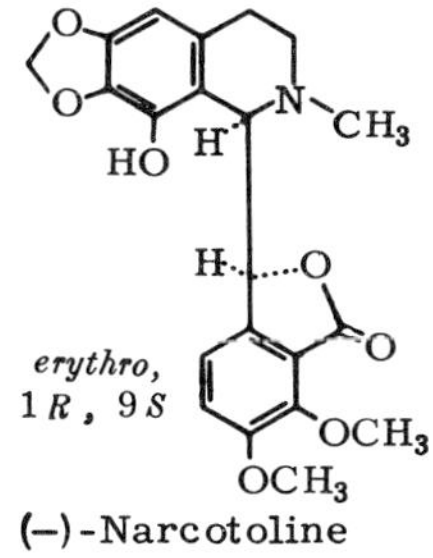

(−)-Narcotoline

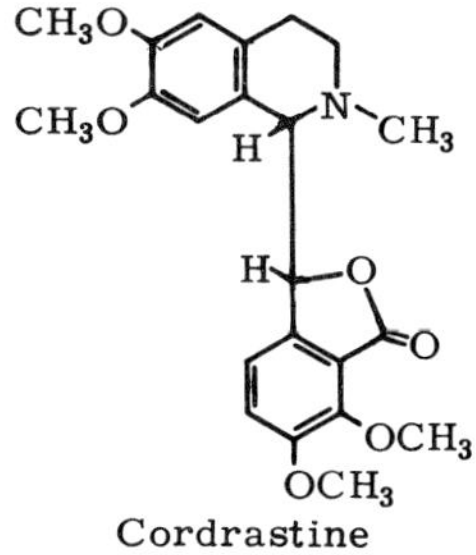

Cordrastine

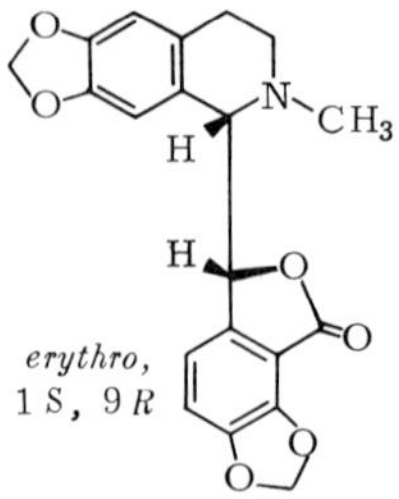

(+)-Bicuculline
(enantiomer also naturally occurring)

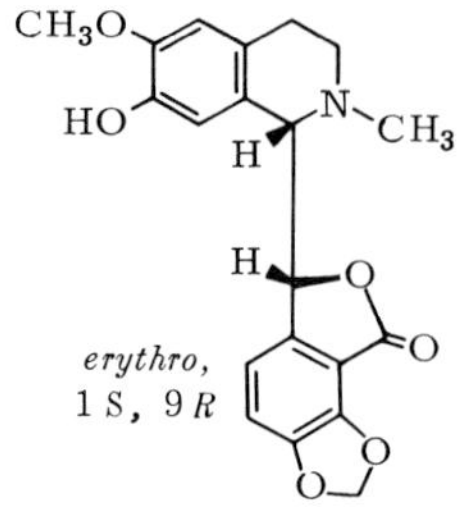

(+)-Corlumidine

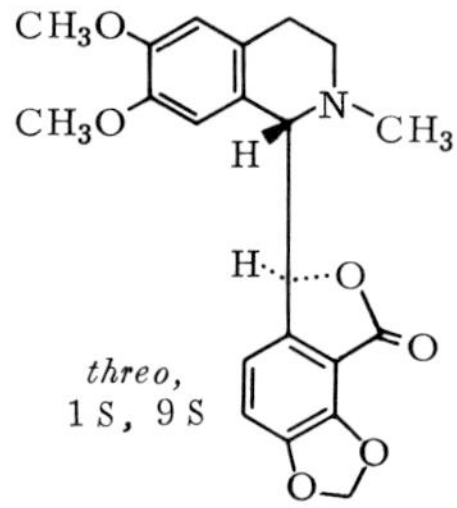

(+)-Adlumine
(enantiomer also naturally occurring)

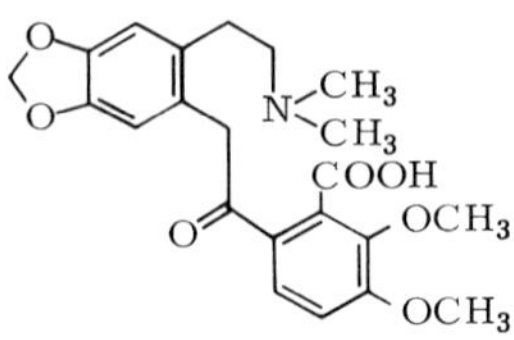

Narceine

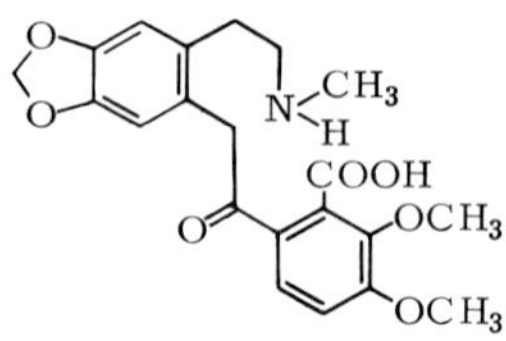

Nornarceine

I. Introduction

The phthalideisoquinoline bases usually possess a tetracyclic nucleus incorporating a γ-lactone ring. They differ from each other either in the nature and position of the aromatic substituents or in the stereochemistry at the asymmetric centers at C-1 and C-9. In two instances, namely, narceine and nornarceine, the γ-lactone ring has been replaced by a ketone and a carboxyl group.[1,1a]

II. Structural Elucidation

The structural elucidation of the phthalideisoquinoline alkaloids was carried out in laboratories in England and Germany over a period of several years but mostly during the time preceding and immediately following World War I. Names associated with this effort are those of Wohler, Freund, Fritsch, Roser, Pyman, Salway, Decker, W. H. Perkin, Jr., and Sir Robert Robinson.

A. Narcotine

The alkaloid $(-)$-α-narcotine, $C_{22}H_{23}O_7N$, is one of the major bases in *Papaver somniferum* L. (Papaveraceae), the source plant for opium. The molecule can be cleaved very readily into two moieties: with dilute sulfuric acid, cotarnine and opianic acid are generated. Under acidic reducing conditions, e.g., zinc in hydrochloric or sulfuric acid, hydrocotarnine and meconine are formed (Scheme I). Cotarnine may also be obtained from narcotine by oxidation with dilute nitric acid, chromic acid, or potassium permanganate.[2]

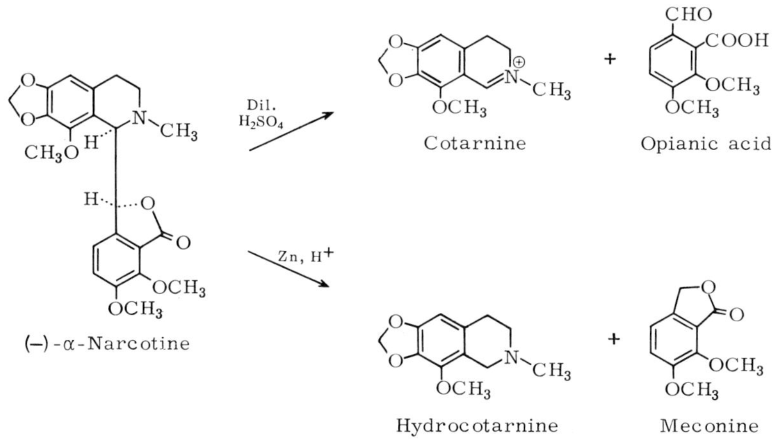

Scheme I

In acid solution, cotarnine exists as an immonium ion, but in alkaline solution the nonionic pseudo-base form predominates (Scheme II). Oxidation of cotarnine leads to oxycotarnine, while reduction gives hydrocotarnine. Disproportionation of cotarnine occurs when the compound is boiled in strong alkali, with formation of oxycotarnine and hydrocotarnine. Nowadays, the reduction of cotarnine to hydrocotarnine is achieved with sodium borohydride, while mercuric acetate is used to carry out the reverse operation (Scheme II).

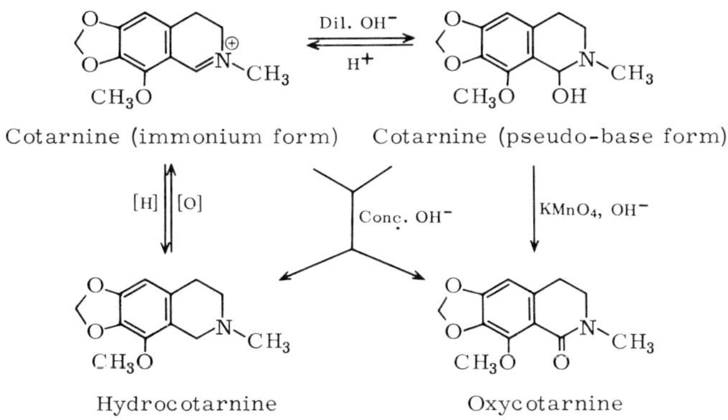

Scheme II

Oxidation of cotarnine with potassium permanganate affords cotarnic acid; alternatively, with hot nitric acid the quaternary salt apophyllenic acid is formed.

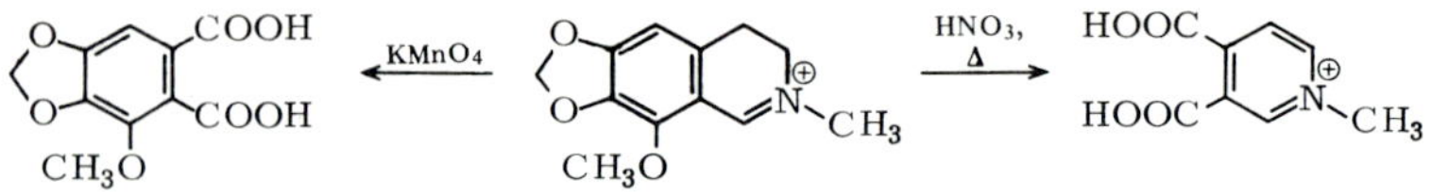

Cotarnic acid Cotarnine Apophyllenic acid

Turning to meconine and opianic acid, the latter compound may be represented either in the aldehydo acid form or the lactol form indicated below. Meconine can be easily oxidized to opianic acid with manganese dioxide. Since meconine has been synthesized by heating 2,3-dimethoxybenzoic acid with formaldehyde and concentrated hydrochloric acid, this approach can also be considered a preparative method for opianic acid.

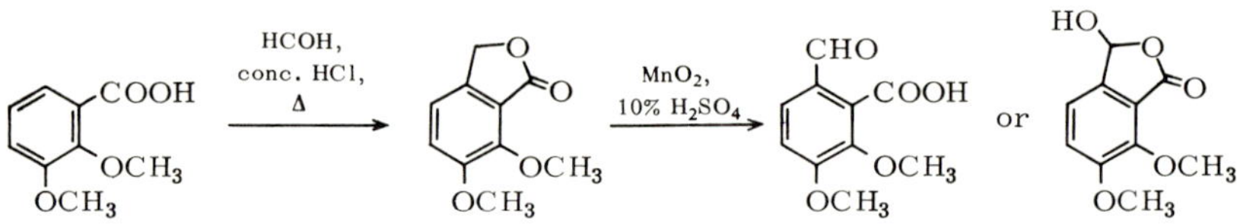

2,3-Dimethoxy-benzoic acid Meconine Opianic acid (aldehydo acid form) Opianic acid (lactol form)

With the structural elucidation of cotarnine, opianic acid, hydrocotarnine, and meconine, and given the presence of a lactone ring in narcotine, the structure of this alkaloid was essentially established.

B. Hydrastine

Several close analogies exist between the chemistry of (−)-β-hydrastine and (−)-α-narcotine. One difference, however, is that although (−)-β-hydrastine can be oxidized to opianic acid and hydrastinine with hot nitric acid, no cleavage of (−)-β-hydrastine takes place when it is treated with zinc or tin in hydrochloric acid (Scheme III). Otherwise, the chemistry of the degradation of hydrastine bears such strong analogy to the narcotine case that the topic need not be covered in detail. A point of special interest is the hydrogenolysis of hydrocotarnine with sodium in ethanol which yields among other products hydrohydrastinine, thus affording an additional link between narcotine and hydrastine (Scheme III).[3]

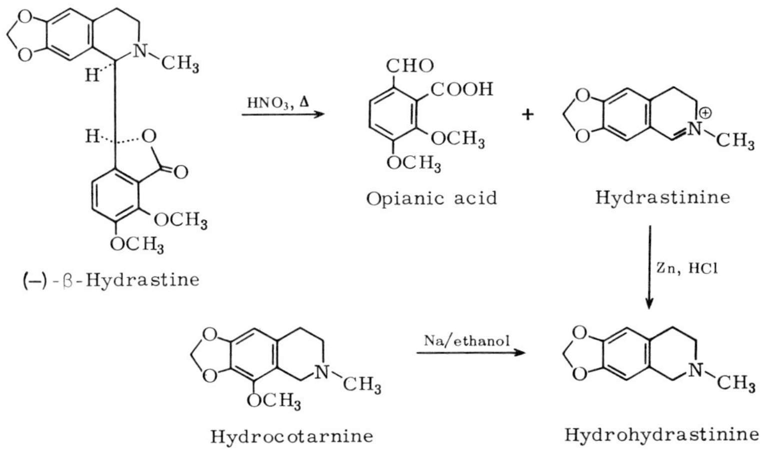

Scheme III

C. Narceine

The structure of the optically inactive and minor opium alkaloid narceine, $C_{23}H_{27}O_8N$, was indicated by the finding that it could also be obtained by heating narcotine methochloride with alkali. The mechanism of this transformation probably involves Hofmann elimination followed by saponification of the resulting enol lactone. Narceine is therefore a highly substituted deoxybenzoin.[4,5,5a]

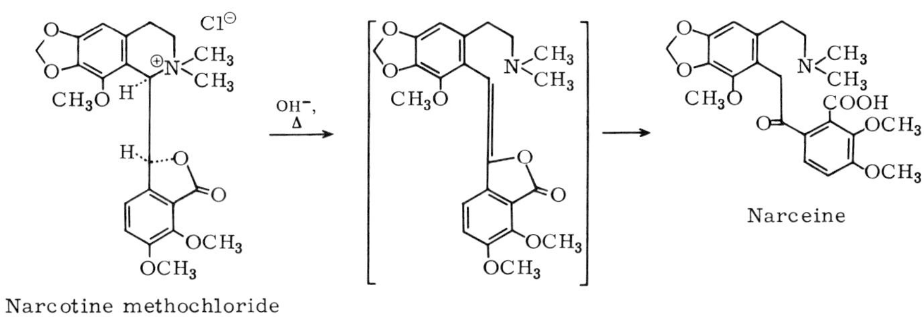

It has also been established that treatment of narceine with ethyl nitrite leads to the oximino derivative **1**, pointing to the presence of a methylene function adjacent to the carbonyl group. Upon heating, the oximino ketone undergoes cleavage to *o*-hemipinic acid and the amino nitrile **2** which affords the known cotarnonitrile when subjected to Hofmann degradation (Scheme IV).[6]

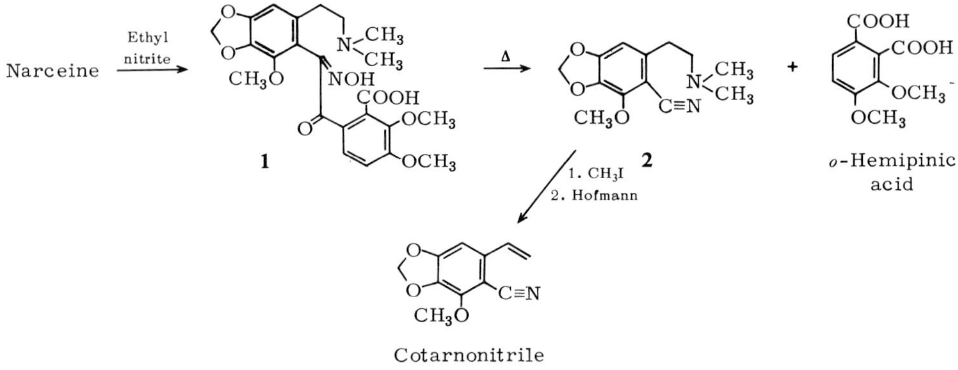

Scheme IV

III. Absolute Configuration and ORD Studies

A. Narcotine

The stereochemistry of (−)-α-narcotine has been considered by Ohta and by Battersby as well as by other groups.

It is known that prolonged action of hot methanolic potassium hydroxide on natural (−)-α-narcotine results in the formation of an equilibrium mixture of the original base and a new optically active diastereoisomer, (−)-β-narcotine, which can be written as shown in Scheme V. Lithium aluminum hydride reduction of the α- and β-narcotines readily affords α-narcotinediol and β-narcotinediol, respectively.[7]

Acetylation of these diols gives rise to the corresponding diacetates **3** and **4**, but subsequent catalytic hydrogenolysis yields one and the same dextrorotatory benzylisoquinoline **5**. The foregoing sequence clearly establishes that α- and β-narcotine must differ from each other only in their stereochemistry at C-9.

The benzylisoquinoline **5** shows a positive Cotton effect near 295 mμ, so that its C-1 hydrogen must be alpha as indicated. It follows that the C-1 hydrogen in (−)-α-narcotine and in (−)-β-narcotine must also be alpha (Scheme V).[7]

Alternatively, α-narcotinediol was cyclized via its monomesylate derivative to the *N*-methotetrahydroprotoberberine salt **6**. This material underwent *N*-demethylation on pyrolysis to yield the protoberberine base **7**. Reductive removal of the hydroxyl group was achieved in ethanolic perchloric acid over a palladium catalyst. The tetrahydroprotoberberine **8** thus obtained showed a strong negative rotation, so that its C-14 hydrogen must be alpha.

The identical sequence was carried out using β-narcotinediol to yield the tetrahydroprotoberberine base **9**. Hydrogenolytic cleavage of this species then provided the same levorotatory tetrahydroprotoberberine **8**. The conclusion is that the C-1 hydrogens in both α- and β-narcotine are alpha (Scheme VI).[8]

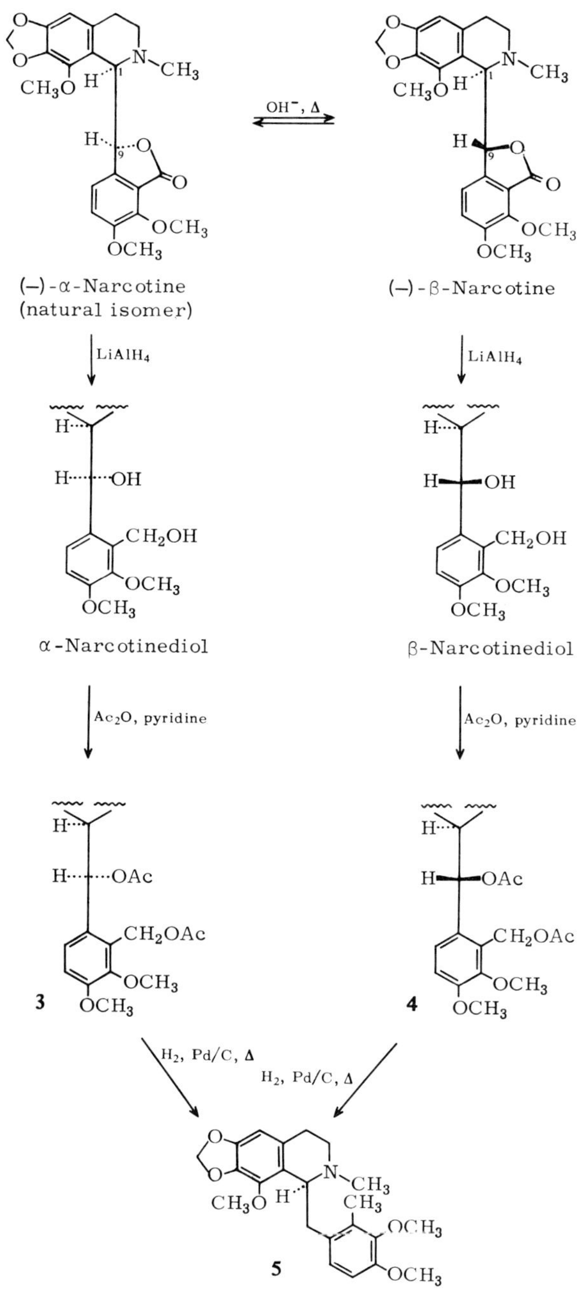
OH⁻, Δ
(—)-α-Narcotine
(natural isomer)
(—)-β-Narcotine
LiAlH4
α-Narcotinediol
β-Narcotinediol
Ac2O, pyridine
3
4
H2, Pd/C, Δ
5

Scheme V

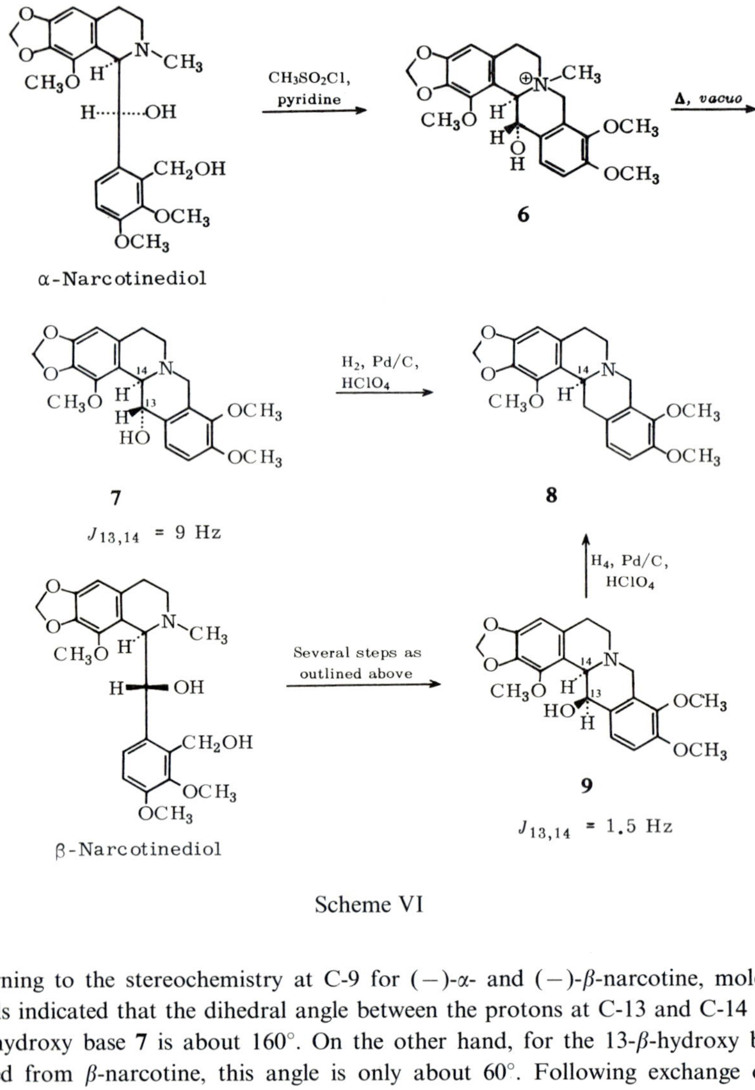

Scheme VI

Turning to the stereochemistry at C-9 for (−)-α- and (−)-β-narcotine, molecular models indicated that the dihedral angle between the protons at C-13 and C-14 of the 13-α-hydroxy base **7** is about 160°. On the other hand, for the 13-β-hydroxy base **9** derived from β-narcotine, this angle is only about 60°. Following exchange of the hydroxylic protons for deuterium, it was determined that the splitting constant $J_{13,14}$ was 9 Hz for species **7**, and only about 1.5 Hz for **9**. The large coupling value of 9 Hz is in accord with a trans arrangement of the C-13,14 hydrogens in **7**, and the small coupling constant of 1.5 Hz argues for a cis relationship in **9**, thus settling the stereochemistry at C-9 for α- and β-narcotine.

Noteworthy is the observation that the hydroxytetrahydroprotoberberine **7** did not exhibit prominent Bohlmann bands between 3.62 and 3.64 μ (2760 to 2750 cm^{-1}), while the diastereoisomeric compound **9** showed bands at 3.56, 3.60, and 3.63 μ (2805, 2774, and 2753 cm^{-1}) due to the presence of a trans-fused B/C quinolizidine system. Species **7** must therefore probably exist in the cis conformation **7a**.[7,*]

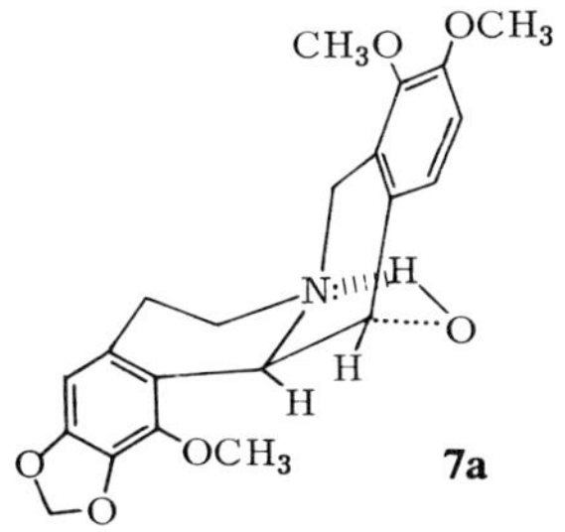

7a

B. Hydrastine

In an analogous sequence, the stereochemistry and absolute configuration of naturally occurring (−)-β-hydrastine and that of its nonnatural diastereoisomer (−)-α-hydrastine have been established by conversion of these bases to (−)-13-epiophiocarpine and the alkaloid (−)-ophiocarpine, respectively. Subsequent hydrogenolysis of the acetate derivatives generated in each case a mixture of (−)- and (±)-canadine (Scheme VII).[9]

It has also been possible for Bláha and his co-workers to derive the relative stereochemistry of (−)-α-narcotine and (−)-β-hydrastine by a careful comparison of appropriate pK_a values, infrared frequencies, and specific rotations.[10]

NMR spectroscopy of the phthalideisoquinolines can also lead to the elucidation of the relative configuration for narcotine. The steric effect of the C-8 methoxyl group is such that the preferred conformation for naturally occurring (−)-α-narcotine can be represented by expression **10**, while the diastereoisomeric (−)-β-narcotine is as in expression **11**.[11]

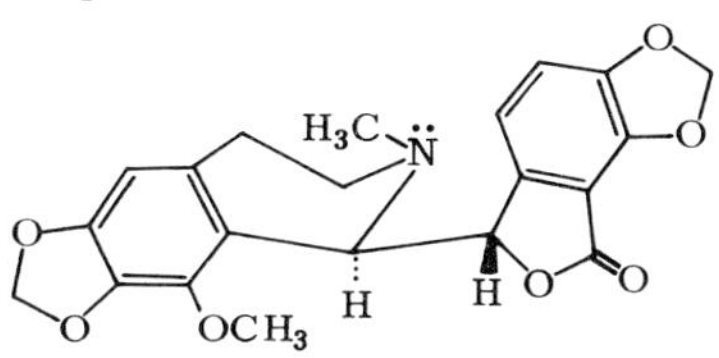

10

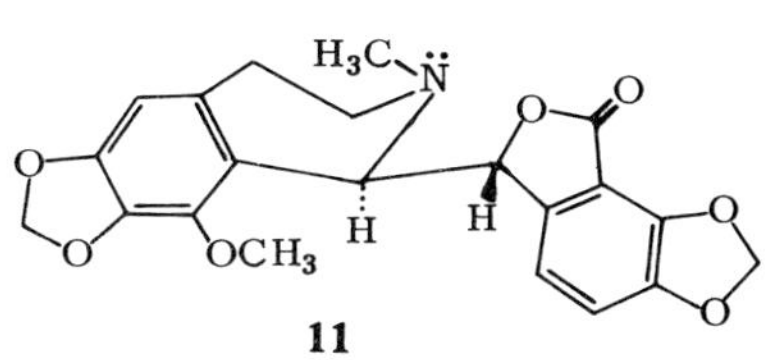

11

The phthalideisoquinolines usually give three apparent ORD Cotton effects from the signs of which the absolute configurations of the two asymmetric centers may be derived.[7,12] The first Cotton effect is between 320 and 335 mμ and reflects the stereochemistry at C-9; it is positive for the 9*R*- and negative for the 9*S*-configuration. This Cotton effect is echoed by a third Cotton effect between 235 and 255 mμ of the same sign and bearing the same relationship to the configuration at C-9.

* For three other possible conformations for the tetrahydroprotoberberines see Chapter 16, Section V, B.

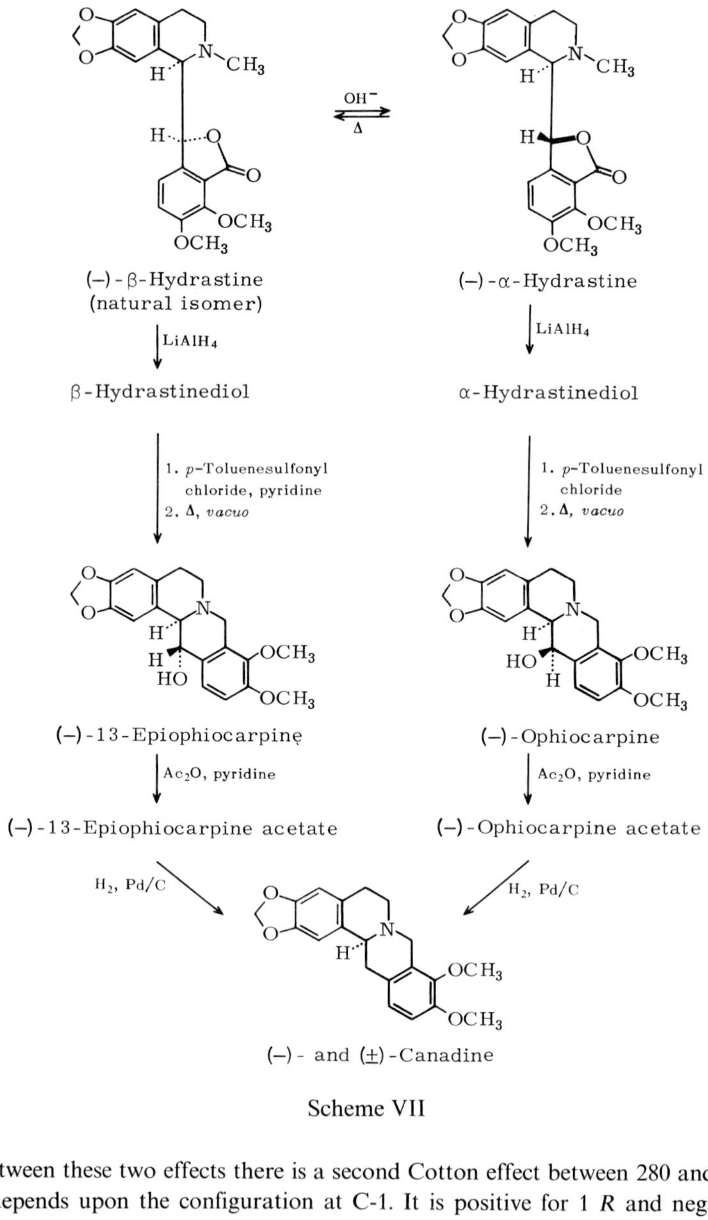

Scheme VII

In between these two effects there is a second Cotton effect between 280 and 300 mμ which depends upon the configuration at C-1. It is positive for 1 *R* and negative for 1 *S*. Where the first and second Cotton effects are of opposite signs, the second extremum of the first Cotton effect may disappear or may be present simply as an inflection.[12] The CD curves of the phthalideisoquinolines are equally useful in settling the stereochemical problem.[12,13]

Simple specific rotations may be of value in preliminary assignments of absolute configuration. Levorotatory free bases usually have the C-1 hydrogen alpha (1 *R*) while the dextro isomers have a beta C-1 hydrogen (1*S*).

IV. Syntheses of Phthalideisoquinolines

A. Narcotine

In the first synthesis of phthalideisoquinoline systems, Perkin and Robinson found that when cotarnine and meconine are simply heated together in alcoholic solution, a small yield of (±)-α-narcotine is obtained. Another diastereoisomer of narcotine had also been expected because of the presence of two asymmetric centers in the alkaloid but was not found. The synthetic (±)-α-narcotine was then resolved, and the (−)-α-narcotine obtained was identical with the natural product.[14]

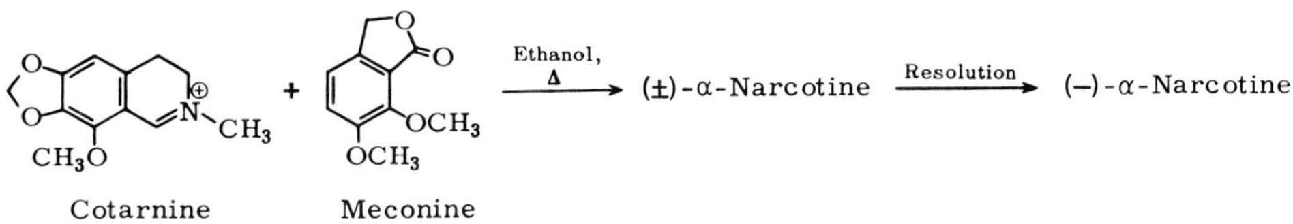

Hope and Robinson then determined that condensation of cotarnine and nitromeconine gave a satisfactory yield of a nitrated narcotine. Elimination of the nitro group via reduction and diazotization produced (±)-*β*-narcotine, the racemic diastereoisomer of natural (−)-α-narcotine, which had not been obtained in the earlier synthesis (Scheme VIII).[15]

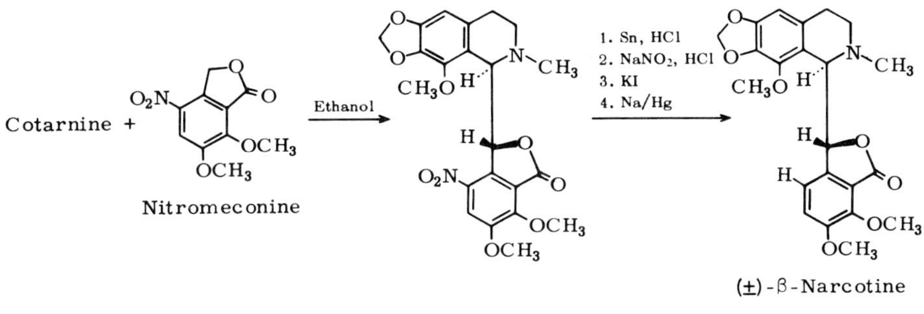

Scheme VIII

That the substituent on the aromatic ring of meconine can definitely affect the stereochemical course of the condensation was further demonstrated by the finding that when cotarnine was condensed with iodomeconine and the adduct reduced, the product was the desired (±)-α-narcotine, corresponding to the natural series.[15]

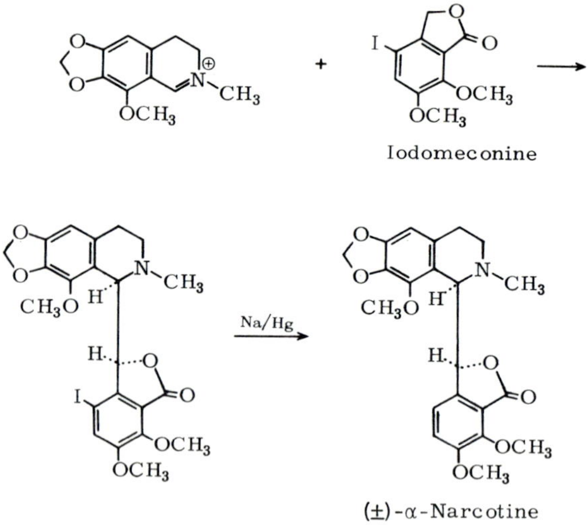

Since it is known that prolonged action of hot methanolic potassium hydroxide on either α- or β-narcotine leads to an equilibrium mixture of the two diastereoisomers, the synthesis of one of these isomers amounts to a preparation of the other.[16]

B. Hydrastine

A synthesis of hydrastine parallel to that of narcotine was first described by Hope and Robinson in 1912. Condensation of nitromeconine with hydrastinine afforded a mixture of diastereoisomeric nitrohydrastines, and the nitro group was subsequently removed to give a mixture of hydrastines. The fact that the initial condensation with nitromeconine was not stereospecific must be due to the absence of a C-8 methoxyl in hydrastinine (Scheme IX).[17,18]

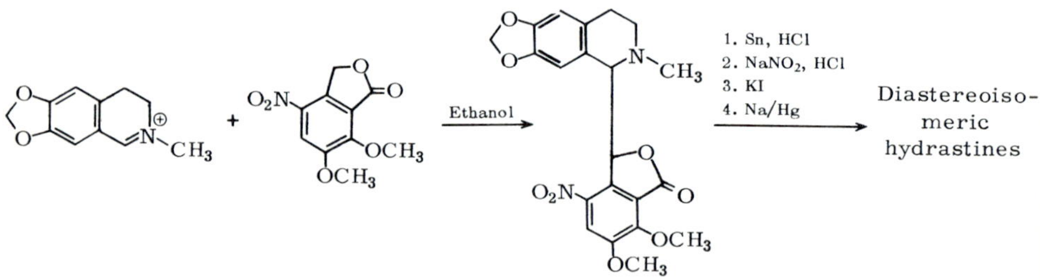

Scheme IX

An improved preparation of hydrastine was carried out by Haworth and Pinder in collaboration with Robinson. The acid chloride of meconine-α-carboxylic acid prepared as shown in Scheme X was used to acylate *N*-methylhomopiperonylamine.

The resulting amide was cyclized using the Bischler–Napieralski procedure and the product catalytically reduced to provide a mixture of diastereoisomeric hydrastines from which (±)-*β*-hydrastine was separated. Resolution then supplied (−)-*β*-hydrastine (Scheme X).[19,20]

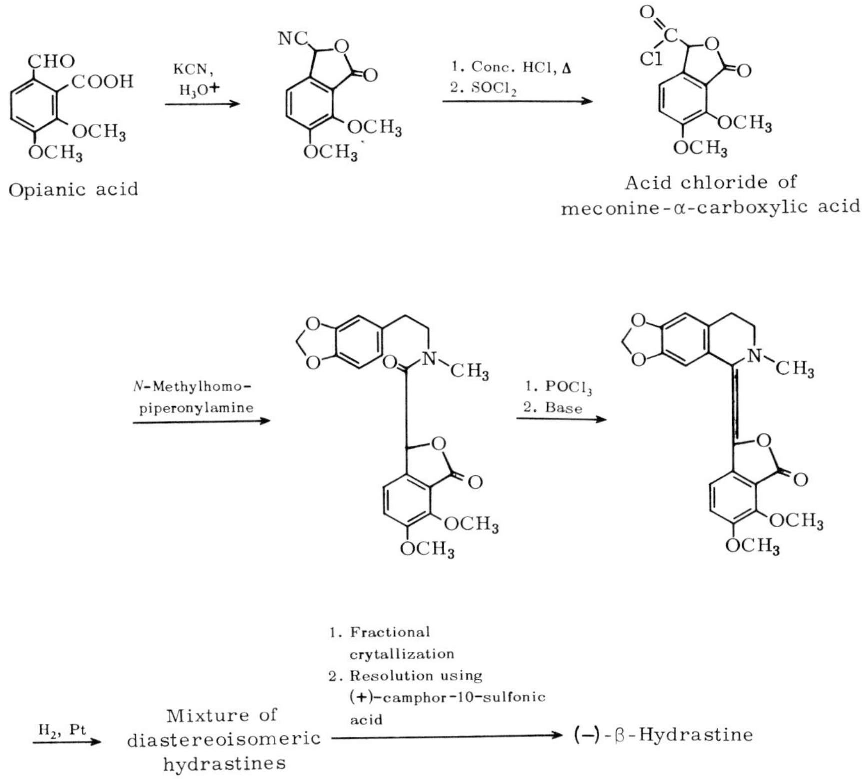

Scheme X

V. Some Reactions of Narcotine and Hydrastine

A. The Conversion of Phthalideisoquinolines to Protoberberines

Robinson and Mirza were the first to carry out the conversion of (−)-*β*-hydrastine to berberine, a transformation which, as was discussed earlier, was to prove of importance in the elucidation of the stereochemistry of the phthalideisoquinolines. Reduction of (−)-*β*-hydrastine with lithium aluminum hydride yielded hydrastinediol, which was cyclized to the dihydroprotoberberine salt **12** by means of thionyl chloride. Pyrolysis of this salt to yield the free base succeeded by oxidation gave the known alkaloid berberine. An item worth noting in this sequence is that the dihydroprotoberberine salt **12** is optically active due to asymmetry at the quaternary nitrogen (Scheme XI).[21]

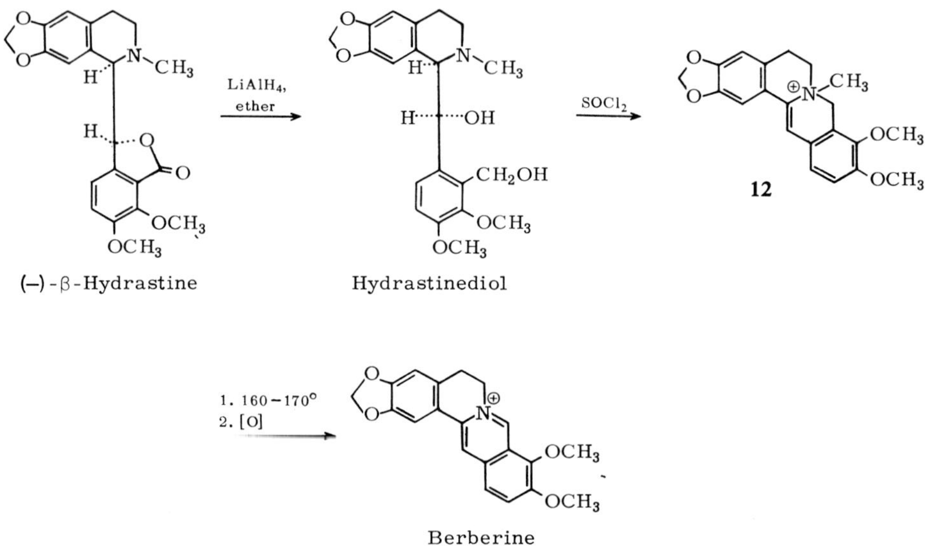

Scheme XI

B. Reduction

The reduction of the lactonic phthalideisoquinolines with mixed metal hydrides can lead to a variety of products depending upon experimental conditions. Treatment of narcotine or hydrastine with lithium aluminum hydride in ether at room temperature gives the corresponding diols,[21] but with lithium aluminum hydride in refluxing tetrahydrofuran, narcotine furnishes the phenolic species **13** resulting from *O*-demethylation at C-8.[22] If the reduction of narcotine is carried out with lithium trimethoxyaluminum hydride in ether at room temperature, two products can be isolated, narcotinediol and the hemiacetal **14**.[23]

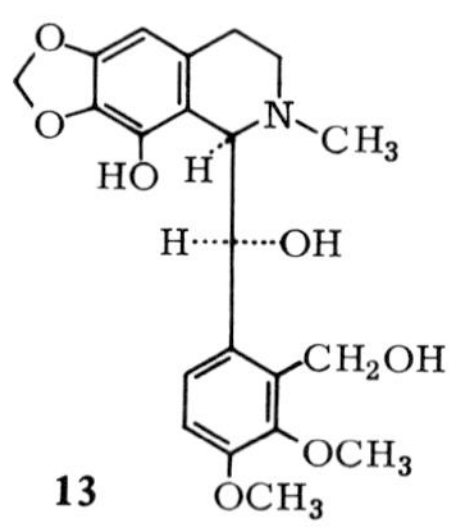

13

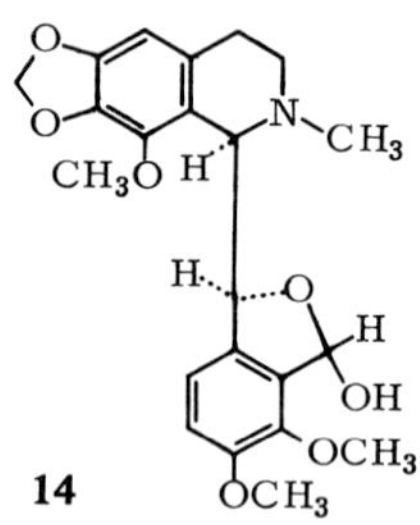

14

C. N-Oxidation and Pyrolysis

When treated with 30% hydrogen peroxide, narcotine and hydrastine yield the corresponding *N*-oxides **15**. Pyrolysis of these oxides *in vacuo* leads to the polycyclic enol lactone **16**.[24]

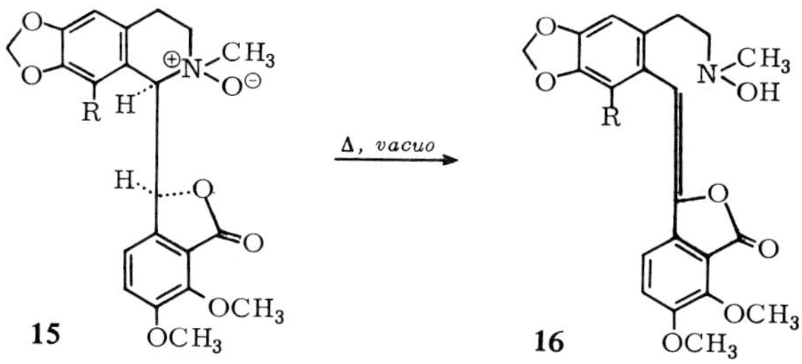

15 16

R = H or CH_3O

D. Amide Formation

Cleavage of the B ring of narcotine has also been achieved by refluxing the alkaloid with benzoyl chloride. The product is the optically inactive benzamide derivative **17**. This reaction has analogy in the aporphine series, in which a similar ring opening of ring B can be carried out with the same reagent.[5]

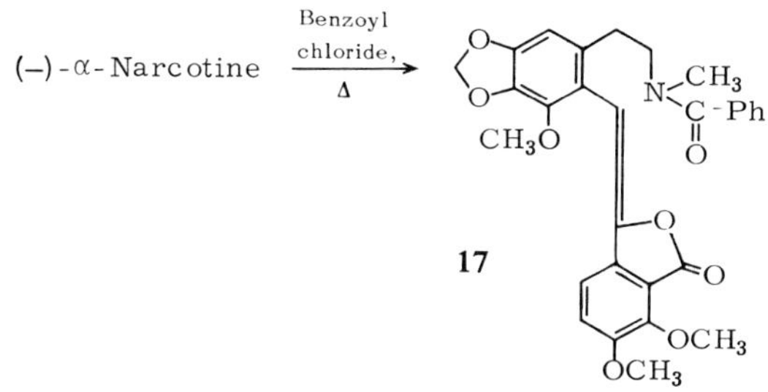

17

E. Reactions with Grignard Reagents

Phenylmagnesium bromide reacts with narcotine to form two hemiketals, **18** and **20**, which are interconvertible in solution via the hydroxy ketone **19**.

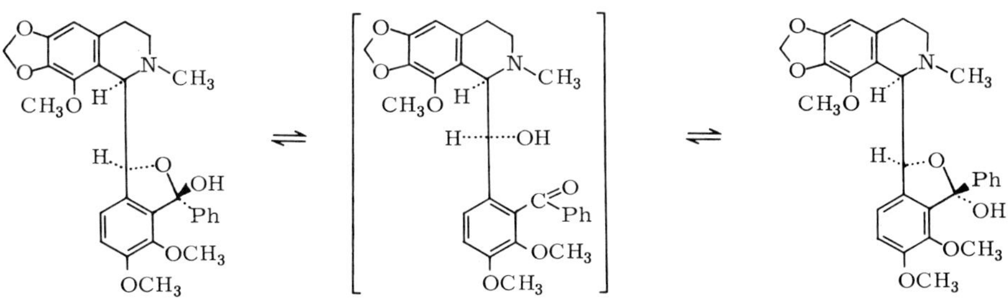

18 19 20

With benzylmagnesium bromide, by contrast, the benzylidene derivative **21** is formed, together with hydrocotarnine.[25,26]

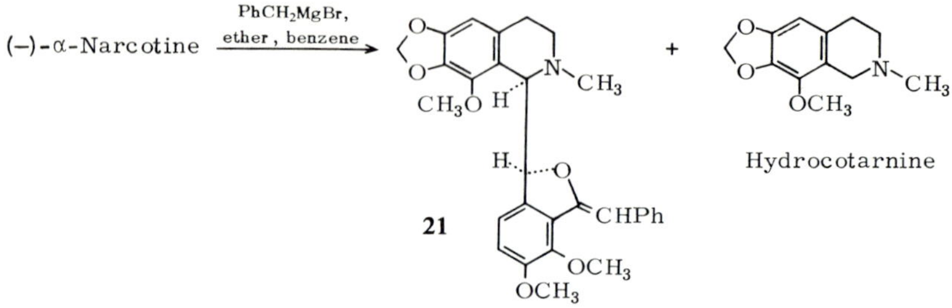

VI. Biosynthesis

It has been postulated that the phthalideisoquinolines are formed in nature by oxidative modification of tetrahydroprotoberberines, and recent work with labeled precursors supports this thesis.[27]

Tyrosine is the precursor for both halves of hydrastine in *Hydrastis canadensis* L. (Ranunculaceae), but dopamine, which is itself derived from tyrosine, can act as the precursor for only one of the two halves of the benzylisoquinoline unit.[28,29]

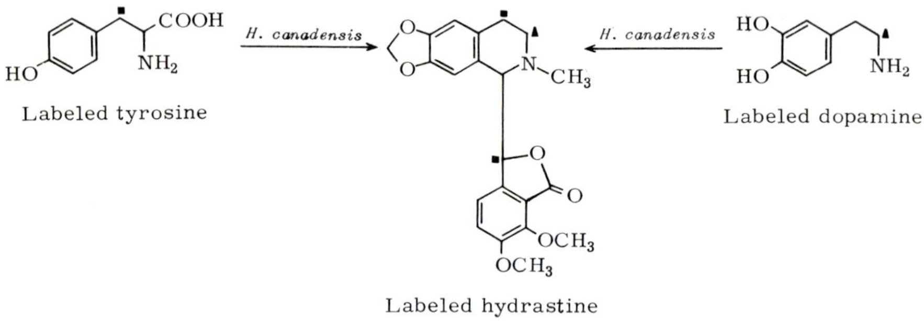

Several feeding experiments have been run to elucidate the biogenesis of narcotine in *Papaver somniferum* L. (Papaveraceae). When labeled (±)-tyrosine was fed to the plant, radioactive narcotine labeled specifically and equally at C-1 and C-3 was obtained. The benzylisoquinoline system of narcotine is thus derived biologically from two Ar−C−C− units which can arise from tyrosine. This result complements the one above using *H. canadensis*.[26,29,30]

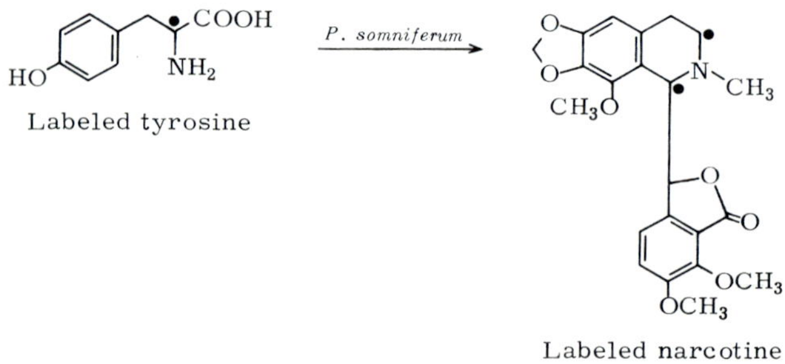

The carbon atoms that arise from the S-methyl of methionine were clearly pinpointed when, after feeding radioactive methionine, narcotine labeled at the lactone carbonyl, the methylenedioxy group, and the N- and O-methyl carbon atoms was obtained.[26,31]

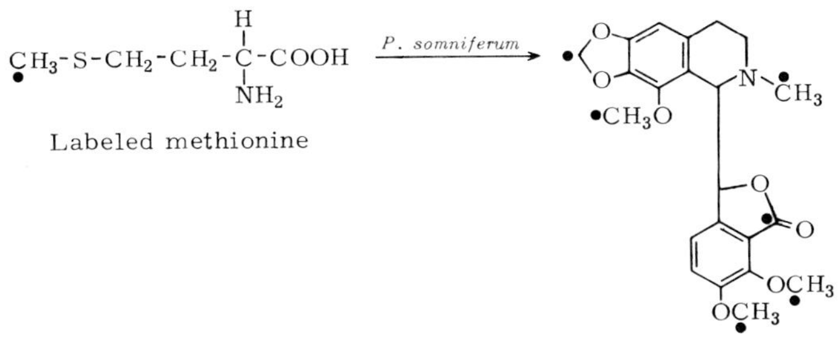

Labeled methionine

Labeled narcotine

Progressing further along the biogenetic locus, the benzylisoquinoline (±)-norlaudanosoline labeled at C-1 led to narcotine also labeled at C-1.[32]

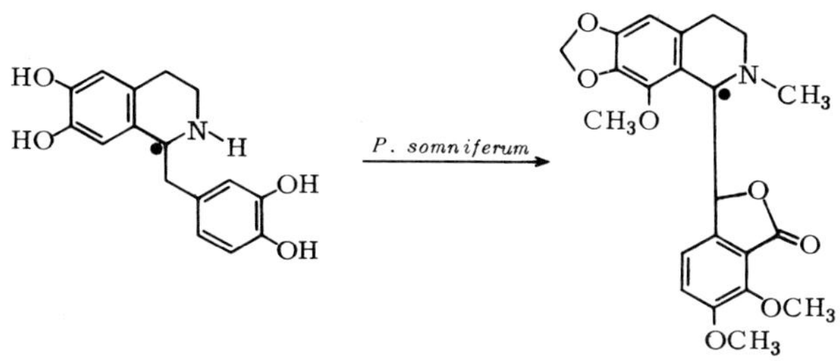

Labeled (±)-norlaudanosoline

Labeled narcotine

Even more significantly, when quadruply labeled (+)- and (−)-reticuline were fed separately to *P. somniferum*, it was found that both enantiomers were incorporated

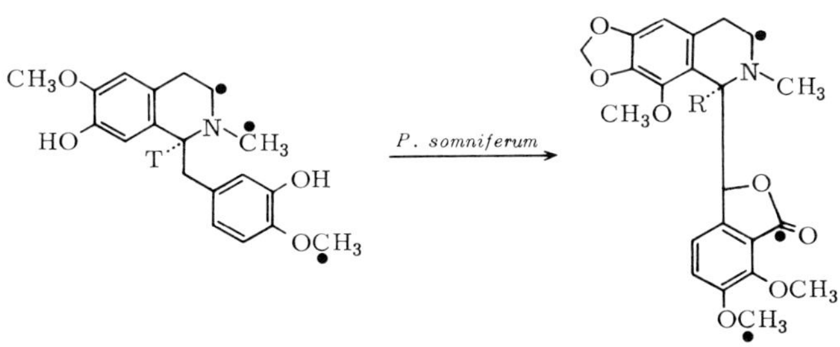

Labeled (+)-reticuline

(Labeled (−)-reticuline somewhat less efficient)

R = H or T

Labeled narcotine (appreciable loss of tritium)

into narcotine, but with the (+)-isomer doing so slightly more efficiently. Evidently epimerization of the wrong benzylisoquinoline precursor must occur, probably by oxidation–reduction at C-1. In keeping with this conclusion considerable loss of tritium occurred in the course of incorporation of both reticulines. Another important observation is that the lactone carbonyl of the phthalideisoquinoline must be derived from the *N*-methyl group of the benzylisoquinoline precursor.[26,32,33]

Finally, it has been found that the feeding of labeled (−)-scoulerine results in the formation of radioactive narcotine. Protoberberines are, therefore, the precursors for the phthalideisoquinolines in plants. Significantly, (−)-scoulerine, which possesses the same absolute configuration as (+)-reticuline and (−)-α-narcotine, was more than one hundred times more efficient than its enantiomer as a precursor for (−)-α-narcotine. The biogenetic sequence in plants is, therefore, benzylisoquinolines → tetrahydroprotoberberines → phthalideisoquinolines.[26,32,33]

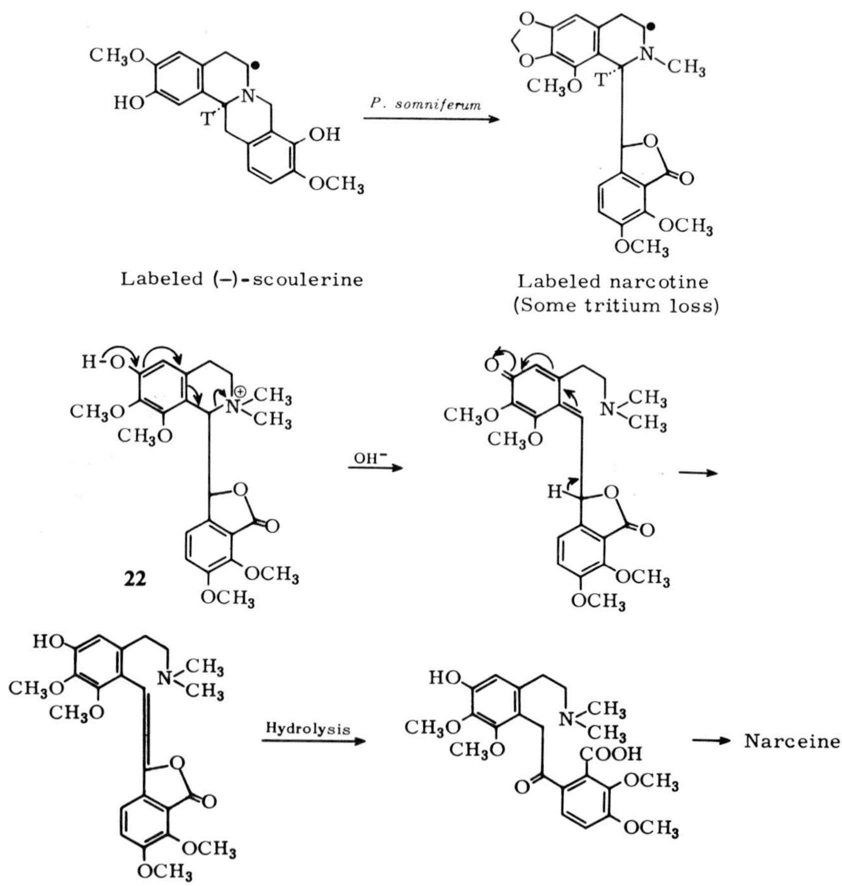

Scheme XII

Little comment has been made in the literature concerning the biogenesis of narceine, and it is generally assumed that this alkaloid is formed in the plant by a Hofmann-type β-elimination on a narcotine *N*-metho salt. A process of this nature is a definite possibility, but is not the only pathway theoretically available. A phenolic salt such as the phthalideisoquinoline **22** could also lead to narceine by the steps indicated (Scheme XII).

VII. Pharmacology

Narcotine has been found to possess antitussive activity, meaning that it may be a valuable agent for the suppression of cough.[34,35] Belying its name, the alkaloid has only a mild narcotic action, and is a much weaker analgesic than morphine or codeine.

Whereas narcotine depresses smooth muscles, hydrastine has a pressor effect.

Esterification of 8-hydroxyhydrastine (≡ narcotoline) with 3,4,5-trimethoxybenzoic anhydride yields the corresponding 8-acyloxyhydrastine, which has been claimed to be a cough-reliever with only mild depressive activity and with no effect upon the respiratory center.[36] Ethyl narceinate (Narcyl) is used as a narcotic, analgetic, and antitussive agent.

VIII. Mass Spectroscopy

The base peak for hydrastine occurs at *m/e* 190. There is no observable molecular ion since the doubly benzylic bond from C-1 to C-9 is extremely weak and breaks with ease.[37]

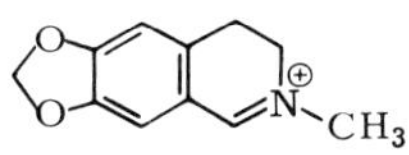

m/e 190

IX. NMR Spectroscopy

The NMR spectra of the phthalideisoquinoline alkaloids hydrastine, bicuculline, and others have been published and interpreted.[38] The chemical shifts for hydrastine are indicated below.[39]

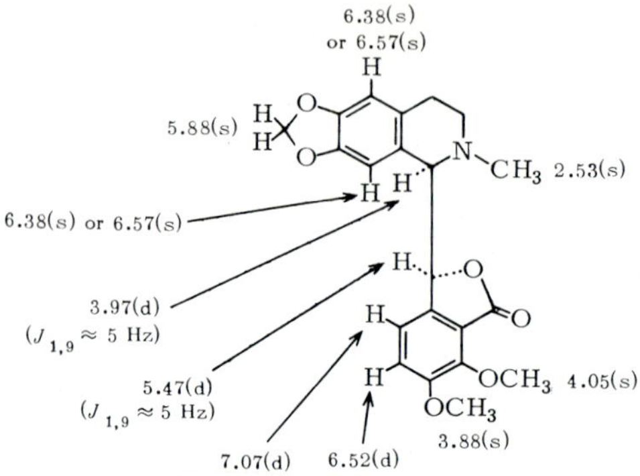

NMR spectral values for hydrastine

X. UV Spectroscopy

The tetracyclic phthalideisoquinolines exhibit three maxima. The central peak is the most characteristic and falls between 290 and 298 mμ.

(−)-α-Narcotine[40]	λ_{max}^{EtOH} 209, 291, and 309−310 mμ (4.86, 3.60, and 3.69)
	λ_{min}^{EtOH} 263 and 296 mμ (3.24 and 3.56)
(−)-β-Narcotine[9]	λ_{max}^{EtOH} 290 and 312 mμ (3.58 and 3.68)
(−)-β-Hydrastine[9]	λ_{max}^{EtOH} 297.5 mμ (3.86)
(−)-α-Hydrastine[9]	λ_{max}^{EtOH} 297.5 mμ (3.86)
(+)-Bicuculline[37]	λ_{max}^{EtOH} 225, 296, and 324 mμ (4.57, 3.81, and 3.77)
Narceine[40]	λ_{max}^{EtOH} 270 mμ (3.98).

REFERENCES

1. For a review on the phthalideisoquinolines, see A. R. Pinder, *Chem. Rev.* **64**, 551 (1964).
1a. Also see J. Stanek, *in* "The Alkaloids" (R. H. F. Manske, ed.), Vol. 9, p. 117. Academic Press, New York, 1967; F. Šantavý, *ibid.* Vol. 12, p. 397.
2. Cleavage with $Hg(OAc)_2$-EDTA gives mixtures of opianic acid and meconine [W. Wiegrebe, E. Roesel, W. D. Sasse, and H. Keppel, *Arch. Pharm.* **302**, 22 (1969)].
3. F. L. Pyman and F. G. P. Remfry, *J. Chem. Soc., London* **101**, 1595 (1912).
4. W. Roser, *Justus Liebigs Ann. Chem.* **247**, 167 (1888).
5. C. R. Addinall and R. T. Major, *J. Amer. Chem. Soc.* **55**, 1202 and 2153 (1933).
5a. For an improved synthesis of nornarceine from (−)-α-narcotine, see W. Klötzer, S. Teitel, and A. Brossi, (in press).
6. M. Freund and P. Oppenheim, *Chem. Ber.* **42**, 1084 (1909).
7. M. Ohta, H. Tani, S. Morozumi, and S. Kodaira, *Chem. Pharm. Bull.* **12**, 1080 (1964).
8, A. R. Battersby and H. Spenser, *J. Chem. Soc.* p. 1087 (1965).
9. M. Ohta, H. Tani, and S. Morozumi, *Chem. Pharm. Bull.* **12**, 1072 (1964); see also Ref. 7.

10. K. Bláha, J. Hrbek, Jr., J. Kovář, L. Pijewska, and F. Šantavý, *Collect. Czech. Chem. Commun.* **29**, 2328 (1964).
11. S. Safe and R. Y. Moir, *Can. J. Chem.* **42**, 160 (1964).
12. G. Snatzke, G. Wollenberg, J. Hrbek, Jr., F. Šantavý, K. Bláha, W. Klyne, and R. J. Swan, *Tetrahedron* **25**, 5059 (1969).
13. T. Kikuchi and T. Nishinaga, *Tetrahedron Lett.* p. 2519 (1969).
14. W. H. Perkin, Jr. and R. Robinson, *J. Chem. Soc., London* **99**, 775 (1911).
15. E. Hope and R. Robinson, *J. Chem. Soc., London* **105**, 2085 (1914).
16. M. A. Marshall, F. L. Pyman, and R. Robinson, *J. Chem. Soc., London* p. 1315 (1934).
17. E. Hope and R. Robinson, *Proc. Chem. Soc., London* **28**, 17 (1912).
18. E. Hope, F. L. Pyman, F. G. P. Remfry, and R. Robinson, *J. Chem. Soc., London* p. 236 (1931).
19. R. D. Haworth and A. R. Pinder, *J. Chem. Soc., London* p. 1776 (1950).
20. R. D. Haworth, A. R. Pinder, and R. Robinson, *Nature (London)* **165**, 529 (1950). See also W. H. Perkin, Jr., J. N. Rây, and R. Robinson, *J. Chem. Soc.* **127**, 740 (1925).
21. R. Mirza and R. Robinson, *Nature (London)* **271** (1950).
22. W. Awe and W. Wiegrebe, *Arch. Pharm. (Weinheim)* **295**, 817 (1962).
23. M. Shamma and J. A. Weiss, unpublished results (1970).
24. K. W. Bentley and A. W. Murray, *J. Chem. Soc., London* p. 2491 (1963).
25. M.-M. Janot and H. Pourat, *Bull. Soc. Chim. Fr.* [6] p. 823 (1955).
26. A. R. Battersby, M. Hirst, D. J. McCaldin, R. Southgate, and J. Staunton, *J. Chem. Soc., C* p. 2163 (1968).
27. R. Robinson, "The Structural Relations of Natural Products." Oxford Univ. Press (Clarendon), London and New York, 1955.
28. I. D. Spenser and J. R. Gear, *J. Amer. Chem. Soc.* , 1059 (1962).
29. G. Kleinschmidt and K. Mothes, *Z. Naturforsch. B* **14**, 52 (1959).
30. J. R. Gear and I. D. Spenser, *Can. J. Chem.* **41**, 783 (1963).
31. R. N. Gupta and I. D. Spenser, *Biochem. Biophys. Res. Commun.* **13**, 115 (1963).
32. A. R. Battersby and M. Hirst, *Tetrahedron Lett.* p. 669 (1965).
33. A. R. Battersby, R. J. Francis, M. Hirst, R. Southgate, and J. Staunton, *Chem. Commun.* p. 602 (1967).
34. M. S. Segal, M. M. Goldstein, and E. O. Attinger, *Dis. Chest.* **32**, 305 (1957).
35. A. Brossi, H. Besendorf, L. A. Pirk, and A. Rheiner, Jr., *Med. Chem.* **5**, 281 (1965).
36. Hungarian Pat. 157,452: *Chem. Abstr.* **73**, 77457u (1970).
37. M. Ohashi, J. M. Wilson, H. Budzikiewicz, M. Shamma, W. A. Slusarchyk, and C. Djerassi, *J. Amer. Chem. Soc.* **85**, 2807 (1963).
38. O. E. Edwards and K. L. Handa, *Can. J. Chem.* **39**, 1801 (1961).
39. "Varian Spectra Catalog," Vol. 1, Spectrum No. 347. Varian Assoc., Palo Alto, California, 1962.
40. P. M. Oertreicher, C. G. Farmilo, and L. Levi, *Bull. Narcotics* **7**, 42 (1954).

Chapter 20/ THE SPIROBENZYLISOQUINOLINES

Occurrence: Fumariaceae
Number: 11
Structures:

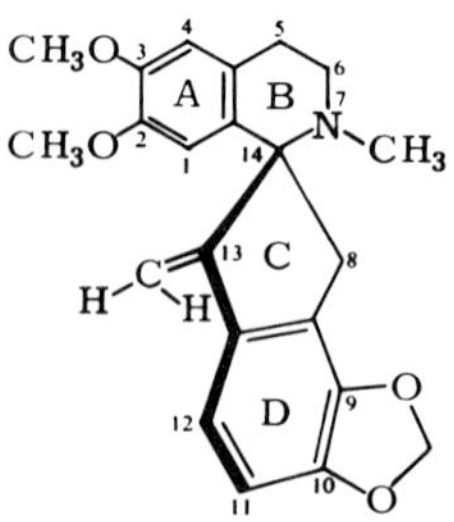

(+)-Ochotensimine[1]

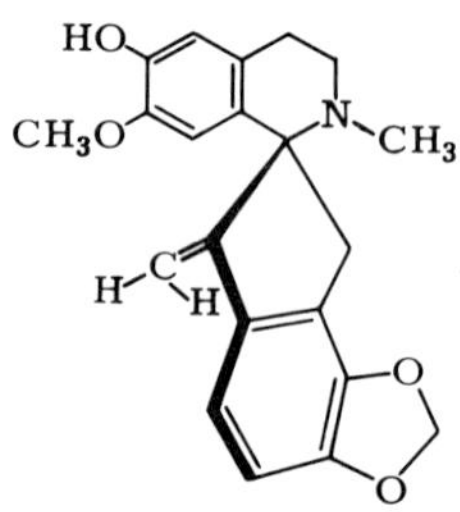

(+)-Ochotensine

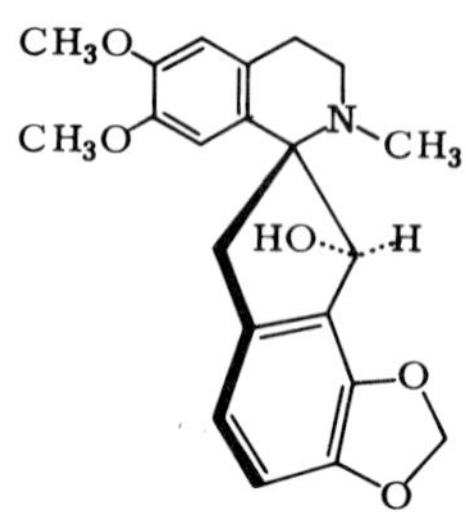

(−)-Fumaricine *

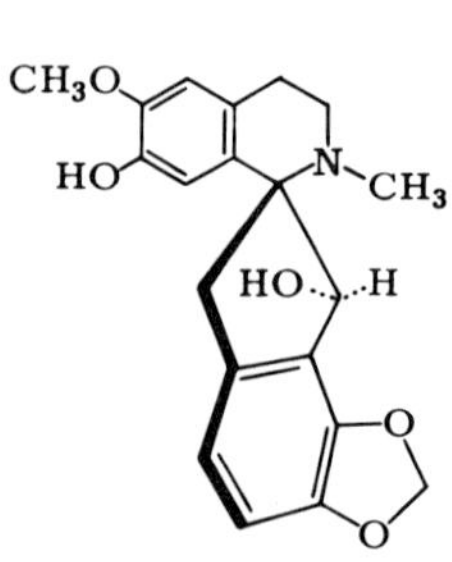

Fumaritine *

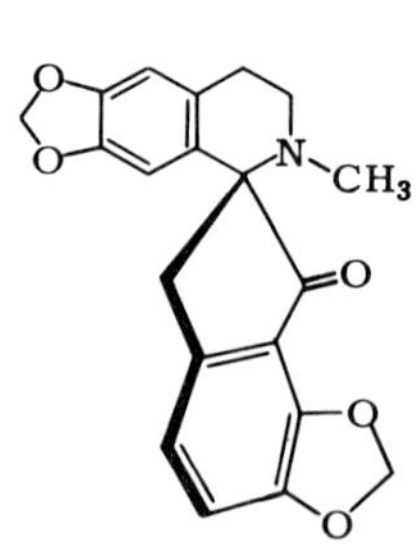

(+)-Fumariline

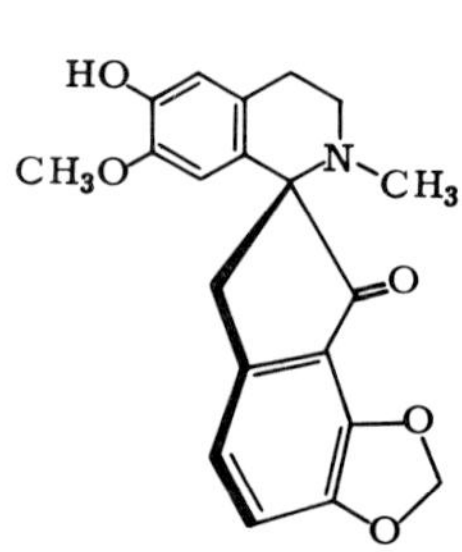

Parfumine *,[2]

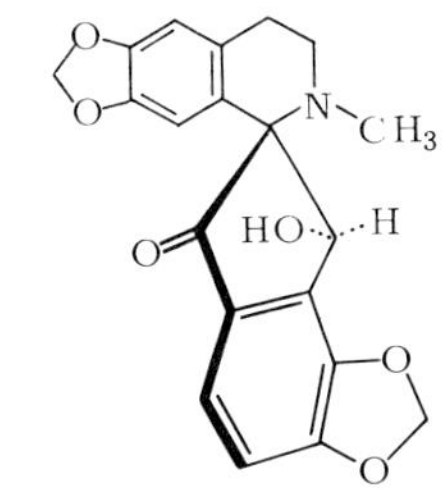

Sibiricine*

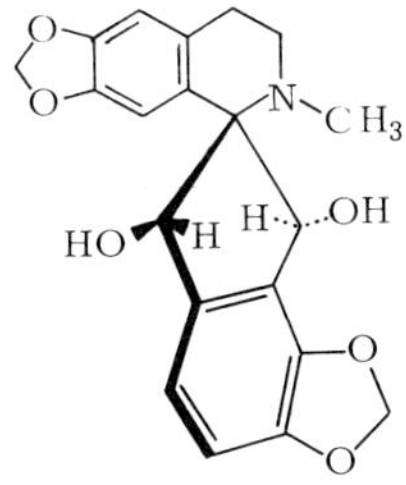

(+)-Ochrobirine‡

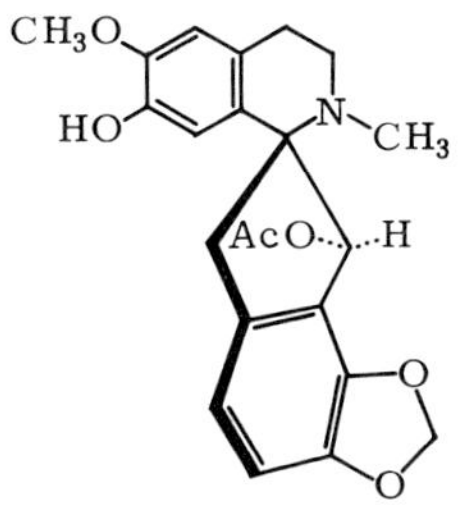

Fumarophycine*

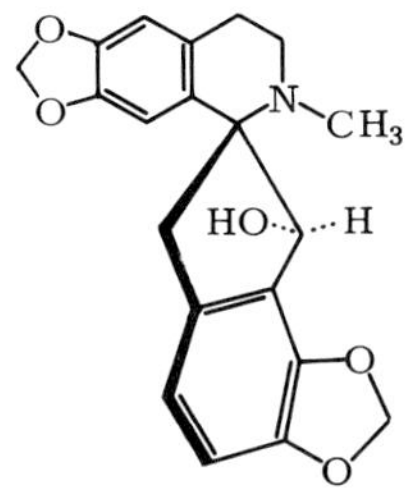

Corydaine[1a,*]

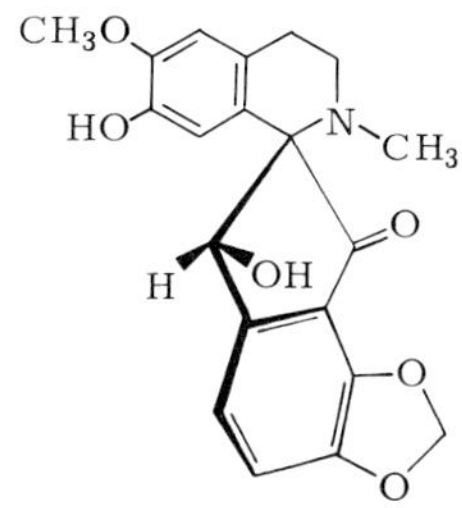

Fumarofine[1b,*]

* Absolute configuration unknown.

‡ There is a protopine alkaloid of the same name.

I. Introduction and Nomenclature[1]

All the known spirobenzylisoquinoline alkaloids originate within the Fumariaceae family, which is botanically very close to the Papaveraceae.

Since the spirobenzylisoquinolines may be derived in nature from protoberberine precursors, the numbering system follows that for the protoberberines. The oxygenated positions in ring D of the spirobenzylisoquinolines are always numbered 9,10 rather than 11,12, in analogy with the protoberberines, most of whom are also C-9,10-oxygenated.

The skeleton **1**, present in all the spirobenzylisoquinolines, can be called the ochotensane system. If the nitrogen is secondary, the molecule is denoted as a norochotensane derivative. The designation of the stereochemistry at C-8 and C-13 can be best described by naming substituents at these sites *syn* if they are on the nitrogen side and *anti* if they are on the same side as ring A. The alkaloid fumaricine as drawn above thus becomes 2,3-dimethoxy-9,10-methylenedioxy-8-*anti*-hydroxyochotensane. Although this may turn out to be a coincidence, all the spirobenzylisoquinoline alkaloids known to date possess a methylenedioxy group in ring D.

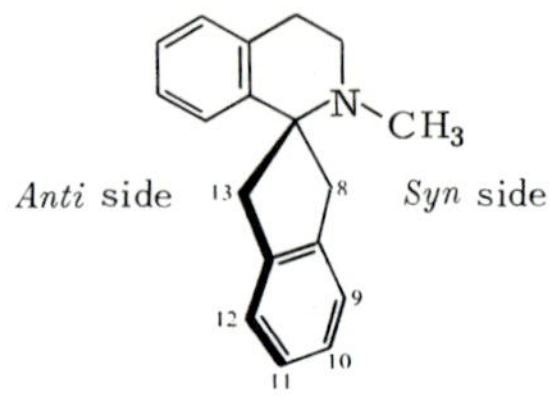

1

II. Structural Elucidation

A. *Ochotensimine and Ochotensine*

Although ochotensine was first isolated in 1936[3] and ochotensimine four years later,[4] the structures of these alkaloids were not determined until 1964, when NMR and mass spectral methods of analysis had become readily available. Both bases were obtained by Manske from *Corydalis* species.

Ochotensimine is the *O*-methyl ether of ochotensine, and the chemical transformations of the former base have been summarized in Scheme I. Hofmann degradation of the alkaloid does not lead to characterizable products. Reduction of ochotensimine methiodide with sodium amalgam, Adams catalyst, or sodium borohydride yields the optically inactive polycyclic base **2**. Catalytic reduction of ochotensimine to dihydroochotensimine, followed by quaternization with methyl iodide and Hofmann degradation, also yields base **2**.

Further Hofmann degradation of the methiodide of base **2** led to a nitrogen-free compound, **3**, and hydrogenation of this material with Adams catalyst reduced only the terminal unsaturation to yield compound **4**. Direct oxidation of ochotensimine methiodide with potassium permanganate produced *m*-hemipinic acid, thus giving some clue to the positions of the two methoxyl groups in the molecule.[5]

The NMR spectrum of ochotensimine proved to be extremely instructive and is summarized in the diagram below. Of importance in the structural elucidation were

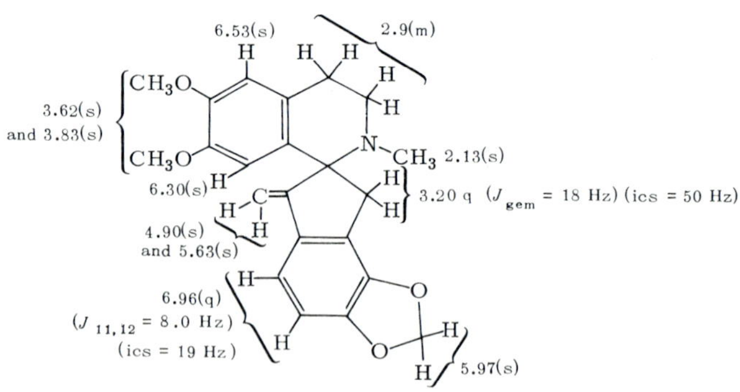

NMR spectral values for ochotensimine

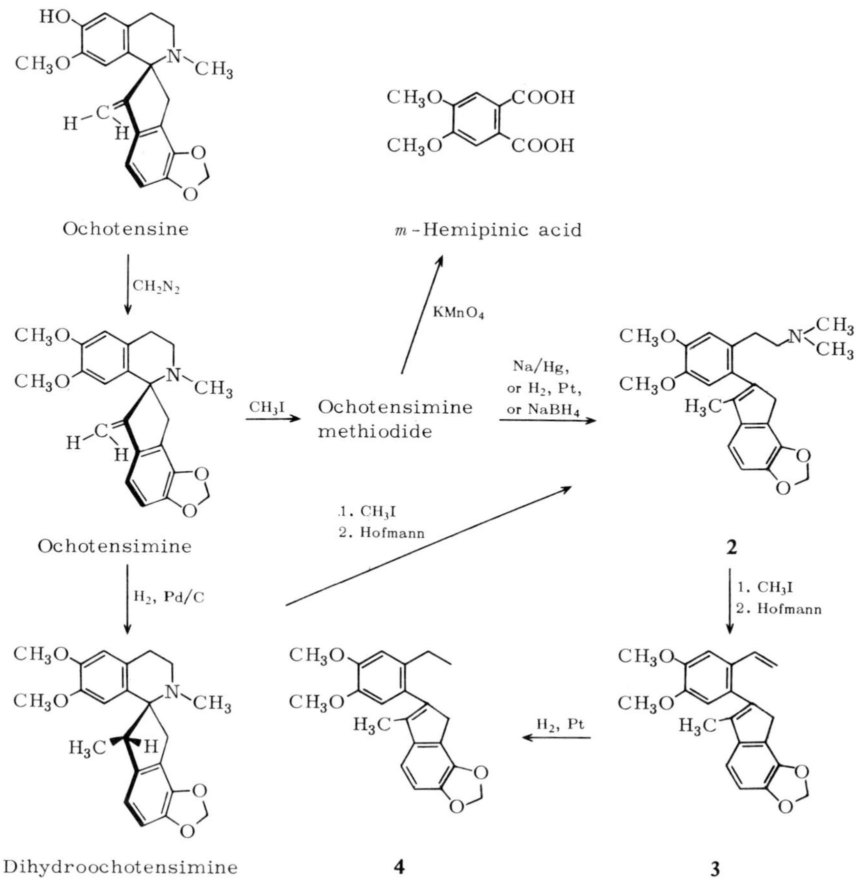

Scheme I

the two vinylic hydrogen singlets at $\delta 4.90$ and 5.63 and the C-8 methylene absorption centered at $\delta 3.20$ and split as a quartet due to the asymmetry at C-14. The appearance of aromatic proton singlets at $\delta 6.30$ and 6.53 and a quartet (2H) at $\delta 6.96$ assisted in fixing the positions of the aromatic substituents.[5]

The NMR spectra of the derivatives of ochotensimine were also useful in settling the structure of the alkaloid. The most outstanding feature in the NMR spectrum of dihydroochotensimine was the presence of a three-proton doublet at $\delta 0.95$ ($J = 7.2$ Hz) due to the C-13 methyl group. The derivative **4** exhibited in its spectrum a triplet at $\delta 1.13$ and a quartet at $\delta 2.52$ ($J = 7.8$ Hz) due to the presence of the ethyl group attached to ring A. This last piece of evidence established unequivocally that ochotensimine must have the $Ar-CH_2-CH_2-N\langle$ grouping common to most benzylisoquinoline alkaloids.

The mass spectrum of ochotensimine was instructive in that the molecular ion peak is also the base peak, underlining the fact that there is no facile pathway by which the molecule can be cleaved into two fragments. The UV spectra of ochotensimine and ochotensine are strongly reminiscent of alkaloidal tetrahydroisoquinoline systems.[5]

Any doubts concerning the structures of ochotensimine and ochotensine were removed in 1966 following the X-ray analysis of ochotensine methiodide.[6]

B. Fumaricine, Fumaritine, and Fumariline

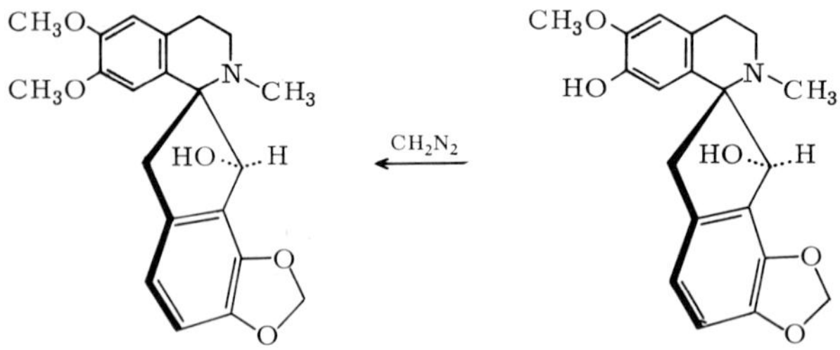

(−)-Fumaricine Fumaritine

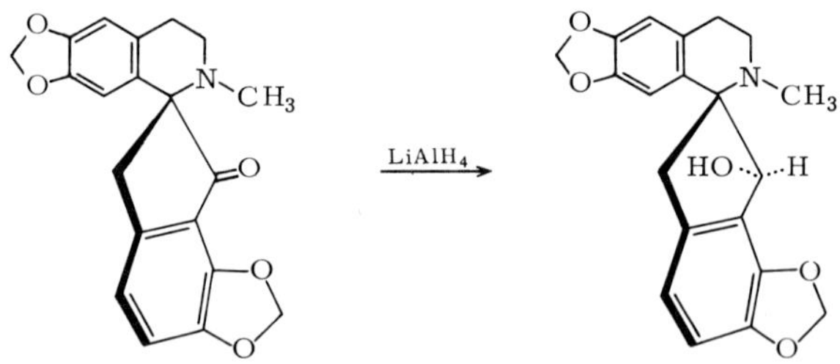

(+)-Fumariline Dihydrofumariline

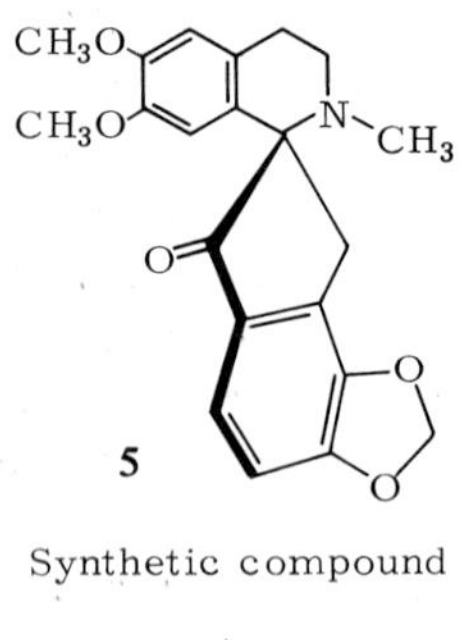

5

Synthetic compound

The three bases fumaricine, fumaritine, and fumariline were isolated from *Fumaria officinalis* L. Because of the paucity of the natural bases and their sensitivity to chemical reagents, it was not possible to carry out chemical degradations. Reliance was placed instead upon spectral methods of analysis.[7,8]

The IR spectrum of fumariline, $C_{20}H_{17}O_5N$, shows a band at 5.85 μ (1709 cm^{-1}), suggesting the presence of a conjugated five-membered cyclic ketone. Furthermore, the UV spectrum of dihydrofumariline formed through sodium borohydride or lithium aluminum hydride reduction of fumariline is virtually superimposable on that of fumaricine, $C_{21}H_{23}O_5N$, but is distinctly different from that of fumariline, again suggesting the presence of a conjugated ketone in fumariline.

The C-8 protons in fumaricine and dihydrofumariline, being singlets, are not coupled to any neighboring aliphatic protons and hence must be adjacent to aromatic rings or quaternary carbons. This conclusion was reinforced by the finding that fumariline did not exchange hydrogen for deuterium in basic or acidic media. Noteworthy also is the observation that the NMR spectrum of fumariline was remarkably similar to that for the synthetic compound **5** obtained by McLean during his synthesis of ochotensimine (see below).

The positions of the aromatic substituents were confirmed through a study of the NOE. In the case of fumariline, there is a 19% NOE for the C-12 aromatic proton when the C-13 protons are saturated. These hydrogens must therefore be close to each other, so that the methylenedioxy group in ring D must be placed at C-9, 10.

Similarly, in the case of fumaricine, the methylenedioxy group must also be at C-9,10, since saturation of the protons at C-13 causes an NOE for the aromatic proton at C-12, and the proton at C-1 shows an NOE when the protons at C-13 are saturated. There is also an observable NOE (14%) for the proton at C-8 when the signal due to the *N*-methyl group is irradiated, so that the C-8 hydrogen and the *N*-methyl group must be on the same side of the spirane system.

As far as fumaritine is concerned, this alkaloid can be converted to fumaricine by treatment with diazomethane. One structural feature of fumaritine that had to be resolved, therefore, was the assignment of the methoxyl group to either C-2 or C-3. Irradiation of the methoxyl group absorption at δ3.85 resulted in a 24% NOE for

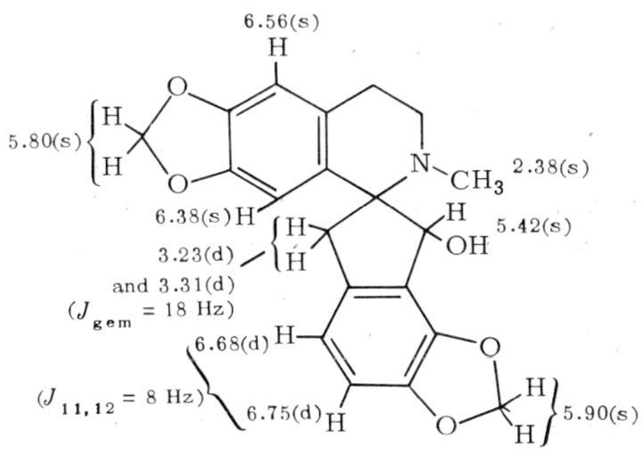

NMR spectral values for dihydrofumariline

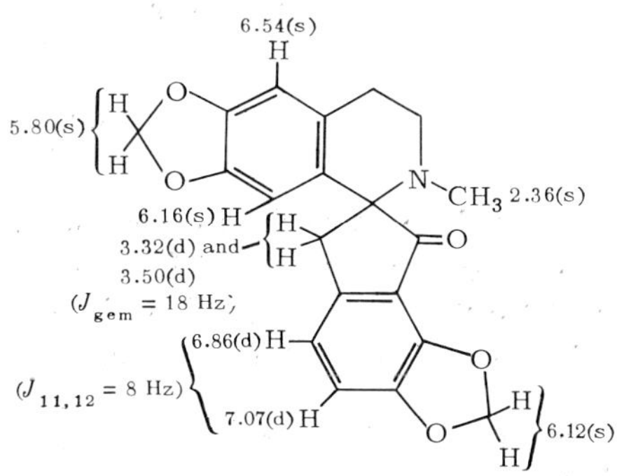

NMR spectral values for fumariline

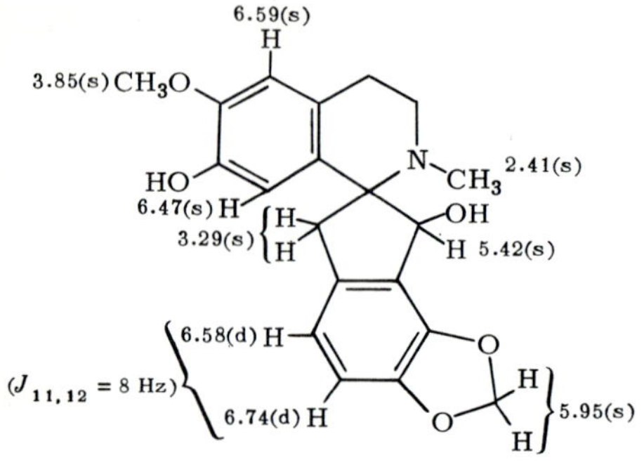

NMR spectral values
for fumaritine

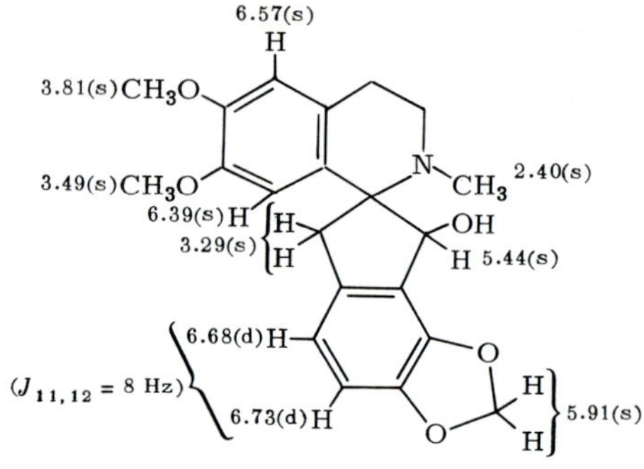

NMR spectral values
for fumaricine

the signal at δ6.59 assigned to the C-4 aromatic proton.[8] It follows that the methoxyl group must be placed at C-3 and the phenolic function at C-2.

C. Sibiricine

Among the alkaloids present in *Corydalis sibirica* L. is the new base sibiricine, $C_{20}H_{17}O_6N$. Other spirobenzylisoquinolines also present in *C. sibirica* are ochotensine and ochrobirine, so that it was immediately suspected that sibricine belonged to the spirobenzylisoquinoline group.

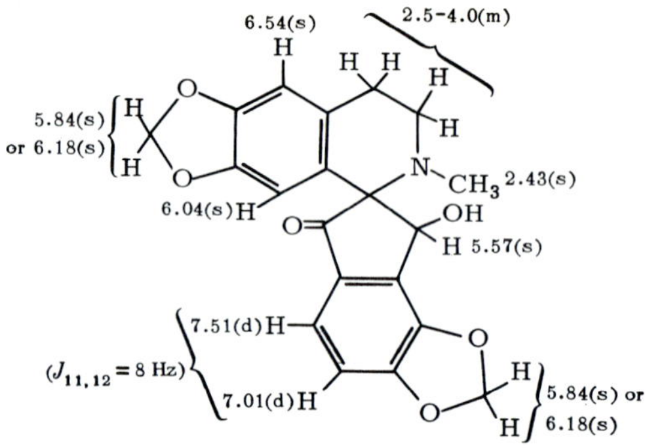

NMR spectral values
for sibiricine

The NMR spectrum of sibiricine bore a direct resemblance to the spectra for fumariline and the synthetic ketone **5**. The configuration of the hydroxyl group at C-8 became apparent from the large NOE of 25% observed at the C-8 proton when the *N*-methyl signal at δ2.43 was irradiated. This effect is possible because these protons must be proximate.

Additionally, the IR spectrum of sibiricine shows a band at 5.85 μ (1710 cm^{-1}) which is virtually at the same frequency as the corresponding carbonyl bands for the synthetic ketone **5** and the alkaloid fumariline.[9]

D. Ochrobirine

The alkaloid (+)-ochrobirine, $C_{20}H_{19}O_6N$, is found in a variety of *Corydalis* species. It is a diol devoid of any carbonyl function, and it readily forms a diacetate. The UV spectrum is reminiscent of that for fumaricine.

The NMR spectrum showed the presence of two methylenedioxy substituents. The hydroxyl groups must be located at C-8 and C-13, since the geminal protons could be observed at $\delta 4.88$ and $\delta 5.42$, respectively.

Irradiation of the *N*-methyl signal at $\delta 2.67$ increased the C-13 proton signal at $\delta 5.42$ by 24% but did not affect the $\delta 4.88$ peak representing the C-8 hydrogen. Additionally, irradiation of the C-8 proton at $\delta 4.88$ produced an increase of 13% in the area of the C-1 proton at $\delta 6.04$. It can be concluded that in ochrobirine the C-13 hydrogen is close to the *N*-methyl group, while the C-8 hydrogen is in the vicinity of the C-1 hydrogen. In a further experiment, it was determined that irradiation of the C-13 proton signal at $\delta 5.42$ increased the area of the C-12 proton signal at $\delta 6.85$, so that the methylenedioxy group must be located at C-9,10.

Finally, oxidation of ochrobirine with dilute nitric acid led to hydrastic acid, thus substantiating the assignment of substituents in ring A.[10]

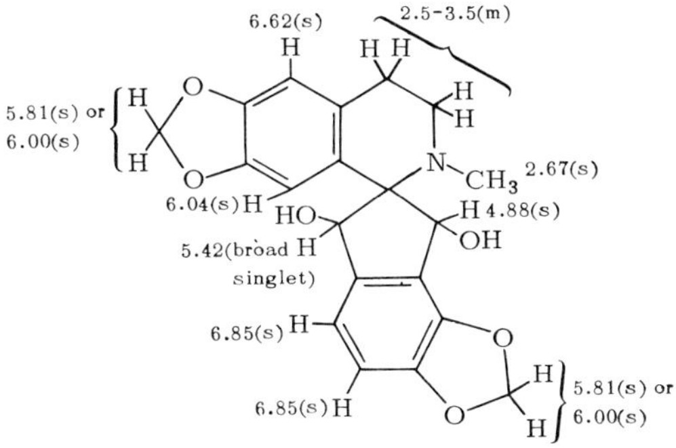

NMR spectral values for ochrobirine

E. Fumarophycine

O-Acetylation of the phenolic alkaloid fumarophycine, derived from *F. officinalis* of Bulgarian origin, gave a product identical to diacetylfumaritine, thus affording a chemical link between these two alkaloids. The structural assignment for fumarophycine was also supported by NMR data.[10a]

III. Syntheses of Spirobenzylisoquinolines

A. The Indanedione Approach

McLean and his co-workers successfully completed a total synthesis of (±)-ochotensimine in 1968.[11] Cyclization of the phenylpropionic acid **6** with phosphorus pentoxide gave the indanone **7**, as shown in Scheme II. Lead tetraacetate was employed to introduce an acetoxyl group adjacent to the carbonyl function, and careful alkaline hydrolysis of the resulting acetate ester afforded the hydroxy ketone **8**. This material was oxidized with Jones reagent to provide the desired 4,5-methylenedioxyindane-1,2-dione (**9**). This compound would not condense with homoveratrylamine, but the more nucleophilic 3-hydroxy-4-methoxyphenylethylamine reacted satisfactorily,* and the tetracyclic base **10** was obtained in good yield. *O*-Methylation of the phenolic function followed by Clarke–Eschweiler *N*-methylation afforded the ketone **11**. The synthesis of ochotensimine was then completed by a Wittig reaction employing methylenetriphenylphosphorane in dimethyl sulfoxide (Scheme II).

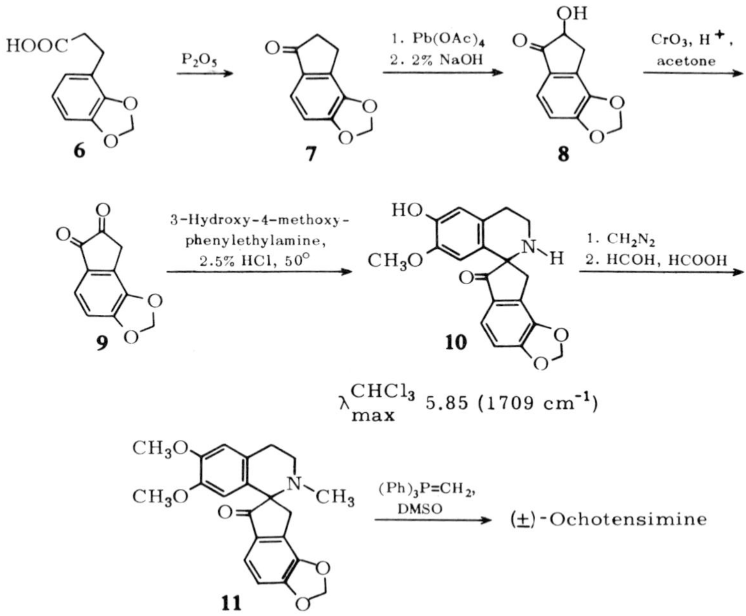

Scheme II

Beckett and Kelly, and Uyeo and his group, have independently carried out syntheses of ochotensimine analogs,[12] as well as of ochotensimine, ochotensine, and fumaricine,[13,14,14a] that parallel the McLean scheme described above.

Manske and Ahmed have also reported on a synthesis of ochrobirine analogs. Condensation of homopiperonylamine with ninhydrin led to the amino dione **12**. *N*-Methylation succeeded by sodium borohydride reduction then furnished a mixture of two diastereoisomeric spirobenzylisoquinoline diols (**13**) (Scheme III).[15]

* *Cf.* Phenolic cyclization, Chapter 1, Section IV, D and Chapter 2 Section VI. E.

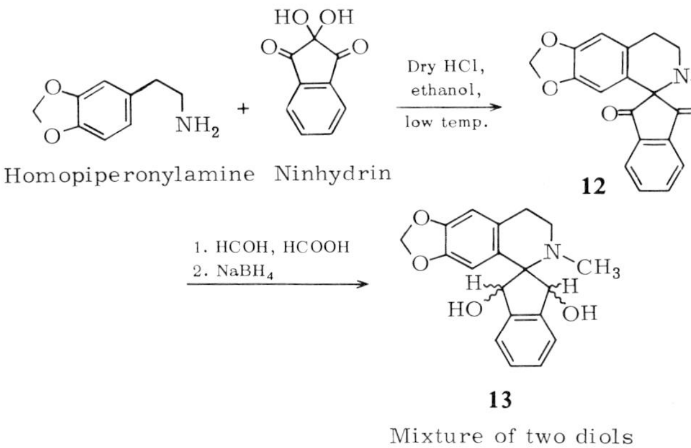

Scheme III

Finally, Kametani and co-workers were able to synthesize (±)-ochrobirine, as shown in Scheme IIIa. 6,7-Methylenedioxyindanone was oxidized to the corresponding methylenedioxyninhydrin using the Kröhnke reaction sequence. Pictet–Spengler condensation with homopiperonylamine, *N*-methylation, and borohydride reduction then furnished (±)-ochrobirine as the sole final product.[15a]

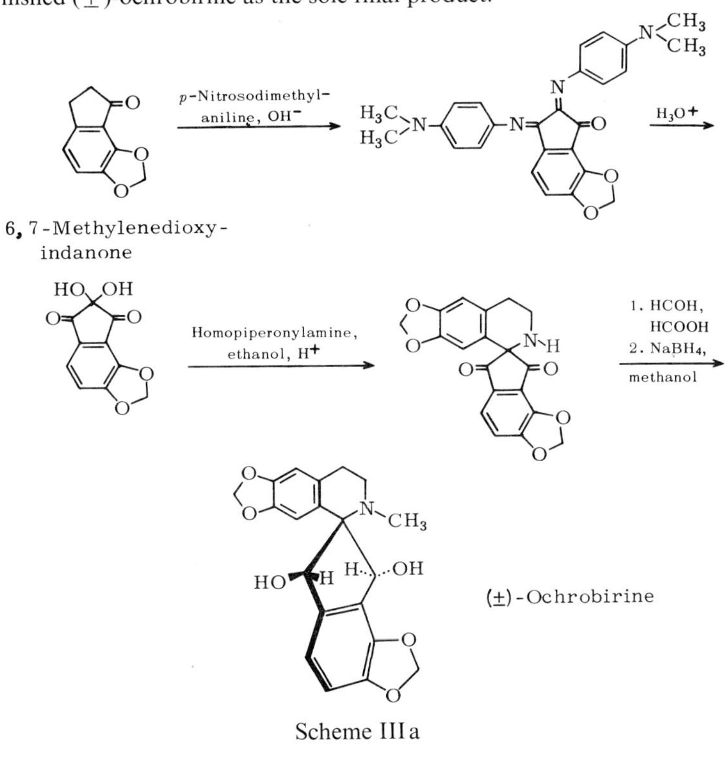

Scheme IIIa

B. The Tetrahydrobenzylisoquinoline Approach

Kametani and his group have succeeded in running an internal Friedel–Crafts condensation on the substituted tetrahydrobenzylisoquinoline **14**. Treatment of the resulting ketone **15** with methylenetriphenylphosphorane in dry ether yielded three products, **16**, **17**, and **18**. The spirobenzylisoquinolines **17** and **18** were then converted to the ochotensimine analog **19**, as described in Scheme IV.[16]

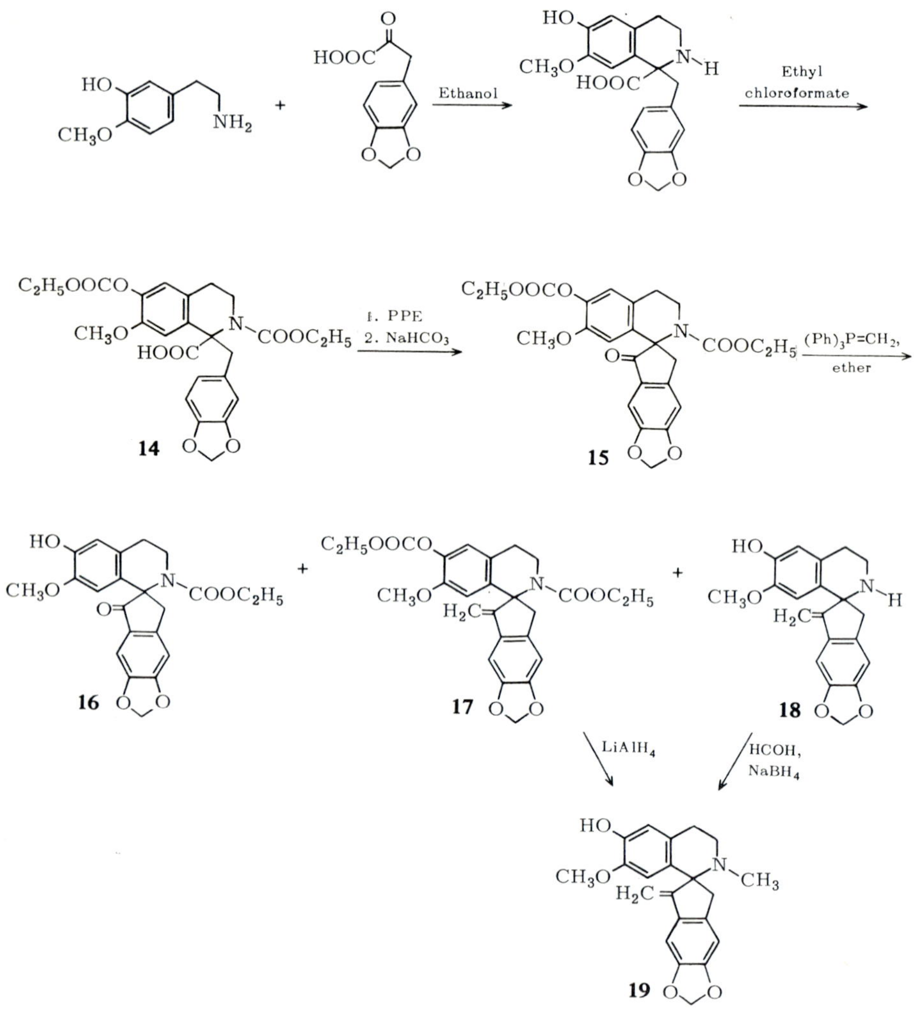

Scheme IV

C. *Approaches using Quinone Methide Intermediates*

1. The Shamma–Jones Synthesis

Any attempt at the formulation of a biogenetic scheme for the spirobenzylisoquinoline alkaloids must proceed from a consideration of the other alkaloids that accompany the spirobenzylisoquinolines in the plant kingdom. These other alkaloids turn out to be of the protoberberine and protopine types. Since protopines are themselves formed from protoberberines, it was possible to adumbrate that protoberberines could also be the biogenetic precursors for the spirobenzylisoquinolines.

Taking ochotensimine as an example, it was logical to assume that this alkaloid still bore some of the scars of its biogenetic history, so that its protoberberine precursor would also have to possess oxygenated functions at C-2, 3, 9, and 10, an *N*-methyl function, and also an unsaturation and a methyl group at C-13. Inclusion of all these requirements into the protoberberine skeleton led to the formulation of the *N*-metho salt **20** as a possible progenitor for ochotensimine.

In a basic medium, this salt can undergo cleavage to the quinonoid intermediate **21** which by the electrocyclic process indicated – essentially a Michael condensation – would form the spirane **22**. A tautomeric shift would then yield the diphenol **23**, which could readily lead to ochotensimine in the plant (Scheme V).

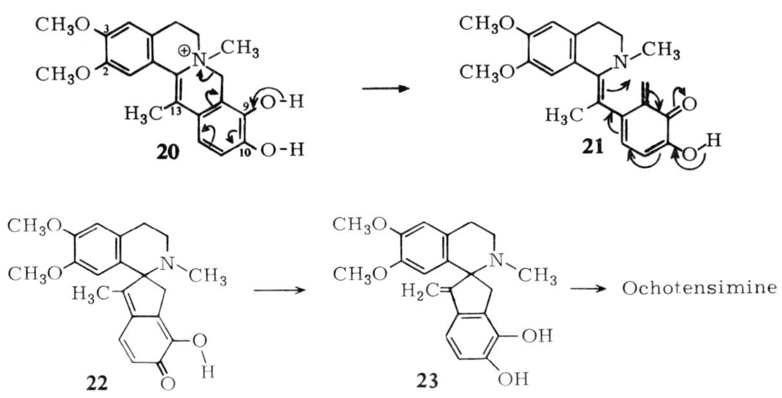

Scheme V

In order to test the above hypothesis, the dibenzyloxyprotoberberine iodide **26** was first synthesized as a precursor to the key diphenolic *N*-metho salt **28**. A 10,11-substituted protoberberine system is much more readily prepared than the 9,10 analog, yet lends itself equally well to the aims of a biogenetic type sequence. Scheme VI describes the steps leading to the preparation of the precursor **26**. The important point to note is that reduction of the imine **24** with sodium borohydride in ethanol at 0° yielded

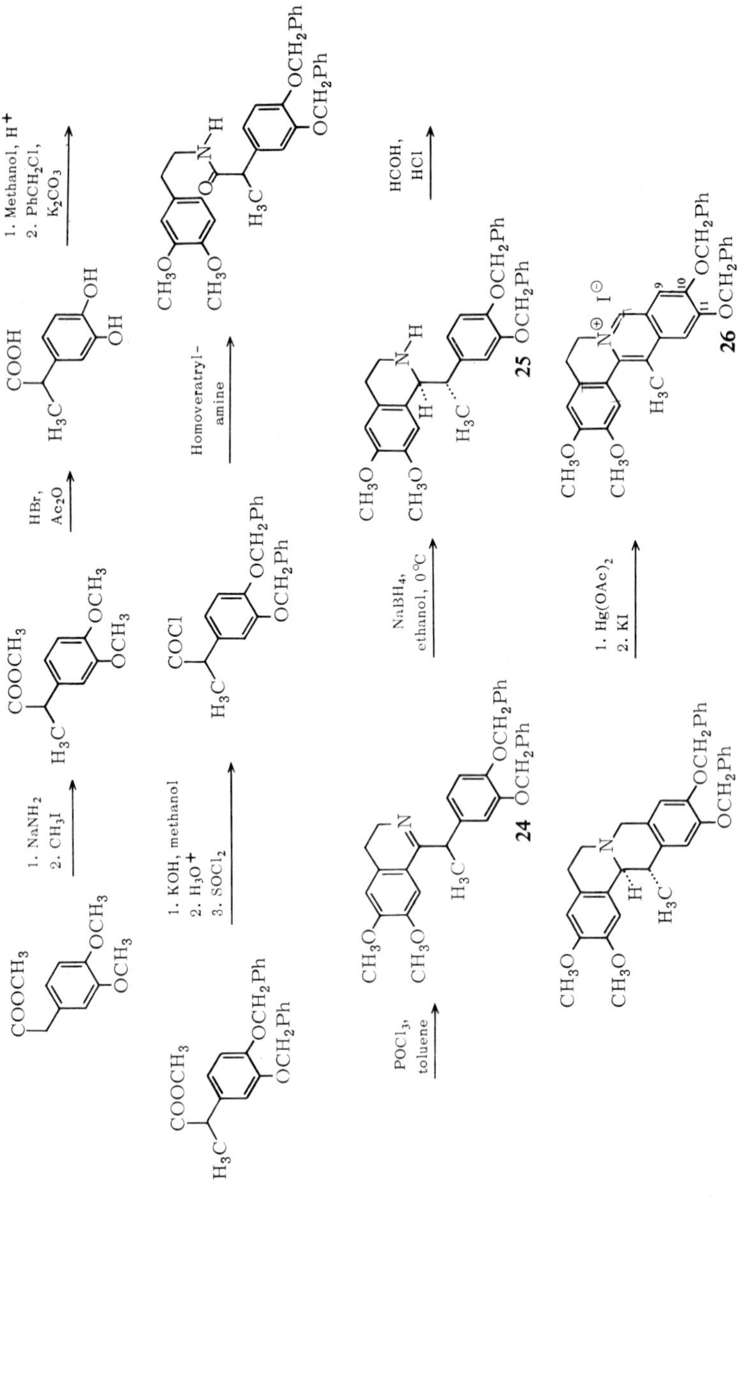

Scheme VI

only one compound, the crystalline benzylisoquinoline **25** produced by hydride addition from the less-hindered side.[17]

The remaining steps in the synthesis are indicated in Scheme VII. Lithium aluminum hydride reduction of the salt **26** followed by immediate treatment with methyl iodide in a dry nitrogen atmosphere furnished the methiodide salt **27**. Selective hydrolysis of the benzyloxy functions was performed with hydrogen bromide in ethanol to give the required diphenolic salt **28**.

When species **28** was refluxed in aqueous ethanolic sodium hydroxide under a nitrogen atmosphere, the quinone methide **29** was formed, characterized by an NMR *C*-methyl singlet at $\delta 1.91$. Addition of DMSO-d_6 (or DMSO) to the quinone methide caused an immediate tautomeric shift with formation of the diphenol **30**. Compound **30** showed the expected vinylic singlets at $\delta 4.60$ and 5.36 (in DMSO-d_6). For further characterization, the critical rearrangement product **30** was diacetylated and reduced with Adams catalyst. The dihydrodiacetate derivative **31** exhibited a C-13-methyl doublet at $\delta 0.95$ ($J = 6.5$ Hz), the corresponding value for dihydroochotensimine being $\delta 0.95$ ($J = 7.2$ Hz). Rates of methiodide formation also showed that in the hydrogenation of the diacetate of **30** the catalyst had approached from the less-hindered side, i.e., the *N*-methyl side, so that the derivative **31** has the stereochemistry indicated. An interesting finding was the fact that in chloroform or ethanol the diphenol **30** tended to tautomerize back to the quinone methide **29**.[17]

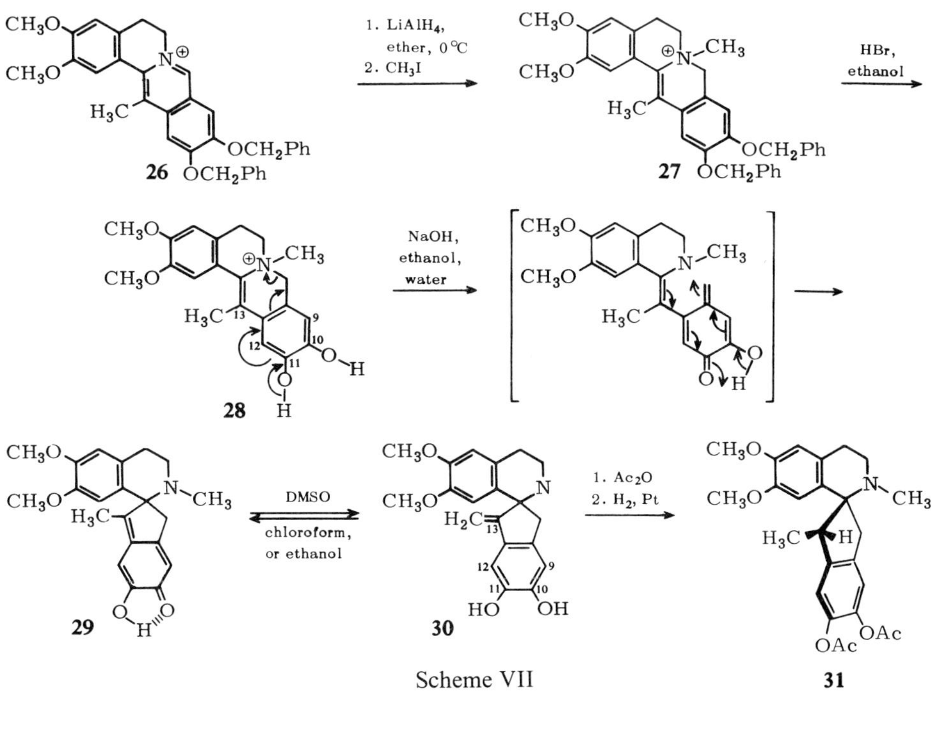

Scheme VII

2. The Shamma–Nugent Syntheses

The Shamma–Nugent syntheses employ as precursors protoberberine *N*-metho salts which are only monohydroxylated in ring D.

The possibility was first considered that the plant does not use an enamine *N*-metho salt such as **20** with two phenolic groups in the same ring as a precursor for ochotensimine, but may prefer instead to proceed through the intermediacy of the diphenolic salt **32** in which the phenolic groups are placed in different rings. Base-induced rearrangement of **32** would then lead to the spirane **33**, which can be transformed into ochotensimine (Scheme VIII).

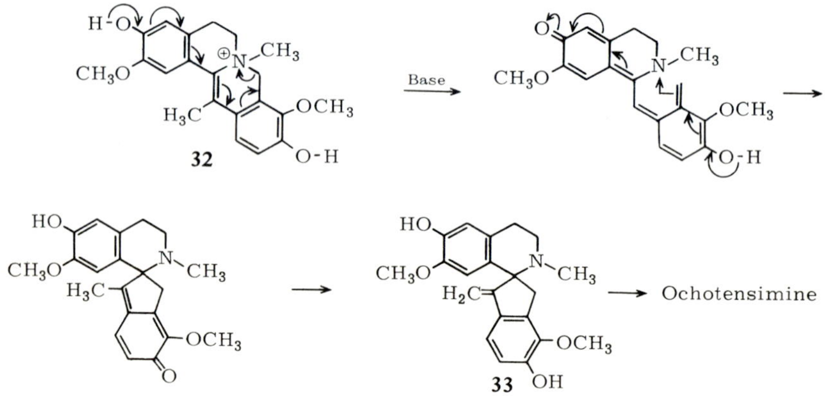

Scheme VIII

To test this thesis, the diphenolic *N*-metho salt **34** was prepared and submitted to prolonged refluxing in aqueous ethanolic sodium hydroxide under a nitrogen blanket. The product was the required spirane **36**, which showed two vinylic proton singlets at δ4.96 and 5.62. The quinone methide **35** could not be isolated because it lacks the stability imparted by internal hydrogen bonding which is present in the analogous compound **29**. Additionally, whereas the salt **28** rearranged to a spiro structure by heating overnight in base, it was necessary for the salt **34** to be refluxed for 4 days to obtain a satisfactory yield of the spiro product **36** (Scheme IX).[18]

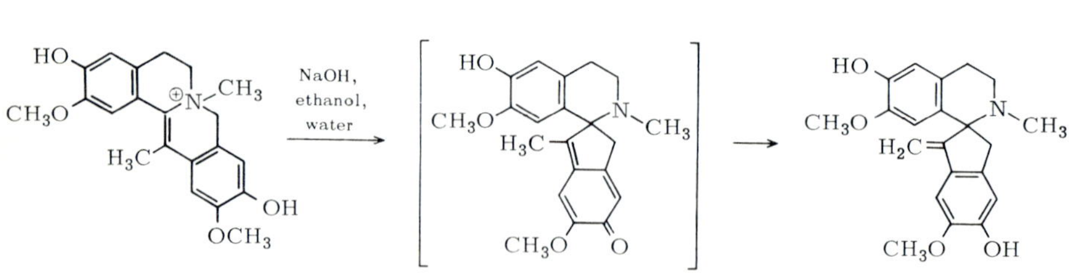

34 **35** **36**

Scheme IX

In the same fashion the monophenolic enamine salt **37** rearranged to the spirobenzylisoquinoline **38** (Scheme X)[19]

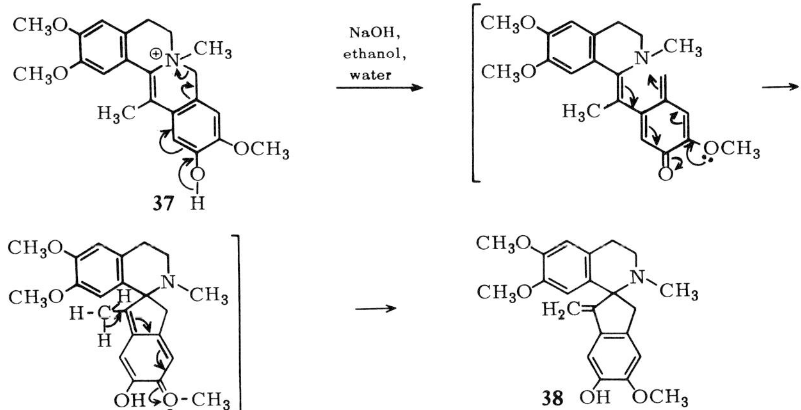

Scheme X

and the monophenolic enamine salt **39** rearranged to the spiro compound **40** (Scheme XI).[19]

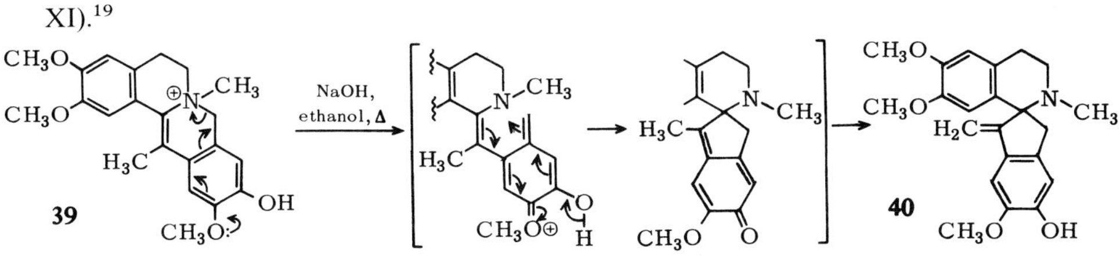

Scheme XI

It should be emphasized that the aforementioned syntheses through quinone methides are no proof for the biogenesis of the spirobenzylisoquinolines. Only experiments with labeled precursors can be conclusive in this respect.

IV. The Protoberberine → Spirobenzylisoquinoline → Dibenzocyclopent[*b*]azepine Rearrangement

Following the above studies on the formation of spirobenzylisoquinolines from phenolic dihydroprotoberberine salts, it was relevant to determine if the monophenol **41** would rearrange in base in the expected manner. Interestingly enough, it was found that the product was the olefinic dibenzocyclopent[*b*]azepine **45** which could be reduced catalytically to a mixture of diastereoisomeric dibenzocyclopent[*b*]-azepines **46** and **47**. In this reaction sequence (Scheme XII) cyclization of the quinone methide **42** leads to the spirobenzylisoquinoline intermediate **43** which possesses a net positive charge spread between rings C and D. This charge induces formation of the aziridinium ion **44** which can readily be transformed to product **45**. A net positive charge is lacking in rings C and D of the spirobenzylisoquinoline intermediates involved in the rearrangements of salts **28**, **34**, **37**, and **39**.[19]

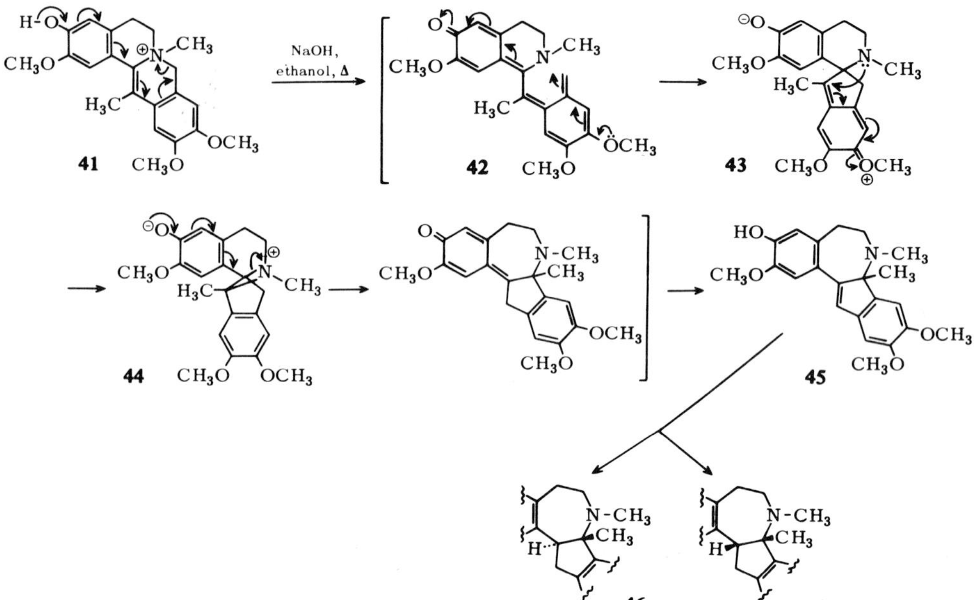

Scheme XII

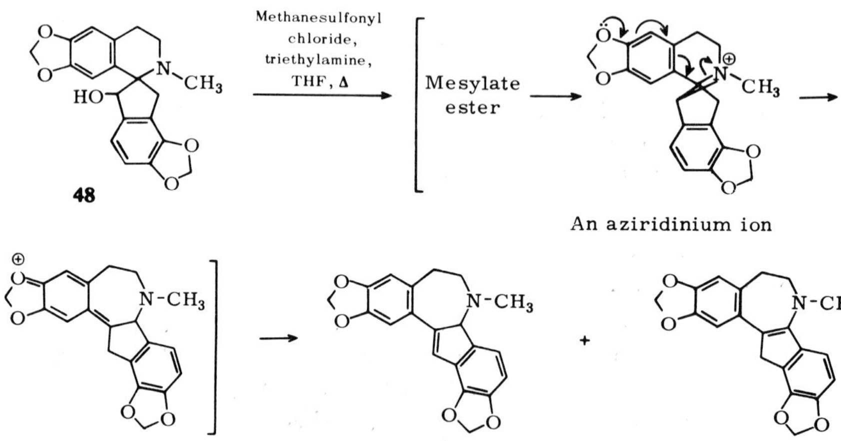

Scheme XIII

It is also relevant to point out that the spirobenzylisoquinoline alcohol **48** can be made to rearrange to two dibenzocyclopent[*b*]azepines when treated with methanesulfonyl chloride in triethylamine,[20] again through the probable intermediacy of an aziridinium ion (Scheme XIII). (See also Chapter 21, Section IX).

V. Mass Spectroscopy

As mentioned earlier, the molecular ion peak for ochotensimine is also the base peak since there is no facile pathway for fragmentation of the molecule.[5]

The mass spectra of fumaricine, fumaritine, and dihydrofumariline show a common fragmentation pattern. The molecular ions are intense but in each case the base peaks are due to the dihydroisoquinoline ion **49** (Scheme XIV).[7,8]

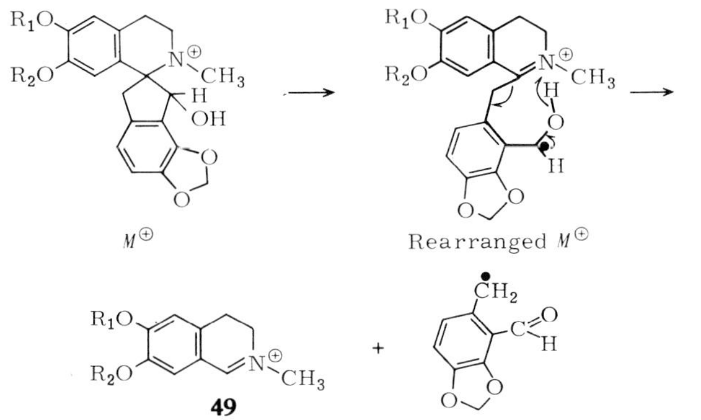

Scheme XIV

The mass spectrum for the ketonic alkaloid fumariline is substantially different from the above. There is again an intense molecular ion peak, but the base peak is at *m/e* 322, corresponding to the loss of the elements of HCO from the molecular ion.[7,8]

The mass spectrum of the keto alcohol sibiricine shows a strong molecular ion peak. Of particular importance is the intense fragment ion at *m/e* 190 resulting from the cleavage typical of ring C-hydroxylated spirobenzylisoquinolines[9]:

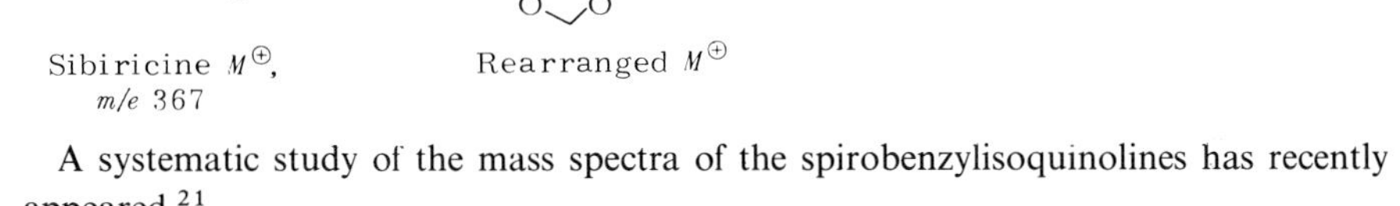

A systematic study of the mass spectra of the spirobenzylisoquinolines has recently appeared.[21]

VI. UV Spectroscopy

UV spectroscopy readily differentiates between ketonic and nonketonic spirobenzylisoquinolines.

Ochotensimine[5]	λ_{max}^{MeOH} 226 and 287 mμ (4.41 and 4.12)
Ochotensine[5]	λ_{max}^{MeOH} 290 mμ (4.26)
Fumaricine[7,8]	λ_{max}^{EtOH} 207, 235, and 288 mμ (4.74, 3.94, and 3.74)
Fumariline[7,8]	λ_{max}^{EtOH} 203, 237, 263, 294, and 355 mμ (4.60, 4.31, 4.05, 3.66, and 3.51)
Sibiricine[9]	λ_{max} 205, 240, 291, and 313 sh mμ (4.80, 3.94, 3.91, and 3.99)
Ochrobirine[10]	λ_{max} 205, 240, and 291 mμ (4.80, 3.94, and 3.91)

REFERENCES

1. For a recent review on the spirobenzylisoquinolines, see M. Shamma, *in* "The Alkaloids" (R. H. F. Manske, ed.), Vol. 13, p. 165. Academic Press, New York, 1971.
1a. D. A. Fesenko and M. E. Perelson, *Khim. Prir. Soedin*, **7**, 166 (1971); *Chem. Abstr.* **75**, 49381n (1971).
1b. C. K.-Yu, J. K. Saunders, D. B. MacLean, and R. H. F. Manske, *Can. J. Chem.* **49**, 3020 (1971).
2. I. A. Israilov, M. S. Yunusov, and S. Yu. Yunusov, *Dokl. Akad. Nauk SSSR* **189**, 1262 (1969); *Chem. Abstr.* **73**, 25715g and 66767u (1970).
3. R. H. F. Manske, *Can. J. Res., Sect. B* **14**, 354 (1936).
4. R. H. F. Manske, *Can. J. Res., Sect. B* **18**, 75 (1940).
5. S. McLean, M.-S. Lin, and R. H. F. Manske, *Can. J. Chem.* **44**, 2449 (1966).
6. S. McLean, M.-S. Lin, A. C. MacDonald, and J. Trotter, *Tetrahedron Lett.* p. 185 (1966).
7. J. K. Saunders, R. A. Bell, C.-Y. Chen, D. B. MacLean, and R. H. F. Manske, *Can. J. Chem.* **46**, 2873 (1968).
8. D. B. MacLean, R. A. Bell, J. K. Saunders, C.-Y. Chen, and R. H. F. Manske, *Can. J. Chem.* **47**, 3593 (1969). For a correlation of intramolecular NOE with internuclear distance, see R. A. Bell and J. K. Saunders, *ibid.* **48**, 1114 (1970).
9. R. H. F. Manske, R. Rodrigo, D. B. MacLean, D. E. F. Gracey, and J. K. Saunders, *Can. J. Chem.* **47**, 3585 (1969).
10. R. H. F. Manske, R. G. A. Rodrigo, D. B. MacLean, D. E. F. Gracey, and J. K. Saunders, *Can. J. Chem.* **47**, 3589 (1969).
10a. M. Castillo, J. K. Saunders, D. B. MacLean, N. M. Mollov, and G. I. Yakimov, *Can. J. Chem.* **49**, 139 (1971).
11. S. McLean, M.-S. Lin, and J. Whelan, *Tetrahedron Lett.* p. 2425 (1968); *Can. J. Chem.* **48**, 948 (1970).
12. B. A. Beckett and R. B. Kelly, *J. Heterocycl. Chem.* **5**, 685 (1968); R. B. Kelly and B. A. Beckett, *Can. J. Chem.* **47**, 2501 (1969).
13. H. Irie, T. Kishimoto, and S. Uyeo, *J. Chem. Soc., C* p. 3051 (1968); p. 1645 (1969).
14. T. Kishimoto and S. Uyeo, *J. Chem. Soc., C* p. 2600 (1969).
14a. T. Kishimoto and S. Uyeo, *J. Chem. Soc., C* p. 1644 (1971).
15. R. H. F. Manske and Q. A. Ahmed, *Can. J. Chem.* **48**, 1280 (1970); T. Kametani, S. Takano, and S. Hibino, *J. Pharm. Soc. Japan* **88**, 1123 (1969).
15a. T. Kametani, S. Hibino, and S. Takano, *Chem. Commun.* p. 925 (1971).
16. T. Kametani, S. Takano, S. Hibino, and T. Terui, *J. Heterocycl. Chem.* **6**, 49 (1969).
17. M. Shamma and C. D. Jones, *J. Amer. Chem. Soc.* **91**, 4009 (1969); **92**, 4943 (1970).
18. M. Shamma and J. F. Nugent, *Tetrahedron Lett.* p. 2625 (1970).
19. M. Shamma and J. F. Nugent, *Chem. Commun.* p. 1642 (1971).
20. H. Irie, S. Tani, and H. Yamane, *Chem. Commun.* p. 1713 (1970).
21. C. K. Yu and D. B. MacLean, *Can. J. Chem.* **49**, 3025 (1971).

Chapter 21 / THE RHOEADINES AND PAPAVERRUBINES

Occurrence: Genus *Papaver* (Papaveraceae)

Approximate Number: 30, about 10 of which have not been isolated from plants, but were obtained through acid-catalyzed isomerization of the natural products

Some Rhoeadines of Interest:

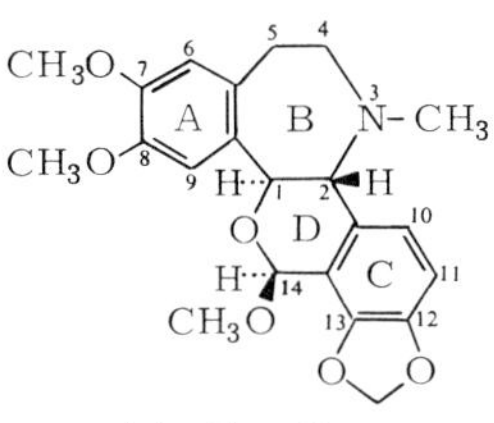

(+)-Glaudine

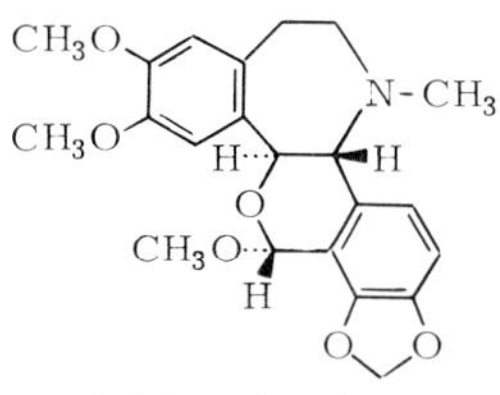

(+)-Epiglaudine

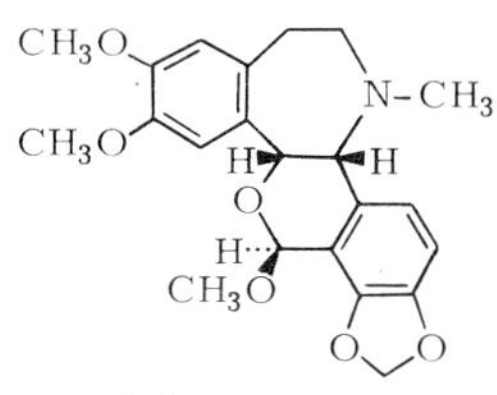

(+)-Oreodine

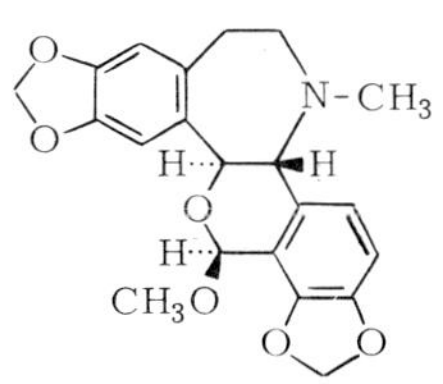

(+)-Isorhoeadine

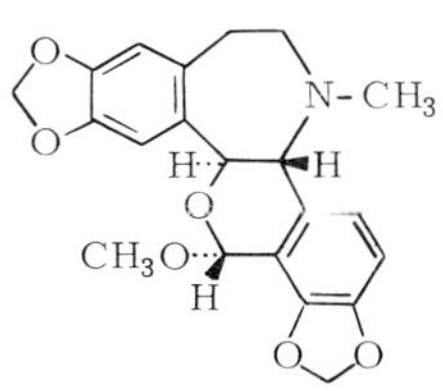

(+)-Epiisorhoeadine

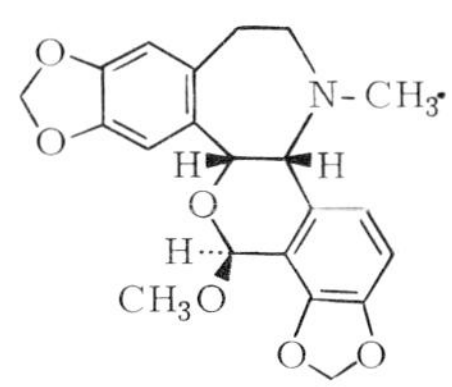

(+)-Rhoeadine

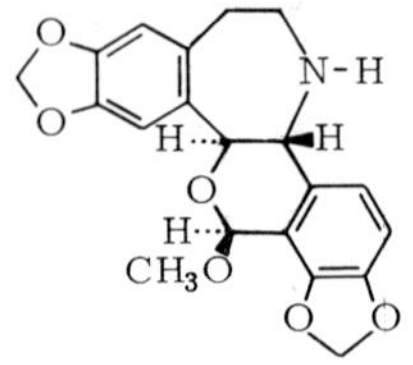

(+)-Papaverrubine A

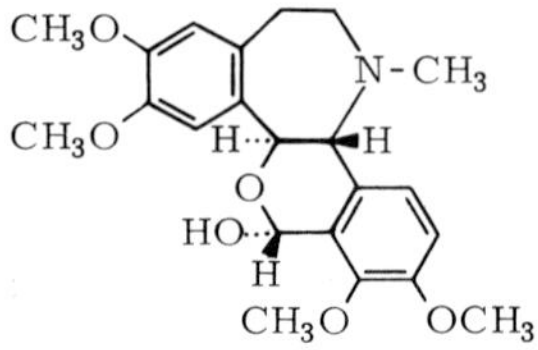

(+)-Alpinigenine

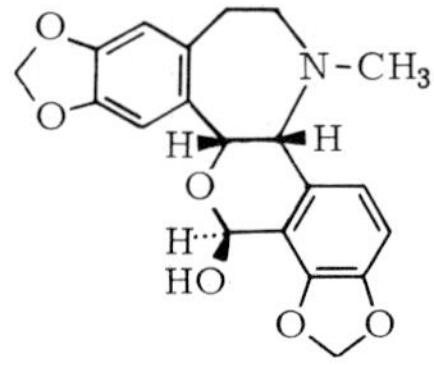

(+)-Rhoeagenine

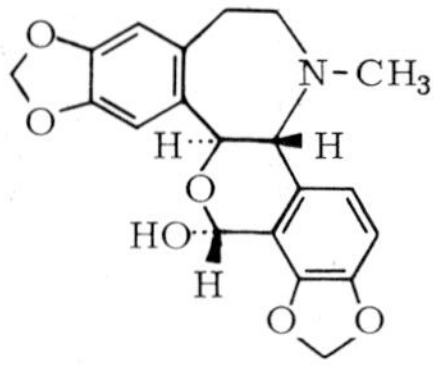

(+)-Isorhoeagenine

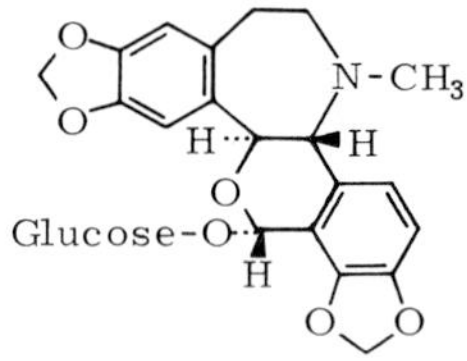

(+)-Isorhoeagenine α-D-glucoside

I. Introduction

The rhoeadines are always present within plants of the genus *Papaver*, family Papaveraceae.[1] They are readily detected on the basis of their characteristic red → brown → green coloration with concentrated sulfuric acid.[2] The outstanding structural feature of the rhoeadines is the presence of a cyclic acetal or hemiacetal system, which was easily recognized in the early structural studies. The seven-membered nitrogen ring is always present, and substituents are inevitably located at C-7, 8, 12, and 13. The papaverrubines are de-*N*-methylrhoeadines and show a red coloration with mineral acids. All the rhoeadines and papaverrubines are dextrorotatory.

The alkaloid rhoeadine, $C_{21}H_{21}O_6N$, was the first rhoeadine-type base to be closely investigated. Early studies indicated the presence of two methylenedioxy groups, one *N*-methyl and one *O*-methyl acetal function. A phthalideisoquinoline structure was first assigned to the alkaloid, but was soon discarded. The correct structure was arrived at by a combination of chemical and physical methods of analysis, especially mass spectroscopy; and these will be discussed in the sections that follow.[3]

II. Reactions of Rhoeadine and Derivatives

Rhoeadine cannot be reduced with lithium aluminum hydride or Adams catalyst. Hofmann degradation of the methiodide salt gave rise to the methine called desrhoeadine and a further Hofmann degradation led to the bismethine **1**. Desrhoeadine and the bismethine **1** were not fully characterized, so that they were of limited value in the structural determination of the alkaloid (Scheme I).

With dilute mineral acid, the hemiacetal rhoeagenine, which is also a natural product was obtained.[4,5] Zinc dust distillation of rhoeagenine gave isoquinoline, while nitric acid oxidation generated hydrastinine.[1,6] Basic permanganate oxidation of rhoeagenine afforded essentially two products, hydrastic acid and 3,4-methylenedioxyphthalic acid, thereby giving a clue to the relative positions of the substituents (Scheme II).[1,7]

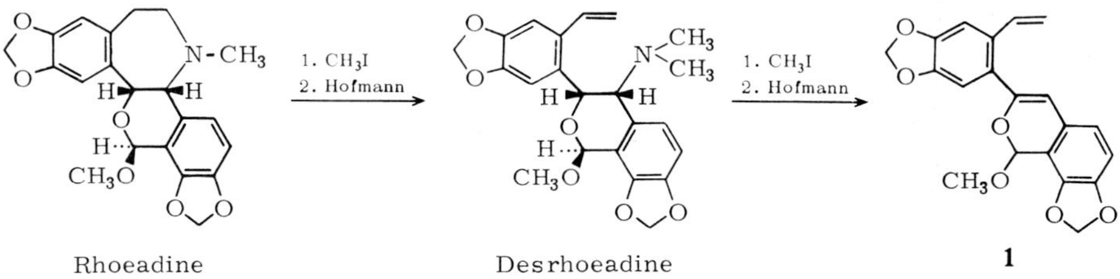

Scheme I

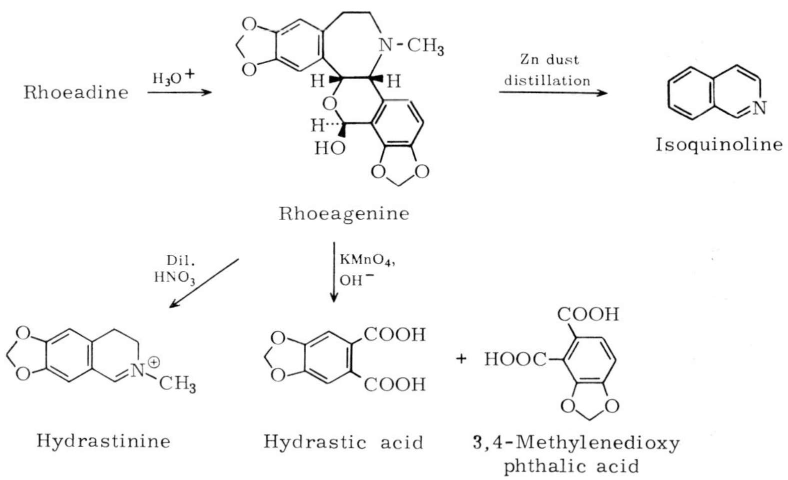

Scheme II

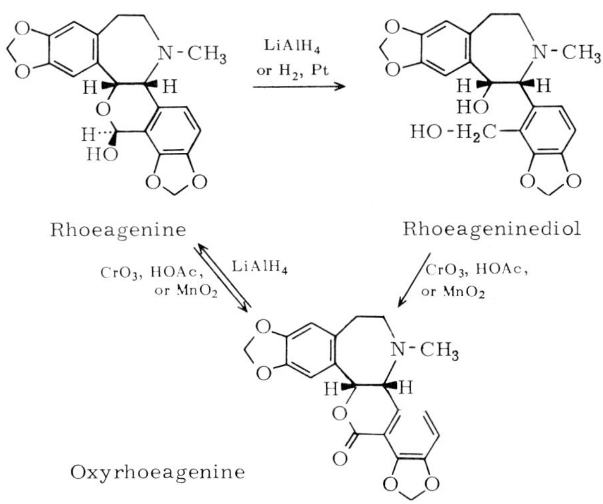

Scheme III

In contrast to rhoeadine, rhoeagenine can be reduced with either lithium aluminum hydride or Adams catalyst to rhoeageninediol, and chromium trioxide or manganese dioxide oxidation of rhoeagenine or rhoeageninediol furnishes the δ-lactone oxyrhoeagenine, which shows a peak in the IR spectrum at 5.80 μ (1725 cm^{-1}). Lithium aluminum hydride reduction of oxyrhoeagenine leads back to rhoeagenine (Scheme III).[3]

Some of the further chemistry of rhoeageninediol has been summarized in Scheme IV. Oxidation with dilute nitric acid furnished hydrastinine and 4,5-methylenedioxyphthalide. Treatment of rhoeageninediol with mineral acid resulted in dehydration, the product being the ether 14-demethoxyrhoeadine.[1,3]

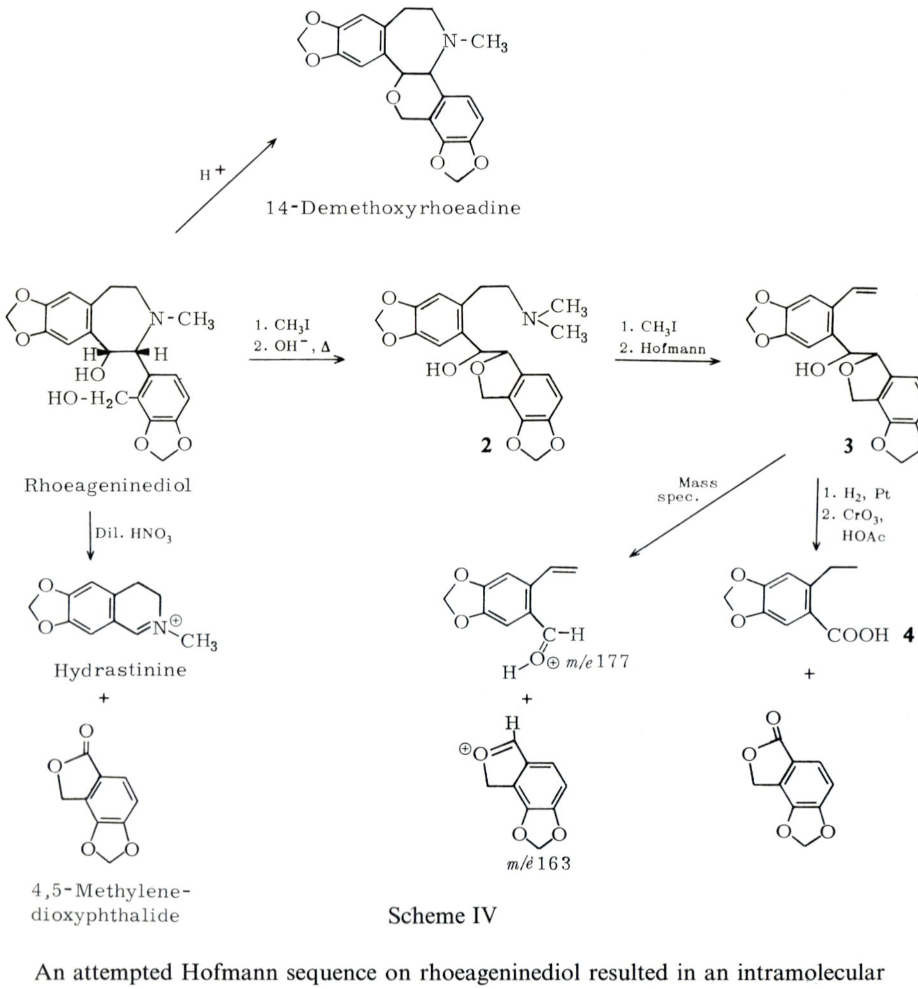

Scheme IV

An attempted Hofmann sequence on rhoeageninediol resulted in an intramolecular S_N2 process with formation of the optically active ether **2**. *N*-Methylation of **2** followed by a Hofmann elimination led to the optically active styrene **3**. The mass spectrum of the styrene **3** was very significant, with intense peaks at *m/e* 177 and 163. Finally, chromium trioxide oxidation of the substituted ethylbenzene obtained through reduction of the styrene **3** gave the acid **4** and 4,5-methylenedioxyphthalide, both of which were fully characterized.[3]

III. An Alternate Degradative Scheme

A new degradation has recently been worked out in connection with some labeling studies. The oxime of the alkaloid alpinigenine was converted to the still optically active cyano acetate derivative **5** by refluxing with acetic anhydride. Attempted Hofmann degradation of the methiodide of **5** produced the bright yellow, optically inactive phthalimidine derivative **6** which is subject to keto–enol tautomerism. Lemieux–Johnson cleavage of **6** yielded two known compounds, the amino aldehyde **7** and hemipinimide (Scheme V).[8,9]

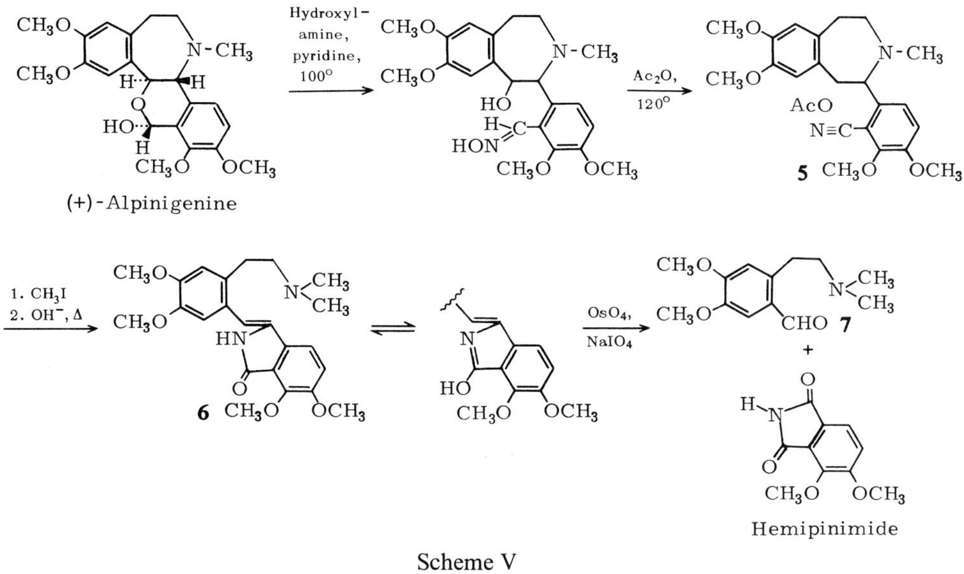

Scheme V

IV. Isomerization of the Rhoeadines and Relative Stereochemistry: NMR Spectroscopy

Differences between the rhoeadine alkaloids may arise because of their relative stereochemistry at the three asymmetric centers, C-1, 2, and 14. In the present discussion, the triad glaudine, epiglaudine, and oreodine will be considered. It is found experimentally that when glaudine is treated with very dilute acid in methanol isomerization at C-14 occurs to give epiglaudine. Stronger acid epimerizes epiglaudine to oreodine, which does not revert back to glaudine (Scheme VI).

The spin–spin coupling constant for the hydrogens at C-1 and 2 is 9 Hz for the B/D trans series of glaudine and epiglaudine and only 2 Hz for the cis-fused analog oreodine. Molecular models show that in the trans series the dihedral angle between the hydrogens in question is in the range of 160 – 180°, thereby accounting for the large coupling constants. The corresponding angle in oreodine is only 75 – 80° so that a low value for $J_{1,2}$ would be predicted and is indeed found.[3,10–13]

Using a combination of kinetic and NMR data, Shamma, Pfeifer and co-workers were able to show that the C-14 methoxyl is beta in glaudine and alpha in epiglaudine, assuming the C-1 hydrogen to be alpha. They also demonstrated that the C-1 and C-2 hydrogens of oreodine must be cis to the C-14 methoxyl group (Scheme VI).[13]

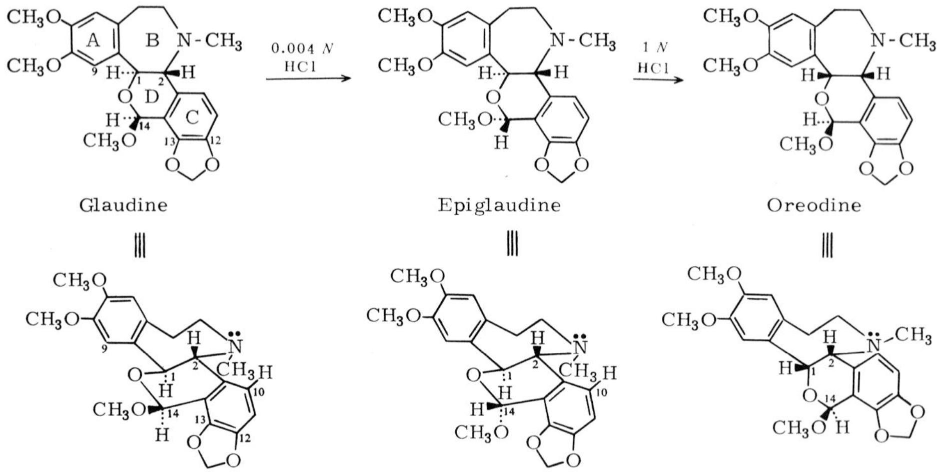

Scheme VI

The reasons for these assignments at C-14 may be summarized as follows[13]:

(1) Molecular models indicate that the above conformations are favored for the pair glaudine–epiglaudine. The C-10 hydrogen is in such close proximity to the nitrogen atom that the N-3 methyl group is forced into an axial position. The very hindered nature of the nitrogen in both glaudine and epiglaudine is reflected in the very slow pseudo-first-order rates of *N*-methylation with methyl iodide: $1.6 \times 10^{-4}\ sec^{-1}$ for glaudine and $2.1 \times 10^{-4}\ sec^{-1}$ for epiglaudine.

(2) The C-1 hydrogen appears at δ5.19 in glaudine and at δ5.57 in epiglaudine. The downfield shift for epiglaudine can be readily accommodated on the basis of the above conformational assignment in which the C-14 methoxyl oxygen is close to the C-1 hydrogen. It is known that anomeric methoxyl protons appear upfield from their equatorial counterparts, and indeed the C-14 methoxyl protons for epiglaudine fall at δ3.56 while the corresponding value for glaudine is δ3.69. No anomaly is involved with the equatorial methoxyl group in glaudine isomerizing in very dilute acid to the

more stable axial position to yield epiglaudine since the "anomeric effect" predicts exactly such a change.[14]*

(3) Oreodine was found to have a quaternization rate with methyl iodide of moderate magnitude, $23.1 \times 10^{-4}\ sec^{-1}$, pointing to only partial hindrance around the nitrogen. The favored conformation above is in accord with such a rate, accessibility to the nitrogen atom being somewhat hindered by ring D. The alternate formulation in which the stereochemistry at C-14 is reversed can be discounted since it would result in inordinate hindrance at the nitrogen by the C-14 methoxyl and a resulting very slow rate of *N*-methylation.

(4) The C-9 hydrogen in glaudine and epiglaudine, being close to the oxygen atom which is incorporated as an integral part of ring D, shows up relatively downfield at $\delta 7.37$ and 7.32, respectively. In oreodine, on the other hand, where no such proximity effect is in force, the C-9 proton appears upfield at $\delta 6.78$.[13,15]

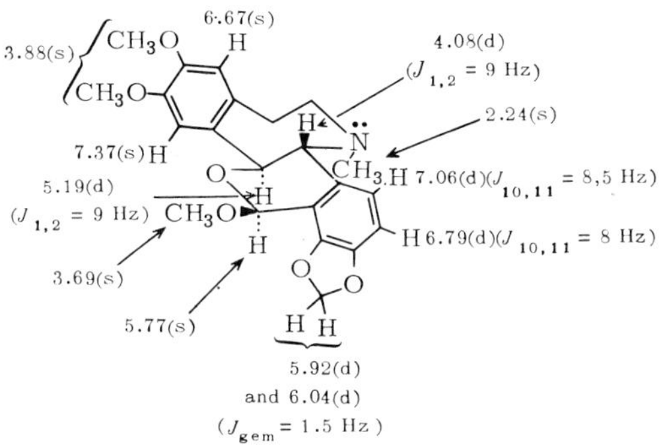

NMR spectral values for glaudine

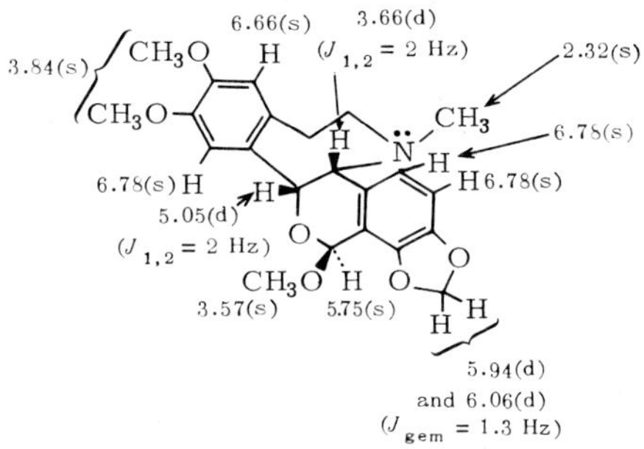

NMR spectral values for oreodine

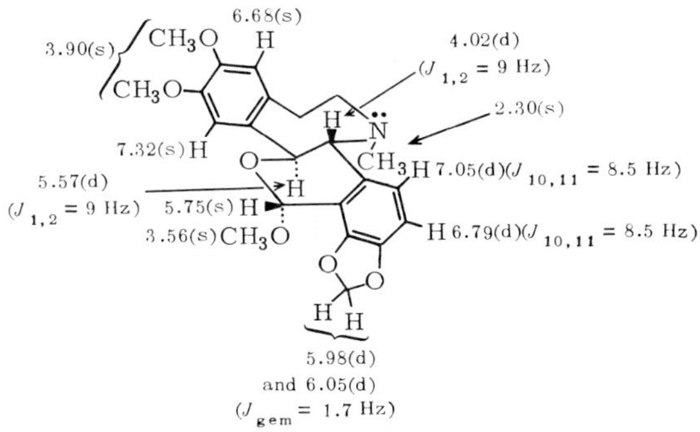

NMR spectral values for epiglaudine

* The chemical shifts quoted here differ very slightly from those in the original paper.

One can easily postulate a mechanism for the isomerization of glaudine to epiglaudine involving opening of ring D followed by recyclization. The irreversible acid-catalyzed isomerization of the trans-B/C-fused epiglaudine to the cis system of oreodine must be due to the latter being thermodynamically more stable.

The preceding diagrams summarize the NMR spectral data for glaudine, epiglaudine, and oreodine. It should be noted that the C-12,13 methylenedioxy hydrogens are in each case split into two doublets because of the asymmetry at C-14.[1]

Whenever a phenolic group is present in ring A, the changes in the chemical shifts of the C-6 and C-9 protons observed upon the addition of sodium deuteroxide to a solution of the alkaloid in DMSO-d_6 can indicate conclusively whether this group is located at C-7 or C-8.[16]

V. (+)-Isorhoeagenine α-D-Glucoside

(+)-Isorhoeagenine α-D-glucoside, $C_{26}H_{29}O_{11}N$, was isolated from *Papaver rhoeas* L. by Němečková and co-workers. Acid hydrolysis yielded rhoeagenine and glucose. NMR spectroscopy indicated, however, that the alkaloidal moiety of the glycoside was isorhoeagenine, since $J_{1,2} = 9.5$ Hz. The known acid-catalyzed isomerization of isorhoeagenine into rhoeagenine had therefore occurred during the hydrolytic process. Application of Klyne's rule of glycoside rotation allowed the assignment of an α-glycosidic linkage to the alkaloid (see Chapter 2, Section V, A).

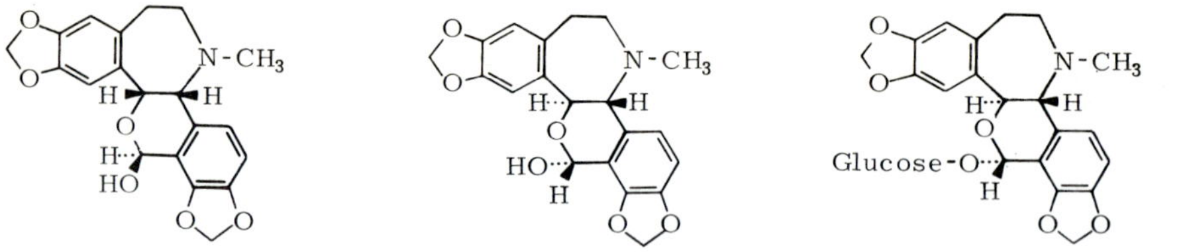

(+)-Rhoeagenine (+)-Isorhoeagenine (+)-Isorhoeagenine α-D-glucoside

The only other glycosidic isoquinoline alkaloids known are the benzylisoquinolines latericine and veronamine and the monoterpenoidal amide ipecoside, which is related to emetine.

VI. Conversion of Rhoeadine to the Protoberberine Alkaloidal Salt Coptisine

The conversion of rhoeadine to rhoeagenine and then to rhoeageninediol has already been considered. Šantavý and co-workers have shown that when rhoeageninediol is boiled for several hours with an excess of thionyl chloride, cyclization with simultaneous dehydrohalogenation occurs to afford the quaternized enamine **8**. *N*-Demethylation with sodium *p*-tolylthiolate was followed either by air oxidation or disproportionation, the product isolated being the known quaternary protoberberine salt coptisine (Scheme VII).[18] Alternatively, hydrogenation of the quaternized enamine iodide **8** with Adams catalyst led to tetrahydrocoptisine methiodide. The known salt **8** had been first obtained by Perkin upon treatment of protopine with phosphorus oxychloride.

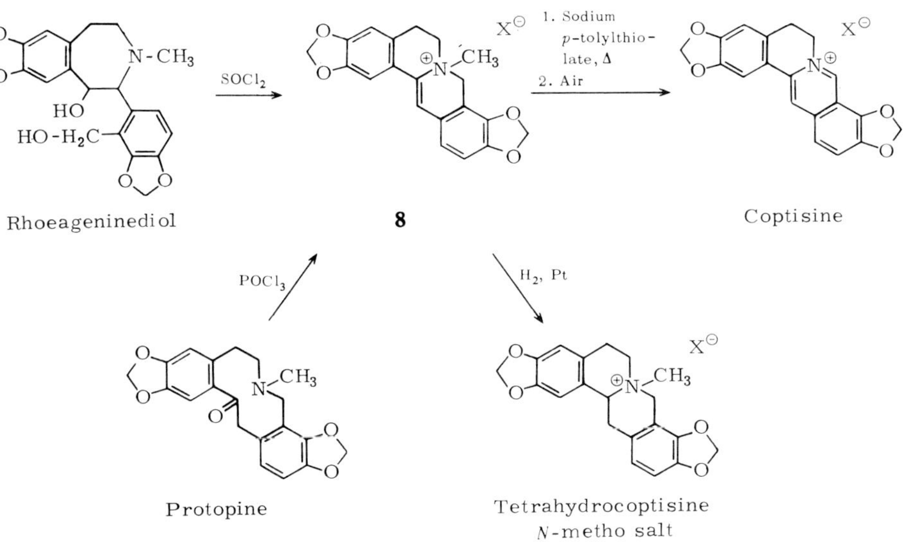

Scheme VII

The mechanism for the transformation of rhoeageninediol to the protoberberine **8** has not been discussed. It may be, however, that rhoeageninediol is first converted to the corresponding dichloride, which through the intermediacy of the aziridinium salt **9** yields the isolated product **8** (Scheme VIII).

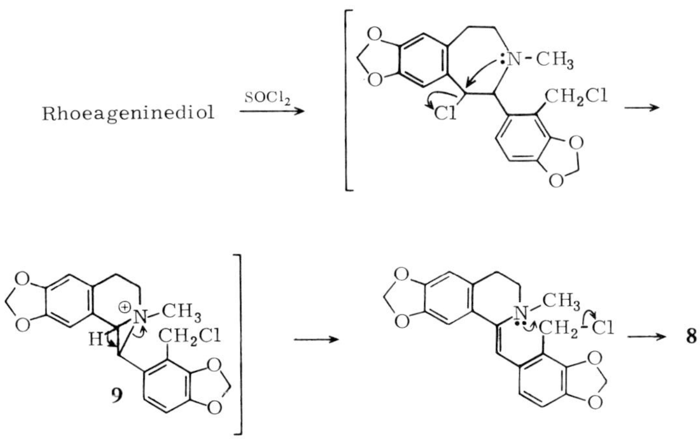

Scheme VIII

Another possible mechanism would proceed by way of protopine:

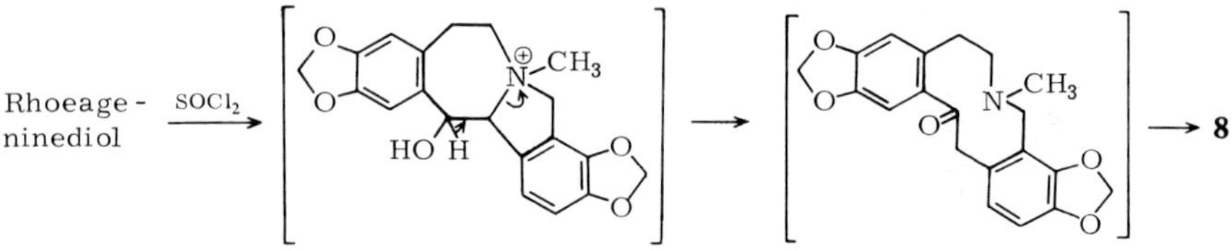

Protopine

VII. Acid-Catalyzed Rearrangement of the Papaverrubines

The papaverrubines are >N–H analogs of the rhoeadines. When the alkaloid papaverrubine A is treated with dilute hydrochloric acid, an intense red salt is formed which has been assigned the quaternary structure **10**. Reduction of this salt affords the colorless tertiary base **11**, which oxidizes readily back to **10**. Oxidation of the tertiary base **11** with mercuric acetate furnishes the red lactam **12** which shows IR absorption at 5.94 μ (1685 cm^{-1}) (Scheme IX).[5,19,20]

Since all the papaverrubines undergo a similar transformation with mineral acids leading to the red salt **10** or one of its analogs, this reaction has become the basis for a color test for this group of alkaloids. The above reaction and color test is widely used in the analysis of samples seized from the illicit opium trade since papaverrubines are always present in opium samples.

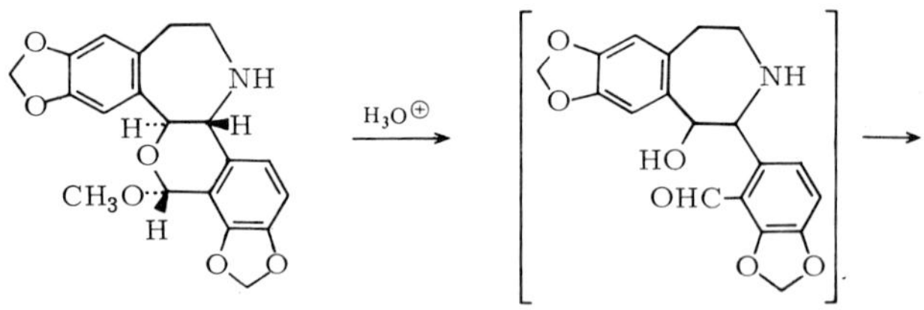

Papaverrubine A

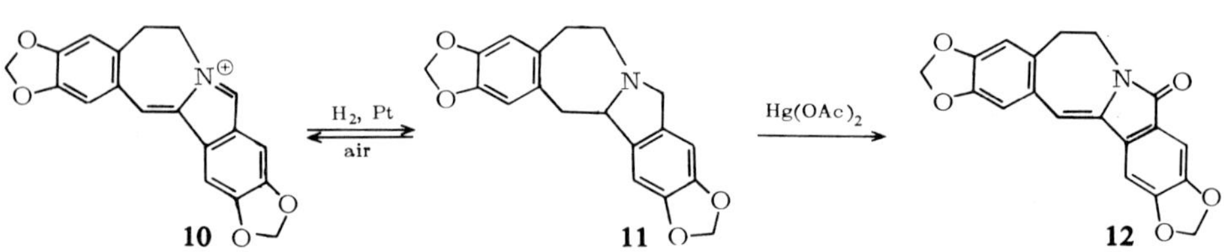

10 **11** **12**

Scheme IX

VIII. Some Early Preparations of Benzazepine Systems

(1) Attempted Pomeranz–Frisch cyclization of the imine acetal **13** in an effort to obtain papaverine afforded mainly the benzazepine **14**, cyclization having occurred in the direction of the more highly activated ring.[21,22]

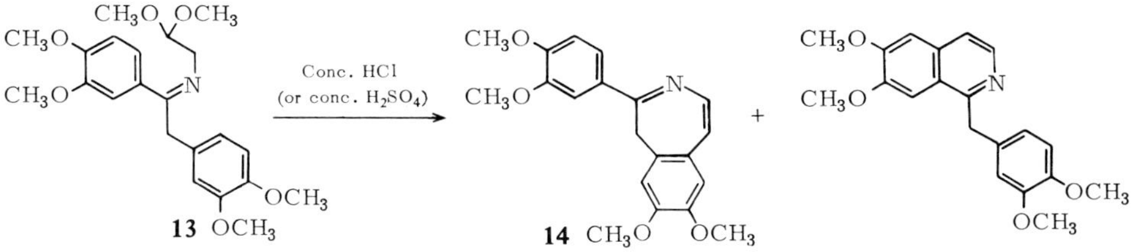

13 14

Papaverine
(minor product)

(2) When either the ketonic immonium salt **15** or the α-amino ketone **16** was reduced with zinc in propionic acid, the benzazepine **17** was isolated.[23]

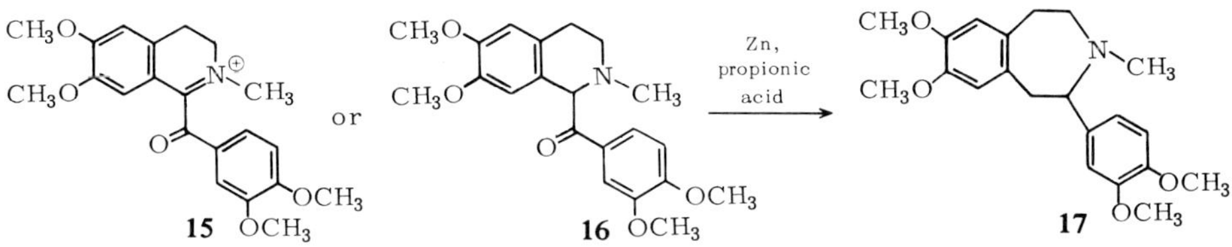

15 16 17

(3) Cyclization of the amino acetal **18** using a Lewis acid has been reported to yield the benzazepine **19**.[24]

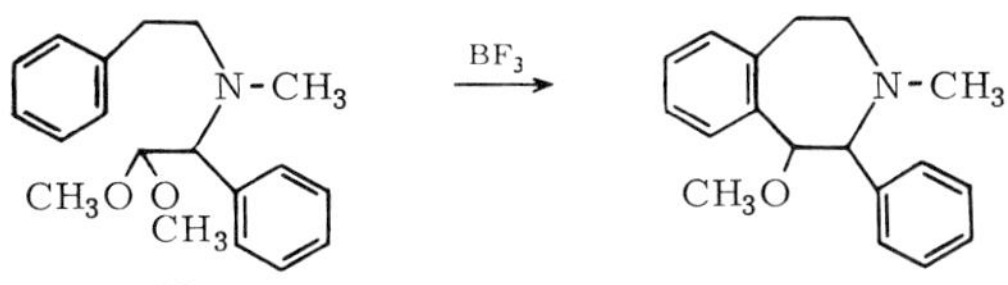

18 19

(4) Treatment of the immonium salt **20** with phenyldiazomethane in methanol furnished the benzazepine **21**.[25]

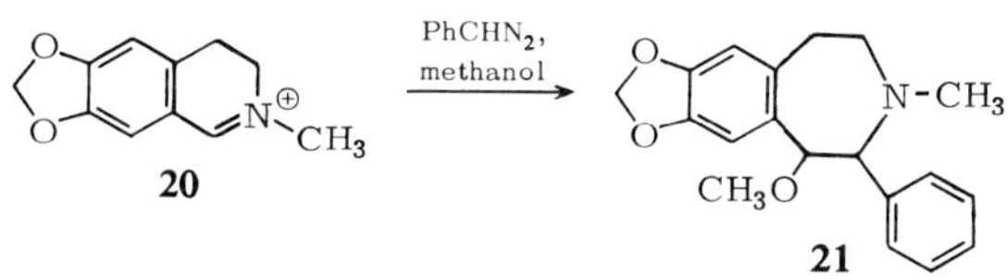

20 21

In a related sequence it was found that treatment of the immonium salt **20** with diazomethane led to the aziridinium salt **22**. Solvolytic ring expansion then generated the benzazepine **23**.[26]

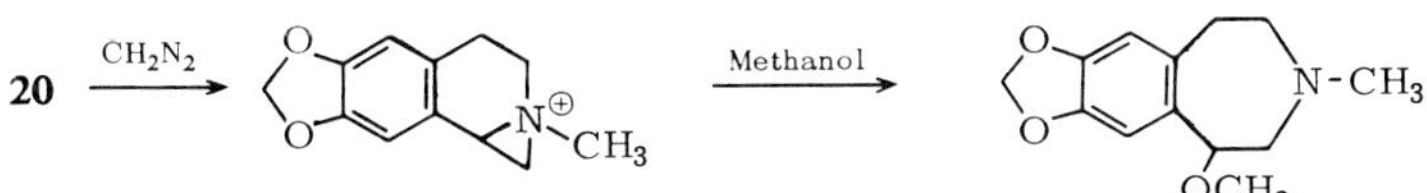

22 23

IX. THE CONVERSION OF A SPIROBENZYLISOQUINOLINE INTO (±)-RHOEAGENINEDIOL

A preparation of (±)-rhoeageninediol has been achieved by Irie and co-workers starting with the synthetic diphenolic ketone **24** by the steps depicted in Scheme IX a.[26a] Lithium aluminum hydride reduction of the keto urethane **25** proceeded from the less hindered side to give the alcohol **26**. When this material was heated with methanesulfonyl chloride in triethylamine and THF, a skeletal rearrangement occurred with formation of two products, **27** and **28**, in a 1 : 1 ratio. Compound **27** was treated with osmium tetroxide, and the resulting glycol **29** could be readily converted to (±)-rhoeageninediol (**30**). (−)-Rhoeageninediol had previously been converted into (+)-rhoeagenine and (+)-rhoeadine[3]; however, the above synthesis of (±)-rhoeageninediol

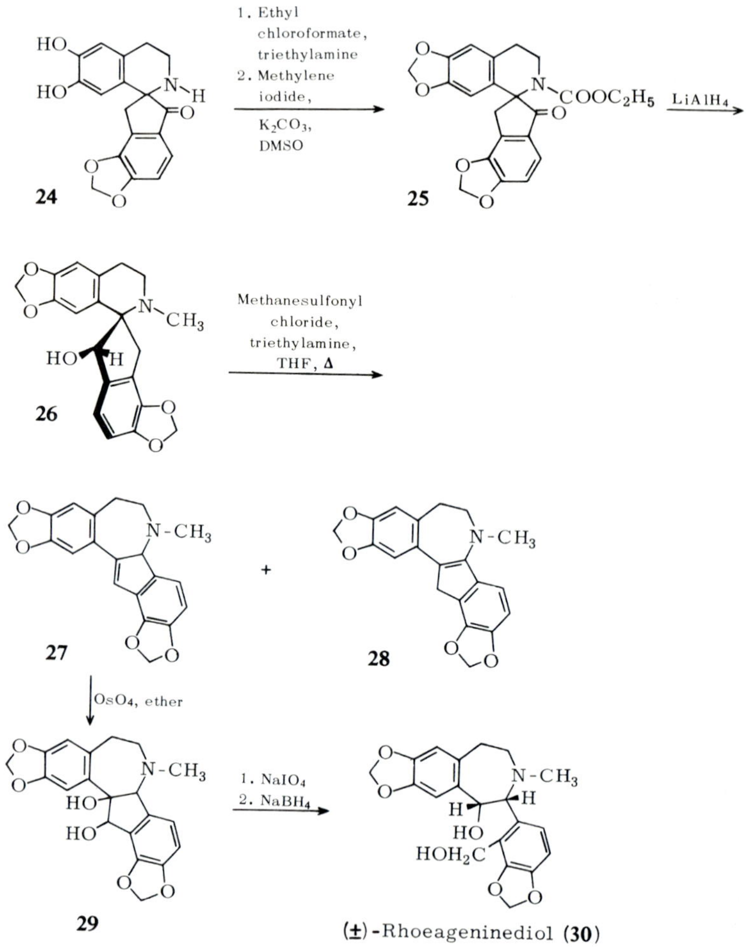

Scheme IX a

cannot be considered a total synthesis of rhoeadine since (±)-rhoeageninediol has not yet been converted into (±)-rhoeadine.

The rearrangement of **26** to the dibenzocyclopent[*b*]azepines **27** and **28** proceeds through the probable intermediacy of an aziridinium ion (see Chapter 20, Section IV).

X. The Roche Nutley Conversions of Phthalideisoquinolines into Rhoeadines

Klötzer and co-workers have carried out the interesting conversion of nornarceine (**31**) into the rhoeadine analogs **37** and **38**. Nornarceine (**31**) derived from naturally

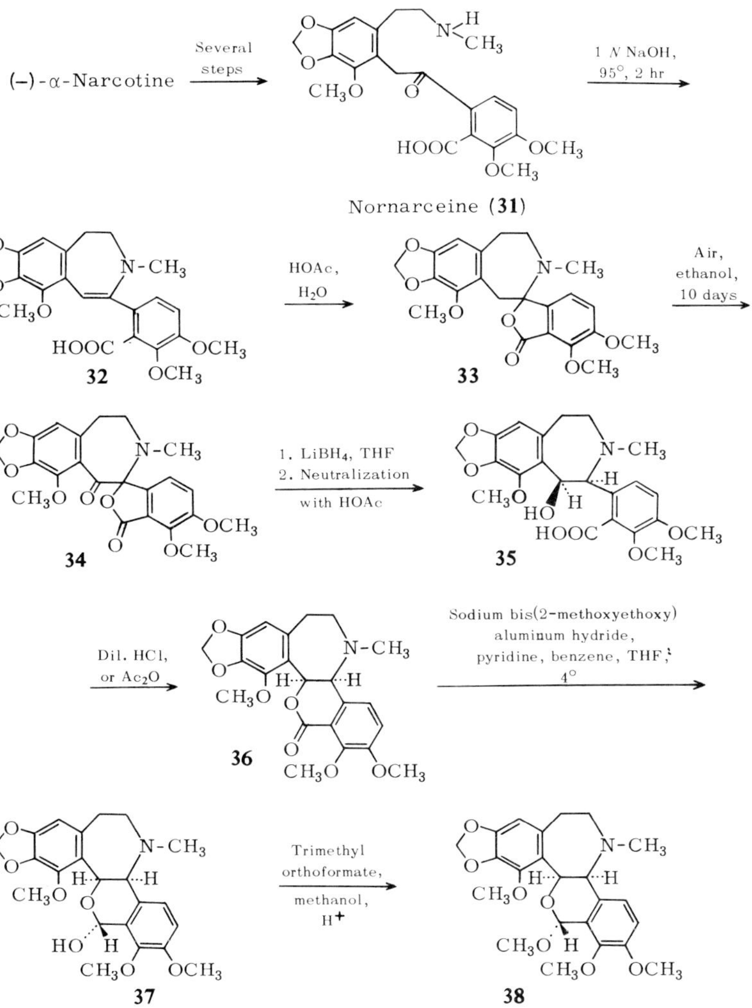

Scheme IXb

occurring (−)-α-narcotine by an improved procedure from that alluded to in Chapter 19, Section II, was first heated in 1 *N* sodium hydroxide to afford the enamine **32** which was readily cyclized in aqueous acetic acid to the amino γ-lactone **33**. Upon standing, **33** oxidized to the ketone **34**. Lithium borohydride reduction followed by neutralization with acetic acid led to the *cis* hydroxy acid **35**. Efforts at cyclization under a variety of experimental conditions always led to the thermodynamically more stable *cis* δ-lactone **36** which was reduced as indicated to the hemiacetal **37**. An X-ray analysis of the methiodide salt of **37** established conclusively the stereochemistry. The methyl acetal **38** could also be derived from **37** (Scheme IX b).[26b]

As an extension of the above work, the phthalideisoquinoline alkaloid (−)-bicuculline was converted into naturally occurring (+)-rhoeadine. Since (−)-bicuculline was obtained from (−)-β-hydrastine whose synthesis had been reported in 1950, this transformation constitutes the first total synthesis of natural (+)-rhoeadine.[26c]

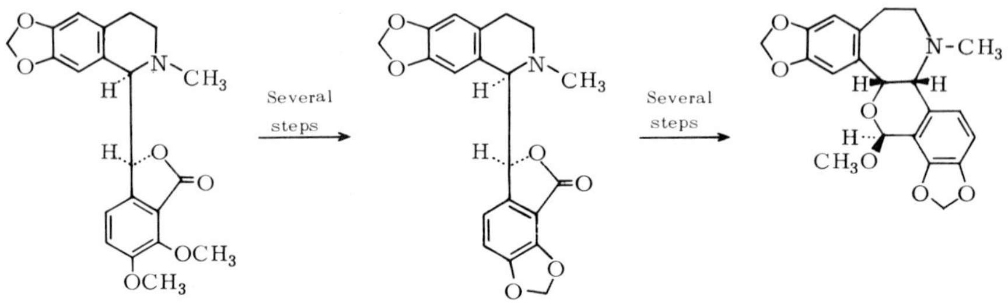

(−)-β-Hydrastine (−)-Bicuculline (+)-Rhoeadine

XI. Optical Activity, ORD, CD, and Absolute Configuration

Šantavý, Hrbek, and Bláha have pointed out that, disregarding the stereochemistry at C-14, the specific rotation at the sodium D line is higher for a rhoeadine type alkaloid with a trans B/D juncture than for its exact analog in the cis series. For example, in chloroform solution, oreodine exhibits $[\alpha]_D + 224°$ and glaudine shows $[\alpha]_D + 455°$.[27] But a study of the ORD and CD curves of a number of rhoeadine alkaloids did not permit a completely unambiguous assignment of absolute configurations.[15,27]

Application of the new aromatic chirality method has led to the determination of the absolute configurations of all the rhoeadine and papaverrubine bases.[27a] A Davydov splitting was observed in the allowed A → B transitions for (+)-glaudine, (+)-epiglaudine, and (+)-oreodine, which were more clearly observed in *n*-hexane than in ethanol solution. The Cotton effect was positive in each of the three cases, so that the two aromatic chromophores interact as depicted in the stereoformulas that follow (positive chirality), thus establishing the absolute configuration for each compound. The chiral center at C-14 can be safely regarded as exerting a negligible influence on the split Cotton effects.

(For other applications of the aromatic chirality method see Chapter 4, Section II, and Chapter 32.)

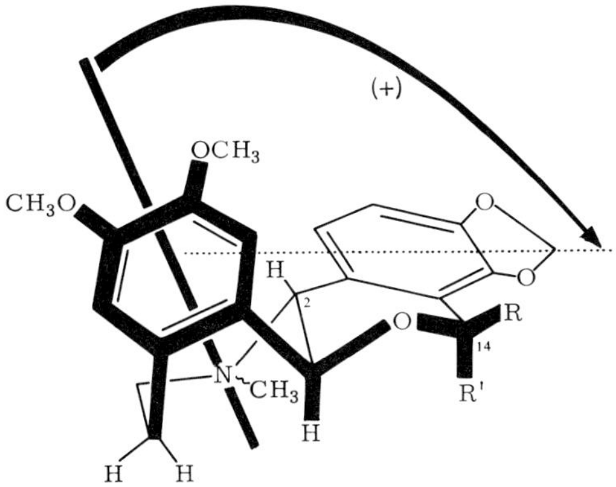

(+)-Glaudine; R = OCH_3, R' = H
(+)-Epiglaudine; R = H, R' = OCH_3

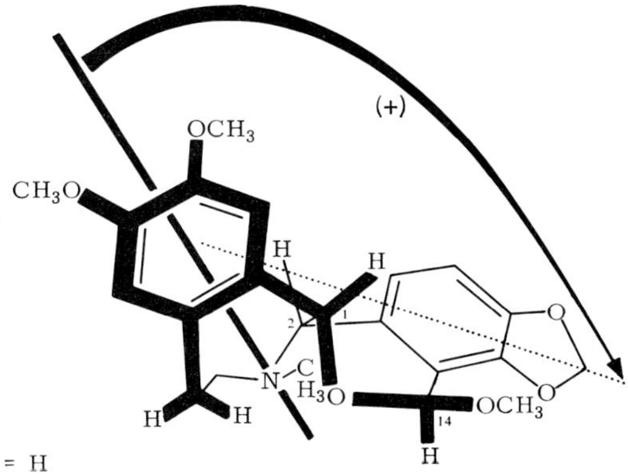

(+)-Oreodine

Since all the rhoeadines and papaverrubines are strongly dextrorotatory and their relative configurations are known, it follows that they must all possess the absolute configurations depicted above. The determination of the configurations of the rhoeadines also clearly indicates that acid-catalyzed transformation of an epiglaudine to an oreodine analog (*trans* → *cis* B/D) proceeds through isomerization at C-1 rather than at C-2.[27a] The following classification can, therefore, be made for the rhoeadine bases:

(a) Glaudine analogs: Isorhoeadine, papaverrubine A, papaverrubine B, papaverrubine D, alpinine, and papaverrubine G.

(b) Epiglaudine analogs: Isorhoeagenine, isorhoeagenine-α-D-glucoside, glaucamine, epipapaverrubine B, *N*-methyl-14-*O*-desmethylepiporphyroxine, papaverrubine C, alpinigenine, and *O*-methylalpinigenine.

(c) Oreodine analogs: Rhoeadine, rhoeagenine, dubirheine, papaverrubine E, oreogenine, and papaverrubine F.

XII. Biogenesis

Because of the resemblance of the rhoeadines to the phthalideisoquinoline alkaloids, it is conceivable that an alkaloid of the narceine type plays a key role in the biogenesis of the rhoeadines. Scheme X has, therefore, been advanced as a possible biogenetic sequence.[3] Its one disadvantage is that it should naturally lead to lactonic oxyrhoeadines, rather than to acetal or hemiacetal species, and no naturally occurring oxyrhoeadines are known. To overcome this difficulty a reductive step was postulated at the terminal stage.[3]

An alternate possibility is that the precursor is not a lactonic phthalideisoquinoline but rather the acetal or hemiacetal analog of a phthalideisoquinoline which, proceeding through the methine and narceine stages parallel to those outlined in Scheme X, could furnish the rhoeadine base without reduction at the terminal stage.

Studies with labeled precursors have been initiated. Following the degradative scheme described in Section III, it was found that when [3-^{14}C]-(±)-tyrosine was

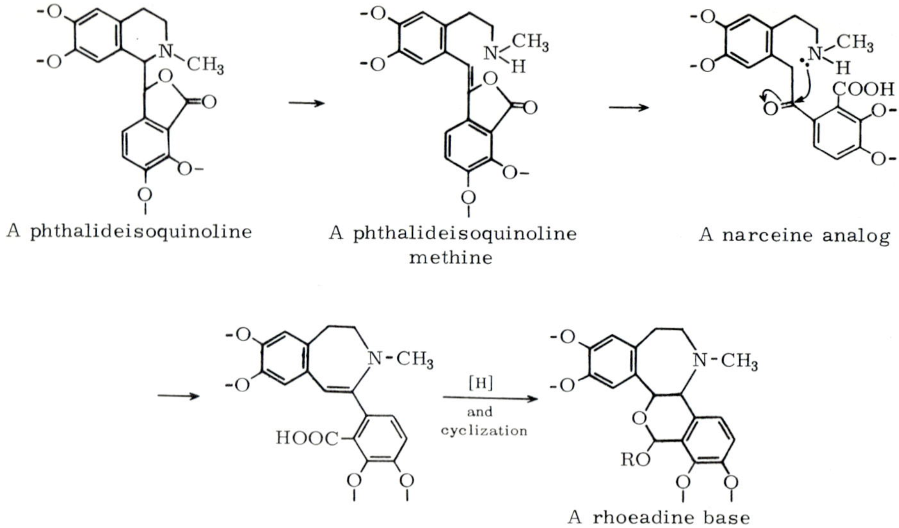

Scheme X

fed to *Papaver bracteatum* Lindl. both the amino aldehyde **7** and the hemipinimide obtained were labeled. These data indicate that as expected 2 moles of tyrosine are involved in the biogenesis of the rhoeadine alkaloids.[8,9]

XIII. Mass Spectroscopy

Dolejš and Hanuš have reported on a comparative study of the mass spectra of the rhoeadines, as a result of which two conclusions could be drawn[28]: (1) The fragmentation pattern for the acetal series is substantially different from that for the hemiacetals, and (2) stereochemical differences do not lead to significantly different spectra.

The mass spectrum for the acetal alkaloid rhoeadine possessing a cis B/D fusion is almost identical to that for its diastereoisomer isorhoeadine with a trans B/D fusion. Both spectra are characterized by three very intense peaks belonging to the molecular ion and ions of *m*/*e* 368 $(M - 15)^+$ and *m*/*e* 177 (base) (Scheme XI).

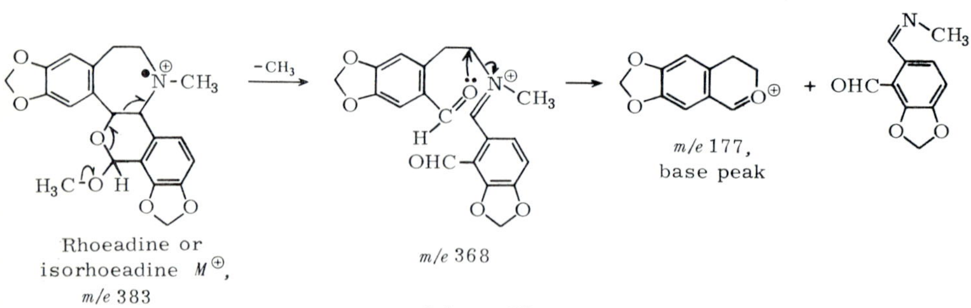

Scheme XI

The hemiacetal alkaloid rhoeagenine, on the other hand, behaves in a completely different fashion under electron impact. The spectrum contains three intense peaks, at *m/e* 206, 192, and 163, formed as indicated in Scheme XII.

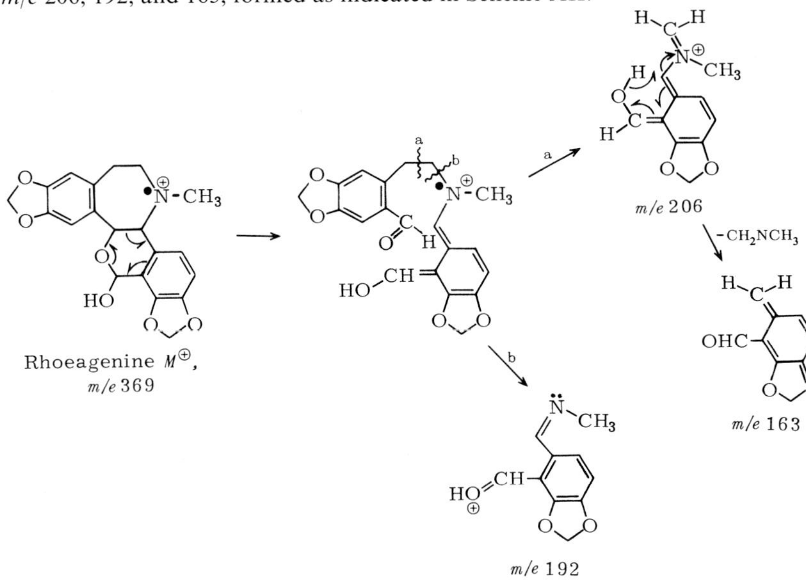

Scheme XII

The mass spectrum of desrhoeadine, the methine base from rhoeadine, is instructive in that it too is composed of essentially three intense peaks at *m/e* 221, 206, and 163. These can be explained by a sequence of cleavages initiated by a retrograde diene condensation of the acetal ring (Scheme XIII).

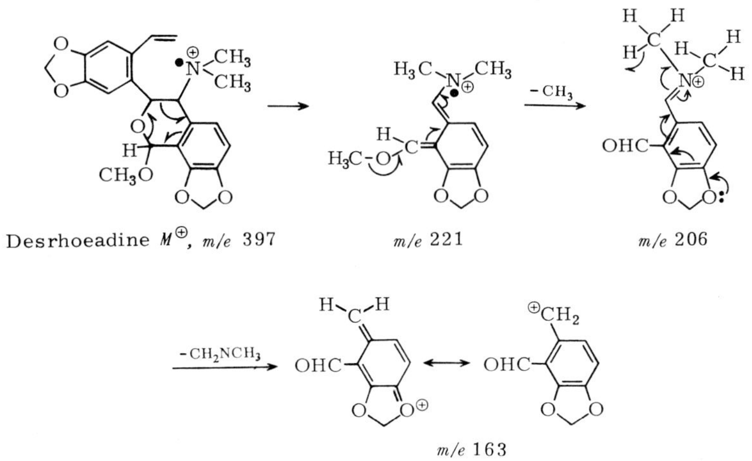

Scheme XIII

XIV. UV Spectroscopy

The rhoeadines show characteristic maxima between 230 and 245 mμ and also between 284 and 294 mμ.

Rhoeadine[1]	λ_{max}^{EtOH} 205, 240, and 292 mμ (4.91, 3.96, and 3.94)
	λ_{min}^{EtOH} 229 and 263 mμ (3.90 and 3.28)
Rhoeagenine[29]	λ_{max}^{MeOH} 243 and 290 mμ (3.97 and 3.96)
Oreodine[29]	λ_{max}^{MeOH} 235 and 285 mμ (4.19 and 3.94)
	λ_{min}^{MeOH} 225 and 260 mμ (4.14 and 3.28)
Alpinigenine[12]	λ_{max}^{MeOH} 230 and 284 mμ (4.19 and 3.79)
	λ_{min}^{MeOH} 260 mμ (3.26)

References

1. For a review, see F. Šantavý, *in* "The Alkaloids" (R. F. H. Manske, ed.), Vol. 12, p. 398. Academic Press, New York, 1970; for a discussion of the *Papaver* alkaloids, which include the rhoeadines and papaverrubines, see L. Kühn, D. Thomas, and S. Pfeifer, *Wiss. Humboldt-Univ. Berlin, Math.-Nat.* **19**, 81 (1970); and S. Pfeifer, *Pharmazie* **26**, 328 (1971).
2. S. Pfeifer, *Pharmazie* **17**, 298 (1962).
3. F. Šantavý, J. L. Kaul, L. Hruban, L. Dolejš, V. Hanuš, K. Bláha, and A. D. Cross, *Collect. Czech. Chem. Commun.* **30**, 335 and 3479 (1965); J. Hrbek, Jr., F. Šantavý, and L. Dolejš, *Collect. Czech. Chem. Commun.* **35**, 3712 (1970).
4. O. Hesse, *Ann. Chem. Pharm.* **149**, 35 (1969).
5. J. Slavík, *Chem. Listy* **52**, 1957 (1958); *Collect. Czech. Chem. Commun.* **24**, 2506 (1959).
6. W. Awe and W. Winkler, *Arch. Pharm.* (*Weinheim*) **290**, 367 (1957).
7. F. Šantavý, M. Maturová, A. Němečkova, and M. Horák, *Collect. Czech. Chem. Commun.* **24**, 3493 (1959); **25**, 1901 (1960).
8. H. Böhm and H. Rönsch, *Z. Naturforsch. B* **23**, 1552 (1968).
9. H. Rönsch, *Tetrahedron Lett.* p. 5121 (1969).
10. A. D. Cross, I. Mann, and S. Pfeifer, *Pharmazie* **21**, 181 (1966).
11. A. Guggisberg, M. Hesse, H. Schmid, H. Böhm, H. Rönsch, and K. Mothes, *Helv. Chim. Acta* **50**, 621 (1967).
12. M. Maturová, H. Potěšilová, F. Šantavý, A. D. Cross, V. Hanuš, and L. Dolejš, *Collect. Czech. Chem. Commun.* **32**, 419 (1967).
13. M. Shamma, J. A. Weiss, S. Pfeifer, and H. Döhnert, *Chem. Commun.* p. 212 (1968).
14. J. T. Edwards and I. Puskas, *Can. J. Chem.* **40**, 711 (1962).
15. See also E. Brochmann-Hanssen and K. Hirai, *J. Pharm. Sci.* **56**, 1658 (1967).
16. E. Brochmann-Hanssen, K. Hirai, B. Nielsen, S. Pfeifer, I. Mann, and L. Kühn, *J. Pharm. Sci.* **57**, 30 (1968).
17. A. Němečková, A. D. Cross, and F. Šantavý, *Naturwissenschaften* **54**, 45 (1967); A. Němečková, F. Šantavý, and D. Walterova, *Collect. Czech. Chem. Commun.* **35**, 1733 (1970).
18. A. Klásek, V. Šimánek, and F. Šantavý, *Tetrahedron Lett.* p. 4549 (1968).
19. D. Walterová and F. Šantavý, *Collect. Czech. Chem. Commun.* **33**, 1623 (1968).
20. S. Pfeifer and I. Mann, *Pharmazie* **20**, 643 (1965).
21. D. W. Brown, S. F. Dyke, G. Hardy, and M. Sainsbury, *Tetrahedron Lett.* p. 2609 (1968).
22. M. Sainsbury, D. W. Brown, S. F. Dyke, and G. Hardy, *Tetrahedron* **25**, 1881 (1969).
23. C. Reby, J. Likforman, and J. Gardent, *C. R. Acad. Sci., Ser. C* **269**, 45 (1969).

24. J. Likforman and J. Gardent, *C. R. Acad. Sci., Ser. C* **268**, 2340 (1969).
25. B. Göber and G. Engelhardt, *Pharmazie* **24**, 423 (1969).
26. V. Snieckus and H. O. Bernhard, *Amer. Chem. Soc., Div. Org. Chem., Joint Conf., Org. Chem. Div., C.I.C.*, No. 32 (1970); and *Tetrahedron* **27**, 2091 (1971).
26a. H. Irie, S. Tani, and H. Yamane, *Chem. Commun.* p. 1713 (1970).
26b. W. Klötzer, S. Teitel, J. F. Blount, and A. Brossi, *J. Amer. Chem. Soc.* **93**, 4321 (1971). For an improved synthesis of nornarceine, see K. Klötzer, S. Teitel, and A. Brossi, (in press).
26c. W. Klötzer, S. Teitel, and A. Brossi, *Helv. Chim. Acta* **54**, 2057 (1971).
27. F. Šantavý, J. Hrbek, Jr., and K. Bláha, *Collect. Czech. Chem. Commun.* **32**, 4452 (1967).
27a. M. Shamma, J. L. Moniot, W. K. Chan, and K. Nakanishi, *Tetrahedron Lett.* p. 4207 (1971). For an X-ray study of rhoeagenine methiodide see C. S. Huber, (in press).
28. L. Dolejš and V. Hanuš, *Tetrahedron* **23**, 2997 (1967).
29. S. Pfeifer and I. Mann, *Pharmazie* **23**, 82 (1968).

Chapter 22 / PROTOSTEPHANINE

Occurrence: Menispermaceae
Structure:

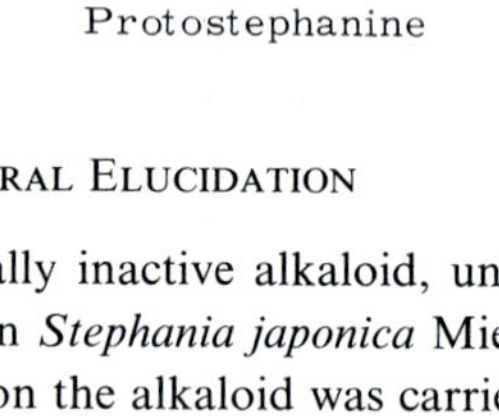

Protostephanine

I. Introduction and Structural Elucidation

Protostephanine is an optically inactive alkaloid, unique in its structural features, found by Kondo and Sanada in *Stephania japonica* Miers. (Menispermaceae).[1]

Extensive degradative work on the alkaloid was carried out in the 1950's by Kondo, Watanabe, and Takeda, and the more important aspects of their studies have been outlined in Schemes I and II. As a result, protostephanine was assigned in 1956 the tricyclic structure shown which incorporates two unusual features, viz., a nine-membered nitrogenous ring and a benzenoid ring with two methoxyl substituents in a meta relationship to each other.

Scheme I delineates the steps through which the positions of the substituents were located. Protostephanine on oxidation with potassium permanganate yielded *m*-hemipinic acid. Upon zinc dust distillation of the alkaloid, a crystalline compound

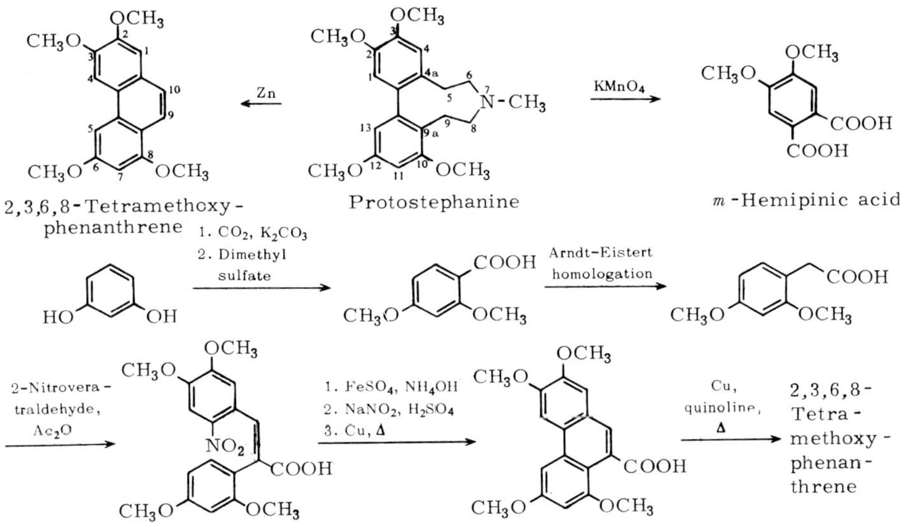

Scheme I

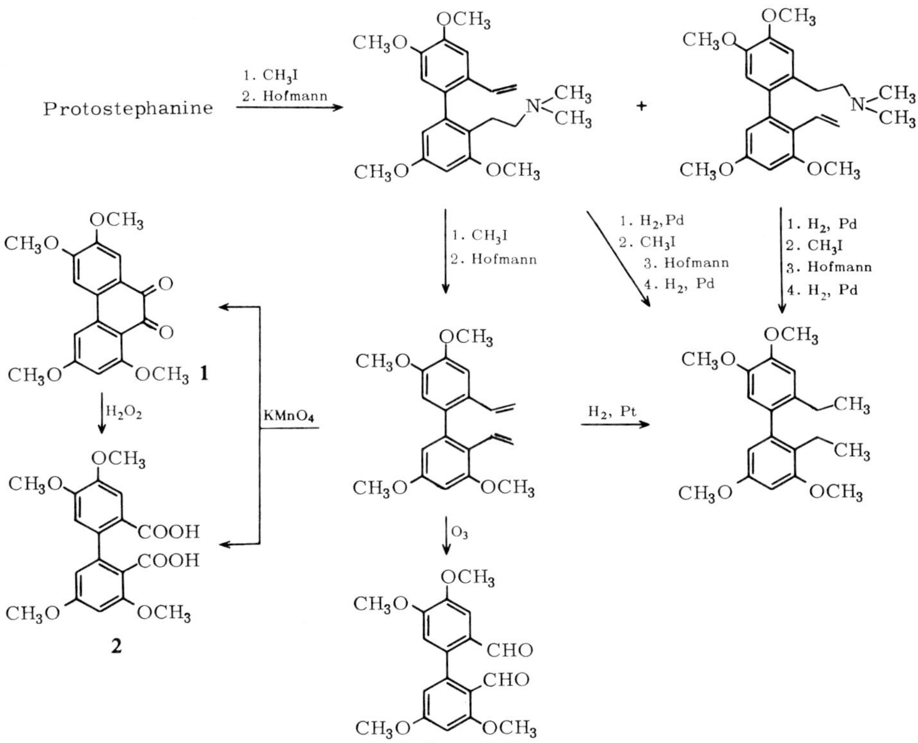

Scheme II

was obtained which was shown by an unequivocal synthesis to be 2,3,6,8-tetramethoxyphenanthrene.

Scheme II is a summary of the Hofmann degradations and subsequent oxidations or reductions that were run starting with protostephanine. These reaction sequences determined that the alkaloid possesses a nine-membered ring which includes a tertiary nitrogen atom. Furthermore, formation of the dione **1** and the diacid **2** established that the aliphatic part of the molecule must be bonded to the biphenyl system in protostephanine at positions C-4a and C-9a.[1]

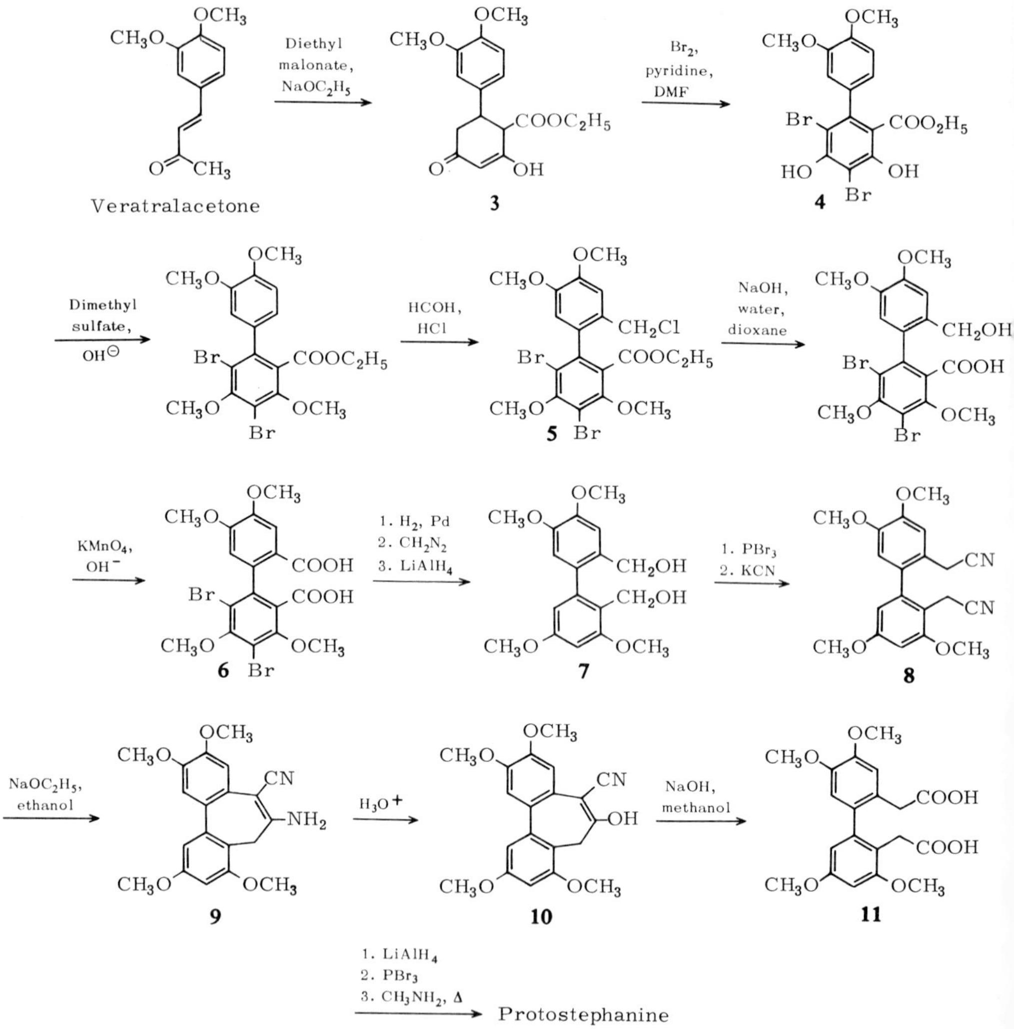

Scheme III

II. The Pecherer–Brossi Synthesis of Protostephanine

A practical synthesis of protostephanine has been carried out by Pecherer and Brossi at the Roche Nutley laboratories. The starting material was veratralacetone, which was condensed with malonic ester to yield the crystalline keto ester **3**. Bromination of this material was accompanied by aromatization so that the dibromoester **4** was obtained. *O*-Methylation followed by chloromethylation afforded the crystalline halogenated ester **5**. The intermediate **6** was obtained by alkaline hydrolysis succeeded by oxidation with permanganate. Debromination was accomplished using palladium and hydrogen. The resulting diacid was then reduced via its dimethyl ester to the diol **7**.

In order to homologate the diol **7** to the important diacid **11**, which incorporates all the carbon atoms of the protostephanine nucleus, the material was converted to the corresponding dinitrile **8** which suffered cyclization in base to the amino nitrile **9** rather than simple hydrolysis. Acid treatment of **9** yielded the corresponding hydroxy nitrile **10**. When this product was refluxed in methanolic sodium hydroxide, ring cleavage and saponification of the cyano group occurred with formation of the desired homologated diacid **11** (Scheme III).

The remainder of the synthesis is straightforward and self-explanatory. It should be noted that in the ultimate step cyclization of the dibromide by means of methylamine proceeded in 34% yield, denoting that the formation of the nine-membered ring in the protostephanine series is not as difficult as might have been anticipated.[2]

Takeda has published the outlines of a synthesis of protostephanine, the details of which are quite similar to the Hoffmann–La Roche approach.[3] But the latter synthesis is the only one presently available which can lead to workable amounts of the natural product.

III. A Suggested Biogenetic Pathway for the Formation of Protostephanine

It is tempting to postulate at first sight that protostephanine is formed in the plant by a simple phenolic oxidative coupling as shown, and in fact such proposals have been recorded in the literature.

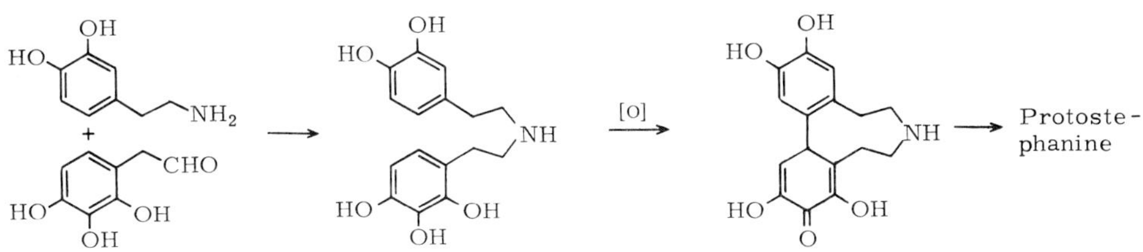

In 1964, however, Barton made the significant suggestion that protostephanine is probably formed in *S. japonica* starting with a benzylisoquinoline base such as **12** which can first undergo internal phenolic oxidative coupling and reduction to the tetracyclic species **13**. Acid-catalyzed dienol–benzene rearrangement then yields the key tetracyclic species **14**, which readily undergoes ring fission to afford the immonium salt **15** with the desired protostephanine skeleton. Reduction to the tertiary amine and *O*-methylation then furnish protostephanine (Scheme IV).[4]

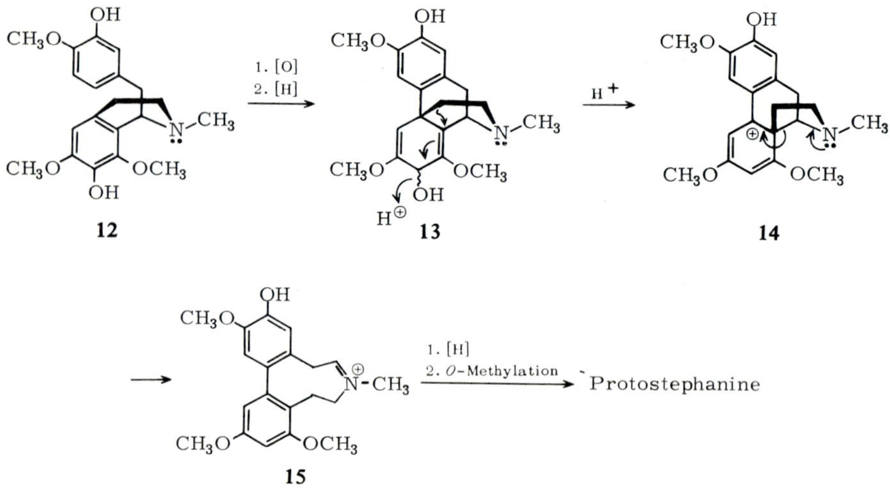

Scheme IV

IV. The Synthesis of Protostephanine by a Route Related to the Biogenetic Pathway

Battersby and his co-workers have carried out successfully a synthesis of protostephanine based on the proposed biogenetic path (Scheme V).

Phenolic oxidative coupling of the same tetrahydrobenzylisoquinoline base **12** mentioned by Barton gave in low yield the dienone **16** which was *O*-methylated to **17**, usually called protostephanone. Protostephanone could more readily be prepared from the pentamethoxytetrahydrobenzylisoquinoline **18** via diazotization and Pschorr cyclization.

When protostephanone (**17**) was reduced with borohydride, a mixture of two diastereoisomeric diols was obtained. This mixture, without separation, underwent an acid-induced rearrangement to afford the dienone **19**. In the subsequent step, which involved borohydride reduction, a mixture of alcohols was again obtained. Careful oxidation with chromium trioxide gave a mixture of ketones **20**. The stage was now set for the rearrangement to the protostephanine skeleton. Cleavage of the critical C–C bond was realized by means of magnesium iodide, which acted as an acid. Lithium aluminum hydride reduction of the resulting unisolated imine produced a mixture of two phenolic compounds with the desired skeleton. Finally, *O*-methylation of the mixture gave protostephanine (Scheme V).[5]

The Barton biogenetic scheme has received further confirmation from the fact that tracer experiments have established the incorporation of the dienone **16** into protostephanine in *S. japonica* with 2.9% incorporation.[5]

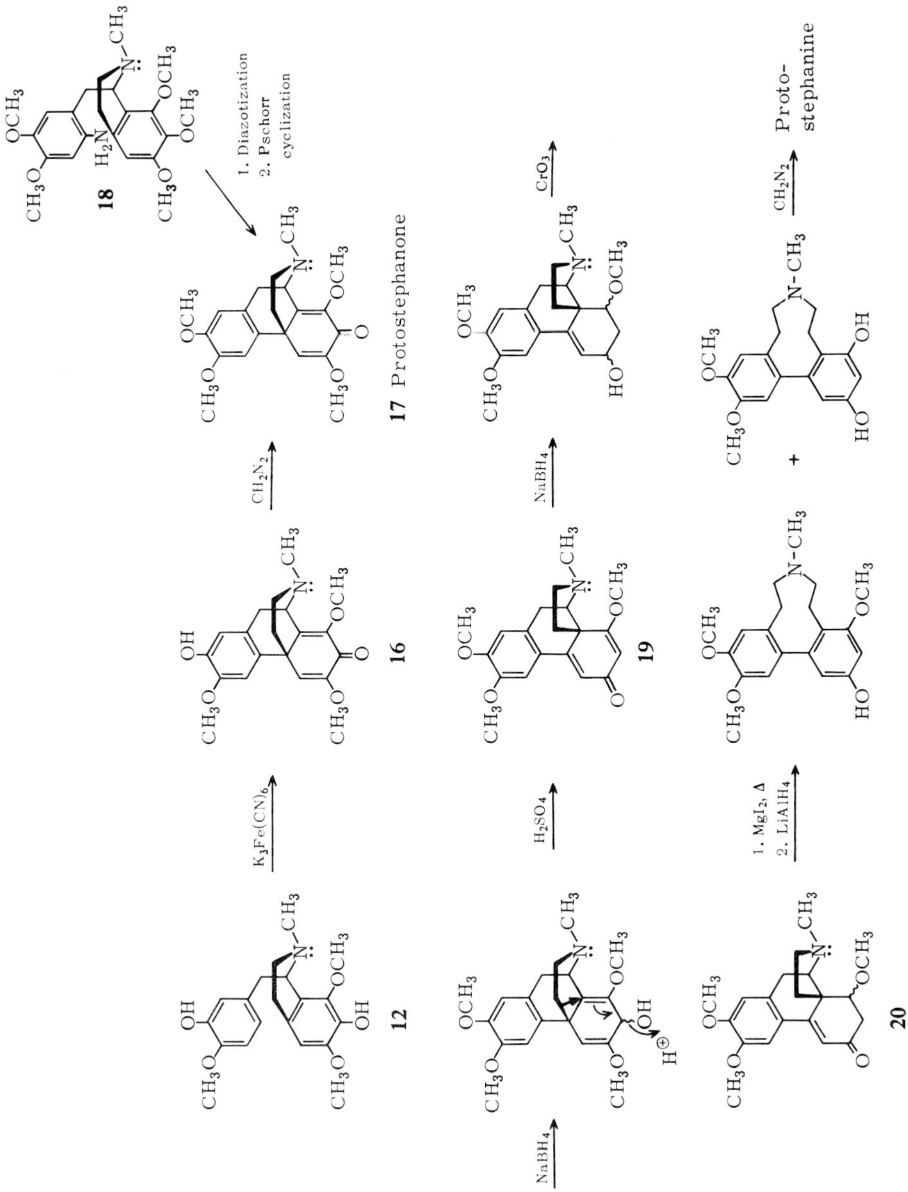
1. Diazotization
2. Pschorr cyclization
18
$K_3Fe(CN)_6$
CH_2N_2
12
16
17 Protostephanone
NaBH4
H2SO4
19
CrO3
1. MgI2, Δ
2. LiAlH4
20
CH2N2
Proto-stephanine

Scheme V

V. The Preparation of Protostephanine Analogs from Thebaine

Upon treatment of the opium alkaloid thebaine with phenylmagnesium bromide, a phenolic base is obtained, labeled phenyldihydrothebaine, which has been conclusively shown through extensive degradative work carried out by Bentley and Robinson and published as early as 1952 to have the structure indicated in Scheme VI. The mechanism for the formation of this nine-membered ring compound bears distinct similarities to the Barton biogenetic proposal for protostephanine, and it is assumed that it is magnesium bromide that triggers the rearrangement.[6] This transformation is the linchpin upon which the postulation of the biogenetic scheme for protostephanine was based.

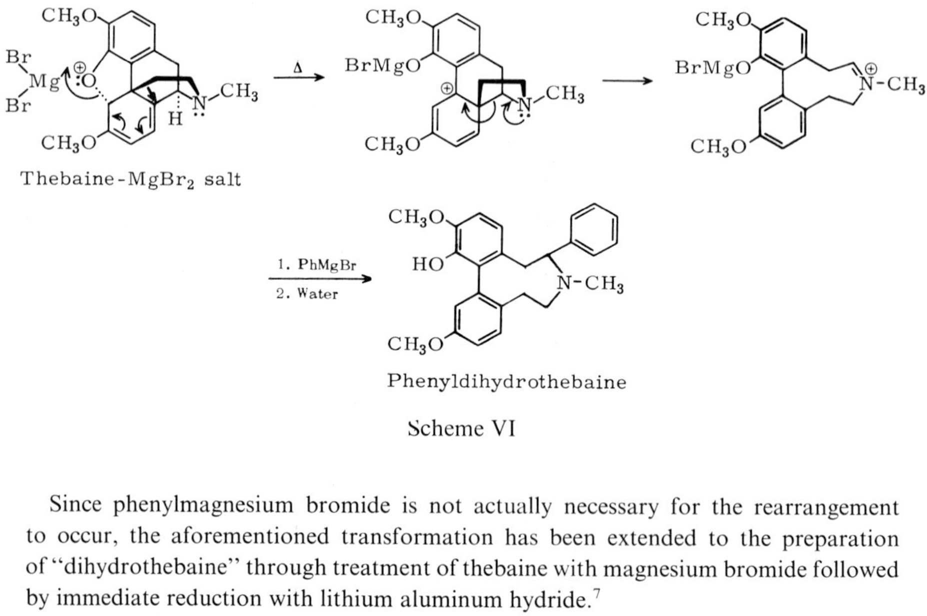

Scheme VI

Since phenylmagnesium bromide is not actually necessary for the rearrangement to occur, the aforementioned transformation has been extended to the preparation of "dihydrothebaine" through treatment of thebaine with magnesium bromide followed by immediate reduction with lithium aluminum hydride.[7]

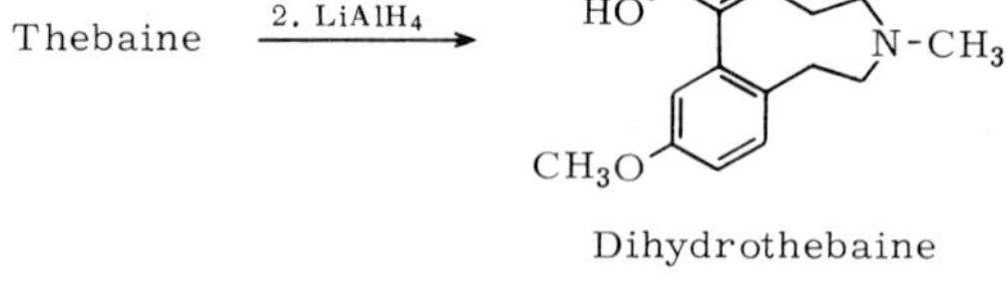

VI. Pharmacology

Protostephanine exerts a moderately strong and persistent hypotensive effect.[8]

VII. UV Spectroscopy

Protostephanine[9] λ_{max}^{MeOH} 283 mμ (3.83)
O-Methyldihydrothebaine methobromide[7] λ_{max}^{EtOH} 225 sh and 280.5 mμ (– and 3.62)

VIII. Erybidine, An Erythrina Alkaloid Related to Protostephanine

The new optically inactive alkaloid erybidine, λ_{max}^{EtOH} 216 and 284 mμ (4.50 and 3.92), recently obtained from *Erythrina xbidwilli* Lindl. (Leguminosae), bears some striking structural similarities to protostephanine, but the biogenetic pathway to erybidine must be different from that followed for protostephanine.[10]

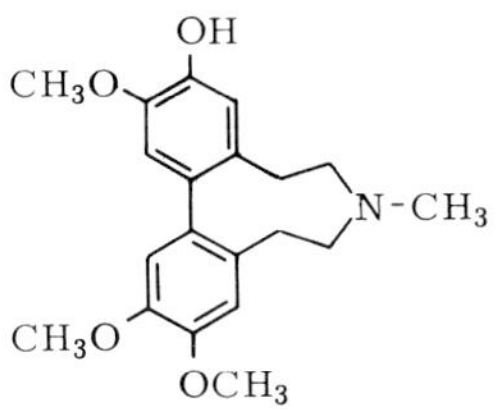

Erybidine

REFERENCES

1. For a lead reference to the structural work on protostephanine, see H. Kondo and K. Takeda, *Annu. Rep. Itsuu Lab.* **9**, 78 (1958); *Chem. Abstr.* **54**, 1580 (1960).
2. B. Pecherer and A. Brossi, *Helv. Chim. Acta* **49**, 2261 (1966); *J. Org. Chem.* **32**, 1053 (1967).
3. K. Takeda, *Annu. Rep. Itsuu Lab.* **13**, 45 (1963); *Chem. Abstr.* **60**, 5570f (1964).
4. D. H. R. Barton, *Pure Appl. Chem.* **9**, 35 (1964).
5. A. R. Battersby, A. K. Bhatnagar, P. Hackett, C. W. Thornber, and J. Staunton, *Chem. Commun.* p. 1214 (1968).
6. K. W. Bentley and R. Robinson, *J. Chem. Soc., London* p. 947 (1952). Phenyldihydrothebaine is actually a mixture of diastereoisomers; see J. A. Berson and M. A. Greenbaum, *J. Amer. Chem. Soc.* **80**, 445 (1958).
7. D. M. Hall and W. W. T. Manser, *Chem. Commun.* p. 112 (1967).
8. B. Pecherer, R. C. Sunbury, and A. Brossi, *J. Med. Chem.* **12**, 149 (1969).
9. A. Brossi and B. Pecherer, private communication (1971).
10. K. Ito, H. Furukawa, and H. Tanaka, *Chem. Pharm. Bull.* **19**, 1509 (1971).

Chapter 23/ EMETINE AND RELATED ALKALOIDS

Occurrence: Alangiaceae and Rubiaceae
Number: 14
Structures:

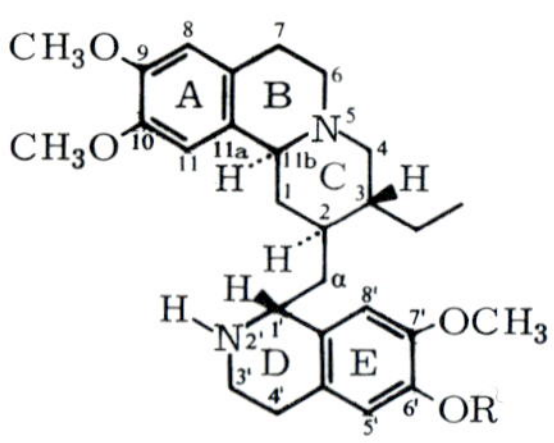

(–)-Emetine[1], R = CH_3
(–)-Cephaeline, R = H
(–)-Alangamide[2] = N-2'-CO-NH-CH_3 emetine

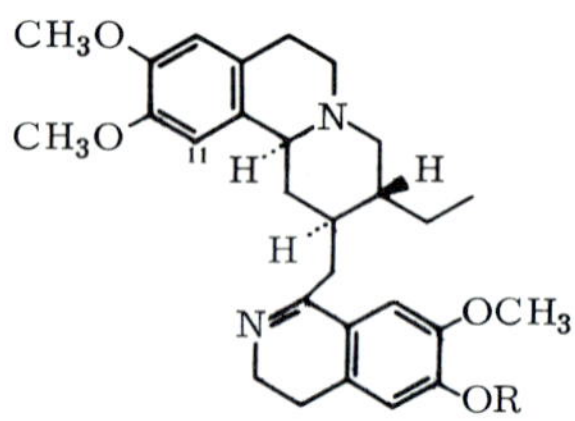

(+)-Psychotrine, R = H
(+)-*O*-Methylpsychotrine, R = CH_3[3]
(+)-Alangicine* ≡ 11-Hydroxypsychotrine[4]

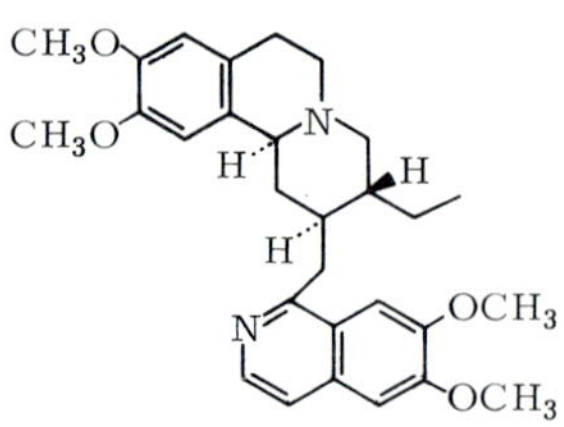

(+)-Emetamine

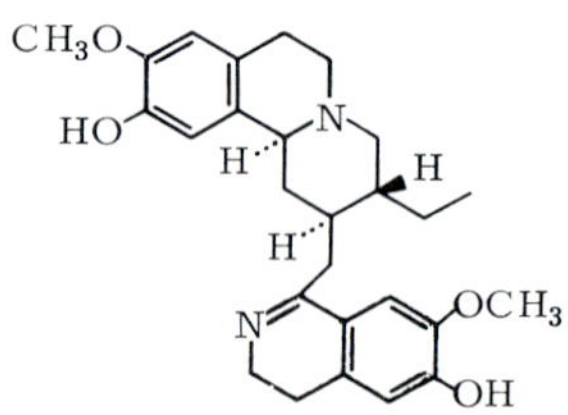

(+)-Demethylpsychotrine[4]

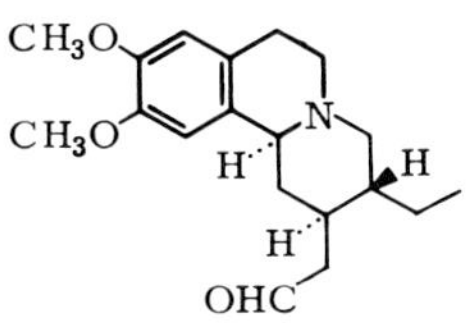

(−)-Protoemetine
Protoemetinol
(CHO replaced by CH_2OH)

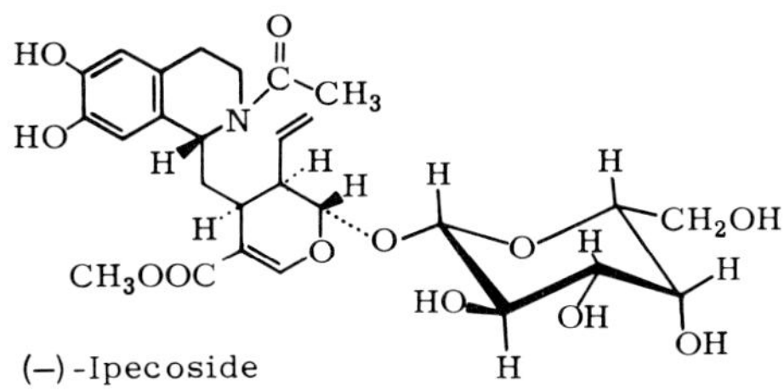

(−)-Ipecoside

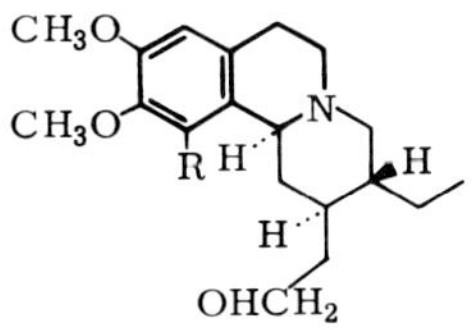

(−)-Dihydroprotoemetine, R = H[5]
(−)-Ankorine,* R = OH[5,6]

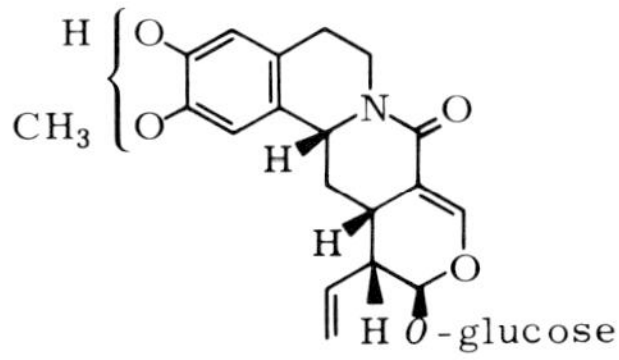

(−)-Alangiside

* Stereochemistry not yet established.

I. Introduction

The original or classical emetine alkaloids consist of five bases: emetine, cephaeline, psychotrine, *O*-methylpsychotrine, and emetamine, obtained mainly from ipecac or ipecacuanha – the roots of the South American straggling bush *Cephaelis ipecacuanha* (Brotero) Rich, or the Central American *Psychotria granadensis* Benth, ex Oerst (Rubiaceae).[1] To this group have been added protoemetine and ipecoside, both of which are also found in ipecac. The alkaloids of ipecac, especially emetine, are important because of their antiamebic activity.

The Indian plant *Alangium lamarckii* Thw. (Alangiaceae) has recently been found to contain emetine, cephaeline, and psychotrine, as well as the new alkaloids alangicine, desmethylpsychotrine, alangamide, dihydroprotoemetine and ankorine.

II. Chemical Interrelationships between the Ipecac Bases

Emetine is levorotatory and has four methoxyl groups. Although the alkaloid contains one tertiary and one secondary nitrogen, no *N*-methyl function is present in it or in any of its immediately related bases. The chemical relationship between emetine and the remaining original ipecac alkaloids have been summarized in Scheme I.

In the catalytic reductions of psychotrine to cephaeline and of *O*-methylpsychotrine to emetine, isocephaeline and isoemetine are also produced. These two reduction products, not found in natural sources and not shown in Scheme I, are diastereoisomeric at C-1′ with cephaeline and emetine, respectively.

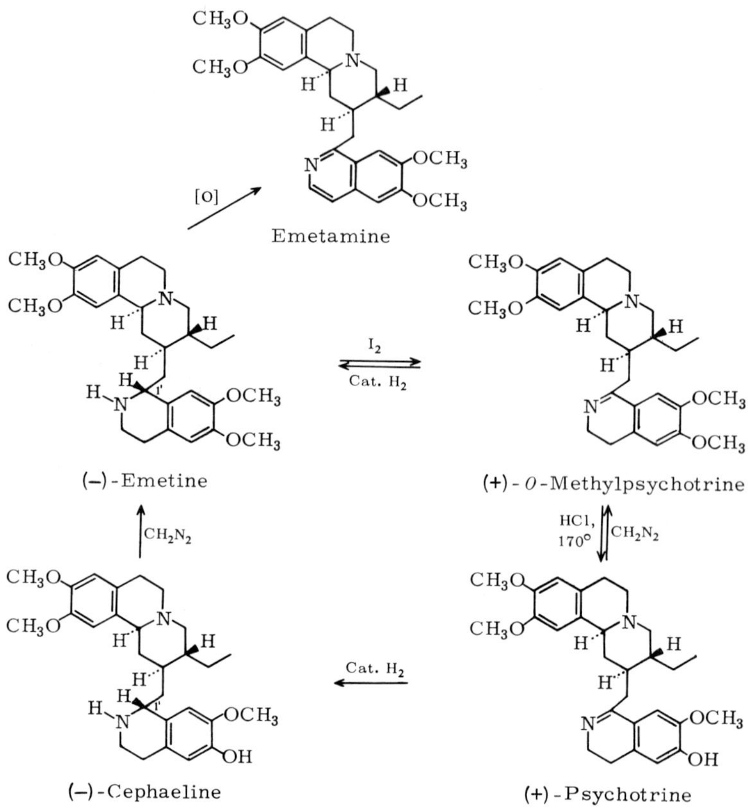

Scheme I

III. Degradation of Emetine

Crude emetine was first isolated in 1817 by Pelletier, but the correct empirical formula of $C_{29}H_{40}O_4N_2$ was not established until 1914 due to insufficient purification. Despite its early discovery, emetine was not obtained in crystalline form until 1953.

Initial studies on the oxidation of emetine with potassium permanganate under a variety of experimental conditions led to the isolation of such fragments as **1** to **4**.

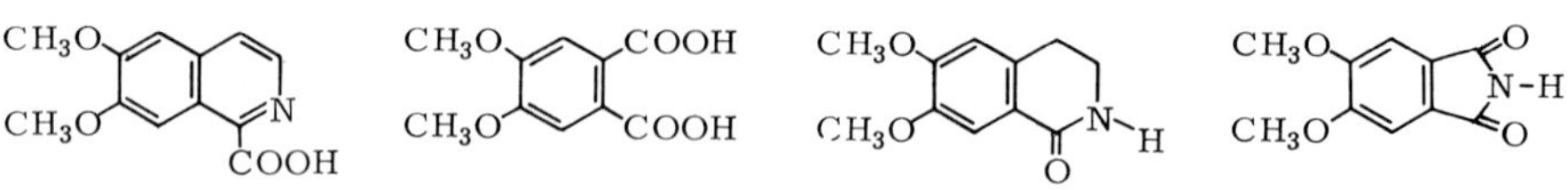

1 **2** *m*-Hemipinic acid **3** **4**

All of these compounds pointed to the presence of the 6,7-dimethoxyisoquinoline system in emetine. But the extremely high yield (96%) of *m*-hemipinic acid (**2**) obtained

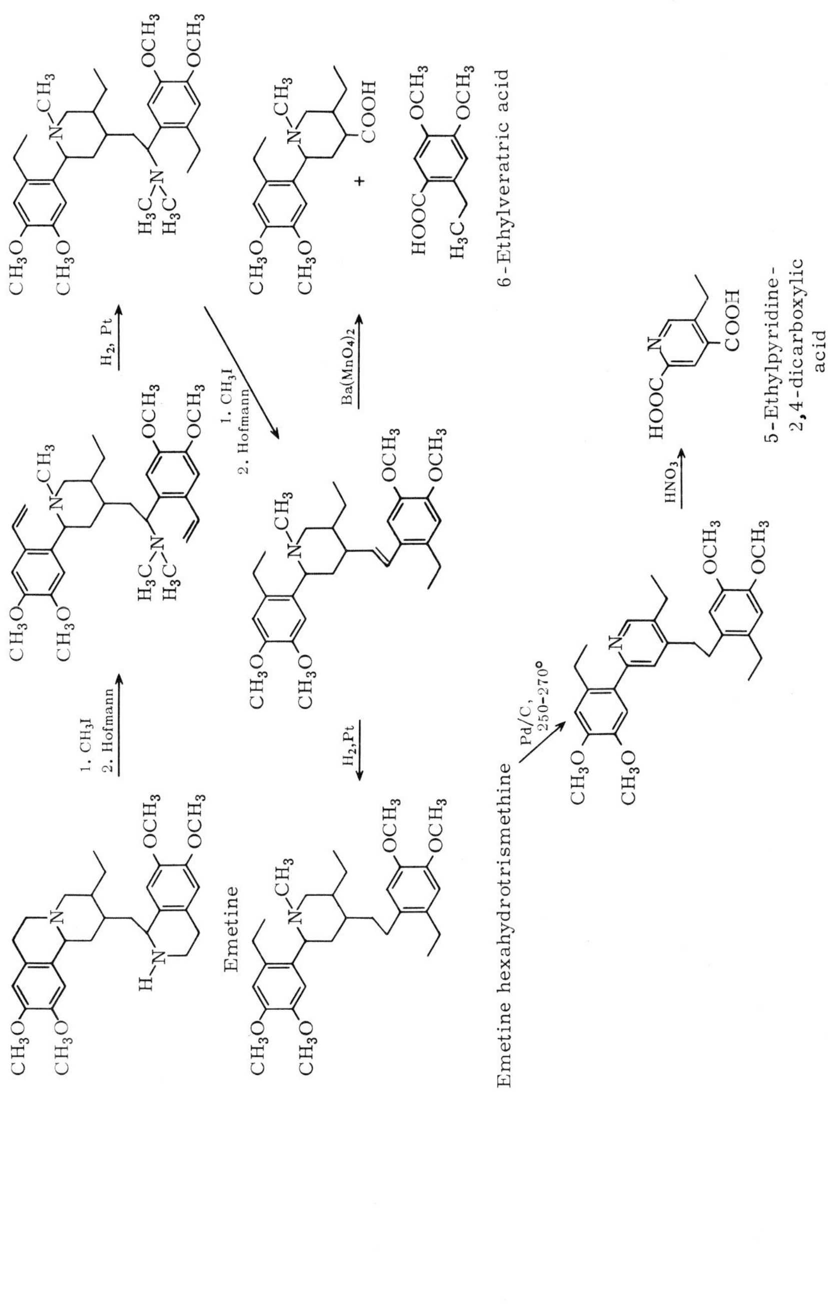
Emetine
1. CH3I
2. Hofmann
H2, Pt
1. CH3I
2. Hofmann
Ba(MnO4)2
6-Ethylveratric acid
H2,Pt
Emetine hexahydrotrismethine
Pd/C, 250-270°
HNO3
5-Ethylpyridine-2,4-dicarboxylic acid

Scheme II

in one of the oxidations indicated that more than one moiety capable of yielding this product was present in the molecule.

Hesse, Pyman, Karrer, Späth, Reichstein, Pailer, and others carried out extensive Hofmann and Emde degradations on emetine. From such work it became obvious that the secondary nitrogen in the alkaloid is monocyclic, while the tertiary nitrogen belongs to two rings. The correct structure for emetine, first tentatively suggested in 1948 by Robinson on biogenetic grounds,[7] was confirmed independently in 1949 by Battersby and Openshaw[8] and by Pailer and Porschinski.[9]

The main lines of the extensive degradative work carried out by Battersby and Openshaw is presented in Schemes II and III. Although certain compounds in these schemes were not completely characterized, the isolation of such shards as 6-ethylveratric acid (Schemes II and III), 5-ethylpyridine-2,4-dicarboxylic acid (Scheme II), formaldehyde and the conjugated diene **5** (Scheme III), all served to define the correct structure for emetine.[8]

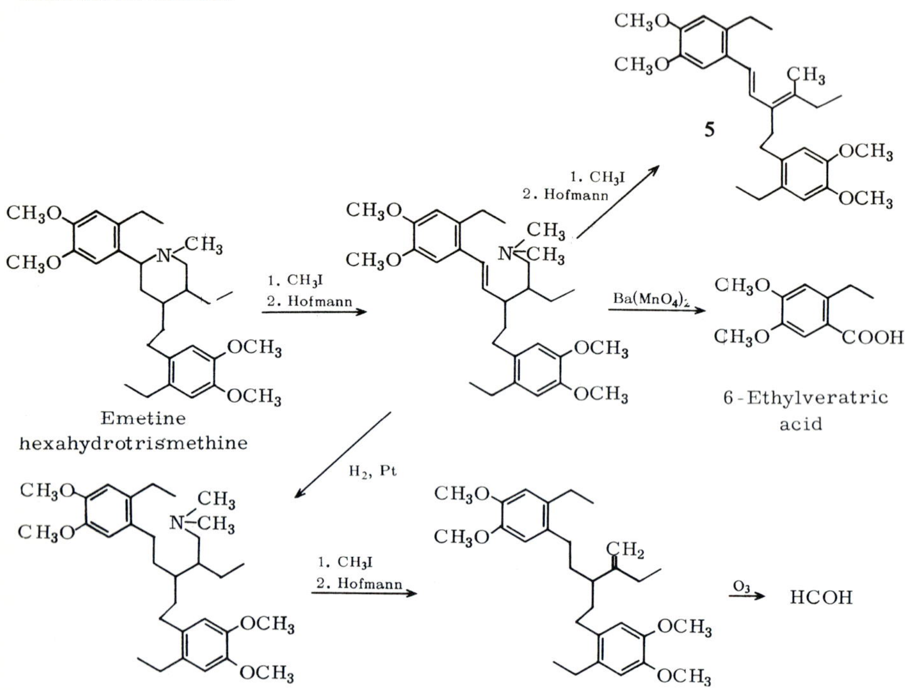

Scheme III

IV. The Position of the Phenolic Function in Cephaeline

The position of the phenolic function in cephaeline was settled conclusively when Pailer and Porschinski converted *O*-ethylcephaeline into the intermediate **6** by a series of Hofmann degradations and catalytic reduction. Ozonization of this salt then gave

2-ethyl-4-ethoxy-5-methoxybenzaldehyde, so that cephaeline must have the phenolic function at C-6′.[9]

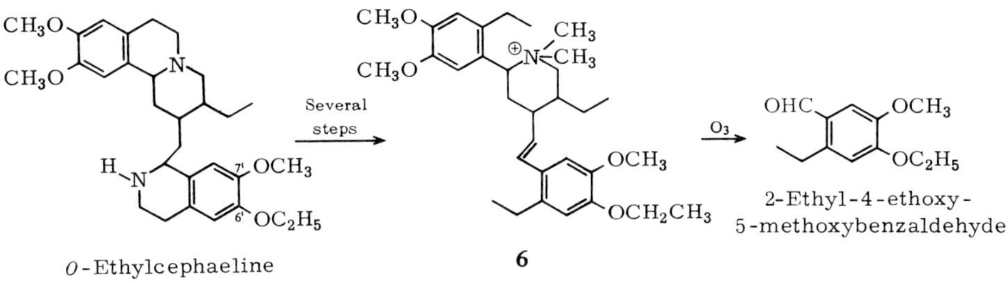

V. PROTOEMETINE

(−)-Protoemetine, $C_{19}H_{27}O_3N$, an ipecac alkaloid of biogenetic interest, was isolated in 1957. The compound is tricyclic and bears two methoxyl groups as well as a single tertiary nitrogen function and an aldehyde group. Its structural elucidation rests mainly upon its conversion to the naturally occurring *O*-methylpsychotrine (Scheme IV).[10]

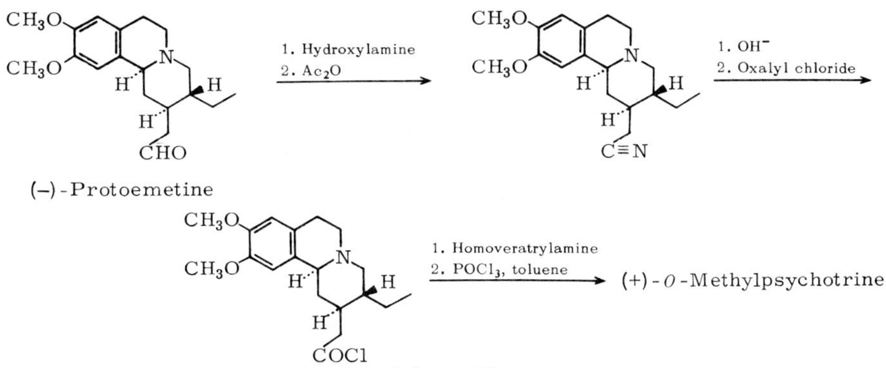

Scheme IV

Since *O*-methylpsychotrine has been reduced to emetine, the above conversion also constitutes a partial synthesis of emetine. Additionally, protoemetine has been condensed with 3-hydroxy-4-methoxyphenylethylamine to produce cephaeline and some isocephaeline. The former product can be *O*-methylated to emetine.[10-12]

VI. THE STEREOCHEMISTRY OF EMETINE

Emetine possesses the thermodynamically stable *trans*-quinolizidine arrangement delineated in which all the substituents are equatorially situated. The more salient reasons for this assignment of configuration at the four asymmetric centers may be summarized as follows:

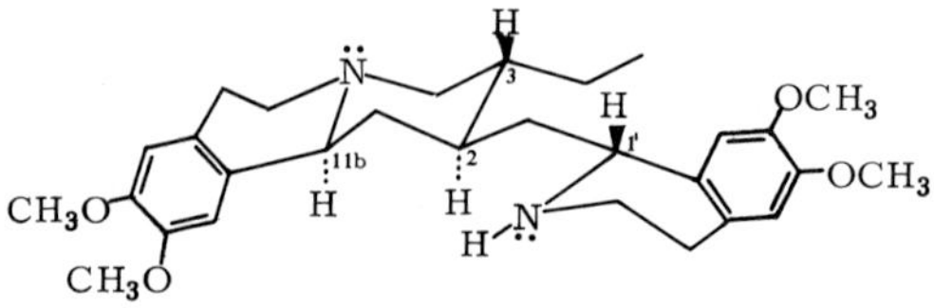

(−)-Emetine

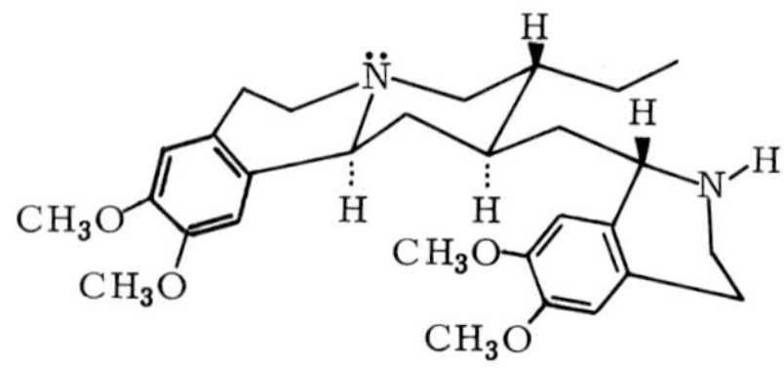

Isoemetine[13]

The ORD curves of emetine and its salts depend primarily upon the benzylic centers of asymmetry at C-11 b and at C-1′. Van Tamelen pointed out that these two centers *must be antipodal to one another*, since emetine hydrobromide exhibits negligible rotational change in the 300 to 700 mμ range, indicating that the contributions of the two

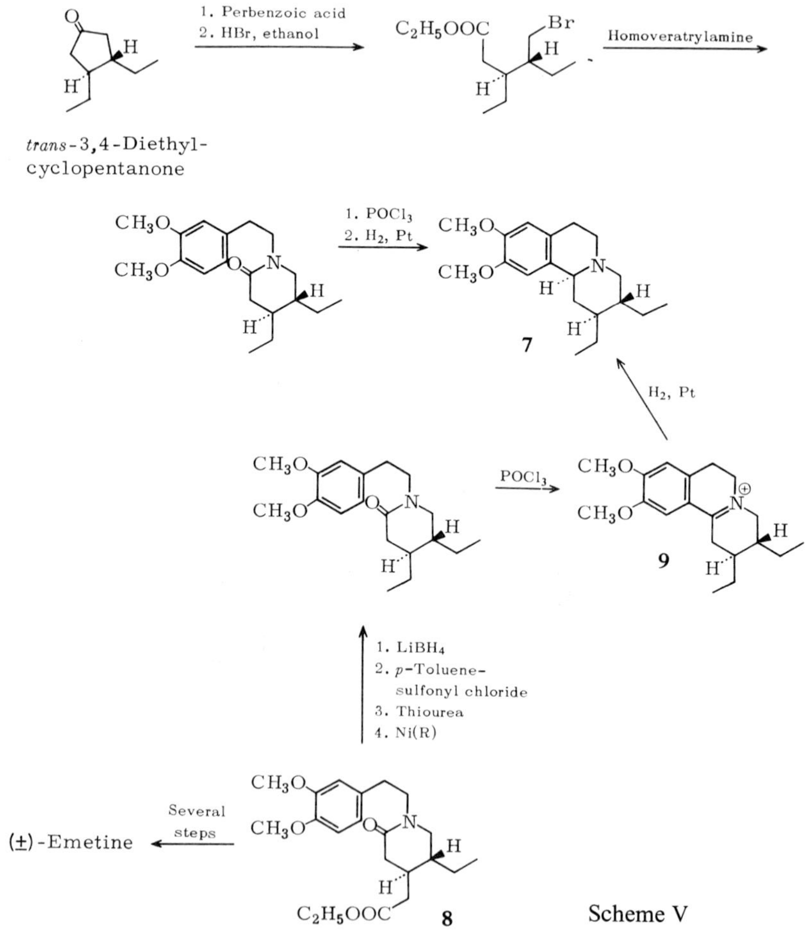

Scheme V

centers essentially cancel each other. On the other hand, isoemetine hydrobromide, which is epimeric with emetine at C-1′, exhibits increasing rotation down to 300 mμ because of optical reinforcement of the two asymmetric centers.[14–16]

The hydrogen atoms at C-2 and C-3 must be trans to each other, and the C-11 b hydrogen must be cis to that at C-2. This conclusion was reached after the synthesis of the tricyclic base **7** of known relative stereochemistry, starting with *trans*-3,4-diethylcyclopentanone. Base **7** proved to be identical with material derived from the lactam ester **8** through the immonium salt **9** by steps not involving any change in stereochemistry. Since the lactam ester **8**, as will be seen in Section VII, A, had previously been converted by Preobrazenskii *et al.* into racemic emetine, the relative configurations at C-2, C-3, and C-11 b in emetine were established (Scheme V).[15–17]

The final step in the preparation of the tricyclic base **7** from the lactam ester **8** by reduction of the immonium salt **9** under a variety of experimental conditions including catalytic hydrogenation, mixed metal hydrides, or sodium in alcohol, always resulted in the formation of only one isomer of **7**. This fact implies that the C-11 b hydrogen in emetine is axial, by analogy with similar reductions in the yohimbine series in which, for example, the immonium salt **10** is reduced to yohimbane.[18] If the thermodynamically less stable isomer of the tricyclic base **7** is desired, it may be obtained together with **7** by reduction of the salt **9** with either zinc or tin in hydrochloric acid.

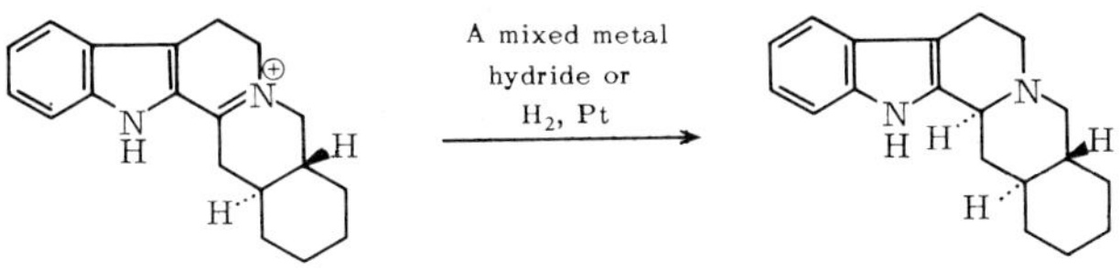

10 Yohimbane

The foregoing conclusions regarding the stereochemistry of emetine are reinforced by the presence in 2′-benzoylemetine of IR Bohlmann bands around 3.65 mμ (2740 cm^{-1}) characteristic of *trans*-quinolizidine systems, indicating that the pair of electrons on N-5 and the hydrogen atom at 11 b in emetine exist primarily in a trans-diaxial relationship.[19]

The absolute configuration of emetine at C-11 b was settled by means of a comparative study of the molecular rotations of *N*-2′-benzoyl- and *N*-2′-acetylemetine versus that for the corresponding 5-11 b dehydroemetinium salts.[20] It was observed that the axial hydrogen at 11 b in natural (−)-emetine makes a large negative contribution: *N*-2′-acetylemetine exhibits $[M]_D^{Py}$ − 401.4°, and the corresponding dehydro (immonium) salt lacking asymmetry at C-11 b has $[M]_D^{Py}$ +625.5°. This situation is analogous to that prevailing in the (−)-tetrahydroprotoberberine series, represented here by (−)-norcoralydine, in which, due to the presence of the 14-α-H, the molecular rotation values usually lie near −1000°. Therefore the corresponding asymmetric centers in the two alkaloidal series must be identical, and *the C-11 b hydrogen in emetine is alpha.*

Battersby and co-workers employed the diene **11**, obtained by sequential Hofmann degradation of *N*-acetylemetine, to study the absolute configuration at C-1′. Compound

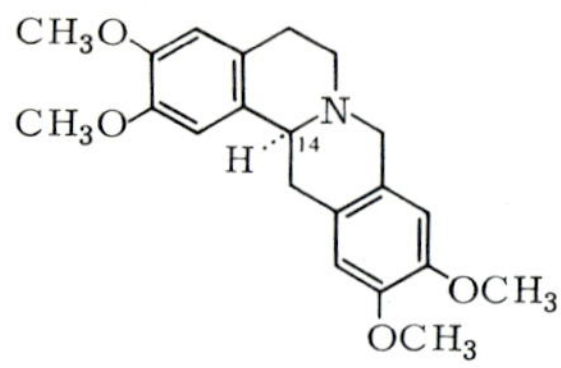

(−)-Norcoralydine

11 is actually related to the diene **5** obtained during the degradative studies on emetine. Oxidation of **11** first with ozone and then either with potassium permanganate or peracetic acid furnished, besides methyl ethyl ketone, the bicyclic (−)-acid amide **12**. The latter was then chemically interrelated with the alkaloid (+)-calycotomine of known absolute configuration.[21]

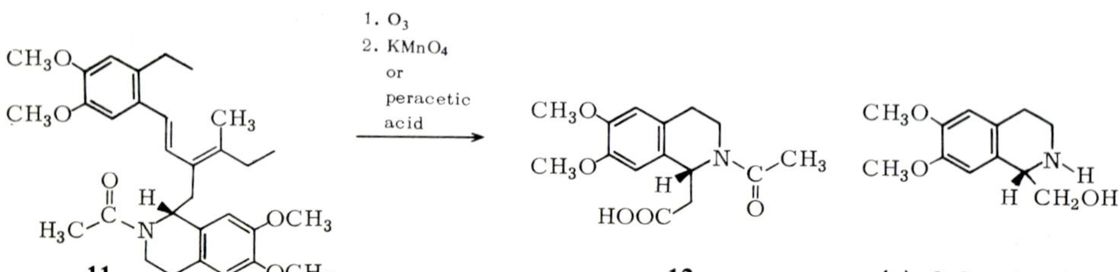

11 **12** (+)-Calycotomine

The stereochemistry of the remaining classical ipecac alkaloids follows from that of emetine. The ORD curve of psychotrine has been recorded[12]; and a chemical correlation has been established between the ipecac and indole alkaloids.[22]

VII. Syntheses of Emetine

More than a dozen syntheses of emetine are known, which make this compound the most synthesized alkaloid known. The study of these syntheses constitutes an interesting and worthwhile intellectual exercise in the laboratory-induced transformation of organic materials, so that most of these preparations will be considered here.

A. The Preobrazhenskii Synthesis

The first synthesis of emetine was reported by Preobrazhenskii and his group in 1950. The sequence was later expanded and modified, and the Russian work in its final form may be summarized as shown in Scheme VI.[23]* Since several steps in the synthesis, including the initial Michael condensation with diethyl glutaconate, lacked stereochemical control, diastereoisomeric intermediates were obtained which had to be separated in the course of the synthesis.

* The structural assignments made for some of the intermediates obtained in this synthesis had to be modified.[24,25]

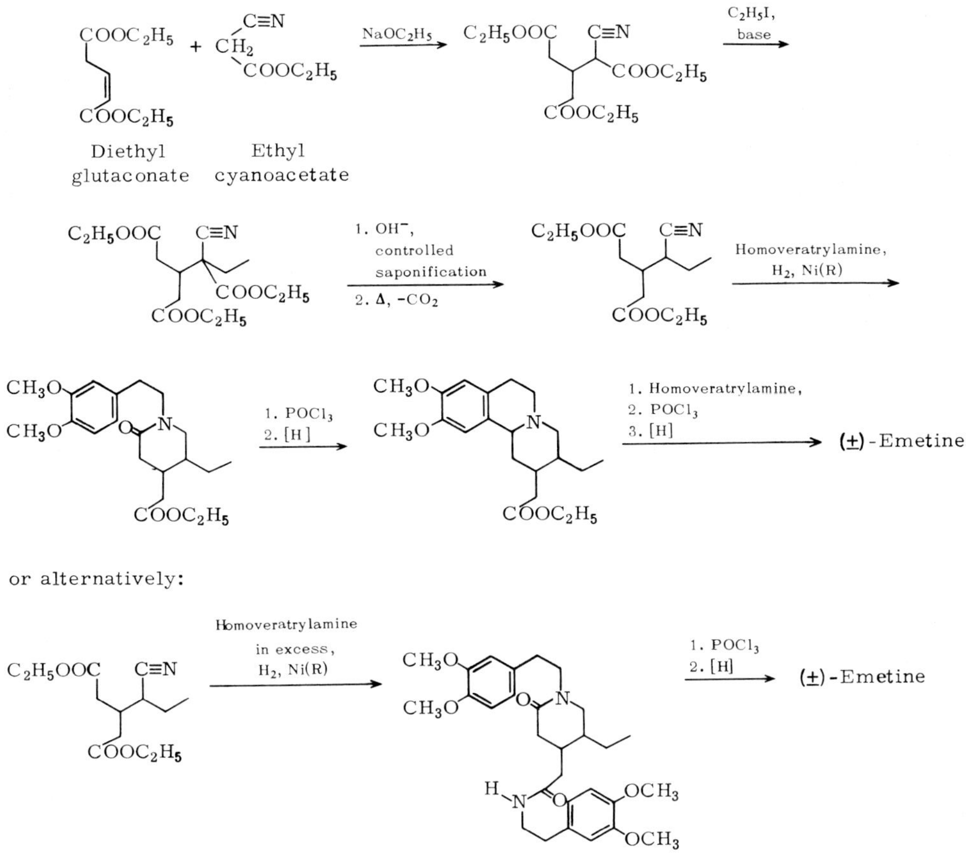

Scheme VI

B. The Battersby–Turner Synthesis

A clean stereospecific synthesis was devised by Battersby and Turner utilizing the β-ketolactam **13** originally prepared by Ban.[26] Michael condensation of the unsaturated lactam **14** with malonic ester gave after work-up the thermodynamically favored trans compound **15**, and subsequent catalytic reduction to the tricyclic ester **16** also proceeded to yield the more stable isomer. The tricyclic ester **16** was found to be identical with that derived from protoemetine, except for the optical activity, so that the remainder of the synthesis essentially followed the conversion of protoemetine to emetine (Scheme VII).[27]

C. The Burgstahler–Bithos Synthesis

A synthesis of emetine reminiscent of the van Tamelen yohimbine synthesis[28] has been carried out by Burgstahler and Bithos utilizing the common reagent gallic acid

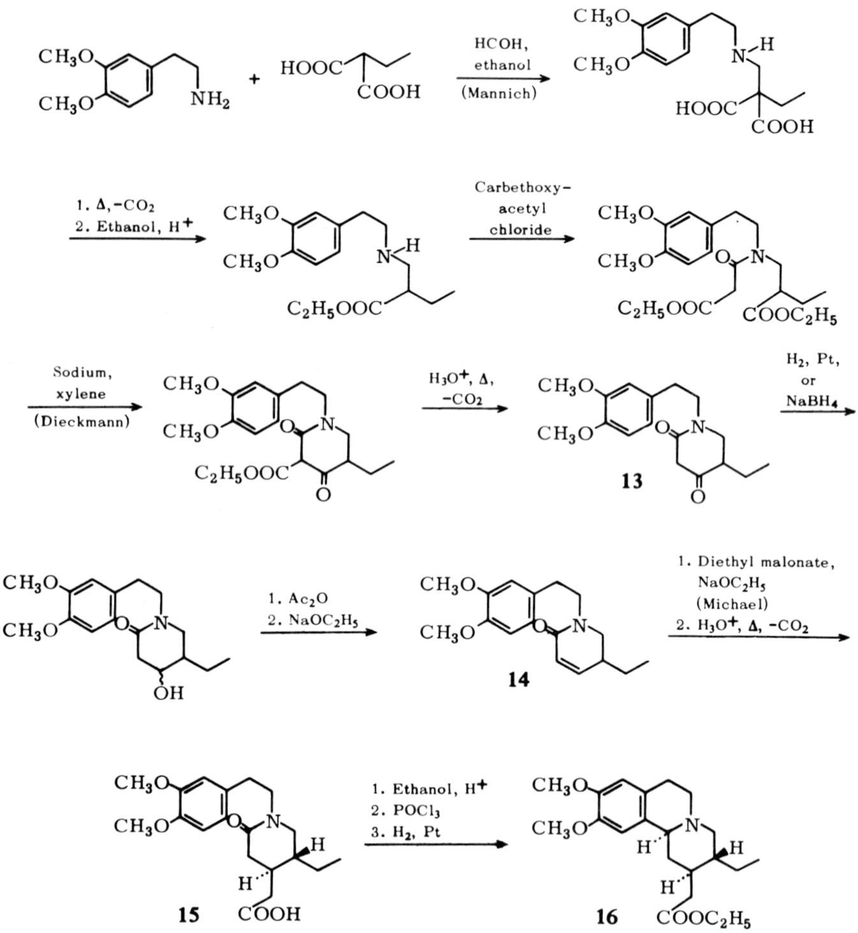

Scheme VII

as starting material. This approach is nonstereospecific and diastereoisomeric compounds had to be separated at proper intervals. Cleavage of the triol **17** with periodic acid produced a mixture of the lactol **18** and the corresponding hydroxyaldehyde, which, without separation, was cyclized to the lactam aldehyde **19** (Scheme VIII).[29]

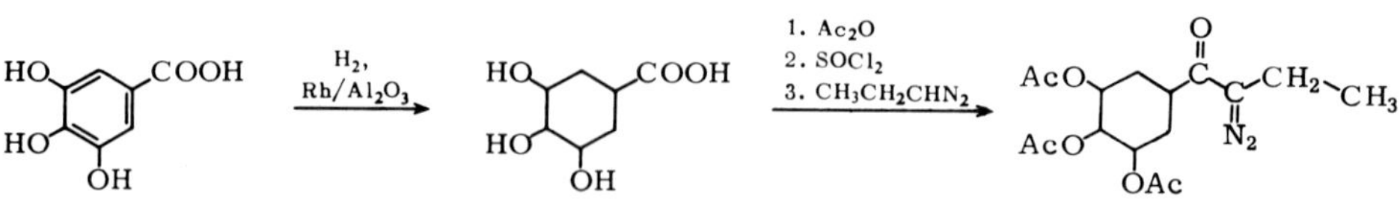

Gallic acid All-*cis* isomer

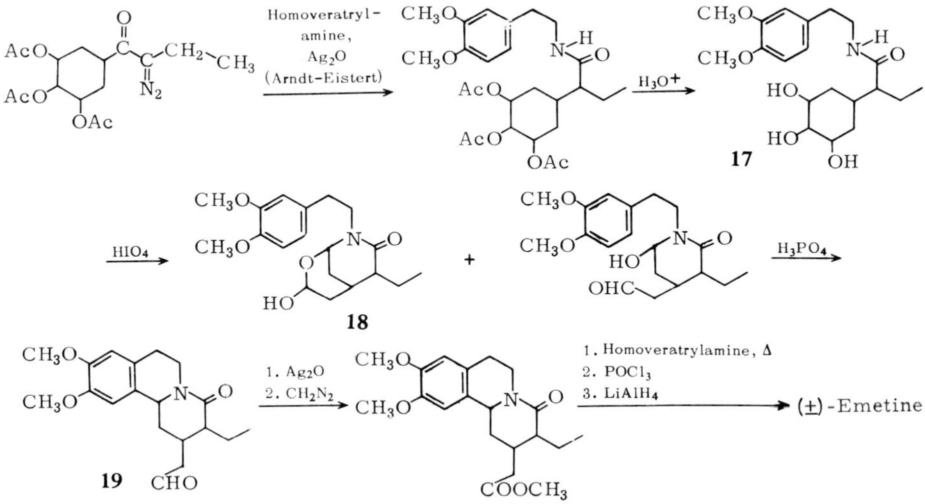

Scheme VIII

D. The van Tamelen Synthesis

In the synthesis of emetine achieved by van Tamelen, Schiemenz, and Arons, the keto triester **20** was first obtained by Michael condensation of methyl acetoacetate with dimethyl glutaconate. Mannich-type condensation of **20** with homoveratrylamine and formaldehyde then provided the key intermediate **21**. The remainder of the synthesis follows straightforward lines, especially since the tricyclic ester **22** had already been converted to emetine by Preobrazhenskii and other workers (Scheme IX).[30]

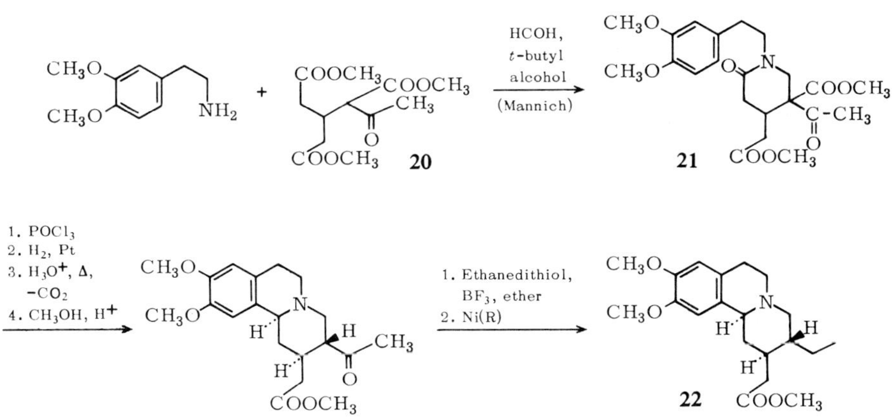

Scheme IX

E. The Glaxo Synthesis

Çlark, Meredith, Ritchie, and Walker have carried out a stereospecific preparation of emetine that follows unusual lines. Condensation of the imine **23** with acetonedicarboxylic acid in dilute aqueous sulfuric acid was found to yield the required *meso* diamino ketone **24**, while if the condensation were carried out in aqueous pyridine the undesired *racemic* diastereoisomer of **24** resulted. One amino function of **24** was protected through monobenzylation, and a Michael-aldol condensation with methyl vinyl ketone gave rise, after acid-catalyzed dehydration, to the critical intermediate **25**. Reduction of the conjugated double bond with lithium in liquid ammonia followed by equilibration with acid gave the more stable trans-fused ketone **26**. Following removal of the ketonic function, reductive debenzylation yielded emetine. The main advantage of this approach is that no isoemetine is obtained as a by-product (Scheme X).[31]

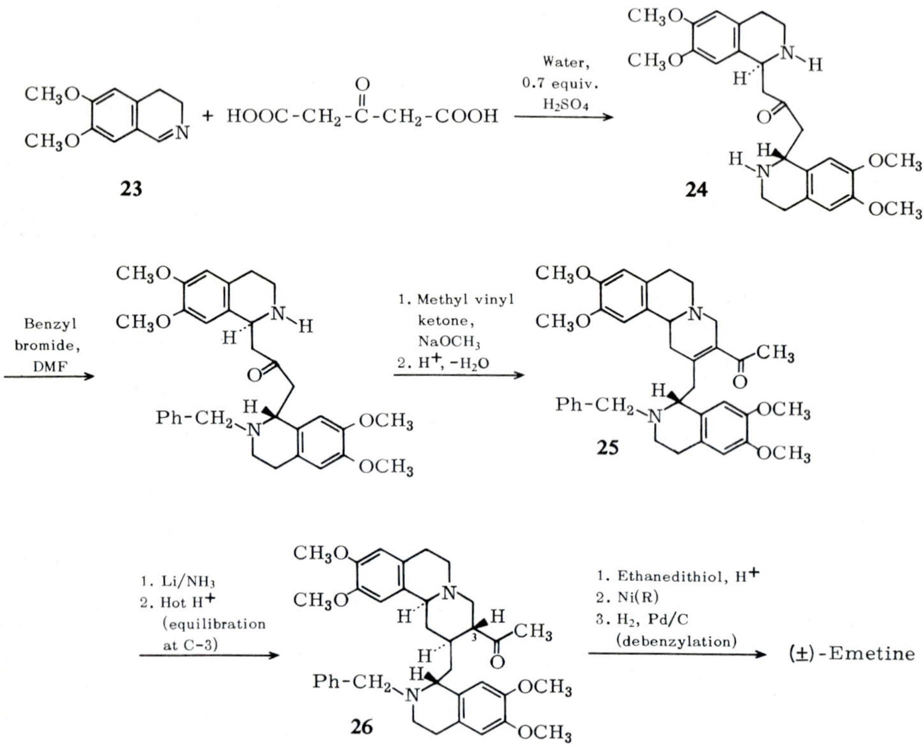

Scheme X

F. The First Roche Basle Synthesis

The aim of this approach was to prepare the tricyclic ketone **29**,[32] which after homologation to the acid **30** would be convertible into emetine (Scheme XI).[13]

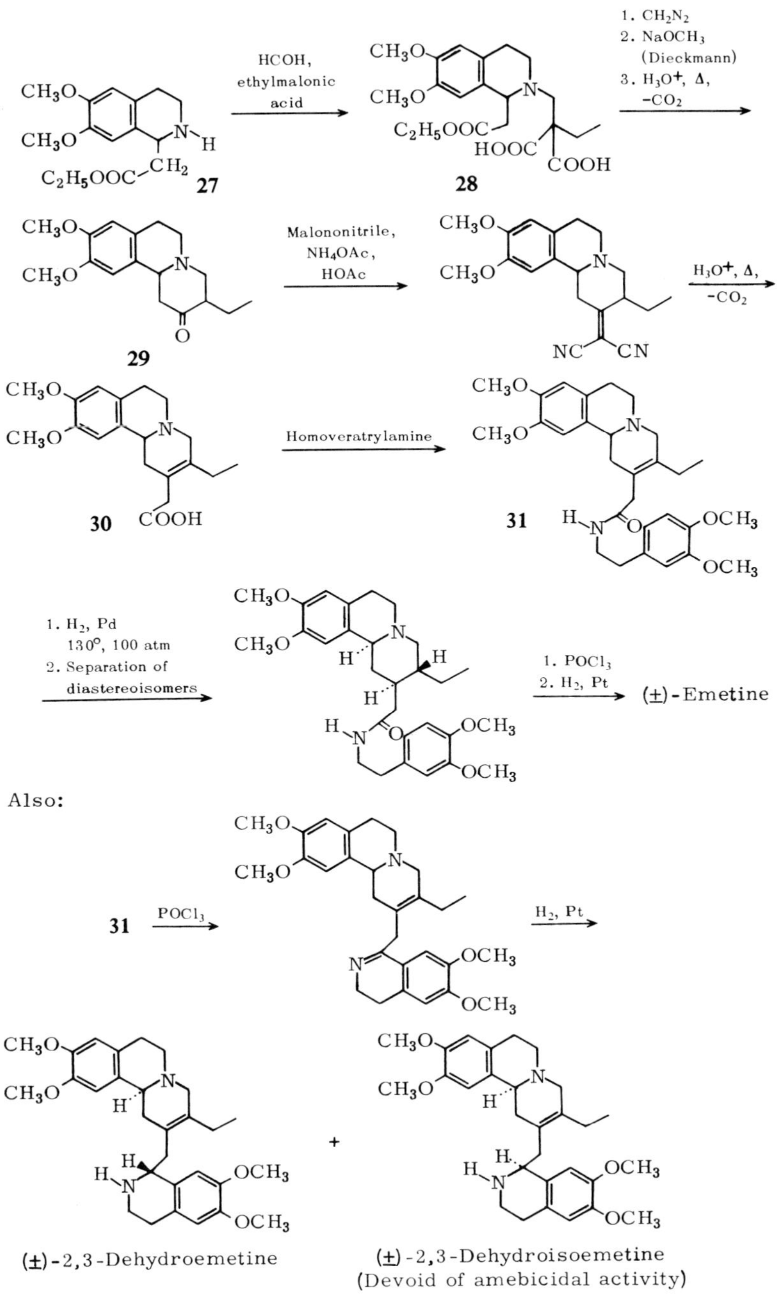
HCOH, ethylmalonic acid
1. CH_2N_2
2. $NaOCH_3$ (Dieckmann)
3. H_3O^+, Δ, $-CO_2$
27
28
29
Malononitrile, NH_4OAc, HOAc
H_3O^+, Δ, $-CO_2$
30
Homoveratrylamine
31
1. H_2, Pd 130°, 100 atm
2. Separation of diastereoisomers
1. $POCl_3$
2. H_2, Pt
(±)-Emetine
Also:
31
$POCl_3$
H_2, Pt
(±)-2,3-Dehydroemetine
(±)-2,3-Dehydroisoemetine (Devoid of amebicidal activity)

Scheme XI

The amino ester **27** yielded the diacid ester **28** through a Mannich-type reaction with formaldehyde and ethylmalonic acid. Esterification, Dieckmann cyclization, hydrolysis, and then decarboxylation produced the required tricyclic ketone **29**. Homologation to **30** was accomplished by condensation with malononitrile followed by hydrolysis and decarboxylation.

The acid **30** was condensed with homoveratrylamine and the resulting amide reduced catalytically to a separable mixture of cis and trans isomers. The trans species after Bischler–Napieralski cyclization and reduction afforded emetine.

Alternatively, the amide **31** could be cyclized to (±)-2,3-dehydro-*O*-methylpsychotrine, and succeeding reduction afforded a separable mixture of (±)-dehydroemetine and (±)-dehydroisoemetine. Dehydroemetine is comparable to emetine in amebicidal activity, so that this synthesis is of commercial importance.[13,33]

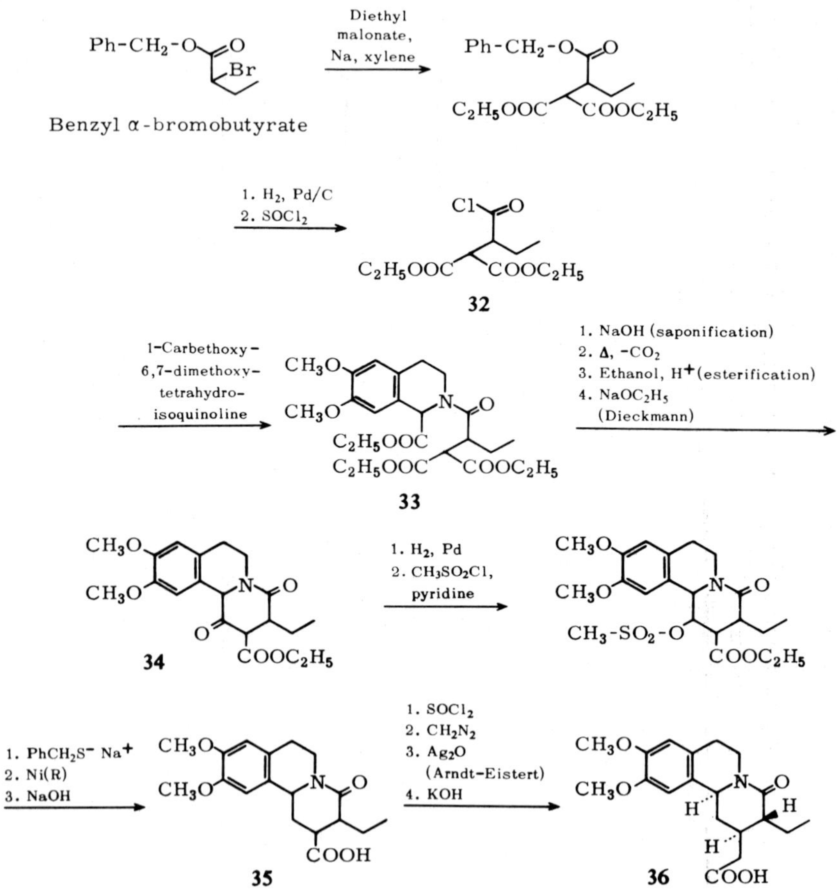

Scheme XII

G. The Second Roche Basle Synthesis

The acid chloride of α-ethyl-β,β-dicarbethoxypropionic acid (**32**) was first prepared by the condensation of benzyl α-bromobutyrate with malonic ester, followed by debenzylation over palladium and treatment with thionyl chloride. This acid chloride was then used to acylate 1-carbethoxy-6,7-dimethoxy-1,2,3,4-tetrahydroisoquinoline to form the amide **33**. The latter material was converted to the keto lactam **34** by the sequence indicated, and further transformation provided the tricyclic acid **35** as a pair of diastereoisomers. One isomer, upon homologation, gave the key homotricyclic lactam acid **36**, which was converted to emetine by the usual sequence (Scheme XII).[34]

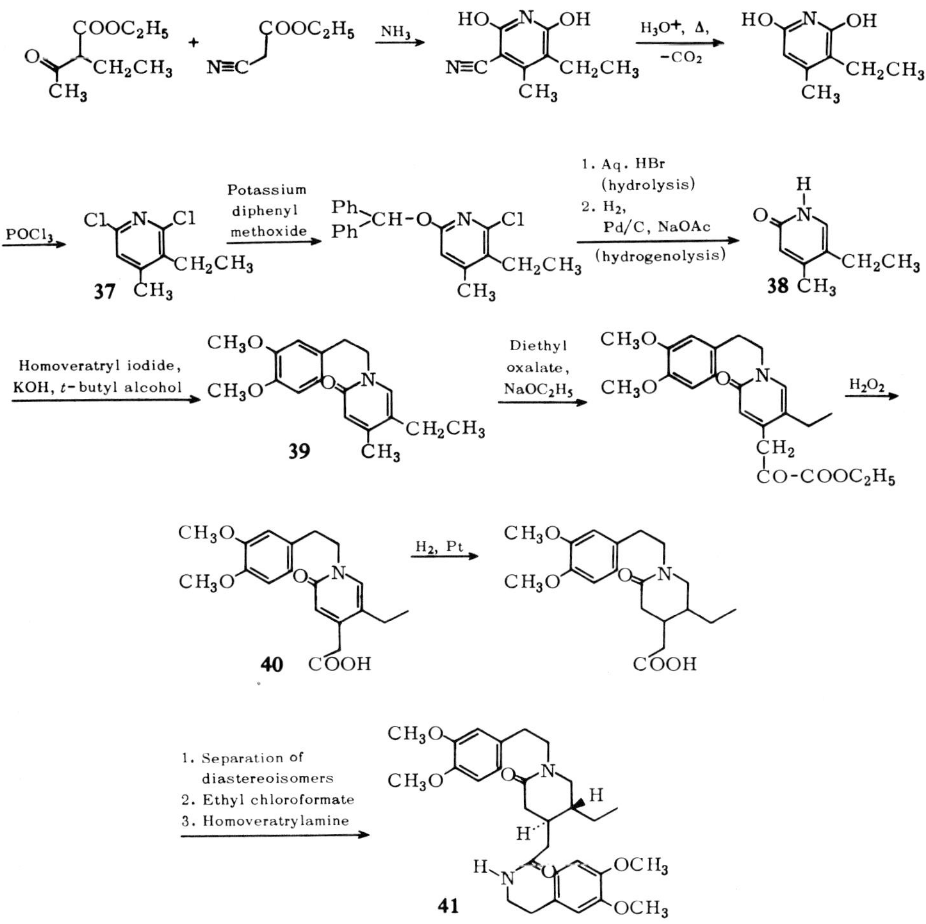

Scheme XIII

H. The Roche Welwyn Synthesis

The application of pyridone chemistry to the synthesis of emetine has been successfully exploited by Barasch, Osbond, and Wickens. The known dichloropyridine **37** was converted to the pyridone **38**, and *N*-alkylation with homoveratryl iodide led to the derivative **39**. Since the *C*-methyl group on the pyridone system is activated through conjugation with the carbonyl group, it was possible to homologate **39** to the acid **40**. Catalytic reduction of **40** produced a separable pair of diastereoisomers, one of which was treated first with ethyl chloroformate and then with homoveratrylamine to give rise to the lactam amide **41**, which was taken to emetine via Bischler–Napieralski cyclization and subsequent reduction (Scheme XIII).[35]

I. The Burroughs–Wellcome Synthesis

Openshaw and Whittaker have developed a commercially applicable synthesis of emetine in which the required optically active intermediate (−)-**44** is obtained by a highly efficient and unusual process (Scheme XIV).[36-38]

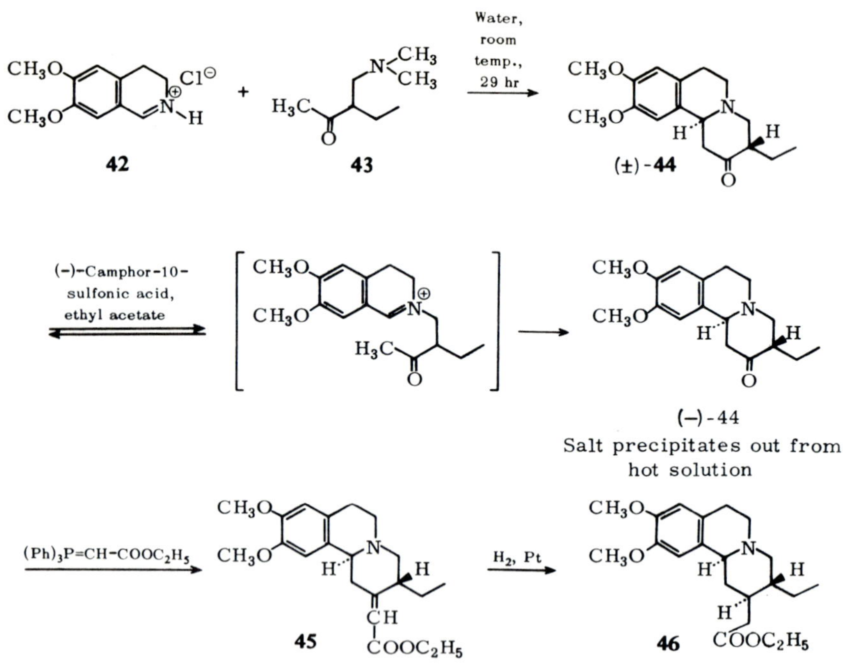

Scheme XIV

Condensation of the imine hydrochloride **42** with the Mannich base **43** afforded the amino ketone **44** in almost quantitative yield. Resolution of this product with (−)-camphor-10-sulfonic acid resulted in crystallization of the salt of the desired levorotatory base. Since ring C of the amino ketone **44** can be readily opened in the manner shown in Scheme XIV, the dextrorotatory amino ketone **44** kept being racemized while the levorotatory enantiomer continuously precipitated out as the salt. The net result was an almost total conversion of the racemic amino ketone **44** into its levorotatory form, epimerization also having occurred at the site of attachment of the ethyl side chain.

Homologation with ethoxycarbonylmethylenetriphenylphosphorane yielded only one geometric isomer of **45** and subsequent catalytic reduction provided the optically active ester **46** which can readily be taken to emetine.

Whenever the intermediate **46** is prepared in any of the aforementioned syntheses in either the levorotatory or the racemic form, condensation with homoveratrylamine succeeded by Bischler–Napieralski cyclization and reduction always leads to a mixture of emetine and isoemetine. Isoemetine is inactive as an amebicide, but it can be *N*-chlorinated on the secondary nitrogen using sodium hypochlorite, and treatment with base leads to *O*-methylpsychotrine by loss of hydrogen chloride. *O*-Methylpsychotrine may then be reduced again to a mixture of emetine and isoemetine.[36,37]

A significant observation of Openshaw and Whittaker is that condensation of the ester **46** with homoveratrylamine is appreciably accelerated by the presence of 2-hydroxypyridine, which acts as a bifunctional catalyst in the condensation of strongly basic amines with esters. From the practical point of view, 2-hydroxypyridine can easily be removed from the resulting amide because of its high solubility in water.[39]

The Openshaw and Whittaker synthesis can also be modified so as to produce the effective amebicides (±)- as well as (−)-2,3-dehydroemetine.[40]

J. The Zymalkowski–Frahm Synthesis

The key δ-lactone carboxylic acid **50** required for a synthesis of emetine was prepared as shown in Scheme XV. The aziridine obtained by the condensation of the acid chloride of **47** with ethylenimine was not isolated, but was forthwith reduced with lithium aluminum hydride to the aldehyde **48**. Homologation using ethoxycarbonylmethylenetriphenylphosphorane yielded in this instance a mixture of geometric isomers which was condensed in a Michael reaction with malonic ester. Hydrolysis, decarboxylation, and lactonization subsequently gave the desired product **50**.[41]

Finally, condensation of the lactone **50** with 2 moles of homoveratrylamine followed by hydrogenation and isomer separation yielded (±)-emetine.

VIII. Ipecoside, a Monoterpenoidal Isoquinoline Glucoside

The neutral glucoside ipecoside, $C_{27}H_{35}NO_{12}$, was obtained in 1952 from ipecac root, and its chemistry has been described so far mainly in communication form.[42]

The glucoside consists of an *N*-acetylated tetrahydroisoquinoline moiety, a C-10 terpenoidal unit, and glucose. Since the term "alkaloid" has never been rigorously

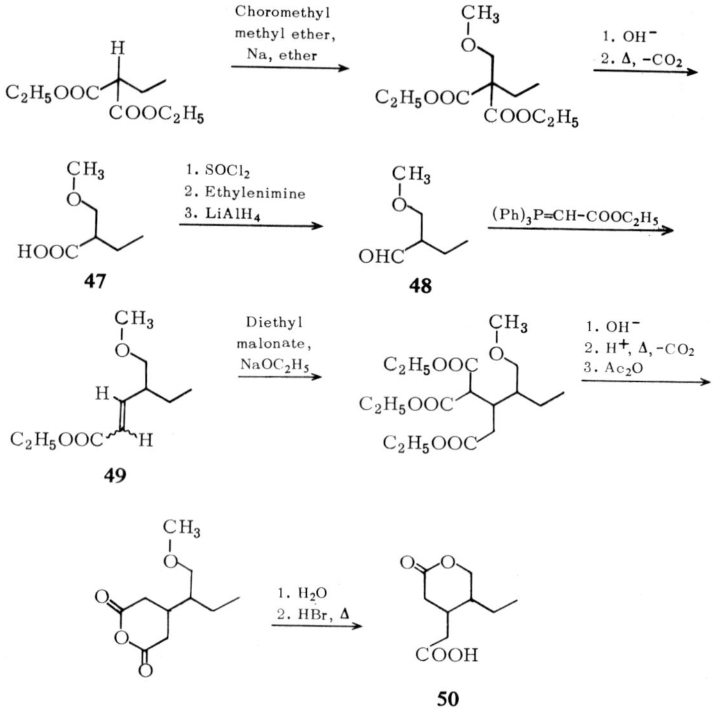

Scheme XV

defined, ipecoside can be considered to be alkaloidal even though it is a neutral compound.

Ipecoside exhibits an amide carbonyl at 6.13 μ (1630 cm^{-1}) and a band at 5.92 μ (1690 cm^{-1}) which was assigned to the $CH_3OOC-\overset{\backslash}{C}=\overset{/}{C}-O-$ system which had been shown to be present in several of the indole alkaloids. The UV data were also in accord with a tetrahydroisoquinoline component, λ_{max} 285 mμ with a bathochromic shift in base due to the presence of the phenolic groups, and also with the presence of the $CH_3OOC-\overset{\backslash}{C}=\overset{/}{C}-O-$ chromophore with λ_{max} 238 mμ.

Catalytic hydrogenation gave dihydroipecoside, which showed unchanged UV and IR spectra (carbonyl region) since only the terminal double bond had been reduced. Mild acid hydrolysis of ipecoside yielded D-glucose, while acetic acid was liberated under conditions sufficiently vigorous to cleave the amide linkage (Scheme XVI).

Ipecoside gave only acetic acid upon Kuhn–Roth oxidation, whereas dihydroipecoside afforded propionic and acetic acids.

The enzyme β-glucosidase hydrolyzes ipecoside so that a β-glucosidic linkage was assigned to the anomeric center. The enzyme also hydrolyzes *O,O*-dimethyldihydroipecoside, though more slowly, to yield an aglycone, probably **51**, which undergoes isomerization with acid to a second, more stable, diastereoisomeric aglycone which

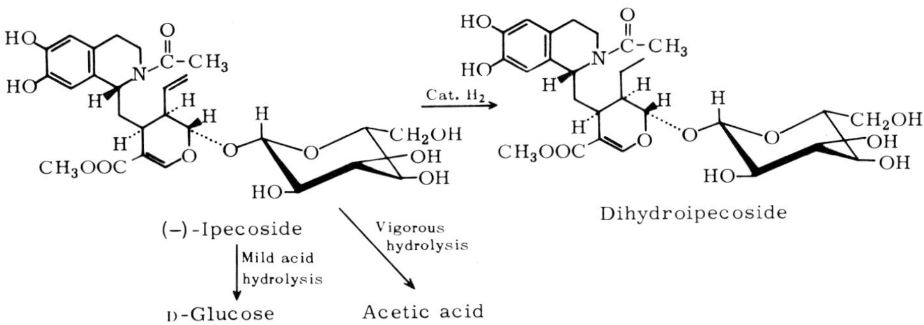

Scheme XVI

has been formulated as **52**. Oxidation of the aglycone **52** furnished the enol δ-lactones **53** and **54**, both of which exhibit carbonyl absorption at 5.64 mμ (1773 cm^{-1}) (Scheme XVII).

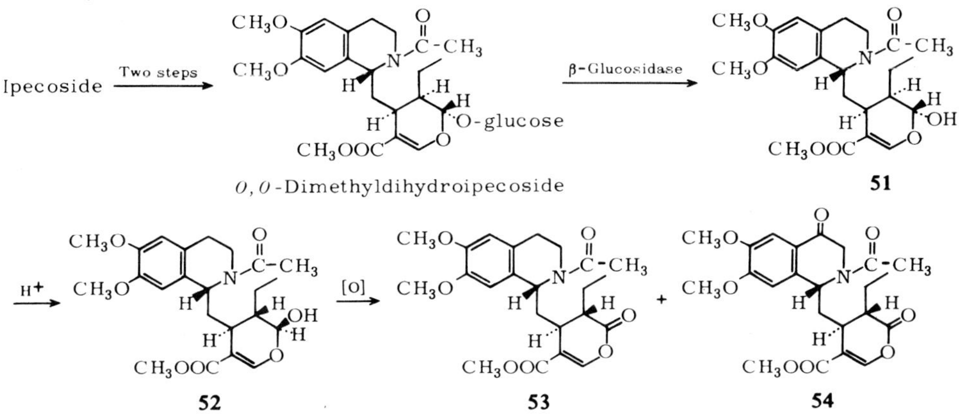

Scheme XVII

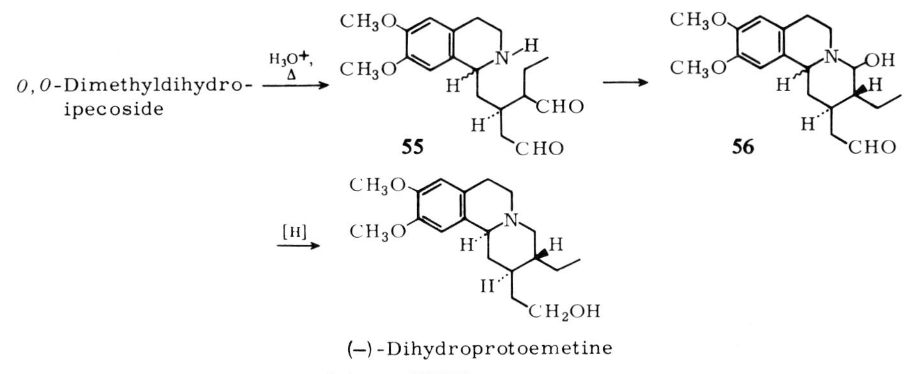

Scheme XVIII

Final proof of structure came from the vigorous acid hydrolysis of *O,O*-dimethyldihydroipecoside, which yielded the aldehydes **55** and **56**. Reduction of the latter product gave (−)-dihydroprotoemetine, identical with authentic material (Scheme XVIII).

IX. Alangiside

(−)-Alangiside is a monoterpenoidal lactam recently isolated from *A. lamarckii*. The molecular formula, $C_{25}H_{31}O_{10}N$, as well as spectral data seemed to indicate that alangiside was closely related to desacetylipecoside. The enzyme β-glucosidase cleaved alangiside to D-glucose and the aglycone $C_{19}H_{21}O_5N$. Additionally, alangiside could undergo the following changes:

(a) Alangiside $\xrightarrow{\text{O-Acetylation}}$ Alangiside penta-*O*-acetate

(b) Alangiside $\xrightarrow{\text{Cat.[H]}}$ Dihydroalangiside $\xrightarrow{\text{O-Acetylation}}$ Dihydroalangiside penta-*O*-acetate

(c) Alangiside $\xrightarrow{CH_2N_2}$ *O*-Methylalangiside $\xrightarrow{\text{O-Acetylation}}$ *O*-Methylalangiside tetra-*O*-acetate

The actual correlation with desacetylipecoside was carried out by treating desacetylipecoside hydrochloride with a weak base to induce lactamization. *O*-Methylation then provided *O*-methylalangiside (Scheme XVIIIa).[42a] The exact position of the phenolic function in alangiside has not yet been settled.

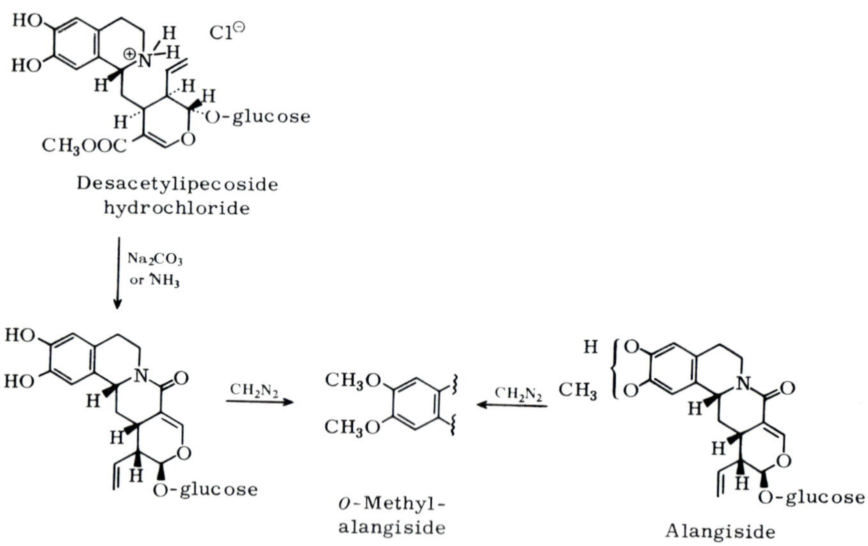

Scheme XVIIIa

X. Biosynthesis

Using feeding experiments, Battersby and Gregory have firmly established that in *Cephaelis ipecacuanha* geraniol is converted into loganin which can then act as a precursor for ipecoside and cephaeline. The C_9 unit of the ipecac alkaloids and the C_{10} unit of ipecoside are, therefore, of monoterpenoid origin.[43,44]

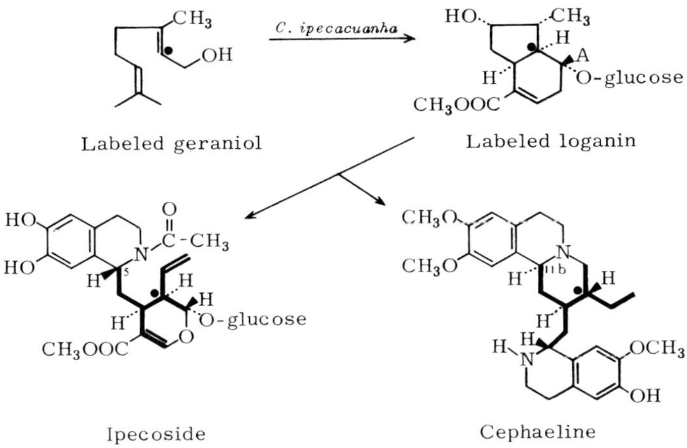

Further studies by Arigoni, Battersby, and their co-workers using mostly *Menyanthes trifoliata* Linn. (Gentianeae), *Vinca rosea* Linn. (Apocynaceae), and *Rauwolfia serpentina* Benth. (Apocynaceae), the last two yielding a variety of indolic alkaloids, have demonstrated that[45-49]:

(1) Geraniol is first oxidized to 10-hydroxygeraniol within the biogenetic locus.

(2) Nerol and 10-hydroxynerol are almost as efficient as geraniol and 10-hydroxygeraniol as precursors for loganin.

(3) Carbon atoms 9 and 10 in 10-hydroxygeraniol and 10-hydroxynerol must equilibrate and become essentially identical by introduction of oxygen at both centers.

(4) 10-Hydrogeraniol or 10-hydroxynerol go to deoxyloganin, which is then oxidized to loganin.

(5) Loganin is cleaved to secologanin, which is a pivotal intermediate. Depending upon the amino acids and the enzymes available in the plant, secologanin can either act as a precursor for desacetylipecoside so as to yield ipecoside and the emetine-type bases, or, alternatively, it can condense to give vincoside, which is a precursor of the indolic iboga, corynanthe, and aspidosperma alkaloids (Scheme XIX).

All of the intermediates mentioned here have been isolated as natural products from plants, some only in trace amounts.

Turning again to *C. ipecacuanha*, it has recently been possible to demonstrate that loganin is changed into secologanin which undergoes conversion to desacetylipecoside and desacetylisoipecoside. It is desacetylipecoside and *not* desacetylisoipecoside which

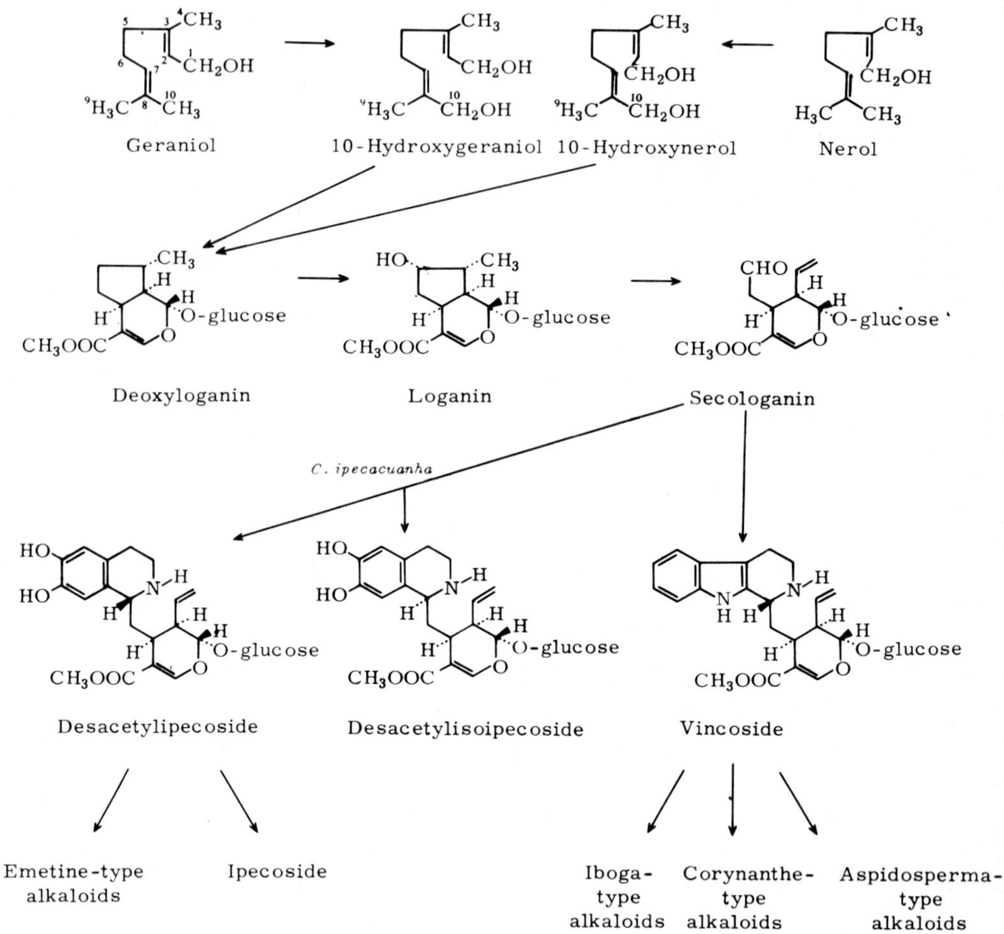

Scheme XIX

can be converted biologically into ipecoside, cephaeline, and emetine. The mechanism of the interesting C-5 → C-11b inversion in going from desacetylipecoside to cephaeline remains to be established.[49a]

Finally, experiments with *C. ipecacuanha* using labeled glycine and sodium acetate have shown that glycine can act as a specific two-carbon precursor of the C_9 unit of cephaeline, delivering 15 – 18% of the activity incorporated to C-15 and none to C-14. On the other hand, radioactive acetic acid did not label this unit with any specificity or efficiency but was incorporated instead into the sterol β-sitosterol that is also present in the plant. The oxidation state of a two-carbon compound may, therefore, exert a major effect upon its utilization in biosynthesis (Scheme XX).[50]

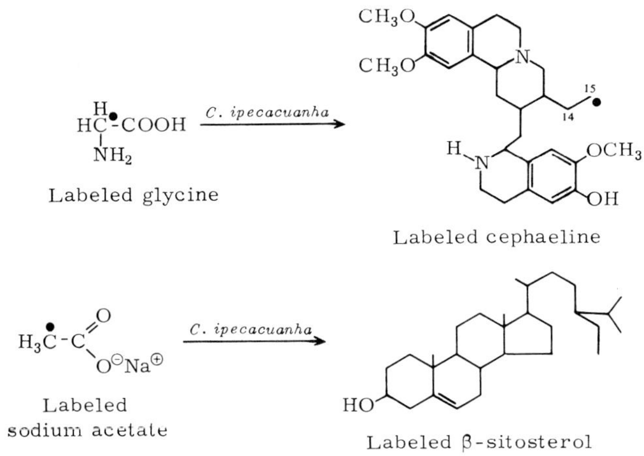

Scheme XX

XI. The Partial Synthesis of Secologanin and Ipecoside

Controlled mild alkaline hydrolysis of the interesting naturally occurring glucoside menthiafolin, obtained from *Menyanthes trifoliata* Linn.,[51] followed by methylation

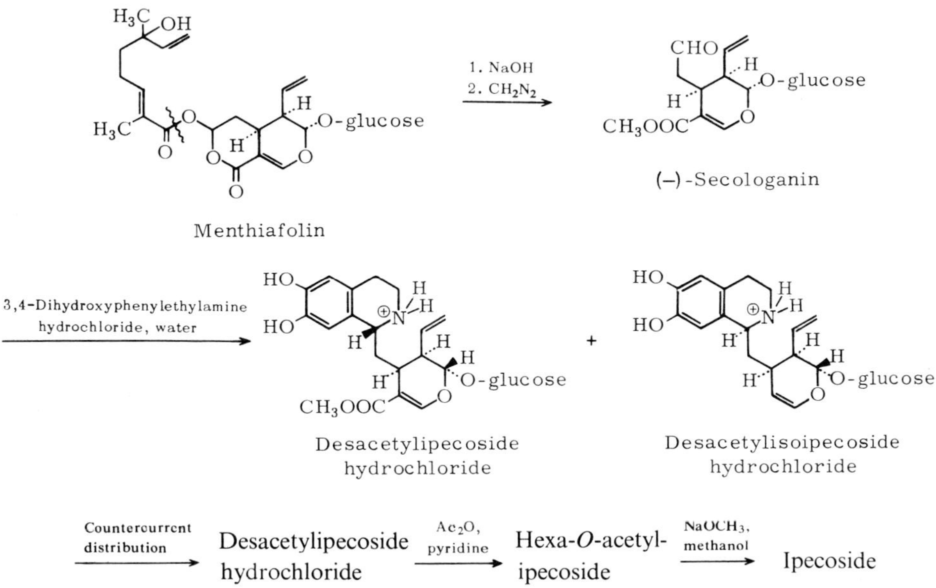

Scheme XXI

with diazomethane, afforded secologanin. Subsequent Mannich condensation with 3,4-dihydroxyphenylethylamine hydrochloride yielded two products, desacetylipecoside and desacetylisoipecoside hydrochlorides, which were separated by countercurrent distribution. Desacetylipecoside was then polyacetylated, and Zemplén deacetylation furnished (−)-ipecoside, identical with the natural product (Scheme XXI).[45,52]

XII. Rubremetine Salts

Mild oxidation of emetine with ferric chloride, bromine, iodine, or mercuric acetate gives reddish, optically active compounds known as rubremetine salts. As a result of this transformation one nitrogen atom loses its basic character while the other becomes quaternary. The structure assigned to the rubremetine cation incorporates a nonbasic nitrogen within a pyrrole ring, a quaternary nitrogen as part of the original isoquinoline system, and asymmetry about C-3 where the ethyl group is attached (Scheme XXII).[53–55]

Catalytic reduction of rubremetine chloride readily furnishes two basic and colorless diastereoisomers, A- and B-dihydrorubremetine, which are epimeric at C-11b. The starting material as well as the products exhibit positive color tests for pyrrole, and spectral data indicate that both A- and B-dihydrorubremetines possess cis-fused quinolizidine systems.[56,57]

The dihydrorubremetines can be hydrogenated further very slowly, but the reaction is appreciably catalyzed by acid. The product from either isomer is optically active tetrahydrorubremetine which still shows pyrrole reactions and contains no N−H function. Tetrahydrorubremetine is formed through the intermediacy of the immonium salt **57** whose formation from the dihydrorubremetines is acid-catalyzed. In line with this reasoning, the dihydrorubremetines undergo mutarotation in acid solution via the salt **57** (Scheme XXII).[56,57]

Several syntheses of the rubremetine cation are available.

XIII. Pharmacology[33]

Ipecac was the standard treatment of the South American Indians before the advent of the white man for what is now recognized as amebic dysentery caused by the microorganism *Entamoeba histolytica*. In 1912 it was recognized that the alkaloid responsible for this drug action is emetine, which is less toxic and more effective than the accompanying base cephaeline.

Emetine is effective only against amebic and not bacillary infections, and the drug is still useful in efforts at eradication of the extraintestinal trophozoites and in the treatment of acute amebic intestinal or extraintestinal infections.

Emetine is toxic and cumulative, and it must not be given beyond a 10-day period. Poisoning by emetine is characterized by muscular tremors and weakness and pains, especially in the extremities. The alkaloid has also been used as a mild emetic, but this is not a common treatment.

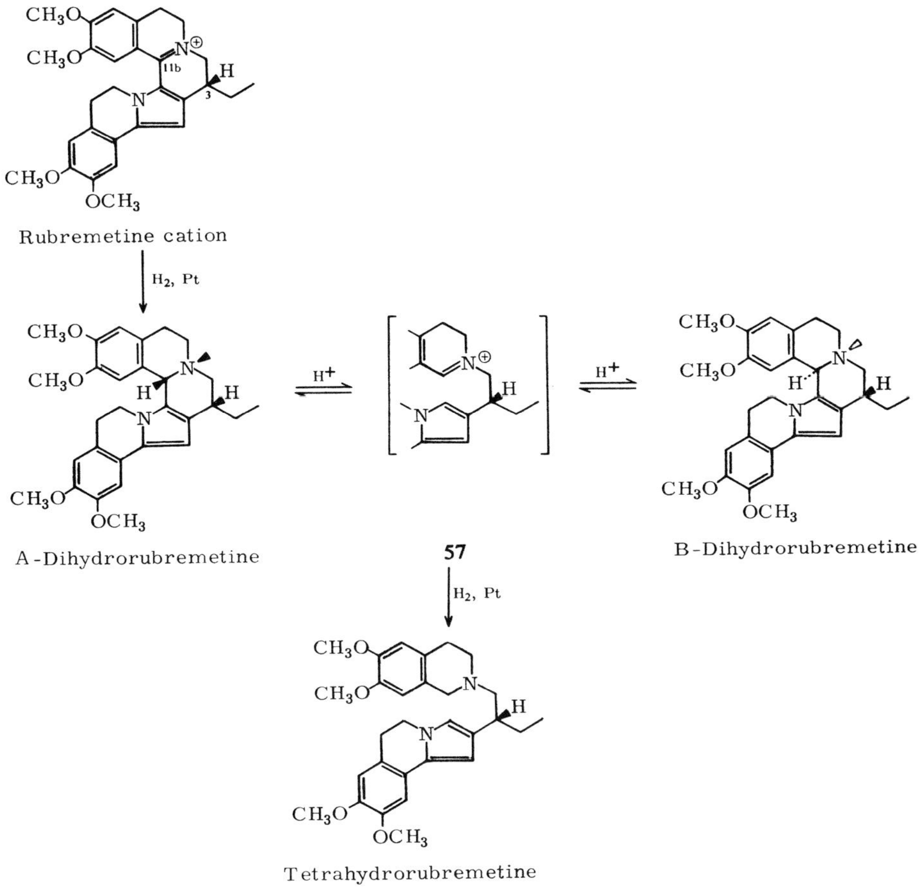

Scheme XXII

(±)-2,3-Dehydroemetine is almost equal to emetine in amebicidal activity, the (−)-enantiomer being the biologically active component. The racemic compound is usually used and it has been found that it is better tolerated and more readily excreted than (−)-emetine. The absence of the C-3 ethyl group of emetine results in a partial loss of antiamebic activity. The pharmacological usefulness of emetine and 2,3-dehydroemetine may be related to the fact that they inhibit protein synthesis in mammalian cells. Emetine also has antiparasitic activity in certain lung diseases and in sheep infected with *Fasciola hepatica*.[1]

As pointed out earlier, (−)-emetine, identical with the natural product, is produced commercially by Burroughs Wellcome & Co. in England, while (±)-2,3-dehydroemetine is manufactured at Hoffmann–La Roche in Switzerland. Since an optical resulution is not involved in the production of (±)-2,3-dehydroemetine, it is more readily available than synthetic (−)-emetine.

Emetine has also shown potential in initial tests as an anticancer drug and is currently undergoing clinical evaluation at the National Institutes of Health.[58]

The benzoquinolizidine derivative tetrabenazine (Nitoman, Hoffmann–La Roche) has been extensively tested because of its psychotropic activity. It exerts a reserpine-like sedative action and has a beneficial effect on various symptoms of schizophrenia as well as in Huntington's chorea. Like reserpine, tetrabenazine decreases the 5-hydroxytryptamine and catechol amine content in the brain. It is less potent than reserpine but of shorter duration and faster onset of action.[59,59a] It is presently in use in some of the Scandinavian countries.

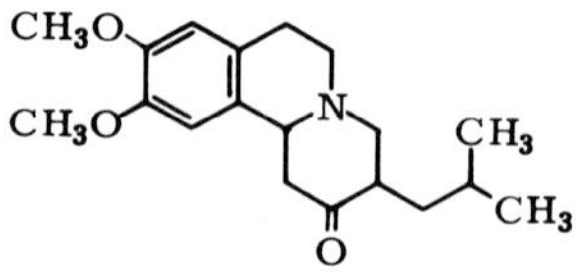

Tetrabenazine

When *N*-2′-benzyloxycarbonylemetine was quaternized with 3,3-ethylenedioxy-1-*p*-toluenesulfonate and the salt reduced with lithium in liquid ammonia, fission of the N-5 to C-11b bond was accompanied by cleavage of the benzyloxycarbonyl group, so that the product was the secoemetine **58** which was de-*N*-alkylated to the derivative **59**. Compound **59** is less potent as an amebicide than the parent alkaloid.[60]

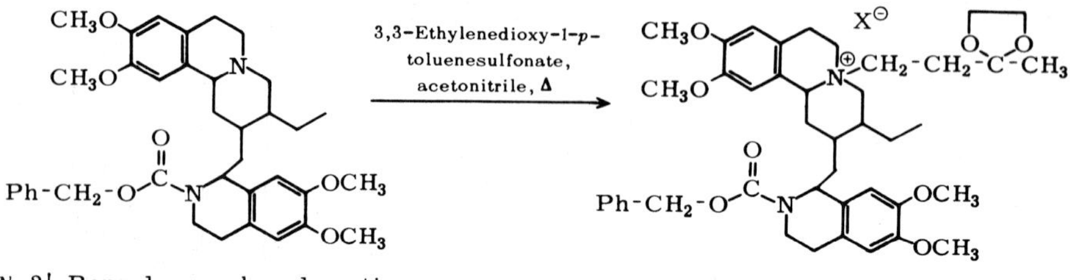

N-2′-Benzyloxycarbonylemetine

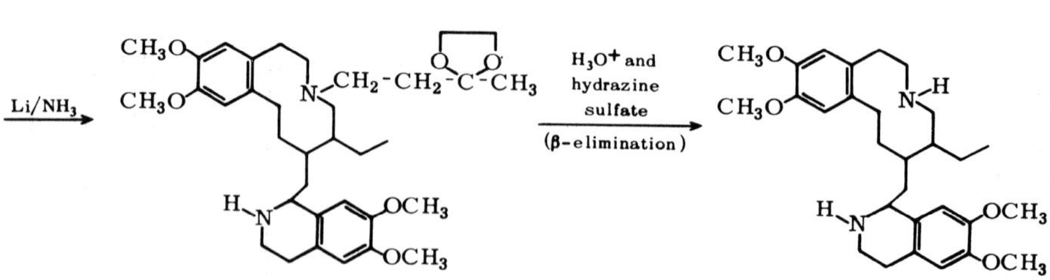

58 **59**

XIV. Mass Spectroscopy

The mass spectral cleavage pattern of the emetine-type alkaloids is dependent on the degree of saturation of ring D. The two most intense peaks in the spectrum of cephaeline are at *m/e* 192 and 178 and are due to ions **60**–**62**.

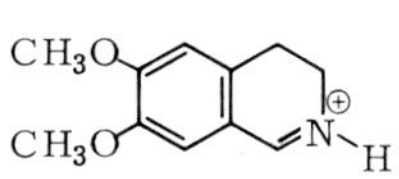

60
m/e 192

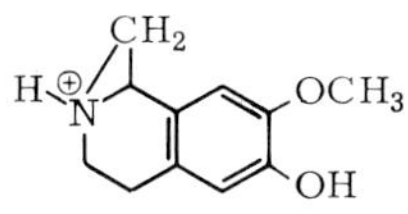

61
m/e 192

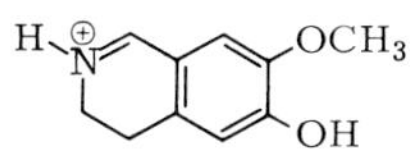

62
m/e 178

O-Methylpsychotrine, which incorporates an imine function in ring D, exhibits a totally different cleavage pattern, with two discernible modes of fragmentation (Scheme XXIII). In type (i) cleavage, the very intense peak at *m/e* 273 and the strong *m/e* 205 peak are produced through cleavage of the allylic C-2 to C-α bond. Type (ii) fission affects primarily ring C and produces the stable ions of *m/e* 190 and 230.[3]

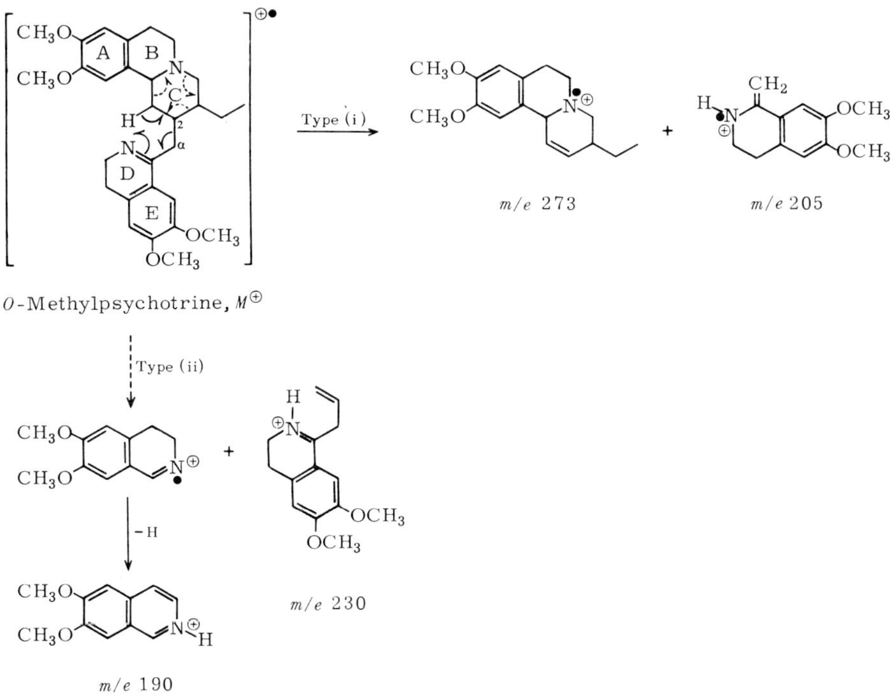

Scheme XXIII

In the spectrum of ankorine, the base peak at *m/e* 334 corresponds to the $(M - 1)$ ion. Other intense peaks are seen at *m/e* 320, 318, 262, 221, 207, and 192, and these are interpreted in Scheme XXIV.[5]

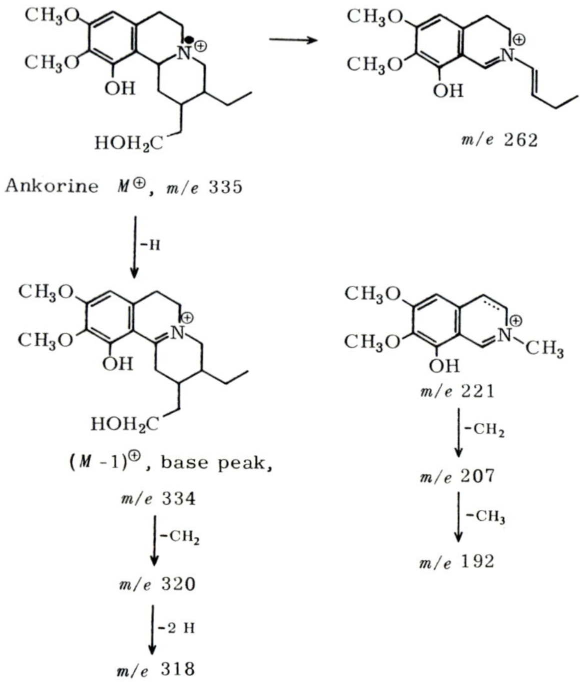

Scheme XXIV

XV. NMR Spectroscopy

It is difficult to make specific assignments for the proton chemical shifts of the classical emetine bases.[12] The NMR spectrum for the simpler alkaloid ankorine is more readily interpreted and the chemical shifts are quoted below.[5]

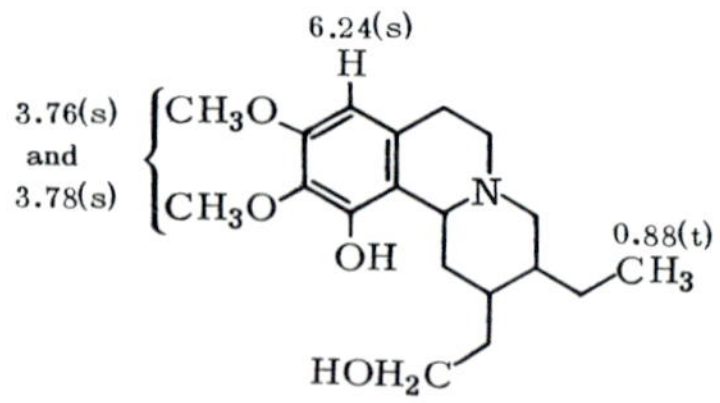

NMR spectral values for ankorine

XVI. UV Spectroscopy

Emetine dihydrochloride[13,61]	λ_{max}^{EtOH} 230 and 283 mμ (4.23 and 3.87)
	λ_{min}^{EtOH} 256 mμ (3.11)
Psychotrine[12]	$\lambda_{max}^{0.1\,N\,HCl}$ 240, 288, 306, and 356 mμ (4.14, 3.76, 3.80, and 3.83)
Desmethylpsychotrine[4]	λ_{max}^{EtOH} 223, 277, 310, and 410 mμ (3.95, 3.83, 3.34, and 3.96)
Alangicine[4]	λ_{max}^{EtOH} 275, 312, and 408 mμ (3.84, 3,42, and 4.09)
Emetamine[62]	λ_{max}^{EtOH} 236 and 283 mμ (4.85 and 3.86)
	λ_{min}^{EtOH} 217, 262, 303, and 321 mμ (4.28, 3.72, 3.36, and 3.48)
Protoemetine perchlorate[10]	λ_{max}^{EtOH} 232 and 283 mμ (3.92 and 3.61)
	λ_{min}^{EtOH} 220 and 254 mμ (3.85 and 2.68)
Ankorine[5]	λ_{max}^{EtOH} 272 mμ (2.96)
Rubremetine chloride[61] (not a natural product)	$\lambda_{max}^{H_2O}$ 257, 286, 300, and 437 mμ (4.28, 4.24, 4.26, and 4.47).

REFERENCES

1. For reviews on the ipecacuanha alkaloids, see H. T. Openshaw, *in* "Chemistry of the Alkaloids" (S. W. Pelletier, ed.), p. 85. Van Nostrand, Princeton, New Jersey, 1970; and A. Brossi, G. V. Parry, and S. Teitel, *in* "The Alkaloids" (R. F. H. Manske, ed.), Vol. 13, p. 189. Academic Press, New York, 1971; and Cs. Szántay in Recent Developments in the Chemistry of Natural Carbon Compounds, Vol. 2, 63 (1967), Akadémiai Kiadó, Budapest.
2. S. C. Pakrashi and E. Ali, *Indian J. Chem.* 7, 635 (1969).
3. H. Budzikiewicz, S. C. Pakrashi, and H. Vorbrüggen, *Tetrahedron* **20**, 399 (1964).
4. S. C. Pakrashi and E. Ali, *Tetrahedron Lett.* p. 2143 (1967).
5. A. R. Battersby, R. S. Kapil, D. S. Bhakuni, S. P. Popli, J. R. Merchant, and S. S. Salgar, *Tetrahedron Lett.* p. 4965 (1966).
6. B. Dasgupta, *J. Pharm. Sci.* **54**, 481 (1965).
7. R. Robinson, *Nature* (*London*) **162**, 524 (1948).
8. A. R. Battersby and H. T. Openshaw, *J. Chem. Soc., London* pp. 3207 and S 59 (1949).
9. M. Pailer and K. Porschinski, *Monatsh. Chem.* **80**, 94 and 101 (1949).
10. A. R. Battersby and B. J. T. Harper, *J. Chem. Soc., London* p. 1748 (1959).
11. Cs. Szantay, L. Töke, and P. Kolonits, *J. Org. Chem.* **31**, 1447 (1966).
12. S. Teitel and A. Brossi, *J. Amer. Chem. Soc.* **88**, 4068 (1966).
13. A. Brossi, M. Baumann, and O. Schnider, *Helv. Chim. Acta* **42**, 1515 (1959).
14. E. E. van Tamelen, P. E. Aldrich, and J. B. Hester, Jr., *J. Amer. Chem. Soc.* **79**, 4817 (1957).
15. E. E. van Tamelen and J. B. Hester, Jr., *J. Amer. Chem. Soc.* **81**, 507 (1959).
16. E. E. van Tamelen, P. E. Aldrich, and J. B. Hester, Jr., *J. Amer. Chem. Soc.* **81**, 6214 (1959).
17. For a closely related sequence, see A. R. Battersby and S. Garratt, *J. Chem. Soc., London* p. 3512 (1959).
18. E. E. van Tamelen and M. Shamma, *J. Amer. Chem. Soc.* **76**, 950 (1954).
19. Y. Ban, M. Terashima, and O. Yonemitsu, *Chem. Ind.* (*London*) pp. 568 and 569 (1959).

20. M. Terashima, *Chem. Pharm. Bull.* **8**, 517 (1960).
21. A. R. Battersby, R. Binks, and T. P. Edwards, *J. Chem. Soc., London* p. 3474 (1960).
22. A. R. Battersby, S. W. Breuer, and S. Garratt, *J. Chem. Soc., C* p. 2467 (1968).
23. R. P. Evstigneeva, R. S. Livshits, L. I. Zakharin, M. S. Bainova, and N. A. Preobrazhenskii, *Dokl. Akad. Nauk SSSR* **75**, 539 (1950); R. P. Evstigneeva and N. A. Preobrazhenskii, *Tetrahedron* **4**, 223 (1958). See also L. I. Zakharin and N. A. Preobrazhenskii, *Z. Obshch. Khim.* **22**, 1890 (1952); **23**, 153 (1953).
24. J. A. Weisbach, J. L. Kirkpatrick, E. L. Anderson, K. Williams, B. Douglas, and H. Rapoport, *J. Amer. Chem. Soc.* **87**, 4221 (1965).
25. A. W. Frahm, *Arch. Pharm.* (*Weinheim*) **301**, 621 (1968).
26. Y. Ban, *Pharm. Bull.* **3**, 53 (1955).
27. A. R. Battersby and J. C. Turner, *Chem. Ind.* (*London*) p. 1324 (1958); *J. Chem. Soc., London* p. 717 (1960).
28. E. E. van Tamelen, M. Shamma, A. W. Burgstahler, J. Wolinsky, R. Tamm, and P. E. Aldrich, *J. Amer. Chem. Soc.* **80**, 5006 (1958); **91**, 7315 (1969).
29. A. W. Burgstahler and Z. J. Bithos, *J. Amer. Chem. Soc.* **82**, 5466 (1960).
30. E. E. van Tamelen, G. P. Schiemenz, and H. L. Arons, *Tetrahedron Lett.* p. 1005 (1963); E. E. van Tamelen, C. Placeway, G. P. Schiemenz, and I. G. Wright, *J. Amer. Chem. Soc.* **91**, 7359 (1969).
31. D. E. Clark, R. F. K. Meredith, A. C. Ritchie, and T. Walker, *J. Chem. Soc., London* p. 2490 (1962).
32. A. R. Battersby, H. T. Openshaw, and H. C. S. Wood, *J. Chem. Soc., London* p. 2463 (1953).
33. A. Brossi, *Pure Appl. Chem.* **19**, 171 (1969).
34. A. Grüssner, E. Jaeger, J. Hellerbach, and O. Schnider, *Helv. Chim. Acta* **42**, 2431 (1959).
35. M. Barasch, J. M. Osbond, and J. C. Wickens, *J. Chem. Soc., London* p. 3530 (1959).
36. H. T. Openshaw and N. Whittaker, *J. Chem. Soc., London* pp. 1449 and 1461 (1963).
37. N. Whittaker, *J. Chem. Soc., C* p. 85 (1969).
38. For a related approach, see Szántay *et al.*[11]
39. H. T. Openshaw and N. Whittaker, *J. Chem. Soc., C* p. 89 (1969).
40. N. Whittaker, *J. Chem. Soc., C* p. 94 (1969).
41. F. Zymalkowski and A. W. Frahm, *Arch. Pharm.* (*Weinheim*) **297**, 219 (1964); *Chem. Abstr.* **60**, 15727b (1964).
42. A. R. Battersby, B. Gregory, H. Spencer, J. C. Turner, M.-M. Janot, P. Potier, P. François, and J. Levisalles, *Chem. Commun.* p. 219 (1967). For the X-ray determination of *O,O*-dimethylipecoside sesquihydrate see O. Kennard, P. J. Roberts, N. W. Isaacs, F. H. Allen, W. D. S. Motherwell, K. H. Gibson, and A. R. Battersby, *Chem. Commun.* p. 899 (1971).
42a. R. S. Kapil, A. Shoeb, S. P. Popli, A. R. Burnett, G. D. Knowles, and A. R. Battersby, *Chem. Commun.* p. 904 (1971).
43. A. R. Battersby and B. Gregory, *Chem. Commun.* p. 134 (1968); see also A. R. Battersby, R. S. Kapil, and R. Southgate, *ibid.* p. 131.
44. For related experiments, see S. Brechbühler-Bader, C. J. Coscia, P. Loew, C. von Szczepanski, and D. Arigoni, *Chem. Commun.* p. 136 (1968); also P. Loew and D. Arigoni, *ibid.* p. 137.
45. A. R. Battersby, A. R. Burnett, and P. G. Parsons, *J. Chem. Soc., C* p. 1187 (1969).
46. A. R. Battersby, A. R. Burnett, and P. G. Parsons, *J. Chem. Soc., C* p. 1193 (1969).
47. S. Escher, P. Loew, and D. Arigoni, *Chem. Commun.* p. 823 (1970).
48. A. R. Battersby, A. R. Burnett, and P. G. Parsons, *Chem. Commun.* p. 826 (1970).
49. A. R. Battersby, S. H. Brown, and T. G. Payne, *Chem. Commun.* p. 827 (1970).
49a. A. R. Battersby and R. J. Parry, *Chem. Commun.* p. 901 (1971).
50. A. K. Garg and J. R. Gear, *Tetrahedron Lett.* p. 4377 (1969).
51. A. R. Battersby, A. R. Burnett, G. D. Knowles, and P. G. Parsons, *Chem. Commun.* p. 1277 (1968).

52. A. R. Battersby, A. R. Burnett, and P. G. Parsons, *Chem. Commun.* p. 1280 (1968).
53. A. R. Battersby, H. T. Openshaw, and H. C. S. Wood, *Experientia* **5**, 114 (1949).
54. H. T. Openshaw and H. C. S. Wood, *J. Chem. Soc., London* p. 391 (1952).
55. A. Brossi, M. Gerecke, A. R. Battersby, R. S. Kapil, Y. Ban, and M. Terashima, *Experientia* **22**, 134 (1966).
56. K.-A. Kovar, H. Ehrhardt, and H. Auterhoff, *Arch. Pharm.* (*Weinheim*) **302**, 220 (1969).
57. K.-A. Kovar, *Arch. Pharm.* (*Weinheim*) **303**, 579 (1970).
58. R. E. Perdue, Jr. and J. L. Hartwell, *Morris Arbor. Bull.* **20**, 35 (1969).
59. A. Pletscher, A. Brossi, and K. F. Gey, *Int. Rev. Neurobiol.* **4**, 275 (1962).
59a. D. E. Schwartz, H. Bruderer, J. Rieder, and A. Brossi, *Biochem. Pharmacol.* **15**, 645 (1966).
60. J. P. Yardley, R. W. Rees, and H. Smith, *J. Med. Chem.* **10**, 1088 (1967).
61. A. Brossi and O. Schnider, *Helv. Chim. Acta* **45**, 1899 (1962).
62. A. R. Battersby, G. C. Davidson, and J. C. Turner, *J. Chem. Soc., London* p. 3899 (1961).

Chapter 24 / AUTUMNALINE: A PHENETHYLISO-QUINOLINE

Occurrence: Liliaceae

Structure:

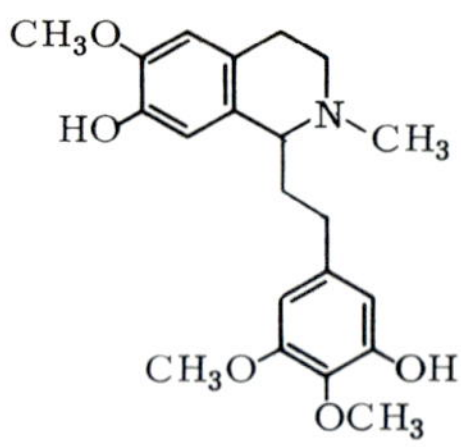

Autumnaline

I. Introduction

The homologated alkaloids can be classified into eight categories: androcymbine, colchicine, homomorphine, homoerythrina, phenethylisoquinoline, bisphenethylisoquinoline, homoaporphine, and homoproaporphine. Examples of the first four are shown below, but a detailed discussion of their chemistry is outside the scope of this book.

This chapter will be devoted to a discussion of the phenethylisoquinoline autumnaline and the physiologically active synthetic compound methopholine. The bisphenethylisoquinoline, homoaporphine, and homoproaporphine bases will be covered in the next two chapters. The isolation and characterization work on these four classes of homologated isoquinoline alkaloids was performed mainly by Battersby, Šantavý, and co-workers.

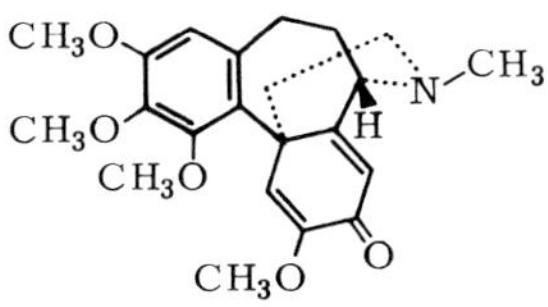

Androcymbine[1]

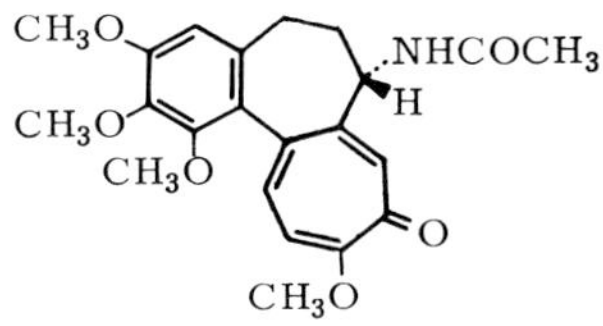

Colchicine[2]

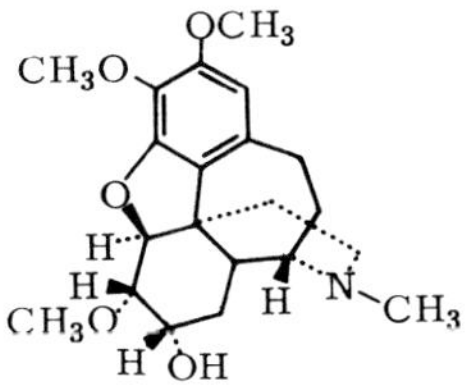

(+)-Kreysiginine[3,4]
(a homomorphine alkaloid; levorotatory enantiomer also known)

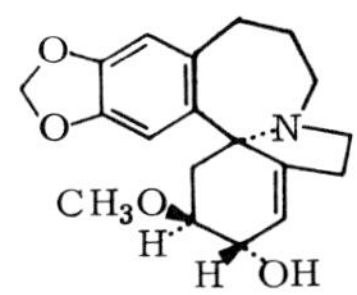

(+)-Schelhammerine[5]
(a homoerythrina alkaloid)

II. AUTUMNALINE

Autumnaline is the only phenethylisoquinoline alkaloid for which a structure has been presented. It was isolated from *Colchicum cornigerum* (Schweinf.) Tackh et Drar. (Liliaceae, subfamily Wurmbaeoidea) and appears to be slightly levorotatory in chloroform, $[\alpha]_D^{22}$ $-5° \pm 3°$, while optically inactive in ethanol.[6,7] The absolute configuration, if the alkaloid is indeed optically active, has not yet been indicated.

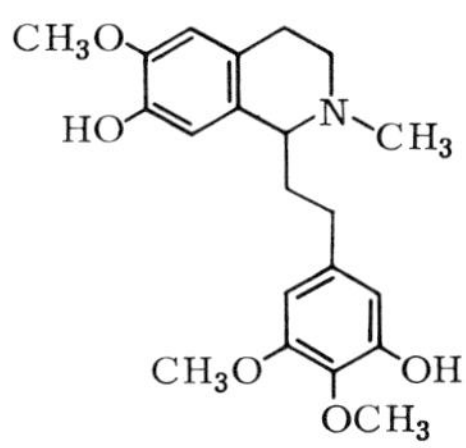

Autumnaline

III. SYNTHESES OF PHENETHYLISOQUINOLINES

The methods available for the preparation of phenethylisoquinolines are best exemplified by the three approaches used to synthesize the drug methopholine.

A. The Acetophenone Approach

The readily available keto amide **1** was condensed with *p*-chlorobenzaldehyde in an aldol reaction to provide the chalcone derivative **2**. The intensely colored imine **3** obtained by hydrolysis of the amide function was then reduced catalytically and the product *N*-methylated to supply methopholine (Scheme I).[8–10]

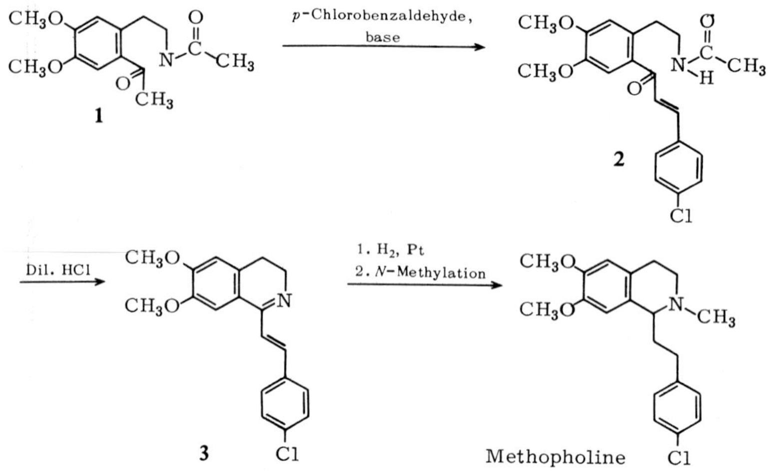

Scheme I

B. Starting from a Quaternary 1-Methyl-3,4-dihydroisoquinoline Salt

The immonium salt **4**, which possesses an activated methyl group, was condensed with *p*-chlorobenzaldehyde using piperidine as the base. The product was the colored salt **5**, which was reduced to methopholine (Scheme II).[11]

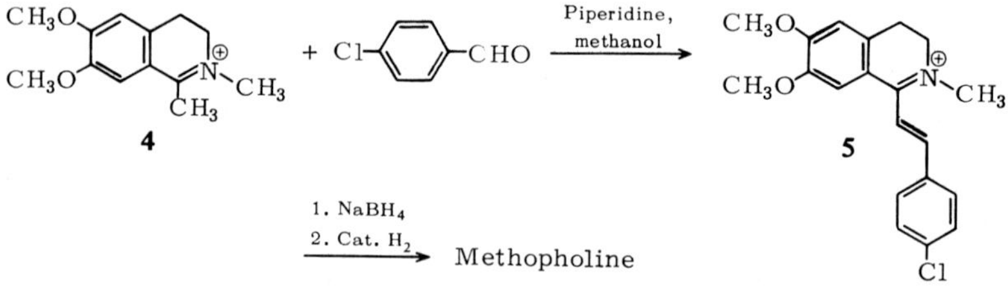

Scheme II

C. Bischler–Napieralski Procedure

This is the most convenient approach. The amide **6** was cyclized to the imine, and *N*-methylation followed by a reductive step furnished methopholine.[11]

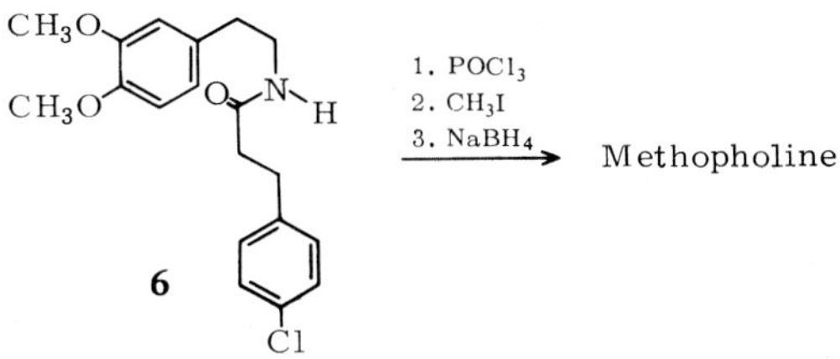

IV. Absolute Configuration

Resolution of the racemic tetrahydroisoquinoline **7** yielded (+)-**7** and (−)-**7**. The dextrorotatory enantiomer was catalytically reduced to the dextrorotary tetrahydrophenethylisoquinoline **8**, whose *N*-methylation afforded (−)-methopholine. The absolute configuration of (+)-**7** was determined by its transformation into (−)-*N*-acetylcalycotomine, identical with the same material derived from (+)-calycotomine of known absolute configuration (Scheme III).[12]

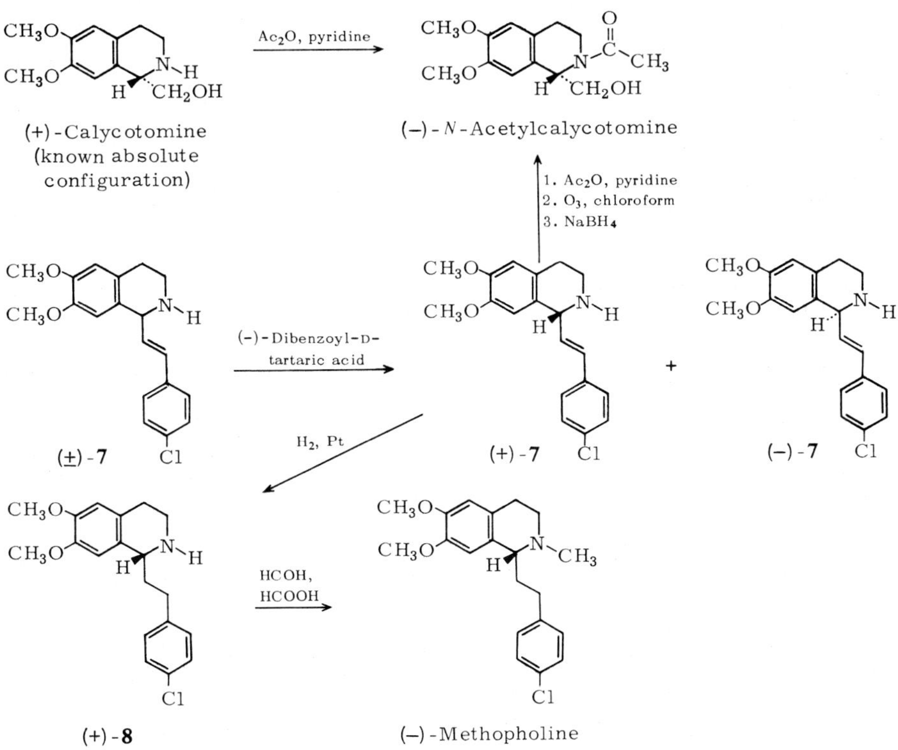

Scheme III

The ORD and CD curves for a number of optically active phenethylisoquinolines have been recorded. When the C-1 hydrogen is *beta* (*R* configuration), the ORD curves are generally negative with minima near 240 and 290 mμ.[13-16]

V. Biosynthesis

No biogenetic studies have been carried out on autumnaline. It has been demonstrated, however, that in *Colchicum autumnale* Linn. (Liliaceae), which produces androcymbine as well as colchicine, phenylalanine is first converted into cinnamic acid.[17-19] Subsequent condensation with an amine such as dopamine yields a phenethylisoquinoline, which can become the precursor for androcymbine analogs, and these in turn can be converted into colchicine (Scheme IV).[17-19]

It follows that cinnamic acid or substituted cinnamic acids may be implicated in the biosynthesis of phenethylisoquinoline and bisphenethylisoquinoline alkaloids, as well as in the biosynthesis of the homoaporphines and the homoproaporphines.

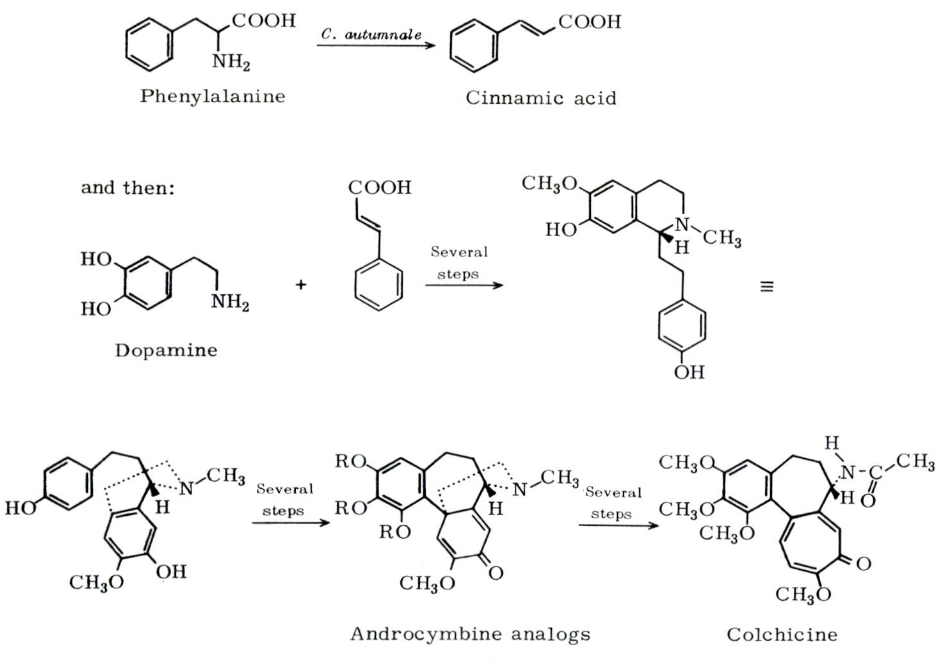

Scheme IV

VI. Pharmacology

The synthetic phenethylisoquinoline methopholine (Versidyne) prepared in the Hoffmann–La Roche laboratories is a safe analgetic, well suited for the treatment of mild to moderately severe pain. When given orally, the analgetic effectiveness of methopholine is equal to that of codeine. But unlike the latter, methopholine does not lead to physical addiction, sedation, or constipation. The levorotatory base with a beta hydrogen at C-1 (*R* configuration) is by far the more active enantiomer.[11]

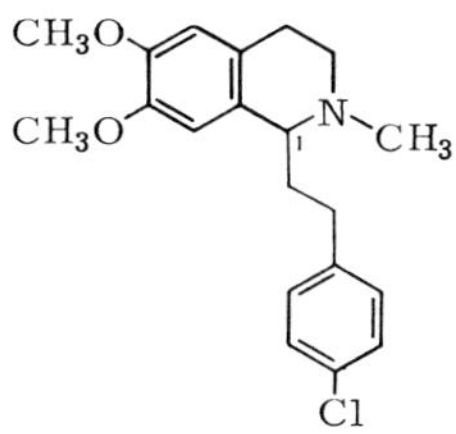

Methopholine

VII. NMR Spectroscopy

The chemical shifts for the synthetic phenethylisoquinoline **9**,[14] **10**, and **11**[20] are listed below.

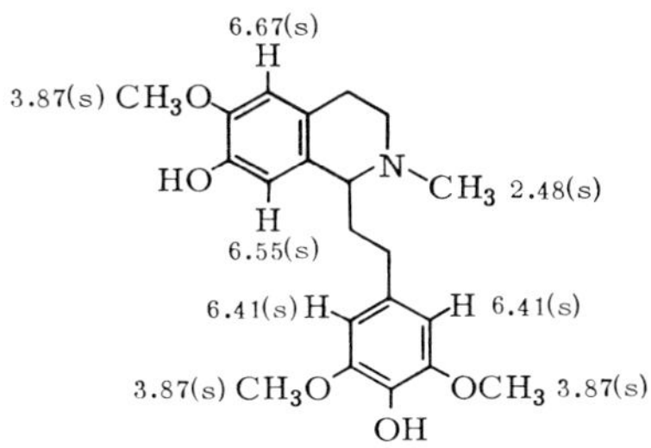

9

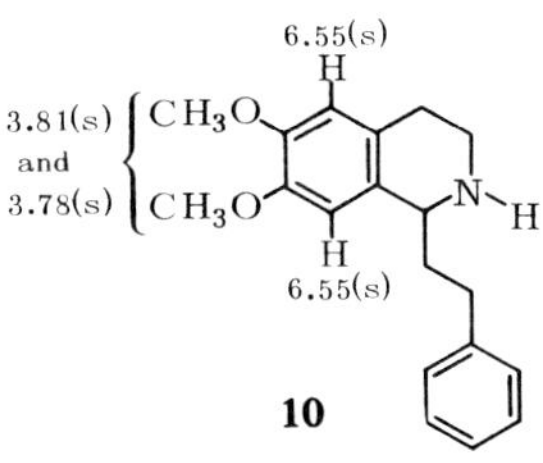

10

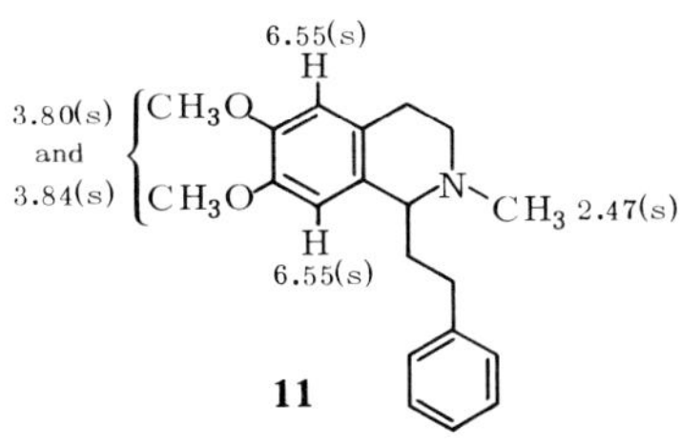

11

VIII. UV Spectroscopy

Autumnaline[6]	λ_{max}^{EtOH} 206, 225 sh, and 285 mμ (4.82, 4.25, and 3.62)
	λ_{min}^{EtOH} 252 mμ (2.88)
Compound **9**[14]	λ_{max}^{MeOH} 230 sh, 283, and 294 sh mμ (4.17, 3.71, and 3.54)

REFERENCES

1. A. R. Battersby, R. B. Herbert, L. Pijewska, and F. Šantavý, *Chem. Commun.* p. 228 (1965).
2. J. Fridrichsons, M. F. Mackay, and A. McL. Mathieson, *Tetrahedron Lett.* p. 2887 (1968).
3. N. K. Hart, S. R. Johns, J. A. Lamberton, and J. K. Saunders, *Tetrahedron Lett.* p. 2891 (1968).
4. A. R. Battersby, M. H. G. Munro, R. B. Bradbury, and F. Šantavý, *Chem. Commun.* p. 695 (1968).
5. J. S. Fitzgerald, S. R. Johns, J. A. Lamberton, and A. A. Sioumis, *Aust. J. Chem.* **22**, 2187 (1969).
6. H. Potěšilová, J. Šantavý, A. El-Hamidi, and F. Šantavý, *Collect. Czech. Chem. Commun.* **34**, 3540 (1969).
7. A. R. Battersby, P. Böhler, M. H. G. Munro, and R. Ramage, *Chem. Commun.* p. 1066 (1969).
8. A. Brossi, J. Würsch, and O. Schnider, *Chimia* **12**, 114 (1958).
9. A. Brossi, H. Besendorf, B. Pellmont, M. Walter, and O. Schnider, *Helv. Chim. Acta* **43**, 1459 (1960).
10. Hoffmann–La Roche Inc., U.S.Pat. 3,067,203 (1962).
11. A. Brossi, H. Besendorf, L. A. Pirk, and A. Rheiner, *Med. Chem.* **5**, 281 (1965).
12. A. Brossi and F. Burkhardt, *Helv. Chim. Acta* **44**, 1558 (1961).
13. A. Rheiner, Jr. and A. Brossi, *Experientia* **20**, 488 (1964).
14. A. Brossi, J. O'Brien, and S. Teitel, *Helv. Chim. Acta* **52**, 678 (1969).
15. A. Brossi and S. Teitel, *Helv. Chim. Acta* **52**, 1228 (1969).
16. T. Kametani, F. Satoh, H. Yagi, and K. Fukumoto, *J. Chem. Soc., C* p. 382 (1970)
17. A. R. Battersby, R. Binks, J. J. Reynolds, and D. A. Yeowell, *J. Chem. Soc., London* p. 4257 (1964).
18. E. Leete, *Tetrahedron Lett.* p. 333 (1965). *Colchicum byzantium* was used for the labeling experiments.
19. A. R. Battersby, *Pure Appl. Chem.* **14**, 117 (1967).
20. K. Ito, H. Furukawa, H. Tanaka, and S. Kato, *J. Pharm. Soc. Japan* **90**, 1169 (1970).

Chapter 25 / MELANTHIOIDINE: A BISPHENETHYLISO-QUINOLINE

Occurrence: Liliaceae
Structure:

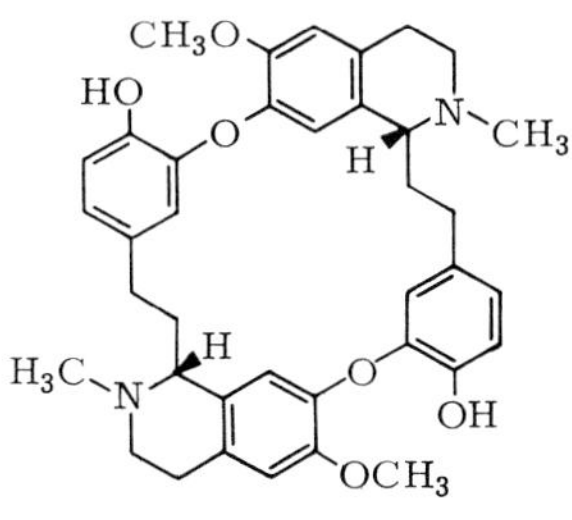

(−)-Melanthioidine

I. Structural Elucidation of Melanthioidine

(−)-Melanthioidine, found in *Androcymbium melanthioides* Willd. (Liliaceae, subfamily Wurmbaeoideae), $C_{38}H_{42}O_6N_2$, λ_{max} 283 mμ (3.34), has two phenolic groups, two methylimino groups, and two methoxyls. NMR spectroscopy indicated the presence of 10 aromatic and 18 nonaromatic protons. The 6 *N*-methyl protons showed up as a singlet at δ2.44, and the 6 *O*-methyl protons also appeared as a singlet at δ3.79, thus underlining the symmetry of the molecule.

Reductive cleavage of *O*,*O*-dimethylmelanthioidine with sodium in liquid ammonia afforded almost exclusively the levorotatory phenethylisoquinoline **1**, whose structure was confirmed by an unambiguous synthesis. Two very minor products were the phenethylisoquinolines **2** and **3**, which indicated that the starting dimer possessed the head-to-tail arrangement (Scheme I).[1,2]

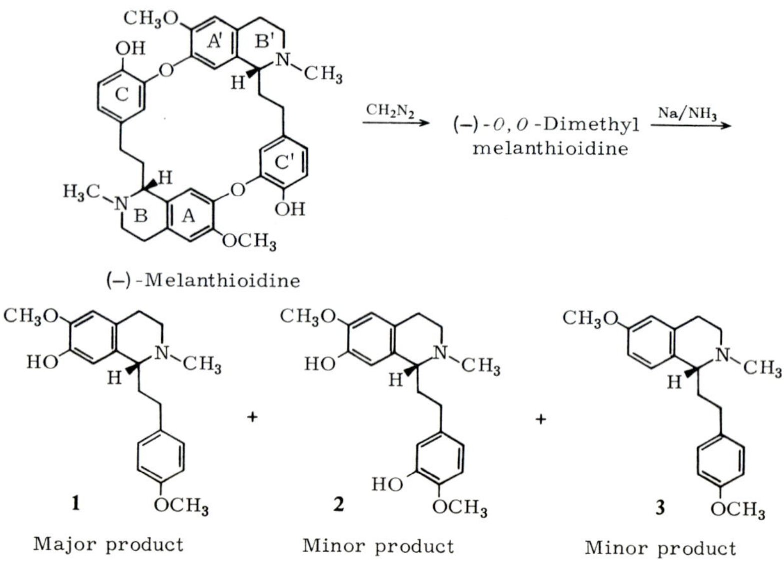

Scheme I

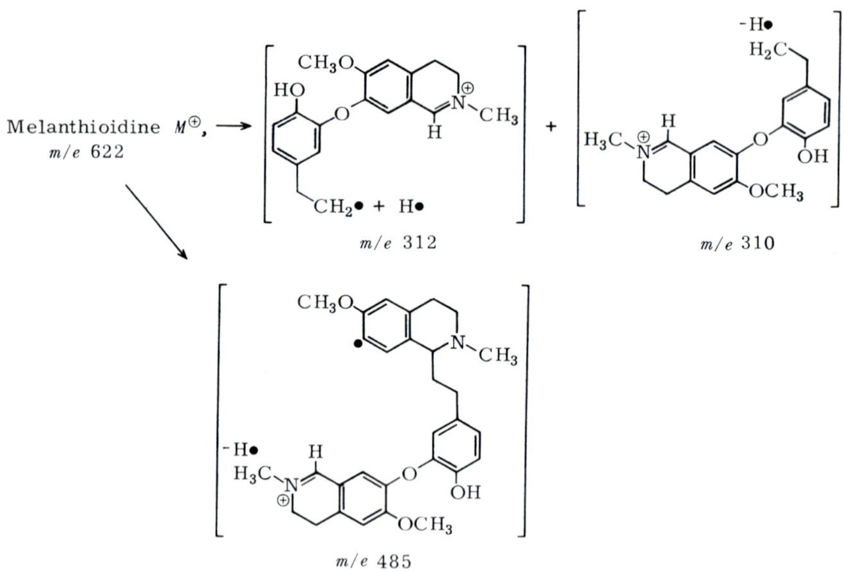

Scheme II

The mass spectrum of melanthioidine showed major fragment ions at *m/e* 312 and 310 corresponding to the cleavage shown in Scheme II. Another important fragment was at *m/e* 485. It should be noted that each cleavage was accompanied by hydrogen transfer.

The fact that the *m/e* 485 peak in the spectrum of melanthioidine was replaced by an *m/e* 499 peak in the spectrum of *O*,*O*-dimethylmelanthioidine, corresponding to an increment of 14 mass units, furnished proof that the phenolic groups in melanthioidine must be located on rings C and C′.

The structural assignment for melanthioidine was then confirmed by a total synthesis of the alkaloid described in the next section.[2]

II. The Battersby Synthesis of Melanthioidine

The synthesis of (±)-melanthioidine was first carried out using a double Ullmann reaction to construct the 20-membered ring. The racemic bromophenethylisoquinoline **4** when treated under Ullmann conditions with copper powder in dry pyridine afforded two diastereoisomeric dibenzylmelanthioidines, which were separated by chromatography. One of these isomers upon catalytic debenzylation was found to yield material corresponding to (±)-melanthioidine (Scheme III).[2,3]

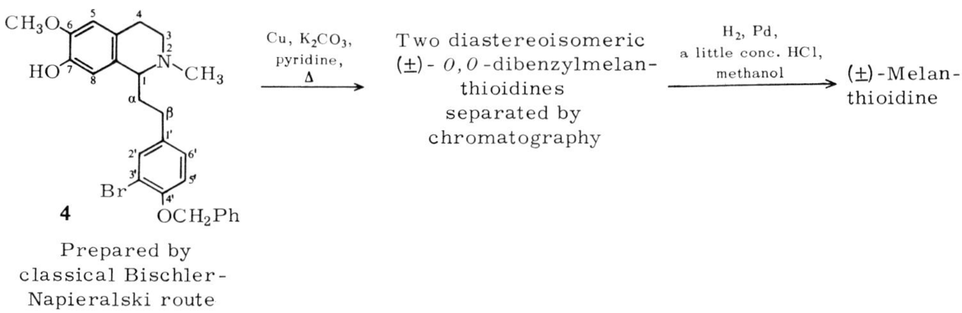

Scheme III

The synthesis was then repeated starting with optically active (−)-**4** which possesses a beta C-1 hydrogen and terminating with (−)-melanthioidine identical in all respects with the natural product.[2]

III. Enzyme-Induced Syntheses of Analogs of Melanthioidine

Two attempts have been made to emulate the biogenetic process leading to melanthioidine using homogenized plant extracts. These efforts proved to be only fractionally successful.

In the first attempt, treatment of the racemic phenethylisoquinoline **5** with hydrogen peroxide and homogenized potato peelings gave a 2% yield of the dimer **6** which involves head-to-tail coupling similar to that prevailing in melanthioidine.[4]

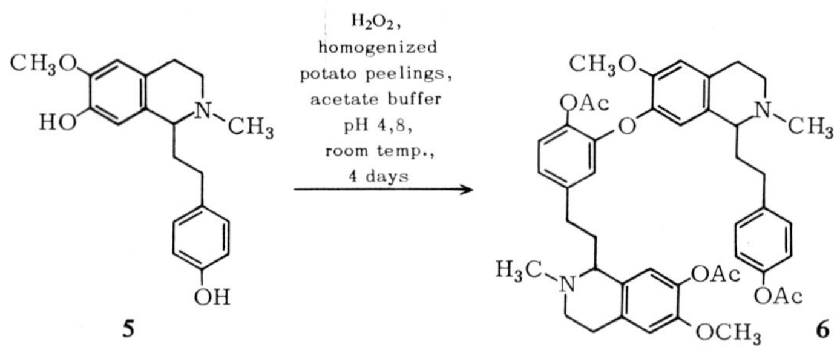

In contrast, the same phenethylisoquinoline **5** with homogenized *Wasabia japonica* Matsumura furnished a small yield of the dimer **7** which is derived from head-to-head coupling.[5]

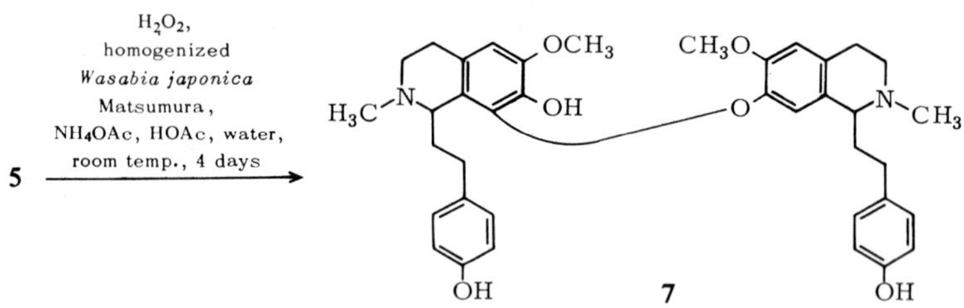

REFERENCES

1. A. R. Battersby, R. B. Herbert, and F. Šantavý, *Chem. Commun.* p. 415 (1965).
2. A. R. Battersby, R. B. Herbert, L. Mo, and F. Šantavý, *J. Chem. Soc., C* p. 1739 (1967).
3. T. Kametani and S. Takano, *Tetrahedron Lett.* p. 121 (1968); T. Kametani, S. Takano, and K. Haga, *Chem. Pharm. Bull.* **16**, 663 (1968).
4. T. Kametani, S. Takano, and T. Kobari, *J. Chem. Soc., C* p. 9 (1969).
5. T. Kametani, S. Takano, and T. Kobari, *J. Chem. Soc., C* p. 2770 (1969).

Chapter 26 / THE HOMOAPORPHINES AND HOMOPROAPORPHINES

Occurrence: Liliaceae

Number: 3 homoaporphines, 1 homoproaporphine, and 2 dihydrohomoproaporphines

Structures:

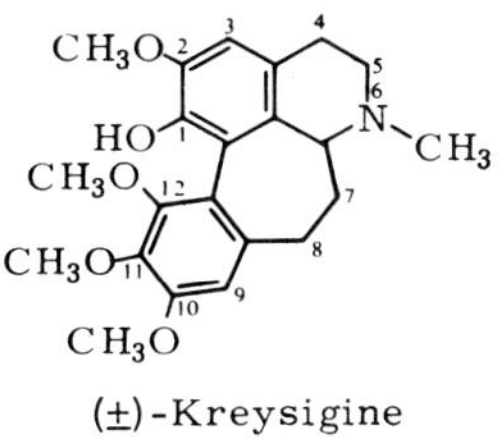

(±)-Kreysigine

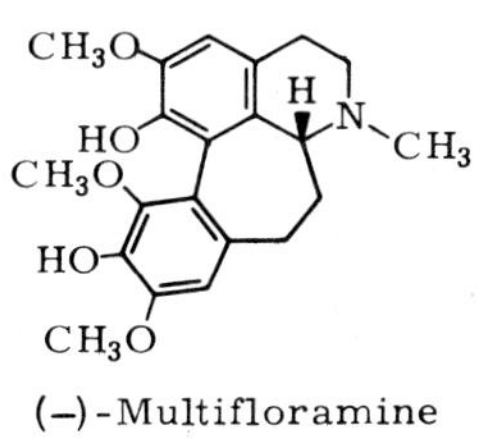

(−)-Multifloramine

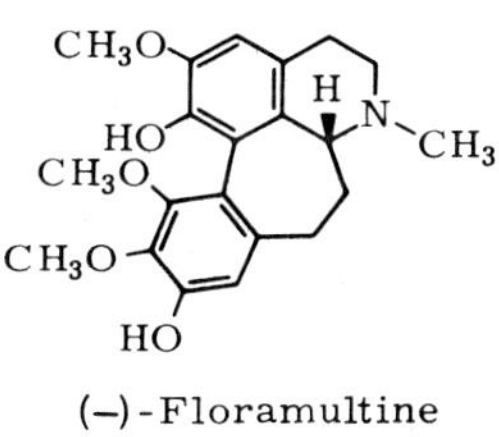

(−)-Floramultine

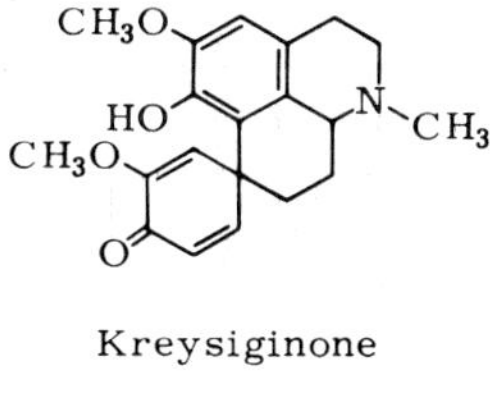

Kreysiginone

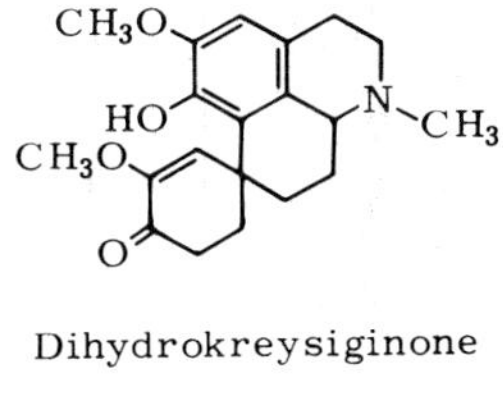

Dihydrokreysiginone

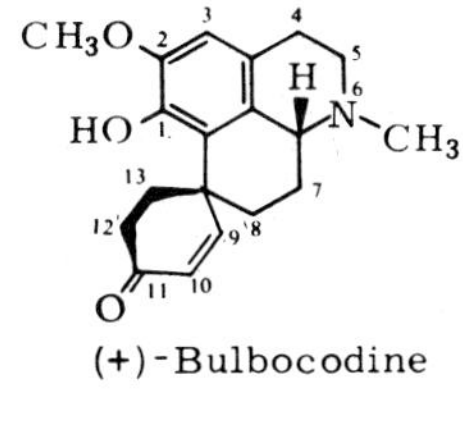

(+)-Bulbocodine

I. Three Homoaporphine Alkaloids: Kreysigine, Multifloramine, and Floramultine

The monophenolic alkaloid kreysigine, $C_{22}H_{27}O_5N$, found in *Kreysigia multiflora* Reichb.[1] in the racemic form and in *Colchicum cornigerum* (Schweinf.) Täckh. et Drav.[2] in the levorotatory form, possesses one *N*-methyl and four *O*-methyl groups.

NMR spectroscopy showed only two aromatic protons as singlets at δ6.54 and 6.59. The signal for one of the methoxyls appeared at higher field, δ3.59, than the others, which were at δ3.83 (3H) and 3.86 (6H), as often observed for the aporphines.[1]

The UV spectrum with maxima at 221, 260, and 293 mμ also bore some resemblance to the aporphines. The structure **5** (Scheme I) was, therefore, considered for the alkaloid on the assumption that the high-field *O*-methyl group in the NMR spectrum was located at C-12. The mass spectrum of kreysigine was also instructive in that the base peak was at $M-17$, indicating the facile loss of the hindered C-1 hydroxyl. This structural assignment was then confirmed by synthesis (Scheme I).[1]

The racemic diphenolic phenethylisoquinoline **1** was oxidized with alkaline ferricyanide in 49% yield to the homoproaporphine **2**. Acid-catalyzed dienone–phenol rearrangement formed the homoaporphine **3**. Finally, *O*-methylation with diazomethane yielded a mixture of two products, **4** and **5**. The former product was identical with *O*-methyl-(±)-kreysigine and the latter with (±)-kreysigine. The intermediate **3** was then found to correspond to the racemic form of the alkaloid (−)-multifloramine found in *K. multiflora* (Scheme I).[1]

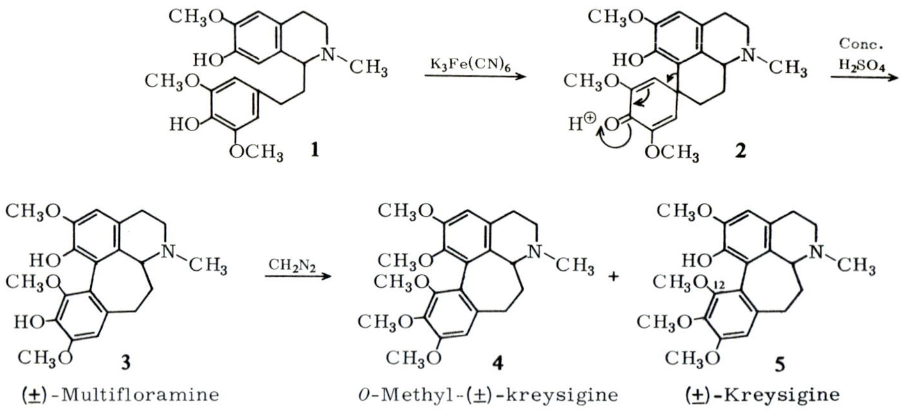

Scheme I

The diphenolic alkaloid (−)-floramultine, $C_{21}H_{25}O_5N$, M^+ *m/e* 371, base peak at $M-17$, is also obtained from *K. multiflora*, and incorporates two phenolic and three *O*-methyl groups. *O*-Methylation with diazomethane produced (−)-kreysigine and (−)-*O*-methylkreysigine, indicating that floramultine must be a de-*O*-methylkreysigine. One of the phenolic functions in floramultine must, therefore, be at C-1 by analogy with kreysigine. Because floramultine shows a high-field C-12 *O*-methyl NMR signal at δ3.55 and exhibits a negative molybdate test for catechol systems, the second phenolic function can be neither at C-12 nor at C-2. It also cannot be at C-11, since floramultine and multifloramine are different compounds. It follows that the second phenolic group in floramultine must be located at C-10. Kreysigine, multifloramine, and floramultine are the first examples of homoaporphine alkaloids.[1]

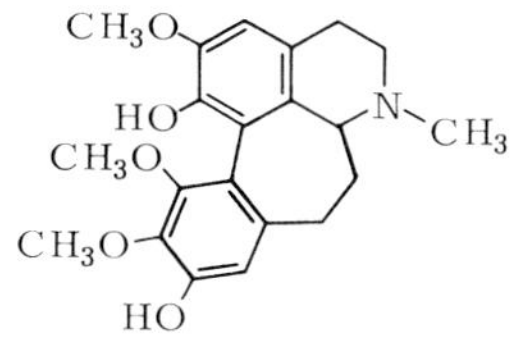

Floramultine

II. The Homoproaporphine Kreysiginone and the Dihydrohomoproaporphine Dihydrokreysiginone

Upon ferricyanide oxidation of the racemic phenethylisoquinoline **6**, Battersby *et al.* isolated the two diastereoisomeric homoproaporphines **7** and **8**. One of them, indicated as **7**, but with no absolute stereochemistry implied, was then found to be present in *K. multiflora*. Also present in the same plant was dihydrokreysiginone, which probably possesses the same stereochemistry as natural kreysiginone.[3]

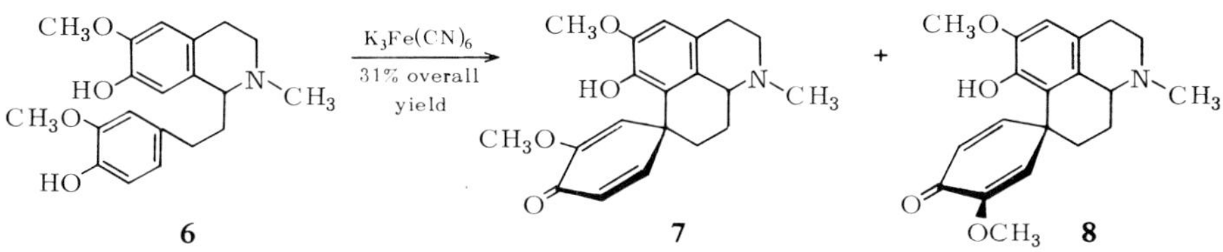

6 **7** **8**

Kreysiginone

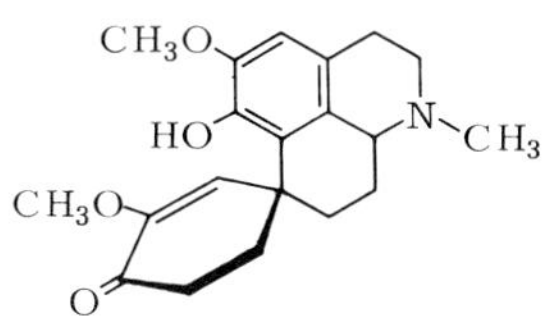

Dihydrokreysiginone

III. Syntheses of Homoaporphines and Homoproaporphines through Phenolic Oxidative Coupling

The syntheses of kreysigine (**5**) and kreysiginone (**7**) derivatives have already been noted in this chapter. Several other instances of phenolic oxidation of phenethylisoquinolines have been recorded and most of these are listed in Scheme II. The structural assignments were in each case supported by spectral analysis. The yields are generally high, and the favored mode of oxidative coupling leads to the formation of homoproaporphines or homoaporphines rather than to species of the androcymbrine type.

In case (b), the tendency toward ortho-ortho coupling of the phenethylisoquinoline **9** to give the modified homoaporphine **10** is strong enough to overcome the steric factor posed by the presence of the C-8 methoxyl group in the starting material, so that even in this instance no ortho-para coupling to an androcymbine analog is observed.

In example (d), the formation of the acetal **12** and the hemiacetal **13** cannot by itself allow for an assignment of relative stereochemistry to the dienone precursor **11**. Assuming the structural assignments for compounds **12** and **13** to be correct, molecular models indicate that these species could have either of the two relative stereochemical arrangements defined by expressions **14** and **15**.

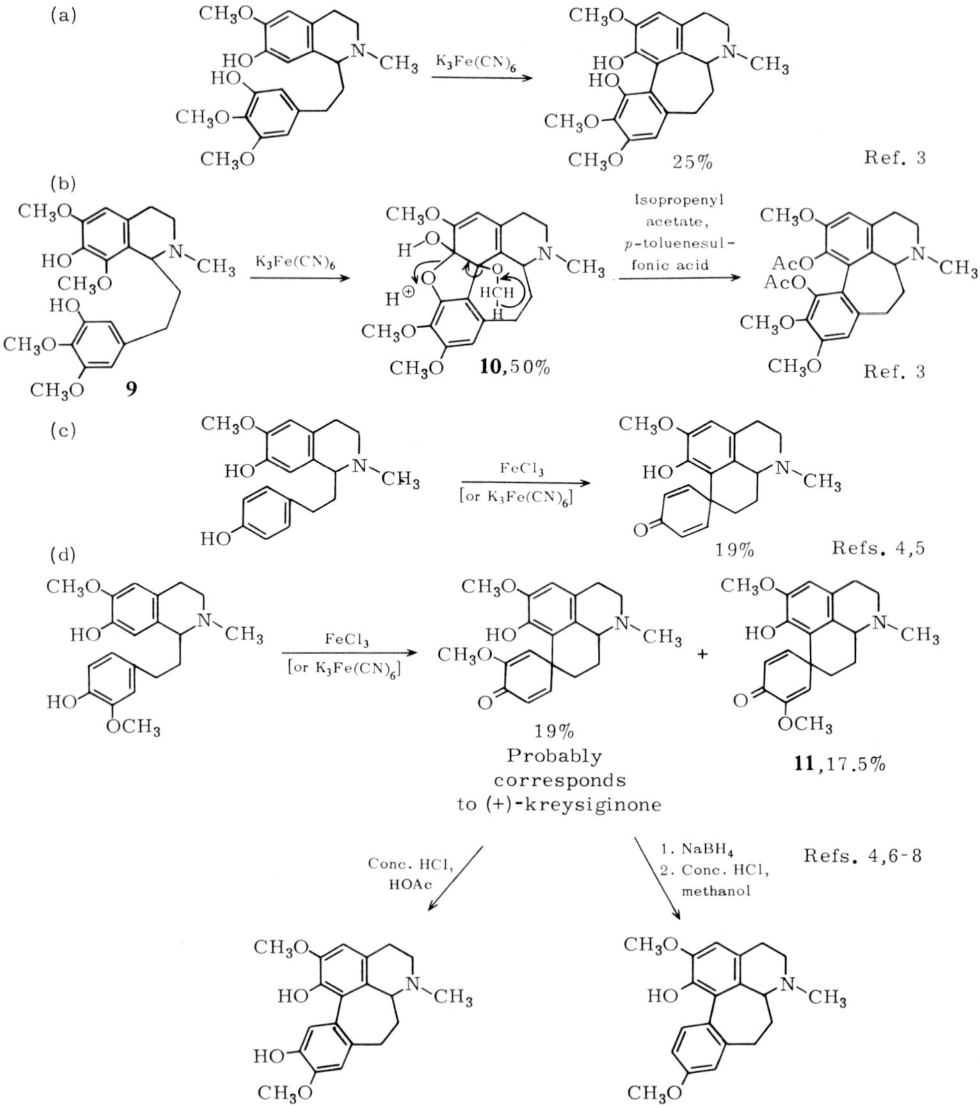

And:

11 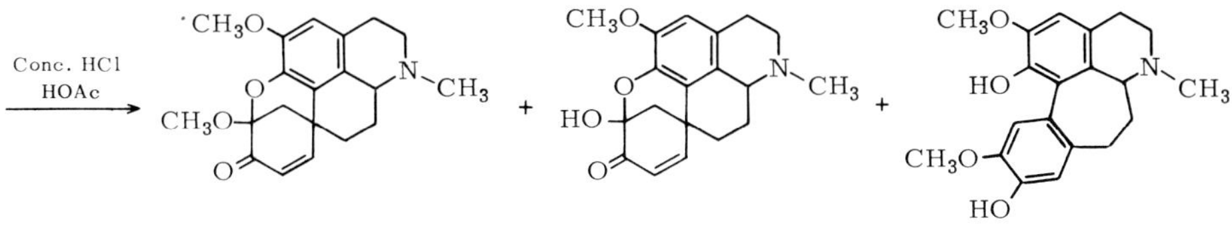

12, 32.5% **13**, 13.5% 4.3%

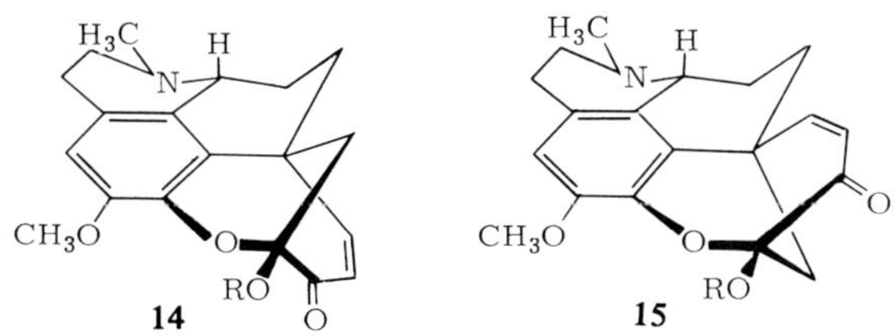

14 **15**

Refs. 4,6,9

(e) 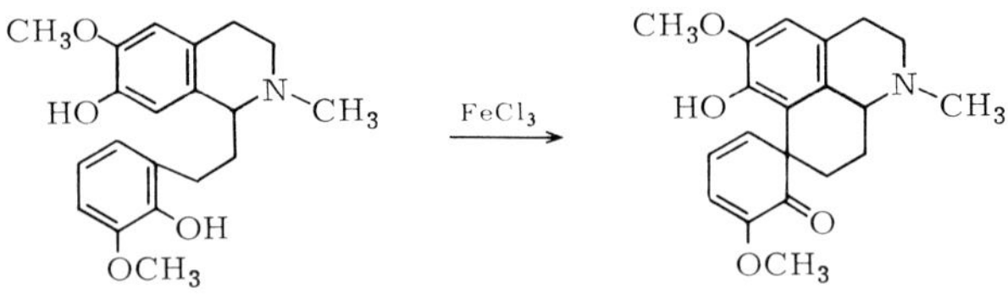

45% Ref. 10

(f) 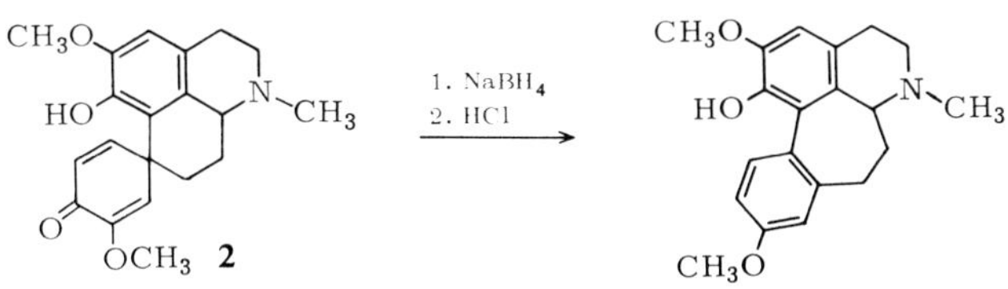

65% from dieniol Refs. 4,9

(g) 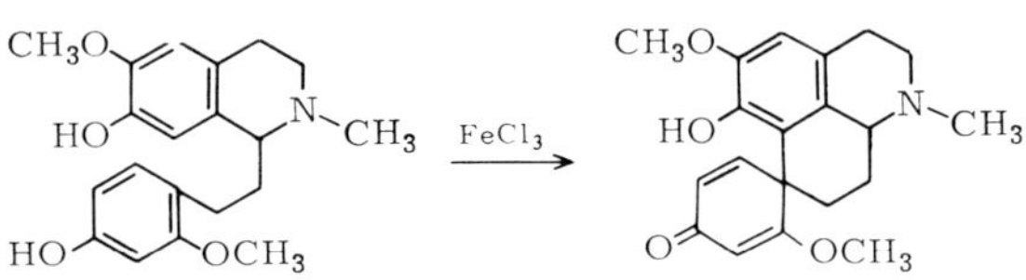

26% Ref. 11

Scheme II

One instance in which an androcymbine analog was obtained is given below, but the yield was on the order of 0.35%, again pointing to the reluctance *in vitro* of the phenethylisoquinolines to form androcymbine-type compounds through para-para coupling.[12]

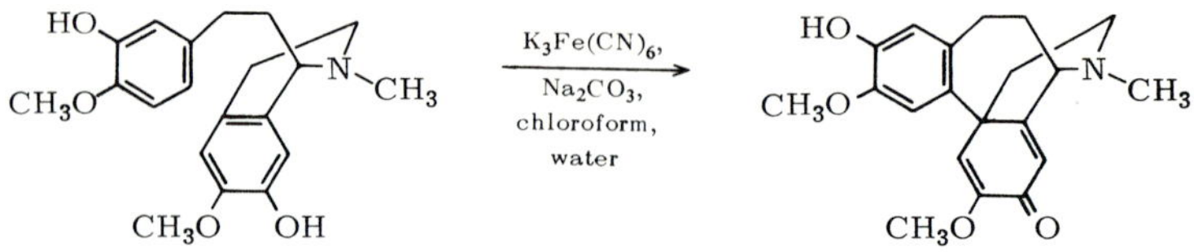

IV. The Synthesis of Reduced Homoproaporphines through Carbonium Ion Intermediates

The amino acetate **16**, obtained as indicated in Scheme III, was *N*-methylated by the Eschweiler–Clarke procedure. Hydrolysis of the acetate function of the resulting tertiary amine gave the amino alcohol **17**, which upon treatment with polyphosphoric acid led to the reduced homoproaporphine **18**.

Alternatively, the amino acetate **16** could be converted to the amide alcohol **19**, which when treated with polyphosphoric acid furnished the reduced homoproaporphine **20**. *N*-Deformylation to the secondary amine **21** was accomplished with ethanolic potassium hydroxide (Scheme III).[13]

V. An Alternate Synthesis of Homoproaporphines through Carbonium Ion Intermediates

Birch reduction of the phenethyltetrahydrobenzylisoquinoline **21a** gave rise to the enol ether **21b**. Treatment of **21b** with hot phosphoric acid led to the reduced homoproaporphine **21c**.[13a]

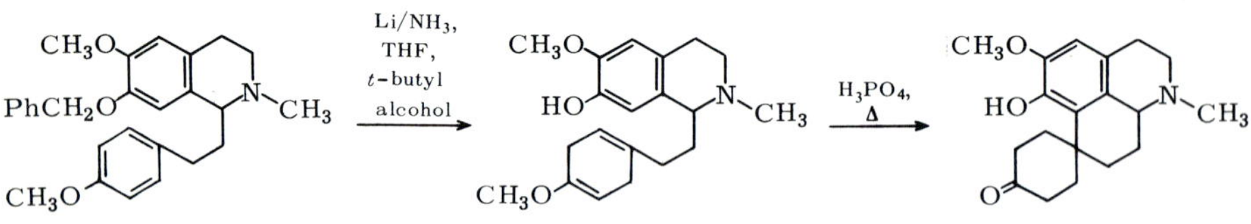

21a **21b** **21c**

This sequence cannot be used to prepare proaporphines since phosphoric acid treatment of the enol ether **21d** gave two compounds presumed to possess the reduced aporphine nucleus **21e**.

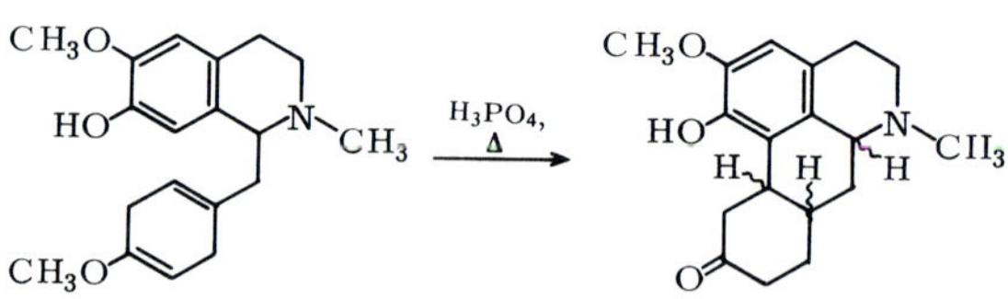

21d **21e**

VI. Photolytic Syntheses of Homoaporphines and Homoproaporphines

Photolysis of the diazonium salt **21f** furnished mostly the phenethylisoquinoline **21g** together with small yields of *O*-methylandrocymbine and kreysigine.[13b]

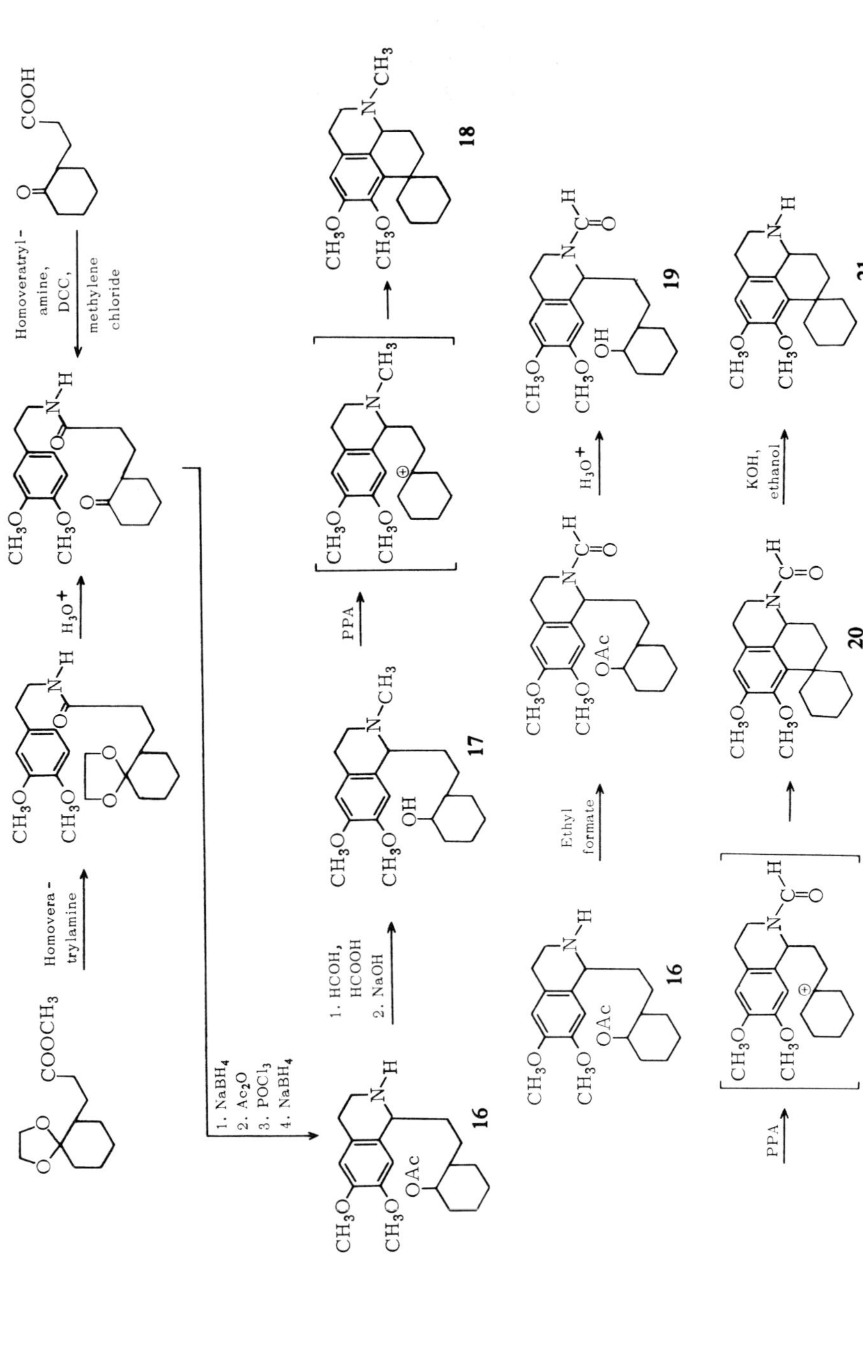
Homovera-trylamine
H_3O^+
Homoveratryl-amine, DCC, methylene chloride
1. $NaBH_4$ 2. Ac_2O 3. $POCl_3$ 4. $NaBH_4$
16
1. HCOH, HCOOH 2. NaOH
17
PPA
18
Ethyl formate
H_3O^+
19
PPA
20
KOH, ethanol
21

Scheme III

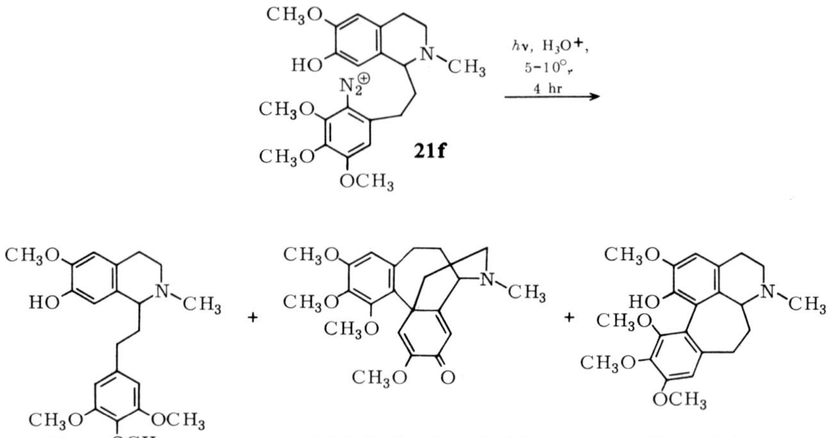

21g O-Methylandrocymbine Kreysigine

Also, irradiation of the brominated phenethyltetrahydroisoquinoline **21h** in alcoholic sodium hydroxide yielded an inseparable mixture of *O*-methylkreysiginone and its diastereoisomer in a 1 : 1 ratio. This type of cyclization has been extended to the preparation of proaporphines.[13c]

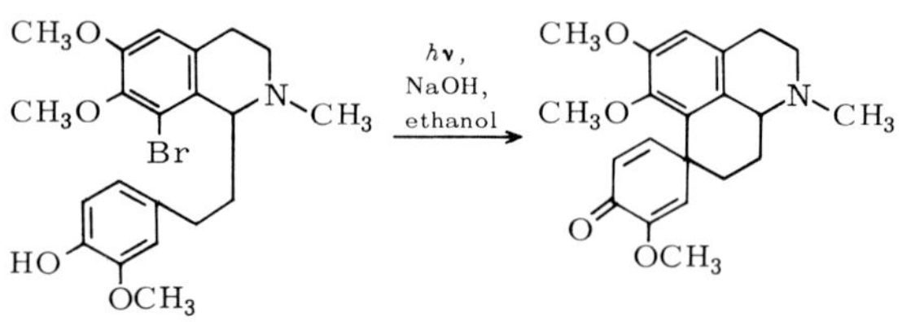

21h O-Methylkreysiginone and diastereoisomer

VII. A Photochemical Reaction of Kreysiginone

In Section III it was seen how treatment of compound **11**, a diastereoisomer of kreysiginone, with concentrated hydrochloric acid yielded mainly the ketal **12** resulting from addition of the C-1 phenol alpha to the ring D carbonyl. When, however, kreysiginone was irradiated with UV light in the absence of any proton source, addition presumably took place beta to the carbonyl to form a material which was formulated as the ether **22**.[14]

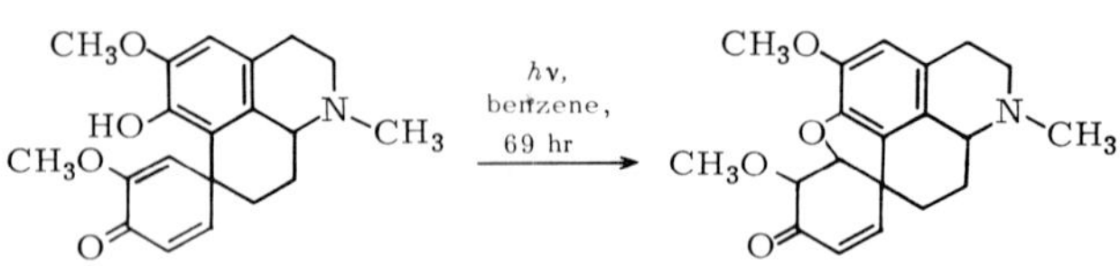

Kreysiginone **22**

VIII. Absolute Configuration

Resolution of the phenethylisoquinoline **23** afforded the antipodes **24** and **25**. *N*-Methylation and debenzylation of the dextrorotatory isomer **24** gave the levorotatory phenolic *N*-methyl base **26**, which upon oxidation with ferric chloride formed the levorotatory homoproaporphine **27**, which had been previously known in the racemic form (see compound **2**). Acid-catalyzed dienone–phenol rearrangement then led to (−)-multifloramine identical with the natural product (Scheme IV). The ORD curves of the homoaporphines and homoproaporphines have been recorded and interpreted.[15]

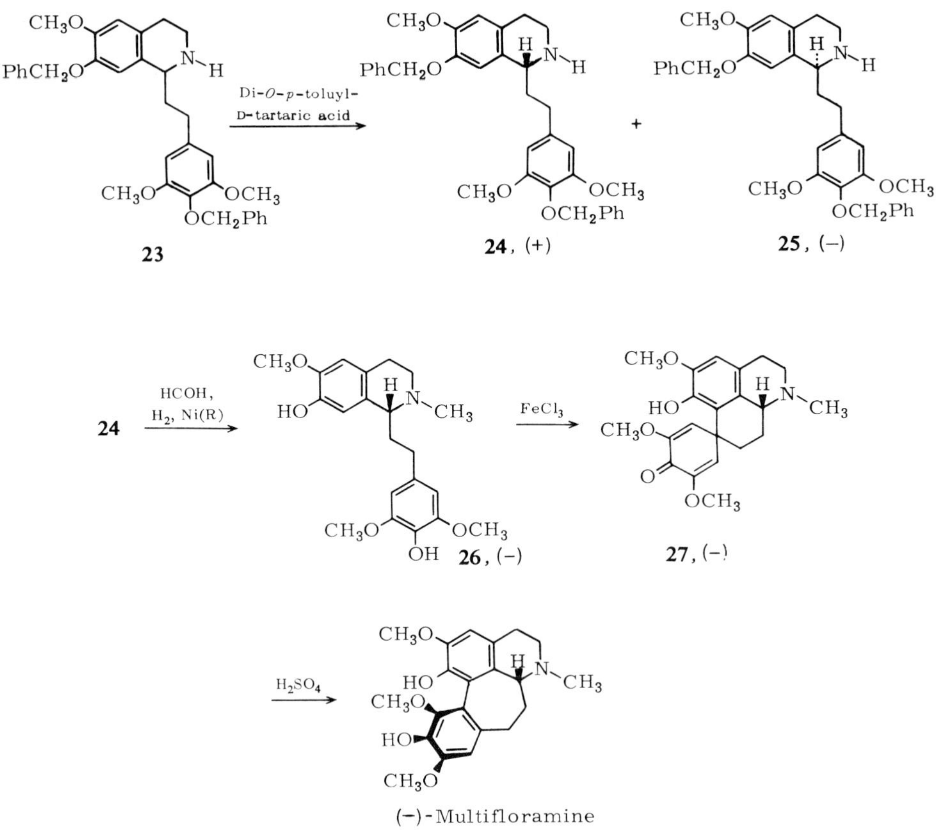

Scheme IV

Naturally occurring (−)-floramultine displays a positive CD curve with a peak at 255 mμ, and previous studies with biphenyl systems have shown this phenomenon to be associated with the absolute configuration displayed here for (−)-floramultine.[16]

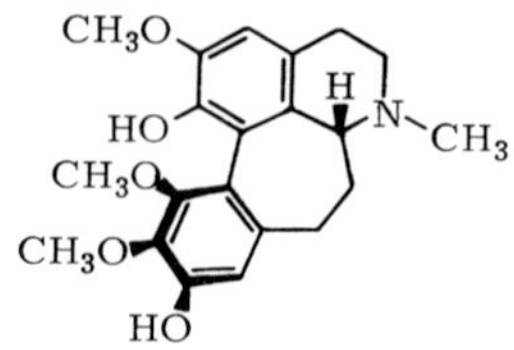

(−)-Floramultine

The ORD and CD curves of the homoproaporphine **27**, of bulbocodine, and of other homoproaporphines have been measured.[9,15,16a]

IX. Biosynthesis of Homoaporphines

In order to find out whether the formation of homoaporphines in plants proceeds by way of homoproaporphines or by direct coupling from a phenethylisoquinoline, Battersby *et al.* fed labeled autumnaline to *K. multiflora*. The total alkaloidal extracts were *O*-methylated with diazomethane to give *O*-methylkreysigine. Good incorporation was registered for autumnaline, but the isomeric phenethylisoquinoline **28** was very poorly incorporated. Autumnaline is therefore transformed into homoaporphines by direct coupling without the intermediacy of homoproaporphines.[17]

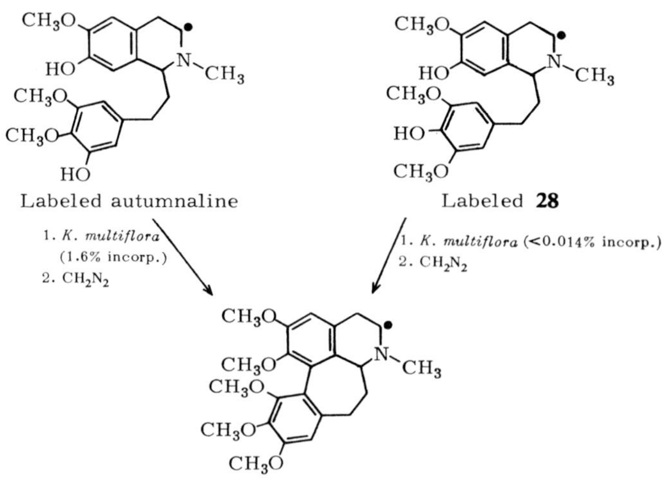

Labeled *O*-methylkreysigine

X. Pharmacology

There is a brief report stating that homoproaporphines may be useful as drugs acting on the central nervous system.[18]

XI. Mass Spectroscopy

The cleavage pattern for the homoproaporphines resembles that of the proaporphines. The principal modes of fission are loss of CH_2NCH_3 from ring B by a retrograde

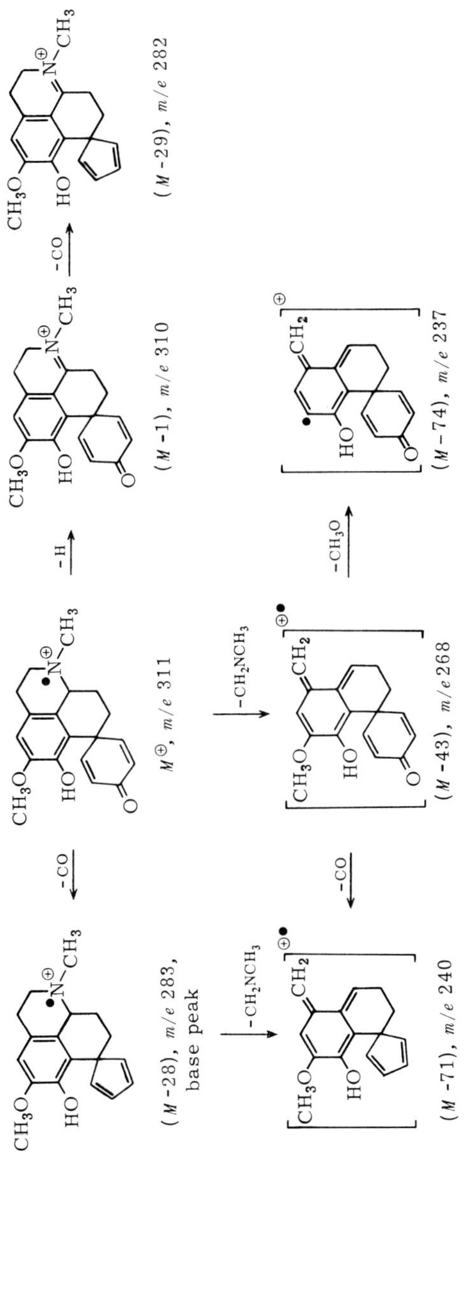

$M^{\oplus}$, m/e 311
-H
(M-1), m/e 310
-CO
(M-29), m/e 282
-CO
(M-28), m/e 283, base peak
-CH_2NCH_3
(M-43), m/e 268
-CH_3O
(M-74), m/e 237
-CO
-CH_2NCH_3
(M-71), m/e 240

Scheme V

Diels–Alder process and loss of the elements of CO from ring D. The example in Scheme V involves the mass spectrum for 1-hydroxy-2-methoxyhomoproaporphine.[5]

XII. NMR Spectroscopy

As with the aporphines, the C-12 proton in the homoaporphine series is found downfield from the remaining aromatic protons, while the C-3 proton is usually located at high field. The C-12 methoxyl is always shifted upfield, again in analogy with the aporphines.

In the homoproaporphine series, a methoxyl group which is part of an enol ether system appears at higher field from the regular aromatic methoxyl groups. The spin–spin couplings of the dienone protons are useful in assigning chemical shifts.

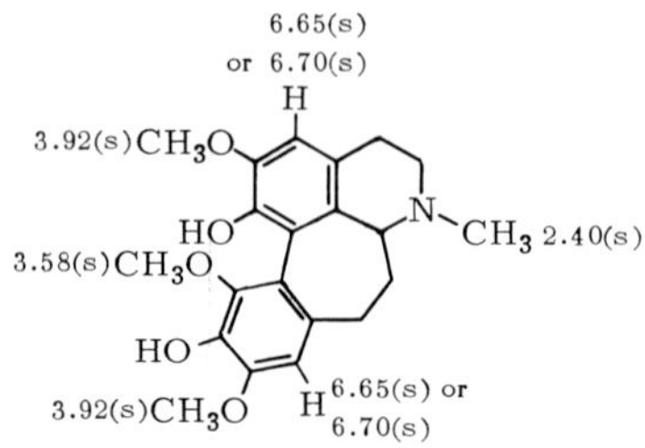

Multifloramine[15]

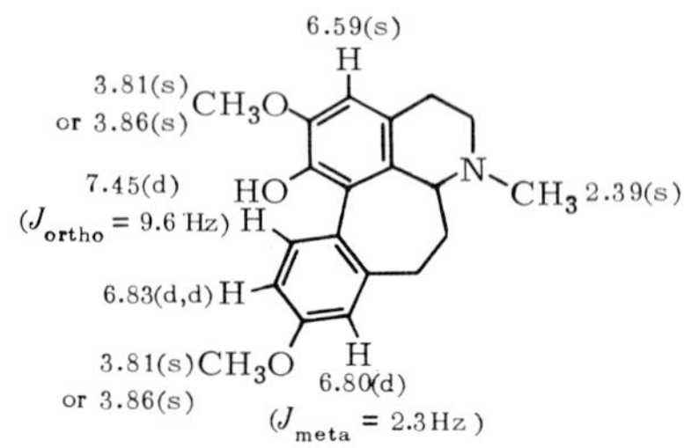

Ref. 8

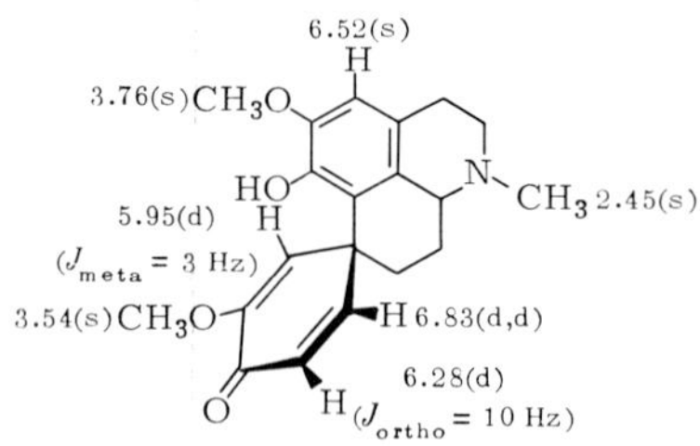

Kreysiginone[3]*

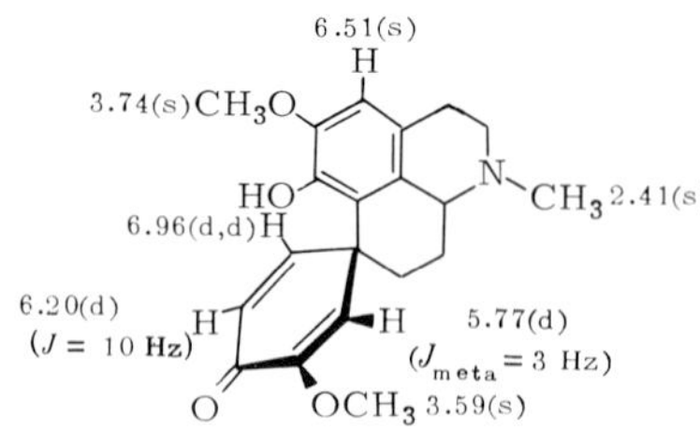

Diastereoisomer of kreysiginone[3]*

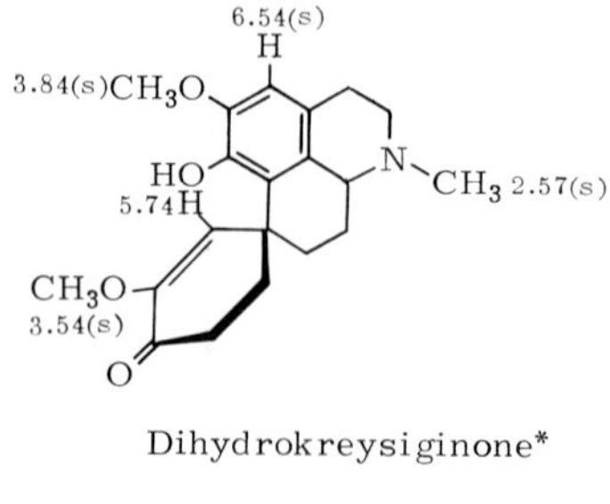

Dihydrokreysiginone*

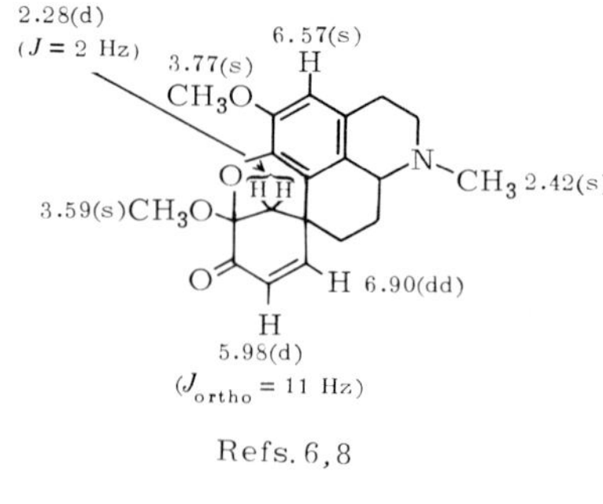

Refs. 6,8

* No absolute stereochemistry implied.

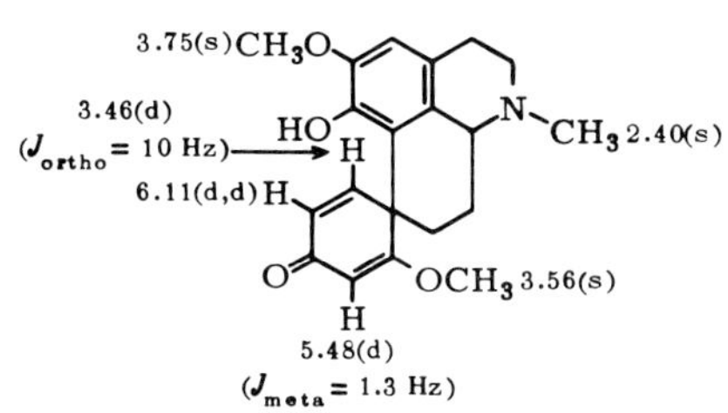

Ref. 11

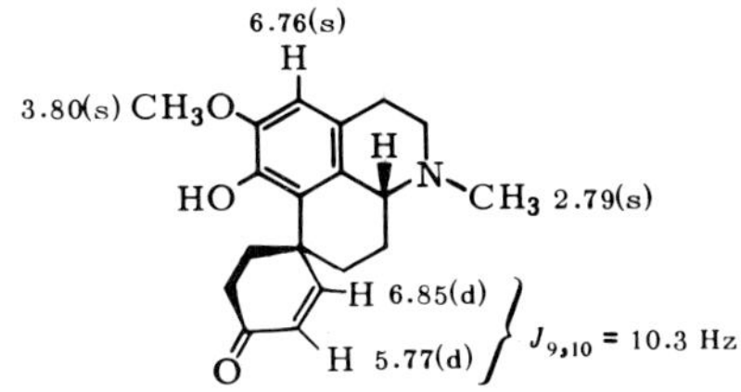

(+)-Bulbocodine[16a]

XIII. UV Spectroscopy

A. Homoaporphines

Homoaporphines with a substituent at C-12 show a central peak at 257 – 258 mμ. When this substituent is missing, the central absorption is in the 260 – 264 mμ range.

Kreysigine[2]	λ_{max}^{EtOH} 218, 257, and 293 mμ (4.62, 4.10, and 3.67) λ_{min}^{EtOH} 246 and 281 mμ (4.05 and 3.64)
Multifloramine[15] (1,11-Dihydroxy-2,10,12-trimethoxyhomoaporphine)	λ_{max}^{MeOH} 216, 257, and 293 mμ (4.66, 4.06, and 3.86)
1,2,10,11,12-Pentamethoxy-homoaporphine hydrochloride[15]	λ_{max}^{MeOH} 218, 258, and 295 sh mμ (4.63, 4.12, and 3.54)
1,11-Dihydroxy-2,10-di-methoxyhomoaporphine[8]	λ_{max} 264 and 291 mμ (4.14 and 4.11)
1-Hydroxy-2,10-dimethoxy-homoaporphine[4]	λ_{max}^{MeOH} 260 and 290 mμ (4.17 and 3.77)
1,10-Dihydroxy-2,11-di-methoxyhomoaporphine[6]	λ_{max} 264 and 293 mμ (4.13 and 4.04)
1-Hydroxy-2,9,11-tri-methoxyhomoaporphine[9]	λ_{max}^{MeOH} 260 and 293 mμ (4.23 and 4.03)

B. Homoproaporphines

The homoproaporphines usually show three maxima in the ultraviolet, with the central absorption between 232 and 243 mμ.

Kreysiginone[3]	λ_{max}^{EtOH} 214, 243, and 287 mμ (4.54, 4.15, and 3.78)
Dihydrokreysiginone[3]	λ_{max}^{EtOH} 220 and 269 mμ
Bulbocodine[16a]	λ_{max}^{EtOH} 232 and 285 mμ (4.12 and 3.45)

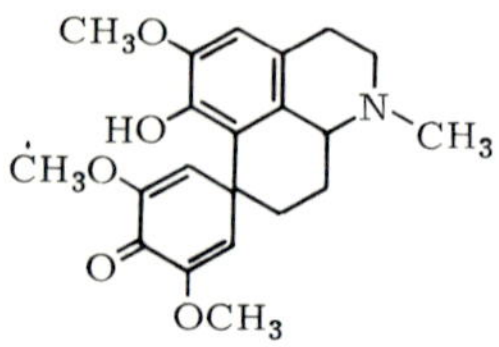

Ref. 15

λ_{max}^{MeOH} 210, 232 sh, and 275 mμ (4.61, 4.06, and 4.14)

1-Hydroxy-2-methoxy-homoproaporphine[5] λ_{max}^{MeOH} 235 and 290 mμ (4.53 and 3.66)

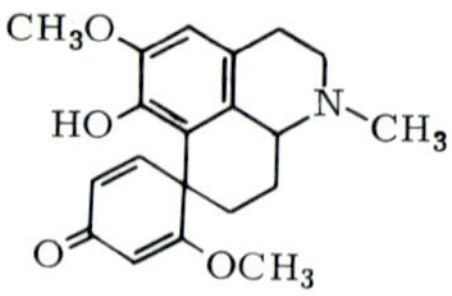

Ref. 11

λ_{mxx}^{MeOH} 235 and 285 mμ (4.19 and 3.87)

REFERENCES

1. A. R. Battersby, R. B. Bradbury, R. B. Herbert, M. H. G. Munro, and R. Ramage, *Chem. Commun.* p. 450 (1967).
2. H. Potěšilova, J. Šantavý, A. El-Hamidi, and F. Šantavý, *Collect. Czech. Chem. Commun.* **34**, 3540 (1969).
3. A. R. Battersby, E. McDonald, M. H. G. Munro, and R. Ramage, *Chem. Commun.* p. 934 (1967).
4. T. Kametani, K. Fukumoto, H. Yagi, and F. Satoh, *Chem. Commun.* p. 878 (1967).
5. T. Kametani, H. Yagi, F. Satoh, and K. Fukomoto, *J. Chem. Soc., C* p. 271 (1968).
6. T. Kametani, F. Satoh, H. Yagi, and K. Fukomoto, *Chem. Commun.* p. 1103 (1967).
7. T. Kametani, F. Satoh, H. Yagi, and K. Fukumoto, *J. Org. Chem.* **33**, 690 (1968).
8. T. Kametani, F. Satoh, H. Yagi, and K. Fukumoto, *J. Chem. Soc., C* p. 1003 (1968).
9. T. Kametani, F. Satoh, H. Yagi, and K. Fukumoto, *J. Chem. Soc., C* p. 382 (1970).
10. T. Kametani, K. Fukumoto, T. Hayasaka, F. Satoh, and K. Kigasawa, *J. Chem. Soc., C* p. 4 (1969).
11. T. Kametani and F. Satoh, *Chem. Pharm. Bull.* **17**, 814 (1969).
12. T. Kametani, K. Fukumoto, M. Koizumi, and A. Kozuka, *J. Chem. Soc., C* p. 1295 (1969).
13. F. Schneider and K. Bernauer, *Helv. Chim. Acta* **53**, 938 (1970).
13a. W. V. Curran, *Chem. Commun.* p. 478 (1971).
13b. T. Kametani, M. Koizumi, K. Shishido, and K. Fukumoto, *J. Chem. Soc., C* p. 1923 (1971).
13c. T. Kametani, T. Sugahara, H. Sugi, S. Shibuya, and K. Fukumoto, *Chem. Commun.* p. 724 (1971). See also T. Kametani, Y. Satoh, S. Shibuya, M. Koizumi, and K. Fukumoto, *J. Org. Chem.* **36**, 3733 (1971); and T. Kametani and M. Koizumi, *J. Chem. Soc., C* p. 3976 (1971).
14. R. E. Harmon, B. L. Jensen, and S. K. Gupta, *160th Meet., Amer. Chem. Soc.* (1970), and R. E. Harmon and B. L. Jensen, *Heterocycl. Chem.* **7**, 1077 (1970).
15. A. Brossi, J. O'Brien, and S. Teitel, *Helv. Chim. Acta* **52**, 678 (1969).
16. A. F. Beecham, N. K. Hart, S. R. Johns, and J. A. Lamberton, *Aust. J. Chem.* **21**, 2829 (1968).
16a. F. Šantavý, P. Sedmera, G. Snatzke, and T. Reichstein, *Helv. Chim. Acta* **54**, 1084 (1971).
17. A. R. Battersby, P. Böhler, M. H. G. Munro, and R. Ramage, *Chem. Commun.* p. 1066 (1969).
18. T. Kametani (Gleran Pharmaceutical Co., Ltd), Jap. Patent 19062 (1970); *Chem. Abstr.* **73**, 56291 v (1970).

Chapter 27 / THE HOMOPROTOBERBERINES

No naturally occurring homoprotoberberines are known. However, by analogy with the homologated alkaloids discussed in the preceding three chapters, there is a fair possibility that such species will eventually be isolated from plant sources.

I. Synthesis

Three methods of synthesis of homoprotoberberines have been developed[1-4]:

A. The Mannich Condensation

Mannich condensation of the phenethylisoquinolines **1** and **3** using formaldehyde and acid leads to the homoprotoberberines **2** and **4**.[1-3] Condensation of **1** involves the less-hindered side of ring C.

In a separate study, reaction of the phenol **5** without the help of an acid catalyst furnished the homoprotoberberine **6** in which cyclization had occurred ortho to the phenolic function. In contrast, cyclization of **5** in the presence of acid proceeded to give the isomeric homoprotoberberine **7** (Scheme I).[4]

B. The Homophthalideisoquinoline Approach: Functionalization at C-13

Condensation of the acid chloride of the known phthalide-3-acetic acid with homoveratrylamine yielded the amide lactone **8**. Bischler–Napieralski cyclization provided the imine **9** which when reduced with Adams catalyst furnished a diastereoisomeric mixture of homophthalideisoquinolines **10** and **11**.

When the mixture of **10** and **11** was treated with methanolic potassium hydroxide, the diastereoisomeric lactams **12** and **13** were obtained and separated. The two products were isolated in the ratio of 7 : 1, respectively (Scheme II).

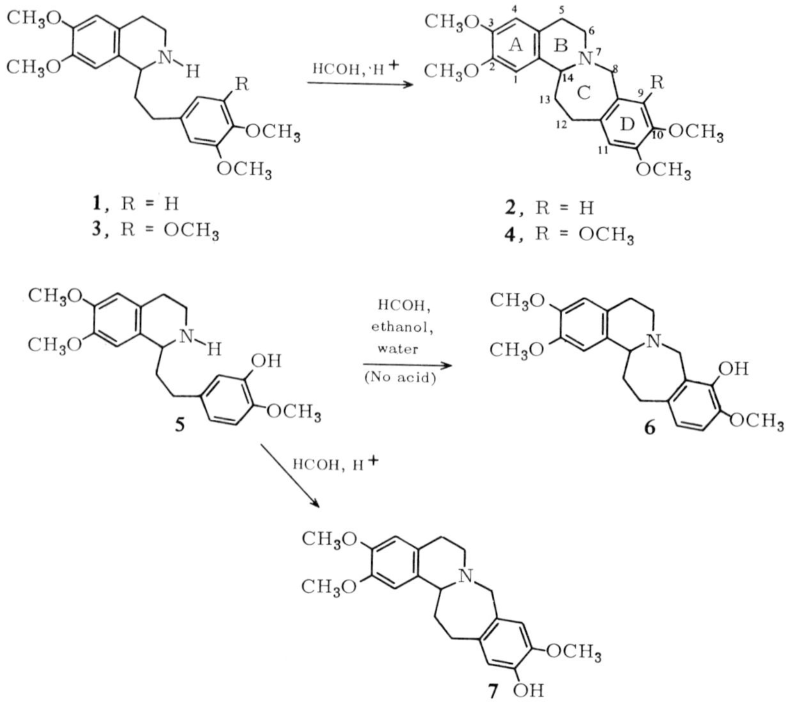

Scheme I

Molecular models clearly indicate that the amino lactone **10** has fewer steric obstacles to overcome in cyclizing to **12** than would the amino lactone **11** in going to **13**, so that the lactam **12** is formed more readily.

Sarett oxidation of the mixture of lactam alcohols **12** and **13** led to the keto lactam **14**. Reduction of this material with sodium borohydride gave the lactam **13** preferentially over **12** in the ratio of at least 10 : 1. Alternatively, hydrogenation of the keto lactam **14** with Adams catalyst afforded a 2 : 3 mixture of **12** and **13** (Scheme II).[2]

The lactam alcohols **12** and **13** were then individually reduced with lithium aluminum hydride to form homoprotoberberines **15** and **16**, respectively. Amine **15** quaternized with methyl iodide at a very fast pseudo-first-order rate of $2.3 \times 10^{-2}\ \text{sec}^{-1}$, while isomer **16** reacted even faster at $3.1 \times 10^{-2}\ \text{sec}^{-1}$, indicating that both compounds are cis-B/C-fused. The absence of Bohlmann bands in the IR spectra added weight to these stereochemical conclusions (Scheme III).[2]

Supporting evidence for the stereochemical assignments was afforded by the NMR spectra of homoprotoberberine alcohols **15** and **16**. The C-8 equatorial hydrogen of species **16** is appreciably more deshielded than the corresponding equatorial hydrogen of **15**. Specifically, in pyridine-d_5, the C-8 methylene protons of **15** appear at $\delta 4.24$ in an AB pattern with an ics of 28 Hz and $J_{a,e} = 14.5$ Hz, whereas for isomer **16** they are

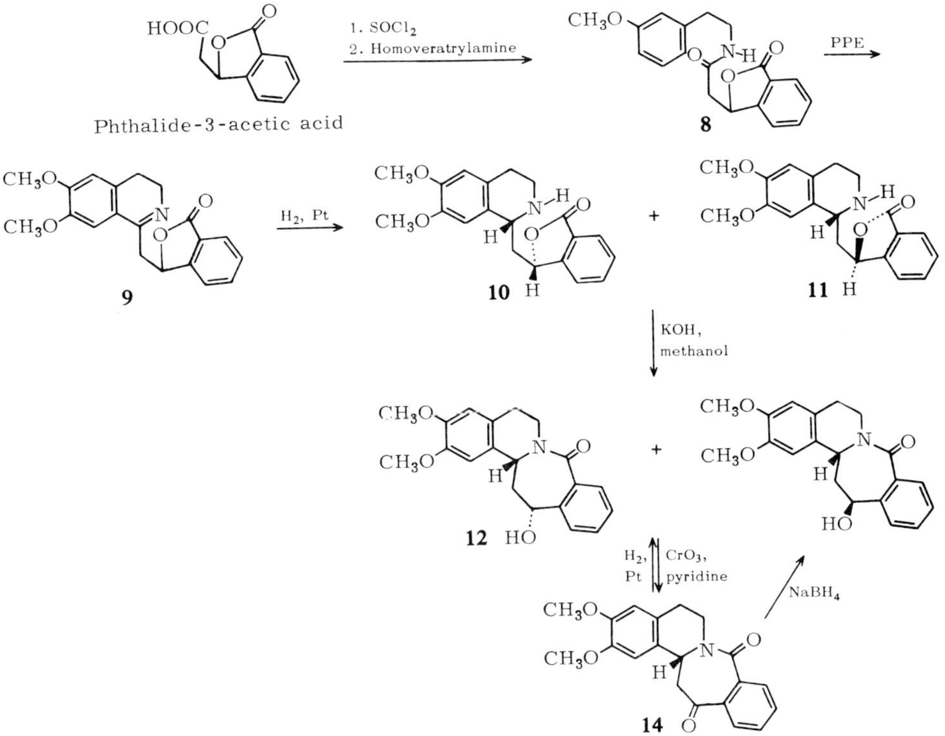

Scheme II

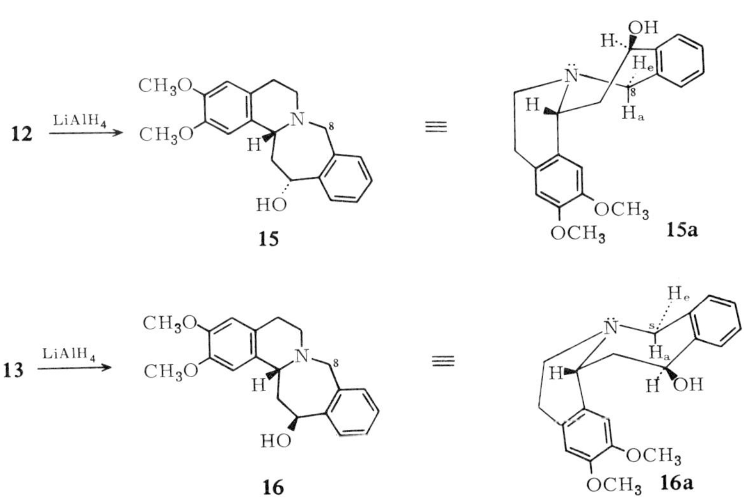

Scheme III

at δ4.81 with an ics of 79 Hz and $J_{a,e} = 14$ Hz. Conformation **16a** is considerably more rigid than **15a**, with the result that the C-8 equatorial hydrogen in **16a** is held more firmly in the plane of the aromatic D ring and is more subject to deshielding (Scheme III).[2]

To obtain the lactam **17**, a mixture of lactam alcohols **12** and **13** was hydrogenolyzed in ethanolic perchloric acid with palladium on carbon. In addition to the lactam **17**, there was a residue of the unhydrogenolyzed lactam alcohol **13**, so that this isomer must suffer hydrogenolysis at a slower rate.

Reduction of the lactam **17** to the crystalline homoprotoberberine **18** was accomplished with lithium aluminum hydride. The rate of methiodide formation for the base **18** was very fast, 3.4×10^{-2} sec^{-1}, pointing again to a cis B/C fusion.[2]

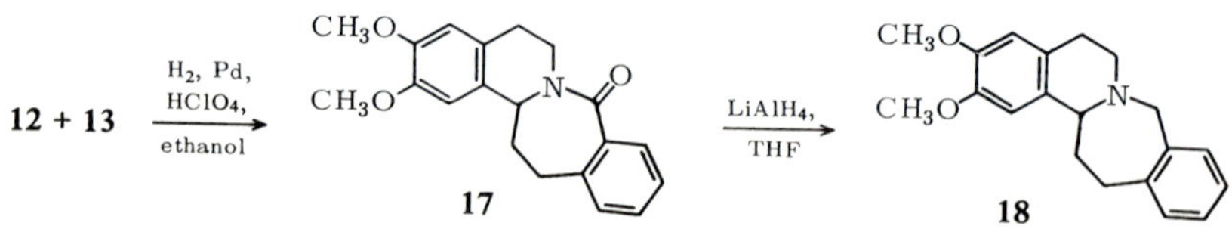

C. The Dieckmann Approach: Functionalization at C-14

Condensation of the amino ester **19** with ethyl 2-bromomethylphenyl acetate (**20**) gave rise to the diester **21**. Dieckmann cyclization followed by hydrolysis and decarboxylation afforded the desired amino ketone **22**. Reduction with Adams catalyst then gave the 14-hydroxyhomoprotoberberine **23**. The pseudo-first-order rate of methiodide formation for the amino alcohol **23** was also very fast, 1.5×10^{-2} sec^{-1}. There is, therefore, a possibility that the favored B/C ring fusion for homoprotoberberines in general is cis (Scheme IV).[2]

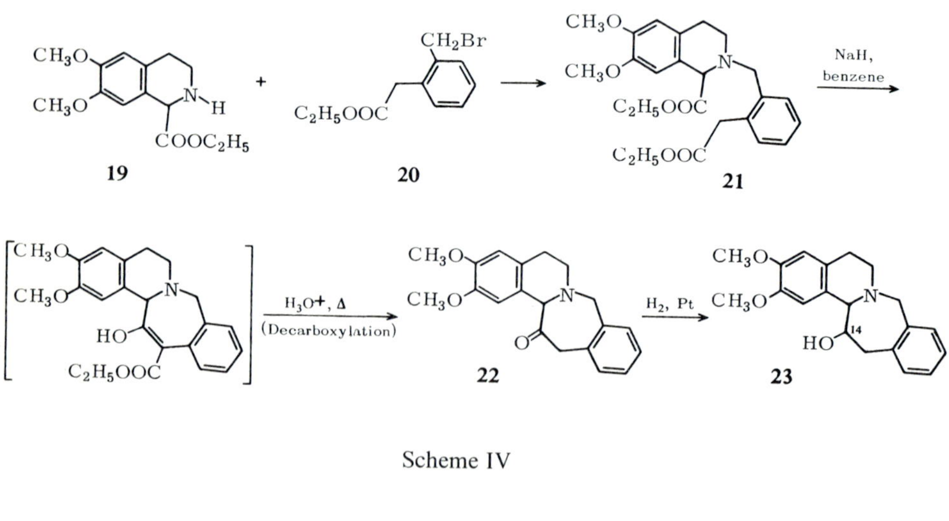

Scheme IV

II. Hydrogenolysis of Homoprotoberberines

When the homoprotoberberine alcohols **15** and **16** or the homoprotoberberine **18** were stirred under a hydrogen atmosphere with 5% palladium on carbon in ethanolic perchloric acid, the tricyclic base **24** was obtained. In like fashion, the 14-hydroxyhomoprotoberberine base **23** could be hydrogenolyzed to the crystalline alcoholic amine **25**. However, the known homoprotoberberines **26** and **4** did not undergo cleavage. It may be argued, therefore, that for steric reasons hydrogenolysis of a homoprotoberberine system will not occur when substituents are present at C-9.[2]

15, 16, or **18**

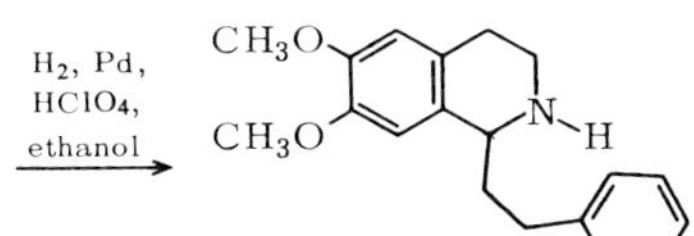

24

23

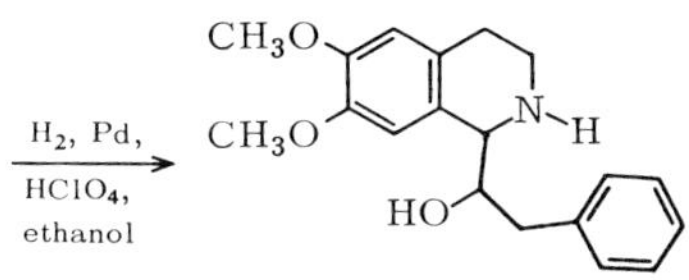

25

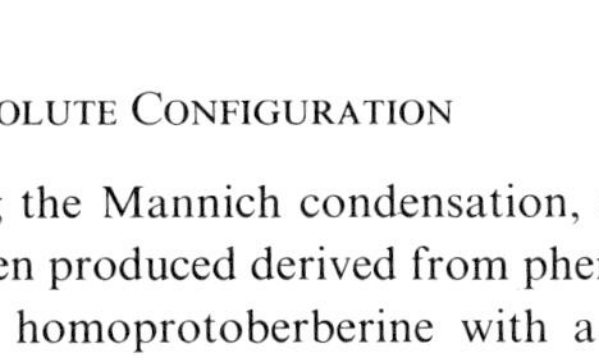

26

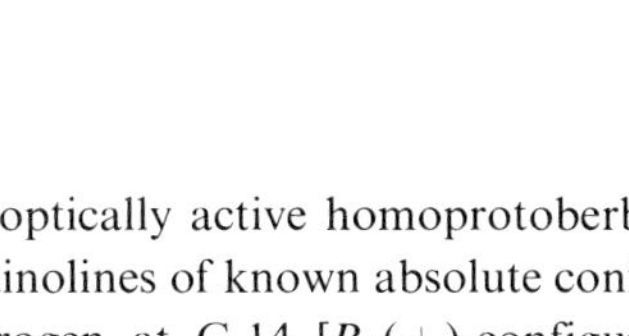

4

III. Absolute Configuration

Using the Mannich condensation, a series of optically active homoprotoberberines have been produced derived from phenethylisoquinolines of known absolute configuration. A homoprotoberberine with a beta hydrogen at C-14 [R-(+)-configuration] shows an ORD curve with a positive Cotton effect, the maximum at or near 247 mμ.[3]

IV. Mass Spectroscopy

The mass spectra of the homoprotoberberines show the expected molecular ion together with the $(M-1)^+$ peak. The main cleavage is along the lines indicated below. The base peak can be either the $(M-1)$ ion or species **28**. Ion **27** can also readily gain aromatization by loss of hydrogen.[1-3]

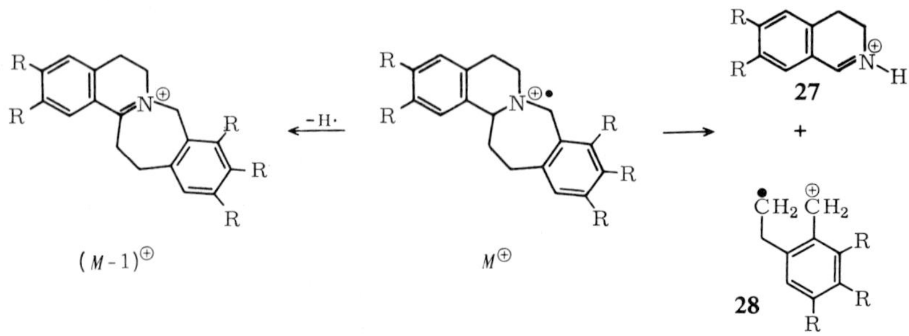

V. NMR Spectroscopy

The usual chemical shifts for the C-8 methylene protons of homoprotoberberines are around δ3.90 and 4.48 with each of the hydrogens appearing as a doublet, $J_{gem} \approx$ 15 Hz.

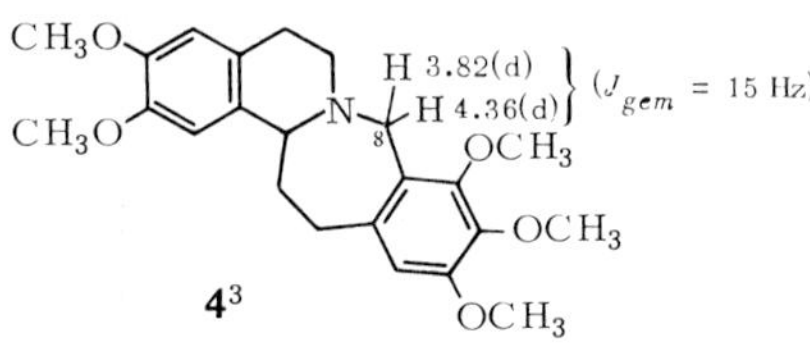

4[3]

Five methoxyl singlets at δ3.70, 3.73, 3.73, 3.77, and 3.77
Three aromatic proton singlets at δ6.58, 6.62, and 6.67

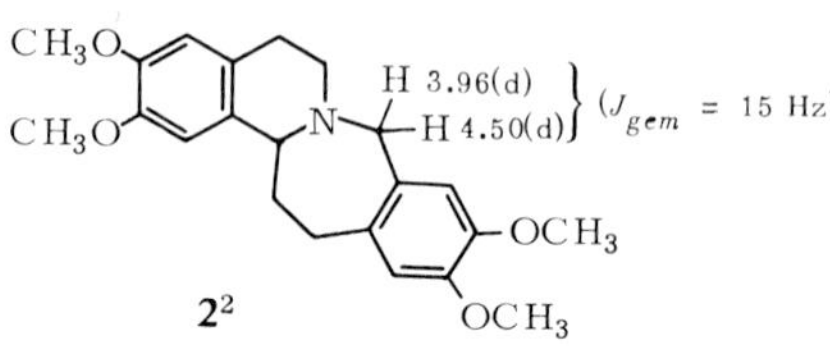

2[2]

Four methoxyl singlets at δ3.83, 3.86, 3.88, and 3.88
Four aromatic proton singlets at δ6.55, 6.57, 6.72, and 6.79

VI. UV Spectroscopy

Ultraviolet spectroscopy is of minimal value in assigning the positions of the substituents in the homoprotoberberines.

2,3,10-Trimethoxy-9-hydroxy-homoprotoberberine[4] λ_{max}^{MeOH} 226 sh and 285 mμ (4.23 and 3.86)

2,3,10-Trimethoxy-11-hydroxyhomoprotoberberine[4]	λ_{max}^{MeOH} 227 sh and 285 mμ (4.27 and 3.85)
2,3,9,10,11-Pentamethoxyhomoprotoberberine (**4**)[3]	λ_{max}^{MeOH} 230 sh and 282 mμ (4.15 and 3.62)

REFERENCES

1. A. Brossi, A. I. Rachlin, S. Teitel, M. Shamma, and M. J. Hillman, *Experientia* **24**, 766 (1968).
2. M. Shamma and M. J. Hillman, *Tetrahedron* **27**, 1363 (1971).
3. A. Brossi and S. Teitel, *Helv. Chim. Acta* **52**, 1228 (1969).
4. T. Kametani, T. Terui, T. Ogino, and K. Fukumoto, *J. Chem. Soc., C* p. 874 (1969).

Chapter 28 / THE PHENYLTETRAHYDRO-ISOQUINOLINES

Occurrence: Orchidaceae
Number: 3
Structures:

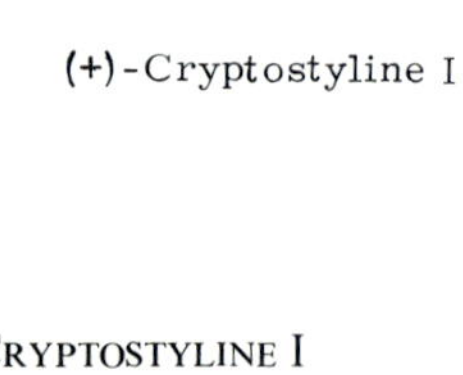

(+)-Cryptostyline I

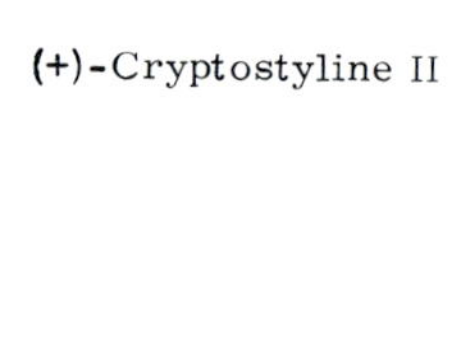

(+)-Cryptostyline II

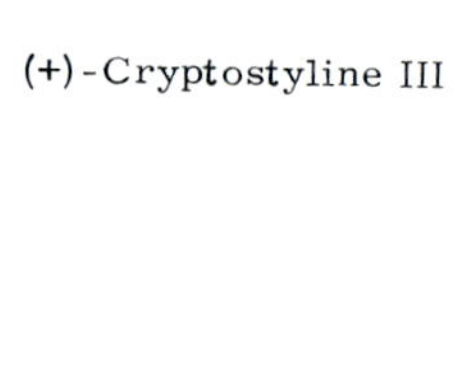

(+)-Cryptostyline III

I. Cryptostyline I

The dextrorotatory alkaloid cryptostyline I, $C_{19}H_{21}O_4N$, was found in *Cryptostylis fulva* Schltr. (Orchidaceae).[1,2] The UV spectrum with λ_{max}^{EtOH} 235 sh and 286.5 mμ (4.37 and 3.89) was clearly indicative of a tetrahydroisoquinoline moiety. The mass spectrum showed the facile loss of a methylenedioxyphenyl residue since the base peak was at *m/e* 206:

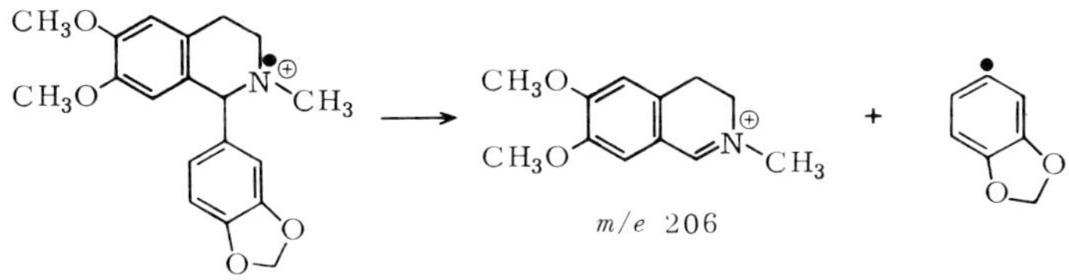

Alkaloid $M^{\oplus}$, m/e 327

In the NMR spectrum, the C-1 proton singlet at δ4.2 was shifted to δ5.9 in the methiodide salt, indicating that the phenyl group is probably attached at C-1. The synthesis of the alkaloid established the structure conclusively (Scheme I).[1]

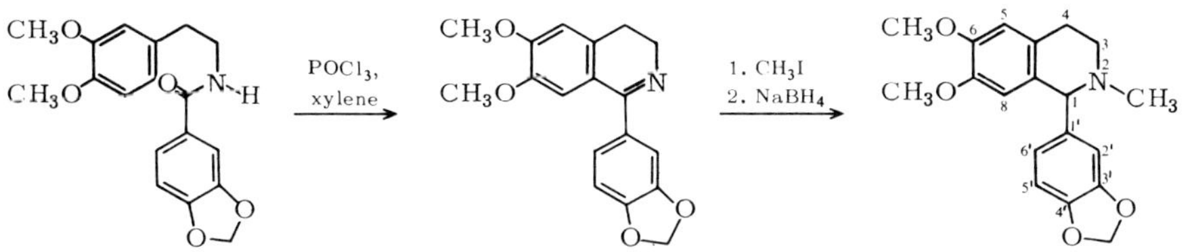

1-(3',4'-Methylenedioxy)-2-methyl-6,7-dimethoxy-tetrahydroiso-quinoline

Scheme I

An example of an alternate method for the preparation of 1-phenyltetrahydroiso-quinolines involves the cyclization of the methyl amide of homoveratric acid with benzaldehyde in the presence of trifluoroacetic acid. The resulting tricyclic lactam can be readily reduced to the tertiary amine (Scheme II).[3]

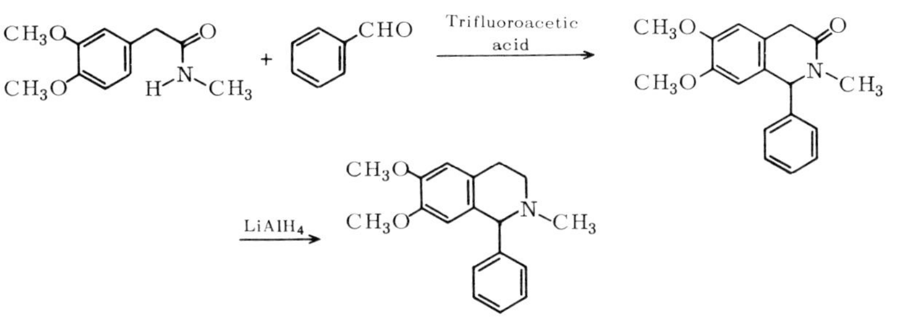

Scheme II

Cryptostyline I, II, and III were obtained from the same plant source.

II. X-Ray Analysis and Absolute Configuration

The ($\pm$)-*N*-norcryptostylines I, II, and III, synthesized by the Bischler–Napieralski approach as described in Scheme I, were resolved mainly through the use of ($-$)-diacetone-2-keto-L-gulonic acid. Reductive *N*-methylation of the three (−)-*N*-norbases **1a**, **b**, **c** afforded the tertiary (+)-bases **2a**, **b**, **c**, and similarly the three (+)-norbases gave rise to three tertiary (−)-bases.[4]

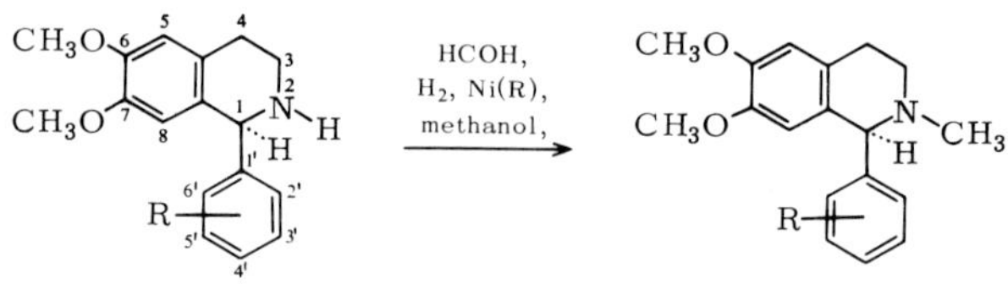

(−)-**1a**, R = 3',4'-methylenedioxy
(−)-**1b**, R = 3',4'-dimethoxy
(−)-**1c**, R = 3',4',5'-trimethoxy

(+)-**2a,b,c**
(Natural series, *S* configuration)

The ORD and CD spectra of (+)-**2a**, **b**, **c** have been described in detail.[4,4a]

The physical and optical data obtained for synthetic crystalline (+)-**2a** and (+)-**2b** were in good agreement with those reported for cryptostyline I and II, respectively, whereas the specific rotation of (+)-**2c** was significantly higher than that given for cryptostyline III. This discrepancy indicates that natural cryptostyline III is optically impure, being made up of about five parts of the dextro isomer and one part of the levo.[4]

A single crystal X-ray analysis of the hydrobromide of unnatural (−)-cryptostyline II (**3**. HBr), besides indicating conclusively the absolute configuration, showed that the aromatic ring of the C-1 substituent lies below and almost perpendicular to the mean plane of the molecule, and that the methoxyl groups at C-3′ and C-4′ are directed toward the far (down) side. This arrangement creates a sterically favorable arrangement for the whole molecule. Based on the above analysis, it follows that unnatural cryptostyline II (**3**) possesses the *R* configuration, and natural (+)-cryptostylines I, II, and III possess the *S* configuration.[4]

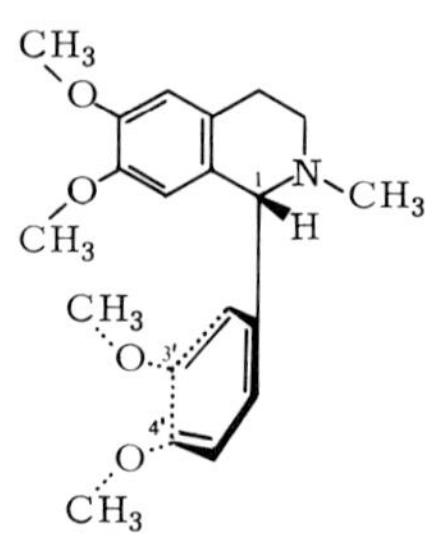

3, unnatural *R* configuration

III. Pharmacology

The synthetic isoquinoline Octaverine has been used as an antispasmodic.[5]

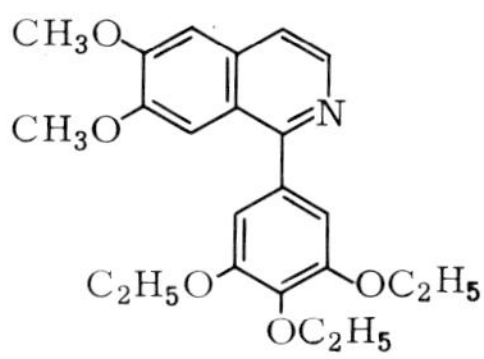

Octaverine

IV. NMR Spectroscopy

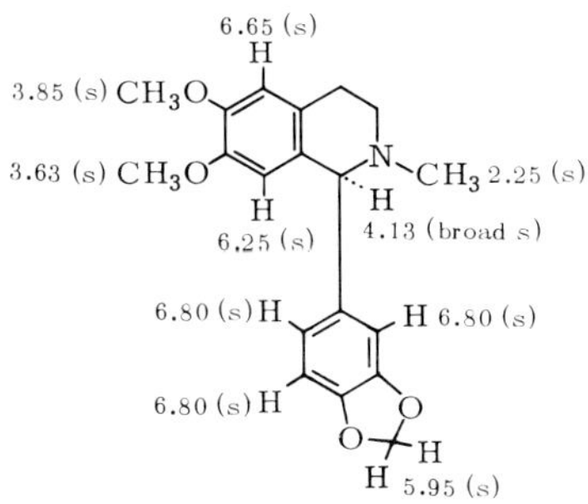

NMR spectral values for cryptostyline I[2,6]

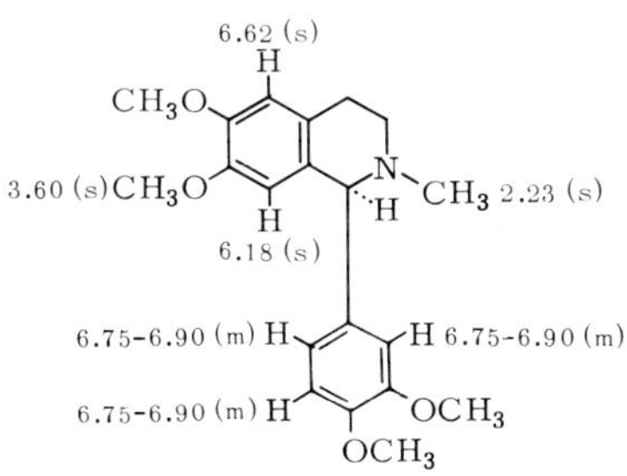

Three methoxyl singlets at δ 3.82, 3.85, and 3.88

NMR spectral values for cryptostyline II[2,6]

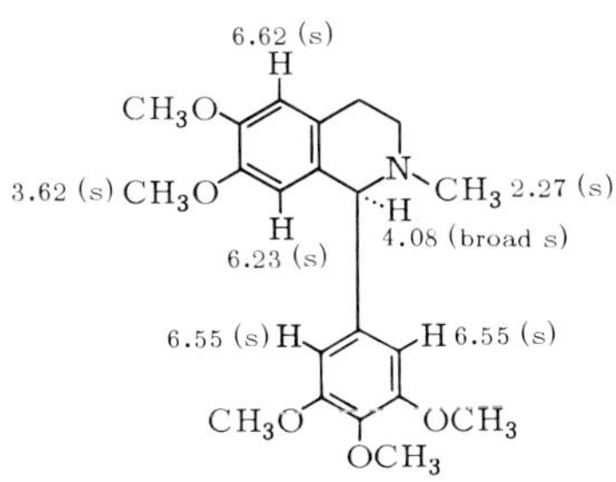

Four methoxyl singlets between δ 3.76 and 3.96

NMR spectral values for cryptostyline III[2,6]

The C-8 proton always appears upfield from the other aromatic protons because of shielding by the C-1 aromatic substituent.[6]

V. UV Spectroscopy

Cryptostyline I[2]: λ_{max}^{EtOH} 235 sh and 286.5 mμ (4.37 and 3.89)
Cryptostyline II[2]: λ_{max}^{EtOH} 228 sh and 281 mμ (4.20 and 3.68)
Cryptostyline III[2]: λ_{max}^{EtOH} 281 mμ (3.57)

References

1. K. Leander and B. Lüning, *Tetrahedron Lett.* p. 1393 (1968).
2. K. Leander, B. Lüning, and E. Ruusa, *Acta Chem. Scand.* **23**, 244 (1969).
3. A. Brossi, *Trans. N.Y. Acad. Sci., Ser. II* **28**, 685 (1966).
4. A. Brossi and S. Teitel, *Helv. Chim. Acta* **54**, 1564 (1971).
4a. The ORD and CD curves of several other 1-phenyltetrahydroisoquinolines have also been recorded: T. Kametani, H. Sugi, H. Yagi, and S. Shibuya, *J. Chem. Soc., C* p. 2213 (1970); but some of the conclusions drawn in this paper may be inconclusive.
5. Patent to ASTA AG Chemische Fabrik, Germany.
6. Some of the assignments for the NMR chemical shifts are by the author.

Chapter 29 / THE *N*-BENZYLTETRAHYDRO-ISOQUINOLINES

Occurrence: Fumariaceae
Number: 2
Structures:

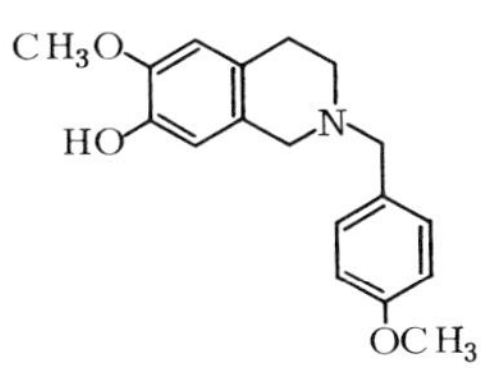

Sendaverine

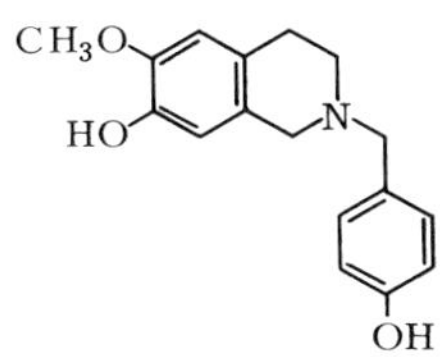

Corgoine

I. Sendaverine

The optically inactive phenolic alkaloid sendaverine, $C_{18}H_{21}O_3N$, was isolated from *Corydalis aurea* Willd. (Fumariaceae) by Manske in 1938, who also showed

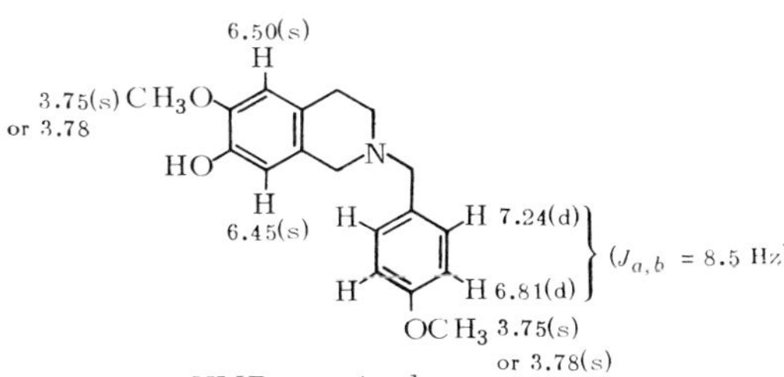

NMR spectral values for sendaverine

that mild oxidation gave *p*-anisic acid.[1] NMR spectroscopy revealed the presence of two methoxyl groups and the absence of an *N*-methyl group even though the nitrogen is tertiary.[2]

The molecular ion in the mass spectrum was at *m/e* 299 and the base peak at *m/e* 121 representing the *p*-methoxybenzyl or methoxycycloheptatrienyl cation. The UV spectrum showed λ_{max}^{EtOH} 255 (?) and 283.5 mμ (4.33 and 3.70).

Final proof of structure was provided by synthesis. *N*-Alkylation of the imine **1** with 4-methoxybenzyl chloride followed by reduction and debenzylation with tin in hydrochloric acid furnished material identical with the natural product (Scheme I).[2]

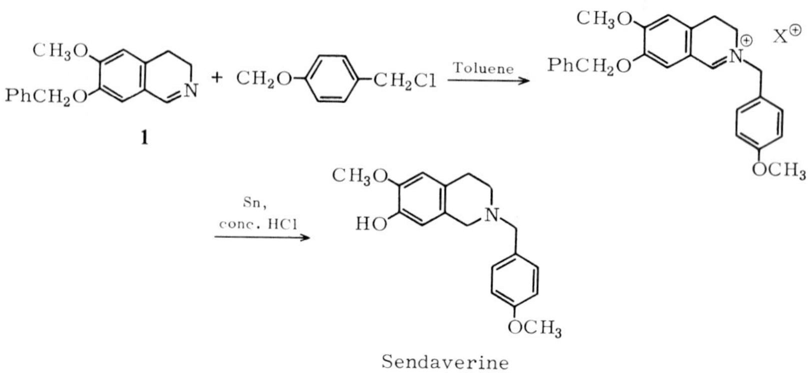

Scheme I

II. Corgoine

Corgoine, $C_{17}H_{19}O_3N$, has recently been isolated from *Corydalis gortschakovii*, and has been shown to have the structure indicated above. Selective *O*-methylation with diazomethane produced sendaverine.[3]

REFERENCES

1. R. H. F. Manske, *Can. J. Res. Sect. B* **16**, 81 (1938); *J. Amer. Chem. Soc.* **74**, 2864 (1952).
2. T. Kametani and K. Ohkubo, *Tetrahedron Lett.* p. 4317 (1965).
3. M. U. Ibragimova, M. S. Yunusov, and S. Yu. Yunusov, *Khim. Prir. Soedin.* **6**, 638 (1970); *Chem. Abstr.* **74**, 54046r (1971).

Chapter 30 / CHERYLLINE

Occurrence: Amaryllidaceae
Structure:

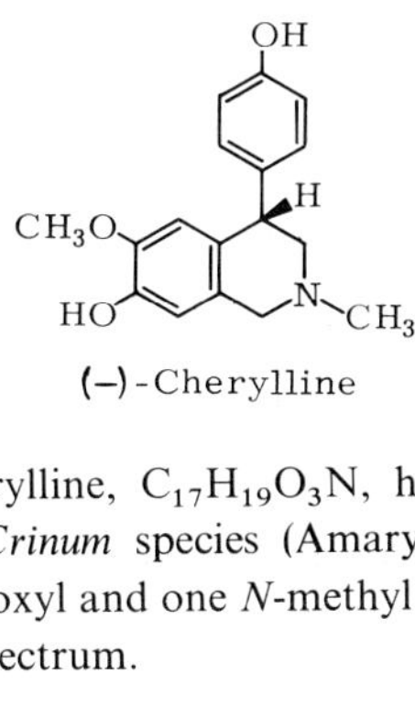

(−)-Cherylline

The phenolic base (−)-cherylline, $C_{17}H_{19}O_3N$, has been obtained from *Crinum powellii* var. *alba* and other *Crinum* species (Amaryllidaceae). The NMR spectrum in DMSO-d_6 showed one methoxyl and one *N*-methyl group and has been summarized below, together with the UV spectrum.

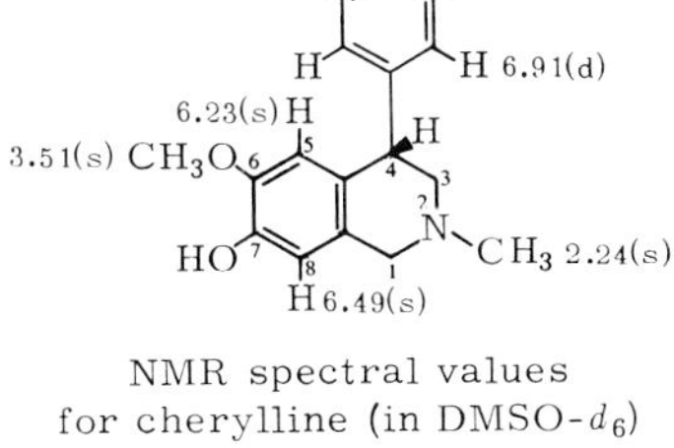

NMR spectral values
for cherylline (in DMSO-d_6)

λ_{max}^{EtOH} 225 sh, 280, 285, and 294 sh mμ (4.18, 3.60, 3.61, and 3.40)

Of the two aromatic singlets, the one at higher field, δ6.23, was assigned to the C-5 position since it must be shielded by the aromatic ring at C-4. Consequently, the 6.49 peak must be due to the C-8 proton. Upon the addition of a drop of NaOD, both singlets shifted upfield to δ6.06, representing a shift of 10 and 26 Hz, respectively. Since it is the C-8 proton that undergoes the larger shift, the phenolic function must be ortho to it and is therefore at C-7.[1]

The foregoing evidence alone is insufficient for a structural assignment, but the synthesis of the alkaloid in its optically active form by Brossi and Teitel settled the structural problem conclusively (Scheme I). The interesting use of selective methoxyl hydrolysis made in the course of the synthesis should be noted. Of the two methoxyl groups in the intermediate **1**, the one that does not hydrolyze is at C-6, a site conjugated to the deactivating imino group.[2]

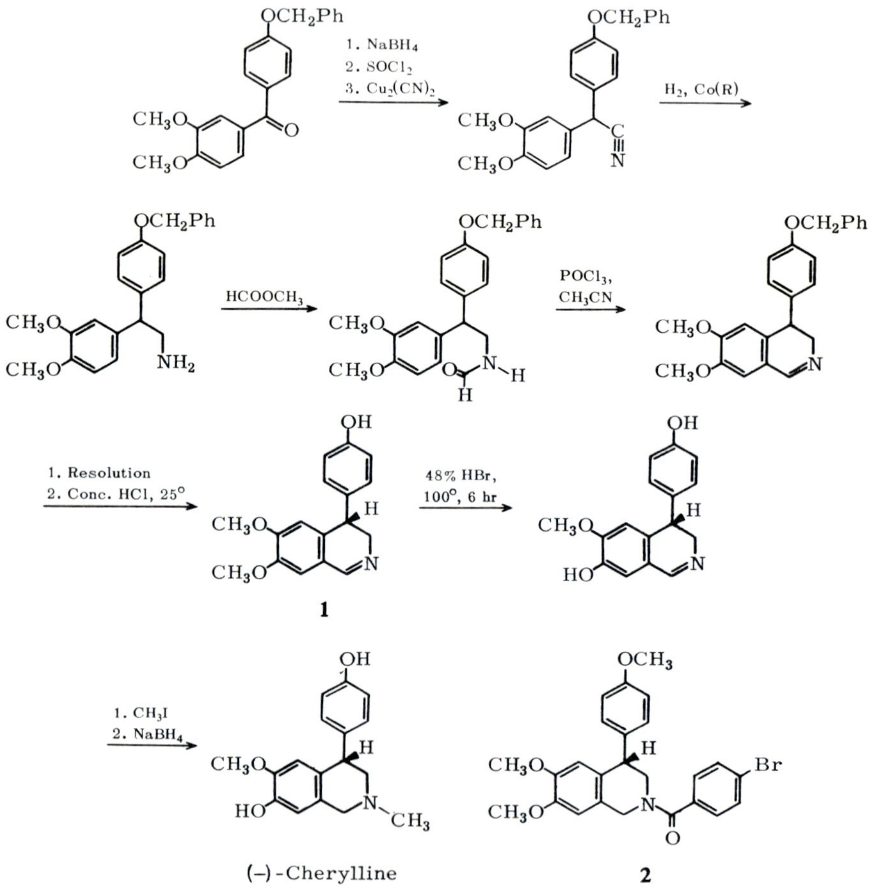

Scheme I

The ORD and CD curves of cherylline have been recorded, and the absolute configuration was assigned by analogy with the ORD curves of 4-aryltetralins.[1,2] The absolute configuration was further confirmed by an X-ray crystallographic study of the *O,O*-dimethyl-*p*-bromobenzamide derivative **2**.[1]

A possible biogenetic pathway for cherylline involves the sequence[3]:

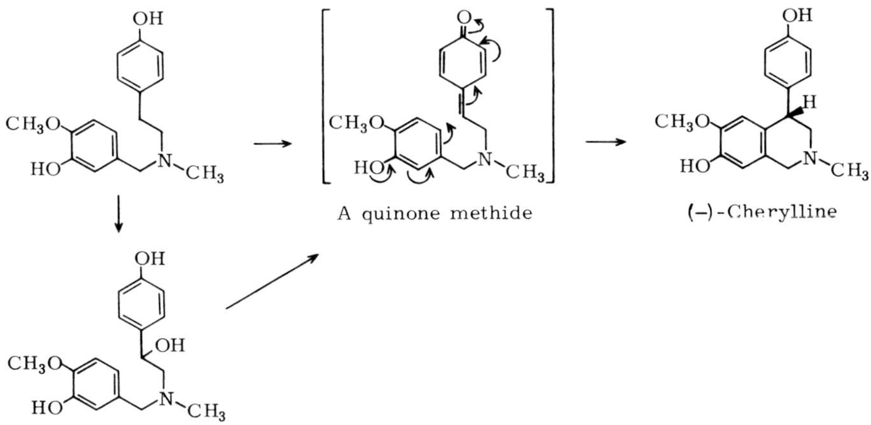

Schwartz and Scott, basing themselves on this sequence, have worked out a biogenetically patterned synthesis of (±)-cherylline (Scheme II).[3]

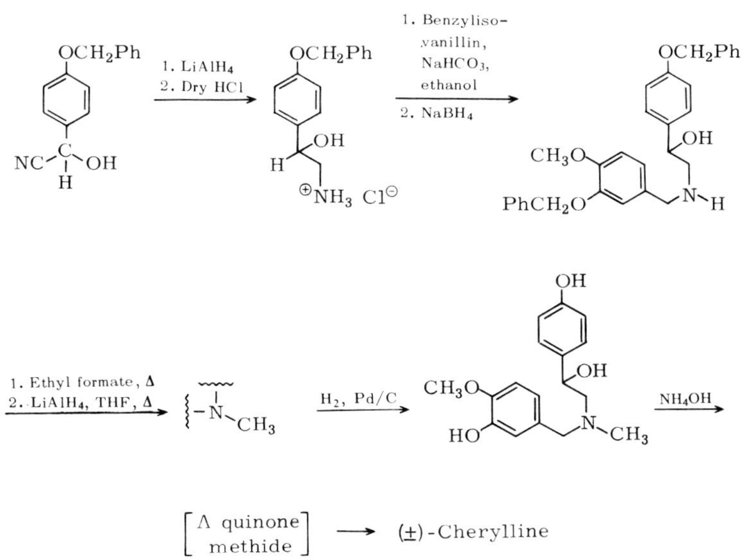

Scheme II

REFERENCES

1. A. Brossi, G. Grethe, S. Teitel, W. C. Wildman, and D. T. Bailey, *J. Org. Chem.* **35**, 1100 (1970).
2. A. Brossi and S. Teitel, *J. Org. Chem.* **35**, 3559 (1970); *Tetrahedron Lett.* p. 417 (1970).
3. M. A. Schwartz and S. W. Scott, *J. Org. Chem.* **36**, 1827 (1971).

Chapter 31 / THE NAPHTHALENOISOQUINOLINES

Occurrence: Ancistrocladaceae
Number: 3
Structures:

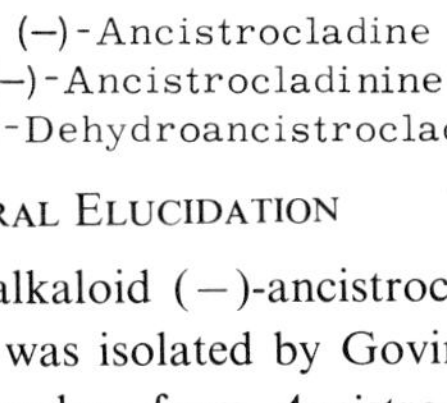

(−)-Ancistrocladine

[(−)-Ancistrocladinine ≡ 1,2-Dehydroancistrocladine]

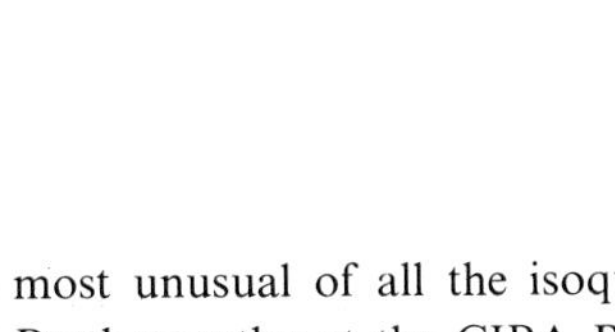

(−)-Ancistrocladisine

I. Structural Elucidation

The new alkaloid (−)-ancistrocladine is the most unusual of all the isoquinoline alkaloids. It was isolated by Govindachari and Parthasarathy at the CIBA Research Centre in Bombay from *Ancistrocladus heyneanus* Wall., the only representative of the family Ancistrocladaceae in India.[1] The alkaloid analyzes for $C_{25}H_{29}O_4N$ and possesses three methoxyls, one aromatic methyl, and two secondary methyl groups as well as one cryptophenolic hydroxyl and one secondary amino group. The UV spectrum, λ_{max} 230, 290, 305, 320, and 335 mμ (4.79, 4.00, 4.04, 3.95, and 3.87), denotes a highly aromatic system, and the IR spectrum shows absorption at 2.91 and 3.01 μ (3440 and 3330 cm^{-1}) due to the −OH and NH functions, respectively.[1]

The isolation of the hitherto unknown methyl 2-methyl-4,5-dimethoxy-1-naphthoate

(1) from the same plant gave a strong clue of the partial composition of the accompanying alkaloid.[1]

In an exhaustive paper on the chemistry of ancistrocladine, the tetracyclic structure shown below was presented for the alkaloid. The structural conclusions were based on detailed chemical degradations (Schemes I–III), accompanied by appropriate spectral data.[2]

Mild permanganate oxidation of the hydrochloride of the alkaloid followed by esterification with diazomethane yielded the methyl ester **1** whose structure was confirmed by synthesis. Noteworthy is the formation of the trialkylated derivative **2**, the γ-lactone **3**, and the completely aromatic isoquinoline **4** (Schemes I and II).

A further degradative sequence involving the Claisen rearrangement of the allyl ether **5**, which eventually terminated in the formation of the furanoid derivative **6**, is shown in Scheme III.[2]

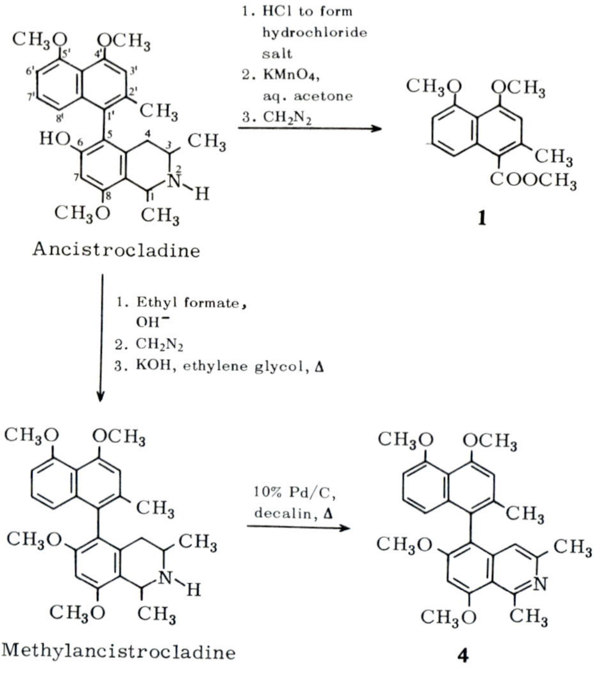

Scheme I

Sodium borohydride reduction of ancistrocladinine, an accompanying alkaloid, gave isoancistrocladine. NMR spectroscopy has shown the C-1 and C-3 methyl groups to be trans in ancistrocladine and cis in isoancistrocladine. Spectral and chemical evidence has also indicated that in ancistrocladisine, found in the same plant, the naphthalene linkage is at C-7 on the dihydroisoquinoline ring.[3] The absolute configuration of the naphthalenoisoquinolines has yet to be established.

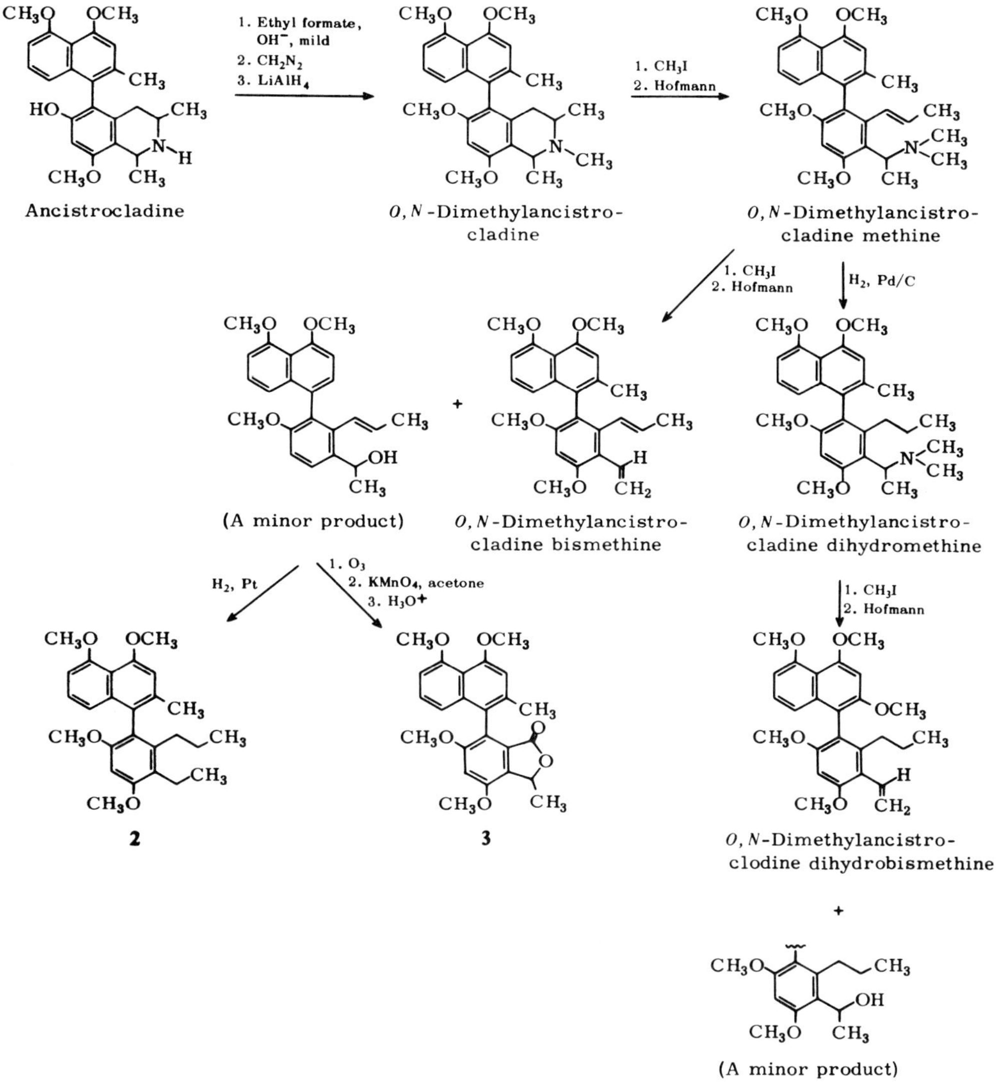

Scheme II

II. Biosynthesis

Ancistrocladine is the first isoquinoline alkaloid found to possess a methyl group at C-3, and its biogenetic origin must be very different from that of the other iso-

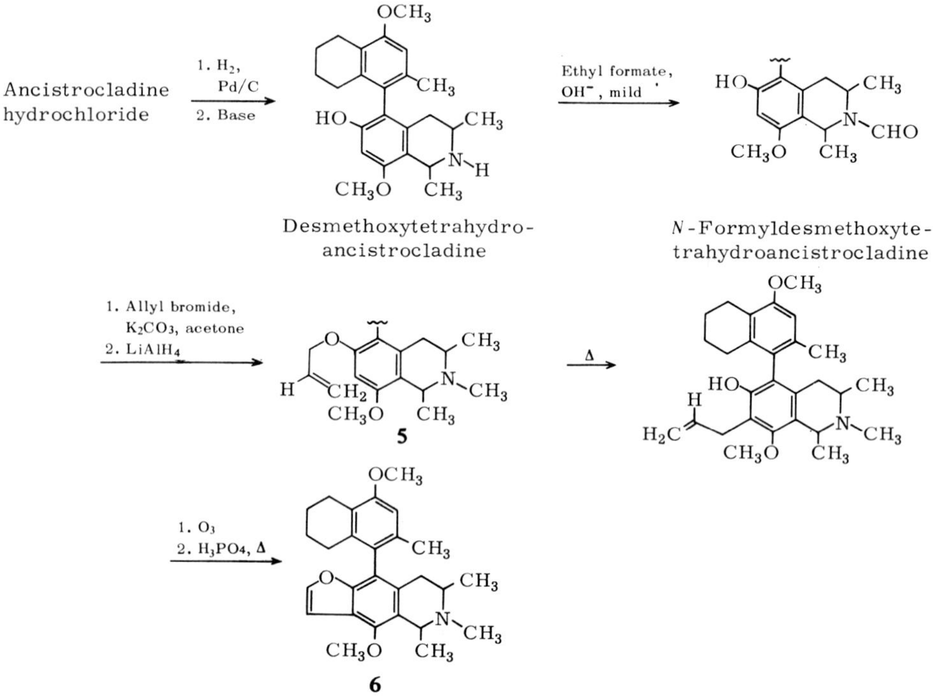

Scheme III

quinoline alkaloids. It is possible that cyclization of polyketide units is followed by phenolic oxidative coupling.[2]

III. NMR Spectroscopy

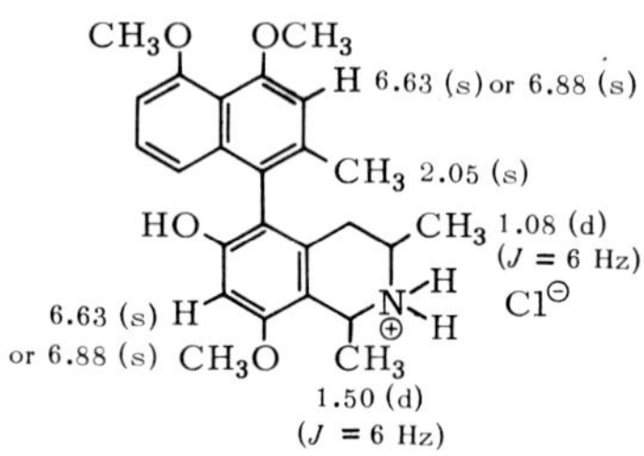

NMR spectral values for ancistrocladine hydrochloride in DMSO-d_6[2]

Three methoxyl singlets at δ 3.85, 3.87, and 3.90

REFERENCES

1. T. R. Govindachari and P. C. Parthasarathy, *Indian J. Chem.* **8**, 567 (1970).
2. T. R. Govindachari and P. C. Parthasarathy, *Tetrahedron* **27**, 1013 (1971).
3. T. R. Govindachari, P. C. Parthasarathy, K. Nagarajan, and H. K. Desai, private communication.

Chapter 32 / ADDENDA

CHAPTER 1/THE SIMPLE TETRAHYDROISOQUINOLINES

Section VIII

Through the use of labeled precursors, it has now been shown conclusively that the biosynthesis of the cactus alkaloid anhalonidine proceeds as described below*:

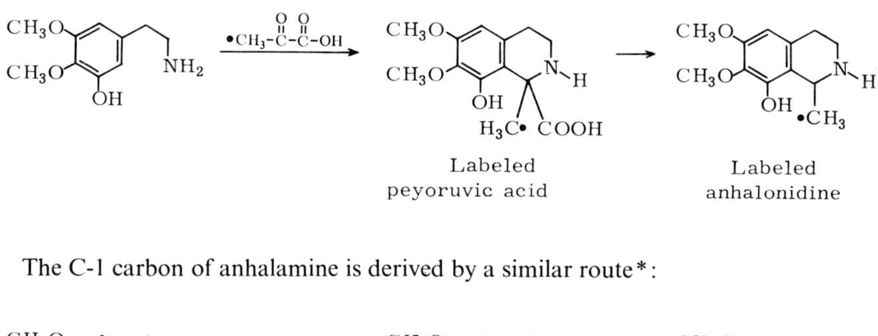

Labeled peyoruvic acid

Labeled anhalonidine

The C-1 carbon of anhalamine is derived by a similar route*:

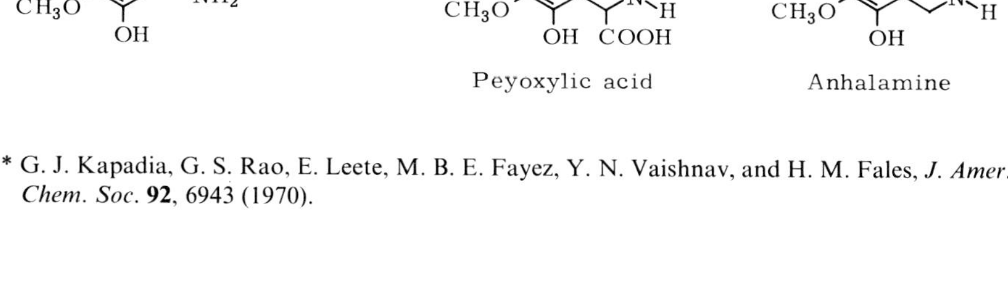

Peyoxylic acid

Anhalamine

* G. J. Kapadia, G. S. Rao, E. Leete, M. B. E. Fayez, Y. N. Vaishnav, and H. M. Fales, *J. Amer. Chem. Soc.* **92**, 6943 (1970).

CHAPTER 2/THE BENZYLISOQUINOLINES

Section IV

A new procedure for the elaboration of benzylisoquinolines has been developed by E. C. Taylor and S. F. Martin. Treatment of 6,7-dimethoxy-1-chloroisoquinoline with the Wittig reagent prepared from veratryl chloride and tri(*n*-butyl)phosphine provided, after work-up, a 74% yield of papaverine.[1]

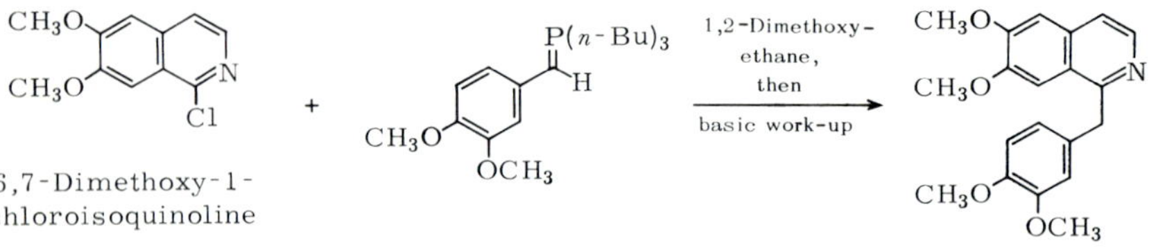

6,7-Dimethoxy-1-chloroisoquinoline

Papaverine

Section VII

Miller, Stermitz, and Falk have found that the electroxidative coupling of the non-phenolic base laudanosine yields a surprisingly high yield (52%) of the morphinandienone *O*-methylflavinanthine.[2]

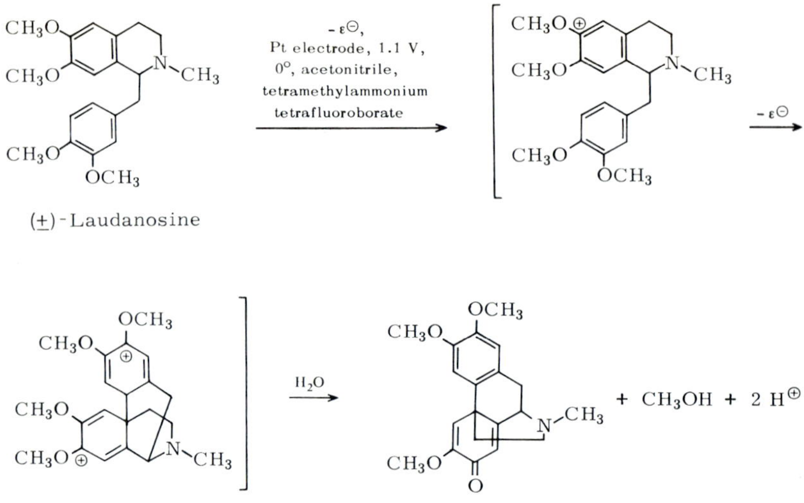

(±)-Laudanosine

O-Methylflavinanthine

A simple and interesting conversion of the tetrahydrobenzylisoquinoline reticuline to an analog of the hasubanan type alkaloid cepharamine has been achieved by Kametani, Kobari, and Fukumoto. The trifluoroacetyl methine **1** was reduced with Adams catalyst, and the product oxidized to the dienone **2** in 20% yield. Hydrolysis of the trifluoroacetyl group followed by treatment with methanolic hydrogen chloride gave rise to the cepharamine analog **3** as yellow needles.[3]

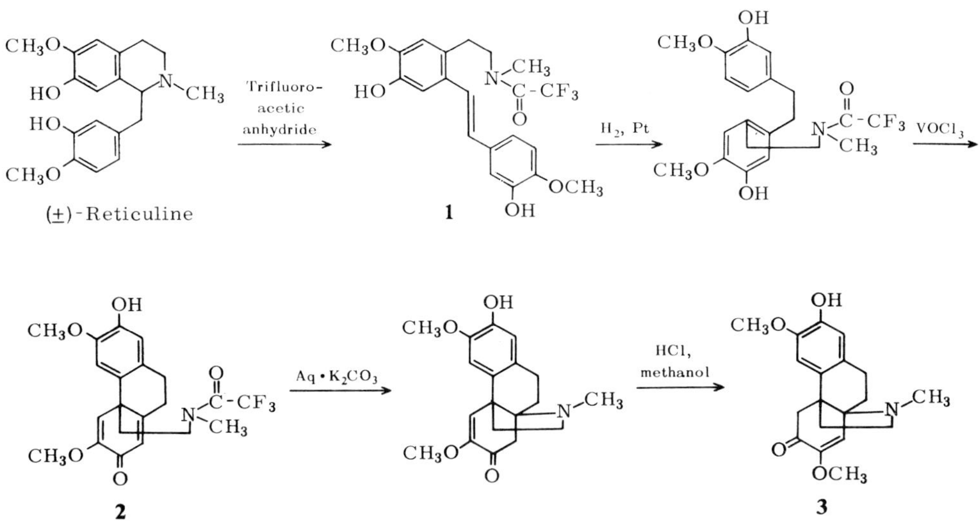

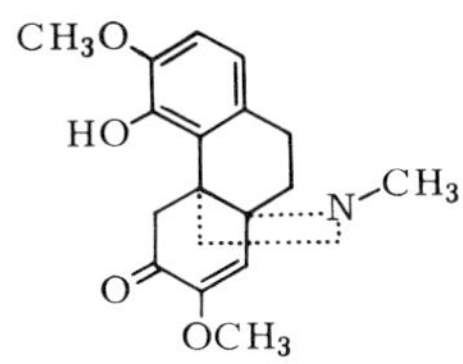
Cepharamine

Section IX

Radioactively labeled (±)-norreticuline was incorporated into papaverine in *Papaver somniferum* L. to a high extent (5%). This finding has led to a proposed biosynthetic pathway for the benzylisoquinolines of that plant.[4]

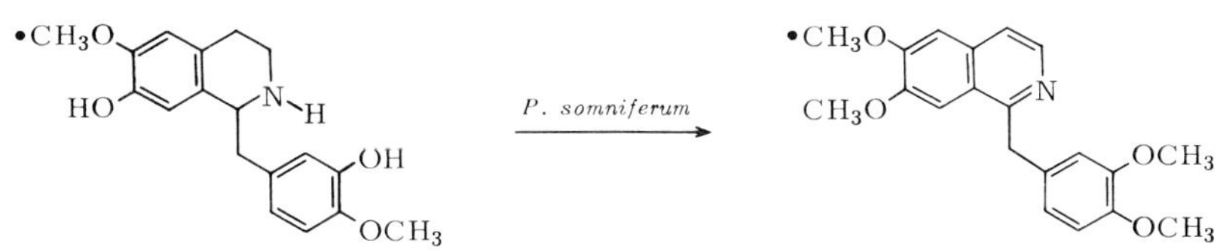

(±)-Norreticuline Papaverine

Section X

Papaverine blocks the action of Ca^{2+} ions in depolarized *taenia coli* of the guinea pig. The stimulatory action of Ca^{2+} ions was reduced because papaverine seems to inhibit the flow of these ions across the membrane.[5]

CHAPTER 4/THE PAVINES AND ISOPAVINES

Section I

A little known synthesis of the pavine skeleton involves the electrophilic molecular rearrangement of the tetracyclic dibenzopyrrocoline *N*-metho salt shown below.[6]

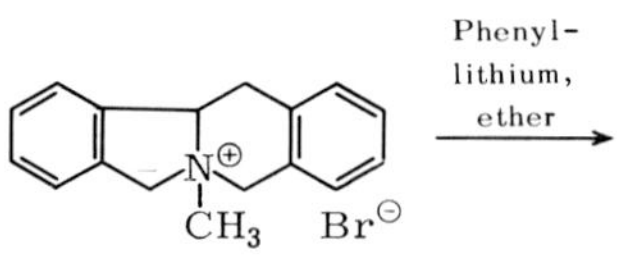

O-Methylthalisopavine has recently been prepared through hydration of 2-methyl-1,2-dihydropapaverine followed by acid-catalyzed cyclization of the resulting C-4 hydroxylated tetrahydrobenzylisoquinoline.[7]

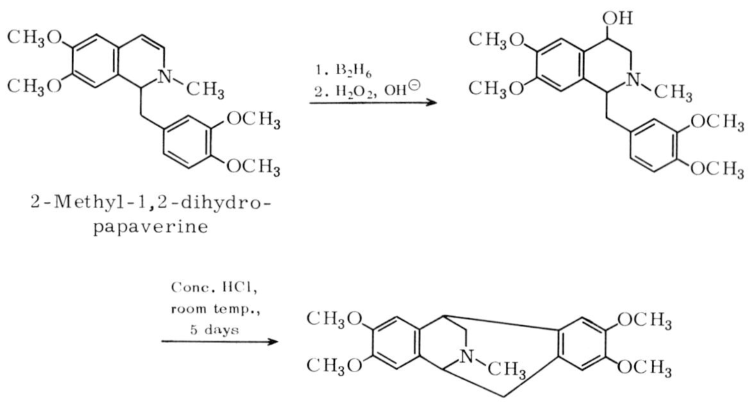

2-Methyl-1,2-dihydro-papaverine

O-Methylthalisopavine

Section II

The features required for a successful application of the aromatic chirality method include:

(a) Two separate chromophoric groups which are close enough in space that their electronic transitions can interact.
(b) The presence of a chiral center close to the chromophoric groups.

(c) The proximity and fair intensity of the absorption bands of the two chromophores.
(d) A knowledge of the directions of the transition moments of the two interacting absorption bands.

If the two chromophoric groups interact, the single Cotton effect observed in the CD curve of the isolated chromophore will split into two Cotton effects of opposite signs centered around the original wavelength; this is the Davydov split.

Numerous models of known structures and also nonempirical calculations have shown that the sign of the split Cotton effect at longer wavelengths is in agreement with the sign or handedness of the spiral or twist involved. A positive split is associated with a positive spiral or twist, and vice versa. Since the intensities of the split Cotton effect are usually distinctly stronger than ordinary Cotton effects, and because the split occurs at about the original UV absorption maximum, it can be readily recognized.

Several isoquinoline alkaloids show UV absorption maxima near 285, 240 sh, and 207 mμ. These bands are usually designated $^1A \rightarrow {}^1L_b$, $^1A \rightarrow {}^1L_a$, and $^1A \rightarrow {}^1B$, respectively, although the nomenclature favored by spectroscopists is $^1A_{1g} \rightarrow {}^1B_{2u}$, $^1A_{1g} \rightarrow {}^1B_{1u}$, and $^1A_{1g} \rightarrow {}^1E_{1u}$, in the same order. The splittings observed in the CD curves correspond to one or more of these bands.

In order to determine the twist of an asymmetric molecule such as amurensine, place strips of tape across the benzenoid rings of a molecular model as shown (▬) and color the ends indicated here by a thick dot (●).

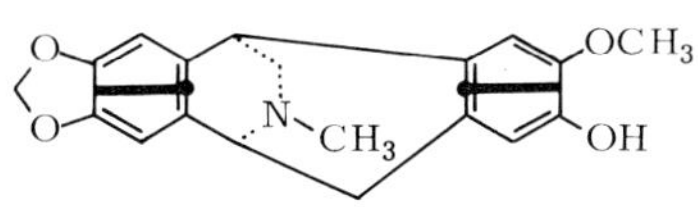

(–)-Amurensine

Turn the molecular model around so that your eye and the two dots are in a straight line. It does not matter from which end of the molecule one is looking. The dihedral angle formed by the two strips of tape can be either of type "A" or "B," in which the thickly dotted line is closer to the eye, and the finely dotted line further away. This angle in amurensine is found to be of type "B," which is negative. The negative twist is reflected by two clearly defined split CD curves centered at 285 and 240 mμ, each negative in sign.

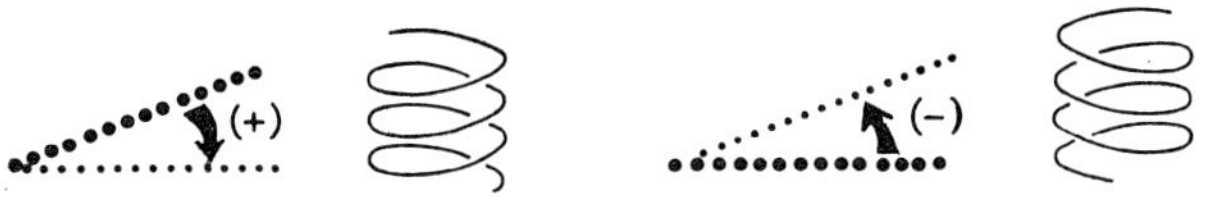

Positive spiral, Type A   Negative spiral, Type B

The strips of tape placed across the aromatic rings in the model shown above are in the direction of the moment for the $^1A \rightarrow {}^1L_b$ transition. The moment for the $^1A \rightarrow {}^1L_a$ transition is at right angle to the $^1A \rightarrow {}^1L_b$ transition.

CHAPTER 5/THE BISBENZYLISOQUINOLINES

Section II

Irradiation of a bisbenzylisoquinoline with UV light in the presence of oxygen results in oxidative cleavage at the C-1 and C-1′ benzylic centers. Isotetrandrine gave as major products the lactam amino alcohol **4** (30%) and the dialdehyde **5** (50%).[8] The presence of phenolic groups complicates the reaction and diminishes the yields; but this oxidative transformation may develop as a useful alternative to the Hofmann degradation in the bisbenzylisoquinoline series.

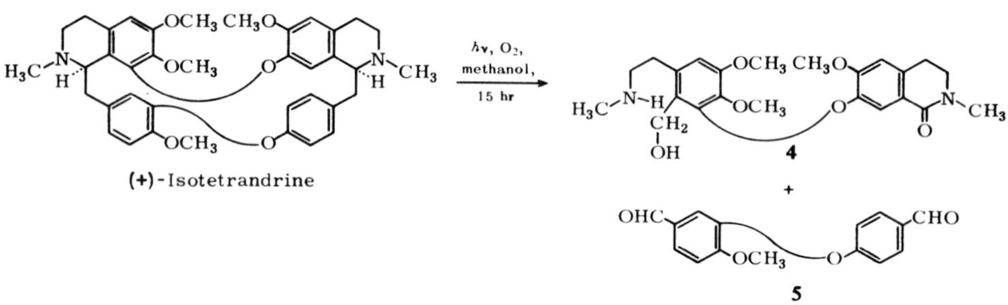

Section III

Thalisopine has now been shown to have the structure indicated below.[9]

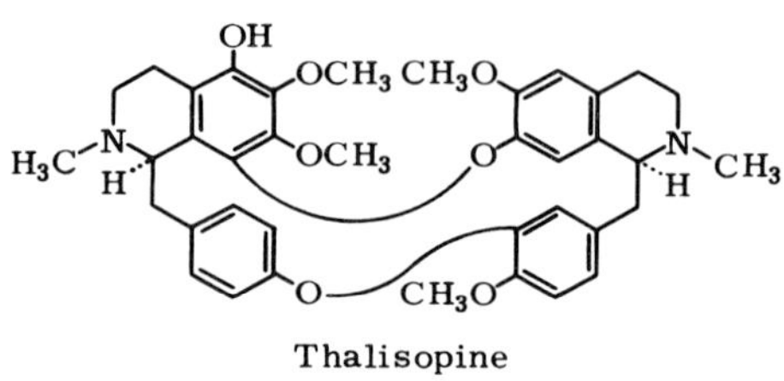

A convenient method for the determination of the position of attachment of a diphenyl ether linkage in a bisbenzylisoquinoline consists in treatment of the *O*-methylated dimer with 3% DCl in D_2O at 120° in a sealed tube for 144 hr. Under these conditions only protons ortho to methoxyl groups are exchanged for deuterium. Subsequent sodium in liquid ammonia cleavage of the *O*-methylated dimer, and characterization of the resulting tetrahydrobenzylisoquinolines, shows that an aromatic hydrogen ortho to a methoxyl group indicates the original site of attachment of the diphenyl ether bond. This reaction sequence was applied to *O*-methyloxyacanthine.[10]

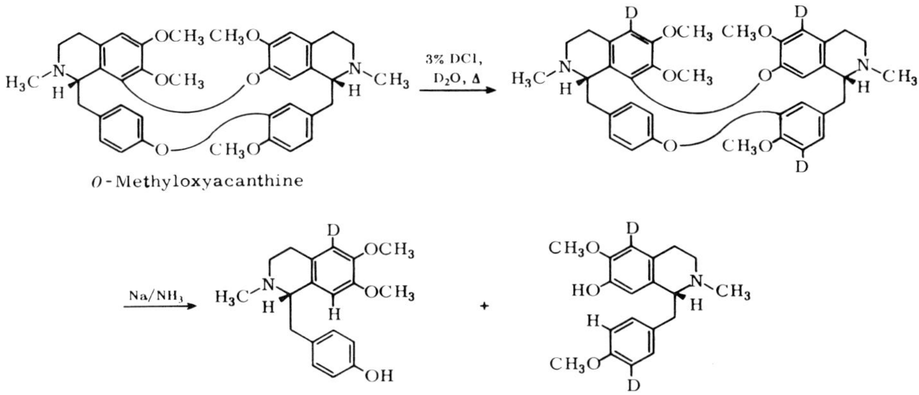

The above method is a very useful supplement to the sodium in liquid deuterated ammonia cleavage of *O*-methylated bisbenzylisoquinolines, which achieves the same purpose.[11]

Section VIII

A mild reagent for *O*-demethylation of ethers is the lithium salt of diphenylphosphine.[12]

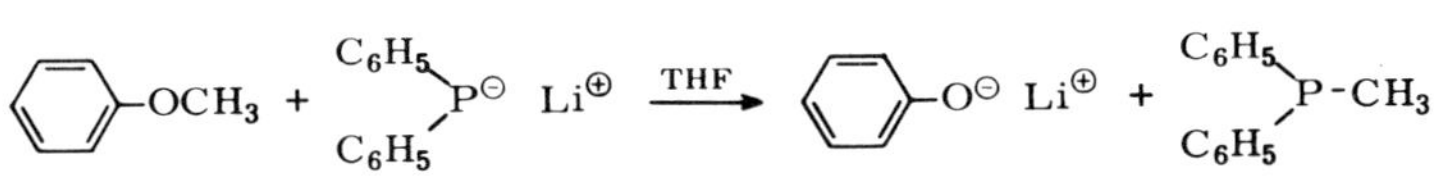

CHAPTER 6/THE CULARINES

Section III

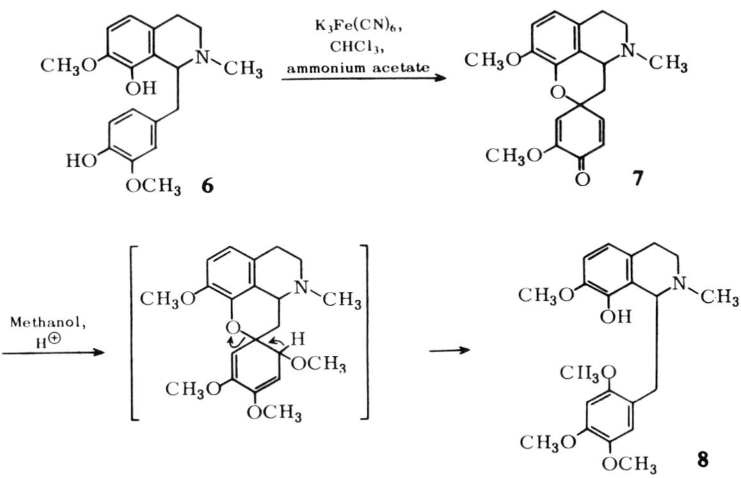

Ferricyanide oxidation of the tetrahydrobenzylisoquinoline **6** afforded a mixture of diastereoisomeric dienones **7**. Attempts to effect the rearrangement of this mixture to a cularine analog under a variety of acidic conditions were to no avail. However, using hydrogen chloride or sulfuric acid in methanol, a product was obtained which was shown to be the benzylisoquinoline **8**. The net result of these transformations is the selective oxidation of ring C of the tetrahydrobenzylisoquinoline.[13]

In a separate study, it was found that rearrangement of the dienone **7** in concentrated sulfuric acid at 5°, followed by *O*-methylation, furnished 1,9,10-trimethoxyaporphine.[14]

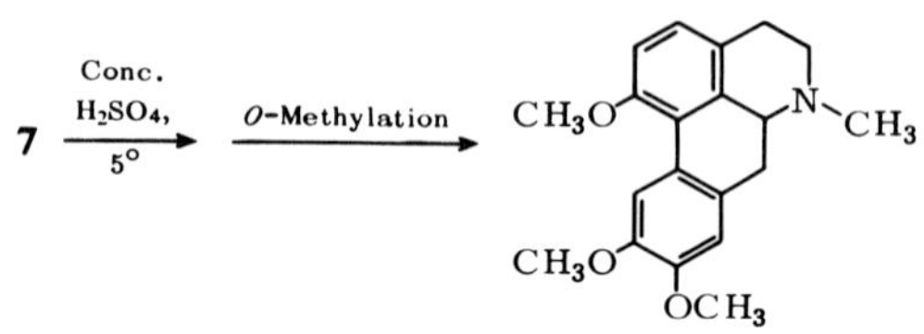

1,9,10-Trimethoxyaporphine

CHAPTER 10/THE APORPHINES

Section II

The novel and unusual quaternary aporphine alkaloid (+)-thalphenine chloride, $C_{21}H_{22}O_4N^+Cl^-$, λ_{max}^{EtOH} 221, 230 sh, 280 sh, 288, 317, and 328 mμ (4.32, 4.21, 3.69, 3.83, 3.97, and 3.87), has been found in *Thalictrum polygamum* Muhl. (Ranunculaceae).[15] The structure was derived from spectral studies as well as from a complete X-ray analysis of thalphenine iodide. Thalphenine is the first aporphine found to possess a methylenoxy bridge.[15] Hofmann degradation of thalphenine gave the optically inactive thalphenine methine (≡ thaliglucine), $C_{21}H_{21}O_4N$, λ_{max}^{EtOH} 221, 250, 260, 272 sh, 287 sh, 317, 350, and 370 mμ (4.26, 4.41, 4.36, 4.13, 3.81, 3.36, and 3.36), also found as a natural product in *Thalictrum* species.[15,16]

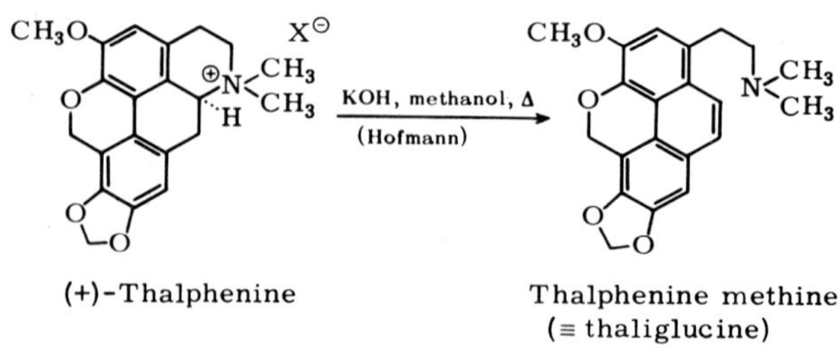

(+)-Thalphenine

Thalphenine methine
(≡ thaliglucine)

Section IV

The intramolecular photocyclization of 7-hydroxylated tetrahydrobenzylisoquinolines monobrominated in ring C has provided a convenient synthesis of aporphines.[17] Irradiation of a solution of the bromophenol **8** and sodium hydroxide in aqueous methanol led to a 52% yield of thaliporphine.

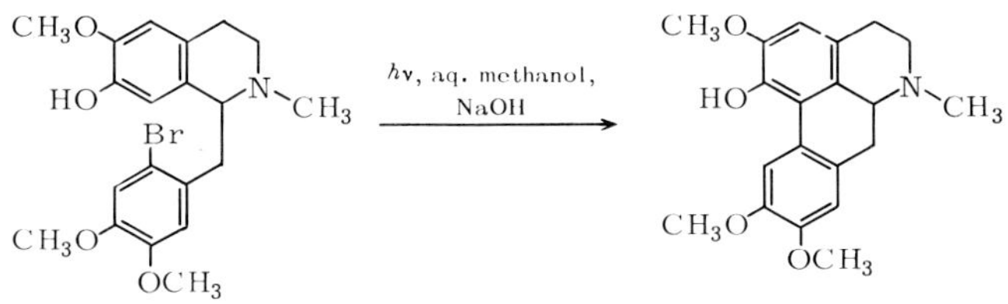

8 Thaliporphine

Another efficient photochemical route to the aporphines involves irradiation of the bromo urethane **9** in the presence of potassium *t*-butoxide.[18] The *N*-carbethoxydehydroaporphine **10** was thus obtained in 72% yield. This method is a distinct improvement over the original use of calcium carbonate in place of potassium *t*-butoxide as the acid scavenger (Chapter 10, Section IV, B).

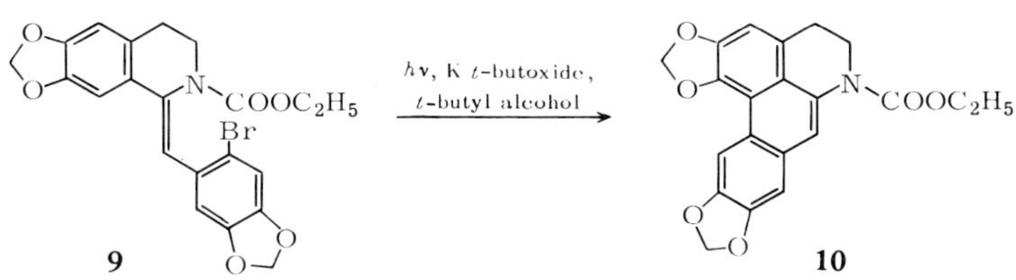

9 **10**

O-Acetylthaliporphine has also been synthesized through the acid-catalyzed rearrangement of a *p*-quinol acetate, in what amounts to a novel and unusual route to the aporphines. Reaction of the tetrahydrobenzylisoquinoline codamine with lead tetraacetate in acetic acid yielded an amorphous product, presumably the tricyclic base **10a**, which was converted to a mixture of *O*-acetylthaliporphine and 4-acetoxy-*O*-acetylthaliporphine through treatment with acetic anhydride and sulfuric acid. The mechanism of formation of the interesting by-product 4-acetoxy-*O*-acetylthaliporphine is unclear. This oxidation using lead tetraacetate should be compared with the rearrangement of quinol acetates leading to C-4 hydroxylation of tetrahydroisoquinolines (Chapter 1, Section VI, A).[19]

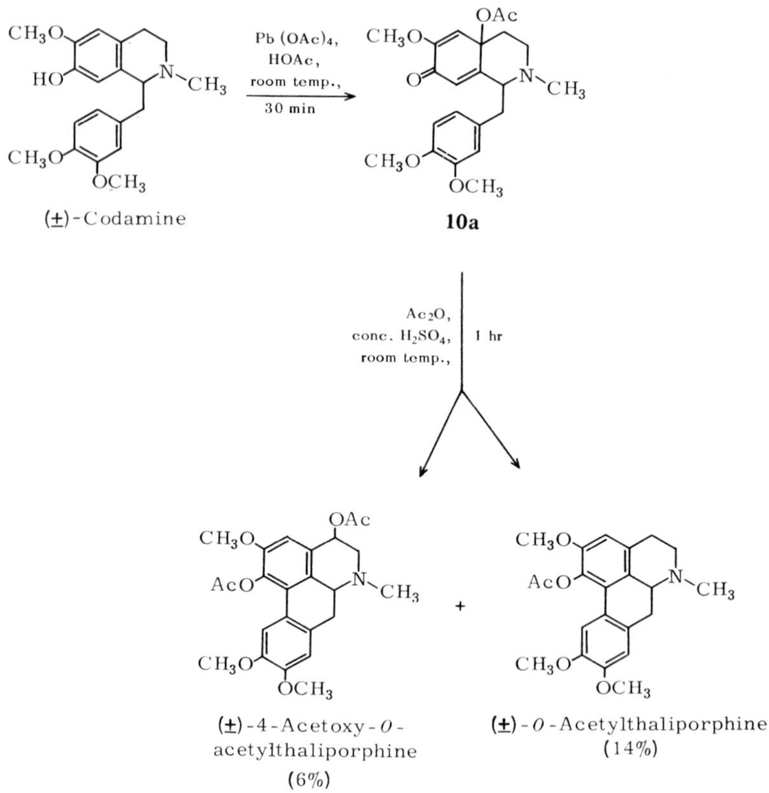

(±)-Codamine

10a

(±)-4-Acetoxy-*O*-acetylthaliporphine (6%)

(±)-*O*-Acetylthaliporphine (14%)

Chapter 14/The Phenanthrene Alkaloids

Section I

Two interesting new phenanthrene alkaloids are thaliglucine (≡ thalphenine methine), which has been discussed above, and thaliglucinone.

Thaliglucine was isolated independently from *Thalictrum rugosum* Ait.[16] and from *T. polygamum* Muhl. (Ranunculaceae).[15] The structure was established by spectral means,[15,16] as well as by a direct chemical correlation with the aporphine (+)-thalphenine, the biogenetic precursor, which was also found in *T. polygamum*.[15]

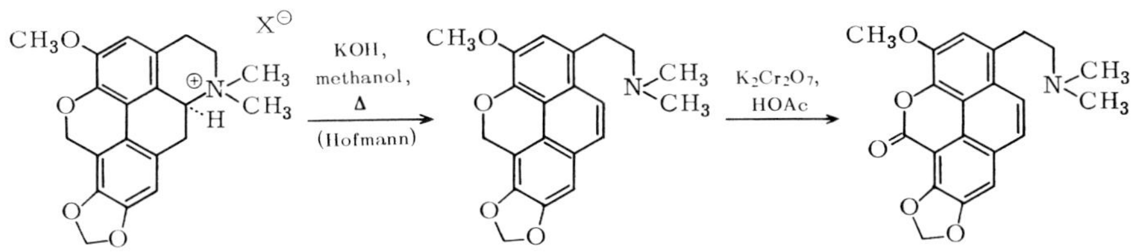

(+)-Thalphenine Thaliglucine (≡ thalphenine methine) Thaliglucinone

The δ-lactone thaliglucinone was found in *T. rugosum* and was obtained as a light yellow crystalline substance. Dichromate oxidation of thaliglucine (≡ thalphenine methine) yielded thaliglucinone.[16]

Chapter 16/The Protoberberines and Retroprotoberberines

Section I

The first dimeric protoberberine alkaloid, bisjatrorrhizine chloride, has been isolated from *Jatrorrhiza palmata* (Lam.) Miers. A partial synthesis was then carried out through phenolic oxidative coupling of jatrorrhizine.[20]

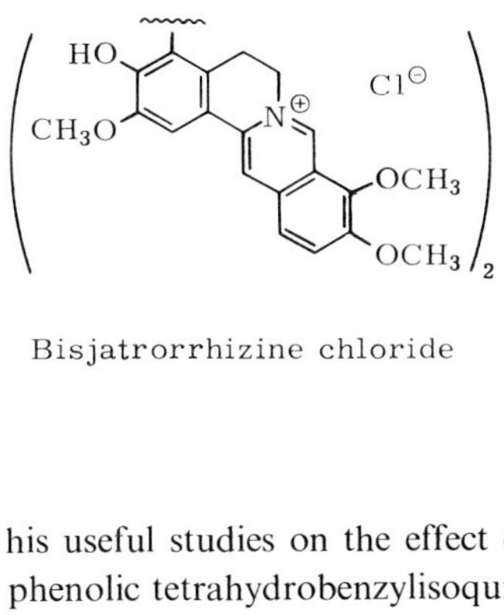

Bisjatrorrhizine chloride

Section III

Kametani has continued his useful studies on the effect of change of pH upon the direction of cyclization of a phenolic tetrahydrobenzylisoquinoline with formaldehyde. Condensation of the tetrahydrobenzylisoquinoline **11** with formaldehyde in methanol at very low pH leads mostly to the 9,10-substituted tetrahydroprotoberberine nandinine **(12).** But the percentage of this product decreases with increasing pH.[21]

On the other hand, Mannich cyclization of the benzyloxytetrahydrobenzylisoquinoline **15** occurred preferentially *ortho* to the phenolic function, regardless of the exact pH, indicating that steric factors are important in this instance. The final product, after *O*-methylation and *O*-debenzylation, was (±)-capaurimine.[22]

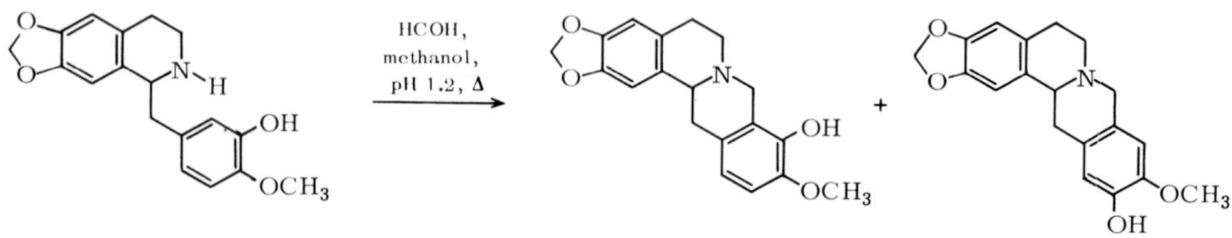

11 Nandinine (**12**) (71%) **13** (18%)

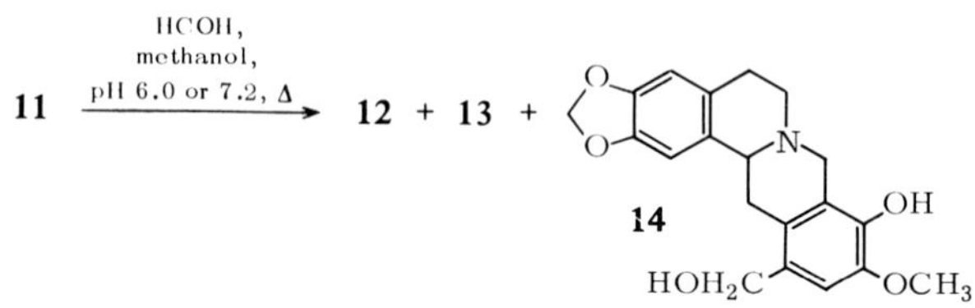

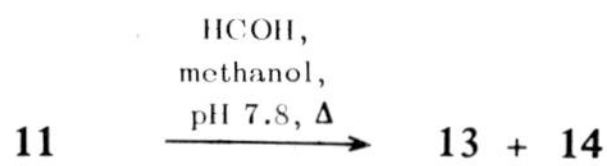

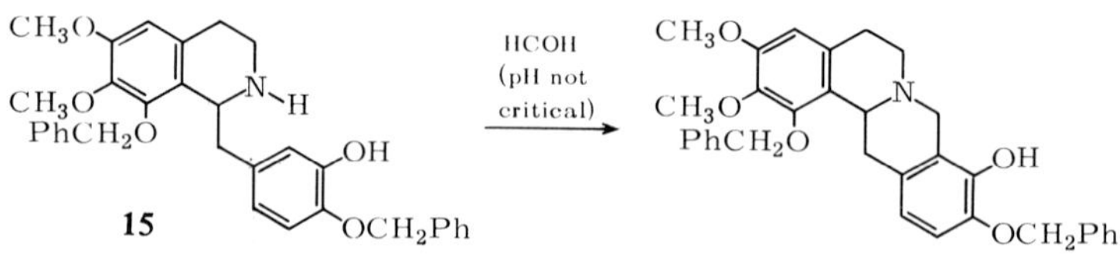

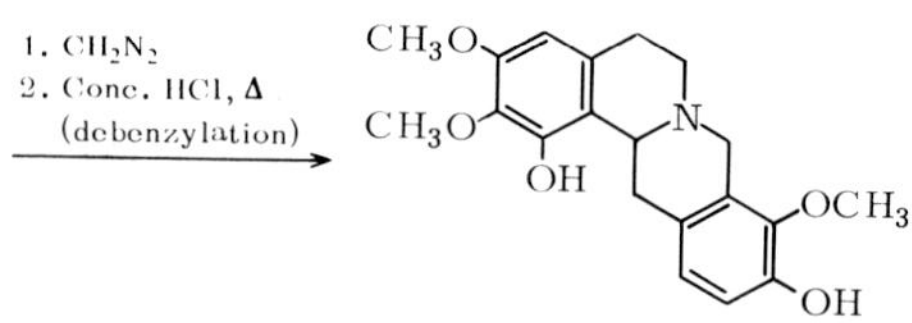

(±)-Capaurimine

The unusually substituted tetrahydroprotoberberine alkaloid caseadine was synthesized, as the racemate, starting with the brominated phenolic amide **16**. Bischler–Napieralski cyclization followed by sodium borohydride reduction and Mannich cyclization furnished the tetrahydroprotoberberine **17** which was debrominated to

(±)-caseadine.[23] *O*-Methylcaseadine had previously been prepared using the ethoxy-carbamido substituent (Chapter 16, Section III, A).

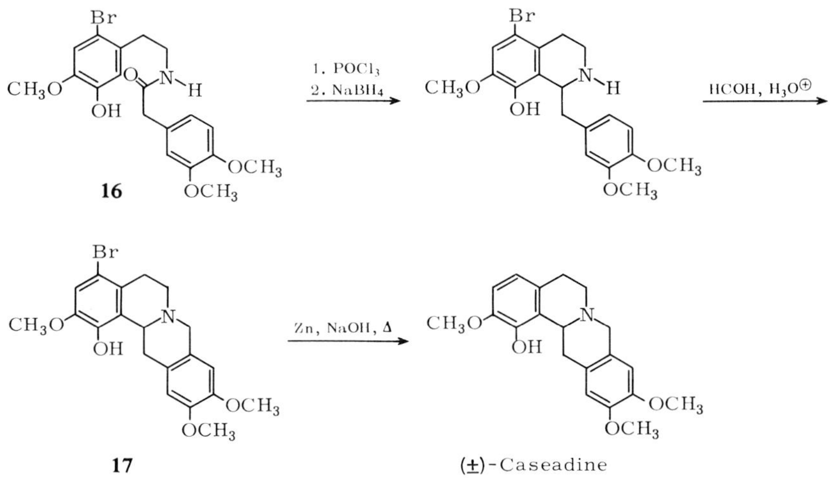

The Pomeranz–Fritsch approach to protoberberines has been further extended through the use of isocoumarin intermediates. Treatment of the isocoumarin **18** with aminoacetaldehyde dimethyl acetal yielded the pyridone **19** which could lead either to the tetrahydroprotoberberine **20** or to the 5-hydroxylated tetrahydroprotoberberine **21**.[24]

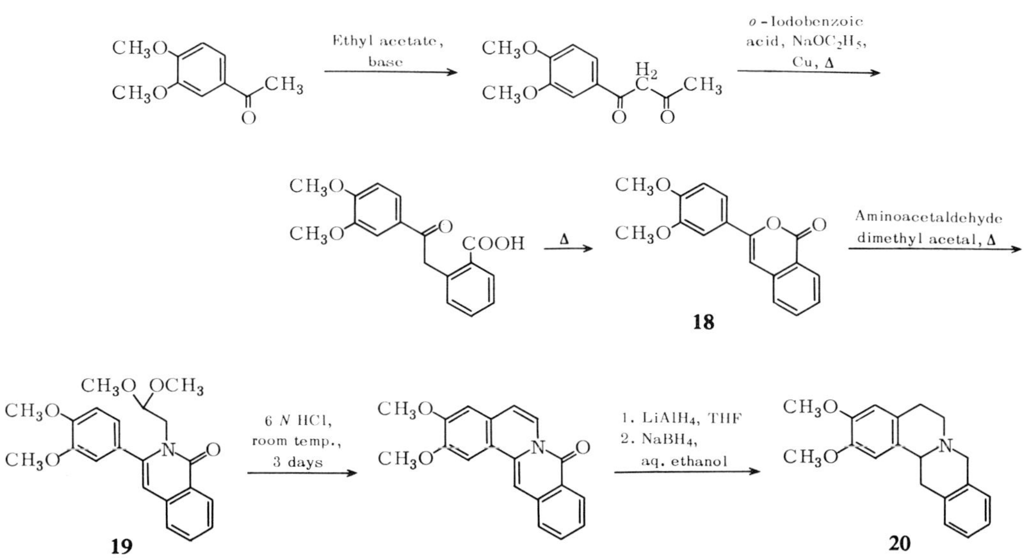

or:

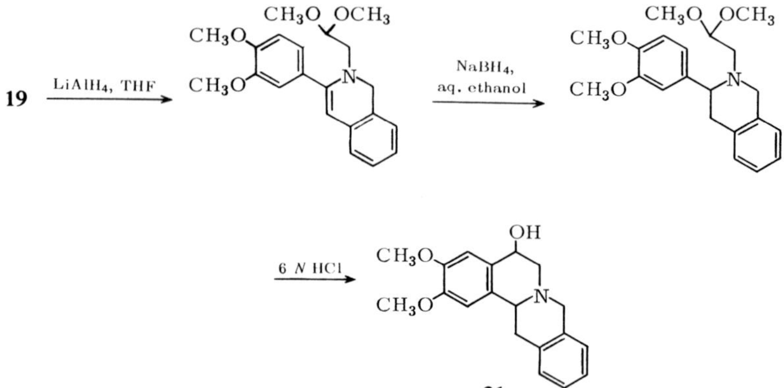

21

CHAPTER 19/THE PHTHALIDEISOQUINOLINES

Section V

The chemistry of α-narcotinediol has been further investigated by Šimanek and Klásek.[25] When this compound was treated with an equivalent amount of *p*-toluene-

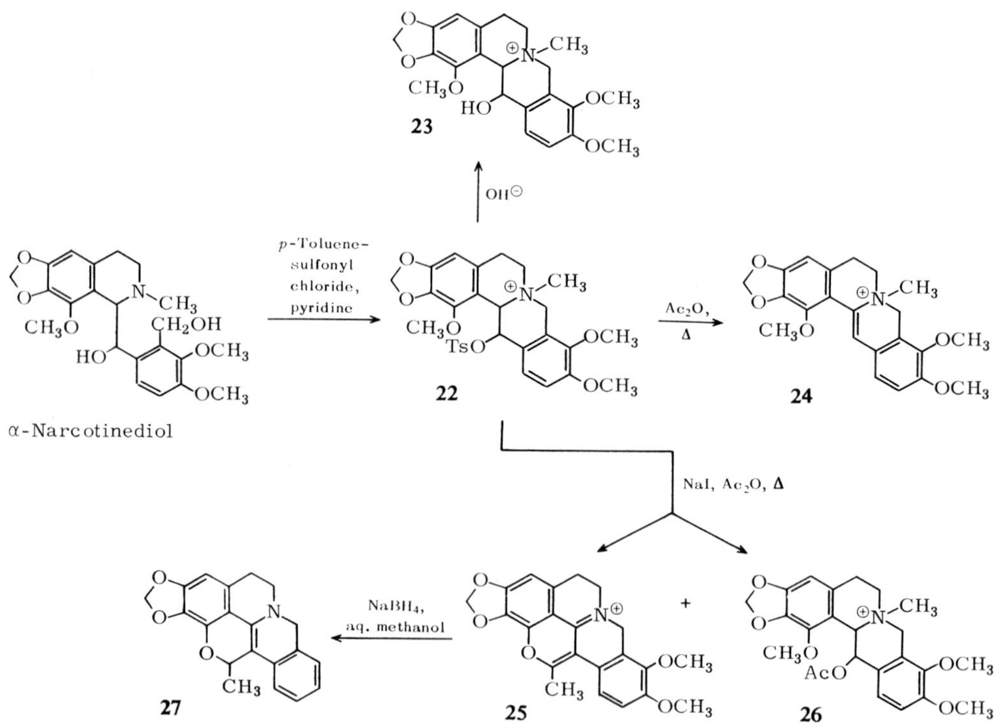

sulfonyl chloride in pyridine, the quaternary salt **22** was obtained, which could be converted to the 13-hydroxylated tetrahydroprotoberberine salt **23**. Otherwise, when the salt **22** was heated with acetic anhydride, the 13-dehydrotetrahydroprotoberberine salt **24** was generated.

If, on the other hand, the salt **22** was refluxed with sodium iodide in acetic anhydride, two products were obtained, namely the yellow immonium salt **25** and the ammonium salt **26**, the product **25** being formed through the intermediacy of the 13-dehydrotetrahydroprotoberberine salt **24**. Finally, sodium borohydride reduction of the immonium salt **25** led to the enamine **27**.

Chapter 20/The Spirobenzylisoquinolines

Section VII

The absolute configuration of (+)-ochotensine, (+)-ochrobirine, dihydrofumariline, and by extension (+)-ochotensimine and (+)-fumariline, has been established by means of the aromatic chirality method. The CD spectra of the first three compounds showed positive Davydov split extrema centered between 278 and 288 mμ, corresponding to the $^1A \rightarrow {}^1L_b$ band. The positive chirality indicates that the 9,10-methylenedioxy group of ring D must be situated in each case below the mean plane of the molecule.[26]

A New Class of Isoquinoline Alkaloids: The Azafluoranthenes

Cava and co-workers have isolated two novel and completely aromatic yellow alkaloids from *Abuta imene*. These are imeluteine and rufescine, to which the azafluoranthene structures indicated below were assigned.[27]

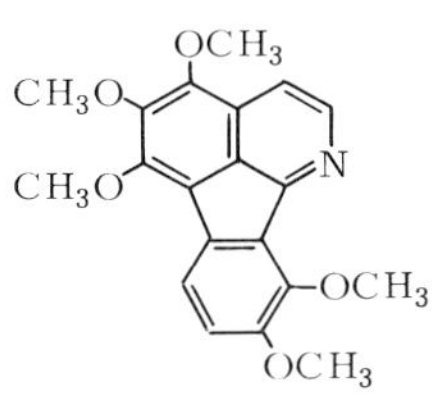

Imeluteine

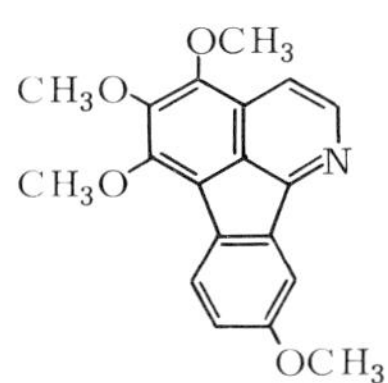

Rufescine

REFERENCES

1. E. C. Taylor and S. F. Martin, *J. Amer. Chem. Soc.* **94**, 2874 (1972).
2. L. L. Miller, F. R. Stermitz, and J. R. Falk, *J. Amer. Chem. Soc.* **93**, 5941 (1971).
3. T. Kametani, T. Kobari, and K. Fukumoto, *Chem. Commun.* p. 288 (1972).
4. E. Brochmann–Hanssen, C.-C. Fu, A. Y. Leung, and G. Zanati, *J. Pharm. Sci.* **60**, 1672 (1971).
5. O. Kadlec and V. Bauer, *Experientia*, **27**, 815 (1971).

6. G. Wittig, H. Tenhaeff, W. Schoch, and G. Koenig, *Justus Liebigs Ann. Chem.* **572**, 1 (1951).
7. S. F. Dyke and A. C. Ellis, *Tetrahedron*, **27**, 3803 (1971).
8. I. R. C. Bick, J. B. Brenner, and P. Wiriyachitra, *Tetrahedron Lett.* p. 4795 (1971).
9. Kh. G. Pulatova, S. Kh. Maekh, Z. F. Ismailov, and Yu. Yunusov, *Chem. Nat. Compds.* **4**, 336 (1968).
10. Y. Inubushi, T. Kikuchi, T. Ikuba, and I. Saji, *Tetrahedron Lett.*, p. 423 (1972).
11. Chapter 5, Ref. 9a, p. 150.
12. F. G. Mann and M. J. Pragnell, *Chem. Ind. (London)* p. 1386 (1964).
13. A. H. Jackson and G. W. Stewart, *Tetrahedron Lett.* p. 4941 (1971).
14. Chapter 6, Ref. 9b, p. 164.
15. M. Shamma, J. L. Moniot, S. Y. Yao, and J. A. Stanko, *Chem. Commun.* p. 408 (1972).
16. N. M. Mollov, L. N. Thuan, and P. P. Panov, *C. R. Acad. Bulg. Sci.* **24**, 1047 (1971).
17. R. J. Spangler and D. C. Boop, *Tetrahedron Lett.* p. 4851 (1971).
18. M. P. Cava, P. Stern, and K. Wakisaka, (in press).
19. O. Hoshino, T. Toshioka, and B. Umezawa, *Chem. Commun.* p. 1533 (1971).
20. M. L. Carvalhas, *J. Chem. Soc., Perkin Trans.* **1**, 327 (1972).
21. T. Kametani, K. Fukumoto, T. Terui, K. Yamaki, and E. Taguchi, *J. Chem. Soc., C* p. 2709 (1971).
22. T. Kametani, T. Honda, and M. Ihara, *J. Chem. Soc., C* p. 2396 (1971).
23. H. Iida, H.-C. Hsu, H. Miyano, and T. Kikuchi, *J. Pharm. Soc. Jap.* **91**, 795 (1971).
24. D. W. Brown, S. F. Dyke, M. Sainsbury, and G. Hardy, *J. Chem. Soc., C* p. 3219 (1971).
25. V. Šimanek and A. Klásek, *Tetrahedron Lett.* p. 4133 (1971).
26. M. Shamma, J. L. Moniot, R. H. F. Manske, W. K. Chan, and K. Nakanishi, *Chem. Commun.* p. 310 (1972).
27. M. P. Cava, K. T. Buck, and A. I. daRocha, *J. Amer. Chem. Soc.* **94**, 5931 (1972).

Appendix / OCCURRENCE OF ISOQUINOLINE ALKALOIDS BY PLANT FAMILIES

Alangiaceae:	Emetine type
Amaryllidaceae:	Cherylline
Ancistrocladaceae:	Naphthalenoisoquinolines
Anonaceae:	Benzylisoquinolines, bisbenzylisoquinolines, aporphines, oxoaporphines, phenanthrenes, protoberberines
Araceae:	Aporphines, oxoaporphines
Aristolochiaceae:	Aporphines, phenanthrenes
Berberidaceae:	Benzylisoquinolines, bisbenzylisoquinolines, aporphines, proaporphine–benzylisoquinoline dimers, aporphine–benzylisoquinoline dimers, protoberberines, protopines, phthalideisoquinolines, taspine
Cactaceae:	Simple tetrahydroisoquinolines and their trimers
Chenopodiaceae:	Simple tetrahydroisoquinolines
Combretaceae:	Benzylisoquinolines
Convolvulaceae:	Protoberberines
Euphorbiaceae:	Proaporphines, aporphines, taspine
Fumariaceae:	Simple tetrahydroisoquinolines, cularines, cularine–morphine dimers, aporphines, protoberberines, benzophenanthridines, protopines, phthalideisoquinolines, spirobenzylisoquinolines, *N*-benzyltetrahydroisoquinolines
Hernandiaceae:	Benzylisoquinolines, isoquinolines, bisbenzylisoquinolines, aporphines, aporphine–benzylisoquinoline dimers, oxoaporphines

Lauraceae:	Benzylisoquinolines, pavines and isopavines, bisbenzylisoquinolines, proaporphines, aporphines, oxoaporphines, phenanthrenes, protoberberines
Leguminoseae:	Simple tetrahydroisoquinolines
Liliaceae:	Homologated isoquinolines
Magnoliaceae:	Benzylisoquinolines, bisbenzylisoquinolines, oxoaporphines
Menispermaceae:	Benzylisoquinolines, bisbenzylisoquinolines, proaporphines, aporphines, oxoaporphines, protoberberines, protostephanine
Monimiaceae:	Benzylisoquinolines, isoquinolones, bisbenzylisoquinolines, proaporphines, aporphines, oxoaporphines, phenanthrenes
Nymphaceae:	Simple tetrahydroisoquinolines, benzylisoquinolines, bisbenzylisoquinolines, proaporphines
Orchidaceae:	Phenyltetrahydroisoquinolines
Papaveraceae:	Simple tetrahydroisoquinolines, benzylisoquinolines, pavines and isopavines, proaporphines, aporphines, oxoaporphines, protoberberines and retroprotoberberines, benzophenanthridines, protopines, phthalideisoquinolines, rhoeadines and papaverrubines
Ranunculaceae:	Simple tetrahydroisoquinolines, benzylisoquinolines, isoquinolones, pavines and isopavines, bisbenzylisoquinolines, aporphines, aporphine–benzylisoquinoline dimers, oxoaporphines, phenanthrenes, protoberberines, protopines, phthalideisoquinolines
Rhamnaceae:	Benzylisoquinolines, aporphines
Rubiaceae:	Emetine type
Rutacae:	Benzylisoquinolines, protoberberines, benzophenanthridines, protopines

AUTHOR INDEX

Numbers in lightface indicate the text pages on which references are cited. The italic numbers indicate the pages on which complete references are listed. Numbers in parentheses are reference numbers.

H

S

T

SUBJECT INDEX

C

D

E

I

P

S

U

ORGANIC CHEMISTRY

A SERIES OF MONOGRAPHS

EDITORS

ALFRED T. BLOMQUIST
Department of Chemistry
Cornell University
Ithaca, New York

HARRY WASSERMAN
Department of Chemistry
Yale University
New Haven, Connecticut

1. Wolfgang Kirmse. CARBENE CHEMISTRY, 1964; 2nd Edition, 1971
2. Brandes H. Smith. BRIDGED AROMATIC COMPOUNDS, 1964
3. Michael Hanack. CONFORMATION THEORY, 1965
4. Donald J. Cram. FUNDAMENTALS OF CARBANION CHEMISTRY, 1965
5. Kenneth B. Wiberg (Editor). OXIDATION IN ORGANIC CHEMISTRY, PART A, 1965; Walter S. Trahanovsky (Editor). OXIDATION IN ORGANIC CHEMISTRY, PART B, *In preparation*
6. R. F. Hudson. STRUCTURE AND MECHANISM IN ORGANO-PHOSPHORUS CHEMISTRY, 1965
7. A. William Johnson. YLID CHEMISTRY, 1966
8. Jan Hamer (Editor). 1,4-CYCLOADDITION REACTIONS, 1967
9. Henri Ulrich. CYCLOADDITION REACTIONS OF HETEROCUMULENES, 1967
10. M. P. Cava and M. J. Mitchell. CYCLOBUTADIENE AND RELATED COMPOUNDS, 1967
11. Reinhard W. Hoffman. DEHYDROBENZENE AND CYCLOALKYNES, 1967
12. Stanley R. Sandler and Wolf Karo. ORGANIC FUNCTIONAL GROUP PREPARATIONS, VOLUME I, 1968; VOLUME II, 1971; VOLUME III, 1972
13. Robert J. Cotter and Markus Matzner. RING-FORMING POLYMERIZATIONS, PART A, 1969; PART B, 1; B, 2, 1972
14. R. H. DeWolfe. CARBOXYLIC ORTHO ACID DERIVATIVES, 1970
15. R. Foster. ORGANIC CHARGE-TRANSFER COMPLEXES, 1969
16. James P. Snyder (Editor). NONBENZENOID AROMATICS, VOLUME I, 1969; VOLUME II, 1971

17. C. H. Rochester. ACIDITY FUNCTIONS, 1970

18. Richard J. Sundberg. THE CHEMISTRY OF INDOLES, 1970

19. A. R. Katritzky and J. M. Lagowski. CHEMISTRY OF THE HETEROCYCLIC *N*-OXIDES, 1970

20. Ivar Ugi (Editor). ISONITRILE CHEMISTRY, 1971

21. G. Chiurdoglu (Editor). CONFORMATIONAL ANALYSIS, 1971

22. Gottfried Schill. CATENANES, ROTAXANES, AND KNOTS, 1971

23. M. Liler. REACTION MECHANISMS IN SULPHURIC ACID AND OTHER STRONG ACID SOLUTIONS, 1971

24. J. B. Stothers. CARBON-13 NMR SPECTROSCOPY, 1972

25. Maurice Shamma. THE ISOQUINOLINE ALKALOIDS: CHEMISTRY AND PHARMACOLOGY, 1972

In preparation

Samuel P. McManus (Editor). ORGANIC REACTIVE INTERMEDIATES